Optics and Spectroscopy

NINTH EDITION

For B.Sc. (Physics) Students of All Indian Universities

R. MURUGESHAN

Head of the Department of Physics (Retired)
Vivekananda College
Tiruvedakam West (Post)
Madurai District
Tamil Nadu

KIRUTHIGA SIVAPRASATH

S Chand And Company Limited

(ISO 9001 Certified Company)

S Chand And Company Limited

(ISO 9001 Certified Company)

Head Office: D-92, Sector–2, Noida – 201301, U.P. (India), Ph. 91-120-4682700

Registered Office: A-27, 2nd Floor, Mohan Co-operative Industrial Estate, New Delhi – 110 044, Phone: 011-49731800

www.**schandpublishing.com;** e-mail: **info@schandpublishing.com**

Marketing Offices:

Chennai	:	Ph: 23632120; chennai@schandpublishing.com
Guwahati	:	Ph: 2738811, 2735640; guwahati@schandpublishing.com
Hyderabad	:	Ph: 40186018; hyderabad@schandpublishing.com
Jalandhar	:	Ph: 4645630; jalandhar@schandpublishing.com
Kolkata	:	Ph: 23357458, 23353914; kolkata@schandpublishing.com
Lucknow	:	Ph: 4003633; lucknow@schandpublishing.com
Mumbai	:	Ph: 25000297; mumbai@schandpublishing.com
Patna	:	Ph: 2260011; patna@schandpublishing.com

First Edition 1997
Subsequent Editions 1998, 2001, 2003, 2005, 2006, 2008, 2010, 2012
Ninth Edition 2014
Reprints 2014, 2015, 2016 (Twice), 2018, 2019, 2020, 2021, 2023, 2024

Reprint 2025

ISBN: 978-81-219-1441-3 **Product Code:** H6OPT68PHYS10ENAI14O

PRINTED IN INDIA

By Vikas Publishing House Private Limited, Plot 20/4, Site-IV, Industrial Area Sahibabad, Ghaziabad – 201 010 and Published by S Chand And Company Limited, A-27, 2nd Floor, Mohan Co-operative Industrial Estate, New Delhi – 110 044.

PREFACE TO THE NINTH REVISED EDITION

The book has been thoroughly revised keeping in mind the latest syllabi of University Grants Commission. Two new chapters have been added as per the syllabus. The complete course in "Optics", as suggested in the UGC Model Curriculum for B.Sc., Physics (General) and B.Sc., Physics (Honours), has been covered.

AUTHORS

PREFACE TO THE FIRST EDITION

This book has been written for the students of B.Sc., Physics of various Indian Universities. The book covers the syllabi, prescribed by Madras, Bharathiyar, Bharathidhasan, Madurai Kamaraj and Manonmaniam Sundaranar Universities.

SI System of Units has been used throughout the text. Proper care has been taken in dealing with the subject with modern outlook. A large number of questions and problems have been given at the end of each Chapter. Students should attempt to tackle them properly for better insight and understanding of the subject.

All available standard books on the subject have been consulted for the preparation of this book. I acknowledge my indebtedness to the authors and publishers of different books on the subject. I am also thankful to my colleagues, Shri M. Muthusamy, Shri K.S. Srinivasa Nambirajan, Shri A.P. Selvarajah and Shri P.K. Veeran, for their co-operation they extended to me in the preparation of this book.

I am thankful to Srimad Nithyanandaji Maharaj, President, Srimad Guhanadaji Maharaj, Secretary, Srimad Sadanandaji Maharaj, Srimad Bhodhanandaji Maharaj and Dr. Subrahmanyam, Principal of our College for their encouragement and help.

Lastly, I would like to thank my publisher, S. Chand & Company Ltd. I especially thank the Managing Director, Shri Ravindra Kumar Gupta.

AUTHOR

UGC Model Curriculum
B.Sc. Physics (General)
University Grants Commission,
New Delhi, 2001

Course 5 : Waves, Acoustics and Optics

5.1 Waves (11)

Waves in Media : Speed of transverse waves on a uniform string, speed of longitudinal waves in a fluid, energy density and energy transmission in waves, typical measurements. Waves over liquid surface: gravity waves and ripples. Group velocity and phase velocity, their measurements. (5)

Superposition of waves : Linear homogenous equations and the superposition principle, nonlinear superposition and consequences. (3)

Standing waves : Standing waves as normal modes of bounded systems, examples, Harmonics and the quality of sound; examples. Chladni's figures and vibrations of a drum. Production and detection of ultrasonic and infrasonic wave and applications. (4)

5.2 Acoustics (13)

Noise and Music : The human ear and its responses; limits of human audibility, intensity and loudness, bel and decibel, the musical scale, temperament and musical instruments.

Reflection, refraction and diffraction of sound : Acoustic impedance of a medium, percentage reflection and refraction at a boundary, impedance matching for transducers, diffraction of sound principle of a sonar system, sound ranging.

Applied acoustics : Transducers and their characteristics. Recording and reproduction of sounds, various systems, Measurements of frequency, waveform, intensity and velocity. The acoustics of halls, reverberation period, Sabine's formula.

5.3 Geometrical Optics (18)

Fermat's principle : Principle of extremum path, the aplanatic points of a sphere and other applications.

General theory of image formation : Cardinal points of an optical system, general relationships, thick lens and lens combinations. Lagrange equation of magnification, telescopic combinations, telephoto lenses and eyepieces. (5)

Aberration in images : Chromatic aberrations, achromatic combination of lenses in contact and separated lenses. Monochromatic aberrations and their reductions; aspherical mirrors and Schmidt corrector plates, aplanatic points, oil immersion objectives, meniscus lens. (4)

Optical instruments : Entrance and exit pupils, need for a multiple lens eyepiece, common types of eyepieces. (5)

5.4 Physical Optics (36)

Interference of light : The principle of superposition, two-slit interference, coherence requirement for the sources, optical path retardations, lateral shift of fringes, Rayleigh refractometer and other applications. Localised fringes; thin films, applications for precision measurements for displacements. (5)

Haidinger fringes : Fringes of equal inclination. Michelson interferometer, its application for precision determination of wavelength, wavelength difference and the width of spectral lines. Twymann-Green interferometer and its uses. Inensity distribution in multiple beam interference, Tolansky fringes, Fabry-Perot interferometer and etalon. (8)

Fresnel diffraction : Fresnel half-period zones, plates, straight edge, rectilinear propagation. (2)

Fraunhofer diffraction: Diffraction at a slit, half-period zones, phasor diagram and integral calculus methods, the intensity distribution, diffraction at a circular aperture and circular disc, resolution of images, Rayleigh criterion, resolving power of telescope and microscopic system, outline of phase contract microscopy. (9)

Diffraction gratings : Diffraction at N parallel slits, intensity distribution, plane diffraction grating, reflection grating and blazed gratings. Concave grating and different mountings. Resolving power of a grating and comparison with resolving powers of prism and of a Fabry-Perot etalon (6)

Double refraction and optical rotation : Refraction, in uniaxial crystals, its electromagnetic theory. Phase retardation plates, double image prism. Rotation of plane of polarisation, origin of optical rotation in liquids and in crystals.

5.5 Lasers (12)

Laser system : Purity of a spectral line, coherence length and coherence time, spatial coherence of a source, Einstein's A and B coefficients, spontaneous and induced emission, conditions for laser action, population inversion. (4)

Application of lasers : Pulsed lasers and tunable lasers, spatial coherence and directionality, estimates of beam intensity, temporal coherence and spectral energy density. (5)

Lasers and nonlinear optics : Polarization P including higher order terms in E and generation of harmonics, momentum mismatch and choice of the right crystal and direction for compensation. (3)

Text and Reference Books

1. AK Ghatak, "Physical Optics"
2. DP Khandelwal; "Optics and Atomic Physics" (Himalaya Publishing House, Bombay, 1988).
3. F Smith and JH Thomson; "Manchester Physics series; Optics" (English Language Book Society and John Wiley, 1977).
4. Born and Wolf; "Optics"
5. KD Moltev; "Optics" (Oxford University Press)
6. Sears; "Optics"
7. Jenkins and White; "Fundamental of Optics" (McGraw-Hill)
8. BB Laud; Laser and Non-linear Optics (Wiley Eastern 1985)
9. Smith and Thomson; "Optics" (John Wiley and Sons).
10. Berkely Physics Coursse; Vol. III "Waves and Oscillations"
11. IG Main; "Vibrations and Waves" (Cambridge University Press)
12. H J Pain; "The Physics of Vibrations and Wave" (MacMillian 1975)

B.Sc. Physics (Honours)

Course p6: Optics

6.1 Geometrical Optics (15)

Fermat's principle : Principle of extremum path; the aplanatic points of a sphere and other applications.

General theory of image formation : Cardinal points of an optical system; general relationships, thick lens and lens combinations, telephoto lenses.

Aberration in images : Chromatic aberrations; achromatic combination of lenses in contact and separated lenses. Monochromatic aberrations and their reduction; aspherical mirrors and Schmidt corrector plates, oil immersion objectives, meniscus lenses.

Optical instruments: entrance and exit pupils, need for a multiple lens eyepiece, common type of eyepieces.

6.2 Physics Optics I (15)

Interference of light : The principle of superposition; two-slit interference, coherence requirements for the sources, localized fringes in thin films, transition from fringes of equal thickness to those of equal inclination. Michelson interferometer; its uses for determination of wavelength, wavelength difference and standardization of the meter. Intensity distribution in multiple beam interference; Tolansky fringes, Fabry-Perot interferometer and etalon.

Fresnel diffraction : Half-period zones, circular apertures and obstacles, straight edge, explanation of rectilinear propagation.

Fraunhofer diffraction : Diffraction at a slit, a circular aperture and a circular disc. Resolution of images; Rayleigh criterion, resolving power of a telescope and a microscope, outline of phase contrast microscopy.

Diffraction grating : Diffraction at N parallel slits; plane diffraction grating, concave grating, resolving power of gratings and prisms.

6.3 Physical Optics II (15)

Double refraction and optical rotation : Double refraction in uniaxial crystals, explanation in terms of electromagnetic theory, phase retardation plates. Rotation of plane of polarization, origin of optical rotation in liquids and in crystals.

Dispersion and scattering: Theory of dispersion of light; absorption bands and anomalous dispersion. Theory of Rayleigh scattering; scattering of X-rays and determination of Z of an atom.

Laser system : Purity of a spectral line; coherence length and coherence time, spatial coherence of a source; Einstein's A and B coefficients; coherence of induced emissions. Conditions for laser action; existence of a metastable state, population inversion by pumping and cavity.

Application of lasers : Spatial coherence and directionality, estimates of beam intensity. Temporal coherence and spectral energy density.

Lasers and nonlinear optics: Polarization P including higher order terms in E and generation of harmonics. Momentum mismatch and choice of right crystal and direction for compensation

CONTENTS

CHAPTER

1

GEOMETRICAL OPTICS

LENS SYSTEMS

1.1. Convex Lens

A convex lens is a transparent refracting medium bounded by two spherical surfaces.

The line joining the centres of curvature of the two surfaces is called the *principal axis.*

A section of the lens through its principal axis is called its *principal section.*

1.2. Principal Focus and Focal Planes

AB represents the principal axis of a lens (Fig. 1.1).

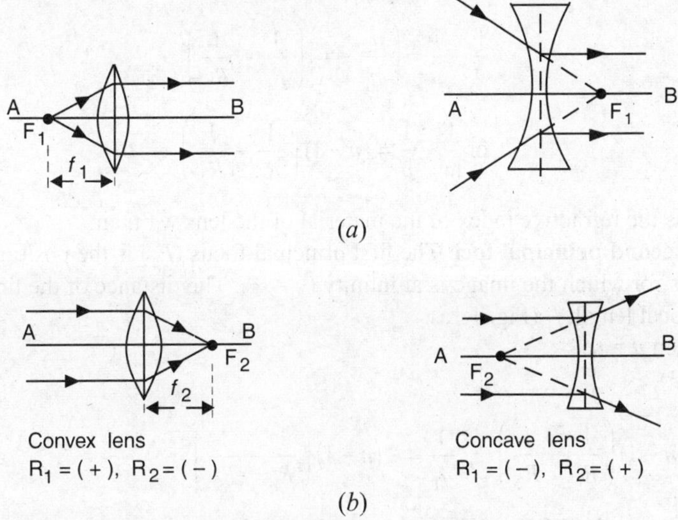

Fig. 1.1.

First Principal Focus (F_1) : *It is that point on the principal axis of the lens, the rays starting from which (convex lens) or appear to converge at which (concave lens) become parallel to principal axis after refraction from the lens.*

The plane passing through F_1 and perpendicular to the principal axis is called *first focal plane.*

Second Principal Focus (F_2) : *Second principal focus is that point on the principal axis at which the rays parallel to principal axis converge (convex lens) or appear to diverge (concave lens) after refraction from the lens.*

The plane passing through F_2 and normal to the principal axis is called the *second focal plane.* The points F_1 and F_2 are called *focal points.*

1.3. Refraction Through a Thin Lens

Consider a thin lens of refractive index n_2 placed in a medium of refractive index n_1 (Fig. 1.2). Let R_1 and R_2 be the radii of the first and second surfaces respectively. Consider a point object O placed on the principal axis. Let I' be the position of the image formed by refraction at the first surface. Let v' be the distance of image I' and u the distance of the object from the pole of the first surface. Then

$$\frac{n_2}{v'} - \frac{n_1}{u} = \frac{n_2 - n_1}{R_1} \quad ...(1)$$

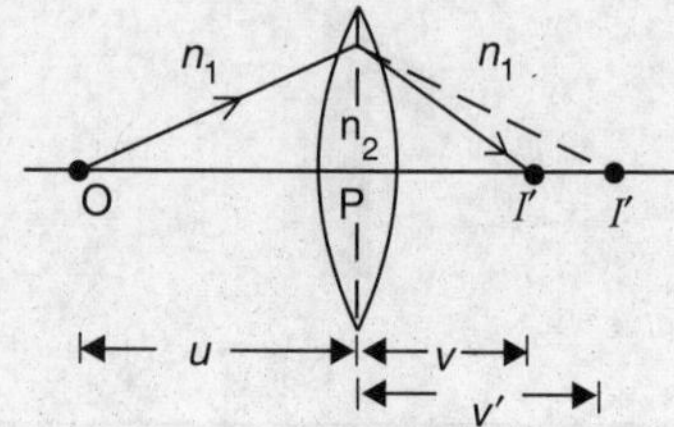

Fig. 1.2.

Now I' acts as the virtual object for the second surface. The final image I is formed at a distance v from the second surface.

Since the rays now pass from a medium of refractive index n_2 into a medium of refractive index n_1, we have

$$\frac{n_1}{v} - \frac{n_2}{v'} = \frac{n_1 - n_2}{R_2} = -\frac{n_2 - n_1}{R_2} \quad ...(2)$$

Adding (1) and (2),
$$\frac{n_1}{v} - \frac{n_1}{u} = (n_2 - n_1)\left(\frac{1}{R_1} - \frac{1}{R_2}\right)$$

Dividing throughout by n_1,

$$\frac{1}{v} - \frac{1}{u} = \left(\frac{n_2}{n_1} - 1\right)\left(\frac{1}{R_1} - \frac{1}{R_2}\right)$$

$$\text{or } \frac{1}{v} - \frac{1}{u} = (n-1)\left(\frac{1}{R_1} - \frac{1}{R_2}\right) \quad ...(3)$$

Here, $n = n_2/n_1$ is the refractive index of the material of the lens w.r.t. air.

First and second principal foci. The first principal focus (F_1) is the position of an object on the principal axis for which the image is at infinity ($v = \infty$). The distance of the first focus from the lens is the first focal length f_1 (Fig. 1.1a).

If $v = \infty$, then $u = f_1$.

From Eq. (3),

$$\frac{1}{\infty} - \frac{1}{f_1} = (n-1)\left(\frac{1}{R_1} - \frac{1}{R_2}\right) \therefore \frac{1}{f_1} = -(n-1)\left(\frac{1}{R_1} - \frac{1}{R_2}\right) \quad ...(4)$$

The second principal focus (F_2) is the position of the image on the principal axis for which the object is at infinity ($u = \infty$). The distance of the second focus from the lens is the second focal length f_2 [Fig. 1.1b]. $\therefore$ $v = f_2$ when $u = \infty$.

From Eq. (3),

$$\frac{1}{f_2} - \frac{1}{\infty} = (n-1)\left(\frac{1}{R_1} - \frac{1}{R_2}\right)$$

$$\therefore \quad \frac{1}{f_2} = (n-1)\left(\frac{1}{R_1} - \frac{1}{R_2}\right) \quad ...(5)$$

Thus the first and second focal lengths of a lens are *numerically* equal when the lens is placed in a uniform medium. The general convention is to call the second principal focal length as the focal length f of the lens.

$$\therefore \qquad \frac{1}{f} = \frac{1}{v} - \frac{1}{u} = (n-1)\left(\frac{1}{R_1} - \frac{1}{R_2}\right) \qquad ...(6)$$

This formula is known as *Lens makers formula.*

Example 1 : *A double convex lens has faces of radii 20 cm each. The index of refraction of glass is 1.5. Compute the focal length of this lens in (i) air and (ii) when immersed in carbon disulphide of refractive index 1.63.*

Solution : *(i)* When the lens is placed in air.

$$\frac{1}{f} = (n-1)\left(\frac{1}{R_1} - \frac{1}{R_2}\right) = (1.5-1)\left[\frac{1}{20} - \frac{1}{(-20)}\right] = \frac{1}{20}$$

or $f = 20$ cm. Thus the focal length of the lens in air is 20 cm.

(ii) When the lens is placed in carbon disulphide.

$$\frac{1}{f} = \frac{n_2 - n_1}{n_1}\left(\frac{1}{R_1} - \frac{1}{R_2}\right) = \frac{1.5 - 1.63}{1.63}\left[\frac{1}{20} - \frac{1}{(-20)}\right] = \frac{-0.13}{16.3}$$

or $$f = -125.4 \text{ cm.}$$

The lens in this case behaves as a diverging or a concave lens.

Example 2 : *Calculate the focal length of a double convex lens for which the radius of curvature of each surface is 0.30 m and refractive index of glass is 1.50.*

Solution : Here, $R_1 = 0.30$ m; $R_2 = -0.30$ m; $n = 1.50$

$$\frac{1}{f} = (n-1)\left(\frac{1}{R_1} - \frac{1}{R_2}\right) = (1.50-1)\left[\frac{1}{0.30} + \frac{1}{0.30}\right] = 3.33$$

$$\therefore \qquad f = 0.30 \text{ m.}$$

Example 3 : *A biconvex lens is made of glass of n = 1.52. If one surface has twice the radius of curvature of the other, and if the focal length is 5 cm, what are the two radii?*

Solution : Here, $n = 1.52$; $f = 5$ cm; $R_2 = -2R_1$.

$$\frac{1}{f} = (n-1)\left(\frac{1}{R_1} - \frac{1}{R_2}\right) = (1.52-1)\left[\frac{1}{R_1} + \frac{1}{2R_1}\right] \text{ or } \frac{1}{5} = \frac{0.78}{R_1}$$

$$\therefore \qquad R_1 = 0.78 \times 5 = 3.90 \text{ cm} ; R_2 = -3.90 \times 2 = -7.80 \text{ cm.}$$

Power of a lens : Power of a lens is its ability to converge or diverge the rays of light. The power of a lens is measured by the reciprocal of its focal length, $P = 1/f$. The unit of power is *dioptre*. One dioptre is the power of a lens whose focal length is 1 metre. Convex lenses have positive power and concave lenses, negative power.

Focal length of combination of two thin lenses in contact. When two thin lenses of focal lengths f_1 and f_2 are placed in contact with each other, then the focal length of the combination is given by $\frac{1}{f} = \frac{1}{f_1} + \frac{1}{f_2}$

Power of the combination = $P = P_1 + P_2$.

1.4. Equivalent Focal Length of Two Thin Lenses Separated by a Distance

Let L_1 and L_2 be two thin lenses of focal lengths f_1 and f_2, placed in air coaxially a distance a apart. Consider a ray PA incident on L_1 parallel to the axis at a height h_1 above the axis [Fig. 1.3]. This ray after refraction through the first lens is directed towards D which is the second principal focus of L_1.

Then, deviation produced by first lens $= \delta_1 = h_1 / f_1$.

The emergent ray from the first lens strikes the lens L_2 at a height h_2. The lens L_2 deviates it further through an angle δ_2. Finally the ray meets the axis at F_2. F_2 is the second principal focus of the lens system.

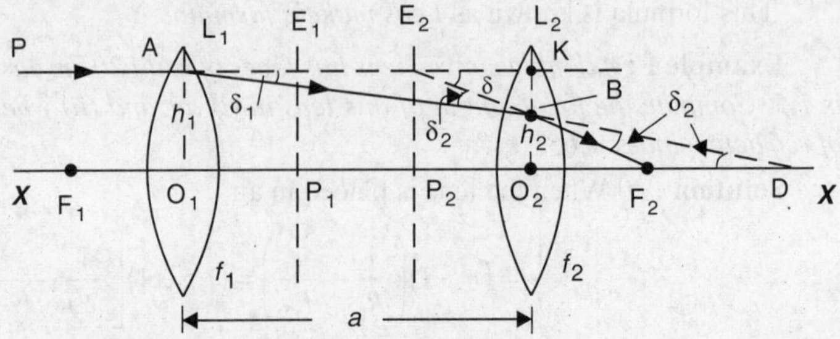

Deviation produced by the second lens $= \delta_2 = h_2 / f_2$.

Fig. 1.3.

PA and F_2B are produced to cut at E_2. Then a single convex lens placed in the position E_2P_2 and having focal length P_2F_2 is *equivalent* to the lens system. Thus $P_2F_2 = f$ is the *equivalent focal length*. Then, deviation produced by equivalent lens $= \delta = h_1 / f$. Now, $\delta = \delta_1 + \delta_2$

$$\therefore \qquad \frac{h_1}{f} = \frac{h_1}{f_1} + \frac{h_2}{f_2} \qquad ...(1)$$

Now, $h_2 = O_2B = O_2K - BK = h_1 - a\delta_1 = h_1 - \dfrac{ah_1}{f_1} = h_1\left(1 - \dfrac{a}{f_1}\right)$

Substituting the value of h_2 in Eq. (1), $\dfrac{h_1}{f} = \dfrac{h_1}{f_1} + \dfrac{h_1}{f_2}\left(1 - \dfrac{a}{f_1}\right)$

$$\text{or} \qquad \frac{1}{f} = \frac{1}{f_1} + \frac{1}{f_2} - \frac{a}{f_1 f_2} \qquad ...(2)$$

$$\therefore \qquad f = \frac{f_1 f_2}{f_1 + f_2 - a} = -\frac{f_1 f_2}{\Delta} \qquad ...(3)$$

Here, $\Delta = a - (f_1 + f_2)$ is known as the *optical separation* or *optical interval* between the two lenses.

Let us find the position of the equivalent lens, *i.e.*, the distance O_2P_2.

The triangles BF_2O_2 and $E_2F_2P_2$ are similar.

$$\therefore \qquad \frac{h_2}{h_1} = \frac{O_2F_2}{P_2F_2} \text{ or } O_2F_2 = \frac{h_2}{h_1} P_2F_2 = \left(1 - \frac{a}{f_1}\right) f \qquad \left(\because \frac{h_2}{h_1} = 1 - \frac{a}{f_1}\right)$$

$$\text{Now,} \qquad O_2P_2 = P_2F_2 - O_2F_2 = f - f\left(1 - \frac{a}{f_1}\right) = \frac{f\,a}{f_1}$$

$$= \left(\frac{f_1 f_2}{f_1 + f_2 - a}\right)\frac{a}{f_1} = \frac{a f_2}{f_1 + f_2 - a}$$

Let $\qquad O_2P_2 = -\beta \qquad$ ($\because P_2$ lies to the left of the lens L_2).

$$\therefore \qquad \beta = \frac{-af_2}{f_1 + f_2 - a} = \frac{-fa}{f_1} \qquad \text{...(4)}$$

Similarly, consider a ray parallel to the axis incident from the right hand side [Fig. 1.4]. Then we can find the position of F_1, the point where the ray intersects the principal axis after refraction through the lens system. E_1P_1 is the first principal plane. P_1 is the first principal point of the lens system. The distance of the first principal point from the first lens is $\alpha = O_1P_1$.

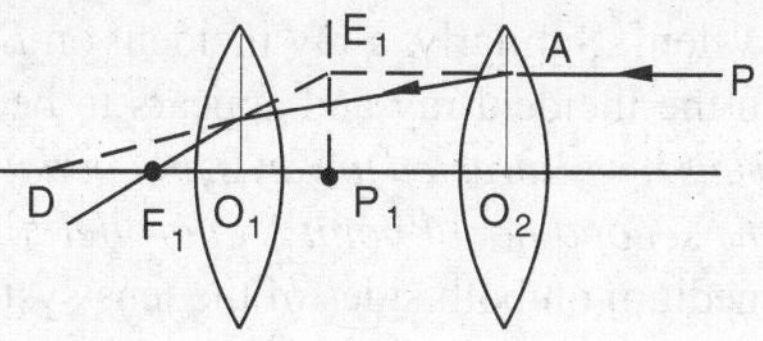

Fig. 1.4.

$$O_1P_1 = \alpha = \frac{af_1}{f_1 + f_2 - a} = \frac{af}{f_2} \qquad \text{...(5)}$$

The first principal focus F_1 is situated at a distance f towards the left of the point P_1.

If P_1 and P_2 are the powers of the component lenses and P the power of the combination, then by Eq. (2),

$$P = P_1 + P_2 - aP_1P_2 \qquad \text{...(6)}$$

1.5. Definitions of Cardinal Points and Respective Planes

For any system of lenses, there are six cardinal points. They are: (i) Two focal points. (ii) Two principal points (iii) Two nodal points.

Consider two convex lenses of focal lengths f_1 and f_2 kept at a distance a apart. In Fig. 1.5 *(a)*, a ray *SB* passing through F_1 emerges from the optical system parallel to the principal axis. *SB* and *PA* produced meet at E_1. A ray incident on a system after passing through a focus becomes parallel to the principal axis. Therefore, F_1 is a focus of the system and its distance is measured from the point P_1.

The first focal point is a point situated on the principal axis in the object space so that rays starting from it become parallel to the principal axis after refraction through the optical system. A plane normal to the principal axis and passing through this point is called the first focal plane.

The vertical plane through E_1, perpendicular to the axis, is called the first principal plane. The point P_1, where the first principal plane cuts the axis, is called the first principal point of the lens. The distance P_1F_1 is called the first principal focal length.

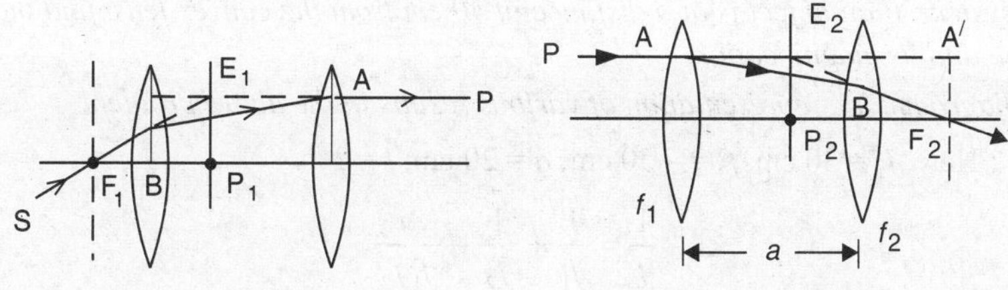

Fig. 1.5(*a*). **Fig. 1.5(*b*).**

A ray *PA* parallel to the principal axis passes as shown in Fig. 1.5*(b)*. It meets the axis at F_2. *The second focal point is a point situated on the principal axis in the image space so that the rays parallel to principal axis from the object space focus at this point. The plane normal to the principal axis and passing through this point is called second focal plane.*

Produce the incident ray (*PA*) forwards and the emergent ray (BF_2) backwards. They meet at E_2. Now draw a plane passing through E_2 and normal to the principal axis. It meets the axis at P_2. The plane passing through E_2P_2 is called the *second principal plane*. The point P_2 is called the *second principal point*. The distance P_2F_2 between the second principal and focal points is called the *second principal focal length.*

If a ray of light $R'S'$ is incident on L_1 such that it is directed to a particular point N_1 on the principal axis, then the emergent ray $T'U'$ will be parallel to $R'S'$ and it appears to be coming from a point N_2 on the principal axis [Fig. 1.5 *(c)*]. The points N_1 and N_2 are called the *nodal points* of the system. Similarly, a ray incident on L_2 such that it is directed towards N_2 emerges from L_1 parallel to the incident ray and appears to be coming from a point N_1. *Thus a ray directed towards first nodal point N_1 emerges through the second nodal point N_2 parallel to the object ray.* When the medium on both sides of the lens system is the same, the nodal points and the principal points coincide. The planes transverse to the principal axis through N_1 and N_2 are called *nodal planes*.

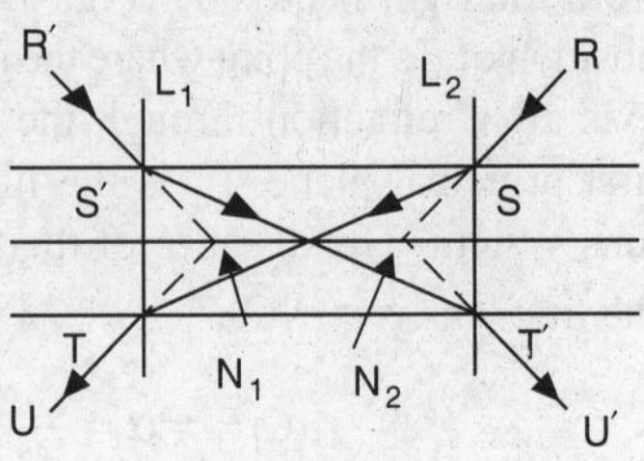

Fig. 1.5(*c*).

We have therefore, for a system of two lenses three pairs of points : (F_1, F_2), (P_1, P_2), (N_1, N_2) with corresponding planes. These points are called the *cardinal points* of the system, and the planes are called the *cardinal planes*.

Example 1 : *Two thin convex lenses of focal lengths 12 and 4 cm are separated by 8 cm. Find the positions of the cardinal points.*

Solution : Here, $f_1 = 12$ cm, $f_2 = 4$ cm, and $a = 8$ cm.

$$\alpha = O_1P_1 = \frac{af_1}{f_1 + f_2 - a} = \frac{8 \times 12}{12 + 4 - 8} = 12 \text{ cm.}$$

$$\beta = O_2P_2 = \frac{-af_2}{f_1 + f_2 - a} = \frac{-8 \times 4}{12 + 4 - 8} = -4 \text{ cm.}$$

$$P_2F_2 = f = \frac{f_1 f_2}{f_1 + f_2 - a} = \frac{12 \times 4}{12 + 4 - 8} = 6 \text{ cm.}$$

$P_1F_1 = -6$ cm.

Fig. 1.6.

N_1 and N_2 coincide with P_1 and P_2. The positions of the cardinal points are plotted in Fig. 1.6.

Example 2 : *A thin convex lens of focal length 30 cm is placed at a distance of 20 cm from a thin concave lens of 50 cm focal length. Calculate the equivalent focal length and locate the cardinal points of the system. If an object is at a distance of 40 cm from the convex lens, find the position of the final image and its magnification.*

Give a diagrammatic representation of various results in the above problem.

Solution : Here, $f_1 = 30$ cm, $f_2 = -50$ cm; $a = 20$ cm, $f = ?$

$$\frac{1}{f} = \frac{1}{f_1} + \frac{1}{f_2} - \frac{a}{f_1 f_2}$$

$$\therefore \quad f = \frac{f_1 f_2}{f_1 + f_2 - a} = \frac{30 \times (-50)}{30 + (-50) - 20} = 37.5 \text{ cm.}$$

Cardinal Points :

Principal points : The distance of the first principal point P_1 from the first (convex) lens is given by

$$\alpha = O_1P_1 = \frac{af}{f_2} = \frac{20 \times 37.5}{-50} = -15$$

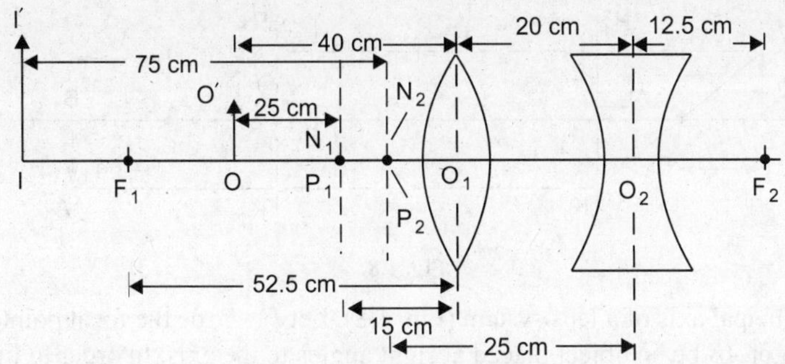

Fig. 1.7.

The first principal point lies at a distance of 15 cm to the left of convex lens O_1.

The distance of second principal point P_2 from the second (concave) lens O_2 is given by

$$\beta = O_2P_2 = \frac{-af}{f_1} = \frac{-20 \times 37.5}{30} = 25 \text{ cm.}$$

The second principal point lies at a distance of 25 cm to the left of concave lens O_2.

Focal points. The distance of first focal point F_1 from the first lens O_1 is given by,

$$O_1F_1 = -f\left(1 - \frac{a}{f_2}\right) = -37.5\left(1 - \frac{20}{-50}\right) = -52.5 \text{ cm.} = -52.5 \text{ cm.}$$

The first focal point lies at a distance of 52.5 cm to the left of convex lens O_1.

The distance of second focal point F_2 from the second lens O_2 is given by,

$$O_2f_2 = f\left(1 - \frac{a}{f_1}\right) = 37.5\left(1 - \frac{20}{30}\right) = 12.5 \text{ cm.}$$

The second focal point lies at a distance of 12.5 cm to the right of concave lens O_2.

Nodal points. As the medium on either side of the lens system is air, the nodal points N_1, N_2 coincide with principal points P_1, P_2 respectively.

The cardinal points P_1, P_2; F_1, F_2; N_1, N_2 are diagrammatically represented in Fig. 1.7.

Position of the final image. The object O is at a distance of 40 cm from lens O_1. Let u and v be the distances of the object and image from P_1 and P_2 respectively. Then

$$u = OP_1 = OO_1 - P_1O_1 = -(40 - 15) = -25 \text{ cm.}$$

From lens formula $\frac{1}{v} - \frac{1}{u} = \frac{1}{f}$, we have

$$\frac{1}{v} - \frac{1}{(-25)} = \frac{1}{37.5} \qquad \text{or } v = -75 \text{ cm.}$$

The image lies at a distance of 75 cm to the left of P_2.

Magnification $m = \frac{v}{u} = \frac{-75}{-25} = 3.$

1.6. Newton's Formula Connecting Conjugate Foci for a Convex Lens System

The knowledge of the cardinal points of an optical system helps to draw graphically the image of an object placed on the principal axis of the system.

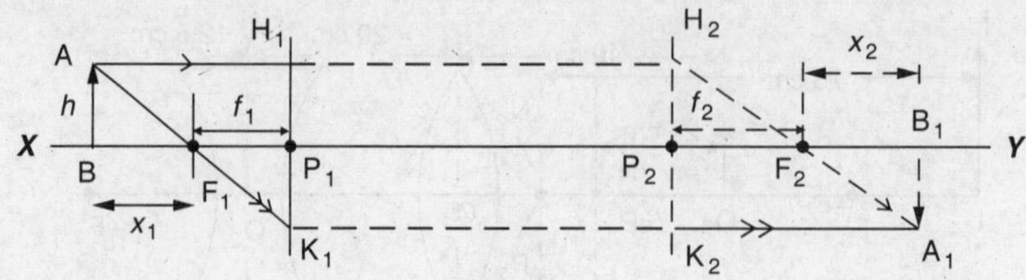

Fig. 1.8.

XY is the principal axis of a lens system (Fig. 1.8). Let F_1, F_2 be the focal points and P_1, P_2 the principal points. Let *AB* be an object placed at right angles to the axis. In order to find the image of the point *A,* make the following construction :

(i) Draw a ray AH_1 parallel to the axis and meeting the first principal plane at H_1. The conjugate ray will proceed from the point H_2 on the second principal plane such that $H_2P_2 = H_1P_1$ and will pass through the second focal point F_2.

(ii) Draw another ray AF_1K_1 passing through the first focal point F_1 and meeting the first principal plane at K_1. This ray, after refraction through the lens system, will emerge parallel to the axis through K_2 such that $K_2P_2 = K_1P_1$.

The point of intersection A_1 of the two emergent rays is the image of *A*. Draw A_1B_1 perpendicular to the axis. A_1B_1 is the image of the object *AB*.

Let f_1 = distance between F_1 and P_1 ; f_2 = distance between F_2 and P_2
x_1 = distance between *B* and F_1 ; x_2 = distance between B_1 and F_2.

$\Delta s\ P_1F_1K_1$ and BF_1A are similar.

$$\therefore \quad \frac{K_1P_1}{AB} = \frac{P_1F_1}{BF_1} \qquad \text{But } K_1P_1 = A_1B_1$$

$$\therefore \quad \frac{A_1B_1}{AB} = \frac{f_1}{x_1} \qquad ...(1)$$

$\Delta s\ A_1B_1F_2$ and $H_2P_2F_2$ are similar.

$$\therefore \quad \frac{A_1B_1}{H_2P_2} = \frac{B_1F_2}{P_2F_2} \qquad \text{But } H_2P_2 = AB$$

$$\therefore \quad \frac{A_1B_1}{AB} = \frac{x_2}{f_2} \qquad ...(2)$$

Comparing Eqs. (1) and (2),

$$\frac{f_1}{x_1} = \frac{x_2}{f_2}$$

$$\therefore \quad x_1x_2 = f_1f_2 \qquad ...(3)$$

This is Newton's formula for a co-axial lens system.

DISPERSION OF LIGHT

1.7. Dispersion Through a Prism

A beam of white light, when it passes through a prism is split up into the constituent colours. This phenomenon is called *dispersion*. The coloured band obtained on the screen is called *spectrum*. It is found that *violet* is deviated more than the *red.* Hence the refractive index of glass is greater for the violet than for the red, *i.e.*, $n_v > n_r$. Thus, as the wavelength increases, the refractive index decreases.

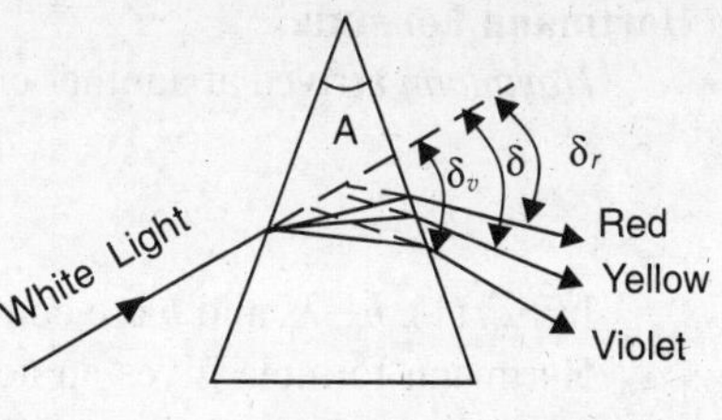

Fig. 1.9.

Angular dispersion. Let δ_v, δ_r and δ be the deviations of violet, red and yellow rays and n_v, n_r and n, the refractive indices of the material of a thin prism for them respectively (Fig. 1.9). The angle of the prism A is small. Then,

$$\delta_v = (n_v - 1)A$$

$$\delta_r = (n_r - 1)A$$

$$\delta = (n - 1)A$$

The difference in deviation for any two colours is called the *angular dispersion* between the two colours. The angular dispersion between violet and red rays is given by,

$$\text{angular dispersion} = \delta_v - \delta_r = (n_v - n_r)A.$$

Dispersive power (ω). *It is defined as the ratio of angular dispersion to the deviation of mean ray.*

∴ The dispersive power of the material of a prism between red and violet rays is given by,

$$\omega = \frac{\delta_v - \delta_r}{\delta} = \frac{(n_v - n_r)A}{(n-1)A} = \frac{n_v - n_r}{n-1} = \frac{dn}{n-1}.$$

Example 1 : *A hollow 60º prism is filled with carbon disulfide whose index of refraction for blue light is 1.652, for red light 1.618. What is the angular dispersion?*

Solution : Given: $A = 60°$, $n_b = 1.652$, $n_r = 1.618$.

∴ Angular dispersion $= (n_b - n_r)A = (1.652–1.618)\,60° = 2.04°$

1.8. Cauchy's Formula

Cauchy showed that the relation between n and λ can be represented by the equation

$$n = A + \frac{B}{\lambda^2} \qquad ...(1)$$

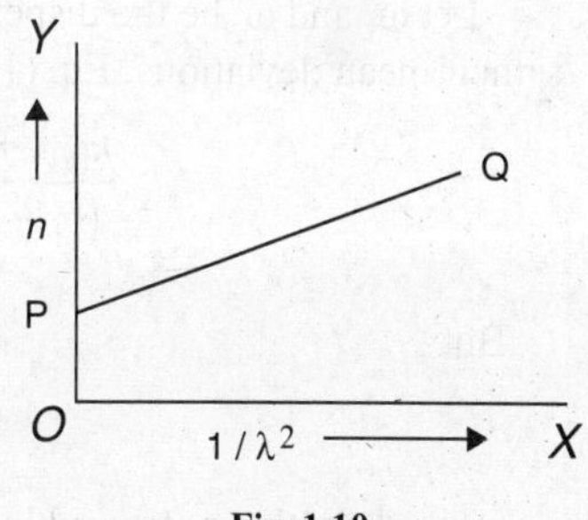

Fig. 1.10.

A and B are called *Cauchy's constants*. The values of A and B depend on the medium. Eq. (1) shows that the refractive index of the medium decreases with increase in wavelength of light.

The constants A and B can be determined by measuring n for two values of λ. If a graph is plotted between n and $1/\lambda^2$, it is a straight line (Fig. 1.10). The intercept OP on the Y-axis gives the value of A. The slope of the line PQ gives the value of B.

Differentiating Eq. (1), w.r.t. λ, $\frac{dn}{d\lambda} = -\frac{2B}{\lambda^3}$ ...(2)

$dn/d\lambda$ is the dispersive power of the medium.

Hence, dispersive power is inversely proportional to the cube of the wavelength.

Example 1: *Calculate the values of Cauchy's constants for crown glass given that*

$$n_C = 1.514,\ n_F = 1.524,\ \lambda_C = 656.3\ nm,\ \lambda_F = 486.2\ nm.$$

Solution : The Cauchy relation is $n = A + \frac{B}{\lambda^2}$.

$$1\cdot514 = A + \frac{B}{\left(656\cdot3 \times 10^{-9}\right)^2} \quad ...(1) \qquad \text{and} \qquad 1.524 = A + \frac{B}{(486.2 \times 10^{-9})^2} \quad (2)$$

Solving Eqs. (1) and (2), we get $A = 1.502$, $B = 5.236 \times 10^{-15}\ \text{m}^2$

Hartmann Formula

Hartmann arrived at another empirical expression for dispersion:

$$n = n_0 + \frac{b}{\lambda - \lambda_0} \qquad \text{...(1)}$$

In Eq. (1), n_0, λ_0 and b are constants to be determined from the observations.

Hartmann formula gives an accurate variation of n with λ, if the range of wavelength is not too large.

1.9. Achromatism in Prisms

Deviation without dispersion. When a beam of white light is passed through a prism, then both deviation and dispersion are produced. A prism that produces deviation without dispersion is called an *achromatic prism*.

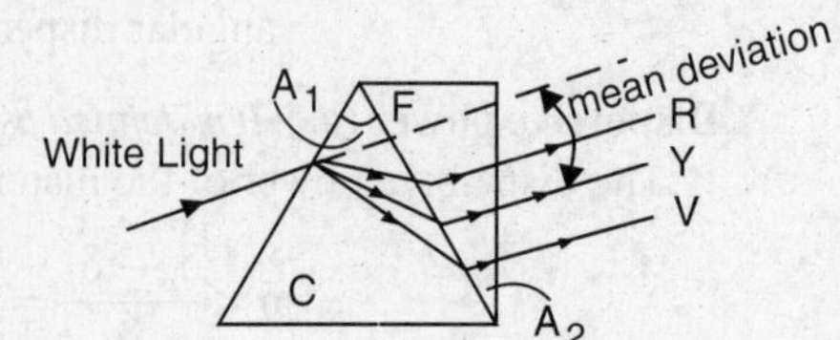

Fig. 1.11.

Consider two prisms, one of crown glass and other of flint glass with angles A_1 and A_2 respectively (Fig. 1.11). Let n_{1v}, n_{1r}, n_1 and n_{2v}, n_{2r}, n_2 be the refractive indices of the two materials for violet, red and the mean yellow rays of light.

Then, angular dispersion produced by crown prism = $(n_{1v} - n_{1r})\, A_1$

Angular dispersion produced by flint prism = $(n_{2v} - n_{2r})\, A_2$.

The condition for deviation without dispersion is

$$(n_{1v} - n_{1r})\, A_1 + (n_{2v} - n_{2r})\, A_2 = 0 \text{ ...(1)} \quad \text{or} \quad \frac{A_2}{A_1} = \frac{(n_{1v} - n_{1r})}{-(n_{2v} - n_2 r)} \qquad \text{...(2)}$$

The *negative* sign shows that the refracting angles of the two prisms are in opposite directions. The rays of different colours emerging from such a combination, are found parallel to each other and hence combine to form white light.

Let ω_1 and ω_2 be the dispersive powers of crown and flint glass prisms respectively and δ_1 and δ_2 their mean deviations. Eq. (1) can be written as

$$\frac{n_{1v} - n_{1r}}{(n_1 - 1)}(n_1 - 1)\, A_1 + \frac{(n_{2v} - n_{2r})}{(n_2 - 1)}(n_2 - 1)\, A_2 = 0$$

But
$$\frac{(n_{1v} - n_{1r})}{(n_1 - 1)} = \omega_1 \text{ and } \frac{n_{2v} - n_{2r}}{n_2 - 1} = \omega_2$$

$$\therefore \qquad \omega_1\delta_1 + \omega_2\delta_2 = 0 \qquad \text{...(3)}$$

The deviation produced by the combination for the mean ray is

$$\delta = \delta_1 + \delta_2 = (n_1 - 1)\, A_1 + (n_2 - 1)\, A_2$$

Substituting for A_2 from Eq. (2),

$$\delta = (n_1 - 1)\, A_1 + (n_2 - 1)\left[-\frac{(n_{1v} - n_{1r})\, A_1}{(n_{2v} - n_{2r})}\right]$$

But
$$(n_1 - 1) = \frac{n_{1v} - n_{1r}}{\omega_1} \text{ and } \frac{n_2 - 1}{n_{2v} - n_{2r}} = \frac{1}{\omega_2}$$

$$\therefore \qquad \delta = (n_{1v} - n_{1r})\, A_1 \left(\frac{1}{\omega_1} - \frac{1}{\omega_2}\right) \qquad \text{...(4)}$$

Since $n_{1v} > n_{1r}$ and $\omega_2 > \omega_1$, δ has a *positive value.*

Example 1 : *Find the angle of a crown glass prism to be combined with a flint glass prism of refracting angle 5° so that the resultant dispersion between C and F lines of the spectrum may be zero. Find also the deviation for the mean ray given that*

	Crown glass	*Flint glass*
n_f	*1.5233*	*1.6385*
n_c	*1.5146*	*1.6224*

Solution : Here, $A_2 = 5°$, $n_{1f} = 1.5233$, $n_{1c} = 1.5146$,

$n_{2f} = 1.6385$. $n_{2c} = 1.6224$; $A_1 = ?$

We have,
$$\frac{A_2}{A_1} = -\frac{(n_{1f} - n_{1c})}{(n_{2f} - n_{2c})}$$

or
$$\frac{5}{A_1} = -\frac{(1.5233 - 1.5146)}{(1.6385 - 1.6224)}$$

or $A_1 = -9.253°$

Now
$$n_1 = \frac{1.5233 + 1.5146}{2} = 1.519$$

$$n_2 = \frac{1.6385 + 1.6224}{2} = 1.630$$

Deviation for the mean ray $= \delta = (n_1 - 1)A_1 + (n_2 - 1)A_2$

$$\delta = (1.519 - 1)(-9.253) + (1.630 - 1)\,5 = -1.652°$$

1.10. Dispersion without Deviation

When white light passes through a prism, it suffers deviation and dispersion. Two prisms of different material may be combined so that the resultant deviation is zero. But there will be still dispersion. Let A_1 and A_2 be the angles of two prisms (Fig. 1.12). Let n_{1v}, n_{1r}, n_1 and n_{2v}, n_{2r}, n_2 be the refractive indices of the two materials for the violet, red and mean rays respectively.

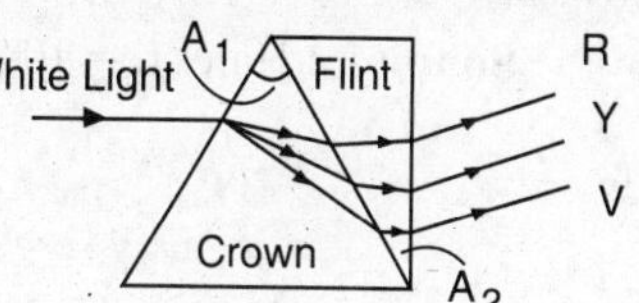

Fig. 1.12.

The deviation produced by the two prisms for the mean ray is,

$$(n_1 - 1)A_1 + (n_2 - 1)A_2 = 0$$

$$\therefore \quad \frac{A_2}{A_1} = -\left[\frac{n_1 - 1}{n_2 - 1}\right]$$

The *negative* sign shows that the refracting angles of the two prisms are in opposite directions. The total dispersion produced by the two prisms is

$$\theta = (n_{1v} - n_{1r})A_1 + (n_{2v} - n_{2r})A_2$$

But
$$A_2 = -\left(\frac{n_1 - 1}{n_2 - 1}\right)A_1$$

$$\therefore \quad \theta = (n_{1v} - n_{1r})A_1 - \frac{(n_{2v} - n_{2r})(n_1 - 1)}{(n_2 - 1)}A_1$$

$$= (n_1 - 1)A_1\left[\frac{n_{1v} - n_{1r}}{(n_1 - 1)} - \frac{(n_{2v} - n_{2r})}{(n_2 - 1)}\right]$$

$$= (n_1 - 1)A_1(\omega_1 - \omega_2)$$

Here, ω_1 and ω_2 are dispersive powers of the materials of the two prisms.

Since $\omega_1 \neq \omega_2$, there will be a resultant dispersion. The final dispersed beam is parallel to the incident beam. This principle is applied in the direct vision spectroscope.

Example 1: *Calculate (a) the refracting angle of a flint glass prism which should be combined with a crown glass prism of refracting angle 9º so that the combination may have no deviation for the D line and (b) the angular separation between the C and F lines, given that the refractive indices of the materials are as follows :*

	C	*D*	*F*
Flint	*1.790*	*1.795*	*1.805*
Crown	*1.527*	*1.530*	*1.535*

Solution : Let A_2 and A_1 be the angles of flint and crown glass prisms.

Here, $A_1 = 9^o$, $n_2 = 1.795$, $n_1 = 1.530$. $A_2 = ?$

We have,
$$\frac{A_2}{A_1} = -\left(\frac{n_1 - 1}{n_2 - 1}\right)$$

$$\frac{A_2}{9} = -\left[\frac{1.530 - 1}{1.795 - 1}\right] \text{ or } A_2 = -6^o$$

The total dispersion produced by the two prisms is,

$$\theta = (n_{1F} - n_{1C})\, A_1 + (n_{2F} - n_{2C})\, A_2$$
$$= (1.535 - 1.527)\, 9 + (1.805 - 1.790)\, (-6)$$
$$= -0.072^o + 0.090^o = -0.018^o$$

The negative sign indicates that the resultant dispersion is in the direction of the deviation produced by the flint glass prism.

Example 2 : *A crown glass prism of angle 10^0 and dispersive power 0.01575 is to be combined with a flint glass prism of dispersive power 0.02764. If the refractive indices of the prisms for sodium D light are respectively 1.511 and 1.621, calculate the angle of the second prism. Also calculate the net dispersion produced by the combination.*

Solution : Here $A_1 = 10^o$, $\omega_1 = 0.01575$, $\omega_2 = 0.02764$, $n_1 = 1.511$, $n_2 = 1.621$, $A_2 = ?$, $\theta = ?$

$$\frac{A_2}{A_1} = -\left[\frac{n_1 - 1}{n_2 - 1}\right] \text{ or } \frac{A_2}{10} = \left[\frac{1.511 - 1}{1.621 - 1}\right]$$

$$\therefore \qquad A_2 = -8.229^o$$

$$\theta = (n_1 - 1)\, A_1\, (\omega_1 - \omega_2) = (1.511 - 1)\, 10\, (0.01575 - 0.02764)$$
$$= -0.0608^o.$$

1.11. Direct Vision Spectroscope

It consists of three crown glass prisms and two flint glass prisms arranged alternately. They are arranged with their refracting angles in opposite directions. The refracting angles of the prisms are such that the total deviation produced for the mean rays is zero. The faces of the prisms are cemented together by canada balsam to reduce loss of intensity of light by reflection at the interfaces of the prisms. The combination is then placed in a metal tube. The tube has a slit and a collimating lens at one end and a telescope at the other end [Fig. 1.13].

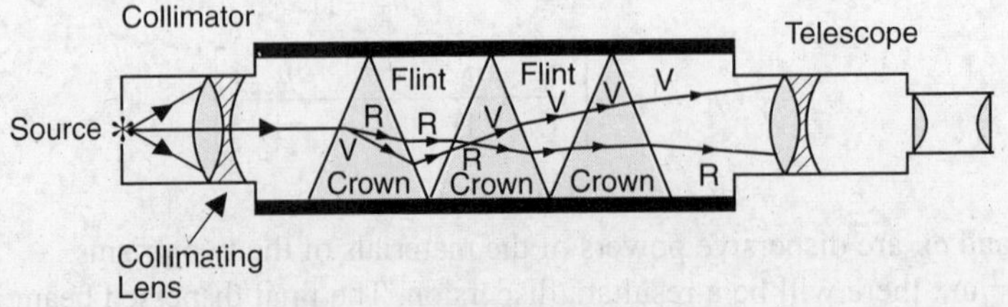

Fig. 1.13.

A narrow beam of white light passing through the slit is rendered parallel by the collimating lens. It then traverses the combination of the prisms, where it is dispersed. The dispersed beam is almost parallel to the incident beam. The paths of the red and violet rays are shown in the figure. The tube is turned directly towards a source of light and its spectrum is observed through the telescope. It cannot be used to make any measurements on the spectra. But it can be used to make a survey on the characteristic features of the spectra.

1.12. Constant Deviation Prism

A constant deviation prism is shown in Fig. 1.14. This single prism *ABCD* is equivalent to two 30° refracting prisms combined with a 45° total reflection prism. The light beam (*EFGHK*) undergoes two refractions which compensate each other. So the deviation of the beam is the 90° change in direction produced by the total reflection. Since there are two refractions, the ray is dispersed and the final image is coloured.

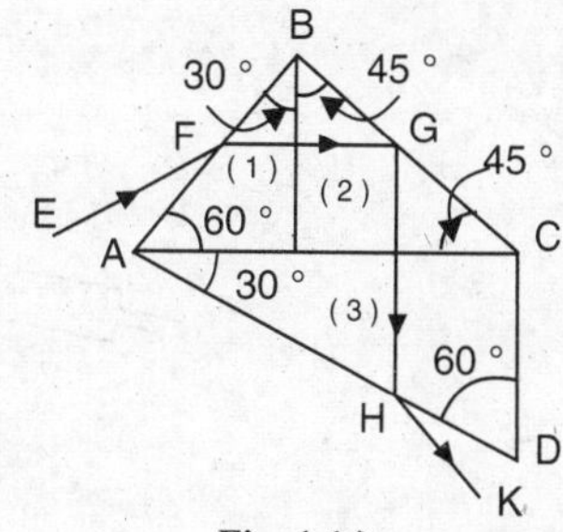

Fig. 1.14.

1.13. Constant Deviation Spectroscope

A spectroscope is an instrument used to produce a spectrum. In a spectroscope, the spectra are observed visually. In a spectrograph, the spectra are photographed. In constant deviation spectroscope, the telescope (*T*) and the collimator (*C*) are fixed at right angles to one another [Fig. 1.15]. A constant deviation prism (*P*) is mounted on the prism table. The prism table can be rotated by means of a micrometer screw carrying a drum (*D*) at one end. The drum is calibrated in terms of wavelengths. When the table is rotated, the lines in the spectrum of source of light appear one by one on the intersection of the cross wires of the telescope. The wavelength of any one line can then be read off from the drum. This instrument is called the Hilger wavelength prism spectrometer. For taking photograph of the lines, the eyepiece in the telescope is replaced by a camera.

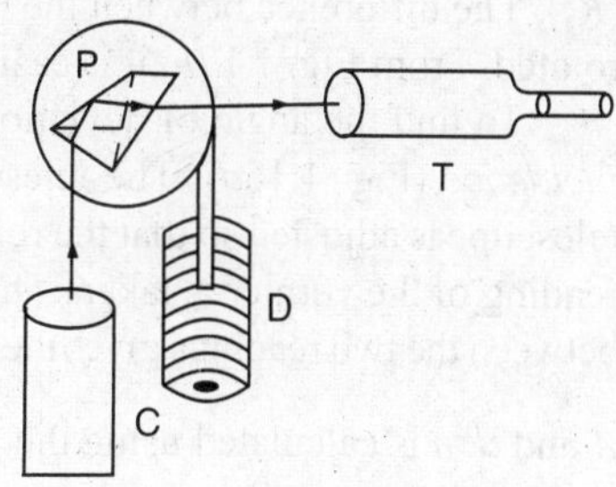

Fig. 1.15.

In order to set the instrument for work, the slit is illuminated with light of known wavelength. The drum is set so that the correct wavelength is indicated on the drum. The prism is placed on the prism table and its position is adjusted till the line is on the intersection of the crosswires. The prism is clamped tight to the table in this position. Any other wavelength can be determined by rotating the drum till the spectral line is on the crosswire and taking the reading of the drum.

1.14. Determination of Refractive Index of the Material of Small Angled Prism

Principle : When a ray is incident normally on one face of the prism, $i_1 = 0$. $\therefore$ $r_1 = 0$ (Fig. 1.16).

Deviation $d = i_1 + i_2 - A = i_2 - A$

$i_2 = A + d,\ r_1 + r_2 = A$

$\therefore \quad r_2 = A$

$$n = \frac{\sin i_2}{\sin r_2} = \frac{\sin (A + d)}{\sin A}$$

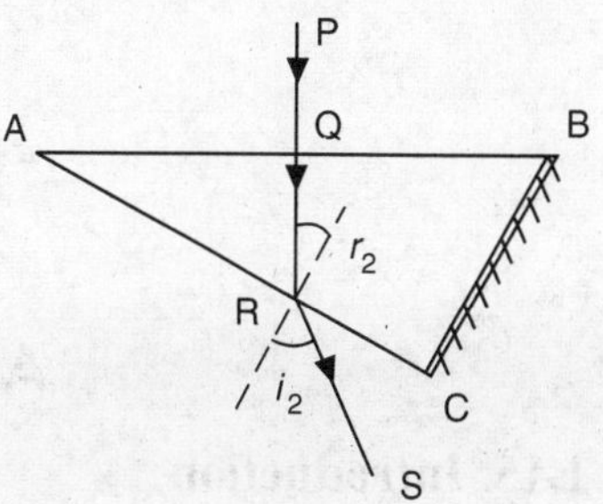

Fig. 1.16.

Here *d* is the angle of deviation at normal incidence. *A* is angle of the prism.

Procedure : The preliminary adjustments of the spectrometer are made. The slit of the spectrometer is illuminated by sodium vapour lamp. The telescope is brought in a line with the collimator and the direct reading is taken. The telescope is rotated through exactly 90° and clamped The small angled prism is mounted centrally on the prism table. The prism table is then carefully rotated so that the image reflected from one of the faces (say *AB*) coincides with the vertical cross-wire (Fig. 1.17*a*). The reading is taken (R_1). Then the prism table is rotated until the image reflected

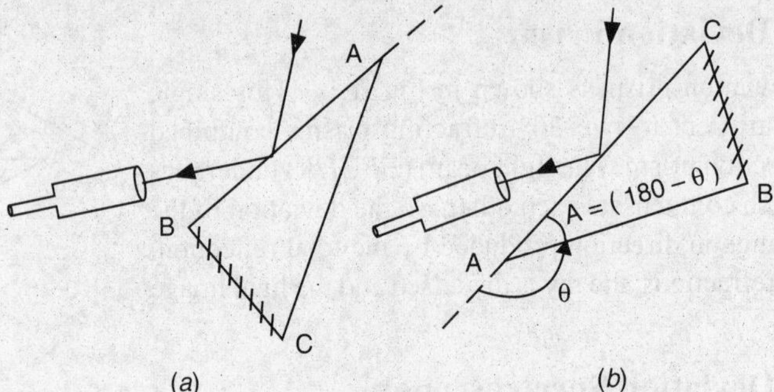

Fig. 1.17.

from the other face *AC* coincides with the vertical cross-wire (Fig. 1.17*b*). Now the reading is taken (R_2). The difference between the two readings gives θ, the angle through which the prism has been rotated. From Fig. 1.17*b*, it is easily seen that the angle of the prism $A = (180 - \theta)$.

To find the angle of deviation corresponding to normal incidence, *the prism is set for normal incidence* (Fig. 1.18*a*). The telescope is adjusted to view the refracted image (Fig. 1.18 *b*). The telescope is adjusted so that the refracted image of the slit coincides with the vertical cross-wire. The reading of the vernier is taken. The prism is removed and the direct reading is taken. The difference between the two readings gives the angle of deviation (*d*) corresponding to normal incidence. Knowing *A* and *d*, *n* is calculated using the formula $n = \dfrac{\sin(A+d)}{\sin A}$.

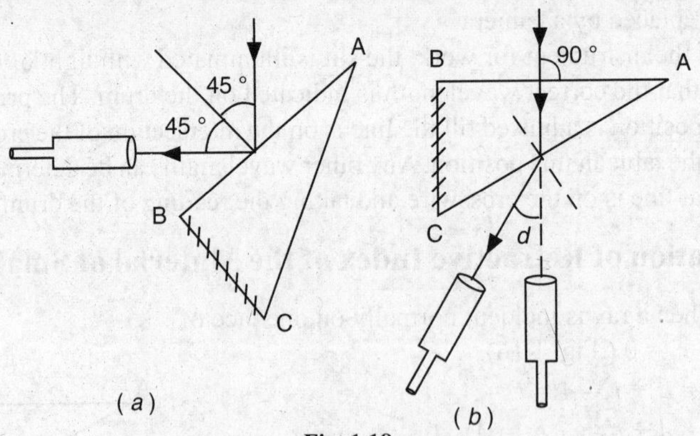

Fig. 1.18.

ABERRATIONS IN LENSES

1.15. Introduction

The deviations in the size, shape, position and colour in the actual images produced by a lens in comparison to the object are called aberrations produced by a lens. Chromatic aberrations are distortions of the image due to the dispersion of light in the lenses of an optical system when white light is used. *The defect of coloured image formed by a lens with white light is called chromatic aberration.* If monochromatic light is used, then such defects are automatically removed. Besides these defects, there are defects which are present even when monochromatic light is used. Such defects are called *monochromatic aberrations*. These aberrations are the result of (i) the large aperture of the optical system, (ii) the large angle subtended by the rays with the principal axis and (iii) the large size of the object. As a result of these aberrations, (i) a point is not imaged as a point, (ii) a plane is not imaged as a plane and (iii) equidistant points are not imaged as equidistant points. Following are the monochromatic aberrations: (i) Spherical aberration, (ii) Astigmatism, (iii) Coma, (iv) Curvature of field and (v) Distortion.

1.16. Spherical Aberration in a Lens

This aberration is due to large aperture of the lenses. The lens of large aperture may be thought to be made up of zones. The marginal and paraxial rays form the images at different places. Fig. 1.19 shows that a monochromatic point source S on the axis is imaged as S_p and S_m. Here, S_m and S_p are the images formed by marginal and paraxial rays respectively. Thus the point object is not imaged as a point. Similarly the focus of marginal and paraxial rays do not coincide. The distance $S_m S_p$ on the axis measures longitudinal spherical aberration.

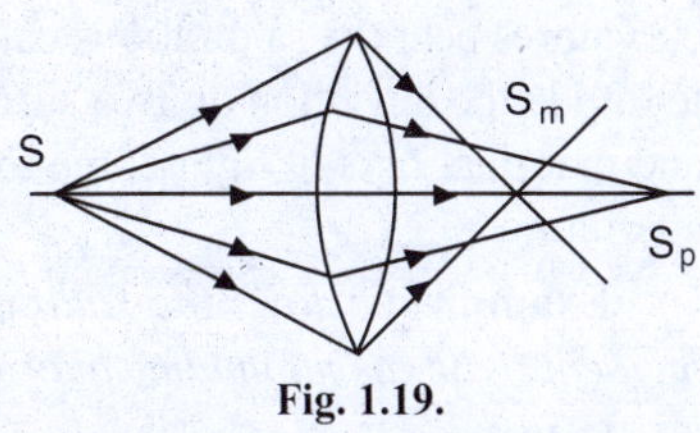

Fig. 1.19.

The failure of a lens to form a point image of a point object on the axis is called spherical aberration.

For rays parallel to principal axis, the distance between the foci of marginal and paraxial rays gives the extent of longitudinal spherical aberration. In Fig. 1.20*a*, F_p and F_m are the foci for the paraxial and the marginal rays respectively. Spherical aberration of a convergent lens is taken to be positive as the distance $(f_p - f_m)$, measured along the axis. The spherical aberration of a diverging lens is negative (Fig. 1.20*b*).

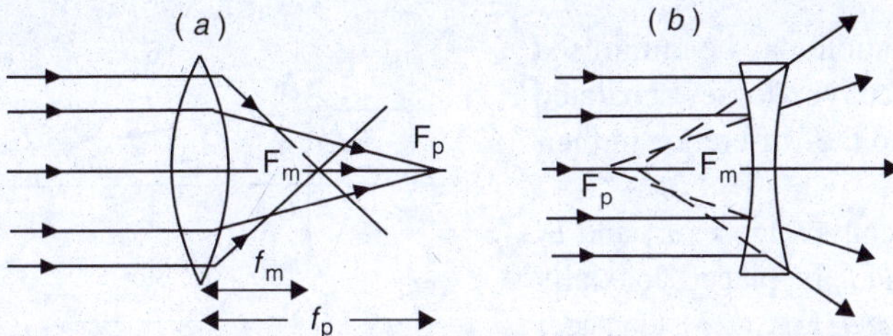

Fig. 1.20.

1.17. Methods of Minimising Spherical Aberration

The following methods are used to reduce spherical aberration.

(i) By using stops : By using stops, we can reduce the lens aperture. We can use either paraxial or marginal rays [Fig. 1.21]. Here, circular discs, called the stops, are used to cut off the unwanted rays. The stop in Fig. 1.21 *(a)* is a disc with a circular hole. It eliminates marginal rays. Fig. 1.21 *(b)* shows a stop to eliminate paraxial rays. But the use of stops reduces the intensity of the image and the resolving power of the instrument.

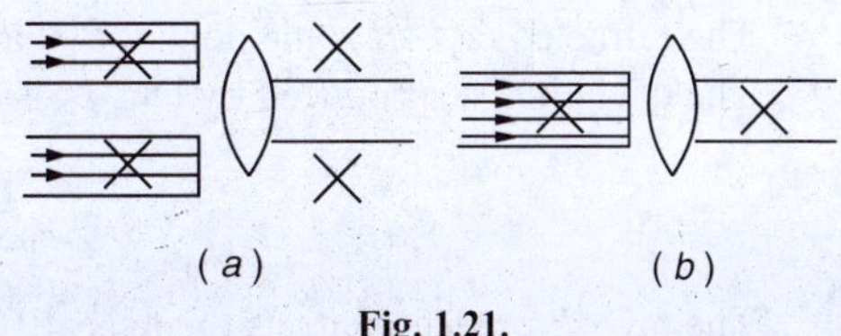

Fig. 1.21.

(ii) By using the two lenses separated by a distance. When two convex lenses separated by a finite distance are used the spherical aberration is minimum when the distance between the lenses is equal to the difference in their focal lengths. In this arrangement, the total deviation is equally shared by the two lenses. Hence the spherical aberration is minimum.

(iii) By using an aplanatic lens

(iv) By using a crossed lens. The radii of curvature R_1, R_2 of a thin lens satisfy the following relation:

$$\frac{1}{f} = (n-1)\left(\frac{1}{R_1} - \frac{1}{R_2}\right)$$

It, therefore, shows that spherical aberration depends upon *(i)* the refractive index of the lens medium (n) and *(ii)* the shape factor β, which is determined by the ratio $\beta = R_1/R_2$. If the refractive index of material of the lens is 1.5, the spherical aberration will be minimum when $\beta = R_1/R_2 = -1/6$. A convex lens whose radii of curvatures bear the said ratio is called as a *crossed lens*. It is essential to divide the deviation on two surfaces equally (Fig. 1.22). The axial and marginal rays of light come to focus with minimum of spherical aberration.

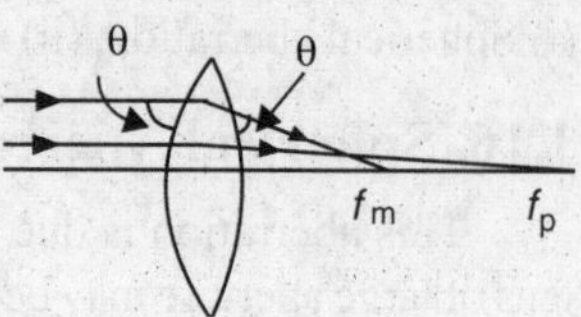

Fig. 1.22.

Example 1 : *Find the radii of curvature for a lens of f = 10 cm, n = 1.5 for which parallel incident light has minimum spherical aberration.*

Solution : Here, $f = 10$ cm; $n = 1.5$.

For minimum spherical aberration, $R_1/R_2 = -1/6$. $\therefore R_2 = -6R_1$.

Now, $$\frac{1}{f} = (n-1)\left(\frac{1}{R_1} - \frac{1}{R_2}\right) \text{ or } \frac{1}{10} = (15-1)\left(\frac{1}{R_1} + \frac{1}{6R_1}\right)$$

$\therefore \quad R_1 = 5.833$ cm; $R_2 = -6R_1 = -35$ cm.

1.18. Condition for Minimum Spherical Aberration of Two Thin Lenses-Separated by a Distance

Spherical aberration may be minimised by using two plano-convex lenses separated by a distance equal to the difference in their focal lengths.

Let two plano-convex lenses L_1 and L_2 of focal lengths f_1 and f_2 be placed coaxially separated by a distance a (Fig. 1.23). Consider a ray OA, parallel to principal axis, incident on lens L_1 at height h_1 above the principal axis.

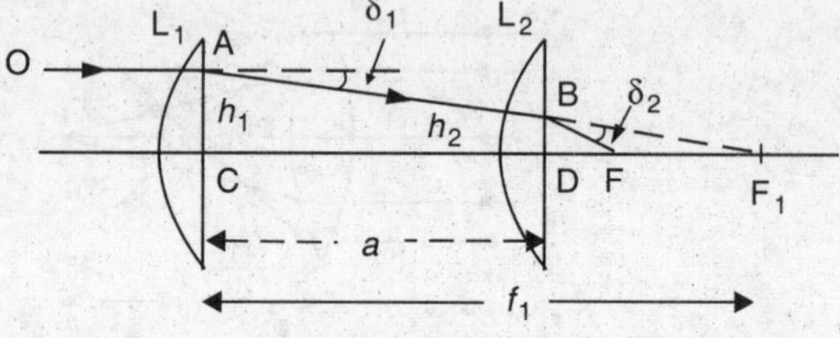

Fig. 1.23.

The deviation δ_1 produced by the lens L_1 is given by

$$\delta_1 = \frac{h_1}{f_1} \quad ...(1)$$

The refracted ray AB is incident at B at a height h_2 from the axis on lens L_2.

The deviation, δ_2 produced by lens L_2 is given by

$$\delta_2 = \frac{h_2}{f_2} \quad ...(2)$$

The ray AB produced meets the axis at F_1 which is the principal focus of lens L_1.

Hence $CF_1 = f_1$.

For minimum spherical aberration, the deviation produced by both the lenses should be equal, i.e., $\delta_1 = \delta_2$

or
$$\frac{h_1}{f_1} = \frac{h_2}{f_2}$$

or
$$\frac{h_1}{h_2} = \frac{f_1}{f_2}. \qquad ...(3)$$

From similar triangles ACF_1 and BDF_1, we get
$$\frac{AC}{BD} = \frac{CF_1}{DF_1} = \frac{CF_1}{CF_1 - CD}$$

or
$$\frac{h_1}{h_2} = \frac{f_1}{f_1 - a} \qquad ...(4)$$

Comparing Eqs. (3) and (4), we get
$$\frac{f_1}{f_2} = \frac{f_1}{f_1 - a} \text{ or } f_2 = f_1 - a$$

$\therefore$
$$a = f_1 - f_2$$

This is the condition for minimum spherical aberration for two lenses separated by a distance.

1.19. Aplanatic Lens

A spherical lens which is free from the defects of spherical aberration and coma is called an *aplanatic lens*. The pair of conjugate points in the lens system free from spherical aberration and coma are called *aplanatic points*.

Aplanatic points are two conjugate points on the axis with respect to a spherical surface such that an object placed at one of these points produces at the second point an image which is free from spherical aberration and coma.

Aplanatic points of a spherical refracting surface : Let C be the centre of curvature of a spherical surface of radius of curvature R and refractive index n [Fig. 1.24]. Consider a point object O placed on the axis at a distance R/n from C. A ray of light OP making an angle θ_1 with the axis after refraction will bend away from the normal along PB. When produced back, it will meet the axis at I making an angle θ_2. The angles of incidence and refraction are i ($\angle OPC$) and r ($\angle NPB$) respectively.

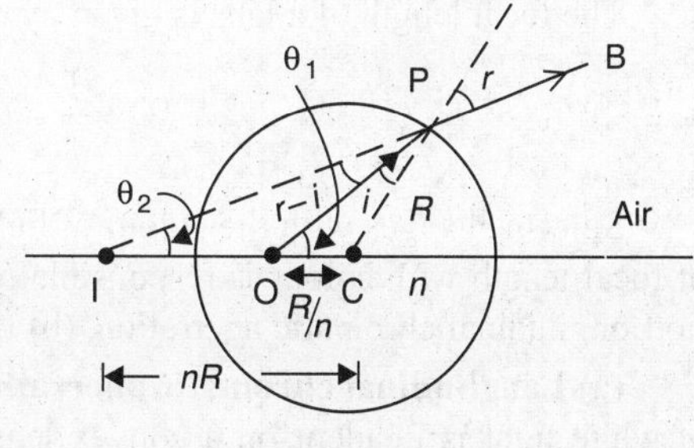

Fig. 1.24.

Then $\sin i / \sin r = 1 / n$

or
$$n \sin i = \sin r \qquad ...(1)$$

$$\Delta\, OPC, \frac{PC}{\sin \theta_1} = \frac{CO}{\sin i} \text{ or } \frac{R}{\sin \theta_1} = \frac{R/n}{\sin i}$$

or
$$n \sin i = \sin \theta_1 \qquad ...(2)$$

From Eqns. (1) and (2), $\sin \theta_1 = \sin r$

or
$$\theta_1 = r \qquad ...(3)$$

In $\Delta\, OPI$, $\theta_1 = \theta_2 + (r - i)$

or
$$\theta_2 = i \qquad ...(4)$$

From similar triangles OCP and PCI, $\dfrac{CI}{CP} = \dfrac{CP}{CO}$ or $CI = \dfrac{CP^2}{CO} = \dfrac{R^2}{R/n}$

$$\therefore \qquad CI = Rn \qquad ...(5)$$

This relation does not contain θ_1 and θ_2. This shows that if the object is placed at a distance R/n from C, then the image is formed at a distance nR irrespective of the slopes θ_1 and θ_2. Hence all the rays starting from O will appear to diverge from I. Therefore, the image formed is free from spherical aberration. O and I are aplanatic points of the spherical lens.

Use of aplanatic lens in high power oil immersion microscope objective. The aplanatic lens is used in high power microscopes to minimise spherical aberration and coma. A typical microscope objective is shown in Fig. 1.25. L_1 is a hemispherical lens of radius R with its plane face directed towards the object O. The object is immersed in cedar wood oil. The refractive index of cedarwood oil and the material of the lens L_1 is the same. The object O is placed in the oil at a distance R/n from the centre of the hemispherical lens. The image is formed at I_1 at a distance of nR from C. Now, I_1 is at the centre of the first surface of the meniscus lens L_2. Therefore, the rays enter the meniscus lens without any deviation. Further, the radius of curvature of the second surface of L_2 is such that I_1 is its first aplanatic point. Thus the rays after refraction through the second surface appear to come from the conjugate aplanatic point I_2. The virtual image I_2 is free from spherical aberration. In this way, a wide beam starting from O is changed into a small angled pencil diverging from I_2. The narrow beam of light from I_2 enters the achromatic combination of lenses which acts as objective of microscope.

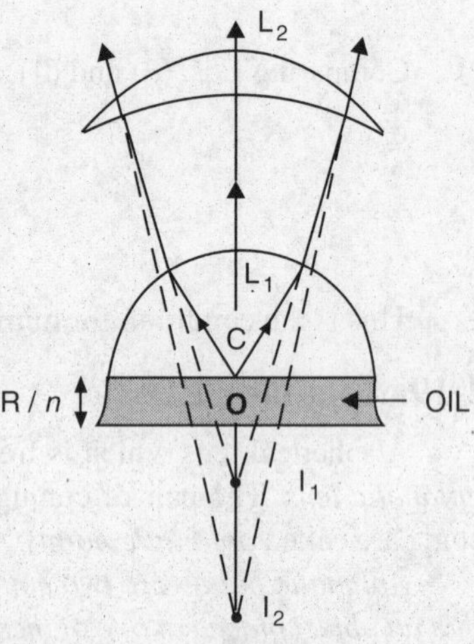

Fig. 1.25.

1.20. Chromatic Aberration in a Lens

The focal length of a lens is given by

$$\frac{1}{f} = (n-1)\left(\frac{1}{R_1} - \frac{1}{R_2}\right)$$

Since n changes with the colour of light, f must be different for different colours. This change of focal length with colour is responsible for chromatic aberration. It is classified into two types : (a) Longitudinal chromatic aberration, (b) Lateral chromatic aberration.

(a) Longitudinal chromatic aberration : A beam of white light is incident on a convex lens parallel to the principal axis [Fig. 1.26]. The dispersion of colours takes place due to prismatic action of the lens. Violet is deviated most and red the least. Red rays are brought to focus at a point farther than the violet rays. Evidently $f_r > f_v$. The difference $f_r - f_v$ is a measure of the axial chromatic aberration of a lens for parallel rays.

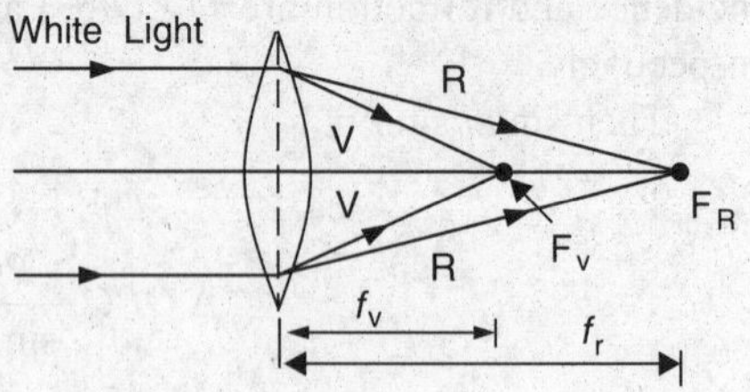

Fig. 1.26.

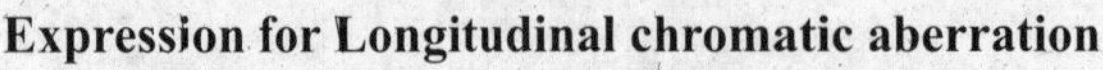

Expression for Longitudinal chromatic aberration

The focal length of a lens is given by

$$\frac{1}{f} = (n-1)\left(\frac{1}{R_1} - \frac{1}{R_2}\right)$$

Let f_v, f_r and f_y be the focal lengths of the lens for violet, red and yellow colours respectively. Also let n_v, n_r and n_y be the respective refractive indices. Then,

$$\frac{1}{f_v} = (n_v - 1)\left(\frac{1}{R_1} - \frac{1}{R_2}\right) \quad ...(1)$$

$$\frac{1}{f_r} = (n_r - 1)\left(\frac{1}{R_1} - \frac{1}{R_2}\right) \quad ...(2)$$

$$\frac{1}{f_y} = (n_y - 1)\left(\frac{1}{R_1} - \frac{1}{R_2}\right) \quad ...(3)$$

Subtracting Eq. (2) from Eq. (1),

$$\frac{1}{f_v} - \frac{1}{f_r} = (n_v - n_r)\left(\frac{1}{R_1} - \frac{1}{R_2}\right)$$

or

$$\frac{f_r - f_v}{f_v\, f_r} = \frac{n_v - n_r}{n_y - 1}(n_y - 1)\left(\frac{1}{R_1} - \frac{1}{R_2}\right)$$

Now, $\omega = (n_v - n_r) / (n_y - 1)$ = dispersive power of the material of the lens; $f_v f_r = f_y^2$.

$$\therefore \qquad f_r - f_v = \omega f_y \quad ...(4)$$

(b) Lateral chromatic aberration : Fig. 1.27 shows a convex lens and an object *AB* placed in front of the lens. The lens forms the image of white object *AB* as B_vA_v and B_rA_r in violet and red colours respectively. The images of other colours lie in between the two. Evidently, the size of red image is greater than the size of violet image ($B_rA_r > B_vA_v$). The difference ($B_rA_r - B_vA_v$) is a measure of lateral or transverse chromatic aberration.

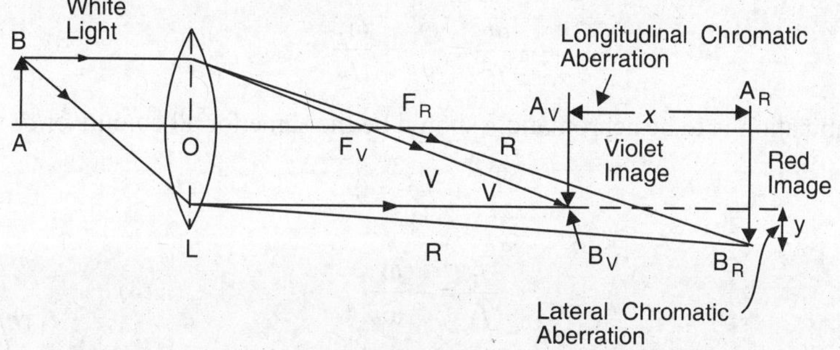

Fig. 1.27.

Chromatic aberration is eliminated by :

(i) keeping two lenses in contact with each other and

(ii) keeping two lenses out of contact.

Achromatic Combination of Lenses

When two or more lenses are combined together in such a way that the combination is free from chromatic aberration, then such a combination is called achromatic combination of lenses.

The minimisation or removal of chromatic aberration is called *achromatisation*. Chromatic aberration cannot be removed completely. Usually achromatism is achieved for two prominent colours.

1.21. Condition for Achromatism of Two Thin Lenses Placed in Contact

The focal length f of a thin lens is given by

$$\frac{1}{f} = (n-1)\left(\frac{1}{R_1} - \frac{1}{R_2}\right) \qquad ...(1)$$

Here, f is the focal length, n the refractive index, R_1 and R_2 the radii of curvature of the two surfaces of the lens. Now we know that f varies with n. Therefore, differentiating Eq. (1),

$$-\frac{df}{f^2} = dn\left(\frac{1}{R_1} - \frac{1}{R_2}\right)$$

$$= \frac{dn}{(n-1)}(n-1)\left(\frac{1}{R_1} - \frac{1}{R_2}\right) = \frac{dn}{(n-1)}\frac{1}{f}$$

or
$$-\frac{df}{f} = \frac{dn}{(n-1)} = \omega \qquad ...(2)$$

Here ω is the dispersive power of the material of the lens.

Let f_1 and f_2 be the focal lengths of the two lenses in contact and ω_1 and ω_2 their dispersive powers. If f is the focal length of the combination, then

$$\frac{1}{f} = \frac{1}{f_1} + \frac{1}{f_2} \qquad ...(3)$$

Differentiating Eq. (3),

$$-\frac{df}{f^2} = -\frac{df_1}{f_1^2} - \frac{df_2}{f_2^2} = -\frac{df_1}{f_1}\frac{1}{f_1} - \frac{df_2}{f_2}\frac{1}{f_2}$$

But
$$-\frac{df}{f_1} = \omega_1 \text{ and } -\frac{df_2}{f_2} = \omega_2.$$

$\therefore$
$$-\frac{df}{f^2} = \frac{\omega_1}{f_1} + \frac{\omega_2}{f_2}$$

If the combination is to be achromatic, f should be the same for all colours or $df = 0$.

$\therefore$
$$\frac{\omega_1}{f_1} + \frac{\omega_2}{f_2} = 0$$

or
$$\frac{f_1}{f_2} = -\frac{\omega_1}{\omega_2} \qquad ...(4)$$

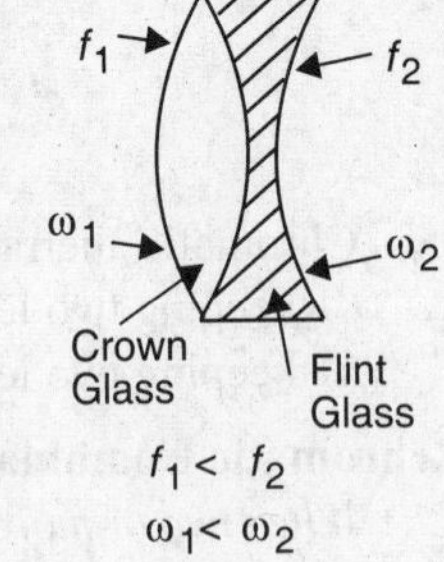

Fig. 1.28.

In order, therefore, to design an achromatic doublet of focal length f, the focal lengths of the constituent lenses must satisfy Eq. (3) and Eq. (4). Knowing ω_1, ω_2 and f, the magnitudes of f_1 and f_2 can be found by solving Eqs. (3) and (4).

Since ω_1 and ω_2 are positive, f_1 and f_2 must be of opposite signs. That is, if one lens is convex, the other should be concave. Since the achromatic doublet is to behave as a converging lens, f_1 must be less than f_2. Consequently, $\omega_1 < \omega_2$. The converging lens is, therefore, made of crown glass (smaller dispersive power) and the diverging lens, of flint glass (larger dispersive power). [Fig. 1.28].

Example 1 : *An achromatic telescope objective of 1.5 m focal length consists of two thin lenses in contact with each other and their dispersive powers are 0.05 and 0.075 respectively. Calculate their focal lengths.*

Solution : Here, $\omega_1 = 0.05$, $\omega_2 = 0.075$ and $f = 1.5$ m.

For an achromat, $\frac{\omega_1}{f_1}+\frac{\omega_2}{f_2}=0$ or $\frac{1}{f_2}=\frac{-\omega_1}{\omega_2 f_1}$

$$\therefore \qquad \frac{1}{f_2}=\frac{-0.05}{0.075\, f_1}=-\frac{2}{3f_1} \qquad ...(1)$$

Also
$$\frac{1}{f}=\frac{1}{f_1}+\frac{1}{f_2}$$

or
$$\frac{1}{1\cdot 5}=\frac{1}{f_1}-\frac{2}{3f_1} \qquad ...(2)$$

$$\therefore \qquad f_1 = 0.5 \text{ m.}$$

From Eq. (1), $\frac{1}{f_2}=-\frac{2}{3\times 0.5}$

$$\therefore \qquad f_2 = -0.75 \text{ m.}$$

Example 2 : *Design an achromatic lens having a focal length of 0.350 m and made of two lenses, one of crown glass of dispersive power 0.0193 and the other of flint glass of dispersive power 0.0271.*

Solution : Here, $\omega_1 = 0.0193$; $\omega_2 = 0.0271$; $f = 0.35$ m.

For achromatism, $\frac{\omega_1}{f_1}+\frac{\omega_2}{f_2}=0$

or
$$\frac{0.0193}{f_1}+\frac{0.0271}{f_2}=0 \qquad ...(1)$$

$$\frac{1}{f}=\frac{1}{f_1}+\frac{1}{f_2};\ \frac{1}{0.35}=\frac{1}{f_1}+\frac{1}{f_2}$$

or
$$\frac{1}{f_1}=\frac{1}{0.35}-\frac{1}{f_2} \qquad ...(2)$$

Substituting the value of $1/f_1$ from Eq. (2) in Eq. (1),

$$0.0193\left(\frac{1}{0.35}-\frac{1}{f_2}\right)+\frac{0.0271}{f_2}=0$$

$$\frac{0.0193}{0.35}=\frac{0.0193}{f_2}-\frac{0.0271}{f_2}=-\frac{0.0078}{f_2}$$

$$\therefore \qquad f_2=-\frac{0.0078\times 0.35}{0.0193}=-0.14\text{m (concave lens)}$$

From Eq. (2), $\frac{1}{f_1}=\frac{1}{0.35}+\frac{1}{0.14}=10$

$\therefore f_1 = 1/10 = 0.1$ m (convex lens)

We should use a convex lens of crown glass of focal length 0.1m and a concave lens (flint) of focal length – 0.14 m.

Example 3 : *A telescope objective is made by cementing together two lenses, the cemented surfaces having the same radii of curvatures. The combination is plano-convex and is of focal length 100 cm. Calculate the radii of curvatures of the component lenses given that,*

	Crown glass	*Flint glass*
Refractive index	*1.50*	*1.60*
Dispersive power	*0.02*	*0.03*

Solution : Let f_1 and f_2 be the focal lengths of the crown glass and flint glass lenses respectively. Then

$$1/f_1 + 1/f_2 = 1/100 \quad ...(1)$$

Further,
$$\frac{\omega_1}{f_1} + \frac{\omega_2}{f_2} = 0$$

or
$$\frac{0.02}{f_1} + \frac{0.03}{f_2} = 0 \quad ...(2)$$

Solving (1) and (2), we get $f_1 = 33.3$ cm. $f_2 = -50$ cm.

Thus, focal length of the plano-concave lens of flint glass = 50 cm.

Focal length of the double convex lens of crown glass = 33.3 cm.

For the plano concave lens, $n = 1.6$, $R_2 = \infty$ and $f = -50$ cm.

Now,
$$\frac{1}{f} = (n-1)\left(\frac{1}{R_1} - \frac{1}{R_2}\right) \text{ or } -\frac{1}{50} = (1.6-1)\left(\frac{1}{R_1} - \frac{1}{\infty}\right)$$

$$\therefore \quad R_1 = -30 \text{ cm.}$$

The cemented surfaces of the concave and convex lens have got the same radii of curvature. Hence the radius of curvature of the second surface of the convex lens is also equal to 30 cm.

Let R' be the radius of curvature of the first surface. Then for the convex lens, $f = 33.3$ cm, $n = 1.5$; $R_2 = -30$ cm; $R' = ?$

$$\frac{1}{33.3} = (1.5-1)\left(\frac{1}{R'} - \frac{1}{-30}\right).$$

$$\therefore \quad R' = 37.45 \text{ cm.}$$

Thus the radii of curvature of the surfaces of the convex lens are 37.45 cm and –30 cm and those of the concave lens are – 30 cm and ∞.

1.22. Condition for Achromatism of Two Thin Lenses Separated by a Finite Distance

Let us consider two convex lenses of focal lengths f_1, f_2 separated by a distance a (Fig. 1.29). The focal length of the combination is

$$\frac{1}{f} = \frac{1}{f_1} + \frac{1}{f_2} - \frac{a}{f_1 f_2} \quad ...(1)$$

Fig. 1.29.

Differentiating Eq. (1),

$$-\frac{df}{f^2} = -\frac{df_1}{f_1^2} - \frac{df_2}{f_2^2} - a\left[-\frac{df_1}{f_1^2 f_2} - \frac{df_2}{f_2^2 f_1}\right]$$

Now,
$$-\frac{df_1}{f_1} = \omega_1 \text{ and } -\frac{df_2}{f_2} = \omega_2$$

$$\therefore \quad -\frac{df}{f^2} = \frac{\omega_1}{f_1} + \frac{\omega_2}{f_2} - a\left[\frac{\omega_1}{f_1 f_2} + \frac{\omega_2}{f_1 f_2}\right]$$

Since both the lenses are of the *same material*, $\omega_1 = \omega_2$.

$$\therefore \quad -\frac{df}{f^2} = \frac{\omega}{f_1} + \frac{\omega}{f_2} - \frac{2a\omega}{f_1 f_2}$$

For an achromatic combination, the focal length f should not change with colour. $\therefore$ $df = 0$. Hence,

$$\frac{\omega}{f_1} + \frac{\omega}{f_2} - \frac{2a\omega}{f_1 f_2} = 0 \text{ or } \frac{1}{f_1} + \frac{1}{f_2} = \frac{2a}{f_1 f_2}$$

$$\therefore \qquad a = \frac{f_1 + f_2}{2} \qquad ...(2)$$

That is, *the distance between the two coaxial lenses must be equal to half the sum of their focal lengths.*

Example 1 : *Two thin converging lenses of focal lengths 10 cm and 20 cm respectively are to form an achromatic combination. Calculate the effective focal length of the combination.*

Solution : Two converging lenses cannot form an achromatic combination if they are in contact. Therefore they should be separated by a distance a. Both the lenses are of the same material. Let their focal lengths be f_1 and f_2. The condition for the combination to be achromatic is

$$a = (f_1 + f_2)/2 \qquad ...(1)$$

The effective focal length f of the combination is

$$\frac{1}{f} = \frac{1}{f_1} + \frac{1}{f_2} - \frac{a}{f_1 f_2} = \frac{1}{f_1} + \frac{1}{f_2} - \frac{f_1 + f_2}{2 f_1 f_2} = \frac{f_1 + f_2}{2 f_1 f_2}$$

$$\therefore \qquad f = \frac{2 f_1 f_2}{f_1 + f_2} = \frac{2 \times 10 \times 20}{10 + 20} = 13.33 \text{ cm.}$$

Example 2 : *A system consisting of two crown glass lenses is to be achromatic for the focal length which is 0.4 m. The focal length of one lens being 0.25 m, find the focal length of the other and the required distance between the lenses.*

Solution : Here, $f = 0.4$ m, $f_1 = 0.25$ m.

$$\frac{1}{f} = \frac{1}{f_1} + \frac{1}{f_2} - \frac{a}{f_1 f_2}$$

or

$$\frac{1}{0.4} = \frac{1}{0.25} + \frac{1}{f_2} - \frac{a}{0.25 f_2} \qquad ...(1)$$

The condition of achromatism is $a = (f_1 + f_2)/2$

$$\therefore \qquad a = (0.25 + f_2)/2 \qquad ...(2)$$

Substituting this value of a in Eq. (1),

$$\frac{1}{0.4} = \frac{1}{0.25} + \frac{1}{f_2} - \frac{(0.25 + f_2)}{2 \times 0.25 f_2}$$

$$\therefore \qquad f_2 = 1m$$

From (2), $a = (0.25 + 1)/2 = 0.625$ m.

1.23. Coma

When a lens is corrected for spherical aberration, it forms a point image of a point object situated on the axis. But *if the point object is situated off the principal axis, the lens, even corrected for spherical aberration, forms a comet-like image in place of point image. This defect in the image is called coma.*

Consider an off axis point A in the object (Fig. 1.30). The rays leaving A and passing through the different zones of the lens such as 11, 22, 33 are brought to focus at different points B_1, B_2, B_3, gradually nearer to the lens. The radius of these circles go on increasing with increase in radius of zone. Thus the resultant image is comet like.

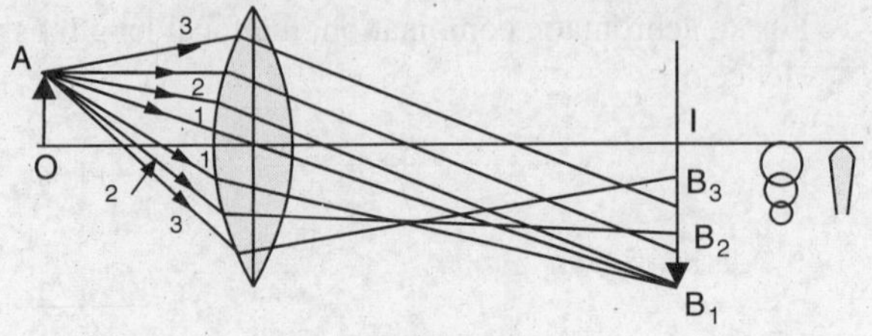

Fig. 1.30.

Removal of coma. The comatic aberration may be eliminated as follows:

1. By using a stop before the lens and so making the outer zones ineffective.

2. By properly choosing the radii of curvature of the lens surfaces. For example, for an object situated at infinity, the comatic aberration may be minimised by taking a lens of $n = 1.5$ and

$$k = \frac{R_1}{R_2} = -\frac{1}{9}.$$

3. **Abbe sine condition.** Abbe showed that coma may be eliminated if each zone of the lens satisfies the *Abbe sine condition*

$$n_1 h_1 \sin \theta_1 = n_2 h_2 \sin \theta_2.$$

Here, n_1 and n_2 are refractive indices of the object and image regions respectively. h_1 and h_2 are the heights of the object and the image. θ_1 and θ_2 are the angles which the incident and the conjugate emergent rays make with the axis (Fig.1.31).

If this condition is satisfied, the lateral magnification

$$\frac{h_2}{h_1} = \frac{n_1 \sin \theta_1}{n_2 \sin \theta_2}$$

will be same for all the rays of light, irrespective of the angles θ_1 and θ_2. Therefore, coma will be eliminated.

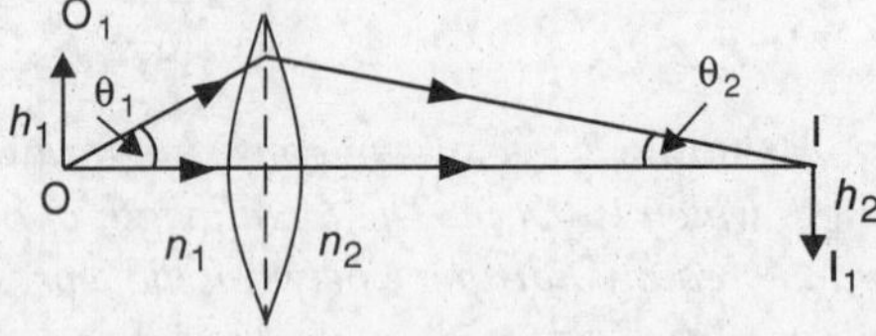

Fig. 1.31.

1.24. Astigmatism and its Minimization

Consider a point B situated off the axis in a line object which is vertically below the axis of the lens. When the cone of rays from B falls on full circumference of the lens, then after refraction all the rays do not meet at a single point (Fig. 1.32).

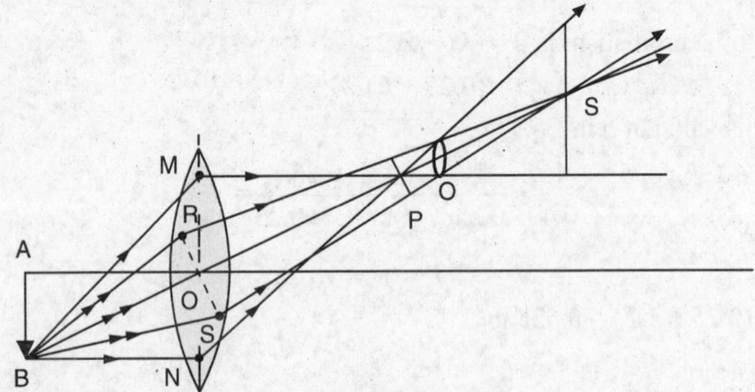

Fig. 1.32.

(a) The rays lying in the vertical plane *BMN* (called *meridional plane*) form the image as a horizontal line *P*.

(b) The rays lying in the horizontal plane *BRS* (called *sagittal plane*) form the image as a vertical line *S*. The *circle of least confusion C* lies in between *P* and S.The best image for the object point

is obtained here. This defect is called *astigmatism.* The distance between P and S is a measure of astigmatism and is called the *astigmatic difference*.

The astigmatic difference in the concave lens is in opposite direction to that produced by a convex lens. Hence astigmatism may be reduced by suitable combination of concave and convex lenses. Such a combination of lenses is called *anastigmatic combination*. It is used in the construction of objective lens in a photographic camera.

EYE-PIECES

1.25. Eye-piece

An *eye-piece* is a combination of lenses designed to magnify the image already formed by the *objective* of a telescope and microscope. An eye-piece consists of two plano-convex lenses. F is called the *field lens* and E the *eye lens* (Fig. 1.33). The field lens has large aperture to increase the field of view. The eye lens mainly magnifies the image. To reduce the spherical aberration , the lenses taken are plano-convex lenses. Further the focal lengths of the two lenses and their separation are selected in such a way as to minimise the chromatic and spherical aberrations.

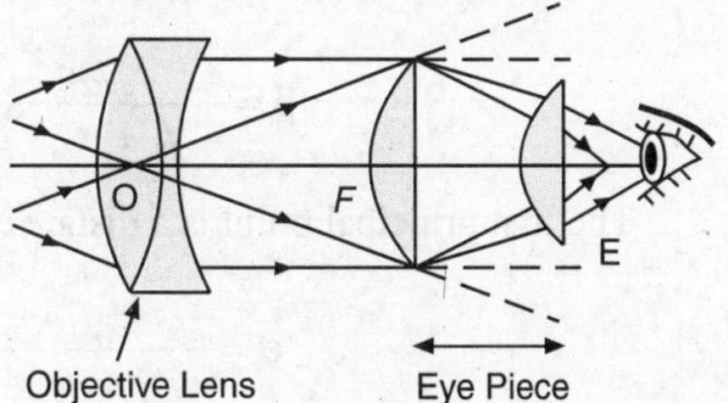

Fig. 1.33.

A combination of lenses is used in an eyepiece in place of a simple lens magnifier for the following reasons :

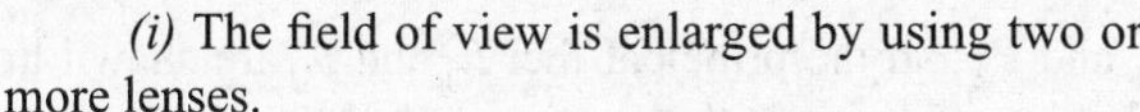

(i) The field of view is enlarged by using two or more lenses.

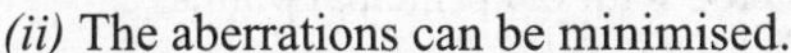

(ii) The aberrations can be minimised.

1.26. Huygens' Eye-piece

Construction: It consists of two plano-convex lenses of focal lengths $3f$ (field lens) and f (eye lens), placed a distance $2f$ apart [Fig. 1.34]. They are arranged with their convex faces towards the incident rays. The eye-piece satisfies the following conditions of minimum spherical and chromatic aberrations.

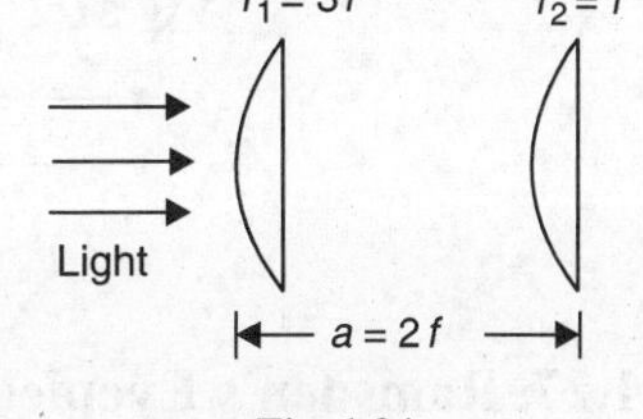

Fig. 1.34.

(i) The distance between the two lenses for minimum spherical aberration is given by $a = f_1 - f_2$. In Huygen's eyepiece, $a = 3f - f = 2f$. Hence this eye-piece satisfies the condition of minimum spherical aberration.

(ii) For chromatic aberration to be minimum $a = (f_1 + f_2)/2$. In Huygens' eyepiece, $a = (3f + f)/2 = 2f$. Hence this eyepiece satisfies the condition of minimum chromatic aberration.

Working : An eye-piece forms the final image at infinity. Thus the field lens forms the image I_2 in the first focal plane of eye-lens, *i.e.*, at a distance f to the left of eye-lens. Now the distance between the field lens and eye-lens is $2f$. Therefore, the image I_2 lies at a distance f to the right of field lens. The image I_1 formed by the objective of microscope or telescope acts as the virtual object for the field lens. Thus we treat I_1 as the virtual object for the field lens, and I_2 as the image of I_1 due to it (Fig. 1.35) or $v = f$, $F = 3f$, $u = ?$ We have

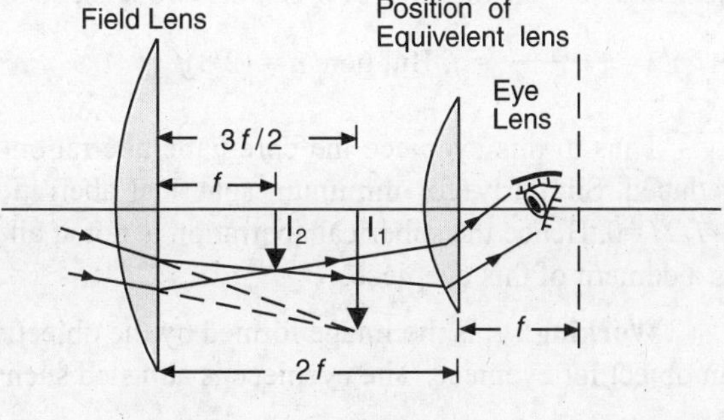

Fig. 1.35.

$$\frac{1}{v}-\frac{1}{u}=\frac{1}{F} \text{ or } \frac{1}{f}-\frac{1}{u}=\frac{1}{3f}$$

$$\therefore \qquad u = 3f/2$$

i.e. I_1 should be formed at a distance (3/2) f from the field lens. Therefore the rays coming from the objective which converge towards I_1, are focussed by the field lens at I_2. The rays starting from I_2 emerge from the eye-lens as a parallel beam.

Cardinal Points of Huygens Eyepiece

The equivalent focal length F of this eyepiece is

$$\frac{1}{F}=\frac{1}{f_1}+\frac{1}{f_2}-\frac{a}{f_1 f_2}=\frac{1}{3f}+\frac{1}{f}-\frac{2f}{3f\times f}=\frac{2}{3f}$$

$$\therefore \qquad F = 3f/2.$$

The second principal point is at a distance β from the eye lens.

$$\beta=-\frac{f_2 a}{f_1+f_2-a}=\frac{-f\times 2f}{3f+f-2f}=\frac{-2f^2}{2f}=-f.$$

The first principal point is a distance α from the field lens.

$$\alpha=+\frac{f_1 a}{f_1+f_2-a}=\frac{3f\times 2f}{3f+f-2f}=\frac{6f^2}{2f}=3f.$$

The position of the principal points P_1 and P_2 and the principal foci F_1 and F_2 are shown in Fig.1.36. Since the system is in air, the nodal points coincide with the principal points.

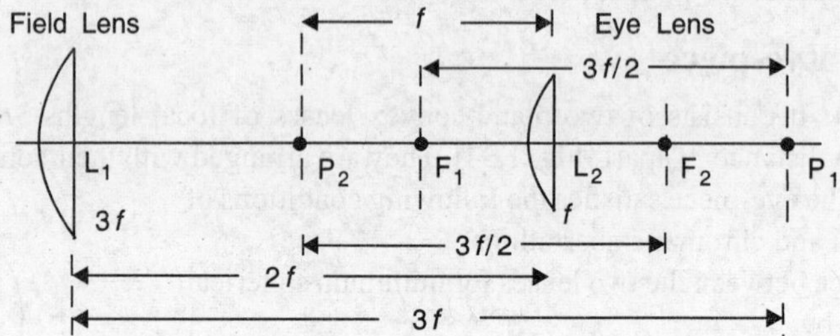

Fig. 1.36.

1.27. Ramsden's Eyepiece

Construction : It consists of two plano convex lenses each of focal length f. The distance between them is (2/3) f [Fig.1.37]. For achromatism, the distance between the two lenses should be $a = (f_1 + f_2)/2 = \frac{(f+f)}{2} = f$. But here $a = (2/3)f$.

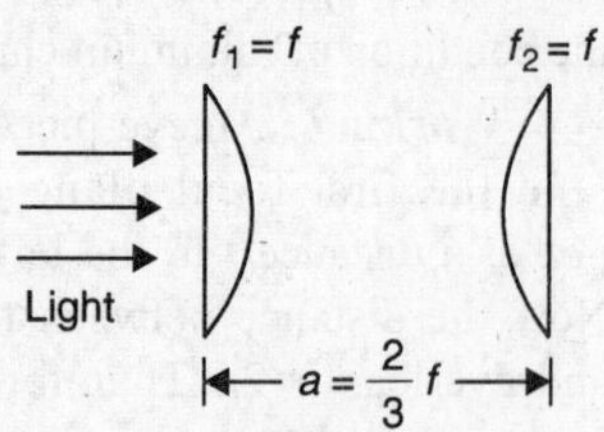

Fig. 1.37.

Thus in this eyepiece the chromatic aberration is only partly reduced. Similarly, for minimum spherical aberration, $a = f_1 - f_2 = f - f = 0$. Hence the spherical aberration is not at all reduced. This is a demerit of this eyepiece.

Working : I_1 is the image formed by the objective of the microscope or telescope. It serves as an object for eyepiece. The eyepiece is adjusted such that the image I_2 formed by the field lens lies in the first focal plane of the eyelens [Fig.1.38]. Then the eye-piece forms the final image at infinity. Since the focal length of the eye lens is f and $a = (2/3) f$, I_2 is at a distance $f/3$ from the field lens. Now, the image I_1 due to objective serves as the object for field lens. I_2 is the image of I_1 due to field lens. Or, $v = -f/3$, $F = f$, $u = ?$

$$\frac{1}{v} - \frac{1}{u} = \frac{1}{F} \text{ or } \frac{1}{-f/3} - \frac{1}{u} = \frac{1}{f} \text{ or } \frac{1}{u} = -\frac{4}{f}$$

$$\therefore \qquad u = -f/4.$$

Thus the eyepiece is so adjusted that the image (I_1) formed by the objective of telescope or microscope lies at a distance $f/4$ towards the left of field lens. The crosswire is placed at I_1. I_1 serves as the object for field lens and its image is formed at I_2.

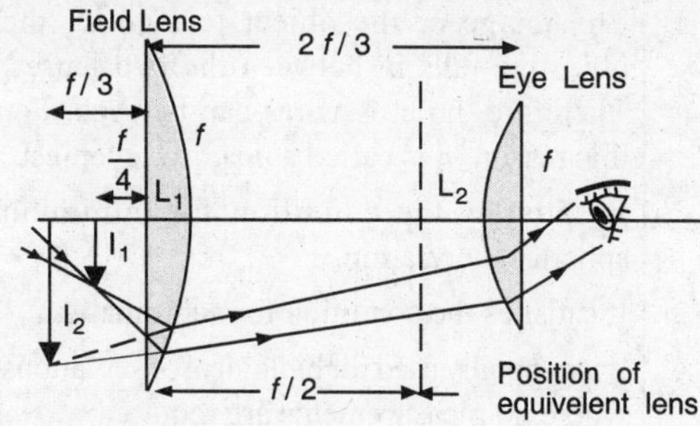

Fig. 1.38.

Cardinal points : The focal length F of the equivalent lens is

$$\frac{1}{F} = \frac{1}{f_1} + \frac{1}{f_2} - \frac{a}{f_1 f_2} = \frac{1}{f} + \frac{1}{f} - \frac{2f/3}{f^2} = \frac{4}{3f}.$$

$$\therefore \qquad F = 3f/4$$

$$\beta = \frac{-f_2 a}{f_1 + f_2 - a} = -\frac{f \times (2f/3)}{2f - (2f/3)} = -\frac{f}{2}$$

$$\alpha = \frac{f_1 a}{f_1 + f_2 - a} = -\frac{f \times (2f/3)}{2f - (2f/3)} = +\frac{f}{2}$$

The positions of the principal points P_1 and P_2 and the principal foci F_1 and F_2 are shown in Fig. 1.39. Since the system is in air, the nodal points coincide with the principal points.

Distance of the first principal focus from the field lens of the eyepiece $= F_1L_1 = F_1P_1 - \alpha = 3f/4 - f/2 = f/4$. Similarly the distance of the second principal focus from the eye lens is $L_2F_2 = P_2F_2 - \beta = 3f/4 - f/2 = f/4$.

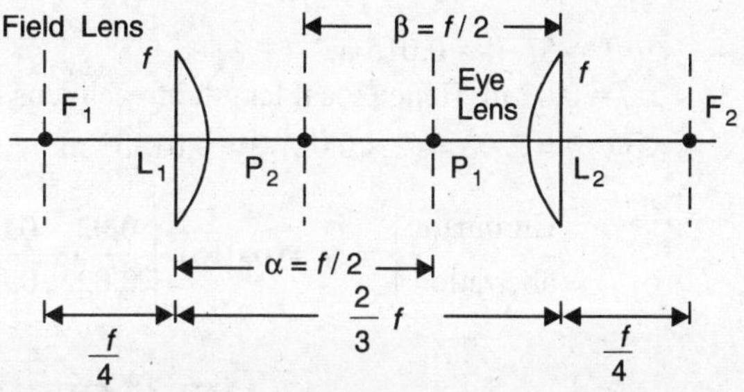

Fig. 1.39.

Example : *A Ramsden's eyepiece is to have an effective focal length of 3 cm. Calculate the focal lengths of the lens components and their distance of separation.*

Solution : We have, $\dfrac{1}{F} = \dfrac{1}{f_1} + \dfrac{1}{f_2} - \dfrac{a}{f_1 f_2}$.

Here, $f_1 = f$, $f_2 = f$, $a = 2f/3$ and $F = 3$cm.

$$\therefore \qquad \frac{1}{3} = \frac{1}{f} + \frac{1}{f} - \frac{2f/3}{f^2}$$

$$\therefore \qquad f = 4 \text{ cm, and } a = 2f/3 = 8/3 \text{ cm.}$$

1.28. Comparison of Eyepieces

	Huygens' eyepiece	Ramsden's eyepiece
1.	The image of the object formed by the objective falls in between the two lenses. Therefore, no crosswires can be used. For this reason, it is called a *negative* eyepiece.	The image of the object formed by the objective lies in front of the field lens. Therefore, crosswires can be used. For this reason, it is called a *positive* eyepiece.
2.	It satisfies the condition for minimum spherical aberration.	It does not satisfy the condition for minimum spherical aberration.
3.	It satisfies the condition for achromatism.	It does not satisfy the condition for achromatism.
4.	It is generally used for biological observations where no measurements are required.	It is used with instruments meant for physical measurements.

Example : *The focal length of an eyepiece should be 0.015 m. Calculate the focal lengths of the two lenses used in the eyepiece and the distance between the two lenses if it is (i) Huygens' eyepiece (ii) Ramsden's eyepiece. If the dispersive power of the material of the lenses used is 0.03, calculate the chromatic aberration produced in each case.*

Solution: *(i)* $F = 3f/2 = 0.015$ m, $\therefore f = 0.01$m, $f_1 = 3f = 0.03$ m;

$$f_2 = f = 0.01 \text{ m and } a = 2f = 0.02\text{m}.$$

$$\left.\begin{array}{l}\text{Chromatic}\\\text{aberration}\end{array}\right\} = df = F^2\left[\frac{\omega_1}{f_1} + \frac{\omega_2}{f_2} - \frac{a(\omega_1 + \omega_2)}{f_1 f_2}\right]$$

$$= (0.015)^2\left[\frac{0.03}{0.03} + \frac{0.03}{0.01} - \frac{0.02 \times (0.03 + 0.03)}{0.03 \times 0.01}\right] = 0.$$

(ii) $F = 3f/4 = 0.015$ m.

$\therefore f = 0.02$ m. Hence focal length of each lens is 0.02 m.

Now $a = 2f/3 = (2 \times 0.02)/3 = 0.01333$ m.

$$\left.\begin{array}{l}\text{Chromatic}\\\text{aberration}\end{array}\right\} = df = (0.015)^2\left[\frac{0.03}{0.02} + \frac{0.03}{0.02} - \frac{0.01333 \times 0.06}{0.02 \times 0.02}\right] = 0.000225 \text{ m}.$$

VELOCITY OF LIGHT

1.29. Importance of Determining Velocity of Light

The velocity of light in vacuum (c) is a constant of nature. Its significance in Physics will be clear from the following:

(*i*) The ratio of the velocity of light in vacuum to that in a medium represents the refractive index of the medium. Hence, from a knowledge of the velocity of light in a medium, its refractive index can be obtained.

(*ii*) According to Einstein, the velocity of light is a universal constant. It is, therefore, adopted as a standard of measurement in Geodetic surveys. In spectroscopy, the wavelengths of spectral lines are determined experimentally. The corresponding frequencies (v) can be calculated from the formula $v = c/\lambda$, if c is accurately known.

(*iii*) It enables the energy and momentum of a given quantity of radiation to be determined. $E = mc^2$. Also $E = hv$. Hence $hv = mc^2$ or $m = hv/c^2$. Momentum = mas × velocity = $(hv/c^2)\, c = hv/c$.

1.30. Houston's Method (Piezo Electric Grating Method)

Principle : In this method *piezo-electric effect* is used for the alternate interruptions of a beam of light. According to this effect, if a quartz crystal is suitably cut and subjected to a high frequency alternating electric field, it is alternately compressed and extended. Thus high frequency oscillations are generated. When these high frequency oscillations are set up in the form of pulses in a quartz crystal, it is crossed by several parallel nodal planes. The intensity and refractive index at nodal planes is different from those at other planes. Thus the crystal acts as a diffraction grating.

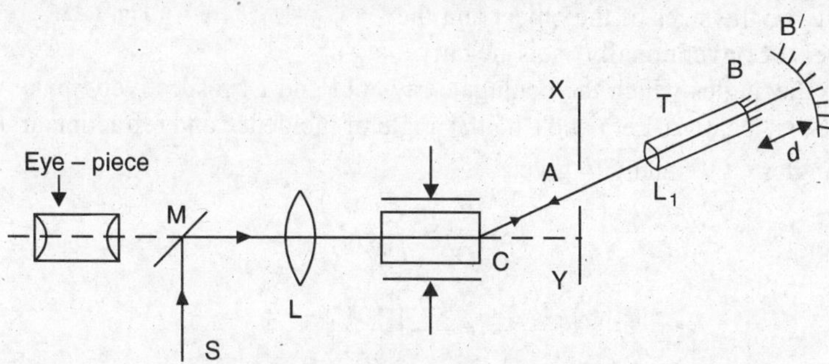

Fig. 1.40.

Experiment : Fig. 1.40 shows Houston's arrangement for determining velocity of light. Monochromatic light from a source *S* after reflection from the glass plate *M* emerges through a convex lens *L* as a parallel beam. Then the light is allowed to fall on the quartz crystal *C*. The quartz crystal acts as a diffraction grating with a frequency double the frequency of oscillations. Then the light passes through the slit *A* of the screen *XY*. Finally the light is focussed by the lens L_1. The convex lens L_1 and the reflecting concave mirror *B* are fixed at the two ends of a tube *T*. The distance between them is the focal length of L_1. Further this distance is also equal to the radius of curvature of the concave mirror *B*.

This light completely retraces its path from *B*. Suppose the returning beam of light finds the vibrating crystal in the same condition as it was when the light went out from it. Then the light emerges in the direction *CM*. Therefore, the eye sees an image of *S*. If there is any change in the condition of the crystal, then no image will be formed at the eyepiece.

Let *f* be the frequency of oscillations of the crystal. Then the rate of formation of the grating is 2*f* per second. Therefore, when the image is seen by the eye, the time taken by light to go from *C* to *B* and back is a simple multiple of $(1/2f)$.

The tube *T* is moved away from the screen along *CB*. Initially, the image disappears. The image will reappear when the tube reaches some position *B'*. This will be so, when the distance *BB*' and back (*i.e.*, 2*d*) is travelled by light in the time $(1/2f)$ second. By taking several such positions, the mean value of *d* is determined.

$$\text{Velocity of light } c = \frac{\text{distance traelled}}{\text{time taken}} = \frac{2d}{1/(2f)} = 4df.$$

The value of the velocity of light in vacuum calculated by Houston was 299, 782 ± 9 km/s. This value agrees very closely with values obtained from other accurate experiments.

1.31. Helmholtz Lagrange Equation for Image Formation for Paraxial Rays

SPS' is a spherical refracting surface separating two media of refractive indices n_1 and n_2 (Fig. 1.41). *P* is the pole, *C* is the centre of curvature of the surface. *OPI* is the optical axis of the system. *OO'* is an object placed on the axis perpendicular to it. An incident ray *OA*, on refraction at *A*, bends towards the normal *CN* and goes in the direction *AI*. Another ray *OP* meets at *I*. *I* is the image of O. To find the image of O', let us take a ray $O'C$. It passes through the centre of curvature C. It strikes the surface normally and goes undeviated. The image of O' lies somewhere on this line. Also it lies on the perpendicular to the axis at I. Hence I', the point of intersection of the two, is the image of O'. II' is the complete image of OO'.

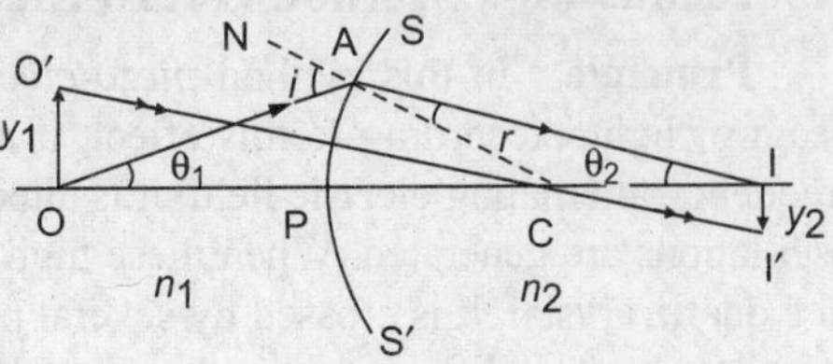

Fig. 1.41.

Let y_1 and y_2 be the sizes of the object and the image respectively. Conventionally y_2 is negative. Let θ_1 and θ_2 be the angles which the conjugate rays OA and IA respectively make with the axis. Conventionally θ_2 is negative. Let i and r be the angle of incidence and refraction at A.

Similar triangles COO' and CII' give:

$$\frac{II'}{OO'} = \frac{CI}{CO}$$

or
$$\frac{-y_2}{y_1} = \frac{CI}{CA}\left(\frac{CA}{CO}\right) \quad \text{...(1)}$$

Using sine law in Δ CIA, we get

$$\frac{CI}{CA} = \frac{\sin r}{\sin(-\theta_2)} = -\frac{\sin r}{\sin \theta_2}$$

Using sine law in Δ COA, we get

$$\frac{CA}{CO} = \frac{\sin \theta_1}{\sin(180° - i)} = \frac{\sin \theta_1}{\sin i}.$$

Substituting these values in Eq. (1), we get

$$\frac{y_2}{y_1} = \frac{\sin r}{\sin \theta_2}\left(\frac{\sin \theta_1}{\sin i}\right).$$

But $\frac{\sin i}{\sin r} = \frac{n_2}{n_1}$ (Snell's law).

$$\therefore \quad \frac{y_2}{y_1} = \frac{n_1 \sin \theta_1}{n_2 \sin \theta_2}$$

or
$$n_1 y_1 \sin \theta_1 = n_2 y_2 \sin \theta_2. \quad \text{...(2)}$$

This equation is called *Abbe's sine condition.*

For paraxial rays, θ_1 and θ_2, are small. So

$\sin \theta_1 = \theta_1 = \tan \theta_1$ and $\sin \theta_2 = \theta_2 = \tan \theta_2$. Then

$$n_1 y_1 \tan \theta_1 = n_2 y_2 \tan \theta_2. \quad \text{...(3)}$$

This relation is called *Langrange equation.*

The ratio y_2/y_1 is the ratio of image to object sizes at right angles to axis and known as *linear transverse magnification* (m_1).

Similarly, the ratio $\tan \theta_2/\tan \theta_1$ is known as *angular magnification* (m_2).

Two are related to each other by Langrange formula, *i.e.*,

$$m_1 m_2 = n_1/n_2 \quad \text{...(4)}$$

Let us now consider a coaxial system having $(n-1)$ refracting surfaces separating n media of refractive indices $n_1, n_2, n_3, \ldots n_n$ respectively.

Then the image formed by the first surface acts as the object for the second and so on. Hence from Eq. (3), we have

$n_1 y_1 \tan \theta_1 = n_2 y_2 \tan \theta_2$ for the I surface

$n_2 y_2 \tan \theta_2 = n_3 y_3 \tan \theta_3$ for the II surface

$n_3 y_3 \tan \theta_3 = n_4 y_4 \tan \theta_4$ for the III surface

....

$n_{n-1}\, y_{n-1} \tan \theta_{n-1} = n_n y_n \tan \theta_n$ for the $(n-1)$th surface

$$\therefore n_1 y_1 \tan \theta_1 = n_n y_n \tan \theta_n. \qquad \ldots(5)$$

This general relation is known as 'Helmholtz's Law of Magnification'.

1.32. Telephoto Lens

The size of the image of a distant object formed by a lens is proportional to the focal length of the lens and the angle subtended by object on the camera lens. The camera size becomes unmanageable if lens of large focal length is used to form the image. To avoid this difficulty a lens combination of two lenses, known as *telephoto lens*, is used.

A telephoto lens combination is shown in Fig. 1.42. It consists of a convex lens L_1 and concave lens L_2 of achromatic type. The position of L_2 is normal lens position in a camera. L_1 is placed ahead of it at a suitable distance. The distance between L_1 and L_2 is less than the focal length of L_1. Also the focal length of L_1 is greater than that of L_2.

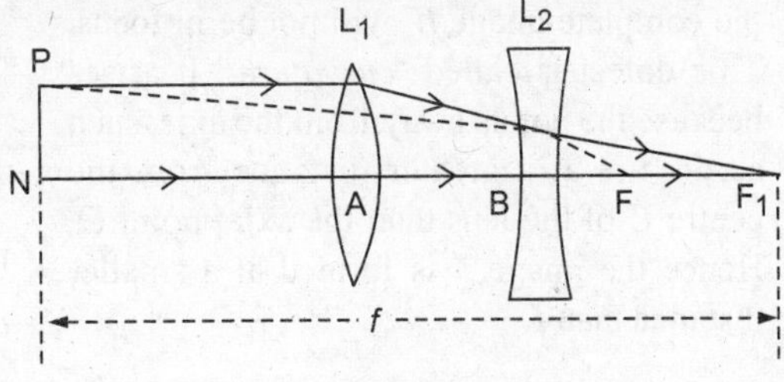

Fig. 1.42.

Parallel rays from a distant object incident on L_1 are brought to focus at F. F acts as virtual object for L_2 which forms its image at F_1. The photographic plate is placed at F_1. The combination has equivalent focal length $NF_1 = f$. Thus telephoto lens acts as a convex lens of focal length NF_1.

If f_1 and f_2 are the focal lengths of the lenses L_1 and L_2 and d is the distance between them, then

$$\frac{1}{f} = \frac{1}{f_1} + \frac{1}{f_2} - \frac{d}{nf_1 f_2}$$

1.33 Entrance Pupil and Exit Pupil

A common arrangement in photographic lenses is to have two separate lens elements with a variable stop between them. Fig. 1.43 is a diagram representing such a combination. L_1 and L_2 are two lenses with a stop $P_0 E_0 L_0$ between them.

For simplicity, the lenses are taken sufficiently large (compared with the diameter of the stop) so that they do not become aperture stops.

By definition, *the entrance pupil of the system is the image of the stop formed by lens L_1*. This image is virtual, erect, and located at *PEL*. By definition, *the exit pupil of the entire system is the image of the stop formed by lens L_2*. This image, located at $P'E'L'$, is also virtual and erect.

The entrance pupil PEL lies in the object space of lens L_1, the stop $P_0E_0L_0$ lies in the image space of L_1 as well as in the object space of L_2, and the exit pupil $P'E'L'$ lies in the image space of L_2.

Points P_0 and P, E_0 and E, and L_0 and L are conjugate pairs of points for L_1. P_0 and P', E_0 and E', and L and L' are conjugate pairs for L_2. This makes points like P and P' conjugate for the whole system. If a point object is located on the axis at O, rays OP and OL limit the bundle that will get through the system. At L_1 these rays are refracted through P_0 and L_0. At L_2 they are again refracted in such directions that they appear to come from P' and L'.

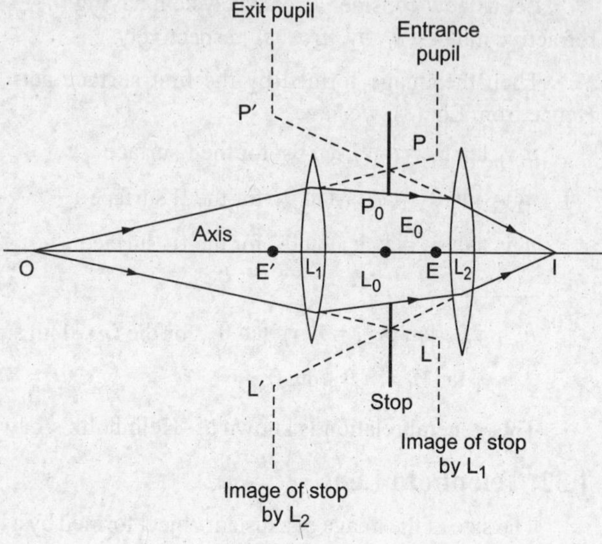

Fig. 1.43.

1.34. Curvature of the Field

Even if a lens is free from spherical aberration, coma and astigmatism, the image of an extended plane object OO' is curved (Fig. 1.44). If a screen is placed at I perpendicular to the axis, the complete image II' will not be in focus. This defect is called '*curvature.*' It arises because the points away from the axis, such as O', are at a greater distance from the centre C of the lens than the axial point O. Hence the image I' is formed at a smaller distance than I.

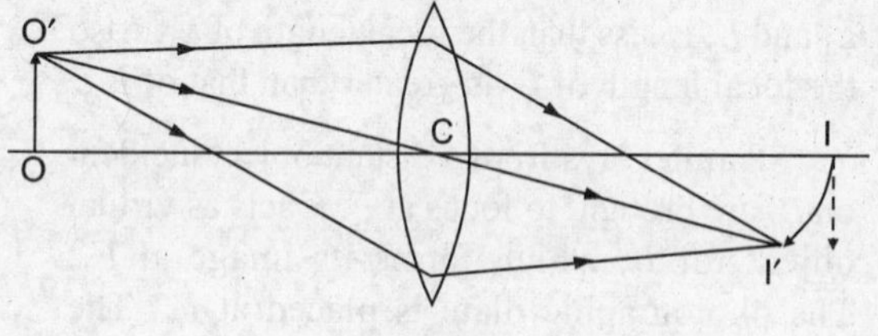

Fig. 1.44.

Removal of Curvature

(*i*) In case of a single lens, curvature can be minimised by using suitable stops.

(*ii*) For a combination of lenses, the condition for absence of curvature is

$$\Sigma \frac{1}{nf} = 0$$

Here n is the refractive index and f is the focal length of a lens.

For two lenses (whether in contact or separated by a distance) the condition reduces to

$$\frac{1}{n_1 f_1} + \frac{1}{n_2 f_2} = 0$$

or
$$n_1 f_1 + n_2 f_2 = 0.$$

This is known as *Petzval's condition.*

Since n_1, n_2 are positive, f_1 and f_2 must have opposite signs. Hence by combining a convex lens of a certain material with a concave lens of the suitable material and focal length, a flat field is obtained.

1.35. Distortion

When a stop is used with a lens to reduce the various aberrations, the image of a plane square-like object placed perpendicular to the axis is not of the same shape as the object. This defect is called '*distortion*'. This is because the chief rays forming images of different points on the object pass through different portions of the lens. Hence different parts of the object suffer different magnifications.

There are two types of distortion : (*i*) barrel-shaped, (*ii*) pin-cushion shaped.

(*i*) When the stop is placed on the object side of the lens (Fig. 1.45*a*), the magnification of the outermost part of the plane object is less than that of the central part, producing barrel-shaped distortion. If the stop is placed on the image side of the lens (Fig. 1.45*b*), then the outermost parts of the object are magnified more than the central parts, producing pin-cushion-shaped distortion.

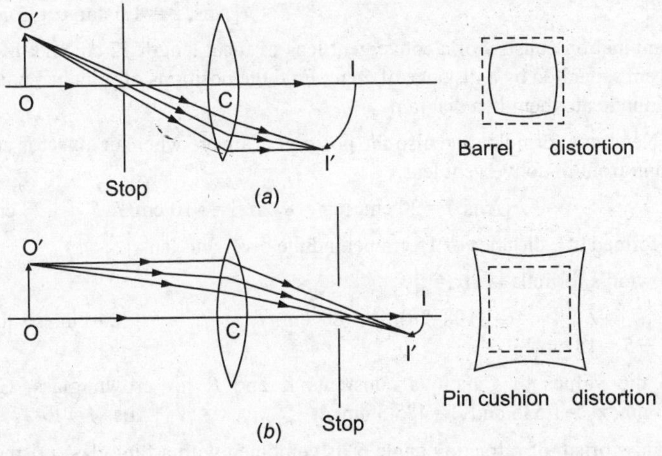

Fig. 1.45.

Removal of Distortion

A combination of two similar meniscus convex lenses, with their concave surfaces facing each other and having an aperture stop in the middle is free from distortion, when the object and image are symmetrically placed (Fig. 1.46). In this manner the pin-cushion-shaped distortion due to the first lens is exactly nullified by the barrel-shaped distortion due to the second lens.

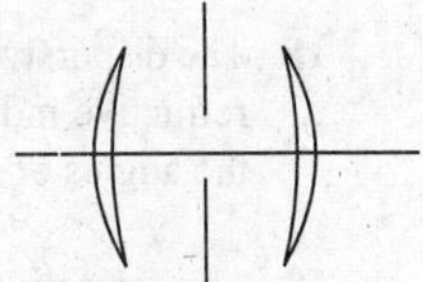

Fig. 1.46.

EXERCISE

1. What are principal points and principal planes?
2. Deduce for a thin lens an equation connecting the focal length, the radii of curvature of the surfaces and the refractive indices of the material of the lens and the surrounding medium.
3. What are the cardinal points of a lens system? Determine the cardinal points in the case of two thin coaxial lenses of focal lengths f_1 and f_2 separated by a distance a.
4. Derive Newton's formula for a convergent system of lenses forming a real image.
5. Define the term "dispersive power". Explain how two narrow angled prisms of different dispersive powers may be combined to produce dispersion without deviation and deviation without dispersion.
6. Give the theory and construction of a direct vision spectroscope.
7. Explain how the refractive index of the material of a small angled prism can be determined with a spectrometer.
8. Explain what is meant by spherical aberration for a lens. How is it caused and what are the ways for its minimisation?
9. Describe the principle of oil immersion objectives.

10. Explain what is meant by chromatic aberration in lenses. Derive the condition for achromatism of two thin lenses separated by a distance.

11. Derive the condition for the achromatism of two lenses mounted coaxially and in contact.

12. Discuss the importance of an eyepiece in an optical instrument. Describe with theory the construction of Ramsden and Huygen eyepieces. What are their relative merits and demerits?

13. Discuss any one method for determining the velocity of light.

14. Two thin convex lenses of focal lengths 5 cm and 2 cm are placed coaxially 3 cm apart. Calculate the equivalent focal length and locate the principal points.

[**Ans.** $\beta = 1.5$ cm, $\alpha = 3.75$ cm, $f = +2.5$ cm.]

15. A lens combination consists of a convergent lens of focal length 12 cm and divergent lens of focal length 15 cm separated by a distance of 6 cm. Find the positions of principal points and focii of lens system and indicate them in a diagram.

By using Newton's formula, find also the position of image when an object is situated at a distance of 108 cm in front of convergent lens.

[**Ans.** $f = 20$ cm, $\alpha = -8$ cm $\beta = -10$ cm, $L_1F_1 = -28$ cm, $L_2F_2 = +10$ cm.]

[image is formed at a distance of 15 cm behind the divergent lens.]

[**Hint**– Newton's formula is $x_1x_2 = f_1f_2$,

Here $x_1 = |L_1O - L_1F_1| = -(108-28) = -80$ cm, $f_2 = -f_1 = f = 20$ cm, so that $x_2 = 5$ cm, and $L_2F_2 = 10 + 5 = 15$ cm.]

16. Calculate the values of Cauchy's constants A and B for crown glass. Given : $n_c = 1.525$; $\lambda_c = 689.3$ nm; $n_b = 1.535$ and $\lambda_f = 485.3$ nm. [**Ans.** $A = 1.427$; $B = 4.6 \times 10^{-15}$ m^2]

17. A crown glass prism of refracting angle 6° is combined with a flint glass prism to obtain deviation without dispersion. If the refractive indices for red and violet rays for crown glass are 1.513 and 1.523 and for the flint glass are 1.645 and 1.665 respectively, find the angle of flint glass prism and mean deviation. [**Ans.** $A_2 = 3°$; $\delta = 1.14°$]

18. The dispersive powers of crown and flint glasses are 0.03 and 0.05 respectively. If the difference in the refractive indices of blue and red colours is 0.015 for crown glass and 0.022 for flint glass, calculate the angles of the two prisms for a deviation of 2° (without dispersion).

[**Ans.** $A_1 = 10°$; $A_2 = 6.82°$]

19. The refractive indices of crown glass for red and blue light are 1.515 and 1.523 respectively ; and those of flint glass for the same two colours are 1.614 and 1.632 respectively. What must be the angle of a flint glass prism which when combined with a crown glass prism of 4° produces (*a*) deviation but no dispersion (*b*) dispersion but no deviation. Calculate also the amount of deviation in (*a*) and the amount of dispersion in (*b*).

[**Ans.** (*a*) $A = -1.778°$, deviation = 0.968°. (*b*) $A = -3.332°$; Dispersion = – 0.028]

20. Spectacle lenses are generally made of crown glass, while spectrometer prisms are made of flint glass. Explain with reasons.

[**Hint :** Crown glass signifies small dispersive power and flint glass signifies large dispersive power].

21. Calculate the focal lengths of two thin lenses separated by 5 cm and satisfying the condition of no chromatic aberration and minimum spherical aberration.

[**Hint :** Let f_1 and f_2 be focal lengths of the two lenses. Since the doublet is corrected for spherical aberration,

$$f_1 - f_2 = a = 5 \quad \text{...(1)}$$

Since the lenses are also corrected for chromatic aberration,

$$a = \frac{f_1 + f_2}{2} = 5. \quad \text{...(2)}$$

or

$$f_1 + f_2 = 10$$

Solving (1) and (2), $f_1 = 7.5$ cm and $f_2 = 2.5$ cm.]

22. Calculate the focal lengths of the crown and flint glass lenses which form an achromatic doublet of focal length 50 cm, when placed in contact. The refractive indices for the C and F lines are given below:

Crown $n_c = 1.515$, $n_f = 1.523$

Flint $n_c = 1.644$, $n_f = 1.664$

[**Ans.** $f_1 = 24.79$ cm (crown glass, convex lens)

$f_2 = -49.19$ cm (flint glass, concave lens)].

23. Calculate the focal length of the components of an achromatic telescope objective having a focal length of 50 cm, made from crown glass and flint glass. Dispersive powers of crown and flint glasses are 0.015 and 0.025 respectively.

[**Ans.** + 20 cm, and – 33.3 cm.]

24. The dispersive powers of crown and flint glass are 0.015 and 0.030 respectively. The refractive indices for the mean rays are 1.52 and 1.65. If one of the surfaces of the flint glass lens is plane, calculate the radii of curvature of the other surfaces of the two lenses which form an achromatic doublet of focal length 80 cm.

[**Ans.** For the flint glass lens, $R_2' = \infty$, $R_1' = -52$ cm.

For the crown glass lens, $R_2 = R_1' = -52$ cm, $R_1 = 34.67$ cm.]

25. Two glasses have dispersive powers in the ratio 2 : 3. These glasses are to be used in the manufacture of an achromatic objective of focal length 20 cm. What are the focal lengths of the lenses?

[**Ans.** $f_1 = 6.67$ cm; $f_2 = -10$ cm].

26. Derive Abbe's sine condition for refraction at spherical surfaces. Write Helmholtz-Langrange equations with its help. Explain different kinds of magnification with their help.

(*Gorakhpur 1980, Delhi 72*)

27. Explain the utility of a telephoto lens in a camera. (*Punjab 76*)

28. Define and explain entrance pupil and exit pupil of an optical system. Locate them for a combination of two lenses having a stop in between. (*Gorakhpur 85*)

29. What are curvature and distortion? What is the cause of each of them? How may they be reduced to a minimum?

30. *The focal length of the more convergent lens of an Huygens' eyepiece is 0.5 cm. Calculate the focal length of the eyepiece.* (*Nagpur University, 2010*)

Ans. The focal lengths of the two lenses of the Huygen's eyepiece are $3f$ and f, and the distance between them is $2f$. In the given eyepiece, the focal length of the more convergent (smaller focal length) lens is 0.5 cm *i.e.*, $f = 0.5$ cm. Thus the focal lengths are 1.5 and 0.5 cm and the distance between them is 1.0 cm.

The focal length F of the eyepiece is given by

$$F = \frac{f_1 f_2}{f_1 + f_2 - a}$$

Here, $f_1 = +1.5$ cm, $f_2 = +0.5$ cm and $a = 1.0$ cm.

$$\therefore \qquad F = \frac{1.5 \times 0.5}{1.5 + 0.5 - 1.0} = +0.75 \text{ cm.}$$

31. What is the optical separation between two thin lenses of focal lengths f_1 and f_2, if the optical system acts as a glass plate? [**Ans.** $d = f_1 + f_2$]

32. *Define optical path. What is the difference between optical path and geometrical path of a ray? How are they related?* (*Meerut U., 2003*)

Ans. The path d of a ray of light in any medium is given by the product *velocity* times *time*:

$$d = vt \qquad ...(1)$$

By definition $\qquad n = c/v \qquad ...(2)$

Here, $\quad n$ = Refractive index of the medium

c = Velocity of light in vacuum or free space

v = Velocity of light in the medium.

$$\therefore \qquad v = c/n \qquad ...(3)$$

Substituting this value of v in Eq. (1), we get

$$d = \frac{c}{n}t \quad \text{or} \quad nd = ct$$

The product nd is called the *optical* path Δ:

$$\Delta = nd \qquad ...(4)$$

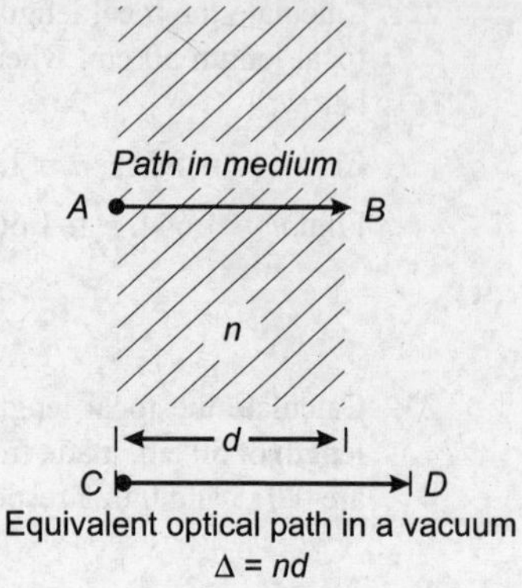

Path in medium

A B

n

d

C D

Equivalent optical path in a vacuum

$\Delta = nd$

Fig. 1.47.

The optical path represents the distance light travels in a vacuum in the same time it travels a distance d in the medium (Fig. 1.47).

The optical path is defined as the product of the geometrical path and the refractive index of the medium.

33. *What is telescopic combination of lenses?* (P. U. 2003)

Ans. If two lenses of focal lengths f_1 and f_2 are placed co-axially at a distance a from each other, then their equivalent focal length is given by

$$\frac{1}{f} = \frac{1}{f_1} + \frac{1}{f_2} - \frac{a}{f_1 f_2} \quad \text{or} \quad f = \frac{f_1 f_2}{f_1 + f_2 - a}$$

If $\qquad f_1 + f_2 = a, f = \infty.$

The focal length of such a combination is infinity. Such a combination is, therefore, called a *telescopic combination of lenses*.

A telescopic combination is used in astronomical telescopes. The incident and emergent rays of such a combination are parallel *i.e.*, the object and the image both lie at infinity.

CHAPTER

2

INTERFERENCE

2.1. Introduction

Interference is the phenomenon of superposition of two coherent waves in the region of superposition. At some points in the medium, the intensity of light is maximum (constructive interference); while at some other points, the intensity is minimum (destructive interference). The positions of maximum intensity are called *maxima*, while those of minimum intensity are called *minima*.

Let a_1 and a_2 be amplitudes of interfering waves and δ the phase difference at a point under consideration. Then,

Condition of maxima : Phase-difference $\delta = 2m\pi$

or Path-difference $\Delta = m\lambda$, $m = 0, 1, 2,.....$, etc.

Maximum amplitude, $A_{max} = a_1 + a_2$

Maximum intensity, $I_{max} = A^2_{max} = (a_1 + a_2)^2$

Condition of minima : Phase-difference $\delta = (2m+1)\pi$

Path-difference $\Delta = (2m+1)\lambda/2$, $m = 0, 1, 2, 3,....$

Minimum amplitude, $A_{min} = (a_1 - a_2)$

Minimum intensity, $I_{min} = (a_1 - a_2)^2$.

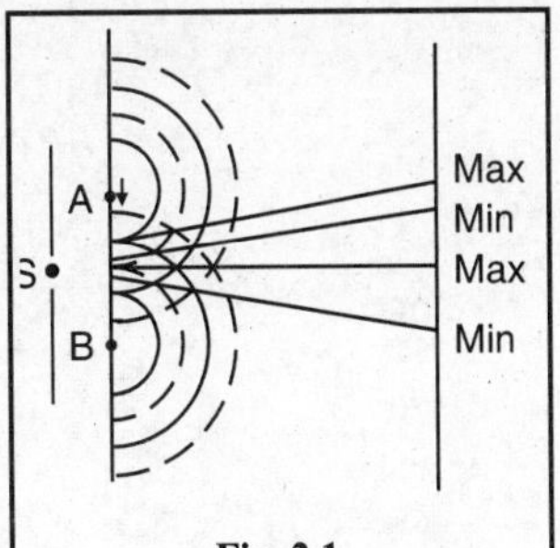

Fig. 2.1.

The relation between the path and phase differences is given by

$$\text{Phase-difference} = \frac{2\pi}{\lambda} \times \text{path difference.}$$

Coherent sources : *Coherent sources are the sources which either have no phase-difference or have always a constant phase-difference between them.*

Conditions for Sustained Interference of Light Waves

(*i*) The two interfering sources must be coherent.

(*ii*) The interfering waves must have equal amplitudes.

(*iii*) The two interfering waves must be propagated along the same line.

(*iv*) The separation between the two sources must be very small.

(*v*) The two sources must be narrow.

(*vi*) *The two interfering sources should emit light of the same frequency or wavelength.*

Young's Experiment

The experimental arrangement is shown in Fig. 2.1. Here *S* is a narrow slit with a source of monochromatic light behind it. *A* and *B* are two narrow slits parallel to *S*. If a white screen is placed in the region beyond the slits, a pattern of bright and dark interference bands (*fringes*) are seen. Two waves that add constructively give a bright fringe. Two waves that add destructively produce a dark fringe. The bright and dark fringes occur alternately at equal distances.

2.2. Theory of Interference Fringes

S is a source of monochromatic light of wavelength λ (Fig. 2.2). A and B are two narrow slits equidistant from S. They act as two virtual coherent sources separated by a distance d. A screen is placed at a distance D from the coherent sources. The point C on the screen is equidistant from A and B. Therefore, the path difference $AC - BC = 0$. Both waves reach C in the same phase. The point C is, therefore, essentially bright.

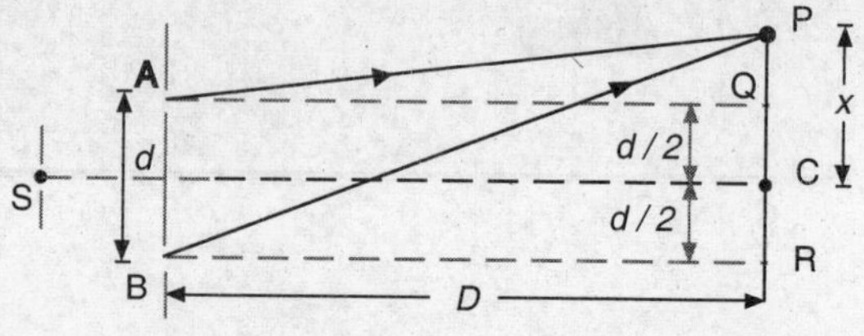

Fig. 2.2.

Path-difference at P : Consider a point P at a distance x from C. Two rays AP and BP starting from A and B reach P and interfere with each other.

The path-difference between the two waves = $BP - AP$

Here, $$PQ = x - \frac{d}{2} \quad \text{and} \quad PR = x + \frac{d}{2}$$

$$(BP)^2 = D^2 + \left(x + \frac{d}{2}\right)^2 \text{ and } (AP)^2 = D^2 + \left(x - \frac{d}{2}\right)^2$$

$$\therefore \quad (BP)^2 - (AP)^2 = \left[D^2 + \left(x + \frac{d}{2}\right)^2\right] - \left[D^2 + \left(x - \frac{d}{2}\right)^2\right] = 2xd$$

or $$(BP - AP) = \frac{2xd}{BP + AP} = \frac{2xd}{2D} \qquad (\because BP \approx AP \approx D)$$

$$\therefore \quad \text{Path-difference} = BP - AP = \frac{xd}{D} \qquad ...(1)$$

$$\text{Phase-difference} = \delta = \frac{2\pi}{\lambda}\left(\frac{xd}{D}\right) \qquad ...(2)$$

Position of bright fringes

If the path-difference is $m\lambda$, the point P is bright. Thus for m^{th} bright fringe, we have

$$\frac{x_m d}{D} = m\lambda \text{ where } m = 0,1,2,3,......$$

or $$x_m = \frac{D\lambda}{d}.m \qquad ...(3)$$

The central bright fringe ($m = 0$) is called the *zeroth - order maximum*.

The bright fringes with $m = 1, 2, 3,.......$ are called the first, the second, the third, etc., order fringes. Their positions are given by

$$x_1 = \frac{D}{d}\lambda,\ x_2 = \frac{D}{d}2\lambda,\ x_3 = \frac{D}{d}3\lambda,...\ x_m = \frac{D}{d}m\lambda,\quad x_{m+1} = \frac{D}{d}(m+1)\lambda$$

The fringe width β is the distance between two consecutive bright (or dark) fringes.

$$\therefore \textit{Fringe-width } \beta = x_{m+1} - x_m = \frac{D}{d}(m+1)\lambda - \frac{D}{d}m\lambda = \frac{\lambda D}{d} \qquad ...(4)$$

Position of dark fringes

If the path difference is an odd multiple of $\lambda/2$, the point P is dark. Thus for m^{th} dark fringe,

$$\frac{x_m d}{D} = (2m+1)\frac{\lambda}{2}, \text{ where } m = 0, 1, 2, 3, \ldots$$

or

$$x_m = (2m+1)\frac{D\lambda}{2d} \quad \ldots(5)$$

The positions of the dark fringes are given by

$$x_0 = \frac{\lambda D}{2d},\ x_1 = \frac{3\lambda D}{2d},\ x_2 = \frac{5\lambda D}{2d},\ x_m = \frac{(2m+1)\lambda D}{2d}, \text{ and } x_{m+1} = \frac{(2m+3)\lambda D}{2d}$$

$$\text{Fringe width} = \beta = x_{m+1} - x_m = \frac{\lambda D}{d} \quad \ldots(6)$$

Since D, d and λ are all constants, fringe-width is constant.

Further, the width of a bright fringe is equal to the width of a dark fringe.

Now, $\beta = \frac{\lambda D}{d}$. That is, the fringe-width β is directly proportional to D and λ and inversely proportional to d. In order to obtain well-separated fringes, β should be large. Since $\beta \propto 1/d$, attempts are made to keep d as small as possible.

2.3. Fresnel's Biprism

It consists of two small angled prisms ABC and ABD cemented along the face AB (Fig. 2.3). $\angle A$ is about 179°. The angles $\angle ACB$ and $\angle ADB$ are about 30′ (α). S is a narrow vertical slit illuminated by monochromatic light of wavelength λ. The beam of light diverging from S falls symmetrically on the faces AC and AD. These beams are deviated towards the base AB, and appear to be diverging from two virtual sources S_1 and S_2. Thus each half of the biprism produces a virtual image. The virtual images S_1 and S_2 act as two coherent sources. In the region MN of the screen, light reaches from both S_1 and S_2. Hence, interference fringes are obtained in the region MN.

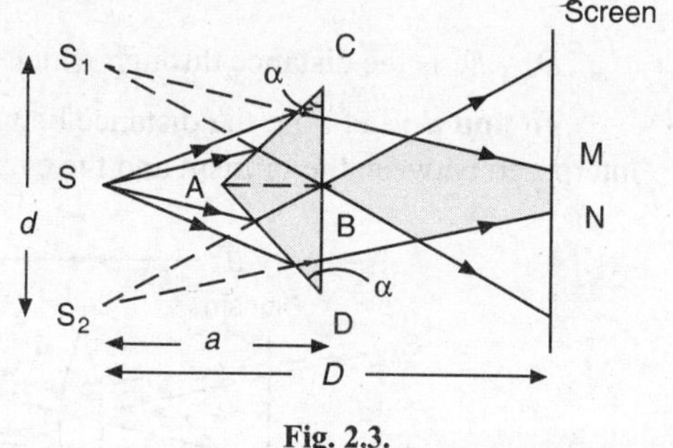

Fig. 2.3.

To find d : Let n = refractive index of the prism material,
a = distance between the slit and the biprism and
α = refracting angle of the biprism.

The deviation produced by each half of the prism = $\delta = (n-1)\alpha$.

∴ Distance between the two virtual sources,

$$d = 2\delta \times a = 2a(n-1)\alpha$$

Here, α is in radians.

Theory : For complete theory, refer to Article 2.2.

The fringe width $\beta = \lambda D/d$

Here, D = distance between source and screen.

Experimental determination of the wavelength of monochromatic light

The experiment is performed on an optical bench. The optical bench is about 1.5 m long. This bench is graduated in millimetres and carries four uprights (Fig.2.4). The first upright carries an adjustable slit, the second a biprism, the third a lens and the fourth a micrometer eyepiece. The heights of uprights are adjustable. The uprights can be made to move along and at right angles to the length of the optical bench.

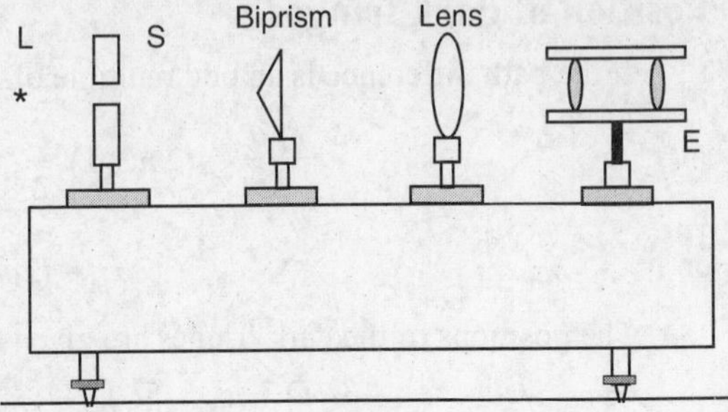

Fig. 2.4.

Illuminate the slit with monochromatic light. Adjust the slit, biprism and the eyepiece to be in the same straight line and at the same height. On making the slit narrow and vertical and looking through the eyepiece, interference fringes will be seen. Rotate the slit by means of the tangent screw so that the fringe system is most clear. In this position, the slit is parallel to the edge of the biprism.

The vertical cross-wire is made to coincide with the centre of a particular bright fringe. The position of the eyepiece is read on the scale (R_1). The eyepiece is moved transversely until the point of intersection of cross-wires moves through a known number (n) of bright fringes. The final reading of the eyepiece is noted (R_2). Then, fringe-width $\beta_1 = (R_1 \sim R_2)/n$.

Let D_1 = distance between the line of the two virtual sources and the focal plane of the eyepiece and d = distance between the virtual sources. Then $\beta_1 = \dfrac{D_1 \lambda}{d}$. But D_1 cannot be measured correctly. Hence the following method is adopted. Without disturbing the slit and the biprism, the eyepiece is moved away through a known distance (say, 20 cm). The band width β_2 in this case is also found out. Now, $\beta_2 = D_2 \lambda / d$.

$$\therefore \qquad \beta_2 - \beta_1 = \frac{\lambda}{d}(D_2 - D_1).$$

$D_2 - D_1$ is the distance through which the eyepiece is moved.

To find d : Let d be the distance between the two virtual sources S_1 and S_2. A convex lens is interposed between the biprism and the eyepiece (Fig. 2.5).

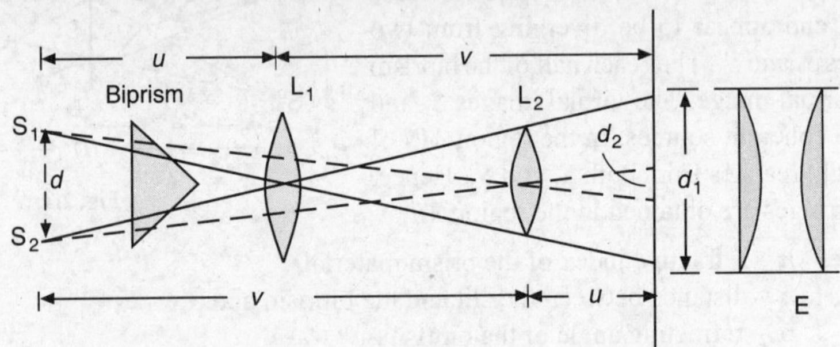

Fig. 2.5.

This lens now forms two real images of the virtual sources S_1 and S_2. The position of the lens is adjusted till the two clear magnified images are seen in the focal plane of the eyepiece. The distance d_1 between these two images is measured (d_1). Let the position of the lens be L_1. Let u and v be the distances of the object and the image respectively from the lens. Then from magnification formula, we have

$$\frac{v}{u} = \frac{d_1}{d} \qquad \text{...(1)}$$

Now the lens alone is moved towards the eyepiece. For the position L_2 of the lens, two clear diminished images of S_1 and S_2 are obtained. The distance d_2 between the images is measured. This is the conjugate position for the lens, *i.e.*, u and v have been interchanged. The magnification in this case is

$$\frac{u}{v} = \frac{d_2}{d} \qquad \text{...(2)}$$

Multiplying equations (1) and (2), $\frac{v}{u} \times \frac{u}{v} = \frac{d_1}{d} \times \frac{d_2}{d}$

or $$d_1 d_2 = d^2 \qquad \therefore d = \sqrt{d_1 d_2}$$

Knowing d_1 and d_2, d is calculated.

λ is calculated using the formula, $\lambda = \frac{(\beta_2 - \beta_1)d}{(D_2 - D_1)}$.

Example 1 : *Calculate the separation between the coherent sources formed by a biprism whose inclined faces make angles of 2° with its base, the slit being 0.1 m away from the biprism. (Refractive index of prism material = 1.5).*

Solution : Distance between virtual sources. $d = 2a\,(n - 1)\,\alpha$

Here, a = distance between the slit and the biprism = 0.1 m.

$$n = 1.5,\ \alpha = 2^\circ = (\pi/180) \times 2 \text{ radians},\ d = ?$$

$$d = 2 \times 0.1 \times (1.5 - 1) \times (\pi/180) \times 2 = 0.003489 \text{ m}.$$

Example 2 : *In a biprism experiment, the obtuse angle of the biprism is 178° and n= 1.524. If the slit to biprism distance is 0.05 m and biprism to screen is 0.65 m, calculate the fringe width for light of λ = 589.3 nm.*

Solution : Let α_1 and α_2 be the refracting angles of the biprism.

Total angular separation of the images is $(n - 1)(\alpha_1 + \alpha_2)$

(*i*) Here, $\alpha_1 + \alpha_2 = (180 - 178) \times (\pi/180) = 2\pi/180$ radians.

a = Distance between the slit and the biprism = 0.05 m.

Linear separation of the images

$$d = (n - 1)(\alpha_1 + \alpha_2)\,a = (1.524 - 1) \times (2\pi/180) \times 0.05$$

$$\therefore \qquad d = 9.141 \times 10^{-4} \text{ m}.$$

(*ii*) Here, $\lambda = 589.3 \times 10^{-9}$ m, $d = 9.141 \times 10^{-4}$ m and

D = distance between the slit and the screen = (0.05 + 0.65) m = 0.7 m

$$\left.\begin{matrix}\text{Fringe}\\ \text{width}\end{matrix}\right\} = \beta = \frac{\lambda D}{d} = \frac{(589.3 \times 10^{-9})(0.7)}{9.141 \times 10^{-4}} = 4.513 \times 10^{-4} \text{ m}.$$

Example 3 : *In a biprism experiment, at a certain position of the eyepiece, the fringe width obtained is 0.2 mm. When the eyepiece is moved away by 50 cm, the fringe width becomes 0.3 mm. If the distance between the two sources is 0.3 cm, find the wavelength of the light used.*

Solution : We have, $\beta_2 - \beta_1 = \frac{\lambda}{d}(D_2 - D_1)$

Here, $\beta_1 = 0.2 \text{ mm} = 2 \times 10^{-4}$ m, $\beta_2 = 0.3 \text{ mm} = 3 \times 10^{-4}$ m,

$d = 0.3 \text{ cm} = 3 \times 10^{-3}$ m and $D_2 - D_1 = 50 \text{ cm} = 0.5 \text{ m} : \lambda = ?$

$$\lambda = \frac{(\beta_2 - \beta_1)d}{(D_2 - D_1)} = \frac{(3 \times 10^{-4} - 2 \times 10^{-4})(3 \times 10^{-3})}{0.5} = 6 \times 10^{-7} \text{ m}.$$

Example 4 : *Determine the wavelength of light used in a biprism experiment making use of the following data: Fringe-width = 0.201 mm, distance between the slit and the cross-wires of the eyepiece = 73.44 cm; distance between the slit images when a convex lens is used in the two conjugate positions respectively = 3.732 mm and 1.242 mm.*

Solution : Here, $\beta = 0.201$ mm $= 2.01 \times 10^{-4}$ m; $D = 0.7344$ m, $d_1 = 3.732 \times 10^{-3}$ m,

$$d_2 = 1.242 \times 10^{-3} \text{ m}, \lambda = ?$$

$$\beta = \lambda D/d \text{ or } \lambda = \beta\, d/D = \beta\sqrt{d_1 d_2}\,/\,D$$

$$\therefore \quad \lambda = \frac{\beta\sqrt{d_1 d_2}}{D} = \frac{(2.01\times10^{-4})\sqrt{(3.732\times10^{-3})\,(1.242\times10^{-3})}}{0.7344}$$

$$= 5.892 \times 10^{-7} \text{ m} = 589.2 \text{ nm}.$$

2.4. Displacement of Fringes

Let A and B be two coherent monochromatic sources (Fig. 2.6). A thin transparent plate of thickness t and refractive index n is introduced in the path of one of the two interfering beams say AP. Light reaching the point P from A has to traverse a distance t in the plate and a distance $(AP-t)$ in air.

Now the path-difference between BP and AP

$$= BP - [(AP - t) + nt] = (BP - AP) - (n-1)\,t$$

The point P is bright if this path difference is $m\lambda$.

i.e.,
$$(BP - AP) - (n-1)\,t = m\lambda$$

$$BP - AP = xd/D$$

$$\frac{xd}{D} - (n-1)\,t = m\lambda$$

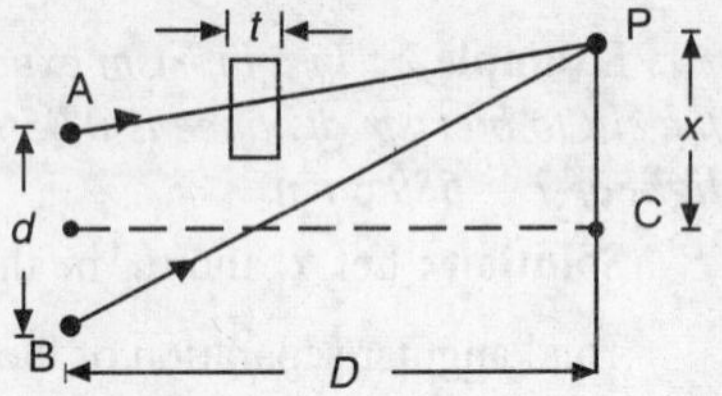

Fig. 2.6.

or
$$x = \frac{D}{d}[(n-1)\,t + m\lambda] \qquad ...(1)$$

The displacement x_0 of the central bright fringe when the glass plate is placed in the path AP is obtained by putting $m = 0$ in Eq. (1).

$$\therefore \quad x_0 = \frac{D}{d}(n-1)\,t \qquad ...(2)$$

To obtain the positions of the first, the second, the third, etc., bright fringes, we put $m = 1, 2, 3, ...$ in Eq. (1).

The fringe-width $= \beta = x_{m+1} - x_m$

$$\therefore \quad \beta = \frac{D}{d}[(n-1)t + (m+1)\lambda] - \frac{D}{d}[(n-1)\,t + m\lambda] = \frac{D}{d}\lambda \qquad ...(3)$$

β is the same for all fringes. This shows that the fringe width is not affected by the introduction of the glass plate. Similarly, it can be shown that the displacement of any dark fringe is also $\frac{D}{d}(n-1)\,t$

Eq.(2) enables us to determine the thickness of extremely thin transparent sheets (like that of mica) by measuring the displacement of the central fringe. From eqn. (3), $D/d = \beta/\lambda$. Substituting for D/d in Eq. (2), we get,

$$x_0 = \frac{\beta}{\lambda}(n-1)\,t \,. \qquad ...(4)$$

To find the thickness of a mica sheet : By measuring the displacement x_0 of the central fringe and the fringe width β, the thickness of the plate can be calculated. When monochromatic light is used, the fringes observed are all alike. Hence it is not possible to measure the shift x_0 of the central bright fringe. With white light, the central fringe is white while all others are coloured. The cross-wire is set on the white fringe and the micrometer reading R_1 is noted. Now the mica sheet is introduced in the path of one of the interfering beams. The whole interference pattern shifts. The cross-wire is again set on the white fringe and the micrometer reading R_2 is noted. Then $(R_1 \sim R_2)$ gives the shift x_0. The fringe-width β is noted as usual, by using monochromatic light of known wavelength λ. n for mica is known. t can be calculated using Eq. (4). or n can be calculated if t is known.

Example 1 : *A thin film of plastic of refractive index 1.45 for light of wavelength 589 nm is inserted normally in the path of one of the interfering beams. The central bright band shifts through 5 fringes. Find the thickness of the film.*

Solution : Here, $n = 1.45$, $\lambda = 589 \times 10^{-9}$ m, x_0 = displacement of the central fringe = $5\,\beta$. $t = ?$

$$x_0 = \frac{\beta}{\lambda}(n-1)\,t \quad \text{or} \quad t = \frac{x_0\,\lambda}{\beta\,(n-1)}$$

$$t = \frac{5\beta \times \lambda}{\beta\,(n-1)} = \frac{5\lambda}{(n-1)} = \frac{5 \times (589 \times 10^{-9})}{1.45 - 1} = 6.544 \times 10^{-6} \text{ m.}$$

Example 2 : *A transparent plate of thickness 10^{-5} m is placed in the path of the interfering beam of a biprism experiment using light of wavelength 500 nm. If the central fringe shifts by a distance equal to the width of ten fringes, calculate the refractive index of the material of the plate.*

Solution : Here, $t = 10^{-5}$ m, $\lambda = 500$ nm $= 500 \times 10^{-9}$ m, $x_0 = 10\,\beta$, $n = ?$

$$x_0 = \frac{\beta}{\lambda}(n-1)\,t \quad \text{or} \quad 10\beta = \frac{\beta}{\lambda}(n-1)\,t$$

$$10 = \frac{(n-1)\,t}{\lambda} \quad \text{or} \quad (n-1)\,t = 10\lambda \quad \text{or} \quad n = \frac{10\lambda}{t} + 1$$

$$\therefore \qquad n = \frac{10 \times (500 \times 10^{-9})}{10^{-5}} + 1 = 1.5$$

Example 3 : *On introducing a thin sheet of glass (n= 1.5) in one of the two interfering beams in a biprism experiment with white light, the shift in central fringe is observed to be 4.5 mm. With sodium light (λ = 589.3 nm) the same arrangement gives fringe width equal to 0.225 mm. Calculate the thickness of the glass plate.*

Solution : Here, $n = 1.5$, $x_0 = 4.5 \times 10^{-3}$ m, $\lambda = 589.3 \times 10^{-9}$m, $\beta = 0.225 \times 10^{-3}$ m; $t = ?$

$$x_0 = \frac{\beta}{\lambda}(n-1)\,t \quad \text{or} \quad t = \frac{x_0\,\lambda}{\beta\,(n-1)}$$

$$\therefore \qquad t = \frac{(4.5 \times 10^{-3}) \times (589.3 \times 10^{-9})}{(0.225 \times 10^{-3})\,(1 \cdot 5 - 1)} = 2.356 \times 10^{-5}\ m.$$

Example 4 : *Two monochromatic beams meeting at a point have a path-difference of 8 wavelengths. A mica sheet of thickness 2.4 microns is introduced in the path of one of the beams. If λ = 600 nm and n=1.5, calculate the path difference now.*

Solution. Refer to Fig. 2.6.

Path-difference before introducing the mica sheet

$$BP - AP = 8\lambda = 8 \times (600 \times 10^{-9})\text{ m} = 4.8 \times 10^{-6}\text{ m.}$$

Path-difference after introducing the mica sheet

$$(BP - AP) - (n - 1)\,t = 4.8 \times 10^{-6}\text{m} - (1{\cdot}5 - 1) \times (2.4 \times 10^{-6})\text{ m} = 3.6 \times 10^{-6}\text{ m}$$

Stoke's law : When a light wave is reflected from the surface of an optically denser medium, it suffers a phase change of π. But it suffers no change in phase when reflected at the surface of optically rarer medium.

2.5. Colours of Thin films

When white light is incident on a thin film (such as soap film or film of oil floating over water) different colours are seen. It can be explained by considering interference between light reflected from the top and the bottom surface of a thin film. Interference in the case of thin films takes place due to (*i*) reflected light and (*ii*) transmitted light.

Interference by reflected light : Consider a thin parallel-sided transparent film of thickness t, refractive index n and surrounded by air (Fig. 2.7). A ray AB of monochromatic light is incident on the upper surface, at an angle i. At B, the light undergoes partial reflection in the direction BC and partial refraction along BD. The angle of refraction is r. At D, it is again partially reflected along DE and partially refracted along DI. The reflected light DE is partially refracted in the direction EF. Let us calculate the path difference between the rays BC and $BDEF$. Let EN and BM be perpendiculars to BC and DE respectively. Let $t = BH =$ thickness of the film. Produce ED and BH to meet at G. Then $\angle HGD = r$.

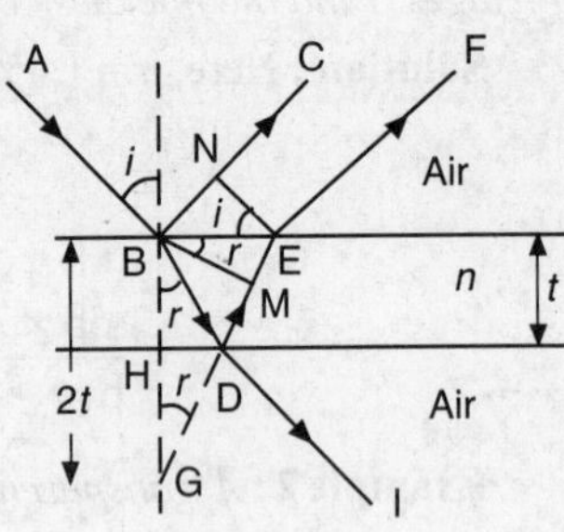

Fig. 2.7.

The path-difference between the two reflected rays $= n\,(BD + DE) - BN$

$$= n\,(BD + DM + ME) - BN \qquad ...(1)$$

Now,
$$n = \frac{\sin i}{\sin r} = \frac{BN/BE}{ME/BE} = \frac{BN}{ME}$$

or
$$BN = n\,ME \qquad ...(2)$$

$\therefore$
$$\text{Path-difference} = n\,(BD + DM + ME) - n\,ME = n\,(BD + DM)$$
$$= n\,(GD + DM) = n\,GM \qquad ...(3)$$

In the Δ BGM, $\cos r = \dfrac{GM}{BG} = \dfrac{GM}{2t}$

or
$$GM = 2\,t\cos r \qquad ...(4)$$

$\therefore$ Path-difference $= n\,GM = 2nt\cos r$ ...(5)

The reflected ray BC suffers a path-difference of $\lambda/2$ on reflection at B.

$\therefore$ Net path-difference between BC and $EF = 2nt\cos r - \lambda/2$.

The film appears *bright* if
$$2nt\cos r - \lambda/2 = m\lambda$$

or
$$2nt\cos r = (2m + 1)\,\frac{\lambda}{2}$$
$$m = 0, 1, 2, 3...... \qquad ...(6)$$

The film appears *dark* if

$$2nt\cos r - \frac{\lambda}{2} = (2m+1)\,\frac{\lambda}{2}. \quad m = 0, 1, 2, \text{... etc}$$

or $$2nt \cos r = (m+1)\,\lambda$$

m is an integer. Therefore, $(m + 1)$ is also an integer.

or $$2nt \cos r = m\,\lambda, \quad m = 0, 1, 2, 3, 4.... \text{ etc.} \quad ...(7)$$

Transmitted system : Consider a thin film of thickness t and refractive index n (Fig.2.8). A ray AB of monochromatic light is incident on the upper surface at an angle i. It is partly refracted along BD. The angle of refraction is r. At D, it is again partially reflected along DE and partially refracted along DI. The ray DE after reflection at E, finally emerges along FK. Let DM and FN be perpendiculars to EF and DI respectively. Produce FE and DH to meet at G. There is no phase change due to reflections at D and E both of which are at a rarer medium.

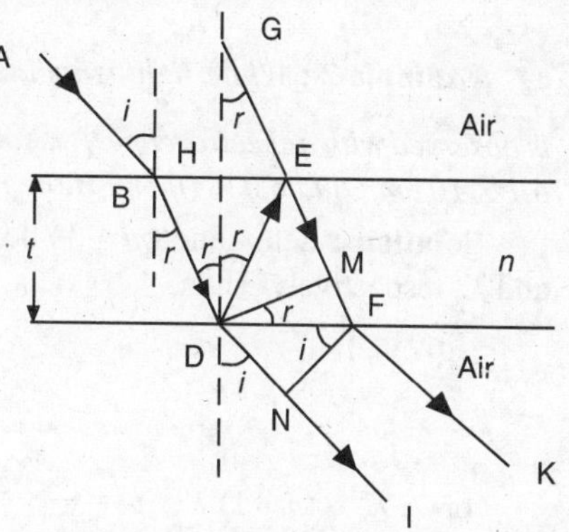

Fig. 2.8.

Therefore, the path difference between the two transmitted rays DI and FK is

$= n\,(DE + EF) - DN = n\,(DE + EM + MF) - DN.$

Now,

$$n = \frac{\sin i}{\sin r} = \frac{DN / DF}{MF / DF} = \frac{DN}{MF} \quad \text{or} \quad DN = n\,MF$$

$\therefore$ Path-difference $= n\,(DE + EM + MF) - n\,MF = n\,(DE + EM)$

$= n\,GM$

In the $\Delta\, DGM$, $\dfrac{GM}{2t} = \cos r$ or $GM = 2t \cos r$

$\therefore$ Path-difference $= n\,GM = 2\,nt \cos r.$ ...(1)

The film appears *bright* if

$$2nt \cos r = m\lambda, \quad m = 0, 1, 2, 3,....\text{etc}, \quad ...(2)$$

The film appears *dark* if

$$2nt \cos r = (2m + 1)\lambda\,/2, \; m = 0, 1, 2, 3,.....\text{etc}. \quad ...(3)$$

These conditions for transmitted light are just opposite to those for the reflected system.

2.6. Production of Colours in Thin Films

When white light is incident on a thin film, the light rays are reflected from the top and bottom surfaces of the film. The condition for brightness or darkness depends upon (*i*) n, (*ii*) t and (*iii*) r. t and r are constant. White light contains a number of wavelengths (colours). n for each colour is different. Condition for brightness in the reflected system is $2nt \cos r = (2m + 1)\,\lambda/2$. The colour for which this condition is satisfied, will be present with maximum intensity. Other colours in its neighbourhood will have lesser intensity. Colours for which the condition of darkness $2nt \cos r = m\lambda$ is satisfied will be absent. Therefore, the *film will appear coloured.*

Example 1 : *A parallel beam of sodium light ($\lambda = 589 \times 10^{-9} m$) is incident on a thin glass plate (n=1.5) such that the angle of refraction into the plate is 60°. Calculate the smallest thickness of the plate which will make it appear dark by reflection.*

Solution : For a dark band in the reflected system

$$2\,n\,t \cos r = m\,\lambda$$

For the least thickness of the plate $m = 1$

$\therefore$ For the film to appear dark, the smallest thickness t is given by

$$2\,n\,t \cos r = \lambda \text{ or } t = \lambda\,/\,(2\,n \cos r)$$

Here, $\lambda = 589 \times 10^{-9}$ *m, n* = 1.5 and $r = 60°$.

$$\therefore \quad t = \frac{\lambda}{2n \cos r} = \frac{589 \times 10^{-9}}{2 \times 1.5 \times \cos 60°} = 392.7 \times 10^{-9} \text{ m.}$$

Example 2 : *White light is incident on a soap film at an angle* $\sin^{-1} \frac{4}{5}$ *and the reflected light is observed with a spectroscope. It is found that two consecutive dark bands correspond to wavelengths* 6.1×10^{-7}*m. and* 6.0×10^{-7}*m. If the refractive index of the film be 4/3, calculate its thickness.*

Solution : Let *m* and $(m + 1)$ be the orders of the consecutive dark bands for wavelengths λ_1 and λ_2 respectively. Then,

$$2nt \cos r = m\,\lambda_1 \quad ...(1)$$

$$2nt \cos r = (m + 1)\,\lambda_2 \quad ...(2)$$

or $m\,\lambda_1 = (m+1)\,\lambda_2 \cdot m = \dfrac{\lambda_2}{\lambda_1 - \lambda_2} = \dfrac{6 \times 10^{-7}}{(6.1 \times 10^{-7} - 6 \times 10^{-7})} = 60$

To find cos r **:** $\sin i = \dfrac{4}{5}; n = \dfrac{4}{3} = \dfrac{\sin i}{\sin r}$. $\therefore \sin r = \dfrac{\sin i}{n} = \dfrac{4/5}{4/3} = 0.6$

$$\cos r = [1 - \sin^2 r]^{1/2} = 0.8$$

Now,
$$2nt \cos r = m\,\lambda_1$$

$$t = \frac{m\lambda_1}{2n \cos r} = \frac{60 \times 6.1 \times 10^{-7}}{2 \times (4/3) \times 0.8} = 1.72 \times 10^{-5} \text{ m.}$$

2.7. Wedge-shaped Film

Consider a wedge-shaped film of refractive index *n* enclosed by two plane surfaces *OP* and *OQ* inclined at an angle θ (Fig.2.9). The thickness of the film increases from *O* to *P*. When the film is illuminated by a parallel beam of monochromatic light, interference occurs between the rays reflected at the upper and lower surfaces of the film. So equidistant alternate dark and bright fringes are observed. The fringes are parallel to the line of intersection of the two surfaces. The interfering rays are *AB* and *DE*, both originating from the same incident ray *SA*.

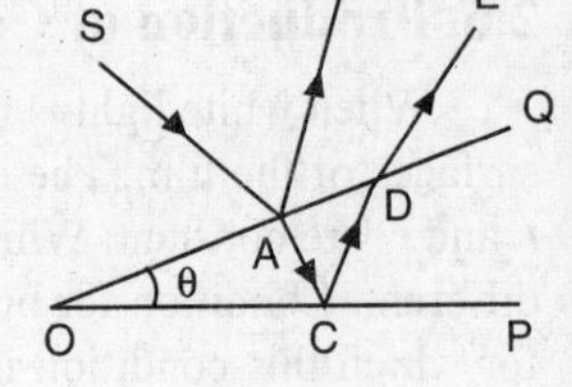

Fig. 2.9.

Expression for the fringe width : The condition for a dark fringe is $2nt \cos r = m\,\lambda$. Here for air $n = 1$. For normal incidence $\cos r = \cos 0 = 1$.

Suppose the *m*th dark fringe is formed where the thickness of the air film is t_m (Fig. 2.10). Then,

$$2 \times 1 \times t_m \times 1 = m\,\lambda$$

or
$$2t_m = m\,\lambda \quad ...(1)$$

Suppose the $(m + 1)$ th dark fringe is formed where the thickness of the air film is t_{m+1}. Then,

$$2t_{m+1} = (m + 1)\,\lambda \quad ...(2)$$

Subtracting (1) from (2), $2(t_{m+1} - t_m) = \lambda$ ....(3)

Let x_{m+1} and x_m be the distances of the $(m + 1)$ th and *m*th dark fringes from *O*.

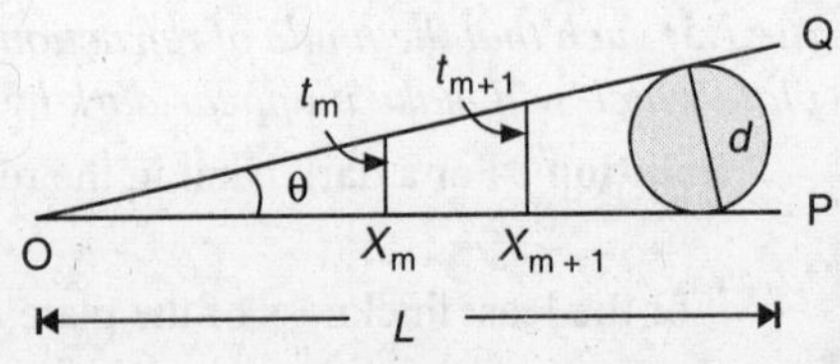

Fig. 2.10.

d = diameter of the wire; L= distance between O and the wire. Then,

$$\frac{t_{m+1}}{x_{m+1}} = \frac{t_m}{x_m} = \frac{d}{L} = \theta$$

$$\therefore \quad t_{m+1} = \frac{d}{L} x_{m+1}; \quad t_m = \frac{d}{L} x_m$$

Substituting these values in Eq. (3), we get

$$2\frac{d}{L}(x_{m+1} - x_m) = \lambda$$

But $x_{m+1} - x_m = \beta$ = fringe width.

or $$2\frac{d}{L}\beta = \lambda$$

$$\therefore \quad \beta = \frac{\lambda L}{2d} = \frac{\lambda}{2\theta}$$

d, λ and L are constants. Therefore, fringe width β is constant. Similarly, if we consider two consecutive bright fringes, the fringe width β will be the same.

Experiment to measure the diameter of a thin wire : An air wedge is formed by inserting the wire between two glass plates. Monochromatic light is reflected vertically downwards on to the wedge by the inclined glass plate G (Fig. 2.11). A travelling microscope M with its axis vertical is placed above G. The microscope is focused to get clear dark and bright fringes. The fringe width (β) is measured. The length (L) of the wedge also is measured. Knowing λ, the diameter (d) of the wire is calculated using the formula,

$$d = \frac{\lambda L}{2\beta}.$$

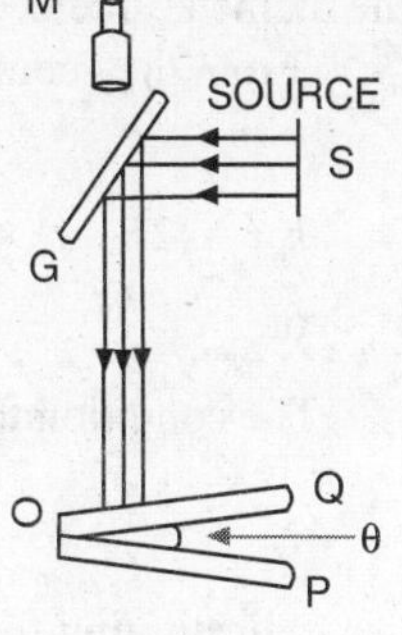

Fig. 2.11.

Testing a surface for planeness : A wedge shaped air film is formed between an optically plane glass plate (OP) and the surface under test (OQ). The fringes will be *straight* if the surface under test is *perfectly plane*. If the surface OQ is not *perfectly plane*, the fringes will be *irregular in shape*. In practice, perfectly plane surfaces are produced by polishing the surfaces and testing them from time to time, until the fringes are straight. In testing for planeness, an extended source of light should be used.

Example 1 : *Light of wavelength 600 nm falls normally on a thin wedge shaped film of refractive index 1.4, forming fringes that are 2mm apart. Find the angle of the wedge.*

Solution : Consider a wedge-shaped film OPQ of refractive index n (Fig.2.10).

$$2nt_m = m\lambda \,;\; 2nt_{m+1} = (m+1)\lambda \qquad \therefore 2n\,(t_{m+1} - t_m) = \lambda$$

$$2n\,(\theta\, x_{m+1} - \theta\, x_m) = \lambda \text{ or } 2n\,\theta\,(x_{m+1} - x_m) = \lambda$$

$$2n\,\theta\,\beta = \lambda \quad \therefore \theta = \frac{\lambda}{2n\,\beta}$$

Here, $\lambda = 600 \times 10^{-9}$ m; $n = 1.4$; $\beta = 2 \times 10^{-3}$ m;

$$\therefore \quad \theta = \frac{600 \times 10^{-9}}{2 \times 1.4 \times (2 \times 10^{-3})} = 1.071 \times 10^{-4} \text{ radian}$$

2.8. Newton's Rings

A plano-convex lens of large radius of curvature is placed with its convex surface in contact with a plane glass plate. Then, an air-film is formed between the lower surface of the lens AOB and the upper surface of the plate POQ (Fig. 2.12). The thickness of the air film is zero at the point of contact O and gradually increases from the point of contact outwards. If monochromatic light is allowed to fall normally on this film, a system of alternate bright and dark concentric rings is formed in the air film. They are called *Newton's rings*. The thickness of air-film remains constant along a circle with its centre at O. Hence, the fringes are in the form of concentric circles.

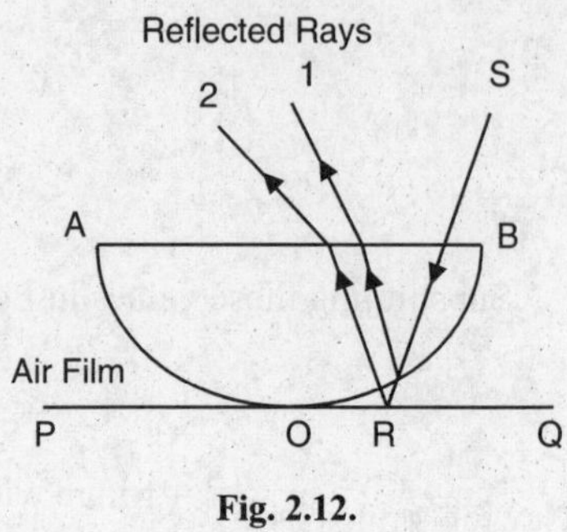

Fig. 2.12.

Newton's rings are formed as a result of interference between the light waves reflected from the upper and lower surface of the air film. 1 and 2 are the interfering rays corresponding to an incident ray SR.

Expression for the radii of the rings : R is the radius of curvature of the lens and A is the centre of curvature (Fig. 2.13). Let there be mth dark ring at point C. The thickness of the air film at point C is $OB = t$. The radius of the mth dark ring is $r_m = BC$. Let DC be the chord and OE the diameter intersecting at right angles at B.

Fig. 2.13.

From the geometry of the circle,

$$DB \times BC = EB \times BO$$

$r_m^2 = (2R - t) \times t$. In practice $t << R$, so that $2R - t \approx 2R$

or $$r_m^2 = 2Rt \qquad ...(1)$$

The condition for a dark ring is $2nt = m\lambda$ where $m = 0, 1, 2, 3, ...$

$$\therefore \qquad t = \frac{m\lambda}{2n} \qquad ... (2)$$

Substituting for t in Eq. (1), we get

$$r_m^2 = \frac{2Rm\lambda}{2n} = \frac{Rm\lambda}{n}$$

$$\therefore \qquad r_m = \sqrt{\frac{mR\lambda}{n}} \qquad ... (3)$$

Thus *the radii of the dark rings are proportional to the square roots of the natural numbers.*

Bright rings : The condition for a bright ring is

$$2\,nt = (2m - 1)\,\lambda/2 \qquad (m = 1, 2, 3, ... \text{ etc.})$$

or $$t = \frac{(2m-1)\,\lambda/2}{2n} \qquad ...(4)$$

Substituting for t in Eq. (1), we get

$$r_m^2 = \frac{2R\,(2m-1)\,\lambda/2}{2n} = \frac{(2m-1)\,R\lambda}{2n}$$

$$\therefore \text{ Radius of } m^{\text{th}} \text{ bright ring} = r_m = \sqrt{\frac{(2m-1)\,R\lambda}{2n}} \qquad ... (5)$$

Thus *the radii of the bright rings are proportional to the square roots of odd natural numbers.* The diameters of mth dark and bright rings are

$$D_m = \sqrt{\frac{4mR\lambda}{n}} \quad \text{Dark ring} \qquad ...(6)$$

$$D_m = \sqrt{\frac{2(2m-1)R\lambda}{n}} \quad \text{Bright ring} \qquad ...(7)$$

2.9. Determination of Wavelength of Sodium Light by Newton's Rings

Experimental arrangement : Fig. 2.14 shows an experimental arrangement for producing Newton's rings by reflected light. S is an extended source of monochromatic light. The light from S is rendered parallel by a convex lens L_1. These horizontal parallel rays fall on a glass plate G at 45°, and are partly reflected from it. This reflected beam falls normally on the lens L placed on the glass plate PQ. Interference occurs between the rays reflected from the upper and lower surfaces of the film. The interference rings are viewed with a microscope M focused on the air film.

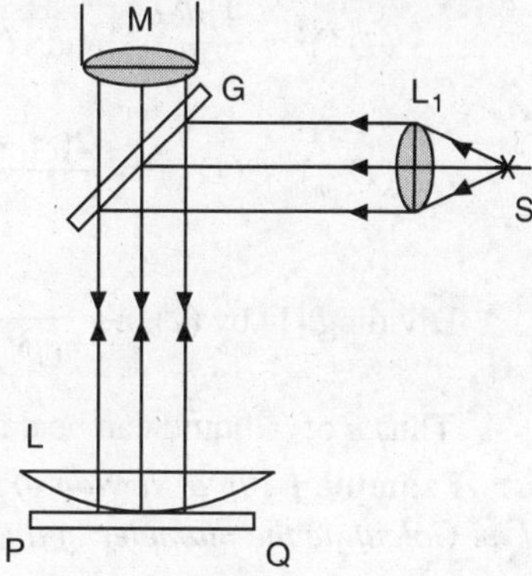

Fig. 2.14.

Procedure : With the help of the travelling microscope the diameters of a number of *dark rings* are measured. The position of the microscope is adjusted to get the centre of Newton's rings at the point of intersection of the cross-wires. The microscope is moved until one cross wire is tangential to the 16th dark ring. The microscope reading is taken. Then the microscope is moved such that the cross-wire is successively tangential to 12th, 8th and 4th dark rings respectively. The readings are noted in each case. Readings corresponding to the same rings are taken on the other side of the centre. The readings are tabulated as follows :

No. of ring	Reading of travelling microscope		Diameter of ring $D = a \sim b$	D^2	$D_m^2 - D_p^2$
	Left (a)	Right (b)			
16					
12					
8					
4					

Average $(D_m^2 - D_p^2) =$

The average value of $(D_m^2 - D_p^2)$ is found.

For an air film $n = 1$.

The diameters of pth and mth dark rings are given by

$$D_p^2 = 4pR\lambda \text{ and } D_m^2 = 4mR\lambda.$$

$$D_m^2 - D_p^2 = 4(m-p)R\lambda$$

$$\therefore \quad \lambda = \frac{D_m^2 - D_p^2}{4(m-p)R}.$$

The radius of curvature R of the lower surface of the lens is found by Boys' method. Substituting this value of R and the average value of $(D_m^2 - D_p^2)$ with $(m-p) = 8$ in the above equation, λ is calculated.

2.10. Determination of Refractive Index of a Liquid by Newton's Rings

The experiment is performed with an air film between the glass plate and the lens as described above. The diameters of pth and mth dark rings with air are given by,

$$D_p^2 = 4pR\lambda \text{ and } D_m^2 = 4mR\lambda$$

$$\therefore \quad D_m^2 - D_p^2 = 4(m-p)R\lambda \qquad ...(1)$$

The transparent liquid of refractive index n is introduced between the lens L and the glass plate G (Fig. 2.15). Then a plano-concave film of liquid is formed between the lens and the glass plate. The diameters of pth and mth dark rings are determined.

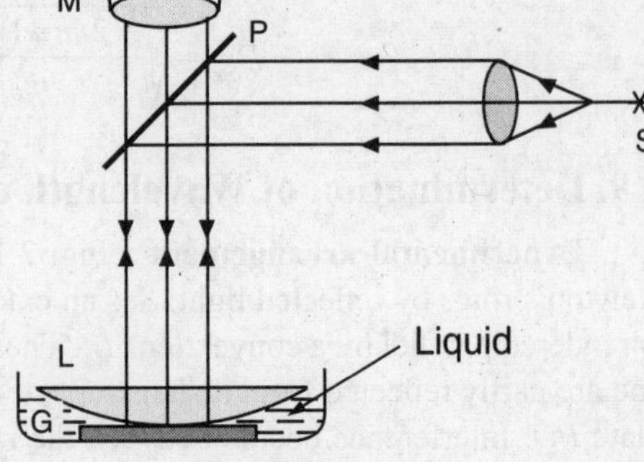

Fig. 2.15.

With liquid film, the diameters of pth and m^{th} dark rings are given by

$$(D_p')^2 = \frac{4pR\lambda}{n} \text{ and } (D_m')^2 \frac{4mR\lambda}{n}$$

$$\therefore \quad (D'_m)^2 - (D'_p)^2 = \frac{2(m-p)Rl}{n} \qquad ...(2)$$

Dividing (1) by (2), $n = \dfrac{D_m^2 - D_p^2}{(D'_m)^2 - (D'_p)^2}$

Thus n of a liquid can be determined by measuring the diameters with and without the liquid.

Example 1 : *In a Newton's ring system, the radius of curvature of the lower face of the lens is 1 m. Calculate the diameter of the 10th bright fringe, if the wavelengtzh of light used is 59×10^{-8} m.*

Solution : Here, $R = 1\text{m}; \lambda = 59 \times 10^{-8}$ m ; $n = 1$, $m = 10$; $D_{10} = ?$

$$D_m = \sqrt{\frac{2(2m-1)R\lambda}{n}}$$

$$\therefore \quad D_{10} = \sqrt{\frac{2 \times (2 \times 10 - 1) \times 1 \times 59 \times 10^{-8}}{1}} = 4.735 \times 10^{-3} \text{ m}$$

Example 2 : *Newton's rings are formed with reflected light of wavelength 5.895×10^{-7} m with a liquid between the plane and the curved surfaces. The diameter of the 5th dark ring is 0.003m and the radius of curvature of the curved surface is 1 metre. Calculate the refractive index of the liquid.*

Solution : $$D_m = \sqrt{\frac{4mR\lambda}{n}} \text{ or } n = \frac{4mR\lambda}{D_m^2}$$

Here, $\lambda = 5.895 \times 10^{-7}$ m, $m = 5$, $D_m = D_5 = 0.003$ m; $R = 1$m; $n = ?$

$$n = \frac{4mR\lambda}{D_m^2} = \frac{4 \times 5 \times 1(5.895 \times 10^{-7})}{(0.003)^2} = 1.31.$$

Example 3 : *Newton's rings are formed with red light of $\lambda = 670$ nm. The radius of the 20th ring is found to be 1.1×10^{-2} m. Find the radius of curvature of the lens and the radius of 30th ring.*

Solution : Let us consider *dark* rings and air-film ($n = 1$).

Here, $\lambda = 670 \times 10^{-9}$m; $m = 20$; $r_m = r_{20} = 1.1 \times 10^{-2}$m $\quad R = ?$

$$r_m = \sqrt{mR\lambda}$$

$$\therefore \qquad R = \frac{r_m^2}{m\lambda} = \frac{(1.1\times10^{-2})^2}{20\times(670\times10^{-9})} = 9\ 29 \text{ m}.$$

$$\text{Radius of } 30^{\text{th}} \text{ ring } r_{30} = \sqrt{mR\lambda} = \sqrt{30\times9.029\times(670\times10^{-9})}$$
$$= 1.347\times10^{-2} \text{ m}.$$

Example 4 : *The diameter of the mth ring changes from 1.2 cm to 1 cm when the air space between the lens and the plate is replaced by a transparent liquid. Find the refractive index of the liquid.*

Solution : $(D_m{}^2)$ liquid $= 4mR\lambda/n$; $(D_m{}^2)_{air} = 4mR\lambda$

$$\therefore \qquad n = \frac{\left(D_m^2\right)_{\text{air}}}{\left(D_m^2\right)_{\text{liquid}}} = \frac{\left(1.2\times10^{-2}\right)^2}{\left(1\times10^{-2}\right)^2} = 1.44.$$

Example 5 : *Explain briefly why the fringes in Newton's rings arrangement are circular and in an air wedge are straight and parallel.*

Ans. In both the arrangements, each fringe is the locus of points of equal thickness of the film. In Newton's rings arrangement, the points of equal thickness of film lie on circles with the point of contact of the lens and the plate as centre. Hence the fringes are concentric circles.

In case of wedge-shaped air-film, the loci of equal thickness are straight lines parallel to the edge of the wedge. Hence the fringes are straight and parallel.

Example 6 : *In a Newton's rings experiment, the diameter of the 5th ring was 0.356 cm and the diameter of 20th ring was 0.618 cm. Find the radius of curvature of the plano convex lens if the wavelength of light used is 589.3 nm.*

Solution. $D_p = D_5 = 0.356\times10^{-2}\ m$; $D_m = D_{20} = 0.618\times10^{-2} m$; $\lambda = 589.3\times10^{-9} m$.

$$R = \frac{D_m^2 - D_p^2}{4(m-p)\lambda} = \frac{(0.618\times10^{-2})^2 - (0.356\times10^{-2})^2}{4\times15\times(589.3\times10^{-9})} = 0.7216\,m.$$

2.11. Michelson's Interferometer

Principle : Here, the two interfering beams are formed by *division of amplitude*. The amplitude of the light beam from an extended source is divided into two parts of equal intensity by partial reflection and refraction. These beams are sent in two perpendicular directions. The two beams are finally brought together after reflection from plane mirrors to produce interference fringes.

Apparatus : M_1 and M_2 are *front silvered plane mirrors* (Fig. 2.16). The two mirrors are mounted vertically on two arms at right angles to each other. The planes of the mirrors can be slightly tilted with the fine screws at their backs. The mirror M_2 is fixed. The mirror M_1 can be moved parallel to itself by means of a very sensitive micrometer screw. G_1 and G_2 are two plane parallel glass plates of equal thickness. The plate G_1 is semi-silvered on the back side. G_1 is a *beam splitter; i.e.*, a beam incident on G_1 is partially reflected and partially transmitted. G_1 is inclined at an angle of 45° to the incident beam. G_2 is called the *compensating plate*. S is a light source.

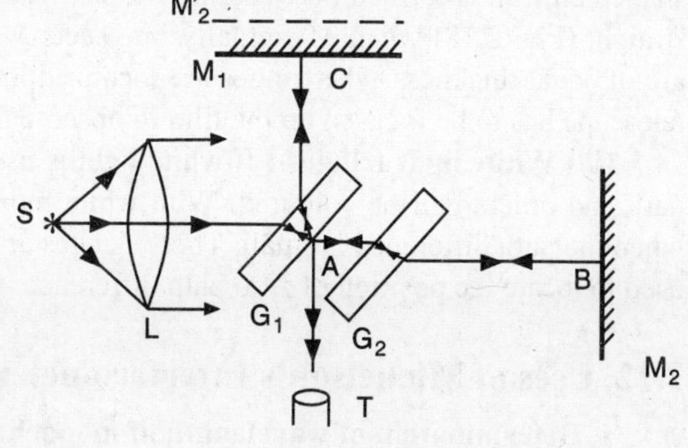

Fig. 2.16.

Working : Light from the source S is rendered parallel by a lens L and falls on the glass plate G_1 at an angle of 45°. At the back surface of G_1, it is partly reflected along AC and partly transmitted along AB. The reflected beam moves towards mirror M_1 and falls normally on it. It is reflected back along the same path and emerges out along AT. The transmitted ray AB falls normally on the mirror M_2. It is reflected along the same path. After reflection at the back surface of G_1, it moves along AT. The two emergent beams have been derived from a single incident beam and are, therefore, *coherent*. The two beams produce interference under suitable conditions.

Function of the compensating plate G_2 : The reflected ray AC passes through G_1 thrice. But the transmitted ray AB passes through G_1 only once. That is why a second plate G_2 of the same thickness and inclination as G_1 is introduced. Thus the function of the plate G_2 is only to equalise the optical paths traversed by both the beams.

Types of Fringes

***(i)* Circular fringes :** Concentric circular fringes are obtained when both the mirrors M_1 and M_2 are mutually perpendicular. The image of M_2 is at M_2' parallel to M_1 (Fig. 2.17). Hence, M_2' and M_1 form the equivalent of a parallel air film. The effective thickness of the air film is varied by moving mirror M_1 **parallel** to itself. Let the eye or the telescope be set along a **direction** making an angle r with the normal to M_1.

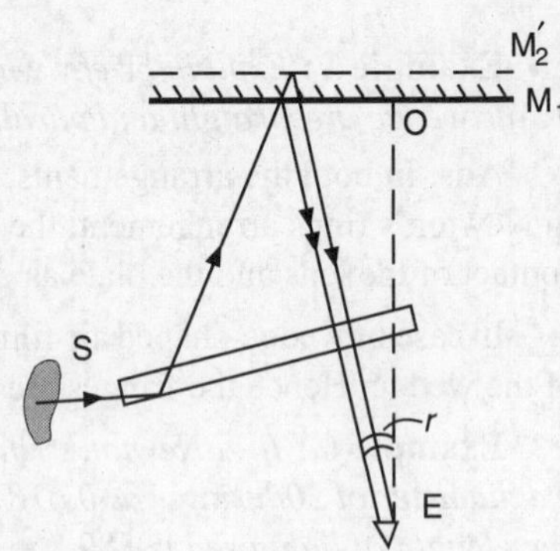

Fig. 2.17.

Then the path difference between the two coherent beams is $2t \cos r$. The condition for a bright ring is $2t \cos r = m\lambda$ where m is an integer.

The condition for a dark ring is

$$2\,t \cos r = (2m - 1)\lambda/2.$$

In either case, r will be constant for given values of t, n and λ. Hence the loci of maxima of intensity will be concentric circles having their centre on the perpendicular from the eye or telescope on M_1. The circular fringes will be situated at infinity. Therefore they can be observed by a telescope focused for infinity. Thus we get *circular fringes of equal inclination* or *Haidinger's fringes*.

If a dark circle appears at the centre of the pattern, the two rays interfere destructively. If the mirror M_1 is then moved by a distance of $\lambda/4$, the path difference changes by $\lambda/2$ (twice the separation between M_1 and M_2'). The two rays will now interfere constructively, giving a bright circle in the middle. As M_1 is moved an additional distance $\lambda/4$, a dark circle will appear once again. Thus, we see that successive dark and bright circles are formed each time M_1 is moved a distance of $\lambda/4$.

***(ii)* Straight fringes :** If M_1 and M_2 are not exactly perpendicular, a wedge shaped air film is formed between M_1 and M_2'. The fringes become practically straight (Fig. 2.18) when M_1 actually intersects M_2' in the middle. The fringes are of equal thickness. The fringes are localized in the airfilm itself. Hence the telescope has to be focused on the film to observe these fringes.

***(iii)* White light fringes :** If white light is used, the central fringe will be dark and others will be coloured. With white light, fringes are observed only when the path difference is small. These fringes are important because they are used to locate the position of zero path difference.

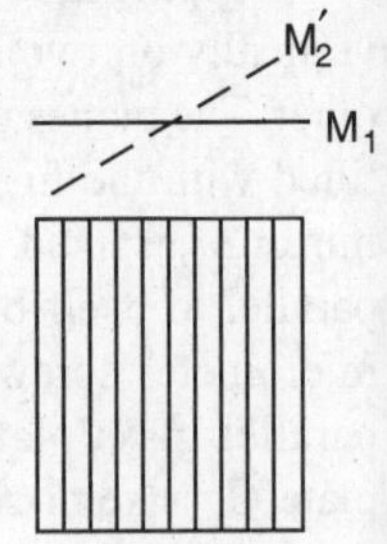

Fig. 2.18.

2.12. Uses of Michelson's Interferometer

1. Determination of wavelength of monochromatic light :

(*i*) Using monochromatic radiation of unknown wavelength λ, the interferometer is adjusted for circular fringes.

(*ii*) With any ring at the centre, the reading of micrometer is noted. Let it be x_1.

(*iii*) Now the mirror M_1 is moved with the help of micrometer screw. The fringes appear to sink or rise due to the change of path difference. Let N fringes move and x_2 be the new reading of the micrometer. When the mirror moves through a distance $\lambda/2$, one fringe shifts. Hence,

$$x_2 - x_1 = x = N \cdot \frac{\lambda}{2} \qquad ...(i)$$

$$\therefore \qquad \lambda = \frac{2(x_2 - x_1)}{N} = \frac{2x}{N} \qquad ..(ii)$$

Example 1: *When the movable mirror of a Michelson interferometer is moved by 0.0589 mm, a shift of 200 fringes is observed. What is the wavelength of light used?*

Solution : When the mirror M_1 is moved through a distance x, the total path difference introduced between the two beams = $2x$. If N fringes move across the field, then $2x = N\lambda$

Here, $x = 0.0589$ mm $= 5.89 \times 10^{-5}$m, $N = 200$; $\lambda = ?$

$$\lambda = \frac{2x}{N} = \frac{2 \times \left(5.89 \times 10^{-5}\right)}{200} = 5.89 \times 10^{-7} \text{ m} = 589 \text{ nm}.$$

Example 2 : *The initial and final readings of a Michelson interferometer screw are 10.7347 mm and 10.6903 mm as 150 fringes pass. Calculate the wavelength of light used.*

Solution : Here, $x = 10.7347 – 10.6903 = 0.0444$ mm $= 4.44 \times 10^{-5}$m; $N = 150$;

$$\lambda = \frac{2x}{N} = \frac{2 \times \left(4.44 \times 10^{-5}\right)}{150} = 5.92 \times 10^{-7} \text{ m} = 592 \text{ nm}.$$

2. Determination of difference in wavelength between two neighbouring lines : Let the source of light emit two close wavelengths λ_1 and λ_2 (like sodium D lines), $\lambda_1 > \lambda_2$. The apparatus is adjusted to form circular rings. Each spectral line produces its own system of rings. We have to consider the superposed fringe-systems. If the bright rings due to λ_1 exactly coincide with bright rings due to λ_2, then the rings are *very distinct* and well defined. This is called *consonance*. If, however, the bright rings due to λ_1 coincide with dark rings due to λ_2, the ring system would disappear producing uniform illumination. This is called *dissonance*. This is the position of maximum *indistinctness*.

Circular rings are formed by adjusting the interferometer using sodium light. The mirror M_1 is gradually moved to obtain dissonance and its position noted. The movement of M_1 is continued in the same direction and successive positions of dissonance are noted. The mean distance x between two successive dissonances is determined. When x is the distance moved by the mirror for two consecutive positions of maximum indistinctness, the path difference is $2x$. During this movement if N is the change in order of the longer wavelength λ_1 at the centre of the field, then $(N + 1)$ will be the change in order of wavelength λ_2 at the centre. Therefore, for dissonance

$$2x = N\lambda_1 = (N + 1)\lambda_2$$

or
$$N = \frac{2x}{\lambda_1} \text{ and } (N+1) = \frac{2x}{\lambda_2} \quad \therefore \frac{2x}{\lambda_2} - \frac{2x}{\lambda_1} = 1$$

or
$$2x\frac{(\lambda_1 - \lambda_2)}{\lambda_1\lambda_2} = 1. \quad \text{or} \quad \lambda_1 - \lambda_2 = \frac{\lambda_1\lambda_2}{2x}$$

Put $\lambda_1 - \lambda_2 = d\lambda$ and $\lambda_1\lambda_2 = \lambda^2$ where λ = mean wavelength.

$$\therefore \qquad d\lambda = \frac{\lambda^2}{2x}$$

From this equation $d\lambda$ can be calculated.

Example 1 : *In an experiment with Michelson's interferometer the scale readings for a pair of positions of maximum indistinctness were found to be 0.6939 mm and 0.9884 mm. If the mean wavelength of the two components of the sodium D-light which was used be 589.3 nm, deduce the difference between the wavelengths of the components.*

Solution : Here, $x = (0.9884 - 0.6939)$ mm $= 0.2945$mm $= 2.945 \times 10^{-4}$ *m*; $\lambda = 589.3 \times 10^{-9}$ *m*; $d\lambda = ?$

$$\therefore \qquad d\lambda = \frac{\lambda^2}{2x} = \frac{\left(589.3 \times 10^{-9}\right)^2}{\left(2 \times 2.945 \times 10^{-4}\right)} \text{ m} = 5.896 \times 10^{-10} \text{ m}.$$

2.13. Jamin's Interferometer

Construction : The essential optical parts are shown in Fig. 2.19. *A* and *B* are two identical optically plane, thick glass plates. The back surfaces of the plates are silvered. The plates are arranged with their faces slightly inclined to each other. *S* is an extended source of monochromatic light, placed at the focal plane of the lens *L*. *A* parallel beam of light is incident on the plate *A*. It gives rise to two beams: (i) R_1 reflected from the front surface and (ii) R_2, reflected from the back surface. Both these reflected beams pass through two identical glass tubes T_1 and T_2. Then the two beams fall on the plate *B*. The beam R_2 is reflected by the front surface. The beam R_1 is reflected from the back surface. The emergent beams have been obtained from the same beam by the division of amplitude. Hence the emergent beams interfere. The straight interference fringes are viewed by a telescope *T*. These are called *Brewster's fringes*.

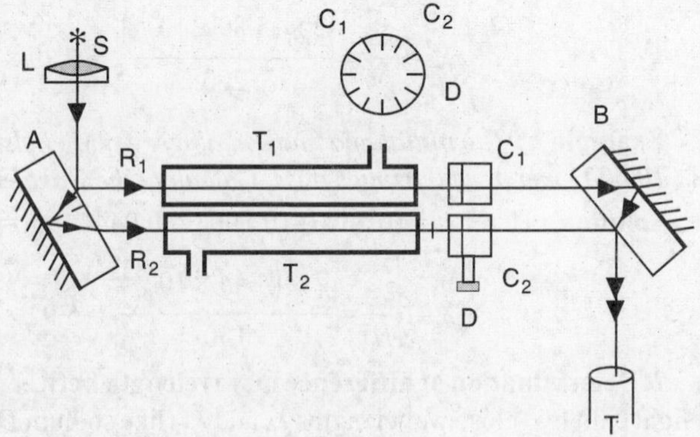

Fig. 2.19.

Measurement of refractive index : This arrangement is used to measure the refractive index of a gas at different pressures. The two tubes T_1 and T_2 are first evacuated. The centre of the cross-wires of the telescope is set on the centre of a bright fringe. Then the gas is allowed to slowly enter one of the tubes. The number of fringes crossing the centre of the field of view is counted. Let it be *m*.

Let n = Refractive index of the gas at final pressure

l = length of each tube.

The path difference between the two beams $= (n - 1)l$

$$\therefore \qquad (n - 1)\, l = m\lambda.$$

Knowing l, λ and m, n is calculated from the relation

$$n = 1 + (m\lambda/l)$$

The experiment is repeated with the gas at different pressures. Let ρ be the density of the gas. It is found that $(n - 1) \propto \rho$ or $(n - 1)/\rho$ is a constant.

This is Gladstone and Dale's law. Since $\rho \propto P$, the pressure of the gas, $(n - 1) \propto P$.

Jamin's compensator : To overcome the difficulty in counting the number of fringes which quickly cross the field of view, a compensator is used. It consists of two glass plates C_1 and C_2 of equal thickness and inclined at a small angle. The plates can be rotated about a common horizontal axis by a calibrated circular disc D. When the disc D is rotated, the interfering beams passing through C_1 and C_2 are affected such that in one case the path increases and in the other case it decreases. The circular disc is calibrated by counting fringes directly with a monochromatic light and is marked in terms of the refractive index and the number of wavelengths.

The tubes T_1 and T_2 are evacuated. Using white light the compensator is adjusted so that the central bright fringe is brought on the cross-wires of the telescope. Then the gas is introduced in one of the tubes at a given pressure and temperature. The fringes will shift. The compensator is then rotated to bring the central white fringe again on the cross-wire. The reading on the calibrated disc directly gives the refractive index of the gas.

Example : *One of the tubes of Jamin's interferometer contains air at room temperature and atmospheric pressure, while the removal of air from the other corresponds to a shift of 120 fringes. If the wavelength of light be 6×10^{-7} m and the tubes are 0.25 m long each, find the refractive index of air.*

Solution : Here, $l = 0.25$ m, $m = 120$, $\lambda = 6 \times 10^{-7}$m.

$$(n-1)\, l = m\lambda$$

$$\therefore \qquad n = 1 + \frac{m\lambda}{l} = 1 + \frac{120 \times \left(6 \times 10^{-7}\right)}{0.25} = 1.000288$$

2.14. Rayleigh's Refractometer

Monochromatic light from a linear source S is made parallel by a lens L_1 and split into two beams by a double slit S_1, S_2 (Fig.2.20). After passing through two exactly similar tubes (T_1,T_2) and the compensating plates (C_1,C_2), the two beams pass through the lens L_2. The two beams interfere in the focal plane of lens L_2. Interference fringes are observed through the eyepiece E.

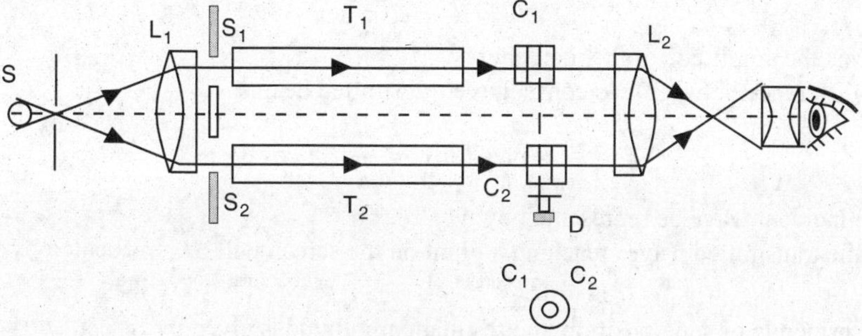

Fig. 2.20.

This instrument is mainly used (*i*) to find the absolute value of ref. index of a gas and (*ii*) to find the variation of ref index of a gas with pressure and temperature.

Initially both the tubes T_1 and T_2 are evacuated. The central white fringe is observed in the field of view of the eyepiece using white light. Now the gas at the given temperature and pressure is introduced in one of the tubes. This brings about a change in the optical path of one interfering beam. So the central white fringe shifts from the field of view. By rotating the circular disc D and thus the plates C_1 and C_2, the central white fringe is brought back to the centre of the field of view. The circular disc D is previously calibrated in terms of ref. index. The reading on the calibrated disc D gives the value of ref. index directly.

2.15. Fabry-Perot Interferometer

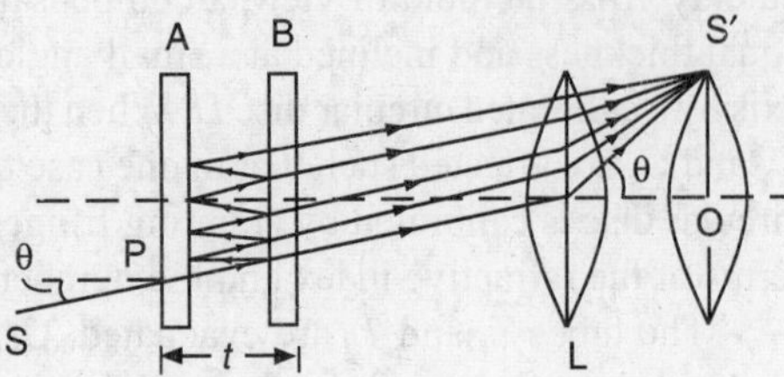

Fig. 2.21.

Construction : This apparatus consists of two glass plates A and B separated by a distance t (Fig.2.21). Their inner surfaces are optically plane, accurately parallel and thinly silvered. Let a beam of monochromatic light SP from an extended source be incident on the glass plate A of the interferometer. The beam suffers multiple reflections in the air film. The parallel transmitted rays are brought to focus at the point S' by the convex lens L.

Let θ be the angle of incidence on the silvered surface of A .

The condition for the rays to produce maxima is

$$2t \cos \theta = m\lambda \quad \text{where } m = 0, 1, 2, 3, \ldots \text{ etc.}$$

The above condition will be satisfied for all points lying on a circle drawn through S' with O as centre. Hence we shall obtain a bright ring through S' . With change in the value of θ, different orders of concentric rings will be produced. Hence the transmitted rays produce concentric dark and bright rings. The phenomenon is called *'multiple-beam interference'*.

In this interferometer, one plate is kept fixed. The other plate can be moved to vary the separation of the plates. The Fabry-Perot interferometer is used to determine wavelengths precisely, to compare two wavelengths, to calibrate the standard metre in terms of wavelengths etc. The Fabry-Perot interferometer produces much sharper fringes than the Michelson interferometer.

2.16. Distribution of Intensity in Fabry-Perot Fringes

Let T and R be the fractions of the incident light intensity which are respectively transmitted and reflected at each silvered surface. Then the fractional transmitted and reflected amplitudes will be $\sqrt{T}$ and $\sqrt{R}$. The amplitudes of the successive rays transmitted through the pair of plates will be

$$aT, aTR, aTR^2, \ldots\ldots.$$

Here, a is the amplitude of the incident ray.

Phase-difference between two consecutive transmitted beams,

$$\delta = \frac{2\pi}{\lambda} \times \text{path-diff.} = \frac{2\pi}{\lambda}(2t \cos \theta) \qquad \ldots(1)$$

Let the incident wave be represented by $y = a\, e^{i\omega t}$

Then, the transmitted waves reaching a point on the screen will be represented by

$$y_1 = aTe^{i\omega t}; \quad y_2 = aTRe^{i(\omega t - \delta)}; \quad y_3 = aTR^2 e^{i(\omega t - 2\delta)}, \ldots\text{etc.}$$

By the principle of superposition, the resultant amplitude is given by

$$A = aT + aTRe^{-i\delta} + aTR^2e^{-2i\delta} + aTR^3e^{-3i\delta} + \ldots$$

$$= aT(1 + Re^{-i\delta} + R^2e^{-2i\delta} + R^3e^{-3i\delta} + \ldots)$$

$$= aT\left(\frac{1}{1 - \text{R}e^{-i\delta}}\right)$$

$$\therefore \quad A^* = aT\left(\frac{1}{1 - \text{R}e^{+i\delta}}\right).$$

Hence, the resultant intensity I is given by

$$I = AA^* = \frac{a^2T^2}{\left(1 - \text{R}e^{-i\delta}\right)\left(1 - \text{R}e^{+i\delta}\right)}$$

$$= \frac{a^2T^2}{1+R^2-R\left(e^{i\delta}+e^{-i\delta}\right)} = \frac{a^2T^2}{1+R^2-2R\cos\delta}$$

$$= \frac{a^2T^2}{(1-R)^2+2R(1-\cos\delta)} = \frac{a^2T^2}{(1-R)^2+4R\sin^2(\delta/2)}$$

$$\therefore \qquad I = \frac{a^2T^2}{(1-R)^2}\left[\frac{1}{1+\dfrac{4R}{(1-R)^2}\sin^2(\delta/2)}\right] \qquad ...(2)$$

Thus the resultant intensity varies with R and δ. Eq. (2) is called *Airy's formula.*

For *maximum intensity* $\sin^2(\delta/2) = 0$, or $\delta = 2m\pi$, $m = 0,1,2,...$

$$\therefore \qquad I_{\max} = \frac{a^2T^2}{(1-R)^2} \qquad ...(3)$$

For *minimum intensity*, $\sin^2(\delta/2) = 1$, or $\delta = (2m+1)\pi$, $m = 0,1,2,..$

$$\therefore \qquad I_{\min} = \frac{a^2T^2}{(1-R)^2}\cdot\frac{1}{1+\dfrac{4R}{(1-R)^2}} = \frac{a^2T^2}{(1+R)^2} \qquad ...(4)$$

Eq. (2) can now be written as

$$I = \frac{I_{\max}}{1+F\sin^2(\delta/2)} \qquad ...(5)$$

Here, $F = 4R/(1-R)^2$ is called the *coefficient of finesse* or *coefficient of sharpness.*

Visibility of fringes : It is given by

$$V = \frac{I_{\max}-I_{\min}}{I_{\max}+I_{\min}} = \frac{2R}{1+R^2} \qquad ...(6)$$

Sharpness of fringes : The sharpness of fringes is determined by the width of the fringes. We define the half width of a fringe as the full width of the fringe corresponding to half of the maximum intensity (Fig. 2.22).

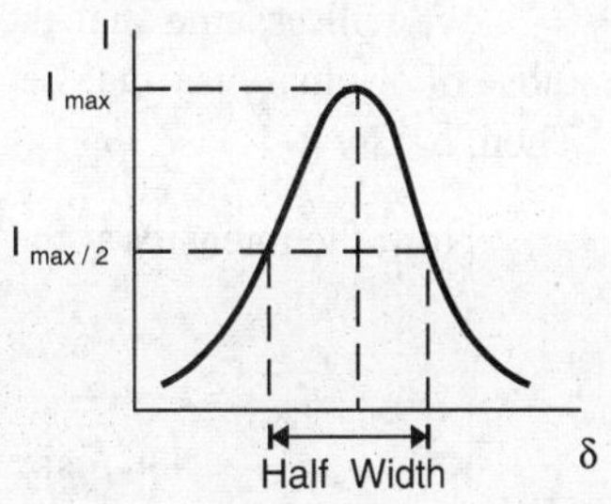

Fig. 2.22.

Now, $$I = \frac{I_{\max}}{1+\dfrac{4R}{(1-R)^2}\sin^2\dfrac{\delta}{2}}$$

When $$\frac{I}{I_{\max}} = \frac{1}{2}, \quad \frac{4R}{(1-R)^2}\sin^2\frac{\delta}{2} = 1$$

or $$\sin\frac{\delta}{2} = \frac{1-R}{2\sqrt{R}}$$

If $\delta/2$ is very small, $\sin\delta/2 \approx \delta/2$. Thus $\delta = \dfrac{1-R}{\sqrt{R}}$

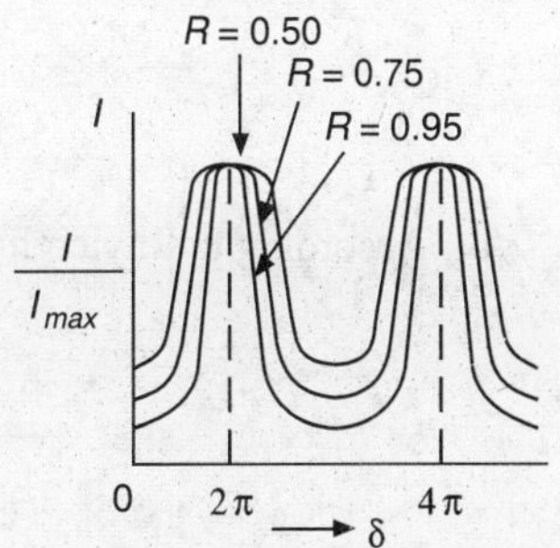

Fig. 2.23.

This gives one-half of the half width and represents the distance from a point of maximum intensity to a point where the intensity has fallen to half the maximum. Hence the half-width is approximately given by

$$\text{Half-width} = \frac{2(1-R)}{\sqrt{R}} \quad \text{...(7)}$$

This expression shows that as reflectance increases, the half-width decreases and the fringes become sharper.

The variation of intensity with phase difference for several values of reflection coefficient is shown in Fig.2.23.

2.17. Resolving Power of Fabry-Perot Interferometer

In F.P. interferometer, the fringe pattern is formed by transmitted light.

The intensity distribution is given by

$$I = \frac{I_0}{1 + F\sin^2(\delta/2)} \quad \text{...(1)}$$

Here, I_0 = incident intensity; $\delta = \frac{4\pi}{\lambda} nt\cos\theta = \frac{4\pi}{\lambda} t\cos\theta \quad (\because n \approx 1)$

A change $\Delta\lambda$ in the wavelength would correspond to a change in δ given by

$$\Delta\delta = \frac{4\pi t\cos\theta}{\lambda^2}\Delta\lambda \text{ (neglecting the negative sign).}$$

Let us assume that there are two closely spaced wavelengths $\lambda + \Delta\lambda$ and λ to be analyzed with a Fabry-Perot interferometer. The total intensity distribution will be (Fig. 2.24)

$$I_{tot} = \frac{I_0}{1 + F\sin^2(\delta/2)} + \frac{I_0}{1 + F\sin^2\left(\frac{\delta - \Delta\delta}{2}\right)}$$

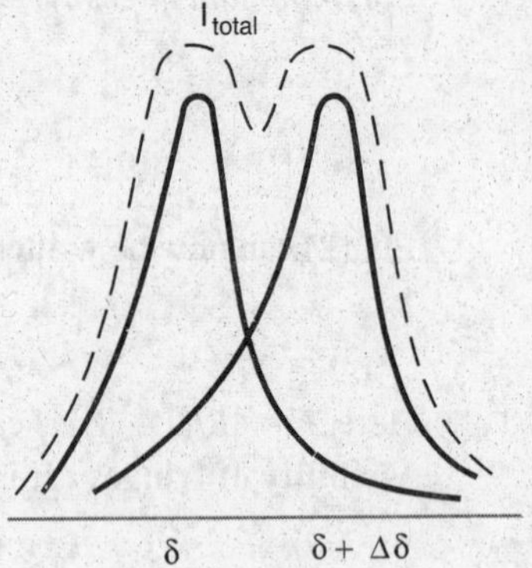

Fig. 2.24.

We will assume that the maxima of I_{tot} to coincide with those of original maxima. Let the order of the fringe for λ be m. Then, $\delta = 2m\pi$.

Now, the intensity at the mid point will correspond to $\delta = 2m\pi + \frac{\Delta\delta}{2}$

$$\therefore \quad I_{mid} = \frac{I_0}{1 + F\sin^2\left(m\pi + \frac{\Delta\delta}{4}\right)} + \frac{I_0}{1 + F\sin^2\left(m\pi - \frac{\Delta\delta}{4}\right)} \approx \frac{2I_0}{1 + F\sin^2\frac{\Delta\delta}{4}}$$

The intensity at maximum will be

$$I_{max} = I_0 + \frac{I_0}{1 + F\sin^2\left(\frac{\Delta\delta}{2}\right)}$$

According to Rayleigh's criterion, the two wavelengths are just resolved if $I_{mid} = 0.811\, I_{max}$

or
$$\frac{2I_0}{1 + F\sin^2\left(\frac{\Delta\delta}{4}\right)} = 0.811\left[I_0 + \frac{I_0}{1 + F\sin^2\left(\frac{\Delta\delta}{2}\right)}\right]$$

$\Delta\delta$ is very small. $\therefore \quad \sin\dfrac{\Delta\delta}{2} \approx \dfrac{\Delta\delta}{2}$. Thus, we get

$$\frac{2}{1+F\dfrac{(\Delta\delta)^2}{16}} = 0.811\left[1+\frac{1}{1+F\dfrac{(\Delta\delta)^2}{4}}\right]$$

or
$$\left[F(\Delta\delta)^2\right]^2 - 15.46\left[F(\Delta\delta)^2\right] - 29.83 = 0 \qquad \text{...(2)}$$

The solution of Eq. (2) is given by

$$\Delta\delta = \frac{4.147}{\sqrt{F}} \qquad \text{...(3)}$$

Thus the resolving power becomes

$$\frac{\lambda}{\Delta\lambda} = \frac{4\pi t\cos\theta}{\lambda\,\Delta\delta} = \frac{4\pi t\cos\theta\sqrt{F}}{4.147\,\lambda} = \frac{2\pi\sqrt{F}}{4.147}m = 1.515\, m\sqrt{F} \qquad \text{...(4)}$$

$$\frac{\lambda}{\Delta\lambda} = m\pi\frac{\sqrt{R}}{(1-R)} \qquad \text{...(5)}$$

Example 1 : *Calculate the visibility of the fringes for a reflection of 80% in multiple-beam interferometry.* (*Allahabad 83*)

Solution : The visibility of fringes, V, in terms of reflectivity R is given by

$$V = \frac{I_{\max} - I_{\min}}{I_{\max} + I_{\min}} = \frac{2R}{1+R^2}.$$

Here, $R = 80\% = 0.8$.

$$\therefore \qquad V = \frac{2\times 0.8}{1+(0.8)^2} = \mathbf{0.976.}$$

Example 2 : *Two F.P. interferometers have equal plate separations but the coefficients of reflection are 0.80 and 0.90. Deduce the relative widths of maxima in the two cases and show it qualitatively with a diagram.* (*Garhwal, 2007; Rohilkhand, 2009; Agra, 2003*)

Solution : In Febry Perot interferometer, the fringe half-width is given by

$$\delta = \frac{1-R}{\sqrt{R}}$$

When $R = 0.8$, $\delta_1 = \dfrac{1-0.8}{\sqrt{0.8}} = \dfrac{0.2}{\sqrt{0.8}} = 0.223$ radians

When $R = 0.9$, $\delta_2 = \dfrac{1-0.9}{\sqrt{0.9}} = \dfrac{0.1}{\sqrt{0.9}} = \dfrac{0.1}{0.94} = 0.106$ radians

$\therefore$ The relative widths of maxima, $\dfrac{\delta_1}{\delta_2} = \dfrac{0.223}{0.106} = 2.1$

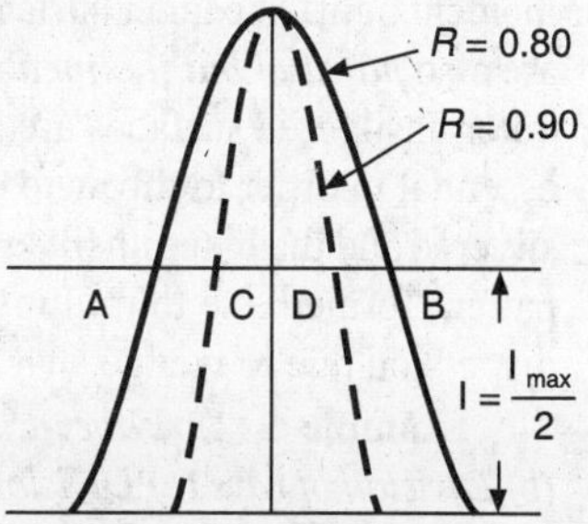

Fig. 2.25

This shows that the width with $R = \mathbf{0.90}$ is even less than half of the width with $R = \mathbf{0.80.}$

The state of affairs is shown diagrammatically in Fig. 2.25 where $\dfrac{AB}{CD} = 2.1$.

Example 3 : *Light having wavelengths 6000Å and 5000Å is incident on a Febry Perot etalon. It is observed that bright fringe is obtained for both colours when the light passes through the etalon in the direction* $\cos^{-1}$ *(0.7) to the normal and that a coincidence is next obtained in the direction* $\cos^{-1}$ *(0.8) to the normal. Deduce the thickness of the air film within the etalon.*

Solution : Let m and n be the orders of the fringes of wave lengths λ_1 and λ_2 which coincide in a direction θ_1. Then, we have

$$2t \cos \theta_1 = m\lambda_1 = n\lambda_2 \qquad ...(1)$$

where t is the thickness of the air film within the etalon.

Let θ_2 be the direction in which fringes again coincide.

Suppose in this position, the order of λ_1 has been increased by p.

Then the order of (smaller) wavelength λ_2 would increase by p +1. Then we have

$$2t \cos \theta_2 = (m+p)\lambda_1 = (n+p+1)\lambda_2 \qquad ...(2)$$

Subtracting equation (1) from (2), we get

$$2t (\cos \theta_2 - \cos \theta_1) = p\lambda_1 = (p+1)\lambda_2 \qquad ...(3)$$

From this:

$$p\lambda_1 = (p+1)\lambda_2$$

or

$$p = \frac{\lambda_2}{\lambda_1 - \lambda_2} \qquad ...(4)$$

Substituting value of p in equation (3), we get

$$2t (\cos \theta_2 - \cos \theta_1) = \frac{\lambda_1 \lambda_2}{\lambda_1 - \lambda_2}$$

or

$$t = \frac{\lambda_1 \lambda_2}{2(\lambda_1 - \lambda_2)(\cos\theta_2 - \cos\theta_1)}$$

Here, λ_1 = 6000Å = 6000×10^{-10} m, λ_2 = 5000Å = 5000×10^{-10} m.

$$\cos \theta_1 = 0.7 \text{ and } \cos \theta_2 = 0.8$$

$$\therefore \qquad t = \frac{(6000 \times 10^{-10}) \times (5000 \times 10^{-10})}{2 \times (6000 - 5000) \times 10^{-10}\ (0.8 - 0.7)}$$

$$= 1.5 \times 10^{-5} \text{ m} = 15\ \mu\text{m}.$$

Example 4 : *Why are the outer surfaces of the two plates of a Fabry-Perot interferometer not exactly parallel to the reflecting surfaces?* (*Himachal Pradesh Univ.*)

Ans. : The inner surfaces of the two plates are *thinly silvered* so as to reflect 80–90% of the incident light. The inner surfaces are *accurately parallel* to each other. The *outer surfaces* of the plates are also *parallel but inclined to their respective inner surfaces*. Each plate is slightly prismatic. The outer unsilvered surfaces are at an angle 1 to 10 min of arc away from the silvered surfaces. This is essential in order to throw to one side the false interference pattern formed by reflections between a silvered and the outer unsilvered surfaces. The false interference pattern is thus not superposed on the pattern formed due to multiple reflections between the inner silvered surfaces. The slight prismatic dispersion due to wedge shaped plates does not materially affect the observed fringe system.

Example 5 : *In Fabry-Perot interferometer, fringes are formed due to (a) Division of wave-front (b) Division of rays (c) Division of amplitude (d) None of these. Mark the correct answer.* (*H.P.U. 2002*)

Ans. : (*c*)

2.18. Construction and Working of F.P. Interferometer

Construction: It consists of two *optically plane* glass plate A, B (Fig. 2.26) placed *accurately parallel* to each other and separated by a distance t.

The inner surfaces of A and B are *thinly silvered* so as to reflect 80–90% of the incident light.

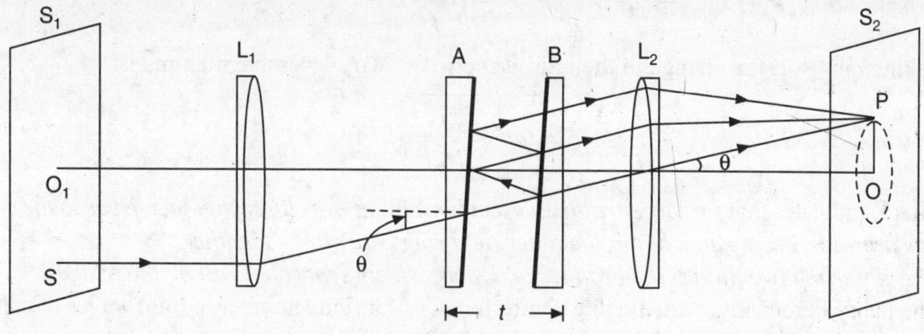

Fig. 2.26

- The plate B facing the observer is fixed. It is provided with screws with which the reflecting surface of B can be made parallel to that of A.
- The plate A is mounted on a carriage. The carriage can be moved in a direction perpendicular to the reflecting surfaces by means of an accurate screw so that the thickness of air film between the coated surfaces of the plates A and B can be varied.
- Light from monochromatic extended source S is rendered parallel by collimating lens.

Working: Monochromatic light from a broad source S_1 is made parallel by the collimating lens L_1. Each parallel ray suffers multiple reflections successively at the two silvered surfaces. At each reflection, a small fraction of light is also transmitted so that each incident ray produces a group of coherent, parallel transmitted rays. There is a constant path difference between any two successive transmitted rays. A telescopic lens L_2 brings these rays to focus at P in its focal plane where they interfere. Thus, the rays from all points of the source produce an interference pattern on a screen S_2 at the *focal plane* of L_2. This is known as ***multiple beam interference***.

2.19. Formation of Circular Fringes

- t is the separation between the plates.
- θ is the inclination of a particular ray with the normal to the silvered surface of A.

For an air film, the optical path difference between two successive transmitted rays corresponding to the incident one is given by

$$\Delta = 2t \cos\theta \qquad ...(1)$$

For the maxima, $\quad 2t\cos\theta = m\lambda \; (m = 0, 1, 2, ...,) \qquad ...(2)$

Here, m is the order of interference and λ is the wavelength of light used.

Fig. 2.27.

The locus of points in the source giving rays of constant inclination θ is a *circle*. Thus, with an extended source, *the interference pattern will be a series of bright concentric rings* (Fig. 2.27) *on a dark background*. Each of the rings will correspond to a particular θ-value.

Linear separation of successive order:

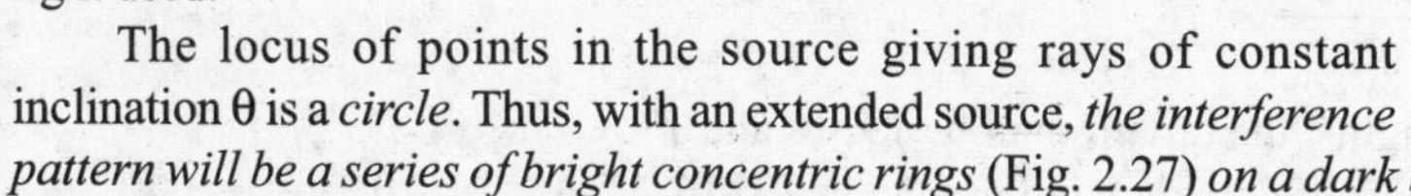

$$m = \frac{2t}{\lambda}\cos\theta = \frac{2t}{\lambda}\left(1 - \frac{\theta^2}{2}\right)$$

f is the focal length of the lens L_2.

$\therefore$ Radius of the mth order bright ring is $r_m = f\theta$.

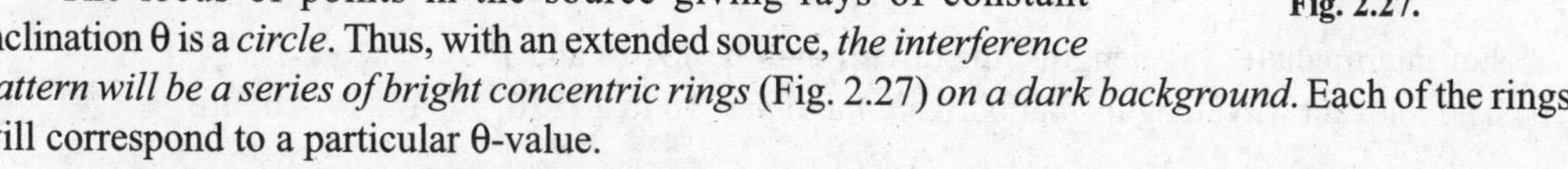

$$\therefore \qquad m = \frac{2t}{\lambda}\left(1 - \frac{\theta^2}{2}\right) = \frac{2t}{\lambda}\left(1 - \frac{r_m^2}{2f^2}\right)$$

$$\Rightarrow \qquad dm = -\frac{2t}{\lambda}\cdot\frac{r_m}{f^2}\,dr$$

Taking $dm = -1$, the change in the radii dr between two successive maxima is

$$\boxed{dr = \frac{\lambda f^2}{2r_m t}} \qquad ...(3)$$

Eq. (3) indicates that *for larger radii, consecutive circles are closer together. Near to the centre, the rings are widely separated but in the outer field rings are closer together.*

The closer the two plates, the broader and more widely separated will be the fringes.

The Fabry Perot fringes arising due to interference of infinite number of light waves of constant inclination to the axis are called *Haidinger fringes*. Path difference of several centimetres may be used without loss of visibility of fringes. Hence very high order of rings may be examined.

Example 1 : *In a Fabry-Perot interferometer, the separation between the plates is 4×10^{-4} cm. Light of wavelength 5000 Å falls normally on the plates. Find the order of the maximum at the centre.*

(*Nagpur University, 2010*)

Solution : In a Fabry Perot interferometer, $2t = m\lambda$.

The order of the maximum at the centre of the interference pattern, is given by

$$m_0 = \frac{2t}{\lambda} = \frac{2\times(4\times10^{-6})}{5000\times10^{-10}} = 16$$

Example 2 : *White light is incident normally on a Fabry-Perot interferometer with plate separation of 4×10^{-6} m. Calculate the wavelengths for which there are interference maxima in the transmitted beam in the range 4000Å to 5000Å.* (*Delhi 2006, Kanpur 80*)

Solution : For a Fabry-Perot interferometer, the condition of maxima in the transmitted beam is

$$2t\cos\theta = m\lambda,$$

where t is plate separation. For normal incidence $\theta = 0°$, so that

$$2t = m\lambda.$$

$$\therefore \qquad \lambda = \frac{2t}{m} = \frac{2\times(4\times10^{-6})}{m}\ \text{m}$$

For 4000 Å (4×10^{-7} m) wavelength, the order at the centre is

$$m = \frac{2\times(4\times10^{-6})}{4\times10^{-7}} = 20.$$

For 5000 Å, $m = \dfrac{2\times(4\times10^{-6})}{5\times10^{-7}} = 16.$

For intermediate wavelengths, the orders shall be 19, 18 and 17.

The relevant wavelengths that correspond to $m = 16$ to 20 (16, 17, 18, 19, 20) are:

$$\lambda_1 = \frac{2\times(4\times10^{-6})}{16} = 5\times10^{-7}\ \text{m} = 5000\ \text{Å}$$

$$\lambda_2 = \frac{2\times(4\times10^{-6})}{17} = 4.706\times10^{-7}\ \text{m} = 4706\ \text{Å}$$

$$\lambda_3 = \frac{2\times(4\times10^{-6})}{18} = 4.444\times10^{-7}\ \text{m} = 4444\ \text{Å}$$

$$\lambda_4 = \frac{2\times(4\times10^{-6})}{19} = 4.211\times10^{-7}\ \text{m} = 4211\ \text{Å}$$

$$\lambda_5 = \frac{2 \times (4 \times 10^{-6})}{20} = 4 \times 10^{-7} \text{ m} = 4000 \text{ Å}$$

∴ The required wavelengths are 4000 Å, 4211 Å, 4444 Å, 4706 Å and 5000 Å.

2.20. Determination of Wavelength

- The Fabry-Perot interferometer is adjusted to produce concentric circular fringes of the monochromatic light of wavelength λ, which we have to determine. For this, the reflecting surfaces of A and B must be parallel.
- Let m be the order of bright fringe at the centre of the fringe system. As at the centre $\theta = 0$, we have

$$2t = m\lambda.$$

If the movable plate is moved a distance $\lambda/2$, $2t$ changes by λ. Hence a bright fringe of next order appears at the centre.

- The movable mirror is moved from one position corresponding to micrometer reading, say, x_1, when there is a bright fringe in the centre to another position corresponding to screw reading, say x_2, when there is again a bright fringe in the centre. The number N of bright fringes which cross the centre of field in this process is counted.

$$\therefore \qquad N \cdot \frac{\lambda}{2} = x_2 - x_1$$

or

$$\lambda = \frac{2(x_2 - x_1)}{N}$$

From this relation, we can determine the value of λ.

Example 1 : *A shift of 100 fringes is observed when movable mirror of Fabry-Perot interferometer is shifted through 0.0295 mm. Calculate the wave length of light used.* (P.U. 2005)

Solution : Here, $x_2 - x_1 = 0.0295 \text{ mm} = 0.0295 \times 10^{-3}$ m; $N = 100$;

$$\therefore \qquad \lambda = \frac{2(x_2 - x_1)}{N} = \frac{2 \times (0.0295 \times 10^{-3})}{100} = 5900 \times 10^{-10} \text{ m} = 5900 \text{ Å}$$

2.21. Etalon and Interferometer

The F.P. instruments are made of two types.

(*i*) In one type, the *separation between two plates is kept fixed.* **It is called F. P. etalon.** Etalons with definite spacing are available in market. Etalons are supplied with a variety of spacers of lengths ranging from 1 to 200 mm for use in the investigation of hyperfine structure of spectral lines. The *etalon* is now invariably used for research.

Construction: In the *etalon*, two semi-silvered plates are mounted in a framework (Fig. 2.28).

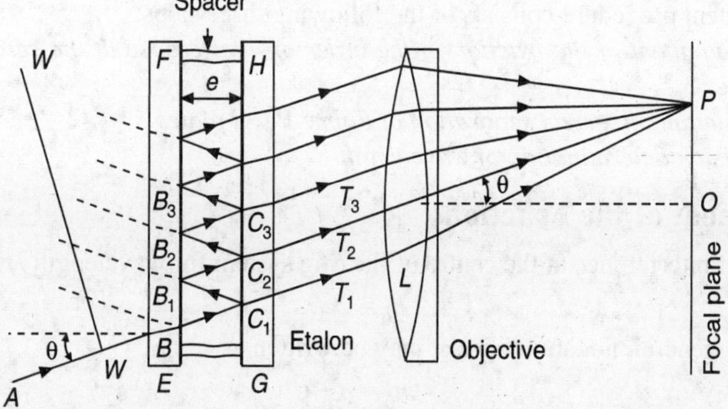

Fig. 2.28.

Their coated faces are kept in constant parallel position, a fixed distance apart, by a fixed spacer. The spacer is commonly a hollow cylinder of invar or silica with three projecting studs at each end. The plates are kept in place slightly pressed against the spacers by the pressure of adjustable springs attached in a proper way to the etalon housing.

e is the separation between the F. P. plates.

Working: Fig. 2.28 shows *formation of multiple reflection fringes by Fabry Perot etalon.* EF and GH are two plane parallel lightly silvered surfaces of the Fabry Perot etalon. Light from a broad source of monochromatic light is incident on the plates at all angles. Consider a plane wave travelling along AB and incident on EF at an angle θ to the normal. The series of parallel transmitted waves C_1T_1, C_2T_2, C_3T_3 ... arise from the same incident wave by its internal multiple partial reflections and refractions between EF and GH. These waves being coherent interfere when brought to a focus at P in the focal plane of an achromatic lens L. *The interference pattern will appear in the form of concentric circles (or rings) where each ring will correspond to one particular value of* θ.

(*ii*) In 2nd, a screw is provided with either plate by which separation between plates can be changed. It is then called F. P. interferometer.

2.22. Determination of Wavelength using Fabry-Perot Etalon-Benoit's Method of Exact Fractions

(*i*) We select at least three accurately known other standard wavelengths for forming Fabry-Perot fringes.

(*ii*) It is also essential to know before-hand, say by grating, the approximate wavelengths to at least 0.1 Å of all the lines whose precision wavelengths are to be determined.

(*iii*) The rings of three known wavelengths and that of the unknown wavelength are photographed on the same plate with a fixed separation of Fabry-Perot plates.

(*iv*) We employ Fabry-Perot etalon "crossed" with the prism spectrograph (Fig. 2.29).

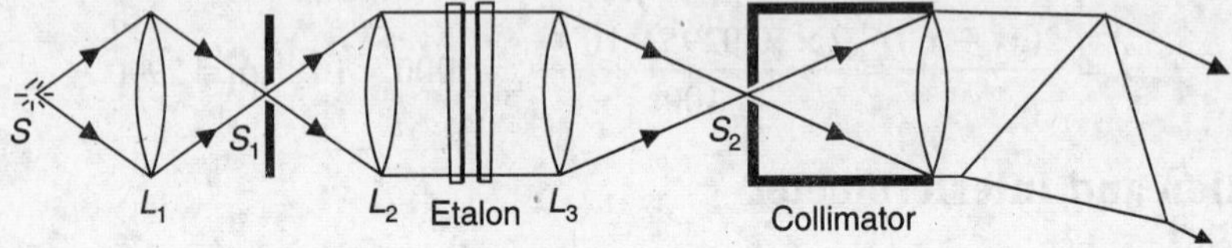

Fig. 2.29.

Light from a wide source slit S_1, which is perpendicular to the plane of the figure, is rendered parallel by the lens L_2 and is incident on the Fabry Perot plates. The well corrected achromatic lens L_3 throws the image of rings on the wide vertical slit S_2 of the prism spectrograph.

The prism spectrograph disperses the lines. The slit image of each wavelength is crossed with a diametrical section of its circular rings, the width of the section being limited by the image width.

The subsequent procedure consists of the following three steps:

(*a*) *Measurement of the fraction in the order of interference at the centre of Fabry-Perot fringes.*

(*b*) *Evaluation of correct separation of Fabry-Perot plates.*

(*c*) *Precision determination of wavelength.*

(*a*) Measurement of the Fraction

The order of interference at the centre of the ring system for wavelength λ is given by

$$n_0 = 2e/\lambda \qquad ...(1)$$

But n_0 is, in general, not an integer. It may be written as

$$n_0 = n + \varepsilon \qquad ...(2)$$

Here, n = order of the first bright ring from the centre and is an integer and
ε = a small fraction less than 1.

$\therefore$ $$2e = (n + \varepsilon)\,\lambda \qquad ...(3)$$

The maximum value of the order is at the centre of the ring system.

The orders of rings decrease as their radii increase.

The order of second bright ring is $(n - 1)$, of the third $(n - 2)$ ...

The order of pth ring from the centre is $(n - p + 1)$.

Let θ_p = Angular radius of the pth bright ring before magnification. Then,

$$(2e/\lambda)\cos\theta_p = (n - p + 1) \qquad ...(4)$$

or $$(n + \varepsilon)\cos\theta_p = (n - p + 1)$$

Since θ_p is small, $$\cos\theta_p = 1 - \tfrac{1}{2}\theta_p^2$$

$\therefore$ $$\tfrac{1}{2}(n + \varepsilon)\,\theta_p^2 = p + \varepsilon - 1 \qquad ...(5)$$

From the photograph of the fringes of λ, ring diameters are measured with a photo-mesuring micrometer.

Let f = focal length of the camera lens and
M = linear magnifying power of the diameter measuring instrument.

Then, the measured linear diameter of the pth ring is given by

$$D_p = 2\theta_p f M \qquad ...(6)$$

Substituting the value of θ_p in Eq. (5), we get

$$\frac{(n + \varepsilon)}{8}\,\frac{D_p^{\,2}}{f^2 M^2} = p + \varepsilon - 1 \qquad ...(7)$$

When $D_p = 0, p = 1 - \varepsilon$.

The squares of ring diameters $(D_p^{\,2})$ are plotted on a graph paper against the corresponding ring numbers (p). The graph is a straight line (Fig. 2.30). The intercept on the ring number axis is $(1 - \varepsilon)$.

Thus the fraction ε of the equation $n_0 = n + \varepsilon$ is experimentally determined.

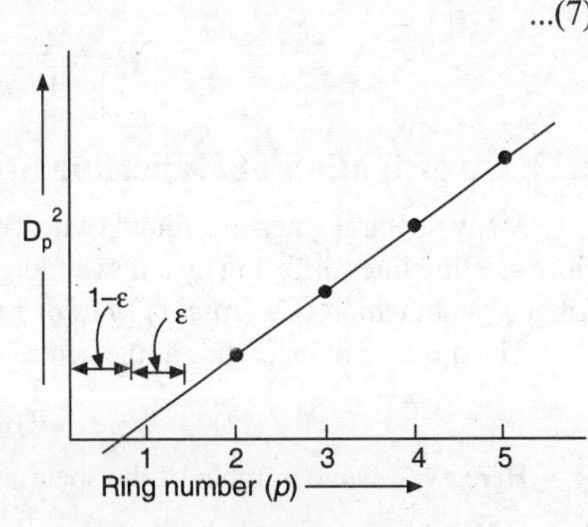

Fig. 2.30.

The procedure outlined above is followed to evaluate the fractions $\varepsilon_1, \varepsilon_2, \varepsilon_3$ in the orders at the centre of respective ring systems of three accurately known wavelengths λ_1, λ_2 and λ_3. Thus in the equation

$$2e = (n_1 + \varepsilon_1)\,\lambda_1 = (n_2 + \varepsilon_2)\,\lambda_2 = (n_3 + \varepsilon_3)\,\lambda_3 \qquad ...(8)$$

we know ε_1, ε_2 and ε_3 and also λ_1, λ_2 and λ_3. But, e is not accurately known. So the accurate values of n_1, n_2 and n_3 are unknown.

(*b*) Precision evaluation of *e*

The approximate separation e' of the Fabry-Perot plates is first obtained by measuring the *spacer* with the precision micrometer. Then, the approximate order in integral number at the centre of the fringe system of λ_1 is given by

$$n_1' = 2e'/\lambda_1 \qquad ...(9)$$

Let x_1 represent the uncertainty in the central order due to not knowing the separation correctly. Corresponding to correct separation e, the order at the centre has one of the following possible values:

$$(n_0)_1 = (n_1' \pm x_1) + \varepsilon_1 \qquad ...(10)$$

x_1 = a small integer = 1, 2, 3

We should now determine the correct value of $(n_0)_1$.

Different possible values of $(n_0)_1$ are substituted successively in the relation

$$(n_0)_1 \times \lambda_1 = (n_0)_2 \times \lambda_2 = (n_0)_3 \times \lambda_3 \quad \text{...(11)}$$

The corresponding values of $(n_0)_2$, the possible order at the centre of fringes of λ_2, is obtained and the corresponding fraction part ε_2' is known. Similar calculations may be done for $(n_0)_3$ and the corresponding fraction ε_3' of λ_3 is known.

It is found that for one particular value of $(n_0)_1$, say $(\boldsymbol{n}_0)_1$, the calculated fractions ε_2' and ε_3' agree with the experimentally observed values ε_2 and ε_3 respectively.

If at the same time $(\boldsymbol{n}_0)_1\, \lambda_1 = 2e'$, we should conclude that this particular value $(\boldsymbol{n}_0)_1$ is the correct order $(n_1 + \varepsilon_1)$ at the centre of the fringe system of λ_1.

Now the correct separation e between the etalon plates is calculated from the relation

$$2e = (n_1 + \varepsilon_1)\, \lambda_1 = (\boldsymbol{n}_0)_1 \times \lambda_1 \quad \text{...(12)}$$

since λ_1 and $(\boldsymbol{n}_0)_1$ are accurately known.

(c) Precision evaluation of wavelength

- Fabry-Perot fringes formed by the unknown wavelength λ_x are analysed. The fractional part ε_x of the order at centre is determined.
- The wavelength λ_x must be previously determined with a grating correct to well within 0.1 Å in order to determine the integral part n_x of the order at the centre without any ambiguity. Then, the correct value of λ_x, is

$$\lambda_x = \frac{2e}{(n_x + \varepsilon_x)} = \frac{(n_1 + \varepsilon_1)\, \lambda_1}{(n_x + \varepsilon_x)}$$

2.23. Investigation of Hyperfine Structure of Spectral Lines

Many spectral lines are found, when examined with this instrument, to consist of extremely close satellite lines differing by a few hundredth of an angstrom. Hyperfine structure means multiple fringe system which is *a group of lines of very close wavelengths*.

The order of interference at the centre of ring system is given by

$$n_0 = 2e/\lambda = 2e\,\bar{\nu} \quad \text{...(1)}$$

Here, $\bar{\nu}$ = wave number of the spectral line which is inverse of wavelength λ.

Differentiating Eq. (1), we get

$$d\,\bar{\nu} = dn_0/2e \quad \text{...(2)}$$

Eq. (2) gives the wave number difference $d\,\bar{\nu}$ between the chief line λ and the satellite $\lambda - d\lambda$.

If ε_1 and ε_2 are respectively the fractional parts of the orders at the centre of close component lines, then $dn_0 = \varepsilon_1 - \varepsilon_2$.

$$\therefore \quad d\,\bar{\nu} = (\varepsilon_1 - \varepsilon_2)/2e \quad \text{...(3)}$$

If the orders are overlapped, then

$$dn_0 = (n_1 + \varepsilon_1) - (n_2 + \varepsilon_2) = (n_1 - n_2) + (\varepsilon_1 - \varepsilon_2)$$

$$\therefore \quad d\,\bar{\nu} = \frac{(n_1 - n_2)}{2e} + \frac{(\varepsilon_1 - \varepsilon_2)}{2e} \quad \text{...(4)}$$

Tolansky has proposed the following method for directly determining $(\varepsilon_1 - \varepsilon_2)$ from the measurement of diameters of the component rings.

The square of measured linear diameter D_p of the *p*th ring from the centre of rings due to the chief line is given by

$$D_p^2 = \frac{8f^2M^2}{(n_0)_1}(p + \varepsilon_1 - 1) \quad ...(5)$$

and

$$D_{p+1}^2 = \frac{8f^2M^2}{(n_0)_1}(p + \varepsilon_1) \quad ...(6)$$

$$\therefore \quad D_{p+1}^2 - D_p^2 = 8f^2M^2/(n_0)_1 = \Delta_1$$

Here, $(n_0)_1$ is the order of interference at the centre of the ring system.

The difference of squares of diameters of any two consecutive rings is, therefore, a constant. From Eqs. (5) and (6), we get

$$\varepsilon_1 = \frac{D_{p+1}^2}{D_{p+1}^2 - D_p^2} - p = \frac{D_{p+1}^2}{\Delta_1} - p \quad ...(7)$$

If measurements are made on more than two rings, a mean value of ε_1 can be found by the method of least squares.

Similarly, for the rings of the satellite line, we have

$$d_{p+1}^2 - d_p^2 = 8f^2M^2/(n_0)_2 = \Delta_2$$

Here, $(n_0)_2$ is the order at the centre for the satellite line and $\varepsilon_2 = \dfrac{d_{p+1}^2}{\Delta_2} - p$.

Hence

$$\varepsilon_1 - \varepsilon_2 = \frac{D_{p+1}^2}{\Delta_1} - \frac{d_{p+1}^2}{\Delta_2} \quad ...(8)$$

Tolansky takes $(n_0)_1 = (n_0)_2$ in all practical work without any appreciable error. Then $\Delta_1 = \Delta_2 = \Delta$.

$$\therefore \quad \varepsilon_1 - \varepsilon_2 = (D_{p+1}^2 - d_{p+1}^2)/\Delta \quad ...(9)$$

and

$$d\bar{\nu} = \frac{1}{2e}\frac{D_{p+1}^2 - d_{p+1}^2}{\Delta} = \frac{1}{2e}\left(\frac{D_{p+1}^2 - d_{p+1}^2}{D_{p+1}^2 - D_p^2}\right) \quad ...(10)$$

If the orders are overlapped, then

$$d\bar{\nu} = \frac{(n_1 - n_2)}{2e} + \frac{1}{2e}\frac{\left(D_{p+1}^2 - d_{p+1}^2\right)}{\left(D_{p+1}^2 - D_p^2\right)} \quad ...(11)$$

Experimental Set Up: Fig. 2.31 (i) shows the experimental arrangement.

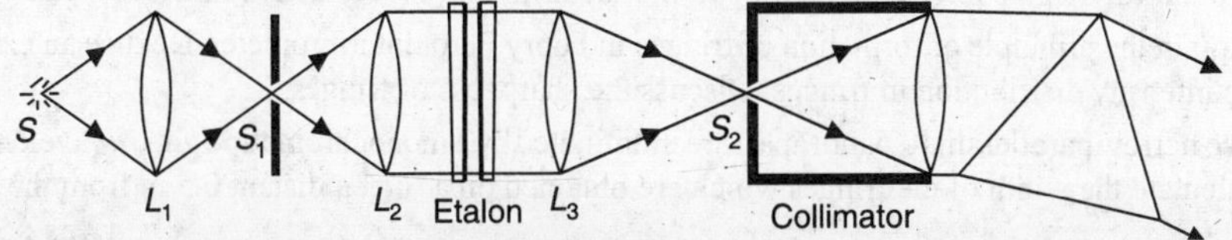

Fig. 2.31 (i)

Light from a wide source slit S_1 is rendered parallel by the lens L_2 and is incident on the F. P. etalon. The lens L_3 throws the image of rings on the wide vertical slit S_2 of the prism spectrograph. The spectrograph separates the spectral components. In the focal plane of the telescope of camera objective, we observe images of the slit, each crossed by fringes as shown in Fig. 2.31 (*ii*).

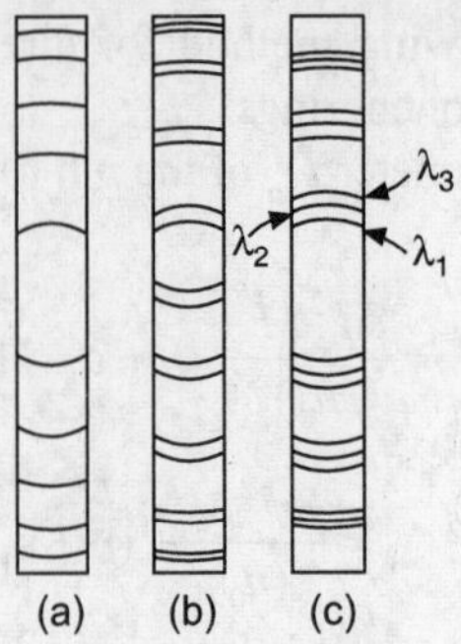

Fig. 2.31 (ii)

Fig. 2.31 (*ii*) (*a*) No structure, (*b*) Structure revealing main line and one satellite, (*c*) Structure revealing main line and two satellites.

EXERCISE

1. Describe, with necessary theory, the method of determining the wavelength of monochromatic light using a Fresnel's biprism.

2. Briefly discuss the effect of introducing a thin plate in the path of one of the interfering beams in a biprism experiment. Deduce an expression for the displacement of the fringes. Show how this method is used for finding the thickness of a mica sheet.

3. Give the theory of interference in thin films and explain the colours of thin films.

4. Discuss the formation of colours in thin films and show that with monochromatic light the interference patterns of reflected and transmitted light are complementary.

5. Explain the formation of interference fringes by an air-wedge. Derive an expression for the fringe width. How can the above method be used to measure the diameter of a thin wire accurately?

6. Explain how the phenomenon of interference is used in testing the optical planeness of a glass surface.

7. What are Newton's rings? How are they formed? Derive an expression for the radius of the *m*th dark ring formed by reflection.

Describe with necessary theory how you would determine the wavelength of monochromatic light using Newton's rings.

8. How can Newton's rings be used to determine the refractive index of a liquid? Derive the formula used.

9. Describe the construction of Michelson's interferometer and explain its working. Discuss the important applications of this interferometer.

10. Describe Michelson's interferometer with the help of line diagram. How are circular fringes formed in it ? How is the wavelength of monochromatic light measured with their help?

11. Explain how Jamin's refractometer is used to study the refractive index of a gas at different pressures.

12. Explain Rayleigh's interferometer method of determining the refractive index of a gas.

13. Explain the principle of formation of fringes in Fabry Perot interferometer. Deduce an expression for the intensity distribution in fringes. Discuss the sharpness of fringes.

14. Two narrow parallel slits 0.5 mm apart are illuminated by a monochromatic light of wavelength 589 nm. Calculate the width of the fringes which are obtained on a screen distant 0.5 m from the slits.

[**Ans :** 5.89×10^{-4}m]

15. In a Young's experiment the separation of bright bands with monochromatic light is 0.54 mm and the separation of the slits is 1mm. If the distance of the slits from the focal plane of the eyepiece is 1 m, calculate the wavelength of light. [**Ans :** 540 nm]

16. In an experiment with a biprism using monochromatic light, the fringe width was found to be 2×10^{-4} m when the distance between the slit and first focal plane of the eyepiece was 1m. A convex lens was kept at a distance of 0.3 m from the slit so as to give an image of the two virtual sources at a distance of 1.2 m from the lens. The distance between the images was found to be 1.18×10^{-2} m. Calculate the wavelength of light. **[Ans :** 5.9×10^{-7}m]

17. In a biprism experiment, the distance between two points were covered by 11 fringes when the wavelength of the light used is 546 nm. When the light is changed, the distance between the same points is covered by 12 fringes. Find the wavelength of that light. **[Ans:** 500.5 nm]

18. When a thin sheet of transparent material of thickness 6.3×10^{-4} cm is introduced in the path of one of the interfering beams, the central fringe shifts to a position occupied by the sixth bright fringe. If $\lambda = 546$ nm, find the refractive index of the sheet. **[Ans:** 1.52]

19. A transparent plate ($n = 1.5$) is placed in the path of one of the interfering beams of a biprism experiment using light of wavelength 500 nm. If the central fringe shifts by a distance equal to the width of ten fringes, calculate the thickness of the plate. **[Ans:** 1×10^{-5}m].

20. White light is reflected from an oil film of thickness 0.01 mm and refractive index 1.4 at an angle of 45° to the vertical. If the reflected light falls on the slit of a spectrometer, calculate the number of dark bands seen between wavelengths 400 and 500 nm. **[Ans:** 12]

21. A soap film of refractive index 4/3 and of thickness 1.5×10^{-6} m is illuminated by white light incident at an angle of 45°. The light reflected by it is examined by a spectroscope in which is found a dark band corresponding to a wavelength of 5×10^{-7} m. Calculate the order of the interference band. **[Ans :** 7]

22. Monochromatic light emitted by a broad source of wavelength 6×10^{-7} m falls normally on two plates of glass which enclose a wedge-shaped film. The plates touch at one end and are separated at a point 15 cm from that end by a wire of 0.05 mm diameter. Find the distance between any two bright fringes. **[Ans :** 9×10^{-4}m]

23. An air wedge is formed by using two plane glass plates inclined at an angle of 0.5° between them. A parallel beam of monochromatic light of wavelength 584 nm is incident normally on it.

(*i*) Find the band width obtained.

(*ii*) If a liquid of refractive index 1.32 is introduced in between the two plates, what will be the band width?

(*iii*) Also find the thickness of the film where the 6th bright band is obtained.

[Ans : (*i*) 3.346×10^{-5} m; (*ii*) 2.535×10^{-5}m; (*iii*) 1.438×10^{-6} m].

24. Newton's rings are obtained by the reflected light. The diameter of the *n*th ring is 2.18 mm and that of the $(n + 10)$ th ring is 4.51 mm. If the wavelength of the light used is 546 nm, calculate the radius of curvature of the surface of the lens. **[Ans :** 0.7137m]

25. In a Newton's ring experiment, the diameters of 4th and 12th dark rings are 0.004 m and 0.007 m respectively. Deduce the diameter of the 20th dark ring. **[Ans :** 9.055×10^{-3}m]

26. A Newton's ring arrangement is used with a source emitting two wavelengths : $\lambda_1 = 6.0 \times 10^{-7}$m and $\lambda_2 = 4.5 \times 10^{-7}$m and it is found that the *n*th dark ring due to n_1 coincides with $(n + 1)$ th dark ring due to λ_2. If the radius of curvature of the curved surface of the lens is 90 cm, find the diameter of *n*th dark ring of λ_1. **[Ans :** $[D_n]\ \lambda_1 = 0.254$ cm.]

27. In moving the mirror of the Michelson's interferometer by 0.0844 mm, the number of fringes counted is 300. Calculate the wavelength. **[Ans :** 5.6×10^{-7}m]

28. What happens to the ring system if a plane polished mirror is used instead of a glass plate in Newton's ring arrangement ?

Ans. The rings disappear and we get uniform illumination. The reason is that the amplitude of the ray reflected from the plane mirror is considerably greater than the amplitude of the ray reflected from the upper surface of air film. Due to the large difference between these two amplitudes, the contrast between dark and bright rings becomes very weak. So the ring system cannot be distinguished.

29. Explain what will happen if white light is used instead of monochromatic light in Newton's ring experiment ?

Ans. The diameter of a ring depends upon the wavelength of light used and isproportional to the square root of the wavelength. When white light is used, there will be superposition of the rings formed by different wavelengths. So few colours rings near the centre are formed. The central spot is dark.

30. Give two examples of each of interference of division of wavefront and division of amplitude.

Ans. Division of wavefront : (*i*) Young's double slit experiment (*ii*) Biprism.

Division of amplitude : (*i*) Colours of thin films (*ii*) Michelson's interferometer.

CHAPTER 3

DIFFRACTION

3.1. Introduction

If an opaque object is placed between a source of light and a screen, a sharp shadow is obtained on the screen. This shows that light travels in straight lines. If, however, the size of the obstacle is *small*, (comparable to the wavelength of light), there is a departure from straight line propagation and the light bends into the geometrical shadow. *This phenomenon of bending of light waves around corners and their spreading into the geometrical shadow of an object is called diffraction.*

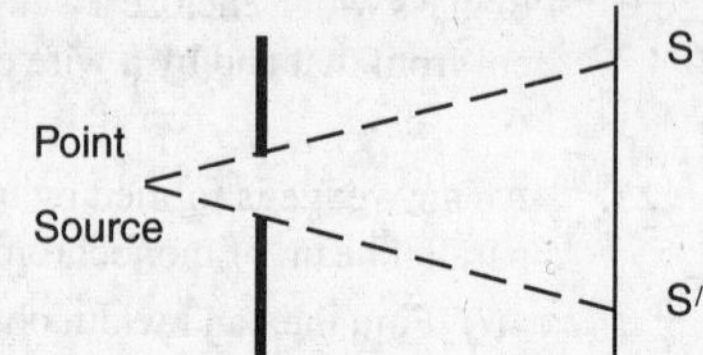

Fig. 3.1.

There are two kinds of diffraction : (i) Fresnel diffraction and (ii) Fraunhofer diffraction.

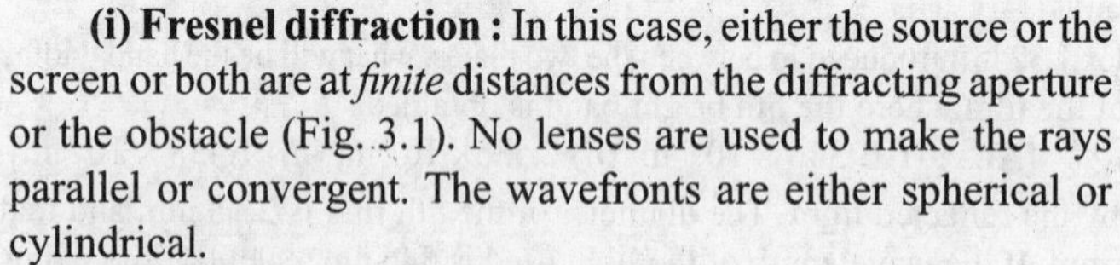

(i) Fresnel diffraction : In this case, either the source or the screen or both are at *finite* distances from the diffracting aperture or the obstacle (Fig. 3.1). No lenses are used to make the rays parallel or convergent. The wavefronts are either spherical or cylindrical.

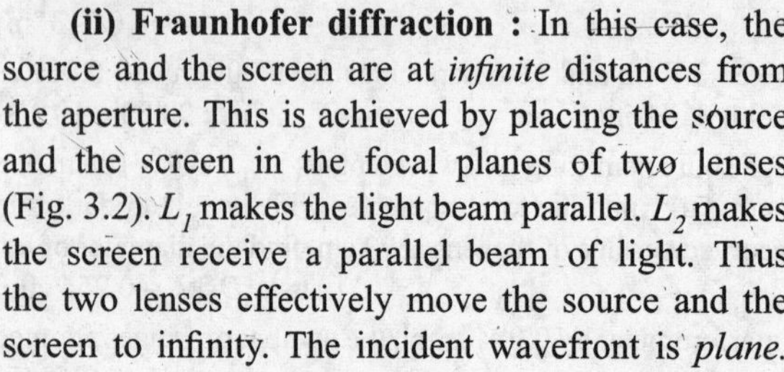

(ii) Fraunhofer diffraction : In this case, the source and the screen are at *infinite* distances from the aperture. This is achieved by placing the source and the screen in the focal planes of two lenses (Fig. 3.2). L_1 makes the light beam parallel. L_2 makes the screen receive a parallel beam of light. Thus the two lenses effectively move the source and the screen to infinity. The incident wavefront is *plane.*

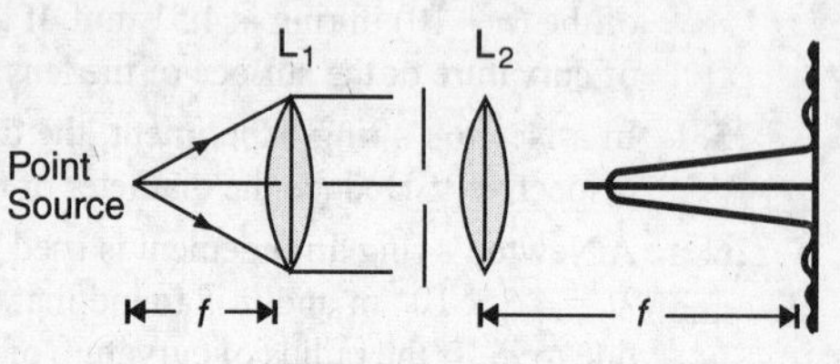

Fig. 3.2.

3.2. Fresnel's Explanation of Rectilinear Propagation of Light

Construction of half-period zones : Let *ABCD* be a plane wavefront of monochromatic light of wavelength λ (Fig. 3.3). It is travelling from *O* to *P*. We want to find the effect of the wavefront at an external point *P*. Fresnel subdivided the wavefront *ABCD* into a number of h.p. zones. From the point *P*, drop a perpendicular *PO* on the wavefront. Let *PO* = *b*. *O* is known as *pole* w.r.t. point *P*. With *P* as centre and radii

$$b+\frac{\lambda}{2}, b+\frac{2\lambda}{2}, b+\frac{3\lambda}{2}, ... b+\frac{n\lambda}{2}$$

draw concentric spheres. These spheres intersect the wavefront $ABCD$ along concentric circles with O as centre. OM_1, OM_2, OM_3, OM_{n-1}, OM_n etc., are the radii of the circles. The area enclosed by the first circle of radius OM_1 is called the first h.p. zone. The annular area between the first circle and second circle is called the second h.p. zone and so on.

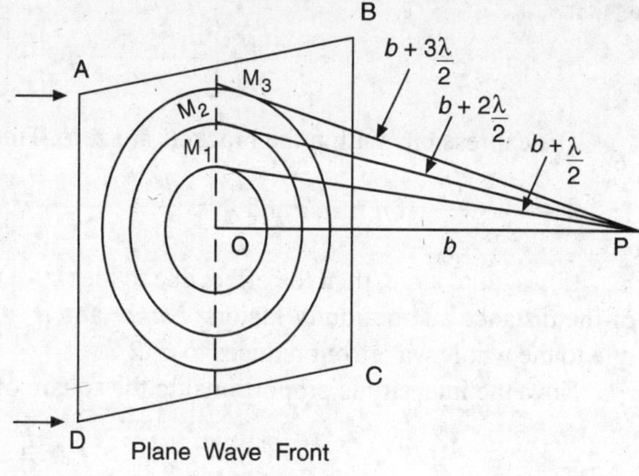

Fig. 3.3.

Radii of h.p. zones : The radius of the nth h.p. zone is

$$OM_n = r_n = \sqrt{\left(b + \frac{n\lambda}{2}\right)^2 - b^2} \approx \sqrt{nb\lambda} \quad ...(1)$$

Thus we see that the *radii of half period zones are proportional to the square roots of the natural numbers.*

Area of a h.p. zone : Area of the nth h.p. zone

$$= \pi(OM_n)^2 - \pi(OM_{n-1})^2 = \pi(nb\lambda) - \pi(n-1)b\lambda = \pi b\lambda$$

The area of the nth zone is independent of n. Thus the area of each h.p. zone is approximately the same.

The amplitude of the disturbance at P due to a given zone is :

(i) directly proportional to the area of the zone,

(ii) inversely proportional to the distance of the point P from the given zone and

(iii) directly proportional to the obliquity factor $(1 + \cos\theta)$.

The resultant amplitude at P due to the whole wavefront :

The resultant amplitude at the point P is

$$A = d_1 - d_2 + d_3 - d_4 + d_5 - \ldots.. d_n \quad ...(2)$$

Here, $d_1, d_2, d_3, \ldots$ are the displacements at P due to the waves from first, second, third,...h.p. zones respectively. The alternate negative and positive signs represent the fact that resultant disturbances produced by two consecutive zones are π out of phase w.r.t. each other.

Here the areas of the h.p. zones are all equal. The distances of the zones from P increase with the order of the zones; displacement $\propto$ 1/distance. Hence d_1, d_2, d_3, ..are in descending order of magnitude. Further as we go from one h.p. zone to the next, the obliquity θ goes on increasing. Hence due to the obliquity factor $(1 + \cos\theta)$, the amplitude due to each successive zone goes on decreasing. Because of these two factors, the amplitudes d_1, d_2, d_3,..decrease monotonically (Fig. 3.4).

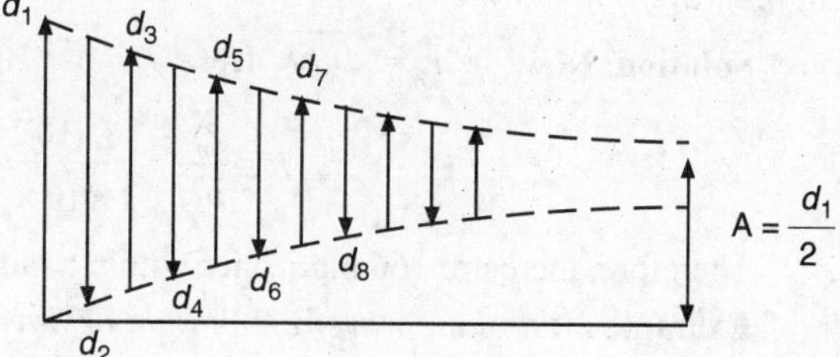

Fig. 3.4.

$$\therefore \quad d_1 > d_2 > d_3 > d_4 \ldots$$

Now, $$d_2 = \frac{d_1 + d_3}{2};\ d_4 = \frac{d_3 + d_5}{2} \text{ and so on.}$$

$$\therefore \quad A = \frac{d_1}{2} + \left[\frac{d_1}{2} - d_2 + \frac{d_3}{2}\right] + \left[\frac{d_3}{2} - d_4 + \frac{d_5}{2}\right] + ..$$

$$= \frac{d_1}{2} + \left[\frac{d_1 + d_3}{2} - d_2\right] + \left[\frac{d_3 + d_5}{2} - d_4\right] + ...$$

The expressions within the brackets are zero. Therefore,

$$A = \frac{d_1}{2} + \frac{d_n}{2} \text{ if } n \text{ is } odd \text{ and } A = \frac{d_1}{2} + \frac{d_{n-1}}{2} - d_n, \text{ if } n \text{ is } even.$$

If n is quite large, then the effect due to the $(n-1)$th or nth zone is almost negligible on account of the distance and obliquity factors. *i.e.*, d_n and $d_{n-1} \to 0$. Therefore, the resultant amplitude at P due to the whole wavefront reduces to $d_1/2$.

Now, the intensity is proportional to the square of the amplitude.

$$\therefore \qquad I = \frac{d_1^2}{4}$$

Thus *the intensity at P due to the whole wavefront is only one fourth of that due to the first h.p. zone alone.*

Explanation of Rectilinear Propagation of Light

Fig. 3.5 shows the variation in the intensity of light at P with the number of exposed zones. From this graph it is obvious that only a few central zones are effective and the intensity at P due to them is practically the same as due to the entire wavefront. If the obstacle is large enough to obstruct the effective central zones, the entire wavefront would become ineffective.

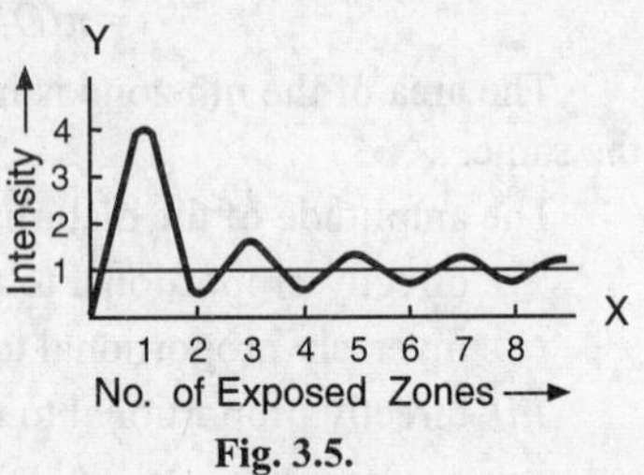

Fig. 3.5.

Consider an opaque body interposed between the wavefront and the point P. Since λ is very small, the area $\pi b \lambda$ of each h.p. zone is very small. Hence even a small obstacle cuts off a considerable number of central effective h.p. zones. So no light will be received at P. The point P is thus practically dark. This explains the darkness observed at P. It is not due to light travelling strictly in straight lines. Thus Fresnel's wave theory accounts for the approximate rectilinear propagation of light.

Example 1 : *How many half period elements are there in a circular portion of 10^{-2} m radius of a plane wavefront given that the wavelength is 6×10^{-7} m and the distance of the point of observation of the wavefront is 1 metre ?*

Solution. Now $\quad r_n = \sqrt{nb\lambda}$ Here, $\lambda = 6 \times 10^{-7}$ m; $b = 1$ m; $r_n = 10^{-2}$ m

$$\therefore \qquad n = \frac{r_n^2}{b\lambda} = \frac{(10^{-2})^2}{1 \times (6 \times 10^{-7})} = 166.7$$

Therefore, there are 166 h.p. zones within a radius of 10^{-2} *m.*

Example 2 : *A plane wavefront of light of wavelength 5×10^{-7} m falls on an aperture and the diffraction pattern is observed in an eyepiece at a distance of 1 metre from the aperture. Find the radius of the 100th h.p. element and the area of a h.p. zone.*

Solution. Now $r_n = \sqrt{nb\lambda}$; Here, $\lambda = 5 \times 10^{-7}$ m; $b = 1$ m; $n = 100$

$$r_{100} = \sqrt{100 \times 1 \times 5 \times 10^{-7}} = 7.07 \times 10^{-3} \text{ m.}$$

$$\left.\begin{array}{r}\text{Area of a}\\\text{h.p. zone}\end{array}\right\} = \pi b \lambda = \pi \times 1 \times (5 \times 10^{-7}) = 1.57 \times 10^{-6} \text{ m}^2.$$

3.3. The Zone-plate

Principle : It is a specially constructed diffraction screen such that light from every alternate zone is cut off. The resultant amplitude at a point P is

$$A = d_1 - d_2 + d_3 - d_4 + d_5 - \ldots\ldots\ldots$$

Here, d_1, d_2, d_3 etc., are the amplitudes at P due to first, second, third,...... h.p. zones.

Now if the *light is obstructed from even h.p. zones*, then the resultant amplitude at P is

$$A = d_1 + d_3 + d_5 + \ldots\ldots\ldots$$

If the *light is obstructed from odd h.p. zones,*

$$A = -(d_2 + d_4 + d_6 + \ldots)$$

In either case, the amplitude and hence the intensity at P is enormously increased.

Now, the radius of nth h.p. zone is $r_n = \sqrt{nb\lambda}$.

Therefore the radii of successive h.p. zones are in the ratio $1 : \sqrt{2} : \sqrt{3} : \sqrt{4}$ etc. This principle is used in the construction of zone plate. It is a simple device for focusing light rays by diffraction.

Construction : A large number of concentric circles with radii proportional to the square roots of natural numbers are drawn on a sheet of white paper. The odd numbered zones are painted black. A highly reduced size photograph of this pattern is taken on a thin glass plate. In the negative of the photograph, the odd zones, which were painted black, appear transparent and the even zones appear black (Fig. 3.6). The resulting glass negative is called *positive zone plate.*

Fig. 3.6.

Theory : S is a point source of monochromatic light giving out spherical waves of wavelength λ (Fig. 3.7). AB represents a positive zone plate. The source is placed on the axis at a distance a from the centre O of the zone plate. P is a point on the screen at a distance b from O.

Let us find the intensity of light at P due to the wave front coming from S. With O as centre and radii $OM_1 = r_1$, $OM_2 = r_2$.......$OM_n = r_n$, divide the plate AB into h.p. zones. From one zone to the next, there is an increasing path difference of $\lambda/2$. Hence,

$$SM_1 + M_1P = SO + OP + \lambda/2$$

$$SM_2 + M_2P = SO + OP + 2(\lambda/2)$$

....

$$SM_n + M_nP = SO + OP + n(\lambda/2) \qquad \ldots(1)$$

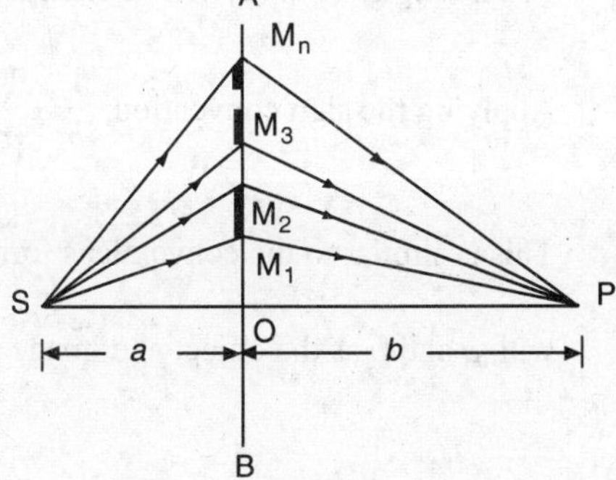

Fig. 3.7.

To find the radius r_n of the nth zone : Let $SO = a$ and $OP = b$.

$$SM_n = \sqrt{(SO^2 + OM_n^2)} = (a^2 + r_n^2)^{\frac{1}{2}} = a\left(1 + \frac{r_n^2}{a^2}\right)^{\frac{1}{2}}$$

$$\therefore \qquad SM_n = a + \frac{r_n^2}{2a} \qquad (\because r_n \ll a)$$

Similarly,

$$M_nP = b + \frac{r_n^2}{2b}$$

Substituting these values in Eq. (1), namely

$$SM_n + M_n P = SO + OP + n\,(\lambda/2),$$

we have
$$a + \frac{r_n^2}{2a} + b + \frac{r_n^2}{2b} = a + b + n\frac{\lambda}{2}$$

or
$$r_n^2\left(\frac{1}{a} + \frac{1}{b}\right) = n\lambda$$

or
$$r_n^2 = \left(\frac{ab\lambda}{a+b}\right)n \qquad ...(2)$$

$\therefore$
$$r_n \propto \sqrt{n} \qquad (\because\ a,\ b \text{ and } \lambda \text{ are constants})$$

Thus *the radii of the various zones are proportional to the square roots of natural numbers.*

The area of the *n*th zone

$$= \pi r_n^2 - \pi r_{n-1}^2 = \pi\left[\frac{n\lambda ab}{a+b} - \frac{(n-1)\lambda ab}{a+b}\right] = \pi\frac{\lambda ab}{a+b} \qquad ...(3)$$

The area is independent of *n*. Hence the area of all the zones is the same. But the distance of the zone from *P* and the obliquity increase as the order of the zone increases. Hence, the amplitude at *P* due to a zone decreases as the order of the zone increases. Let d_1, d_2, d_3 ...... be the displacements at *P* due to the first, second, third etc., zones. Then, the resultant amplitude at *P* is

$$A = d_1 + d_3 + d_5 + ... \qquad ...(4)$$

This displacement *A* is enormously greater than $d_1/2$, the resultant due to all the zones. Hence the point *P* is extremely bright. *P* can be said to be the image of *S*. This explains the *focusing* action of a zone plate. It thus behaves like a convex lens.

Focal length : Eq. (2) can be written as $\frac{1}{a} + \frac{1}{b} = \frac{n\lambda}{r_n^2}$

Applying the sign convention, $\frac{1}{b} - \frac{1}{a} = \frac{n\lambda}{r_n^2}$ ...(5)

This is similar to the convex lens formula $\frac{1}{v} - \frac{1}{u} = \frac{1}{f}$ ...(6)

with *a* and *b* as the object and image distances.

$\therefore$
$$f = \frac{r_n^2}{n\lambda} \qquad ...(7)$$

Here, *f* is called the *primary* or *first order focal length* of the zone plate.

Thus *the zone plate acts as a convergent lens.*

3.4. Multiple Foci in a Zone Plate

For a given wavelength λ, a lens has a single focal length *f*. But a zone plate has a number of foci.

If the point source *S* is at infinity (*i.e.,* if $a = \infty$), the area of each zone $= \pi b\lambda$ from Eq. (3). In this case, the image *P* is formed at a distance $b = r_n^2/(n\lambda)$ = principal focal length.

Now consider a point P_3 along the axis of the zone plate at a distance $b/3$ from the zone plate. Then the area of each h.p. zone with respect to P_3 will be $1/3\ \pi b\lambda$. Hence each zone on the zone plate will contain three h.p. elements corresponding to P_3. The resultant amplitude at P_3 will be

$$A = (m_1 - m_2 + m_3) + (m_7 - m_8 + m_9) + (m_{13} - m_{14} + m_{15}) + \ldots\ldots\ldots$$

Here $m_1, m_2, m_3, \ldots\ldots\ldots$ are roughly one third of d_1, d_2, d_3, *etc.* $m_4, m_5, m_6, m_{10}, m_{11}, m_{12} \ldots$ *are* absent as the respective zones are blackened.

$$\therefore \qquad A = \left(m_1 - \frac{m_1 + m_3}{2} + m_3 \right) + \left(m_7 - \frac{m_7 + m_9}{2} + m_9 \right) + \ldots$$

$$= \frac{1}{2}(m_1 + m_3 + m_7 + m_9 + m_{13} + m_{15} + \ldots)$$

Thus point P_3 is sufficiently bright. P_3 represents the position of the second focal point. Its distance from O is the *second focal length*. It is given by $f_2 = \dfrac{r_n^2}{3n\lambda}$.

The image obtained in this case is however less in intensity than the image at P. Similarly other foci occur on the axis at distances $b/5$, $b/7$,.... from the zone plate. But the intensity of successive foci decreases gradually. There is no image formation corresponding to $b/2$, $b/4$, etc.

Thus the various focal lengths are given by,

$$f_p = \frac{r_n^2}{(2p-1)n\lambda} \qquad \text{where} \qquad p = 1, 2, 3\ldots$$

$$\therefore \qquad f_1 = \frac{r_n^2}{1n\lambda};\ f_2 = \frac{r_n^2}{3n\lambda} = \frac{f_1}{3};\ f_5 = \frac{r_n^2}{5n\lambda} = \frac{f_1}{5}\ldots$$

3.5. Comparison of a Zone Plate with a Convex Lens

Similarities : (*i*) The distances of the object and image are connected together by similar formulae in both the cases.

Convex lens : $\dfrac{1}{v} - \dfrac{1}{u} = \dfrac{1}{f}$; Zone plate : $\dfrac{1}{b} - \dfrac{1}{a} = \dfrac{n\lambda}{r_n^2}$

(*ii*) Focal length of both varies with the wavelength λ. Hence both show chromatic aberration.

Differences

(*i*) A convex lens has only one focus. But a zone plate has several foci.

(*ii*) The rays are brought to focus by refraction in a convex lens. But the image is formed by diffraction in a zone plate.

(*iii*) The image due to a convex lens is more intense than that due to a zone plate.

(*iv*) In a zone plate $f_r < f_v$ since $f = \dfrac{r_n^2}{n\lambda}$ and $\lambda_r > \lambda_v$.

But in case of a lens $f_r > f_v$.

(*v*) In the case of a convex lens, all the waves meet in phase at the image, after traversing the same optical path. But in the zone plate, the waves travel unequal optical paths. The rays from two successive transparent zones differ in path by λ.

Example 1 : *For a wavelength of light* $\lambda = 6 \times 10^{-7}$ *m and radius of the first h.p. zone* $r_1 = 6 \times 10^{-4}$ *m , a zone plate brings rays to focus at its brightest spot. Find the focal length of the equivalent lens.*

Solution. $f = \dfrac{r_n^2}{n\lambda} = \dfrac{r_1^2}{\lambda} = \dfrac{(6 \times 10^{-4})^2}{(6 \times 10^{-7})} = 0.6$ m

Example 2 : *Find the radii of the first three transparent zones of a zone plate behaving like a convex lens of focal length 1m for light of wavelength λ= 589.3 nm.*

Solution. If r_n is the radius of the nth h.p. zone, then

$$f = \frac{r_n^2}{v\lambda} \text{ or } r_n = \sqrt{n\lambda f}$$

Here, $f = 1$ m and $\lambda = 589.3$ nm $= 5.893 \times 10^{-7}$ m.

If we suppose that the even number of zones are opaque, then the first three transparent zones will correspond to $n = 1$, 3 and 5 respectively. Hence the radii of the first three transparent zones are given by

$$r_1 = \sqrt{1 \times (5.893 \times 10^{-7}) \times 1} = 7.67 \times 10^{-4} \text{ m}$$

$$r_2 = \sqrt{3 \times (5.893 \times 10^{-7}) \times 1} = 1.339 \times 10^{-3} \text{ m}$$

$$r_3 = \sqrt{5 \times (5.893 \times 10^{-7}) \times 1} = 1.716 \times 10^{-3} \text{ m.}$$

Example 3 : *A zone plate gives a series of images of a point source on its axis. If the first strongest and second strongest images are at distances of 30 cm and 6 cm respectively from the zone plate, both on the same side remote from the source, calculate the distance of the source from the zone plate.*

Solution. Let a and b be the distances of the object and the image from the zone plate. Let r_n be the radius of the nth zone. Then

$$\frac{1}{f_p} = \frac{1}{b} - \frac{1}{a} = \frac{(2p-1)n\lambda}{r_n^2}$$

For the first strongest primary image, $p = 1$ and $b = 0.3$ m.

$$\therefore \qquad \frac{1}{0.3} - \frac{1}{a} = \frac{n\lambda}{r_n^2} \qquad \text{...(1)}$$

For the second strongest image, $p = 2$ and $b = 0.06$ m.

$$\therefore \qquad \frac{1}{0.06} - \frac{1}{a} = \frac{3n\lambda}{r_n^2} \qquad \text{...(2)}$$

Dividing (2) by (1), $\dfrac{\dfrac{1}{0.06} - \dfrac{1}{a}}{\dfrac{1}{0.3} - \dfrac{1}{a}} = 3$; Simplifying, $a = -0.2998$ m.

Therefore, the distance of the source from the zone plate = 0.2998 m.

Negative sign shows that the source is to the left of the zone plate.

Example 4 : *A zone plate is made by arranging the radii of the circles of various zones equal to the radii of dark rings formed when a plano-convex lens of radius of curvature 1.2 m kept on a plane glass plate and a monochromatic light of wavelength 6×10^{-7} m is used. Calculate the various focal lengths of the zone plate.*

Solution: In Newton's ring experiment, the radius of the nth dark ring is

$$r_n = \sqrt{nR\lambda} \quad \text{or} \quad r_n^2 = nR\lambda.$$

Let r_n be the radius of the nth zone of the zone plate. Then,

$$\left.\begin{array}{l}\text{principal focal length}\\ \text{of the zone plate}\end{array}\right\} = f = \frac{r_n^2}{n\lambda}.$$

According to the given problem, the radii of the h.p. zones are the same as the radii of Newton's rings.

$$\therefore \qquad f = \frac{r_n^2}{n\lambda} = \frac{nR\lambda}{n\lambda} = R = 1.2 \text{ m.}$$

The principal focal length of the zone plate is 1.2 m.

$$\therefore \qquad f_1 = 1.2 \text{ m}; \; f_3 = \frac{f_1}{3} = 0.4 \text{ m}; \; f_3 = \frac{f_1}{5} = 0.24 \text{ m. etc.}$$

3.6. Diffraction at a Circular Aperture

(i) Intensity at an axial point : Let AB be a small circular aperture (say a pinhole), of centre O and radius r (Fig.3.8). S represents a point source of monochromatic light. XY is a screen perpendicular to the plane of the paper. CD represents the spherical wavefront generated by S. Let the wavefront be divided into h.p. zones with respect to the axial point P. With respect to the point P, O is the pole of the wavefront. If the aperture AB exposes only one h.p. zone, then the amplitude at P is d_1. If the whole wavefront is exposed, then the *resultant* amplitude at P will be $d_1/2$. Thus the resultant amplitude at P due to first zone alone is twice the amplitude due to the entire wavefront. Therefore the resultant intensity at P due to first zone alone is 4 times the intensity due to the entire wavefront. Consequently, the point P is extremely bright and gives the position of maximum intensity.

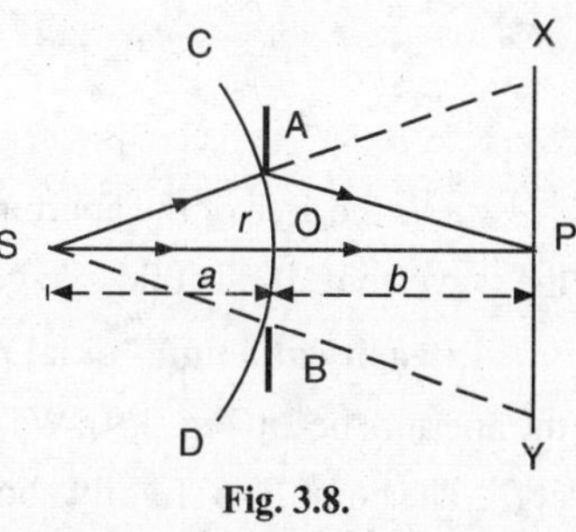

Fig. 3.8.

If the radius of the aperture is such that it accommodates two h.p. zones, then the resultant amplitude at $P = d_1 - d_2$.

The intensity at $P = (d_1 - d_2)^2$.

This intensity has a minimum value since $d_1 \approx d_2$.

If three h.p. zones are contained within the aperture, the resultant intensity at P would be $(d_1 - d_2 + d_3)^2$ (maximum) and so on. *Thus the axial point P is bright or dark according as the aperture allows an odd or even number of h.p. zones.*

Suppose we keep the radius of the aperture fixed and move the screen gradually towards the aperture. Then the number of h. p. zones allowed by the aperture increases gradually. *Thus by continuously moving the screen towards the aperture, the axial point P becomes alternately bright and dark, depending on whether the aperture exposes odd or even number of h.p. zones.*

Mathematical Theory:

Let $SO = a$ = distance of the aperture from the source

$OP = b$ = distance of the aperture from the screen

$OA = OB = r$ = radius of the aperture.

Then, the path difference between the waves reaching P along the paths SAP and SOP is

$$\delta = (SA + AP) - (SO + OP)$$

$$= [(a^2 + r^2)^{1/2} + (b^2 + r^2)^{1/2}] - [a + b]$$

$$= a\left(1 + \frac{r^2}{a^2}\right)^{\frac{1}{2}} + b\left(1 + \frac{r^2}{b^2}\right)^{\frac{1}{2}} - (a + b)$$

$$= a\left(1 + \frac{r^2}{2a^2}\right) + b\left(1 + \frac{r^2}{2b^2}\right) - (a + b) = \frac{r^2}{2}\left(\frac{1}{a} + \frac{1}{b}\right)$$

If the screen is placed at such a distance that the aperture allows n full h.p. zones, then we have

$$\frac{r^2}{2}\left(\frac{1}{a}+\frac{1}{b}\right)=n\frac{\lambda}{2} \quad \text{or} \quad \frac{1}{a}+\frac{1}{b}=\frac{n\lambda}{r^2}.$$

The point P on the screen will have minimum intensity if n is even and maximum intensity if n is odd.

If a, r and λ are kept fixed, the value of b determines the value of n. Hence if we move the screen gradually towards the aperture, P will appear alternately bright and dark. Similar observations can be had by keeping a, b and λ fixed and gradually varying r.

If the source is situated at infinity, then $a = \infty$.

$$\therefore \quad \frac{1}{b}=\frac{1}{f}=\frac{n\lambda}{r^2} \quad \text{or} \quad n=\frac{r^2}{b\lambda}=\frac{\pi r^2}{\pi b\lambda}$$

πr^2 is the area of the aperture. $\pi b\lambda$ is the area of h.p. zone, constructed for a point distant b from the wavefront. If n is odd, we get a maximum at P. If n is even, we get a minimum at P.

Intensity at a non - axial point: Let the diameter AB of the aperture be $2r$ (Fig. 3.9). We want to calculate the intensity at P'. The point P' is slightly above P. Let $PP' = x$ and $OP = b$. If the source S is at a considerable distance from the aperture, then the wavefront striking the aperture is almost plane.

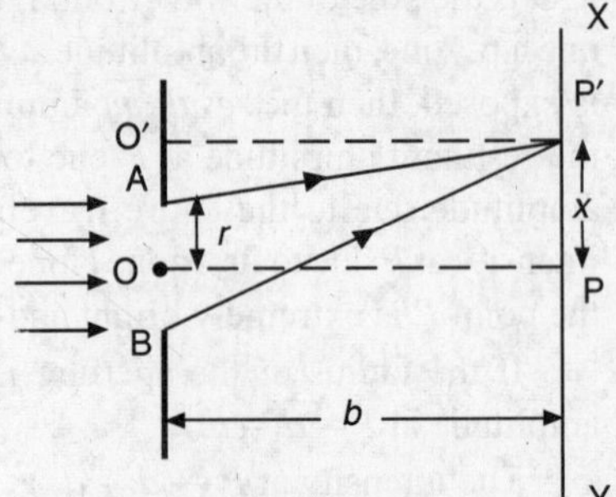

Fig. 3.9.

Hence the path difference between the secondary waves reaching P' from A and B is

$$\delta = BP' - AP'$$

$$= \sqrt{[b^2+(x+r)^2]}-\sqrt{[b^2+(x-r)^2]}$$

$$= \left[b+\frac{(x+r)^2}{2b}\right]-\left[b+\frac{(x-r)^2}{2b}\right]=\frac{2rx}{b}$$

The point P' will be of minimum intensity if this path difference δ is an even multiple of $\lambda/2$. That is,

$$\frac{2rx}{b}=2n\frac{\lambda}{2} \quad (2n \text{ indicates even number of zones})$$

$$\therefore \text{ radii of dark rings are given by } x_n=\frac{nb\lambda}{2r} \qquad ...(1)$$

The point P' will be of maximum intensity if the path difference δ is an odd multiple of $\lambda/2$.

That is, $\frac{2rx}{b}=(2n+1)\frac{\lambda}{2}$

$$\therefore \text{ radii of bright rings are given by } x_n=\frac{(2n+1)b\lambda}{4r} \qquad ...(2)$$

Thus with a circular aperture, the diffraction pattern on a screen consists of concentric bright and dark rings with a central dark or bright spot depending on the distance b (Fig.3.10).

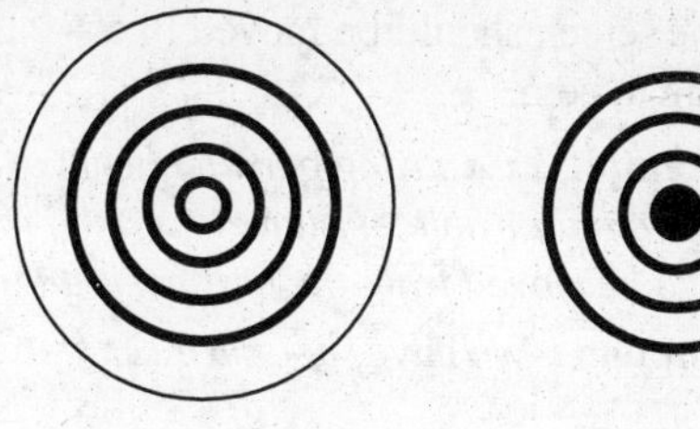

Fig. 3.10.

Importance : Diffraction at a circular aperture is of considerable importance in the formation of images by microscopes and telescopes. When light from a distant star is passed through the objective of a telescope, the incident plane wavefront is limited by the circular aperture of the objective. Hence a diffraction pattern due to a circular aperture is obtained in the focal plane of the telescope. Thus the image of a distant star is seen to consist of a bright central disc surrounded by dark and bright rings.

From Eq. (1), the radius of the first dark ring is $x_1 = \frac{b\lambda}{2r} = \frac{b\lambda}{D}$

For an incident plane wavefront, $b = f$. $\therefore\ x_1 = \frac{f\lambda}{D}$

Here f is the focal length and D the diameter of the aperture of the telescope objective.

Example 1 : *A circular aperture of 1.2×10^{-3} m diameter is illuminated by plane waves of monochromatic light. The diffracted light is received on a distant screen which is gradually moved towards the aperture. If the centre of the circular patch of light first becomes dark when the screen is 0.3 m from the aperture, find the wavelength of light used.*

Solution. For the first darkness on the axis, the aperture should contain first two h.p. zones only.
Area of each h.p. zone for plane waves = $\pi\, b\, \lambda$
Area of the circular aperture = $\pi\, r^2$

$\therefore$ For first darkness, $\frac{\pi r^2}{\pi b\lambda} = 2$ or $\lambda = \frac{r^2}{2b}$

Here, $r = (1.2 \times 10^{-3} / 2)$ m $= 6 \times 10^{-4}$ m and $b = 0.3$ m. $\lambda = ?$

$$\therefore \quad \lambda = \frac{r^2}{2b} = \frac{(6 \times 10^{-4})^2}{2 \times 0.3} = 6 \times 10^{-7} \text{ m} = 600 \text{ nm}.$$

Example 2 : *A parallel beam of wavelength 600 nm is normally incident on a circular opening in an opaque screen. What should be the radius of the opening such that the intensity at a point 0.5m from the screen is 4 times the intensity on it when the screen is absent? Calculate the distance through which the screen should be moved towards the point to reduce the intensity at the point to zero.*

Solution : If the screen is placed at such a distance that the aperture allows n full h.p. zones, then

$$n = \frac{\pi\, r^2}{\pi\, b\lambda} = \frac{r^2}{b\lambda} \text{ or } r = \sqrt{nb\lambda}$$

The intensity in the presence of the opening will be four times the intensity in the absence of the opening if only the first h.p. zone is exposed.

$\therefore n = 1$. Here, $b = 0.5$ m and $\lambda = 600$ nm $= 6 \times 10^{-7}$ m; $r = ?$

$$\therefore\ r = \sqrt{nb\lambda} = \sqrt{1 \times 0.5 \times (6 \times 10^{-7})} = 5.477 \times 10^{-4} \text{ m}.$$

The intensity at the point will be zero when the opening contains two h.p. zones. Here, $n = 2$. Let the new distance of the point from the screen be b_1.

$$\therefore \qquad b_1 = \frac{r^2}{n\lambda} = \frac{(5.477 \times 10^{-4})^2}{2 \times (6 \times 10^{-7})} = 0.25 \text{ m}.$$

The screen should be moved (0.5 – 0.25) = 0.25 m towards the point to reduce the intensity at the point to zero.

Example 3 : *A strong parallel beam of monochromatic light is incident normally on a thin plate having a circular hole of diameter 1 mm. If the wavelength of light used is 5×10^{-7}m, how far is the screen to be moved from the first position where the centre is black to a second similar position?*

Solution : We have, $b = r^2/(n\,\lambda)$.

For first minimum $n = 2$, let $b = b_1$. Then $b_1 = \frac{r^2}{2\lambda}$ ...(1)

For second minimum $n = 4$, let $b = b_2$. Then $b_2 = \frac{r^2}{4\lambda}$ ...(2)

$$\left.\begin{array}{l}\text{Distance between}\\ \text{these two positions}\end{array}\right\} = b_1 - b_2 = \frac{r^2}{2\lambda} - \frac{r^2}{4\lambda} = \frac{r^2}{4\lambda}$$

Here, $r = (10^{-3}/2)$ m $= 5 \times 10^{-4}$ m, $\lambda = 5 \times 10^{-7}$ m, $b_1 - b_2 = ?$

$$\therefore \qquad b_1 - b_2 = \frac{r^2}{4\lambda} = \frac{(5\times 10^{-4})^2}{4\times(5\times 10^{-7})} = 0.125 \text{ m}$$

3.7. Diffraction at a Straight Edge

Let O be a sharp straight edge of an opaque obstacle OD (*e.g.*, an edge of a razor blade). S is a narrow slit illuminated by a monochromatic source of light of wavelength λ (Fig. 3.11). Both are parallel to each other and perpendicular to the plane of the paper. Let XY be a screen parallel to OD. AB is the incident cylindrical wavefront. Join SO and produce it to meet the screen at P so that SOP is perpendicular to the screen. Let us determine the distribution of light intensity on the screen.

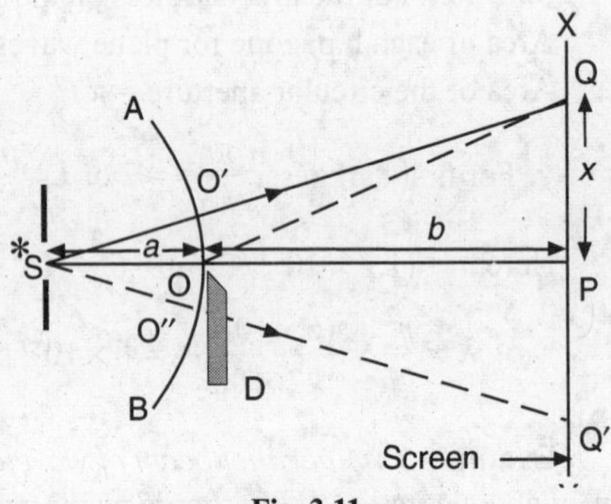

Fig. 3.11.

According to the laws of geometrical optics, we should get uniform illumination above P and complete darkness below P. But in actual practice we observe the following :

1. Just above P, alternate dark and bright bands of unequal width and varying intensity parallel to the straight edge are observed. The bands become closer and less distinct as we move away from P. Ultimately the bands disappear and we get uniform illumination.

2. Inside the geometric shadow, the illumination falls off rapidly. Even at a short distance from the edge of the shadow, the illumination is negligible.

(a) Explanation of diffraction bands: Let Q be a point in the illuminated region. We want to find the intensity at Q. Join SQ cutting the wavefront at O'. O' is the pole of the wavefront with respect to Q. With Q as centre and radii equal to

$$QO' + \frac{\lambda}{2},\ QO' + \frac{2\lambda}{2},\ QO' + \frac{3\lambda}{2}, \ldots QO' + \frac{n\lambda}{2}$$

mark off points on AB. Then through these points draw lines parallel to the length of the slit. The wavefront is thus divided into h.p. strips. There are two similar halves of the wavefront. One ($O'A$) is above O' and the other ($O'B$) below it.

The intensity of light at Q is due to

(i) the entire upper half wavefront beyond O' and

(ii) the exposed portion $O'O$ of the lower half.

The amplitude at Q due to the upper half wavefront $O'A = d_1/2$

If the exposed portion $O'O$ of the lower half wavefront contains one h.p. strip, the resultant amplitude at $Q = \frac{d_1}{2} + d_1 = \frac{3d_1}{2}$ (maximum)

If the exposed portion $O'O$ of the lower half wavefront contains two h.p. strips, the resultant amplitude at $Q = \frac{d_1}{2} + d_1 - d_2 = \frac{3d_1}{2} - d_2$ (minimum)

Similarly we get maximum at Q if the lower half wavefront exposes three, five, seven......h.p. strips. We get minimum at Q if the lower half wavefront exposes four, six,........ h.p. strips. Thus the intensity will be maximum or minimum according as OO' contains odd or even number of h.p. strips. This gives rise to *diffraction bands*. The intensity of a maximum goes on decreasing and that of a minimum goes on increasing. Finally when we move a large distance from P, we get a uniform illumination (Fig. 3.12).

Mathematical theory: Let $SO = a$, $OP = b$ and $PQ = x$ (Fig.3.11).

The path-difference between the vibrations from O' and those from O

$$\delta = OQ - O'Q$$

Now, $$OQ = \sqrt{b^2 + x^2} = b\left(1 + \frac{x^2}{b^2}\right)^{\frac{1}{2}} = b + \frac{x^2}{2b} \qquad (\because x << b)$$

Similarly, $$SQ = \sqrt{(a+b)^2 + x^2} = a + b + \frac{x^2}{2(a+b)}$$

$\therefore$ $$O'Q = SQ - SO' = a + b + \frac{x^2}{2(a+b)} - a = b + \frac{x^2}{2(a+b)}$$

$$\delta = OQ - O'Q = b + \frac{x^2}{2b} - b - \frac{x^2}{2(a+b)} = \frac{x^2}{2}\left(\frac{1}{b} - \frac{1}{a+b}\right) = \frac{ax^2}{2b\,(a+b)}$$

For the displacement to be maximum, δ must be equal to an odd multiple of $\lambda/2$.

$\therefore$ for *maximum* brightness,

$$\frac{ax^2}{2b(a+b)} = (2n+1)\frac{\lambda}{2} \qquad (n = 0, 1, 2, 3...)$$

or $$x_{max} = \sqrt{\frac{b(a+b)(2n+1)\lambda}{a}} \text{ or } x_{\max} \propto \sqrt{(2n+1)}$$

Similarly, for *minimum* brightness, $$\frac{ax^2}{2b(a+b)} = 2n\frac{\lambda}{2}$$

or $$x_{min} = \sqrt{\frac{2b(a+b)n\lambda}{a}} \text{ or } x_{min} \propto \sqrt{2n}$$

Thus $x \propto \sqrt{(2n+1)}$ or $x = k\sqrt{(2n+1)}$ for a bright band.

$$x \propto \sqrt{2n} \text{ or } x = k\sqrt{2n} \text{ for dark band.}$$

Thus the position of a dark or bright band $\propto \sqrt{n}$.

Hence dark and bright bands are *unequally spaced.*

(b) Intensity at the edge of the geometrical shadow : Let us find the intensity at the edge P of the geometrical shadow. The pole of the wavefront with respect to P is O. The wavefront is divided into h.p. strips with respect to P. The lower half of the wavefront is completely blocked by the obstacle. The illumination at P is due to the upper half of the wavefront only. The displacement at P is half that due to the whole wavefront. Thus the intensity at P is *one quarter* of the intensity of that produced by the complete exposed wavefront.

(c) Intensity within the geometrical shadow: Q' is a point within the geometrical shadow. The corresponding pole is O''. Then the entire lower half wavefront $O''B$ and the portion $O''O$ of the upper half are obstructed. Thus the intensity at Q is due to only the portion OA of the upper half wavefront $O''A$. As Q' moves gradually below P inside the geometrical shadow, the first, the first two, the first three etc., h.p. strips of the upper half of the wavefront are successively cut off. So the amplitudes at various points below P are $d_2/2$, $d_3/2$, $d_4/2$, etc. The intensities are $(d_2/2)^2$, $(d_3/2)^2$, $(d_4/2)^2$, etc. d_1, d_2, d_3, etc. decrease in magnitude rapidly. Thus we see that the intensity inside the geometrical shadow decreases rapidly to zero as we move away from P. The intensity becomes zero at a point for which all the effective h.p. strips of the upper half wavefront are cut off.

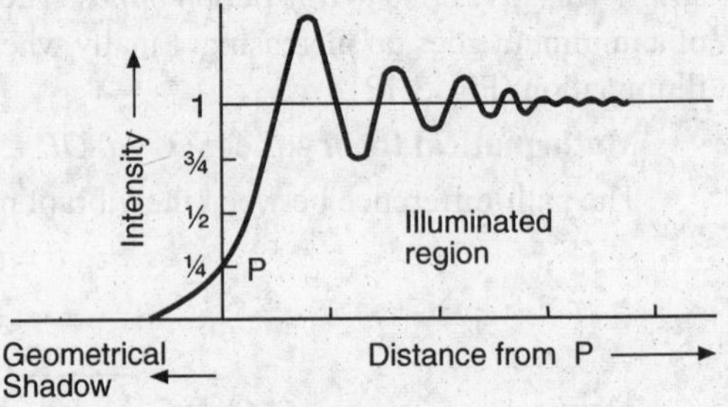

Fig. 3.12.

(d) Intensity distribution curve: The graph shown in Fig. 3.12 depicts the intensity distribution.

Example 1 : *A narrow slit is illuminated by monochromatic light of wavelength 6×10^{-7}m. It is placed at a distance of 0.1m from a straight edge. If measurements are carried out at a distance equal to 1m from the edge, what is the distance between the first and second dark bands?*

Solution.
$$\left.\begin{array}{l}\text{Distance of } n\text{th order dark band} \\ \text{from the point } P \text{ on the screen}\end{array}\right\} = (x_n)_{min} = \sqrt{\frac{2b(a+b)n\lambda}{a}}$$

Here, $\lambda = 6 \times 10^{-7}$ m; $a = 0.1$ m, $b = 1$m: $x_2 - x_1 = ?$

Distance between the first and second dark bands $= x_2 - x_1$

$$= \sqrt{\frac{2b(a+b)2\lambda}{a}} - \sqrt{\frac{2b(a+b)\lambda}{a}} = \sqrt{\frac{2b(a+b)\lambda}{a}}\left(\sqrt{2} - \sqrt{1}\right)$$

$$= \sqrt{\frac{2 \times 1 \times (0.1+1)(6 \times 10^{-7})}{0.1}}\,(1.414 - 1) = 1.504 \times 10^{-3} \text{ m.}$$

Example 2: *A straight edge is illuminated by monochromatic light from a narrow slit parallel to the edge and at a distance of 0.2 m from it. The distances of the first and second dark bands from the edge of the geometric shadow on a screen at a distance of 0.6 m from the straight edge are 1.7×10^{-3} m and 2.4×10^{-3}m. Calculate the wavelength of light.*

Solution : Let x_n be the distance of the nth dark band from the edge of the geometric shadow.

$$\text{Then } x_n = \sqrt{\frac{2b(a+b)n\lambda}{a}} \text{ or } x_n^2 = \frac{2b(a+b)n\lambda}{a}$$

Here , $a = 0.2$ m; $b = 0.6$ m; $x_1 = 1.7 \times 10^{-3}$ m; $x_2 = 2.4 \times 10^{-3}$ m; $\lambda = ?$

$$\therefore \qquad (1.7\times10^{-3})^2 = \frac{2\times0.6\times0.8\times1\times\lambda}{0.2}$$

and
$$(2.4\times10^{-3})^2 = \frac{2\times0.6\times0.8\times2\times\lambda}{0.2}$$

$$(2.4\times10^{-3})^2 - (1.7\times10^{-3})^2 = \frac{2\times0.6\times0.8\times\lambda}{0.2}$$

$$\therefore \qquad \lambda = \frac{[(2.4\times10^{-3})^2 - (1.7\times10^{-3})^2]\times0.2}{2\times0.6\times0.8} = 5.979\times10^{-3}\ \text{m}$$

3.8. Fresnel Diffraction at a Narrow Rectangular Aperture (Narrow slit)

Let S be a narrow rectangular slit source of monochromatic light, placed perpendicular to the plane of the paper (Fig. 3.13). AB is a rectangular aperture held parallel to S . XY is a screen, perpendicular to the plane of the paper. CD is the cylindrical wavefront diverging from S and incident on AB . According to the laws of geometrical optics, we should get uniform illumination in the region EF and complete darkness in the region above E and below F. But actually we get a diffraction pattern depending on the width of the aperture. Now there are three cases :

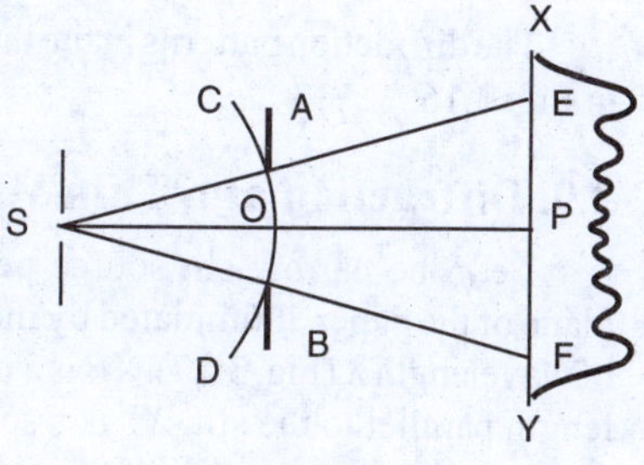

Fig. 3.13.

1. Wide aperture: Suppose the aperture is wide enough to expose a large number of h.p. zones. Then the two edges A and B will behave like two independent straight edges. Hence diffraction bands of unequal width are produced at E and F inside the geometrical bright portion. As we move away from E and F towards P, the midpoint of EF , the bands go on decreasing in width and visibility. The general illumination increases till it becomes uniform near about P. Above E and below F, the intensity rapidly falls to zero.

2. Narrow aperture: Consider the middle point P of the geometrical image EF on the screen (Fig.3.14). Suppose the slit is so narrow that only a few h.p. zones are exposed. If the number of these h.p. zones is odd, P will be of maximum intensity and if even P will be of minimum intensity. Thus P can be *bright* or *dark*.

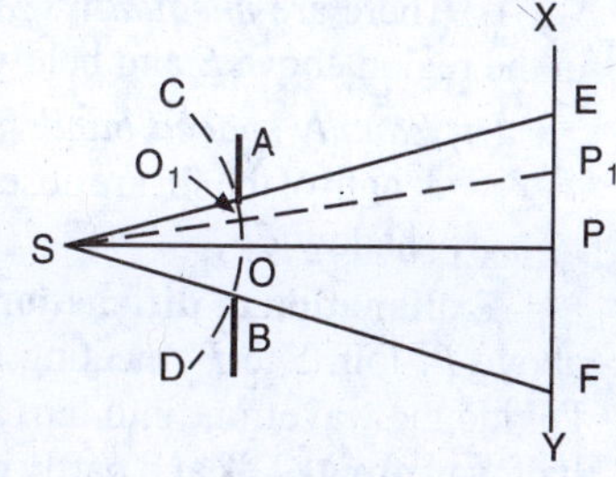

Fig. 3.14.

Consider a point P_1 in the illuminated region EF. O_1 is the pole of the wavefront w.r.t. P_1 . The intensity at P_1 will depend on the number h.p. zones in the upper exposed part AO_1 and lower exposed part O_1B of the wavefront . The point P_1 will be of maximum intensity if an odd number of h.p. zones are contained in the upper and in the lower exposed part of the wavefront. But if their number is even, P_1 will be of minimum intensity. Hence a set of alternately dark and bright diffraction bands will be seen inside EF.

Finally consider a point P_2 within the geometrical shadow (Fig. 3.15). O_2 is the pole of the wavefront

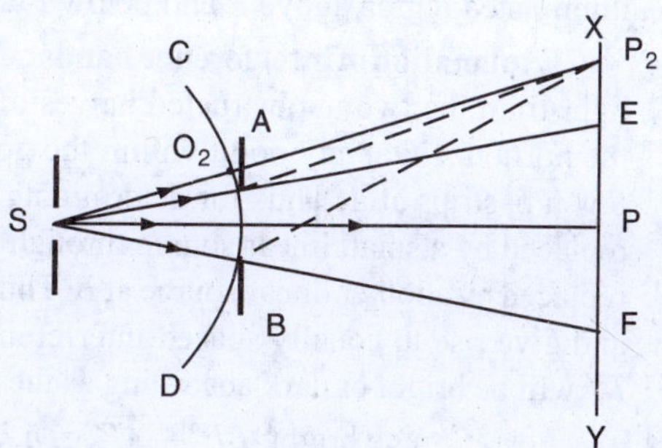

Fig. 3.15.

w.r.t. P_2. Divide the wavefront into h.p. zones. Then all the upper h.p. zones (between O_2 C) are cut off by the obstacle. Only a few h.p. elements from the lower half wavefront O_2D send light to P_2.

The intensity at P_2 will be maximum or minimum according as AB contains an odd or even number of h.p. elements. Hence beyond E and F in the geometrical shadow, diffraction bands of decreasing intensity and width will be seen. If the point P_2 is far away from E, all the effective h.p. zones are cut off by the obstacle and hence maxima and minima become indistinguishable.

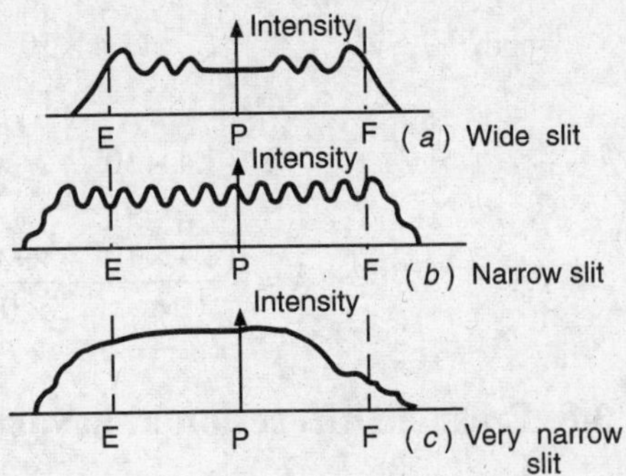

Fig. 3.16.

3. Very narrow aperture : When the aperture comprises one or less than one, h.p. element w.r.t. P, it is always bright. There will be no bands inside the geometrical image EF. There will however be fringes outside the region EF.

The diffraction patterns in the three cases are shown in Fig. 3.16.

3.9. Diffraction at a Thin Wire

Let S be narrow slit source perpendicular to the plane of the paper, illuminated by monochromatic light of wavelength λ (Fig. 3.17). AB is a narrow wire with its length parallel to the slit. XY is a screen perpendicular to the plane of the paper. WW' is a cylindrical wavefront diverging from S. According to geometrical optics, we should get uniform illumination above E and below F, and complete darkness within EF. But actually we observe the following:

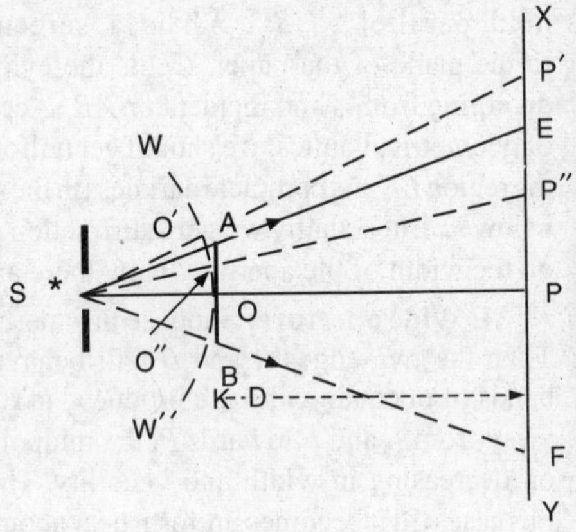

Fig. 3.17.

(*i*) There are *unequally spaced diffraction bands* in the region above E and below F.

(*ii*) *Equally spaced bands* running parallel to the length of the slit are observed in the geometrical shadow EF.

Explanation of diffraction bands: Consider a point P' on the screen in the illuminated region above P. Join S to P' meeting the wavefront at O'. Thus O' is the pole of the wavefront w.r.t. P'. Divide the wavefront into h.p. strips. Then the intensity at P' will be partly due to the entire upper half wavefront O' W and partly due to the part $O'A$. The intensity at P' will be maximum or minimum depending upon whether the number of h.p. strips between O' and A is odd or even. Thus in the illuminated region above E and below F there will be diffraction bands.

Explanation of interference bands : Consider point P'' inside the geometrical shadow. P'' receives light from the two unobstructed halves of the wavefront AW and BW'. We know that the effect of the portion AW at any point within the geometrical shadow can be considered as entirely due to a few h.p. strips at A. Thus for determining the intensity at P'', the part AW of the wavefront may be replaced by a small linear source through A. Similarly, the lower part BW' of the wavefront may be replaced by another linear source at B. Thus the edges A and B of the wire act as two coherent sources and give rise to equally spaced interference bands in the region of geometrical shadow. The point P'' will be bright or dark according as the path difference $BP'' - AP''$ is even or odd multiple of $\lambda/2$.

For P'' to be bright, $BP'' - AP'' = n\lambda$ and

for P'' to be dark, $BP'' - AP'' = (2n-1)\lambda/2$.

The centre P will be always bright. Fringe width $\beta = \frac{D\lambda}{d}$.

Here, D = distance between the screen and the wire and d = diameter of the wire.

Effect of increasing the thickness of the wire : If the thickness of the wire is gradually increased, the width of the interference fringes decreases. If the wire is sufficiently thick, the interference fringes completely disappear. The intensity, however, falls off rapidly as we enter the geometrical shadow from either side. The diffraction pattern in the illuminated region remains unaffected.

To measure the diameter of a wire: A narrow vertical slit, a narrow wire and a micrometer eyepiece are arranged on an optical bench. The centres of these components are arranged to lie in the same horizontal line. When the slit is illuminated with monochromatic light, interference fringes of equal width are seen within the geometrical shadow of the wire. The fringe width β is measured. D, the distance between the slit and the cross-wires is also measured. λ is known.

$\therefore$ The diameter of the wire = $d = D\lambda/\beta$.

Example : *If the diameter of a wire be 2×10^{-4} m, calculate the separation between the fringes formed on a screen placed at 0.5m. from the wire. The light source used has a wavelength of 5×10^{-7}m and is placed at a finite distance from the wire.*

Solution : The two edges of the wire act as if two small coherent sources are situated there. They, therefore, produce equidistant interference bands.

Diameter of the wire = Distance between coherent sources = 2×10^{-4} m :

D = distance between wire and screen = 0.5m; $\lambda = 5 \times 10^{-7}$m; β = ?

$$\beta = \frac{D\lambda}{d} = \frac{0.5 \times (5 \times 10^{-7})}{2 \times 10^{-4}} = 1.25 \times 10^{-3} \text{ m.}$$

3.10. Fraunhofer Diffraction at a Single Slit

Let a parallel beam of monochromatic light of wavelength λ be incident normally upon a narrow slit of width $AB = a$ (Fig. 3.18). The diffracted light is focused by a convex lens on a screen placed in the focal plane of the lens. The diffraction pattern obtained on the screen consists of a central bright band, having alternate dark and weak bright bands of decreasing intensity on both sides.

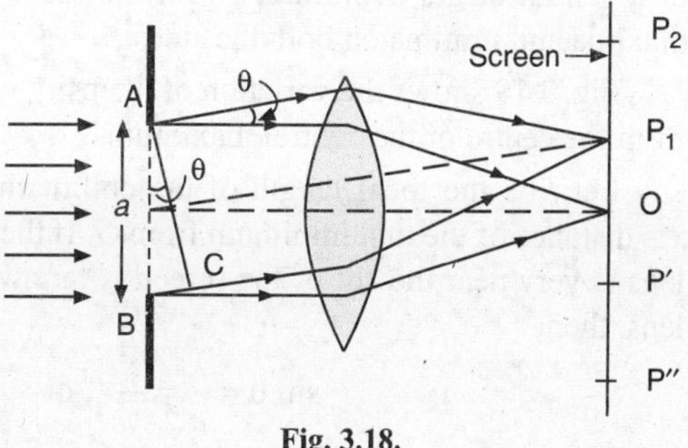

Fig. 3.18.

As a plane wavefront is incident on the slit AB, each point on the wavefront becomes a source of secondary wavelets. The rays diffracted along the direction of incident rays are focused at O. The point O is optically equidistant from all points on the slit AB. Therefore, all the secondary wavelets from AB reach O in the same phase. Hence there is maximum intensity at O.

Let us consider the intensity at any point P_1 above O. The secondary waves travelling at an angle θ with the normal are focused at P_1. Let AC be drawn perpendicular to BC. The optical lengths of all the rays from the plane AC to the focal point P_1 are the same. The path difference between the secondary waves originating from extreme points A and B is

$$BC = AB \sin\theta = a \sin\theta.$$

Let this path difference be one wavelength. If now we imagine the aperture AB to be divided into two equal halves, then the wavelets from *corresponding points* in each half will differ in phase at P_1 by $\lambda/2$. They would mutually interfere and cancel out each other's effect thus producing the *first* minimum at P_1.

$$\therefore \quad BC = a \sin\theta = \lambda$$

or
$$\sin\theta = \frac{\lambda}{a} \text{ or } \theta = \frac{\lambda}{a} \qquad (\because \theta \text{ is very small})$$

Hence the *first minimum on either side of O* will occur in a direction given by

$$\theta = \frac{\lambda}{a}.$$

Suppose for another point P_2, the path difference BC is 2 λ. Now the slit can be supposed to be divided into four equal parts. The rays from corresponding points separated by a distance $a/4$ in the two halves of each half of the slit will have a path difference of $\lambda/2$. These rays will mutually interfere and cancel out each other. A second minimum, therefore, occurs at P_2 in a direction θ given by

$$BC = a \sin\theta = 2\lambda.$$

In general the various secondary minima will occur when the path difference between the extreme rays is an even multiple of $\lambda/2$ or an integral multiple of λ.

$$\therefore \quad \text{For minima, } BC = a \sin\theta = 2n\,(\lambda/2) = n\lambda \qquad ...(1)$$

Besides the *central maximum* at O, there are *secondary maxima* which lie in between the minima on either side of the central (principal) maximum. These are situated in directions in which the path difference BC is an odd multiple of $\lambda/2$.

$\therefore$ For secondary maxima, $BC = a \sin\theta = (2n+1)\,\lambda/2$ ...(2)

The intensity of these secondary maxima is much less and falls off rapidly as we move outwards.

Thus the diffraction pattern due to a single slit consists of a central bright maximum at O followed by secondary maxima and minima on both the sides.

Fig. 3.19 shows the variation of intensity with distance from the centre of the central maximum.

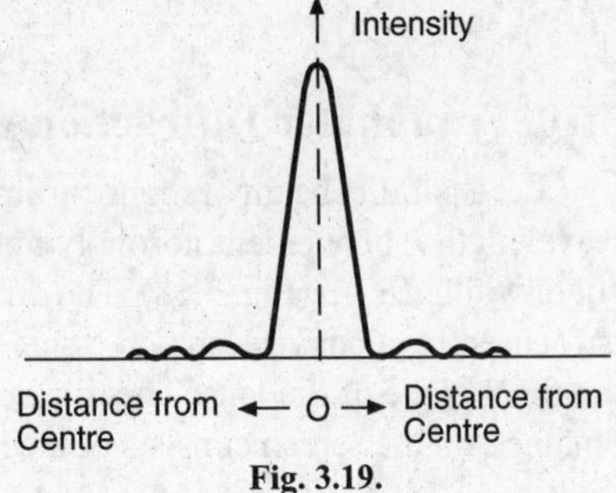

Fig. 3.19.

Let f be the focal length of the collimating lens. Let x = distance of the first minimum from O. If the collimating lens is very near the slit or the screen is far away from the lens, then

$$\sin\theta = \frac{\lambda}{a} = \frac{x}{f} \quad \text{or} \quad x = \frac{f\lambda}{a}$$

$$\therefore \left.\begin{array}{l}\text{Width of the}\\ \text{central maximum}\end{array}\right\} = 2x = \frac{2f\lambda}{a}$$

Hence the width of the central maximum is proportional to the wavelength of light λ.

3.11. Fraunhofer Diffraction at a Double Slit

Let AB and CD be two parallel slits of equal width a and separated by an opaque distance b (Fig. 3.20). Let a plane wave-front be incident normally upon the slits. The light diffracted from these slits is focused by a lens L on a screen XY placed in the focal plane of lens L. The slits and the screen are perpendicular to the plane of the paper.

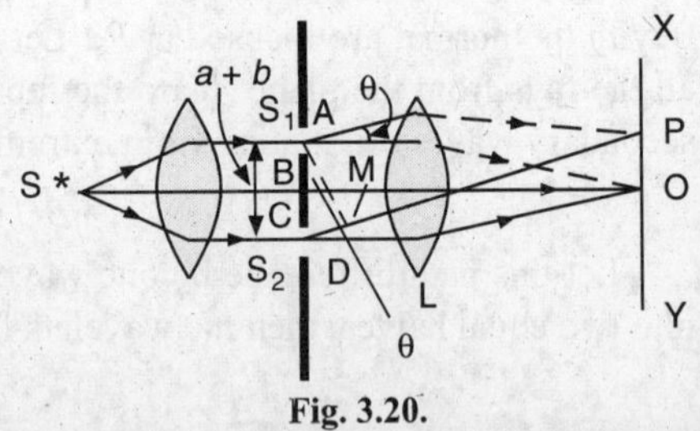

Fig. 3.20.

By Huygen's principle, every point in the slits AB and CD sends out secondary wavelets in all directions. All the secondary waves travelling in the direction of incident light come to focus at O.

From the theory of diffraction at a single slit, the resultant amplitude R due to wavelets diffracted from each slit in a direction θ is

$$R = A\frac{\sin\alpha}{\alpha}$$

Here A is a constant being equal to the amplitude due to a single slit when $\theta = 0$, and $\alpha = \pi a \sin\theta/\lambda$.

We can, therefore, consider the two slits as equivalent to two coherent sources placed at the middle points S_1 and S_2 of the slits and each sending a wavelet of amplitude $A \sin\alpha/\alpha$ in a direction θ to the normal. Hence the resultant amplitude at P will be due to the interference between the two waves of same amplitude R and a phase difference ϕ.

S_1M is the perpendicular drawn from S_1 on S_2P. Then the path difference S_2M between secondary wavelets from S_1 and S_2 is $(a + b)\sin\theta$. The corresponding phase difference is $(2\pi/\lambda)(a + b)\sin\theta$.

Let $\phi = (2\pi/\lambda)(a + b)\sin\theta = 2\beta$

From the triangle of amplitudes (Fig. 3.21), the resultant amplitude at P is

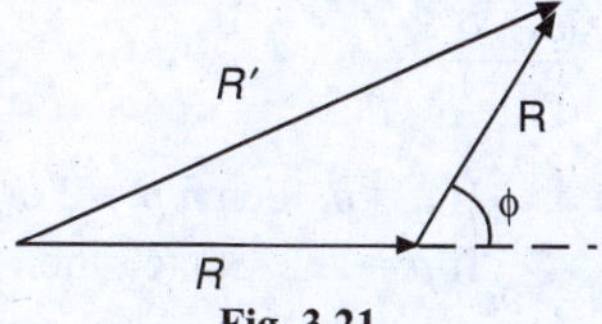

Fig. 3.21.

$$\begin{aligned} R'^2 &= R^2 + R^2 + 2RR\cos\phi \\ &= 2R^2(1 + \cos\phi) = 2R^2.2\cos^2(\phi/2) = 4R^2\cos^2(\phi/2) \end{aligned}$$

Substituting the values of R and ϕ, we get

$$R'^2 = 4A^2\frac{\sin^2\alpha}{\alpha^2}\cos^2\beta$$

Hence intensity at P would be

$$I = 4A^2\frac{\sin^2\alpha}{\alpha^2}\cos^2\beta$$

Thus the resultant intensity at any point depends on two factors :

(i) $\mathbf{(4A^2\sin^2\alpha)/\alpha^2}$ **:** This factor gives the diffraction effects due to a single slit. There is a central maximum in the direction $\alpha = 0$. The central maximum has on either side alternate minima and secondary maxima of diminishing intensity. The angular positions of minima are given by $\alpha = n\pi$ where $n = 1, 2, 3, \ldots$. The positions of secondary maxima are at $\alpha = \pm\frac{3\pi}{2}, \pm\frac{5\pi}{2}, \pm\frac{7\pi}{2}, \ldots$

(ii) The second term $\cos^2\beta$ is due to the combined effect of both the slits and hence is due to interference. The intensity will be maximum when $\cos^2\beta = 1$.

i.e., $\beta = \pm n\pi$ where $n = 0, 1, 2, 3 \ldots$, or $\frac{\pi}{\lambda}(a + b)\sin\theta = \pm n\pi$

or $$(a + b)\sin\theta = \pm n\lambda$$

The intensity will be minimum when $\cos^2\beta = 0$ or $\beta = \pm(2n + 1)\pi/2$

or $$\frac{\pi}{\lambda}(a + b)\sin\theta = \pm(2n + 1)\frac{\pi}{2} \text{ or } (a + b)\sin\theta = \pm(2n + 1)\frac{\lambda}{2}$$

Fig. 3.22 *(a)* and Fig. 3.22 *(b)* represent the intensity distribution determined by the factors $(A^2 \sin^2 \alpha)/\alpha^2$ and $\cos^2 \beta$ respectively. The resultant intensity is given by Fig. 3.22 *(c)*.

Missing orders. Suppose for some value of θ, the following conditions are simultaneously satisfied:

$$(a + b) \sin \theta = n\lambda \text{ interference maxima} \quad ...(1)$$

$$a \sin \theta = p\lambda \text{ diffraction minima} \quad ...(2)$$

Thus the positions of certain interference maxima correspond to the diffraction minima at the same positions on the screen. As a result, the interference pattern will be missing in this direction θ.

Dividing Eq. (1) by Eq. (2), we get

$$\frac{a+b}{a} = \frac{n}{p}$$

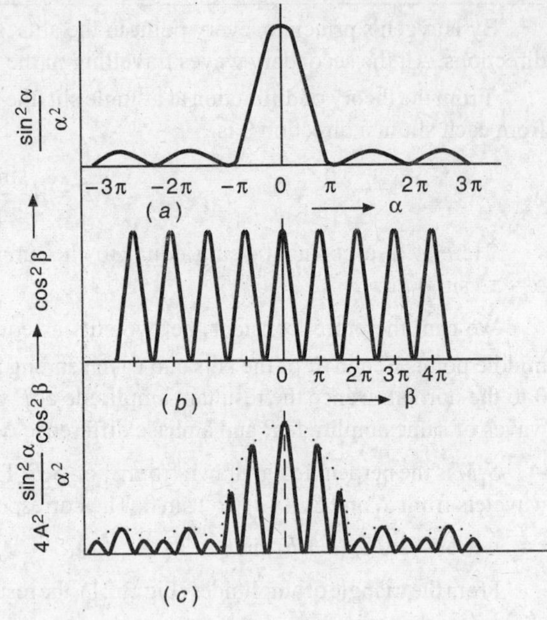

Fig. 3.22.

If $a = b$, then $n/p = 2$ or $n = 2p$.

If $p = 1, 2, 3$, etc., then $n = 2, 4, 6$ etc.

Thus the orders 2, 4, 6 etc., of the interference maxima will be missing. That is, they will coincide with diffraction minima of 1st, 2nd, 3rd, etc., orders.

3.12. Plane Transmission Diffraction Grating

An arrangement consisting of a large number of parallel slits of equal width and separated from one another by equal opaque spaces is called a diffraction grating.

It is constructed by ruling equidistant parallel lines with a fine diamond point on an optically plane glass plate. The ruled lines are opaque to light. These are called *opacities*. The space in between any two lines is transparent to light. The spaces are called the *transparencies*. Such a grating is called *transmission grating*.

Theory : Consider a parallel beam of light incident normally on a grating *XY* (Fig. 3.23). *AB*, *CD*, *EF* ... are the transparent slits. Let the width of each slit be *a* and the width of each opaque portion be *b*. Then the distance $(a + b)$ is called the *grating constant* or *grating element*. The points in the consecutive slits separated by the distance $(a + b)$ are called the '*corresponding points*'. Most of the light issuing from the spaces will go straight on. But as the width of spaces is comparable to the wavelength of light, part of this light spreads out in all directions, on leaving the slits.

(i) Suppose a telescope with its axis normal to the grating is placed in the path of diffracted light. Then the rays issuing out normally are brought to focus at a point *O* lying on the principal axis of the lens *L*. All the rays reaching *O* are in phase with each other. Hence the rays reinforce producing a central bright band (*central maximum*).

(ii) The rays diffracted at an angle θ with the grating normal reach P_1 on passing through the lens in different phases. Draw *AK* perpendicular to the direction of the diffracted light. Then *CN* is the path difference between the rays diffracted from the two corresponding points *A* and *C* at an angle θ.

The path difference $CN = AC \sin \theta = (a + b) \sin \theta$.

If this path difference is an even multiple of $\lambda/2$, then the point *P* will be bright.

Hence for maximum intensity, we have

$$(a + b) \sin \theta = \pm n\lambda$$

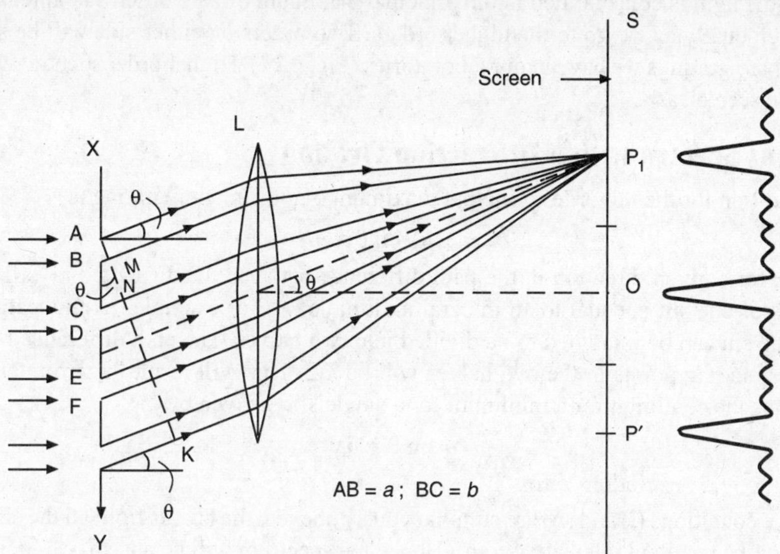

Fig. 3.23.

where n is an integer, 0, 1, 2, 3, etc. n is called the *order* of the interference maximum.

The point P will be dark if $(a + b) \sin \theta = \pm (2n + 1)\lambda/2$.

Thus the diffracted rays from any pair of corresponding points of the slits will produce constructive or destructive interference at a point P according as the path difference is an even or odd multiple of $\lambda/2$. This condition holds true for all the rays from the corresponding points of any pair of adjoining slits in the entire grating surface. We find therefore that brightness and darkness are alternate.

For $n = 0$, we get central maximum at O. When $n = 0$, $\sin \theta = 0$ and $\theta = 0$.

Hence, when there is no diffraction, the light travels straight and is said to be of zero order.

$$\text{For } n = \pm 1, \sin \theta_1 = \pm \frac{\lambda}{(a+b)}$$

This gives the condition for the *first order principal maximum* intensity point on either side of O, *i.e.*, at P_1 and P'. The intensity at P_1 is less than the intensity at O.

$$\text{For } n = \pm 2, \sin \theta_2 = \pm \frac{2\lambda}{(a+b)}.$$

This gives the direction of the *second order principal maxima.*

$$\text{For } \pm 3, \sin \theta_3 = \pm \frac{3\lambda}{(a+b)}.$$

This gives the *III order principal maximum* and so on.

Let monochromatic light be incident normally on a grating. Then there will be a central bright image of the source with bright images on either side corresponding to different orders.

Let white light be incident normally on a grating. Then the central image is white, since for $n = 0$, we have $\theta = 0$ irrespective of λ. When $n = 1$, $\sin \theta_1 = \lambda/(a+b)$. Hence the angle of diffraction will change with the wave length. Since $\lambda_r > \lambda_v$, $\theta_{1r} > \theta_{1v}$. Each wavelength gives rise to its own principal maximum in each order. Hence *each order will consist of a spectrum.*

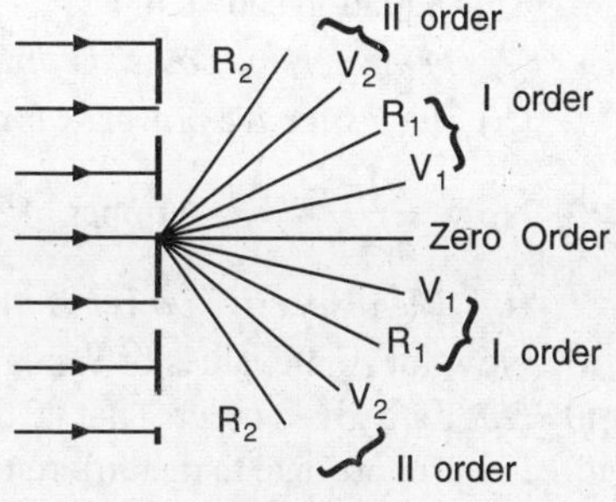

Fig. 3.24.

Most of the light is concentrated in the principal maximum of zero order. The intensity goes on diminishing gradually as we go to the higher orders. Two orders on either side will be seen clearly with the ordinary gratings we have in our laboratories (Fig. 3.24). Higher order spectra will be feeble and cannot be seen clearly.

3.13. Absent Spectra with a Diffraction Grating

The condition for the *n*th order principal maximum of a diffraction grating is

$$(a + b) \sin \theta = n\lambda \qquad ...(i)$$

Suppose for a given direction θ, the path difference between the diffracted rays from the two extreme ends of one slit is equal to an integral multiple of λ. For example, let this path difference be λ. Then the slit can be imagined to be divided into two halves. The path difference between any pair of corresponding points in the two halves will be $\lambda/2$. This will result in zero intensity in that direction. Thus the condition for a minimum for a single slit is given by

$$a \sin \theta = p\lambda \qquad ...(ii)$$

where $p = 1, 2, 3, \ldots$ excluding zero.

When the conditions (i) and (ii) are simultaneously obeyed, the beams from all the slits reinforce each other but the resultant intensity is zero. Hence the spectrum will be absent.

Dividing (*i*) by (*ii*), we get
$$\frac{a+b}{a} = \frac{n}{p} \qquad ...(iii)$$

This is the condition for the *n*th order spectrum to be absent in the diffraction pattern.

(*i*) Let $b = a$. Then from Eq. (*iii*), $n = 2p$.

So the 2nd, 4th, 6th etc., orders of the spectrum will be missing corresponding to the minima due to a single slit given by $p = 1, 2, 3$, etc.

(*ii*) Also when $b = 2a$, $n = 3p$.

So the 3rd, 6th, 9th etc., orders of the spectrum will be missing corresponding to the minima due to a single slit given by $p = 1, 2, 3$, etc.

3.14. Dispersive Power of a Grating

The dispersive power of a diffraction grating is defined as the rate of change of the angle of diffraction with the wavelength of light. If θ and $(\theta + d\theta)$ are the angles of diffraction corresponding to the wavelengths λ and $(\lambda + d\lambda)$, then $d\theta/d\lambda$ is called the dispersive power of the grating.

For a plane transmission grating, we have $(a + b) \sin \theta = n\lambda$ ...(1)

Here, $(a + b)$ is the grating element. θ is the angle of diffraction for *n*th order spectrum.

Differentiating Eq. (1) w.r.t. λ, we get $(a + b) \cos \theta \, (d\theta/d\lambda) = n$

$$\therefore \qquad \textit{Dispersive power} = \frac{d\theta}{d\lambda} = \frac{n}{(a+b)\cos\theta} \qquad ...(2)$$

This equation shows that :

(*i*) *The dispersive power is directly proportional to n, the order of the spectrum.*

(*ii*) *The dispersive power is inversely proportional to the grating element* $(a + b)$.

Now, $\dfrac{1}{a+b} = N$ = number of lines/cm of the grating. $\therefore \; \dfrac{d\theta}{d\lambda} \propto N$.

(*iii*) *The dispersive power is inversely proportional to cos* θ.

Now for small values of θ, $\cos\theta$ is constant. If we neglect the influence of the factor $\cos\theta$, then $d\theta \propto d\lambda$ for a given order. That is, the angular dispersion of two spectral lines in a particular order is directly proportional to the difference in wavelength between the two spectral lines. Such a spectrum is called a *normal spectrum.*

3.15. Overlapping of Spectral Lines

Suppose the light incident on the grating surface consists of a large range of wavelengths. Then, the spectral lines of shorter wavelength and of the higher order overlap on the spectral lines of the longer wavelength and of lower order. Thus the third order red line ($\lambda = 700$ nm), fourth order green line ($\lambda = 525$nm) and fifth order violet line ($\lambda = 420$ nm) will all occur in the same direction θ because

$$(a + b) \sin \theta = 3 \times 700 \times 10^{-9} = 4 \times 525 \times 10^{-9} = 5 \times 420 \times 10^{-9}$$

Obviously, in the visible region of spectrum there is no overlapping. The visible region of the spectrum extends from 400 nm to 720 nm. If a photographic plate is employed for observations, then the spectrum may extend down to 200 nm in the ultra-violet region. Then, a line of 400 nm in the first order will overlap a line of 200 nm in the second order. The overlapping of different wavelengths can be avoided by the use of suitable filters to absorb those undesired wavelengths which may overlap.

3.16. Grating at Oblique Incidence

Consider a parallel beam of light incident obliquely on the grating surface at an angle of incidence i. Let θ be the corresponding angle of diffraction (Fig. 3.25).

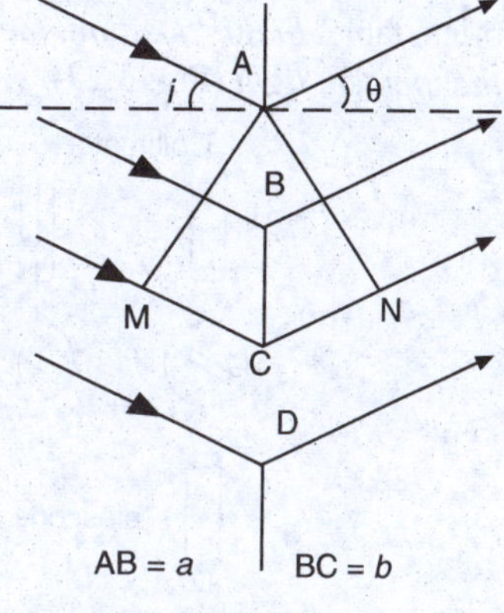

Fig. 3.25.

The path difference between the secondary waves from the corresponding points A and C is

$p = MC + CN = AC \sin i + AC \sin \theta = (AB + BC) (\sin i + \sin \theta)$

Here, $AB = a =$ the width of the slit and $BC = b =$ the width of the opaque portion.

$$\therefore \quad p = (a + b) (\sin \theta + \sin i) \quad ...(1)$$

For nth order principal maximum, $(a + b) (\sin \theta + \sin i) = n\lambda$ ...(2)

$$\text{or} \quad (a + b)\left[2 \sin \frac{\theta + i}{2} \cos \frac{\theta - i}{2}\right] = n\lambda \quad ...(3)$$

$$\text{or} \quad \sin \frac{\theta + i}{2} = \frac{n\lambda}{2(a + b) \cos \{(\theta - i)/2\}}$$

The deviation of the diffracted beam from its original path $= \theta + i$.

Now, $\sin \{(\theta + i)/2\}$ and hence the deviation $(\theta + i)$ would be minimum when $\cos \{(\theta - i)/2\}$ is maximum or $(\theta - i)/2 = 0$ or $\theta = i$.

Thus the rays would suffer minimum deviation when the angle of incidence equals the angle of diffraction.

Angle of minimum deviation $= D = \theta + i = 2\theta = 2i$

$$\therefore \quad \theta = \frac{D}{2} \text{ and } i = \frac{D}{2}$$

Now Eq.(2) is modified as

$$2 (a + b) \sin (D/2) = n\lambda \quad ...(4)$$

Eq. (4) is the grating equation for *minimum deviation position.*

3.17. Determination of Wavelength of Light using Transmission Grating (Normal Incidence)

Principle. If θ is the angle of diffraction for a wavelength λ in the nth order, then

$$(a + b) \sin \theta = n\lambda.$$

Here, $(a+b)$ is called *grating element*. If the grating element $(a+b)$ and the angle of diffraction θ for order n are determined, then λ can be calculated.

(i) Determination of $(a+b)$: On every grating, the number of rulings per inch is written. Let N be the number of rulings per inch. Then $(a+b) = 0.0254/N$ metre.

(ii) Determination of angle of diffraction θ : Adjust a spectrometer (Fig. 3.26) for work. Illuminate the slit with the source whose wavelength is required. Take the direct reading of the telescope. Rotate the telescope by 90° and fix it. Mount the grating on the prism table so that the rules are vertical and the ruled surface is at the centre of the prism table. Rotate the prism table until the image of the slit reflected by the grating coincides with the intersection of the cross-wires in the telescope. Take the reading of the prism table. Rotate the prism table by 45° in the proper direction so that the plane of the grating is normal from the collimator and fix the prism table. *In this position, the grating is normal to the incident light* (Fig. 3.27).

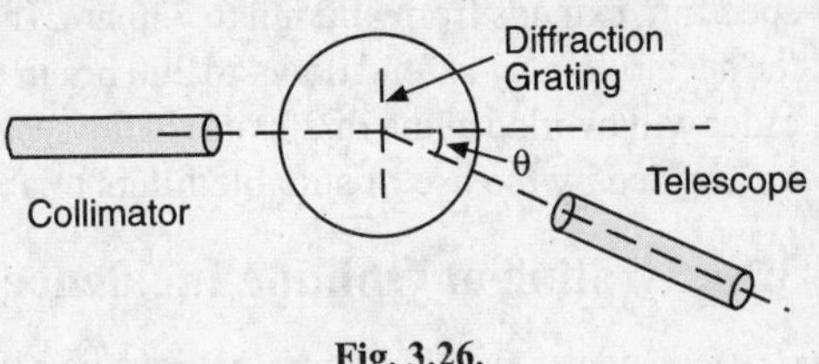

Fig. 3.26.

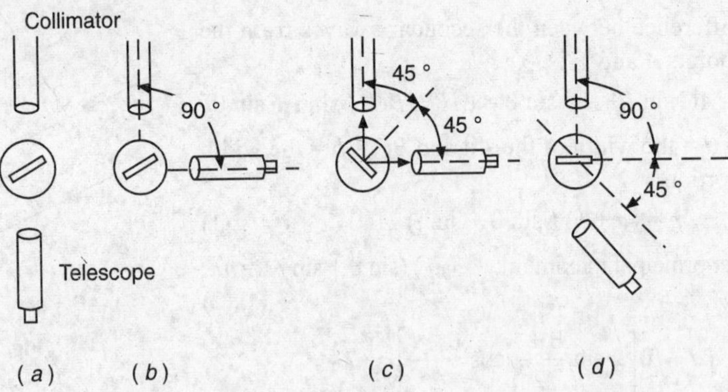

Fig. 3.27.

Bring the telescope to take the direct reading. Get the first order spectrum in the telescope. Make the cross-wire coincide with the various spectral lines one after another. Take the reading in each case. The difference between this and the direct reading gives the angle of diffraction. The angle of diffraction may be found for the spectrum on the other side of the normal as well as for other orders of spectra. The readings are tabulated as shown below:

Order	Colour	Vernier Reading		2θ	θ	λ
		Left	Right			
I						
II						

Let θ be the mean angle of diffraction for a wavelength λ in the nth order. Thus knowing $(a+b)$, θ and n, λ can be calculated.

Minimum deviation method : The telescope is turned and the central bright image of the slit is located. The telescope is turned further away and the first order image is located. The prism table is rotated in such a direction as to shift the image towards the axis of the collimator. This is continued until the image reaches its limiting position. The telescope is turned using the fine motion screw until the point of intersection of the crosswires coincides with the image. The reading R_1 on the circular scale is noted.

The telescope is turned to get the direct image of the slit at the point of intersection of the crosswires. The reading R_2 on the circular scale is noted. The angle of minimum deviation of the first order image is then $D = R_1 \sim R_2$. $(a + b) = 0.0254/N$ metre. The wavelength λ of the light is calculated using the formula

$$2\,(a + b) \sin (D/2) = n\lambda.$$

Example 1: *Light of wavelength 500 nm is incident normally on a plane transmission grating. A second order spectral line is observed at an angle of 30°. Calculate the number of lines per metre on the grating surface.*

Solution : Here, $\lambda = 500 \times 10^{-9}$ m; $n = 2$; $\theta = 30^o$.

We have, $(a + b) \sin \theta = n\lambda$

$$\therefore \quad (a+b) = \frac{n\lambda}{\sin\theta} = \frac{2\times(500\times10^{-9})}{\sin 30°} = 2\times10^{-6} \text{ m.}$$

$$\left.\begin{matrix}\text{No. of lines}\\ \text{per metre}\end{matrix}\right\} = N = \frac{1}{(a+b)} = \frac{1}{2\times10^{-6}} = 5\times10^{5}$$

Example 2: *What is the highest order spectrum which may be seen with monochromatic light of wavelength 600 nm by means of a diffraction grating with 5000 lines/cm.*

Solution : The number of orders visible with a grating is given by

$$n = \frac{(a+b)\sin\theta}{\lambda}$$

n will be maximum when $\sin \theta = 1$.

$$\therefore \quad n_{max} = \frac{(a+b)}{\lambda}$$

Here, $$(a+b) = \frac{1}{5000}\text{ cm} = \frac{1}{500000}\text{ m; } \lambda = 6\times10^{-7}\text{ m.}$$

$$\therefore \quad n_{max} = \frac{1/500000}{6\times10^{-7}} = 3.333 \approx 3.$$

Example 3 : *What will be the angular separation between the two sodium lines of wavelength 589 nm and 589.6 nm, when a parallel beam of light is incident normally at a plane transmission grating of 6 × 10⁵ lines per metre in the second order spectrum?*

Solution : We have, $(a + b) \sin \theta = n\lambda$

For λ_1 : $(a + b) = 1/(6 \times 10^5)$ m; $\lambda_1 = 589 \times 10^{-9}$ m ; $n = 2$; $\theta_1 = ?$

$$\sin\theta_1 = \frac{n\lambda_1}{(a+b)} = \frac{2\times(589\times10^{-9})}{1/(6\times10^{5})} = 0.7068$$

$$\therefore \quad \theta_1 = 44^o\,59'$$

For $$\lambda_2;\ \sin\theta_2 = \frac{n\lambda_2}{(a+b)} = \frac{2\times(589.6\times10^{-9})}{1/(6\times10^{5})} = 0.7075$$

$$\therefore \quad \theta_2 = 45°2'$$

$\therefore$ Angular separation of the lines $= \theta_2 - \theta_1 = 3'$.

Example 4 : *Find the value of λ when the angle of minimum deviation is 30° for the second order and the number of lines per cm is 5000.*

Solution. At the angle of minimum deviation, we have

$$2(a+b)\sin(D/2) = n\lambda$$

Here, $D = 30^\circ$; $D/2 = 15^\circ$ and $\sin D/2 = \sin 15^\circ = 0.2588$, $n = 2$, $(a+b) = 1/5000$ cm, $\lambda = ?$

$$\therefore \quad \lambda = \frac{2(a+b)\sin D/2}{n} = \frac{2\times(1/5000)\times 0.2588}{2} = 5.176\times 10^{-5} \text{ cm.}$$

Example 5 : *A diffraction grating used at normal incidence gives a green line (5400 Å) in a certain order superimposed on the violet line (4050 Å) of the next higher order. If the angle of diffraction is 30º, how many lines per cm are there in the grating?*

Solution : We have, $(a+b)\sin\theta = n\lambda$

Let the green line in the nth order be superimposed on the violet line in the $(n+1)$th order. Then

$$(a+b)\sin\theta = n\,\lambda_1 = (n+1)\,\lambda_2$$

Here, $\lambda_1 = 5400$ Å $= 5400\times 10^{-8}$ cm, $\lambda_2 = 4050\times 10^{-8}$ cm and $\theta = 30^\circ$.

$$(a+b)\sin 30^\circ = n\times(5400\times 10^{-8}) \quad ...(1)$$

and

$$(a+b)\sin 30^\circ = (n+1)\times(4050\times 10^{-8}) \quad ...(2)$$

Dividing (1) by (2), $\dfrac{n\times 5400}{(n+1)\,4050} = 1 \quad \therefore n = 3.$

Substituting the value of n in Eq. (1),

$$(a+b)\sin 30^\circ = 3\times(5400\times 10^{-8})$$

or

$$(a+b) = \frac{3\times 5400\times 10^{-8}}{\sin 30^\circ} = 3.24\times 10^{-4} \text{ cm}$$

$$\therefore \quad \left.\begin{matrix}\text{Number of}\\ \text{lines per cm}\end{matrix}\right\} = N = \frac{1}{(a+b)} = \frac{1}{(3.24\times 10^{-4})} = 3086$$

Example 6 : *A beam of light having wavelengths around 5000 Å is incident normally on a plane diffraction grating. The angle of deviation for a certain order maximum is found to be 45º. Calculate the dispersive power of the grating in this spectral region.*

Solution : For the nth order maximum we get, $(a+b)\sin\theta = n\lambda$.

Differentiating, $(a+b)\cos\theta\,(d\theta/d\lambda) = n \;\therefore\; \cot\theta\,\dfrac{d\theta}{d\lambda} = \dfrac{1}{\lambda}$

$$\frac{d\theta}{d\lambda} = \frac{1}{\lambda\cot\theta} = \frac{1}{(5000\times 10^{-8})\times\cot 45^\circ} = 2\times 10^{4} \text{ rad/cm.}$$

Example 7 : *Show that only the first order spectrum is possible in a grating if the grating element is less than twice the wavelength of light used.*

Solution : The various orders of spectra formed by a grating are given by

$$(a+b)\sin\theta = n\lambda.$$

Here, $(a+b)$ is the grating element, θ the angle of diffraction and n the order of the maximum. The maximum possible value of $\sin\theta$ is 1.

$$\therefore \quad n_{max} = \frac{(a+b)}{\lambda}$$

This expression shows that if the grating element $(a+b) < 2\lambda$, then

$$n_{max} < \frac{2\lambda}{\lambda} < 2.$$

This means that only first order will be obtained.

3.18. Concave Reflection Grating

A concave reflection grating is made by ruling parallel equidistant lines on a polished metallic concave surface. The rulings are equally spaced along the chord AB and not along the arc of the surface as shown in Fig. 3.28. In the Fig. 3.28, the rulings are indicated by dots. The ruled lines behave as obstacles and the clear spaces between them as polished strips, *i.e.*, reflecting elements.

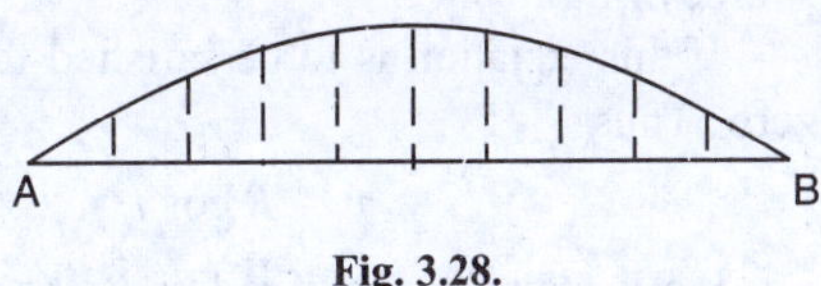

Fig. 3.28.

Let a = width of clear space and b = width of a ruling. Then $(a + b)$ is the *grating element*.

The rulings act as a grating to form spectra. The concave reflecting surface helps in automatically focusing the spectrum. Thus this grating avoids the use of the lenses. This grating can be conveniently used in ultraviolet and infrared regions which are absorbed by glass lenses.

Theory: Let GG' be the grating surface and C its centre of curvature (Fig. 3.29). Let A and B be two corresponding points on the grating surface. So $AB = (a + b)$ is the grating element. Let S be a linear source of light illuminated by monochromatic light of wavelength λ. Light rays SA and SB incident at angles i and $i + di$ respectively are diffracted at the grating at angles θ and $\theta + d\theta$. The two diffracted rays are brought to a focus at S'. With S and S' as centres, draw arcs AM and BN of radii SA and $S'B$ respectively.

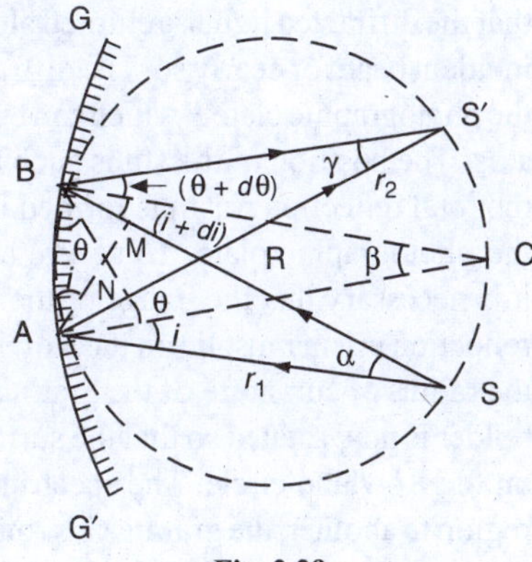

Fig. 3.29.

The path difference between the rays SAS' and SBS'

$$= (SB + BS') - (SA + AS')$$
$$= (SM + MB + BS') - (SA + AN + NS')$$
$$= MB - AN \ (\because SA = SM \text{ and } BS' = NS')$$
$$= AB \sin i - AB \sin\theta = (a + b)(\sin i - \sin\theta)$$

Condition of maxima : The rays will produce maximum intensity at S' if

$$(a + b)(\sin i - \sin\theta) = n\lambda \qquad ...(1)$$

If all the diffracted rays of a given wavelength form the image at S', then the path difference $(a + b)(\sin i - \sin\theta)$ should be constant for any pair of diffracted rays.

Hence differentiating Eq. (1), we get

$$\cos i \, di - \cos\theta \, d\theta = 0 \qquad ...(2)$$

Let α, β and γ be the angles as shown in Fig. 3.29. Then

$$\alpha + i = \beta + i + di \qquad \therefore di = \alpha - \beta$$
$$\beta + \theta = \gamma + \theta + d\theta \qquad \therefore d\theta = \beta - \gamma$$

Hence Eq. (2) becomes

$$\cos i(\alpha - \beta) - \cos\theta(\beta - \gamma) = 0 \qquad ...(3)$$

Let R be the radius of curvature of the grating and $SA = r_1$ and $AS' = r_2$. Then,

$$r_1\alpha = AM = (a + b)\cos i \,;\ R\beta = AB = (a + b) \text{ and } r_2\gamma = BN = (a + b)\cos\theta$$

$$\therefore \qquad \alpha = \frac{(a + b)\cos i}{r_1};\ \beta = \frac{(a + b)}{R};\ \gamma = \frac{(a + b)\cos\theta}{r_2}$$

Substituting these values of α, β and γ in Eq. (3) and simplifying,

$$\therefore \qquad \cos i\left(\frac{\cos i}{r_1} - \frac{1}{R}\right) - \cos\theta\left(\frac{1}{R} - \frac{\cos\theta}{r_2}\right) = 0 \qquad ...(4)$$

This is the general equation which relates the position of diffracted image with the position of source of light.

If this equation is to be satisfied for all values of i and θ, then the bracketed terms should be zero : Thus

$$r_1 = R \cos i \qquad \text{and} \qquad r_2 = R \cos \theta$$

i.e., if S lies on a circle of radius R, then S' also lies on the same circle.

We thus conclude that *the grating, the light source S and the spectral lines lie on the circumference of the same circle*. This circle is called the *Rowland circle*.

Eagle mounting : The plan of Eagle mounting is shown in Fig. 3.30. G is the grating which can be rotated about a vertical axis and moved along a track. S is the slit. Light from the slit is reflected towards the grating by a right angled prism. The angle of diffraction used is nearly equal to the angle of incidence so that the diffracted light is returned along the same path as the incident beam or nearly so. The diffracted light is focused on the photographic plate P which can be rotated about a vertical axis. The position of the slit is such that its virtual image in the total reflection prism is formed just below the centre of the photographic plate. To secure a sharp focused spectra it is necessary that the image of the slit formed by the total reflection prism must lie on the Rowland circle described on the radius of curvature of the grating as diameter. The plate holder is now rotated so that the surface of the plate also lies on the Rowland circle. The spectrum is then nicely focused on the plate. To pass from one spectral region to another, the grating G is rotated by proper amount and the above adjustments for focusing are repeated. The instrument can be mounted in a long box or room where the temperature is held constant.

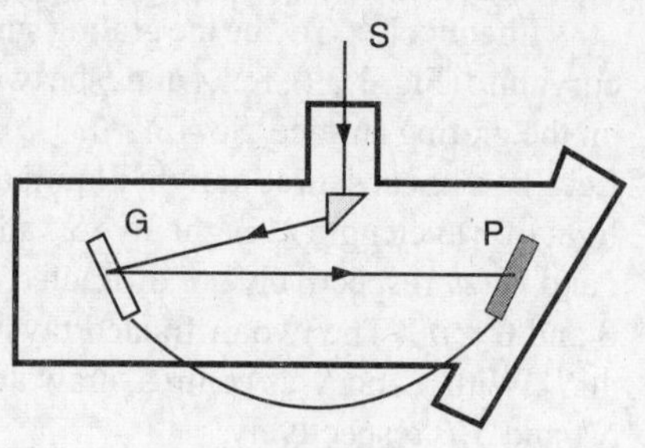

Fig. 3.30.

The advantages of this mounting are that it is cheap and quite compact. Astigmatism is least in this type of mounting. This mounting is commonly used in vacuum spectrography for the investigation of ultraviolet region below 200 *nm*. Since air absorbs these wavelengths, it must be pumped out of the spectrograph.

3.19. Resolving Power of Optical Instruments

The capacity of an instrument to show two close things separately is called '*resolution*'. *The ability of an optical instrument to produce distinctly separate spectral lines of light having two or more wavelengths very close to each other or to resolve the images of two nearby points is called its resolving power.*

Rayleigh's criterion for resolution : Rayleigh proposed the following criterion for resolution :

Two spectral lines of equal intensities are just resolved by an optical instrument when the principal maximum of the diffraction pattern due to one falls on the first minimum of the diffraction pattern of the other. .

Consider the intensity distribution curves of two wavelengths λ and $\lambda + d\lambda$ (Fig. 3.31).

The principal maximum of one coincides with the first minimum of the other. The resultant intensity curve shows

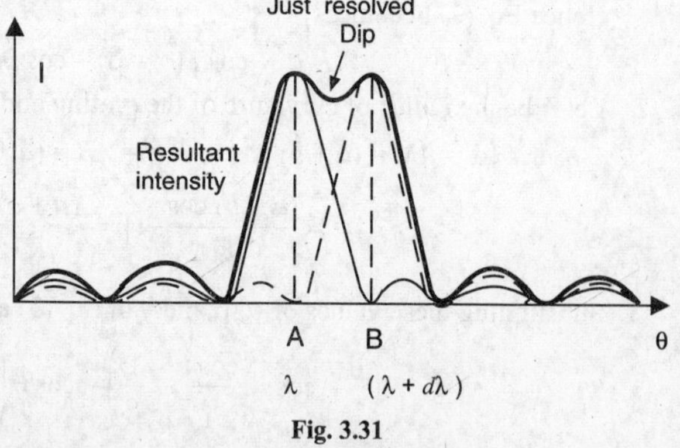

Fig. 3.31

a distinct dip in the middle indicating the presence of two different wavelengths. The lines are said to be 'just' resolved.

A grating or a prism gives us spectral resolution. If an optical instrument 'just' resolves two spectral lines of wavelengths λ and $\lambda + d\lambda$, then $\lambda/d\lambda$ is a measure of the 'resolving power' of the instrument.

3.20. Resolving Power of a Telescope

The telescope is used to see distant objects. Therefore the amount of details which the telescope reveals depends on the angle subtended at its objective by the two point objects.

The *RP* of a telescope is therefore defined as *the reciprocal of the smallest angle subtended at the objective by the two distant object points which can be just seen as separate ones through the telescope.*

Let D be the diameter of the telescope objective AB (Fig. 3.32). Two distant point objects O and O' subtend an angle $d\theta$ at the objective. The image of each object is a diffraction pattern. P_1 and P_2 are the positions of the central maxima of the two images. According to Rayleigh, the two objects O and O' will be just resolved when the central maximum of the diffraction pattern of one coincides with the first minimum of the diffraction pattern of the other.

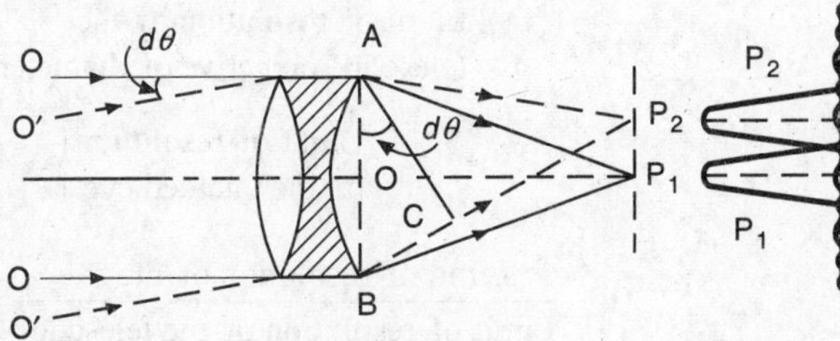

Fig. 3.32.

The path difference between AP_1 and BP_1 is zero. Hence they reinforce with one another at P_1. Thus P_1 corresponds to the position of the central maximum of the first image. Let $\angle P_2 A P_1 = d\theta$.

The path difference between BP_2 and $AP_2 = BC$.

Consider ΔABC.

$BC = AB \sin d\theta = AB.d\theta = D.d\theta$ (for small angles)

If $D.d\theta = \lambda$, then P_2 corresponds to the first minimum of the first image. But P_2 is also the central maximum of the second image. Thus, Rayleigh's condition of resolution is satisfied if

$$D.d\theta = \lambda$$

or
$$d\theta = \lambda/D \quad ...(1)$$

This condition holds good for rectangular apertures.

For circular apertures, Airy showed that

$$d\theta = 1.22\,(\lambda/D) \quad ...(2)$$

The reciprocal of $d\theta$ measures the *RP* of the telescope.

$$\therefore \quad RP = \frac{1}{d\theta} = \frac{D}{1.22\lambda} \quad ...(3)$$

RP depends upon the diameter of the objective and the wavelength of light used.

$RP \propto D$ and $RP \propto 1/\lambda$. Hence *RP* can be increased by using objectives of large diameters.

Let r be the radius of the first dark ring and f, the focal length of the objective. Then,

$$d\theta = \frac{r}{f} = \frac{1.22\,\lambda}{D}$$

or
$$r = \frac{1.22\,f\lambda}{D} \quad ...(4)$$

Eq. (4) shows that if f and λ are small, and D is large, then the radius of the central bright disc is small. Thus the diffraction pattern will appear sharper and the angular separation $d\theta$ between two just resolvable point objects will be smaller. So the *RP* of the telescope will be higher.

3.21. Relation between Magnifying Power and Resolving Power of a Telescope

The magnifying power of a telescope is

$$M = \frac{\text{Diameter of the objective}}{\text{Diameter of the exit pupil}} = \frac{D}{d}$$

The magnifying power is said to be normal if diameter of the exit pupil is equal to the diameter d_e of the pupil of the eye. Hence normal magnifying power $= \dfrac{D}{d_e}$.

$$\left.\begin{array}{r}\text{Limit of resolution of the}\\ \text{telescope objective of diameter } D\end{array}\right\} = d\theta = \frac{1.22\lambda}{D}$$

$$\left.\begin{array}{r}\text{Limit of resolution}\\ \text{of the unaided eye}\end{array}\right\} = d\theta' = \frac{1.22\lambda}{d_e}$$

$$\therefore \quad \frac{\text{Limit of resolution of the eye}}{\text{Limit of resolution of the telescope}} = \frac{d\theta'}{d\theta} = \frac{1.22\lambda / d_e}{1.22\lambda / D} = \frac{D}{d_e}$$

= Normal magnifying power of the telescope.

Thus, (*normal magnifying power of telescope*) × (*its limit of resolution*) = *limit of resolution of the unaided eye.*

Example 1 : *A diffraction limited telescope with a 7.6 cm aperture is aimed at a target 12.5 km away. Assuming light of 500 nm wavelength, calculate the linear resolution of the telescope.*

Solution: Here, $\lambda = 500 \times 10^{-9}$ m; $D = 7.6 \times 10^{-2}$ m.

Limit of resolution of the telescope,

$$d\theta = \frac{1.22\,\lambda}{D} = \frac{1.22 \times (500 \times 10^{-9})}{7.6 \times 10^{-2}} = 8.026 \times 10^{-6} \text{ radian.}$$

Distance of the target from the telescope = 12.5×10^3 m

∴ Linear resolution of the telescope,

$$x = (8.026 \times 10^{-6}) \times (12.5 \times 10^3) = 0.1003 \text{ m.}$$

Example 2 : *Calculate the aperture of the objective of a telescope which may be used to resolve stars separated by 6.1×10^{-6} radians for light of wavelength 589.6 nm.*

Solution : Here, $d\theta = 6.1 \times 10^{-6}$ radians; $\lambda = 589.6 \times 10^{-9}$ m. $D = ?$

$$d\theta = 1.22\,(\lambda/D)$$

$$\therefore \quad D = \frac{1.22\lambda}{d\theta} = \frac{1.22 \times (589.6 \times 10^{-9})}{(6.1 \times 10^{-6})} = 0.1179 \text{ m.}$$

Example 3 : *A telescope of aperture 3×10^{-2} m is focused on a window at 80 m distance fitted with a wiremesh of spacing 2×10^{-3} m. Will the telescope be able to observe the wiremesh ? ($\lambda = 5.5 \times 10^{-7}$ m).*

Solution : Here, $D = 3 \times 10^{-2}$ m; $\lambda = 5.5 \times 10^{-7}$ m ; $d\theta = ?$

$$d\theta = \frac{1.22\lambda}{D} = \frac{1.22 \times (5.5 \times 10^{-7})}{3 \times 10^{-2}} = 2.237 \times 10^{-5} \text{ radian}$$

The telescope will be able to resolve the wiremesh, if the angle α subtended by it at the objective is equal to or greater than $d\theta$, the limit of resolution.

The angle subtended by the wiremesh at the objective is

$$\alpha = \frac{\text{Spacing of wiremesh}}{\text{distance of the objective from the wiremesh}} = \frac{2 \times 10^{-3}}{80} = 2.5 \times 10^{-5} \text{ rad.}$$

$\alpha > d\theta$. Hence the telescope will be able to observe the wiremesh.

3.22. Resolving Power of a Microscope

The resolving power of a microscope represents its ability to form distinctly separate images of two objects lying close together. *It is measured by the smallest distance between two point-objects whose images are just resolved by the objective of the microscope.* The smaller is the distance, the higher is the *RP*.

Consider two point objects O and O' (Fig. 3.33). The linear distance between O and $O' = d$. I and I' are the images of O and O' formed by the objective AB of the microscope. The images are actually Fraunhoffer diffraction patterns consisting of a central bright disc surrounded by a series of alternate dark and bright rings. The centres of the discs lie at I and I'. According to Rayleigh's criterion, O and O' will be just resolved if the position of the central maximum of I also corresponds to the first minimum of I'. Airy has shown that this will happen when the path difference between the extreme rays, $O'BI - O'AI$, is given by

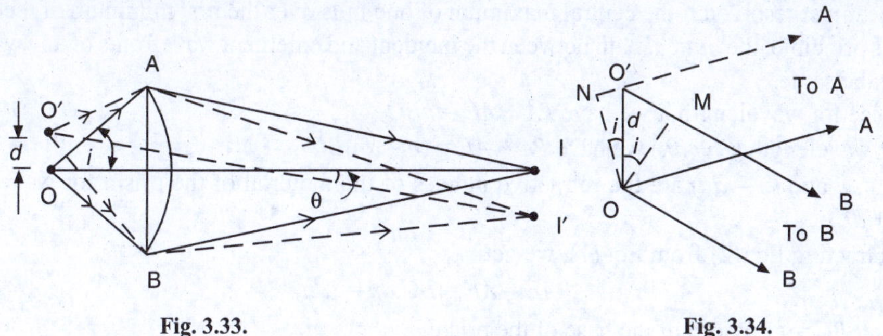

Fig. 3.33. Fig. 3.34.

$$O'BI - O'AI = 1.22\,\lambda.$$

But the paths AI and BI are equal. Hence the above condition becomes

$$O'B - O'A = 1.22\,\lambda \qquad \text{...(1)}$$

O and O' are very close together. Hence, we can take $O'A$ to be parallel to OA, and $O'B$ parallel to OB. From Fig. 3.34,

$$O'B - OB = O'M = d \sin i \qquad \text{...(2)}$$

and

$$OA - O'A = O'N = d \sin i \qquad \text{... (3)}$$

Now, $OA = OB$. Adding Eqs. (2) and (3), $O'B - O'A = 2d \sin i$

Here, i is the semi-vertical angle of the cone of rays received by the objective from O. Now, Eq. (1) becomes,

$$2d \sin i = 1.22\,\lambda \text{ or } d = \frac{1.22\,\lambda}{2 \sin i}$$

If the space between the object and the objective is filled with an oil of refractive index n, then $d = \dfrac{1.22\lambda}{2n \sin i}$

$n \sin i$ is called the *numerical aperture* of the objective of the microscope.

$$\therefore \qquad d = \frac{1.22\,\lambda}{2N.A.}$$

This measures the limit of resolution of the microscope. Its reciprocal is called as resolving power.

3.23. Resolving Power of a Prism

The R.P. of a prism is defined as its capacity to form separate maxima of two wavelengths very close together. *If λ and $(\lambda + d\lambda)$ are the wavelengths of two close lines which are just resolved, then $\lambda/d\lambda$ measures the R.P. of the prism.*

Let ABC be the section of a prism, which is placed in the minimum deviation position (Fig. 3.35). A parallel beam of light consisting of wavelengths λ and $(\lambda + d\lambda)$ is incident on the prism. BP is the incident plane wavefront. CE and CF are the emergent wavefronts for λ and $(\lambda + d\lambda)$. The telescope objective of the spectrometer focuses the wavefront CE at P_1 and CF at P_2. P_1 and P_2 are the positions of the central maxima of diffraction patterns for λ and $\lambda + d\lambda$ respectively. The two spectral lines will appear just resolved if the central maximum of one falls over the first minimum of the other. By Fermat's principle, the *optical* path between the incident and emergent wavefronts of any wavelength must be the same.

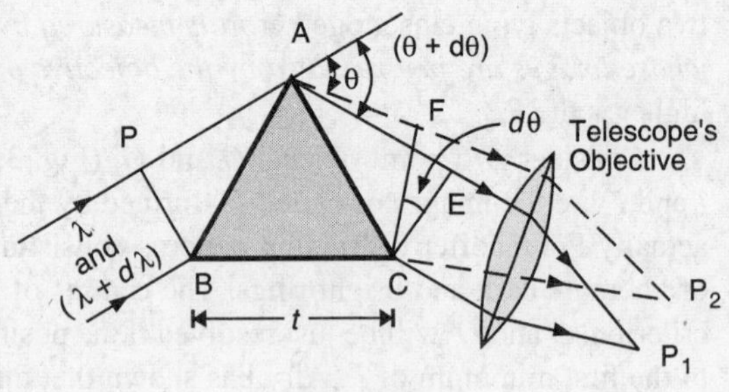

Fig. 3.35.

Hence for wavelength λ we have $PA + AE = nBC$ ...(1)

For wavelength $(\lambda + d\lambda)$ we have $PA + AF = (n - dn)BC$ ...(2)

Here, n and $(n - dn)$ are the refractive indices of the material of the prism for wavelengths λ and $(\lambda + d\lambda)$.

Subtracting Eq. (2) from Eq. (1), we get

$$AE - AF = BC.dn = t\,dn \qquad ...(3)$$

Here, $BC = t$ = width of the base of the prism.

Let θ and $(\theta + d\theta)$ be the angles of deviation for wavelengths $(\lambda + d\lambda)$ and λ respectively. Then the angle between the two emergent wavefronts CE and CF is $d\theta$. Hence from the geometry of Fig. 3.35, we have

$$AE - AF = EF = CE.d\theta = D.d\theta \qquad ...(4)$$

Here, $CE = D$ = the width of the emergent beam = the diameter of the telescope objective.

From Eqs. (3) and (4), $D.d\theta = t.dn$

$$\text{or} \qquad d\theta = t\,dn/D \qquad ...(5)$$

But the limiting condition for resolution in the case of a telescope is

$$d\theta = \lambda/D \qquad ...(6)$$

From Eqs. (5) and (6), $\dfrac{\lambda}{D} = t\dfrac{dn}{D}$ or $\lambda = t\,dn$

$$\therefore \qquad \frac{\lambda}{d\lambda} = t\frac{dn}{d\lambda} \qquad ...(7)$$

That is, *(i)* $RP \propto t$ and *(ii)* $RP \propto (dn/d\lambda)$.

Example 1 : *Calculate the minimum thickness of the base of a prism which will just resolve the D_1 and D_2 lines of sodium. Given n for wavelength 656.3 nm = 1.6545 and for wavelength 527 nm = 1.6635.*

Solution : $\dfrac{dn}{d\lambda} = \dfrac{1.6635 - 1.6545}{(6.563 - 5.270) \times 10^{-7}} = 6.96 \times 10^4 \text{ m}^{-1}$

Now, for sodium light,

λ for $D_1 = 5.89 \times 10^{-7}$ m; λ for $D_2 = 5.896 \times 10^{-7}$ m.

Mean $\lambda = 5.893 \times 10^{-7}$ m ; $d\lambda = 0.006 \times 10^{-7}$ m.

We have, R.P. $= \dfrac{\lambda}{d\lambda} = t\dfrac{dn}{d\lambda}$

$$\therefore \quad t = \frac{\lambda / d\lambda}{dn / d\lambda} = \frac{(5.893 \times 10^{-7}) / (0.006 \times 10^{-7})}{6.96 \times 10^4} = 0.0141 \text{ m.}$$

3.24. Resolving Power of a Plane Diffraction Grating

The R.P. of a grating is defined as its ability to show two neighbouring lines in a spectrum as separate. It is measured by the ratio $\lambda/d\lambda$, where λ is the wavelength of a spectral line and $d\lambda$ is the least difference in the wavelengths of two neighbouring spectral lines which can just be resolved.

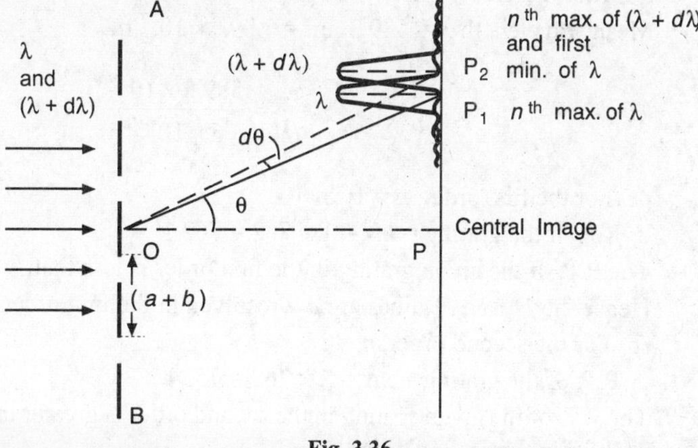

Fig. 3.36.

In Fig. 3.36, light of two wavelengths λ and $(\lambda + d\lambda)$ is incident normally on the surface of a plane transmission grating AB. The light of each wavelength would form a separate diffraction pattern of the slit. P_1 is the nth primary maximum of a spectral line of wavelength λ at an angle of diffraction θ. Then,

$$(a + b) \sin \theta = n\lambda \qquad \text{...(1)}$$

Here, $(a + b)$ is the grating element.

P_2 is the nth primary maximum of a second spectral line of wavelength $(\lambda + d\lambda)$ at an angle of diffraction $(\theta + d\theta)$. Then,

$$(a + b) \sin (\theta + d\theta) = n(\lambda + d\lambda) \qquad \text{...(2)}$$

According to Rayleigh, the two spectral lines will appear just resolved if the principal maximum due to $(\lambda + d\lambda)$ falls on the first minimum of λ or vice versa. Thus, the two lines will appear just resolved if the angle of diffraction $(\theta + d\theta)$ also corresponds to the direction of first secondary minimum after the nth primary maximum at P_1 corresponding to wavelength λ. This is possible if the extra path difference introduced is λ/N. Here, N is the total number of lines in the grating.

$$\therefore \quad (a + b) \sin (\theta + d\theta) = n\lambda + \frac{\lambda}{N}$$

Equating the right hand sides of Eqs. (2) and (3),

$$n(\lambda + d\lambda) = n\lambda + \frac{\lambda}{N} \quad \text{or} \quad n\,d\lambda = \frac{\lambda}{N}$$

$$\therefore \qquad \frac{\lambda}{d\lambda} = nN$$

The R.P. *increases* with

(i) the order n of the spectrum and (ii) the total number of lines N on the grating. The R.P. is independent of the grating element $(a + b)$.

Example 1 : *A diffraction grating has 0.15 m of surface ruled with 6×10^5 lines/metre. What is its R.P. in the first order ?*

Solution : Here, $N = (6 \times 10^5) \times 0.15 = 9 \times 10^4$ and $n = 1$; R.P. = ?

$$\therefore \qquad \left.\begin{array}{l}\text{Resolving power}\\ \text{of the grating}\end{array}\right\} = \frac{\lambda}{d\lambda} = nN = 1 \times (9 \times 10^4) = 9 \times 10^4$$

Example 2 : *Examine if D_1 and D_2 lines of sodium will be clearly separated in the (a) first order and (b) second order by a 1" grating having 300 lines/cm. Wavelength of D_1 and D_2 lines are 589.6 and 589 nm respectively.*

Solution : Here, $N = 2.54 \times 300 = 762$; $d\lambda = 0.6$ nm $= 6 \times 10^{-10}$ m;

Mean wavelength $\lambda = 589.3$ nm $= 5.893 \times 10^{-7}$ m,

$$\therefore \qquad \frac{\lambda}{d\lambda} = \frac{589.3 \times 10^{-9}}{6 \times 10^{-10}} = 982.2$$

(a) For the first order, $n = 1$;

$\therefore$ R.P. of the grating $= nN = 1 \times 762 = 762$

The R.P. of the given grating in the first order is less than $\lambda/d\lambda$.

Hence the D_1 and D_2 lines are *not* resolved in the first order.

(b) For the second order, $n = 2$

$\therefore$ R.P. of the grating $= nN = 2 \times 762 = 1524$.

The R.P. of the given grating in the second order is greater than $\lambda/d\lambda$. Hence the D_1 and D_2 lines are resolved in the second order.

Example 3 : *A plane transmission diffraction grating has 40,000 lines in all with grating element 12.5×10^{-5} cm. Calculate the maximum resolving power for which it can be used in the range of wavelength 5000 Å.*

Solution : For a grating, $(a + b) \sin\theta = n\lambda$

The maximum order (for $\theta = 90°$) is

$$n_{max} = \frac{a+b}{\lambda} = \frac{12.5 \times 10^{-7}}{5000 \times 10^{-10}} = 25$$

Hence the second order is the highest which is observed with this grating. Therefore the maximum resolving power obtained with this grating is

$$R = Nn = 40{,}000 \times 2 = 80{,}000$$

3.25. Comparison of Prism and Grating Spectra

Prism Spectrum	Grating Spectrum
1. A prism produces only one spectrum.	1. A grating gives a number of spectra of different orders on each side of the central maximum.
2. All the light energy emerging from a prism is concentrated in a single spectrum. So the intensities of spectral lines are large.	2. In a grating, most of the light is concentrated in the central maximum and the rest is distributed in the spectra of different orders. So the intensities of spectral lines are very small.
3. The prism spectrum is produced by dispersion. Dispersion is due to different velocities of light through the prism for different wavelengths.	3. The grating spectrum is formed by diffraction of light. For maximum intensity, the angle of diffraction is different for different wave-lengths.
4. In a prism, the deviation is least for red and greatest for violet colour.	4. In a grating, the deviation is least for violet and greatest for red.
5. The prism spectrum depends on the material of the prism.	5. The grating spectra are independent of the material of the grating.
6. The dispersive power of a prism is $\frac{d\delta}{d\lambda} \propto \frac{1}{\lambda^3}$. Thus there is greater dispersion in the violet region (shorter wavelength) than in the red region. The spectrum obtained with a prism is *irrational spectrum.*	6. The dispersive power of a grating is $\frac{d\theta}{d\lambda} = \frac{n}{(a+b)\cos\theta}$ It is a constant for a particular order. Thus the spectral lines are evenly distributed. The spectrum obtained is said to be *normal.*
7. The resolving power of a prism is small. Hence usually the prism spectrum does not show the fine structure of spectral lines.	7. The resolving power of the grating is large. Hence the grating spectrum shows the fine structure of spectral lines.

EXERCISE

1. Distinguish between Fraunhofer and Fresnel class of diffraction.
2. Starting from the division of a plane wave-front into half-period elements, give Fresnel's explanation of rectilinear propagation of light on this theory.
3. Give the theory of the zone plate. Derive an expression for its focal length. Compare its action with that of a convex lens.
4. Describe and explain the Fresnel's diffraction at a circular aperture. What happens if (i) size of aperture is changed (ii) screen is gradually moved.
5. Describe Fresnel's diffraction at a straight edge. Derive an expression for the distance of nth bright band from the edge of geometrical shadow. Sketch the intensity distribution for Fresnel diffraction by a straight edge.
6. Discuss the theory of Fresnel diffraction by a slit.
7. Describe the diffraction pattern given by a narrow wire illuminated by monochromatic light from a narrow slit parallel to wire and explain it theoretically. How do you measure the wire thickness with its help? What changes occur if the wire is thick?
8. Discuss the phenomenon of Fraunhofer diffraction at a single slit.

9. Describe with necessary theory, Fraunhofer diffraction due to a double slit.
10. Discuss the theory of diffraction grating. Describe in detail how you would use a transmission grating to determine wavelength of light.
11. What are absent spectra in a grating? What particular spectra would be absent if the width of opacities be double that of the transparencies in such a grating?
12. Define and explain the term "dispersive power of a grating". Derive an expression for it.
13. Explain the formation of spectra by a plane diffraction grating. What are its chief characteristics? What do you understand by overlapping and absent spectra?
14. Give the theory of concave reflection grating. Explain the Eagle mounting of concave grating.
15. Explain Rayleigh criterion of resolution. Derive an expression for the resolving power of a telescope.
16. Derive an expression for the resolving power of a microscope.
17. Derive an expression for the chromatic resolving power of a prism.
18. Obtain an expression for the resolving power of a plane transmission grating.
19. Distinguish between the spectra obtained with a grating and a prism.
20. A sharp razor blade is held vertically in a beam of white light diverging from a point source. Discuss the nature of fringes obtained on a screen placed behind the blade.

    [Ans. When white light is used, coloured diffraction bands are produced. Since $x \propto \sqrt{\lambda}$, the violet and blue bands lie nearer the edge of the geometrical shadow than the red bands. Only a small number of bands are visible near the edge. Then there is general illumination due to overlapping of bands of different colours].
21. A screen is placed at a distance of 1 metre from a circular hole illuminated by a parallel beam of light of wavelength 640 nm. Compute the radius of the fourth half-period zone.

    [Ans : $r_4 = \sqrt{(4 \times 100 \times 640 \times 10^{-9})} = 1.6 \times 10^{-3}$ m].
22. A point source of light of wavelength 6×10^{-7} m is placed at a distance of 1 m from a zone plate. Its image is obtained at a distance of 2 m on the other side. Calculate (i) the focal length of the zone plate, (ii) the power of an equivalent lens which may replace the zone plate, (iii) the radius of first zone on the plate.

    [Ans. (i) (2/3) m; (ii) 1.5 dioptre; (iii) 6.324×10^{-4} m].
23. In an experiment with a zone plate the wavelength of the light used is 589.3 nm. A point source is kept at a certain position in front of the zone plate. At a distance of 60 cm from the zone plate a clear image is obtained. On moving the eyepiece nearer, next clear image is obtained at a distance of 30 cm from the zone plate. Calculate the principal focal length of the zone plate and the area of each zone. Also calculate the radius of the first zone. **[Ans :** $f = 1.2$ m; Area $= 2.2 \times 10^{-6}$ m^2 ; $r_1 = 8.4 \times 10^{-4}$ m]
24. The diameter of the first ring of a zone plate is 1 mm. If plane waves ($\lambda = 500$ nm) fall on the plate, where should the screen be placed so that light is focused to a brightest spot?

    [Hint : $f = \dfrac{r_1^2}{\lambda} = \dfrac{(0.5 \times 10^{-3})^2}{500 \times 10^{-9}} = 0.5$ m].
25. A parallel beam of microwaves ($\lambda = 0.8$ cm) is incident on a circular aperture of adjustable radius. If a detector is placed on the axis of aperture 80 cm behind it and the radius of aperture is gradually increased, at what value of radius would the response of detector be maximum? **[Ans :** 0.08 m]
26. Parallel monochromatic light ($\lambda = 600$ nm) falls normally on a circular aperture of diameter 0.12 cm and is viewed from the opposite side along a line through the centre of the hole and normal to its plane. Calculate the three largest distances from the hole at which the intensity is (*i*) maximum, (*ii*) zero.

[**Hint :** *(i)* $b = \frac{r^2}{\lambda}, \frac{r^2}{3\lambda}, \frac{r^2}{5\lambda} = 0.6$ m, 0.12 m

(ii) $b = \frac{r^2}{2\lambda}, \frac{r^2}{4\lambda}, \frac{r^2}{6\lambda} = 0.3$ m, 0.15 m, 0.1 m]

27. A narrow slit illuminated by light of wavelength 640 nm is placed at a distance of 3m from a straight edge. If the distance between the straight edge and the screen is 6m, calculate the distance between the first and the 4th dark band. [**Ans :** 4.8×10^{-3} m]

28. A slit and screen are kept 2.5 metres apart and a straight edge is placed half way between them with the edge parallel to the slit. Find the distance of fifth bright band from the end of geometrical shadow if the wavelength of light used be 600 nm. [**Ans :** 1.161×10^{-4} m]

29. Fringes are seen within the shadow of a thin wire held vertically in front of a narrow vertical slit illuminated by monochromatic light. Find the diameter of the wire if the fringe width at a distance of 4 m from the wire is 1.8 mm and the wavelength of light used is 540 nm. [**Ans :** 1.2×10^{-3} m]

30. Calculate the angular separation between the first order minima on either side of the central maximum when the slit is 6×10^{-4} cm wide and light illuminating it has a wavelength of 600 nm.

[**Ans :** $2\theta = 73^\circ 44'$]

31. In a double-slit Fraunhofer diffraction pattern, the screen is 1.6 m away from the slits. The slit widths are 0.08 mm and they are 0.4 mm apart. Calculate the wavelength of light if the fringe-spacing is 0.25 cm. Also deduce the missing orders. [**Ans :** 625 nm. 6th, 12th, 18th,... orders will be missing.]

32. Calculate the angle of diffraction at which the second order maximum occurs for light of wavelength 589.6 nm with a plane diffraction grating having 6000 lines per cm. [**Ans :** $45^\circ 2'$]

33. Monochromatic light of wavelength 6.56×10^{-5} cm falls normally on a grating consisting of parallel wires equidistant from one another. The first order spectrum is observed at 15′ from zero position. Find the grating constant. [**Ans :** 1.015×10^{-2} m]

34. The limits of visible spectrum are approximately 400 mμ to 700 mμ (milli micron). Find the angular breadth of the first order visible spectrum produced by a plane grating, having 15000 lines per inch, when light is incident normally on the grating (Given sin $13^\circ 40' = 0.237$, sin $24^\circ 30' = 0.415$).

[**Ans :** $10^\circ 44'$]

35. Light of wavelength 5000 Å is incident normally on a plane transmission grating. Find the difference in the angles of deviation in the first and third order spectra. The number of lines per cm on the grating surface is 6000. [**Ans :** $46^\circ 42'$]

36. The Mount Wilson Observatory telescope has an objective of diameter 2.54 m. Assuming the mean wavelength of light to be 5.5×10^{-7} m, estimate the smallest angular separation of the two stars which can be resolved. [**Ans :** 2.642×10^{-7} rad.]

37. The mean wavelength effective in visual observation may be treated as 580 nm. What is the smallest angular separation (in seconds) between two point stars which a telescope of 0.2 m diameter objective can resolve? [**Ans :** 0.730 sec.]

38. Find the separation of two points on the moon that can be resolved by a 5 m telescope. The distance of the moon is 3.8×10^8m. The eye is most sensitive to light of wavelength 5.5×10^{-7} m.

[**Ans :** 50.996 m]

39. Calculate the smallest angular separation resolved by the average human eye, given : aperture $= 2.5 \times 10^{-3}$ m, and effective wavelength $= 5.5 \times 10^{-7}$ m. If a scale with millimetre markings is viewed by the unaided eye, what is the largest distance at which the markings will be visible ? [**Ans :** 3.726 m.]

40. The objective of a certain telescope has a diameter of 5.08 m. Calculate the smallest angle resolved by it. What is the minimum magnification required to enable the human eye to utilize such a high resolution? Given effective $\lambda = 5.5 \times 10^{-7}$ m and aperture of human eye $= 2.54 \times 10^{-3}$ m.

[**Ans :** 2.642×10^{-4} rad.; 2000]

41. Calculate the least value of the angular separation of two stars which can be resolved by a telescope of 2m aperture. If the aperture of the human eye be 2×10^{-3} m and the focal length of the eyepiece be 2.54×10^{-2} m, what must be the minimum focal length of the objective if the full resolving power of the telescope is to be utilized? (Assume $\lambda = 6 \times 10^{-7}$m). [**Ans.** 3.66×10^{-7} rad; 25.4 m.]

42. A microscope is used to resolve two self-luminous objects separated by a distance of 5×10^{-7} m and placed in air. What should be the minimum overall magnifying power of the microscope in order to see the resolved images of the objects? The resolving power of the eye is 1.5' and the least distance of distinct vision is 0.25 m.

[**Hint :** $\left.\begin{array}{l}\text{linear limit of}\\ \text{resolution of eye}\end{array}\right\} = x_e = \frac{1.5}{60} \times \frac{\pi}{180} \times 0.25 = 1.09 \times 10^{-4}$ m.

$$\therefore \quad M = \frac{x_e}{x_m} = \frac{1.09 \times 10^{-4}}{5 \times 10^{-7}} = 218$$

43. A prism spectrometer uses a flint glass prism of base 10^{-2} m. For flint glass dispersion $dn/d\lambda$ is 10^5m^{-1} in the range $\lambda = 6 \times 10^{-7}$m. What is the smallest difference of wavelength in this range which this spectrometer may resolve? [**Ans :** 60 nm.]

44. The refractive indices of the material of a prism are 1.624 for 456 nm and 1.586 for 656 nm. Calculate *(i)* R.P. of the prism if the base length is 1.6×10^{-2} m; *(ii)* base length required to resolve the sodium lines 589 nm and 589.6 nm. [**Ans :** 3040; 0.005169 m.]

45. Light is incident normally on a grating 0.5 cm wide with 2500 lines. Find the angular separation of the two sodium lines in the first order spectrum. Can they be seen distinctly? Wavelengths of the lines are 589 and 589.6 nm.

[**Ans :** 3.139×10^{-4} radian; R. P. required = 982.2. Actual R.P. = n N = $1 \times 2500 = 2500$

$\therefore$ R.P. available > R.P. required.

Therefore, the lines will be resolved and will be seen as separate.]

46. Examine if D_1 and D_2 lines of sodium will be clearly resolved in the first order by 2.5 cm grating having 300 lines per cm. (wavelengths of D_1 and D_2 lines are 589·6 and 589 nm respectively).

[Ans : Not resolved]

CHAPTER 4

POLARISATION OF LIGHT

4.1 Introduction

Experiments involving interference and diffraction show the wave nature of light. Experiments involving polarization show the transverse nature of light waves. Light is an electro-magnetic wave, in which electric and magnetic field vectors are varying in time and space at right angles to the direction of propagation of the wave. The electric vector **E** functions as the light vector. In a beam of ordinary light, the vibrations are equally probable in all directions at right angles to the direction of propagation. There is thus perfect symmetry around the direction of propagation (Fig. 4.1). Such light waves are called *un-polarised* waves.

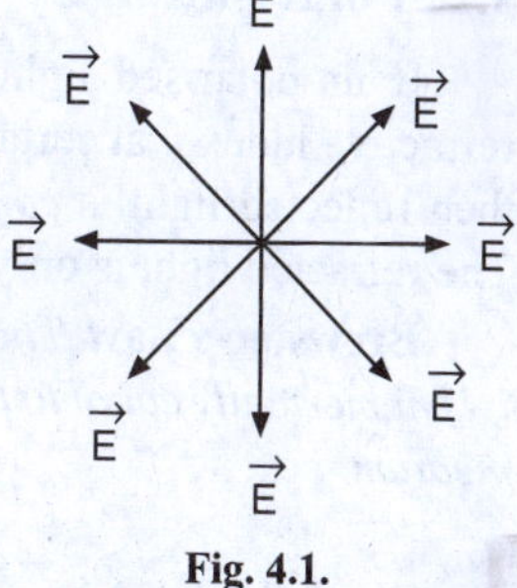

Fig. 4.1.

Polarisation of Light. Suppose ordinary light is incident normally on a pair of parallel tourmaline crystal plates P_1 and P_2 (Fig. 4.2) cut parallel to their crystallographic axis. Then, the emergent light shows a variation in intensity as P_2 is rotated. The intensity is a maximum when the axis of P_2 is parallel to that of P_1 (Fig. 4.2 *a*), and minimum when at right angles (Fig. 4.2 *b*). This shows that the light emerging from P_1 *is not symmetrical about the direction of propagation of light, but its vibrations are confined only to a single line in a plane perpendicular to the direction of propagation.* Such light is called *polarised light. Plane-polarised light is the light in which the light vector vibrates along a fixed straight line in a plane perpendicular to the direction of propagation.*

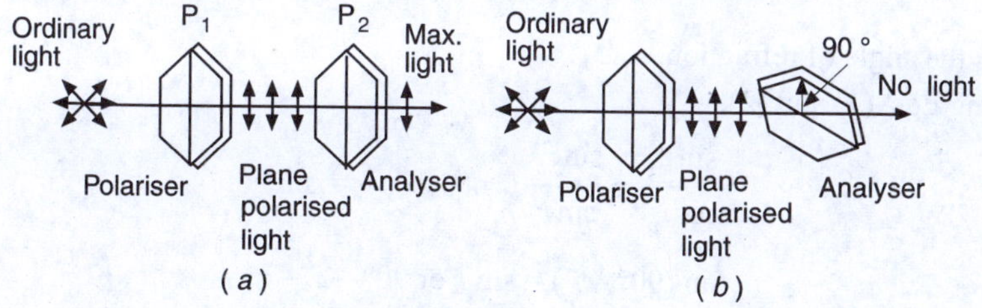

Fig. 4.2.

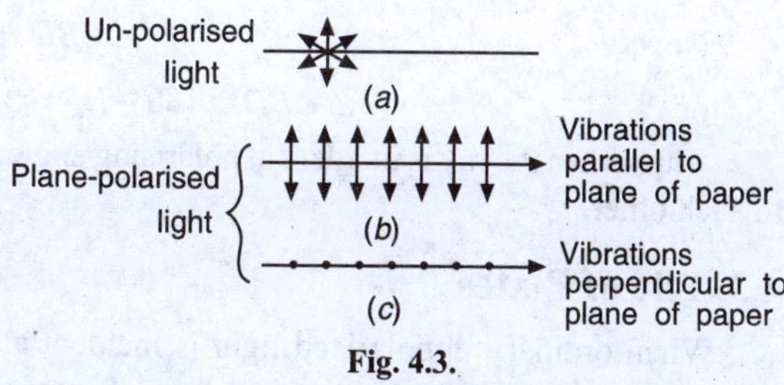

Fig. 4.3.

In a *plane polarised* beam of light, the vibrations are along a single straight line.

If the vibrations are parallel to the plane of the paper, they are represented by *arrows* (Fig. 4.3*b*).

If they are along a straight line perpendicular to the plane of the paper, they are represented by *dots* (Fig. 4.3*c*).

A beam of ordinary light can be considered as consisting of two sets of vibrations– one set vibrating in one plane and the other perpendicular to it. Hence Fig. 4.3 *(a)* can be taken to represent an ordinary beam of light.

The plane in which the vibrations take place (*i.e.,* the plane containing the direction of vibration and the direction of propagation) *is called the plane of vibration.* The plane *ABCD* (Fig. 4.4) is the plane of vibration. *A plane perpendicular to the plane of vibration is called the plane of polarization. Thus plane of polarization is the plane passing through the direction of propagation and containing no vibrations.* The plane *EFGH* is the plane of polarization.

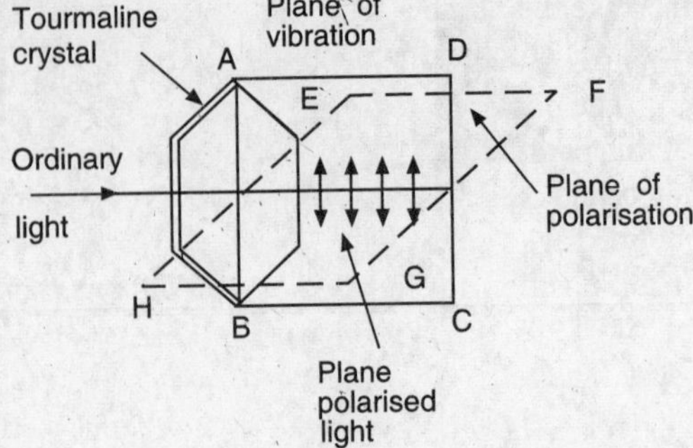

Fig. 4.4.

4.2. Polarisation by Reflection

If un-polarised light falls on a transparent surface of refractive index *n* at certain angle i_p, called polarising angle, then reflected light is *completely plane polarised* (Fig. 4.5). The refracted light is only partially polarised.

Brewster's Law. *The tangent of the angle of polarisation i_p is numerically equal to the refractive index n of the reflecting medium, i.e.,*

$$n = \tan i_p$$

Fig. 4.5.

From Brewster law, $$n = \tan i_p = \frac{\sin i_p}{\cos i_p} \quad ...(1)$$

By Snell's law, $$n = \frac{\sin i}{\sin r} \quad ...(2)$$

Here, *r* is the angle of refraction.

Comparing Eqs. (1) and (2),

$$\frac{\sin i_p}{\cos i_p} = \frac{\sin i_p}{\sin r} \quad \text{or} \quad \cos i_p = \sin r$$

or $$\sin (90° - i_p) = \sin r \text{ or } 90° - i_p = r$$

or $$r + i_p = 90°.$$

Now, $$i_p + \angle CBD + r = 180°$$

or $$\angle CBD = 180 - (i_p + r) = 180° - 90° = 90°$$

Thus when the ray is incident at polarising angle, the reflected and refracted rays are perpendicular to each other.

4.3. Pile of Plates

When ordinary unpolarised light is incident at polarising angle on a glass plate, the reflected light is completely polarised in the plane of incidence. That is, the reflected light consists of only ⊥ vibrations. However, its intensity is low because only 15% of ⊥ vibrations are reflected and the remaining 85% are refracted. The refracted light is not completely plane polarised. It consists of 85% of the ⊥ vibrations and all || vibrations.

By using a pile of plates, more and more of the ⊥ vibrations can be filtered out of the refracted beam by repeated reflections. To increase the intensity of plane polarised reflected beam, we should increase the number of reflections and consequently the number of plates. The refracted light also becomes completely plane-polarised with vibrations in the plane of incidence.

The pile of plates (Fig. 4.6) consists of a number of thin parallel-faced glass plates placed one over the top of the other inside a wooden tube. The plates are inclined at an angle of 32.5° with the axis of the tube. In this position, the light would be incident at an angle of 57.5° (90° – 32.5°) which is the polarising angle for glass. The reflected as well as the transmitted beams of light are plane polarised. In reflected beam, the vibrations are ⊥ to the plane of incidence. In transmitted beam, the vibrations are || to the plane of incidence.

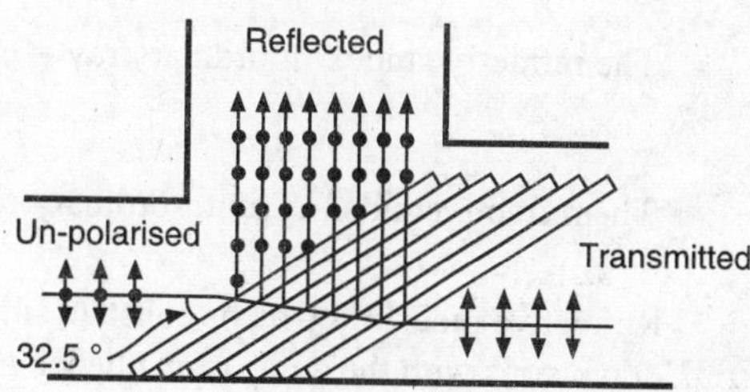

Fig. 4.6.

Let I_p and I_d denote the intensities of components with vibrations || and ⊥ to the planes of incidence respectively in the transmitted light. Then, the degree of polarisation is given by

$$p = \frac{I_p - I_d}{I_p + I_d} = \frac{m}{m + [2n/(1-n^2)]^2}$$

Here, m is the number of plates and n is the refractive index of their material.

4.4. Law of Malus

Statement : *When a 'completely' plane-polarised light beam is incident on an analyser, the intensity of the emergent light varies as the square of the cosine of the angle between the planes of transmission of the analyser and the polariser.*

Proof : Let the angle between the planes of the polariser and the analyser be θ. Let $OP = A$ be the amplitude of vibrations of emergent plane polarised light from the polariser. Resolve the amplitude A into two components, $OB = A \cos \theta$ and $OC = A \sin \theta$ respectively, along and ⊥ to the plane of transmission of the analyser (Fig. 4.7). The parallel component $A \cos \theta$ is transmitted by the analyser while the perpendicular component $A \sin \theta$ is reflected.

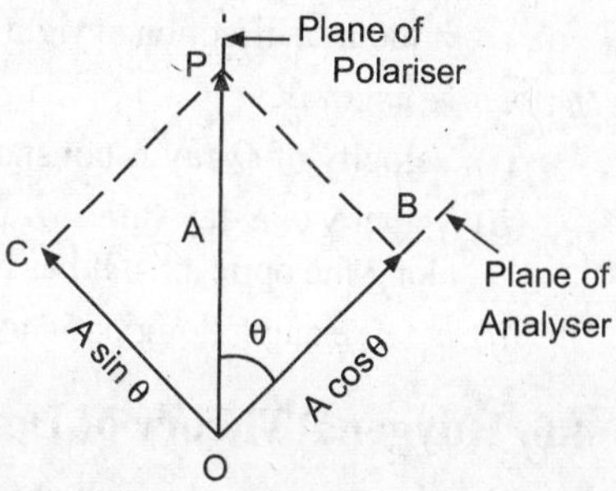

Fig. 4.7.

∴ Intensity of the beam transmitted by the analyser A, $I_\theta = (A \cos \theta)^2 = A^2 \cos^2\theta = I_0 \cos^2\theta$

Here, $I_o = A^2$ is the intensity of the plane polarised light incident on the analyser.

Thus $I_\theta = I_o \cos^2 \theta$. This is the *Malus law.*

4.5. Double Refraction

When a ray of unpolarised light is passed through a calcite crystal, it is split up into two refracted rays. One of the refracted rays follows the ordinary laws of refraction. Hence it is called the *ordinary ray*. The other refracted ray does not follow the ordinary laws of refraction. It is called the *extraordinary ray*.

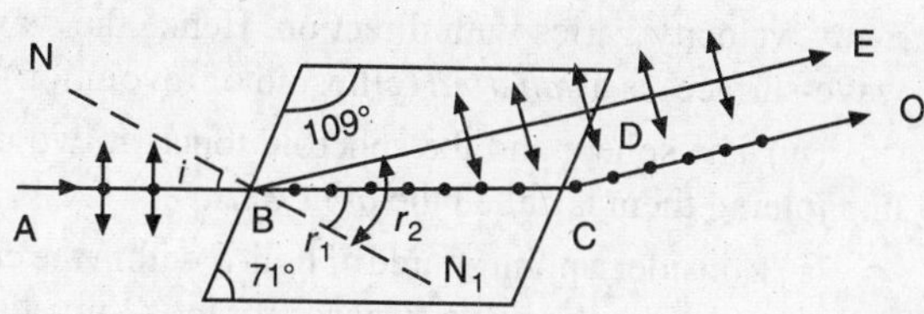

Fig. 4.8.

Therefore, if an object is viewed through such a crystal, two images of the object are observed. One image corresponds to O-ray and the other to E-ray. This phenomenon is called *double refraction.*

Let a ray AB be incident on a calcite crystal at an angle of incidence i [Fig. 4.8.]. It is split up into two rays (O and E) inside the crystal. The O-ray travelling along BC makes an angle of refraction r_1. The E-ray travelling along BD makes an angle of refraction r_2. The refracted rays emerge out along CO and DE which are parallel to each other and also parallel to the incident beam.

$$\text{The refractive index of ordinary ray} = n_o = \frac{\sin i}{\sin r_1}.$$

$$\text{The refractive index of extra-ordinary ray} = n_e = \frac{\sin i}{\sin r_2}$$

Refractive index of O- ray is constant since it obeys ordinary laws of refraction. Refractive index of E- ray varies with the angle of incidence i. The O- ray has the same velocity in all directions. The E- ray has different velocities in different directions. Along the optic axis, the velocity of the E- ray is the same as that of the O- ray.

Uniaxial crystals. Crystals which possess only one optic axis are called *uniaxial crystals*. There are two types of uniaxial crystals : *negative* and *positive*. Calcite is negative. Quartz is positive.

In negative crystals,

(*i*) Velocity of O-ray is constant in all directions.

(*ii*) Velocity of E-ray is different in different directions. It is minimum and equal to the velocity of O-ray along optic axis. It is maximum in a direction at right angles to the direction of the optic axis.

(*iii*) The refractive index of E-ray varies, having *maximum* value along optic axis and *minimum* value in a direction at right angles to it.

In positive crystals,

(*i*) Velocity of O-ray is constant in all directions.

(*ii*) Velocity of E-ray varies with the direction. It is *maximum* and equal to the velocity of the O-ray along the optic axis. It has a *minimum* value in a direction perpendicular to the optic axis.

(*iii*) $n_o < n_e$ except along optic axis where the two are equal.

4.6. Huygens' Theory of Double Refraction in Uniaxial Crystals

Huygens extended his principle of secondary wavelets to explain double refraction. He assumed:

(*i*) When a beam of light strikes the surface of a doubly refracting crystal, each point on the surface becomes the origin of two wavefronts which spread out into the crystal. One wavefront is for O-ray and the other is for E-ray. In other words, the wave-surface is double.

(*ii*) For the O-ray, the crystal is isotropic. So the O-ray travels with the same velocity in all directions. Hence the wavefront (or wave-surface) corresponding to O-ray is *spherical* .

(*iii*) For the E-ray, the crystal is anisotropic. So for the E-ray, velocity varies with direction. Hence the wavefront (or wave-surface) is a *spheroid* (ellipsoid of revolution).

(*iv*) The sphere and the spheroid touch at two points. The line joining them is called the *optic axis*.

(*v*) Consider a point source of light S within the crystal (Fig. 4.9). O and E are the wave surfaces for the O-ray and E-ray. In

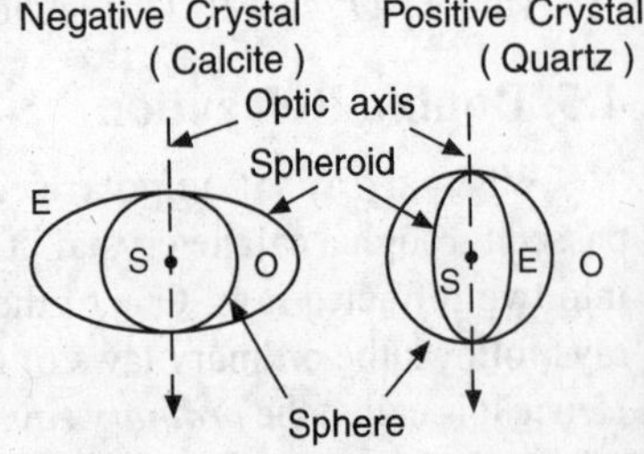

Fig. 4.9.

a negative crystal, the ordinary wave-surface (sphere) lies within the extra-ordinary wave-surface (spheroid). In a positive crystal, the ordinary wave-surface (sphere) lies outside the extraordinary wave-surface (spheroid).

4.7. Huygens' Construction for Double Refraction in Uniaxial Crystals

Optic axis in the plane of incidence and inclined to the refracting surface :

(a) Oblique incidence : Consider a plane wavefront *W* incident on a negative uniaxial crystal, calcite (Fig. 4.10). When the point *A* of the wavefront strikes the surface of the crystal, two secondary wavelets start from that point. The ordinary disturbance at *A* gives rise to an ordinary secondary wavelet, spherical in shape. The extraordinary disturbance gives rise to a spheroidal secondary wavelet. The sphere and the spheroid touch each other along the optic axis through *A*. When the point *B* of the incident wavefront strikes the crystal at *C*, the wavelets have attained the sizes shown in Fig. 4.10. The tangent plane *CD* drawn from *C* to the spherical wavelet gives the ordinary wavefront in the crystal. The tangent plane *CF* drawn from *C* to the spheroidal wavelet gives the extraordinary wavefront in the crystal. The incident ray *PA* has given rise to an ordinary refracted ray *AD* and an extraordinary refracted ray *AF*. In a negative crystal, the *E*-ray is refracted farther from the optic axis than the *O*-ray. The reverse is the case for a positive crystal. The two rays travel in different directions with different speeds.

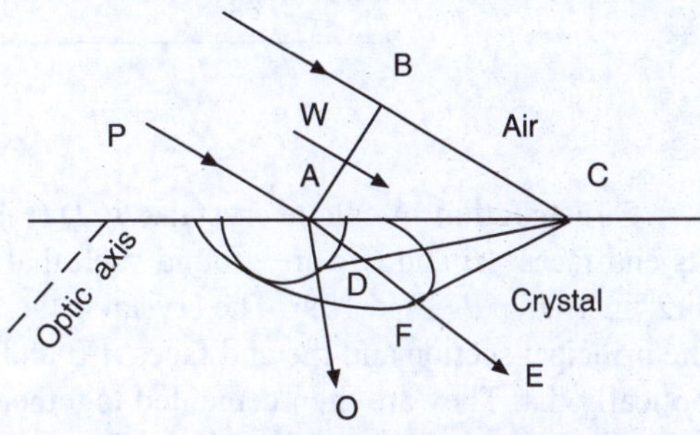

Fig. 4.10.

(b) Normal incidence : Consider the refraction of a plane e.m. wave by a negative uniaxial crystal (calcite). The direction of the optic axis is shown as a dashed line in Fig. 4.11. In order to determine the *O*-ray, we draw a sphere of radius V_o from the point *B*. Similarly we draw another sphere from the point *D*. The common tangent plane to these spheres is shown as *OO'*. *OO'* represents the wavefront corresponding to the ordinary refracted ray. In order to determine the *E*-ray, we draw an ellipse (centred at the point *B*) with its minor axis along the optic axis. The ellipsoid of revolution is obtained by rotating the ellipse about the optic axis. Similarly, we draw another ellipsoid of revolution from the point *D*. The common tangent plane to these ellipsoids is shown as *EE'*.

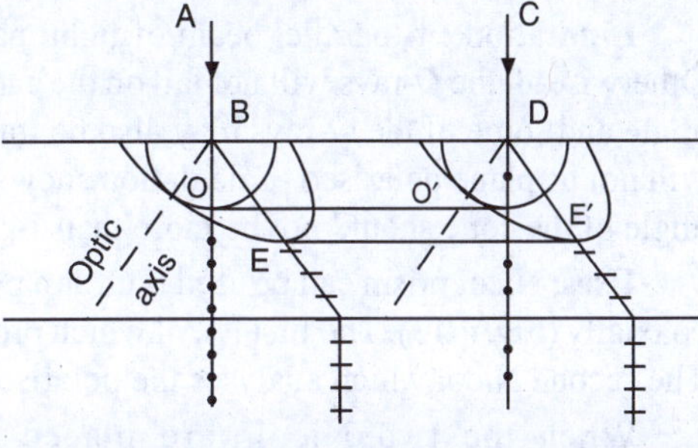

Fig. 4.11.

If we join the point *B* to the point of contact *O*, then corresponding to the incident ray *AB*, the direction of the *O*-ray will be along *BO*. Similarly, join the point *B* to the point of contact *E* (between the ellipsoid of revolution and the tangent plane *EE'*). Then corresponding to the incident ray *AB*, the direction of the *E*-ray will be along *BE*.

4.8. Nicol Prism

Principle : It is an optical device made from a calcite crystal. It is used for producing and analysing plane polarised light. Its action is based on the phenomenon of *double refraction*. When a ray of ordinary unpolarised light is passed through a calcite crystal, it is split up into the *O*-ray and the *E*-ray. Both these rays are plane polarised. In Nicol prism, *O*-ray is eliminated by total internal reflection. Hence only *E*-ray is transmitted through the prism. The *E*-ray is plane polarised and has its vibrations parallel to the principal section.

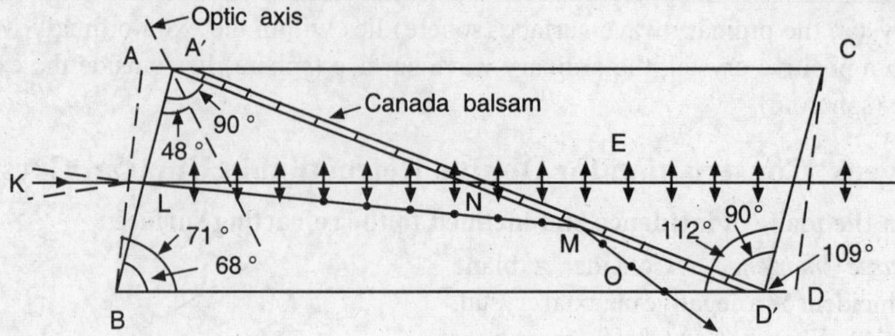

Fig. 4.12.

Construction: A calcite crystal *ABCD* (Fig. 4.12) whose length is 3 times its breadth is taken. Its end faces *AB* and *CD* are ground such that the angles in the principal section become 68° and 112° instead of 71° and 109°. The crystal is then cut apart along the plane *A′D* perpendicular to both the principal section and the end faces *A′B* and *CD′*. The two cut surfaces are ground and polished optically flat. They are then cemented together by Canada balsam which is a transparent liquid of refractive index 1.55 for sodium light. The crystal is then enclosed in a tube blackened inside.

Action : A ray *KL* of unpolarised light nearly parallel to *BD′* is incident on the face *A′B*. The ray is split up into an *O*-ray *LM* and an *E*-ray *LN*. The cutting is such that the *O*-ray is incident on the canada balsam surface at an angle greater than the critical angle and is totally reflected sideways. The sides of the prism are blackened to absorb the *O*-ray. The *E*-ray travels from an optically rarer medium to a denser medium. Therefore, it is transmitted through the prism. This *E*-ray is plane polarised and has vibrations in the principal section parallel to the shorter diagonal of the end face of the crystal. Thus *we get a single beam of plane polarised light*. So the Nicol prism can be used as a polariser.

Limitation : A parallel beam of light, parallel to the longer side of the prism should be used. Otherwise all the *O*-rays will not fall on the canada balsam surface at an angle greater than the critical angle and some of the *O*-rays may also be transmitted. So the light emerging from the Nicol prism will not be plane polarised. Calculation shows that if the beam is convergent or divergent, the semi-angle of the cone should not be more than 14° with the axis of the cone parallel to the longer side.

Uses: Nicol prism can be used both as a polariser and an analyser. Consider two Nicols arranged coaxially (Fig. 4.13). The first Nicol which produces the plane-polarized light is called the *polariser*. The second Nicol which analyses the polarised light is called the *analyser*.

When the two Nicols are placed with their principal sections parallel to each other as in Fig. 4.13 *(a)*, then the *E*-ray transmitted by one is freely transmitted by the other. This position and the other position corresponding to the angle of 180° between the two principal sections is known as *parallel Nicols*. The intensity of emergent light in these settings is maximum. On rotating one of the two Nicols, the intensity of the transmitted light decreases. When the principal sections of the two Nicols are mutually perpendicular, no light is transmitted by the system. It is so because the *E*-ray from the first Nicol forms an *O*-ray, for the second and is, therefore, totally reflected. The two Nicols are said to be *crossed* in this position [Fig. 4.13(b)].

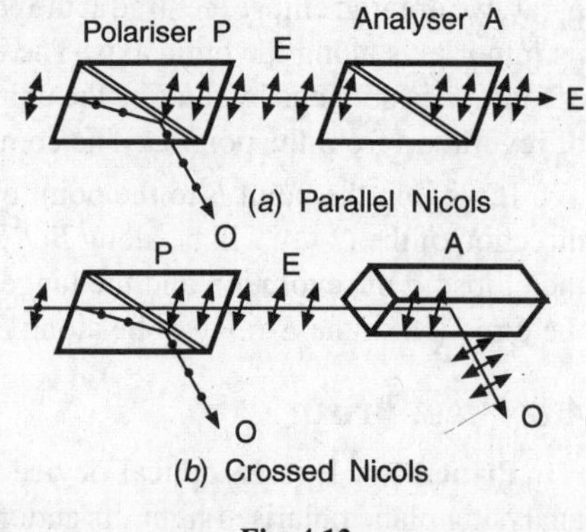

Fig. 4.13.

Let I_0 be the intensity of transmitted beam when the principal sections of the two Nicols are parallel. Let I_θ be the intensity when the principal sections are inclined at an angle θ. Then according to Malus law,

$$I_\theta = I_0 \cos^2\theta.$$

The above facts can be used for detecting plane polarised light. If the given light on examination through a rotating Nicol shows a variation in intensity with minimum intensity zero, the given light is plane polarised.

4.9. Double-image Polarizing Prisms

Nicol prism cannot be used in ultraviolet light because Canada balsam layer absorbs these radiations. Moreover in Nicol prism only the E-ray is transmitted while O-ray is lost by total internal reflection. Sometimes it is desirable to have both the rays in order to obtain two widely separated images. This can be achieved by optical devices known as 'double-image prisms'. There are two such prisms : the Rochon prism and the Wollaston prism.

Rochon prism. It consists of two right-angled prisms ABC and ADC (Fig. 4.14) of quartz cemented together with glycerine or castor oil so as to form a rectangular block. The optic axis of the prism ABC is in the plane of incidence and perpendicular to the refracting edge *i.e.*, parallel to its base BC. The prism ADC has its optic axis perpendicular to the plane of incidence (*i.e.*, perpendicular to the plane of the paper) and parallel to the refracting edge.

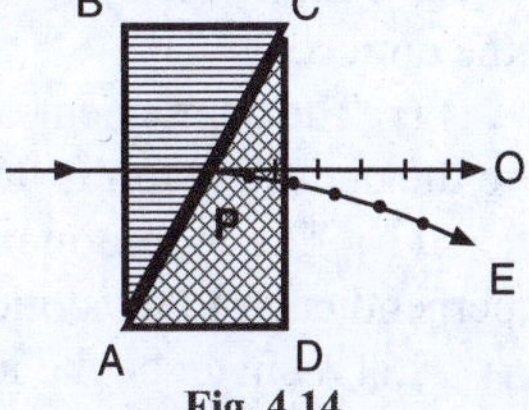

Fig. 4.14.

A ray of light incident normally on the face AB travels undeviated along the optic axis of the first prism. On entering the second prism ADC at P, it splits up into O and E-rays. The O-ray travels undeviated, thus remaining achromatic. The E-ray is both deviated and dispersed. When only one plane-polarised ray is required, the E-ray on account of its deviation can be cut off.

Wollaston prism. It consists of two prisms ABC and ADC of quartz cut in such a way that the optic axis of ABC lies in the plane of incidence and is parallel to the face AB. The optic axis of ADC is perpendicular to the plane of incidence (Fig. 4.15). Both the prisms are cemented together by glycerine or castor oil.

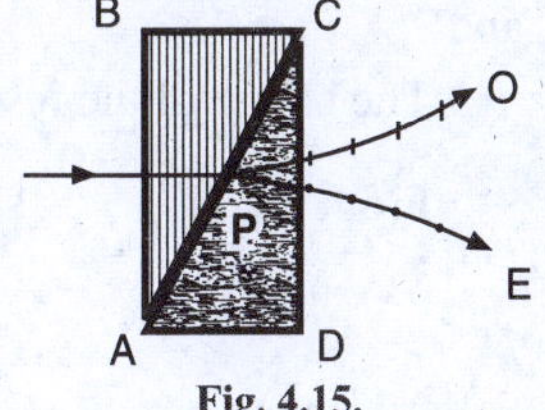

Fig. 4.15.

The incident ray entering normally to the surface AB travels perpendicular to the optic axis. Therefore, the O and E-rays travel in the first prism in the same direction but with different velocities. The second prism is cut with its optic axis perpendicular to that of the first. So the O-ray in the first prism becomes the E-ray in the second and vice-versa. Hence both the rays are deviated and dispersed in the second prism resulting in greater and greater separation of the two rays. The prism is specially useful in determining the percentage of polarisation in a partially polarized beam.

4.10. Plane, Circularly and Elliptically Polarised Light

In the *plane polarized light,* the light vector vibrates simple harmonically along a fixed straight line perpendicular to the direction of propagation *i.e.*, the orientation of light vector remains unchanged while its magnitude undergoes variations.

When two plane polarised waves are superimposed, then under certain conditions, the resultant light vector may rotate. If the magnitude of light vector remains constant while its orientation varies regularly (so that the tip of the vector traces a circle), the light is said to be *circularly polarised.* If, however, both the magnitude and the orientation of the resultant light vector vary, the tip of the vector traces an ellipse. Then, the light is said to be *elliptically polarised.*

4.11. Theory of Production of Elliptically and Circularly Polarised Light

Plane polarised light is incident normally on a calcite crystal cut with its faces parallel to the optic axis (Fig. 4.16). On entering the doubly refracting calcite, the plane polarised beam is split up into O and E rays. Both the O and E components are plane polarised with their planes of vibration at right angles to each other. Both these rays travel in the plate along the same path but with different velocities. For calcite, $V_E > V_o$. Hence the O ray falls behind the E ray. When the rays have travelled through the thickness d of the crystal, a phase difference δ is introduced between them.

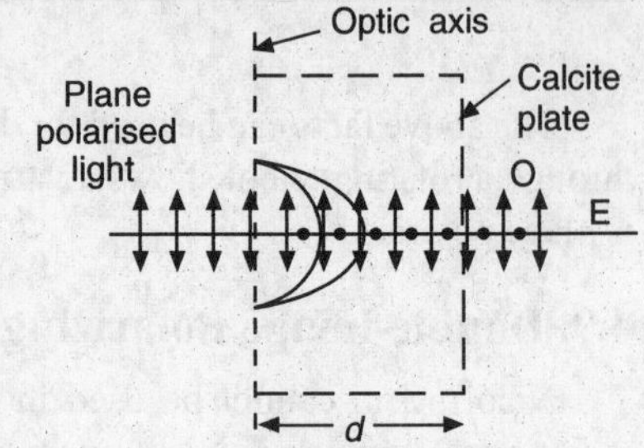

Fig. 4.16.

Theory : Let the linear vibrations in the incident plane polarised light be along MN inclined at an angle θ with the optic axis (Fig. 4.17). Let A be the amplitude of vibrations in the incident light. It can be resolved into two components: $A\cos\theta$ along the optic axis and $A\sin\theta$ perpendicular to the optic axis.

(*i*) The component $A\cos\theta$ having vibrations parallel to the optic axis forms the E wave.

(*ii*) The component $A\sin\theta$ having vibrations perpendicular to the optic axis forms the O wave.

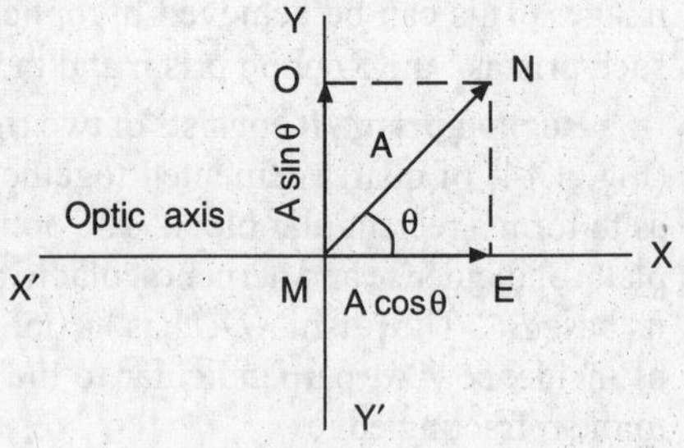

Let $A\sin\omega t$ be the incident wave.

The two emergent plane-polarised E- and O-waves would be represented respectively by

$x = A\cos\theta\sin(\omega t + \delta)$ and $y = A\sin\theta\sin\omega t$.

Fig. 4.17.

Put $A\cos\theta = a$ and $A\sin\theta = b$. Then,

$$x = a\sin(\omega t + \delta) \qquad ...(1)$$

and

$$y = b\sin\omega t. \qquad ...(2)$$

The nature of the resultant vibration can be obtained by eliminating t from Eqs. (1) and (2).

From Eq. (1), $\frac{x}{a} = \sin\omega t\cos\delta + \cos\omega t\sin\delta$

$$= \sin\omega t\cos\delta + \sqrt{(1-\sin^2\omega t)}\sin\delta$$

But from Eq. (2), $\sin\omega t = y/b$.

$$\therefore \quad \frac{x}{a} = \frac{y}{b}\cos\delta + \left\{\sqrt{\left(1-\frac{y^2}{b^2}\right)}\right\}\sin\delta \text{ or } \left(\frac{x}{a}-\frac{y}{b}\cos\delta\right)^2 = \left(1-\frac{y^2}{b^2}\right)\sin^2\delta$$

or

$$\frac{x^2}{a^2}+\frac{y^2}{b^2}-\frac{2xy}{ab}\cos\delta = \sin^2\delta. \qquad ...(3)$$

This equation represents an ellipse. Hence, the emergent light from the crystal is generally *elliptically polarised.* The exact nature of the resultant motion depends on the value of δ, which depends on the thickness d of the crystal plate.

Special Cases

(i) Plane polarised light. When $\delta = 0$, $\sin\delta = 0$ and $\cos\delta = 1$.

From Eq. (3), $\frac{x^2}{a^2}+\frac{y^2}{b^2}-\frac{2xy}{ab} = 0$ or $\left(\frac{x}{a}-\frac{y}{b}\right)^2 = 0$ or $y = \frac{b}{a}x$.

This is the equation of a straight line.

Thus the emergent light is *plane polarised* with vibrations in the same plans as the incident light (Fig. 4.18).

When $\delta = \pi$, $\cos\delta = -1$ and $\sin\delta = 0$.

Eq. (3) gives, $$y = -\frac{b}{a}x$$

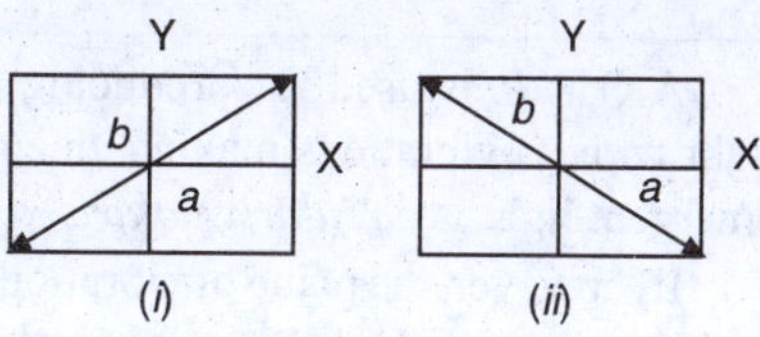

Fig. 4.18.

This is the equation of a straight line passing through the origin with a slope $(-b/a)$.

Hence the emergent light is *plane polarised.* But the direction of vibration now makes an angle $2\tan^{-1}(b/a) = 2\theta$ with that of the incident light. This case is the basis of a *half wave plate.*

(ii) Elliptically polarised light : When $\delta = \pi/2$, $\cos\delta = 0$ and $\sin\delta = 1$.

Eq. (3) gives $\frac{x^2}{a^2} + \frac{y^2}{b^2} = 1$

This is the equation of a symmetrical ellipse. The emergent light is *elliptically polarised* provided $a \neq b$ (Fig. 4.19).

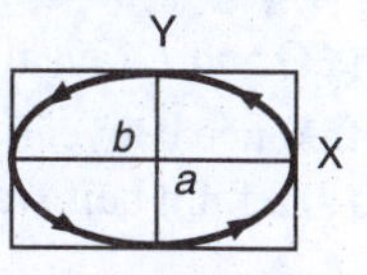

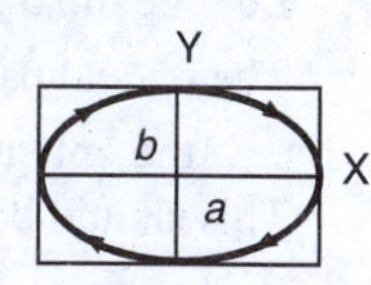

Fig. 4.19.

(iii) Circularly polarised light. Let $a = b$, *i.e.*, the amplitudes of the two vibrations are equal ($\theta = 45°$). Let $\delta = \pi/2$. Then Eq. (3) gives
$$x^2 + y^2 = a^2.$$

This is the equation of a circle of radius a. Thus the emergent light is *circularly polarised* (Fig. 4.20). This case forms the basis of *quarter wave plate.*

Thus we see that the plane-polarised and circularly-polarised lights are the special cases of elliptically-polarised light.

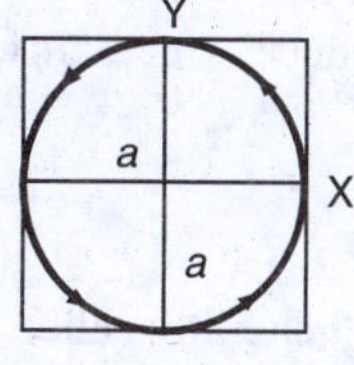

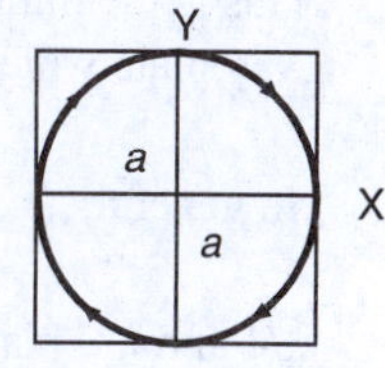

Fig. 4.20.

4.12. Quarter Wave Plate

A plate of doubly refracting uniaxial crystal cut with its optic axis parallel to the refracting faces and capable of producing a path difference of $\lambda/4$ (or a phase difference of $\pi/2$) between the ordinary and extraordinary waves is called a 'quarter wave plate'.

Consider a plate of doubly refracting uniaxial crystal cut with its faces parallel to the optic axis. A beam of mono-chromatic light of wavelength λ is incident normally on the plate (Fig. 4.21). It is broken up into O and E waves inside the plate. Both of these waves travel in the same direction (perpendicular to the faces) but with *different* velocities.

Let t be the thickness of the plate. The optical paths of the O and E waves in the plate are $n_o t$ and $n_e t$.

The path difference between the two waves on emerging $= (n_o \sim n_e)\,t$

If $(n_o \sim n_e)\,t = \lambda/4$, the plate is called a quarter wave plate.

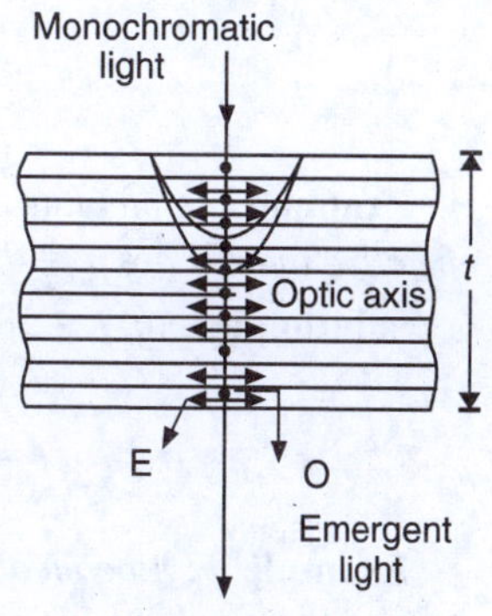

Fig. 4.21.

$$\therefore \qquad (n_o \sim n_e)\,t = \lambda/4 \text{ or } t = \frac{\lambda}{4(n_o \sim n_e)}$$

For a *positive crystal* such as *quartz*, $n_e > n_o$.

For a *negative crystal* such as *calcite*, $n_o > n_e$.

If the thickness of the plate is such that $(n_o \sim n_e)\,t = (2m + 1)\,\lambda/4$ where m is an integer, the plate still acts as a Q.W.P.

A Q.W.P. is used for producing circularly and elliptically polarised lights. If plane polarised light with its vibrations making an angle of 45° with the optic axis is passed through a Q.W.P., the emergent light is *circularly polarised.*

If, however, the plane of vibrations of the incident plane polarised light is not inclined at an angle of 45° to the optic axis, the emergent light is *elliptically polarised* .

4.13. Half Wave Plate

A plate of doubly refracting uniaxial crystal cut with its optic axis parallel to the refracting faces and capable of producing a path difference of λ/2 (or a phase difference of π) between O and E rays is called a 'half wave plate'.

Let t be the thickness of such a plate.

The optical paths for the O and E rays through the crystal are $n_o t$ and $n_e t$.

∴ the optical path difference between the O and E rays = $(n_o \sim n_e)\, t$.

This should be equal to λ/2 for a half wave plate.

$$\therefore \qquad (n_o \sim n_e)\, t = \lambda/2 \text{ or } t = \frac{\lambda}{2(n_o \sim n_e)}$$

This is the minimum thickness.

Any plate which introduces a path difference of $(2m + 1)\, \lambda/2$ will also act as a half wave plate.

In general,
$$t = \frac{(2m+1)\lambda}{2(n_0 \sim n_e)}.$$

λ/4 and λ/2 plates are generally called as *retardation plates* .

Example 1 : *Calculate the plate thickness of a quarter wave plate for light of wavelength 5.9×10^{-7} m. Principal refractive indices are $n_o = 1.544$ and $n_e = 1.553$.*

Solution : Here, $\lambda = 5.9 \times 10^{-7}$ m; $n_e = 1.553$, $n_o = 1.544$.

$$\therefore \qquad t = \frac{\lambda}{4(n_e - n_o)} = \frac{5.9\times10^{-7}}{4(1.553-1.544)} = 1.63\times10^{-5} \text{ m.}$$

Example 2 : *Calculate the thickness of a half-wave plate for light of wavelength 580 nm. Principal refractive indices are $n_o = 1.544$ and $n_e = 1.553$.*

Solution : Here, $\lambda = 580 \times 10^{-9}$ m; $n_e = 1.553$; $n_o = 1.544$. $t = ?$

$$t = \frac{\lambda}{2(n_e - n_o)} = \frac{580\times10^{-9}}{2(1.553-1.554)} = \mathbf{3.22\ 10^{-5}\ m.}$$

Example 3 : *A beam of plane polarised light for which $\lambda = 6 \times 10^{-7}$m, falls normally on a thin piece of quartz whose optic axis lies in the surface. If $n_e = 1.553$ and $n_o = 1.544$, find the thickness of the quartz which will introduce a phase difference of π radians between the E and O rays on emergence.*

Solution : In this case, the quartz plate must act as half-wave plate. Let t be the required thickness. Then,

$$t = \frac{\lambda}{2(n_e - n_o)} = \frac{6\times10^{-7}}{2(1.553-1.544)} = \mathbf{3.3\times10^{-5}\ m.}$$

Example 4: *A phase retardation plate of quartz has thickness 0.1436 mm. For what wavelengths in the visible region (450 nm – 800 nm) will it act as (i) quarter-wave plate, (ii) half-wave plate? Given : $n_o = 1.5443$, $n_e = 1.5533$.*

Solution : (i) The thickness of a λ/4 – plate is given by

$$t = \frac{m\lambda}{4(n_e - n_o)}, \; m = 1,3,5,7,...$$

$$\lambda = \frac{4(n_e - n_o)t}{m} = \frac{4(1.5533 - 1.5443) \times (0.1436 \times 10^{-3})}{m} \text{ metre}$$

$$= \frac{5169.6}{m} \text{ nm.}$$

Put m = 1, 3, 5, 7, 9, 11, 13, ...

λ = 5169.6 nm, 1723.2 nm, 1033.9 nm, 738.5 nm, 574.4 nm, 469.9 nm...

The wavelengths in the visible region are 738.5 nm, 574.4 nm, 469.9 nm.

(*ii*) For λ/2-plate, we shall have

$$\lambda = \frac{2(n_e - n_o)t}{m} \text{ metre} = \frac{2584.8}{m} \text{ nm.}$$

$$= 2584.8 \text{ nm}, 861.6 \text{ nm}, 517 \text{ nm}, 369.3 \text{ nm} ...$$

The wavelength in the visible region is **517 nm.**

4.14. Production and Detection of Plane, Circularly and Elliptically Polarised Light

Production of plane polarised light. Generally a Nicol prism is used for producing plane polarised light. A beam of unpolarised monochromatic light is passed through a Nicol prism. It is split up into *O* and *E* components inside the Nicol. The *O* component is totally reflected at the canada balsam layer and is absorbed by the blackened sides. The *E* component passes through the end face. It is plane polarised with vibrations in the plane of the paper.

Detection : The beam is allowed to pass through a Nicol prism. The Nicol is rotated gradually about the incident beam as axis. If the intensity of the light varies and light is *completely extinguished twice in each rotation* of the Nicol, then the beam is plane polarised.

Production of circularly polarised light. Circularly polarised light is produced if the amplitudes of the *O* and *E* rays are equal and there is a phase difference of π/2 or a path difference of λ/4 between them.

A parallel beam of monochromatic plane polarised light is allowed to fall normally on a Q.W.P. such that the vibrations in the incident plane polarised light make an angle of 45°, with the optic axis of the plate. The light emerging from the Q.W.P., is circularly polarised (Fig. 4.22).

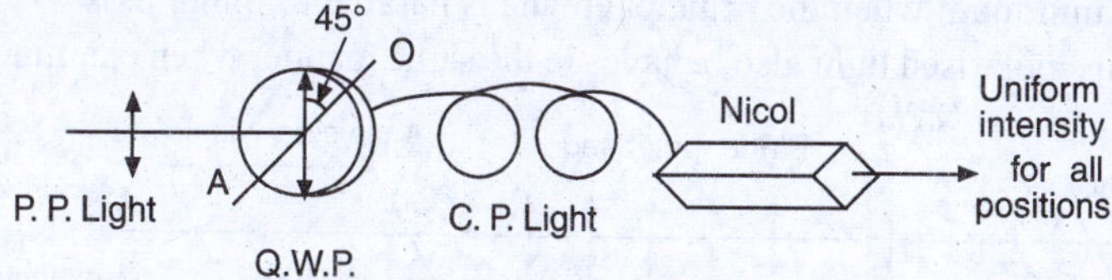

Fig. 4.22

Detection. The circularly polarised light, when observed through a rotating Nicol, shows no variation in intensity. If ordinary unpolarised light is viewed by a rotating Nicol, the intensity here also remains constant. In this respect, circularly polarised light resembles ordinary unpolarised light. We can distinguish between them by using a Q.W.P.

If circularly polarised light is passed through a Q.W.P., it is converted into plane polarised light. This plane polarised light can be extinguished by means of a rotating Nicol twice in a rotation (Fig. 4.23*a*).

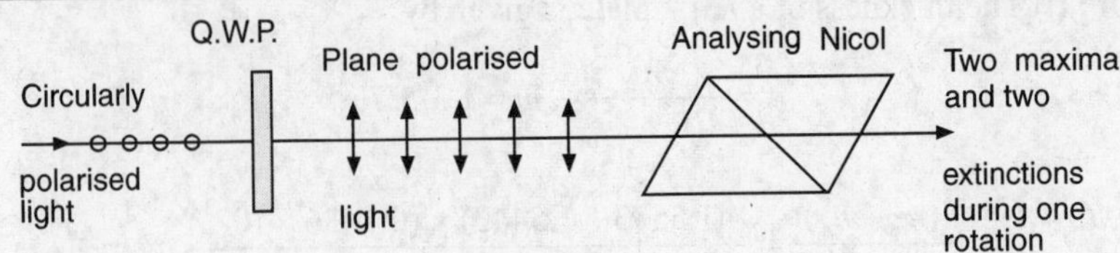

Fig. 4.23 *(a)*

On the other hand, ordinary unpolarised light on passing through a Q.W.P remains ordinary unpolarised light. It cannot be extinguished by a rotating Nicol (Fig. 4.23 *b*).

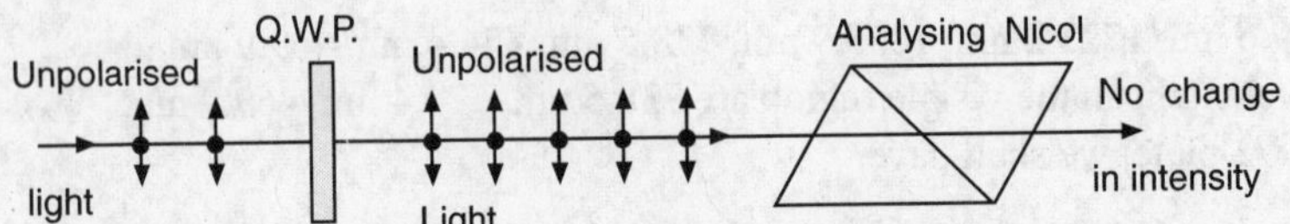

Fig. 4.23 *(b)*

Production of elliptically polarised light. Monochromatic plane polarised light is allowed to fall normally on a Q.W.P, such that the vibrations in the plane polarised incident light make an angle θ ($\theta \neq 0, 45°, 90°$) with the optic axis of the plate. The plane polarised light on entering the Q.W.P. is split up into O and E components having *unequal amplitudes*. On emergence from the Q.W.P. there is a phase difference of $\pi/2$ between the two components. They combine to form elliptically polarised light (Fig. 4.24).

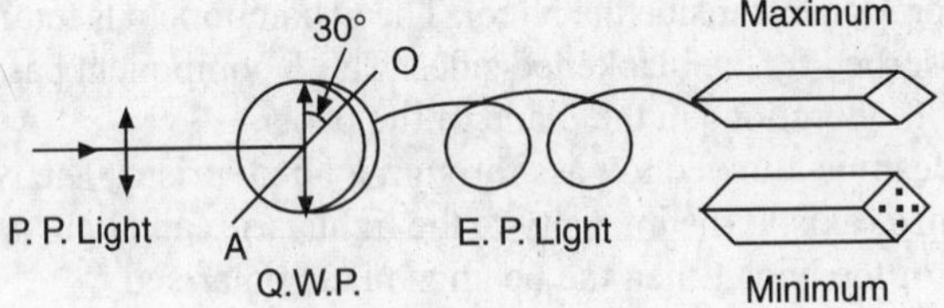

Fig. 4.24.

Detection. If elliptically polarised light is examined using a Nicol, the intensity of the emergent light varies between a maximum and a non-zero minimum as the Nicol is rotated. The intensity will be maximum when the principal plane of the Nicol is parallel to the major axis of the ellipse. The intensity will be minimum when the principal plane is parallel to minor axis.

Partially plane polarised light also behaves in the same manner when examined through a Nicol.

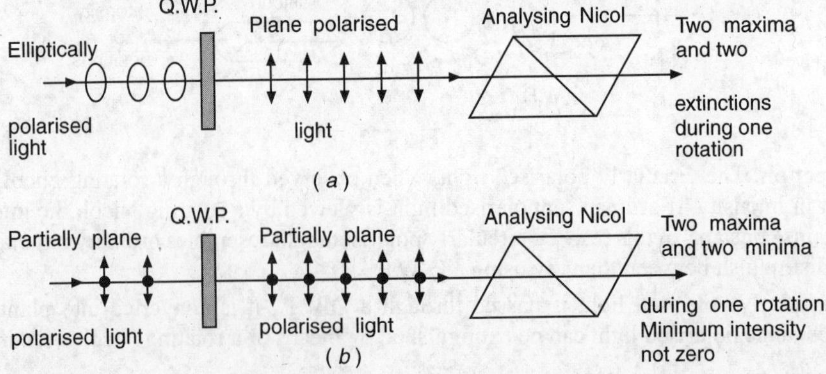

Fig. 4.25 (*a* & *b*).

To distinguish between elliptically polarised and partially plane polarised light, it is passed through a Q.W.P. The Q.W.P. converts the elliptically polarised light into plane polarised light. It will give two maximum and two complete extinctions when observed through a rotating Nicol (Fig. 4.25*a*). On the other hand, if the incident light is partially plane polarised light, it will remain as such when passed through the Q.W.P. If this is examined using a Nicol, the intensity varies between a maximum and non-zero minimum as the Nicol is rotated (Fig. 4.25 *b*).

Analysis of polarised light. Suppose we are supplied with light coming out from a hole and are asked to find its state of polarisation. The light may be (i) unpolarised, (ii) plane polarised, (iii) partially plane polarised, (iv) circularly polarised or (v) elliptically polarised. A Nicol prism and a quarter wave plate are taken and the following tests are applied.

Test 1. The given beam of light is passed through a Nicol prism. The Nicol prism is rotated about the direction of propagation of light as axis. The changes in intensity are noted. There are three possibilities :

(*i*) *The intensity does not vary at all* . Then the given beam of light is either *unpolarised or circularly polarised.*

(*ii*) The intensity shows variations–two maxima and two minima during one rotation but intensity is never zero. Then the given beam of light is either partially plane polarised or elliptically polarised.

(*iii*) The intensity shows variations and *completely extinguished twice in each rotation.* Then the given beam of light is *completely plane polarised* .

Tests 2. To distinguish between circularly polarised light and unpolarised light, *the given beam of light is first passed through a Q.W.P., and then examined through a rotating Nicol* .

(*i*) If the intensity varies with zero minimum, the given light is circularly polarised [Fig. 4.23*(a)*].

(*ii*) If there is no variation in intensity, the given beam of light is unpolarised [Fig. 4.23*b*].

Test 3 : To distinguish between elliptically polarised and partially plane polarised light, the Nicol is first adjusted for maximum intensity. Then a Q.W.P., is inserted between the given light and Nicol so that light falls normally on it and its optic axis is parallel to the principal section of the Nicol.

(*i*) On rotating the Nicol, if the intensity varies with zero minimum, the given light is elliptically polarised [Fig. 4.25*a*].

(*ii*) On rotating the Nicol, if the emergent light shows variation in intensity with a non-zero minimum, the given beam of light is partially plane polarised. (Fig. 4.25b).

OPTICAL ACTIVITY

4.15. Introduction

The property of rotating the plane of vibration of polarised light by certain crystals and other substances is called optical activity.

Substances which rotate the plane of polarisation are known as optically active substances. Substances like cinnabar, sodium chlorate, sugar crystals, turpentine oil, sugar solution, quinine sulphate solution etc., are optically active.

If we take two crossed Nicols N_1 and N_2 , the light incident on N_1 does not pass through N_2. But if we introduce some particular substance (such as sugar solution, quartz crystal etc.) between these crossed nicols, N_1 and N_2 (Fig. 4.26), then some light begins to pass through N_2. The light is, however, again completely cut off if N_2 is rotated through a certain angle. This shows that the light emerging from the quartz crystal is still plane-polarised, but its plane of polarisation has been rotated by the quartz crystal through a certain angle. Thus quartz is optically-active.

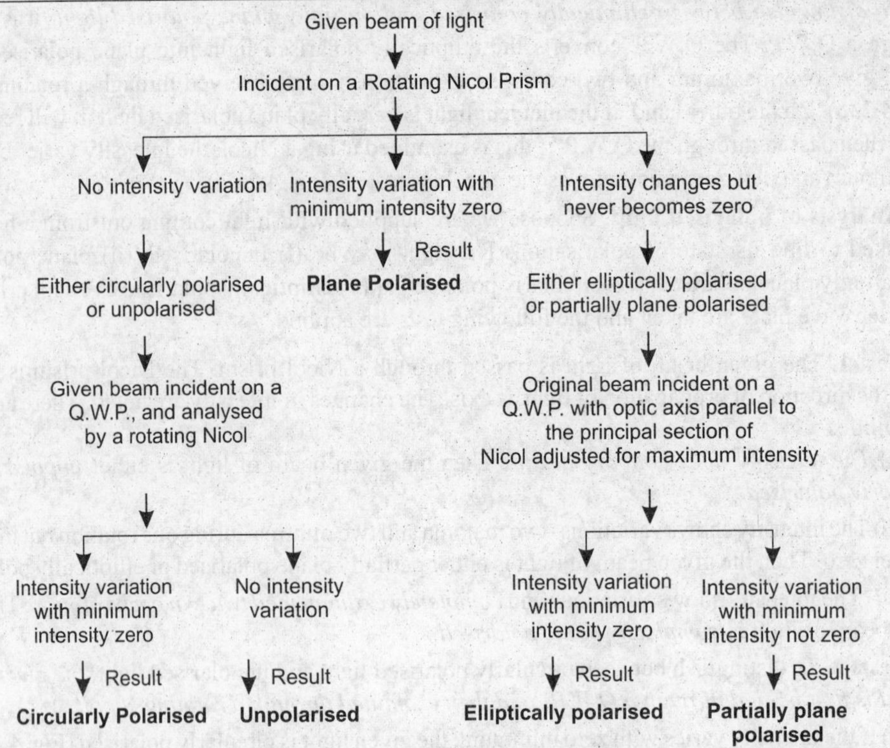

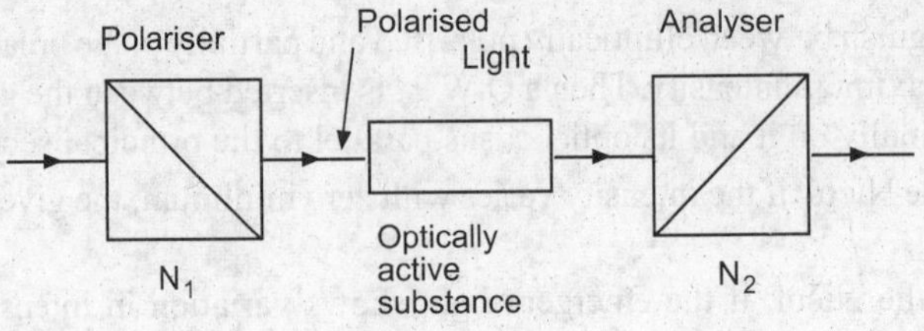

Fig. 4.26.

There are two types of optically-active substances. Those which rotate the plane of polarisation clockwise (looking against the direction of light) are called 'dextrorotatory' or 'right-handed', while those which rotate anti-clockwise are called 'laevorotatory' or 'left-handed'. Quartz occurs in both forms. Cane sugar is dextro whereas fruit sugar is laevo. The angle through which the plane of polarisation is rotated by the substance is called the *angle of rotation* θ.

4.16. Biot's Laws for Rotatory Polarisation

(*i*) The angle of rotation θ of the plane of polarisation, for a given wavelength, is directly proportional to the length (l) of the optically-active substance traversed. *i.e.*, $\theta \propto l$.

(*ii*) For solutions, angle of rotation θ for a given pathlength is proportional to the concentration (C) of the solution, *i.e.*, $\theta \propto C$.

(*iii*) The rotation produced by a number of optically-active substances is equal to the algebraic sum of individual rotations. The anti-clockwise and clock-wise rotations are taken with opposite signs,

$$\theta = \theta_1 + \theta_2 + \theta_3 + ...$$

(*iv*) The angle of rotation is approximately inversely proportional to the square of wavelength of light employed *i.e.*, $\theta \propto 1/\lambda^2$.

4.17. Fresnel's Theory of Optical Rotation

It is based on the fact that any S.H.M., along a straight line, can be considered as a resultant of two opposite, circular motions of the same frequency. Fresnel made the following assumptions:

(i) When a plane polarised beam enters an optically-active substance, it breaks up into two oppositely directed circularly polarised beams—one clockwise and the other anti-clockwise.

(ii) In an optically-active substance, the right-circular and left-circular vibrations travel with different speeds V_R and V_L. In a dextrorotatory substance, $V_R > V_L$. In a laevorotatory substance, $V_L > V_R$. Hence a relative phase-difference is developed between the two beams as they travel ahead.

(*iii*) On emergence from the optically - active substance, the two circular vibrations recombine to give a plane polarised beam. But the plane of polarisation of the emergent beam has been rotated "with respect to that of the incident beam by an angle" which depends on the phase difference between the two beams.

Mathematical treatment. Let a plane polarised beam be incident normally on a quartz crystal plate, cut with its faces perpendicular to the optic axis. Let the vibrations in the incident light be represented by

$$y = 2a \sin \omega t$$

Here, $2a$ is the amplitude of incident vibrations. The vibrations are executed along the *Y*-axis.

On entering the quartz plate, the incident beam breaks up into two oppositely directed circularly polarised beams [Fig. 4.27 *a*].

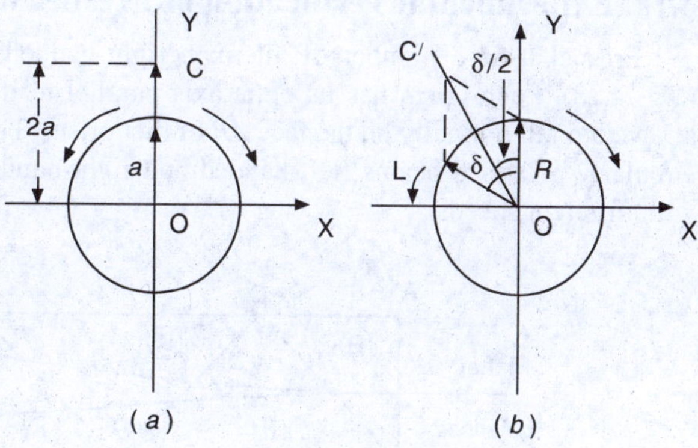

Fig. 4.27.

The circular vibrations are represented by the equations

$$x_1 = a \cos \omega t \qquad y_1 = a \sin \omega t \qquad \text{(right-handed)}$$

and $$x_2 = -a \cos \omega t \qquad y_2 = a \sin \omega t \qquad \text{(left-handed)}$$

The two circular vibrations travel with *different* velocities through the crystal. Therefore, when they emerge from the quartz plate, a phase-difference δ is introduced between them [Fig. 4.27*b*].

The emergent circular vibrations are represented by the equations

$$x_1 = a \cos \omega t, \qquad y_1 = a \sin \omega t$$

and $$x_2 = -a \cos (\omega t + \delta), \qquad y_2 = a \sin (\omega t + \delta)$$

The resultant vibrations along the *X* and *Y* axes are

$$X = x_1 + x_2 = a \cos \omega t - a \cos (\omega t + \delta) = 2a \sin \frac{\delta}{2} \sin \left(\omega t + \frac{\delta}{2} \right)$$

and $$Y = y_1 + y_2 = a \sin \omega t + a \sin (\omega t + \delta) = 2a \cos \frac{\delta}{2} \sin \left(\omega t + \frac{\delta}{2} \right)$$

These resultant vibrations along the *X* axis and *Y* axis are perpendicular to each other and are of the same period and phase. When compounded together, they give a resultant linear vibration along *OC′* which makes an angle δ/2 with the Y axis such that

$$\frac{X}{Y} = \tan \frac{\delta}{2}.$$

Thus on emergence, plane of vibration is along OC' whereas before entering the crystal, it was along OC. In this way, plane of vibration (or plane of polarisation) has been rotated by an angle $\delta/2$, where δ is the phase-difference between the two emergent circularly polarised vibrations.

Let n_R and n_L be the refractive-indices of quartz for clockwise and anti-clockwise vibrations respectively.

Then optical path difference in passing through a thickness l of the quartz plate is $(n_L \sim n_R)l$

$$\left.\begin{array}{l}\text{Phase difference between}\\ \text{the two circular components}\end{array}\right\} = \delta = \frac{2\pi}{\lambda}(n_L \sim n_R)l$$

$$\left.\begin{array}{l}\text{Angle of rotation of}\\ \text{the plane of polarisation}\end{array}\right\} = \theta = \frac{\delta}{2} = \frac{\pi l}{\lambda}(n_L \sim n_R)$$

4.18. Experimental Verification of Fresnel's Theory

Fresnel used a number of alternate right-handed (R) and left-handed (L) quartz prisms (Fig. 4.28). Each prism has its optic axis parallel to its base. A beam of plane polarised light is allowed to fall normally on the face AB of first prism. This plane polarised beam is split up into two circularly polarised beams (left-handed and right-handed) which travel along the same direction with different speeds.

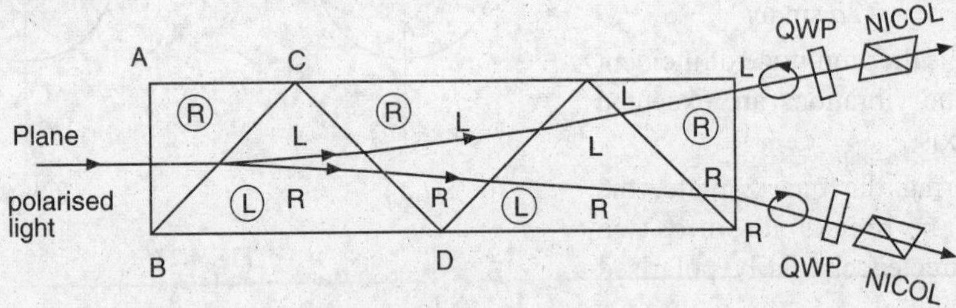

Fig. 4.28.

In the R prism, right-handed circularly polarised beam travels faster than the left-handed beam. When the beam is incident on the oblique surface BC of second prism (L), right handed beam becomes slower than left-hand and *vice versa*. Hence the second prism is optically denser medium for right-handed beam and rarer for left-handed beam. Therefore, in the second prism right-handed beam bends towards the normal (downwards). But the left-handed beam, bends away from the normal (upwards). Applying this rule at successive interfaces, the vibrations separate more and more. When the two beams emerge out, they are widely apart.

When the beams emerging from the system are analysed by a Nicol prism and a quarter wave plate, it is found that both the beams are circularly polarised in opposite directions. Thus, Fresnel's theory is experimentally verified.

4.19. Specific Rotation

(*i*) For crystals, the specific rotation (α) is defined as the rotation in degrees produced by a plate of the crystal, 1 mm thick. $\theta = \alpha l$.

Here, θ is the rotation measured in degrees, l is the path length measured in millimetres. The unit of α is deg mm^{-1}.

(*ii*) In the case of a pure liquid, $\theta = \alpha l \rho$

Here, the path-length l is measured in decimetres and ρ is the density in gm/cc.

(*iii*) *The specific rotation of a solution, at a given temperature and for a given wavelength of light is defined as the rotation (in degrees) produced by 1 decimetre length of the solution when its concentration is 1 gm / cc.*

Unit : deg. dm^{-1} $(gm/cc)^{-1}$. $\theta = \alpha\, Cl$.

Here θ is the rotation produced in degrees, l is the length of the solution in decimetre (10 cm) and C is the concentration in gm/cc.

Molecular rotation. The product of specific rotation and the molecular weight of the substance is called the molecular rotation.

4.20. Laurent's Half- shade Polarimeter

Construction : Its optical parts are shown in Fig. 4.29. Light from a monochromatic source S is rendered parallel by a convex lens L and falls on the polarising Nicol P which converts it into plane polarised light. This light passes through a half-shade device H and then through the tube T containing the solution. The transmitted light passes through the analyser A. The light emerging from the analyser is observed through a telescope G. The analysing Nicol A can be rotated about the axis of the tube and its position can be read on a circular scale S.

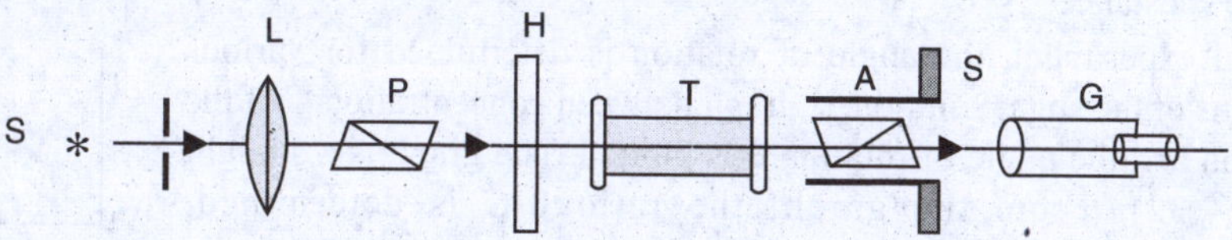

Fig. 4.29.

Working of the half-shade device. It consists of a semicircular glass plate XBY cemented to a semicircular quartz plate XDY (Fig. 4.30). The quartz plate is cut with its optic axis parallel to the line of separation XCY. The thickness of the quartz plate is such that it introduces a phase difference of π between the O and E vibrations, *i.e.*, it is a *half wave plate.* The thickness of the glass plate is such that it absorbs the same amount of light as the quartz plate.

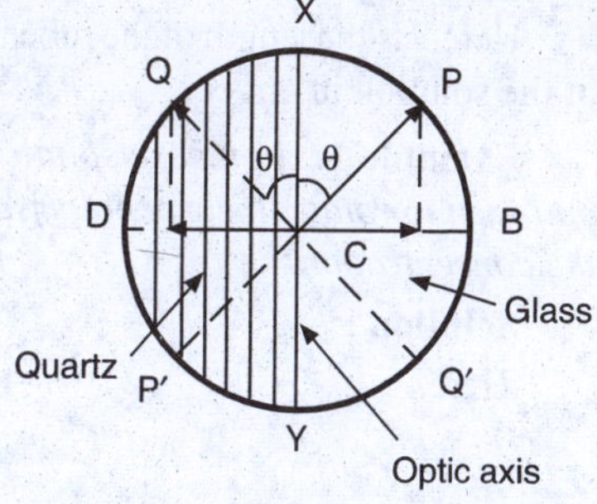

Fig. 4.30.

The light from the polariser (P) is plane polarised and falls normally on the half-shade plate. Let CP be the direction of vibrations in the plane polarised light. CP is inclined at an angle θ to the optic axis XY of the quartz half. On passing through the glass half, the vibrations will remain along CP. But on passing through the quartz half, the vibrations will be split up into E and O components. The vibrations of the O component are along CB and those of E component along CX. O component travels faster than the E component within quartz. Hence, on emergence, the O component gains a phase of π over the E component. Thus, on emergence from the quartz plate, O component has vibrations along CD. The E component has vibrations still along CX. Thus the light emerging from the quartz plate has resultant vibration along CQ such that $\angle PCX = \angle QCX = \theta$. *Thus the effect of quartz plate is to rotate the plane of polarization by an angle* 2θ.

Thus there are two plane polarised beams. One emerges from the glass plate with vibrations in the plane CP, while the other emerges from the quartz plate with vibrations in the plane CQ.

Let the principal plane of the analysing Nicol be parallel to QCQ'. Then light from the quartz plate will pass through the analyser. But the light from the glass plate will be partly stopped by the analyser. Hence *quartz half will be brighter than the glass half* [Fig. 4.31(*i*)].

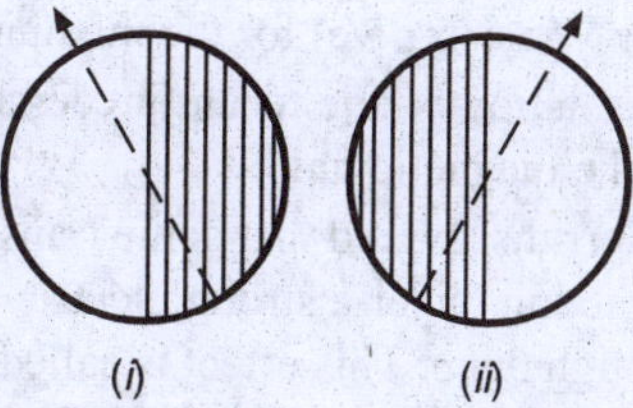

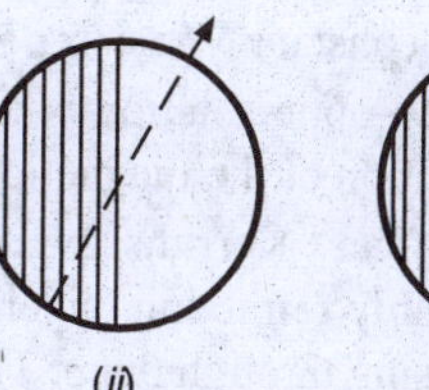

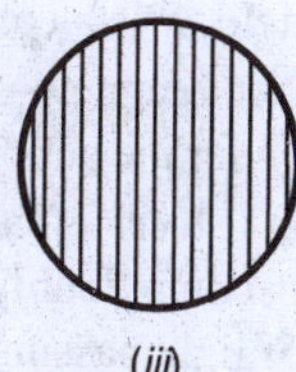

Fig. 4.31.

Let the principal plane of the analyser be parallel to PCP'. Then the light from the glass plate will pass through the analyser. But the light from the quartz plate will be partly stopped. Therefore, *glass half will be brighter than the quartz half* [Fig. 4.31(*ii*)].

When the principal plane of the analyser is parallel to XCY, it is equally inclined to the two plane polarised beams. Hence the two halves of the field appear *equally illuminated* [Fig. 4.31(*iii*)].

4.21. Determination of Specific Rotation of Sugar Solution

The tube is first filled with water. The analyser A is adjusted to obtain the condition of equal brightness of the two halves of the field of view. The position of the analyser is read on the circular scale (R_0).

The tube is filled with the sugar solution of known concentration. Now the analyser is rotated to obtain equally illuminated position of the field of view again. The position of the analyser is again read on the scale (R_1). Now $R_1 \sim R_0 = \theta$ = angle of rotation. The direction of rotation of the analyser gives the sense of rotation of the plane of polarization by the active substance.

In actual experiment, the angle of rotation is determined for various concentrations of the sugar solution. A graph between concentration C of the sugar solution and the angle of rotation θ is plotted. The graph is a straight line (Fig. 4.32). From the graph, the ratio θ/C is determined. The specific rotation of sugar is calculated from the relation, $\alpha = \dfrac{\theta}{lC}$.

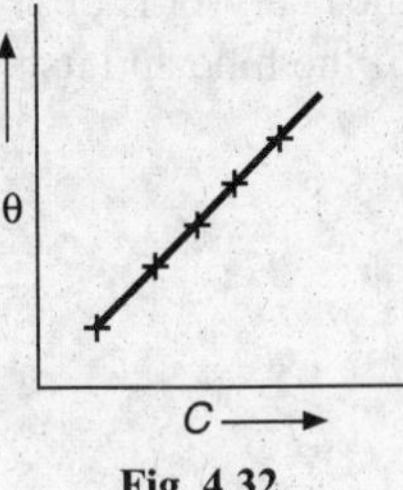

Fig. 4.32.

Here, l is the length of the tube in decimetre, θ is the rotation in degrees and C is the concentration of the solution in gm/cc.

Example 1: *A glucose solution of unknown concentration is contained in a 12 cm long tube and seen to rotate linearly polarised light by 2.5°. If the specific rotation of glucose is 52°, what is the concentration?*

Solution :

Here, $l = 1.2$ dm; $\theta = 2.5°$; $\alpha = 52°$. $C = ?$

$$\alpha = \theta / (lC).$$

$$C = \frac{\theta}{l\alpha} = \frac{2.5}{1.2 \times 52} = 0.0401 \text{ gm/cc}.$$

4.22. Optical Rotation by Magnetic and Electric Fields

Faraday effect (Magneto-optic rotation). If optically inactive substances are subjected to a magnetic field, they acquire the ability to rotate the plane of polarization of light propagated along the direction of the applied field. This is called the *Faraday effect*. The effect is most pronounced in media of high refractive index, and is absent in vacuum. Faraday effect is observed in many solids, liquids and gases.

The angle of rotation of the plane of polarization is given by $\theta = KBl$.

Here l is the path length of the light in the substance, B is the induction of the uniform magnetic field, K is a constant (+ ve or – ve) for the medium and is called *Verdet's constant*.

Magneto-optic rotation is due to the precession of the electron orbits of the atoms produced under the action of the magnetic field.

Kerr effect : Kerr discovered that some transparent substances, which are otherwise isotropic, become doubly refracting under a strong electric field. Hence they rotate the plane of polarization of light passing through them. This effect is called *Kerr effect* . Nitrobenzene shows the effect quite prominently. Owing to the action of the field, the dielectric behaves optically like a uniaxial crystal with its optic axis parallel to the field direction.

EXERCISE

1. Explain polarization by reflection and arrive at Brewster's law.
2. State and explain Malus law.
3. What is meant by 'double refraction'? Give Huygens' theory of double refraction in uniaxial crystals.
4. Describe how a Nicol prism can be used as an analyser.
5. Describe the method of producing linearly polarised light, circularly polarised light and elliptically polarised light. Explain how they can be distinguished from one another.
6. Explain the construction, principle and use of a (i) quarter wave plate and (ii) half wave plate.
7. What is optical activity? Describe Fresnel's explanation for the optical activity.
8. Give Fresnel's explanation for the optical activity. What is the specific rotatory power? Describe Laurent's half-shade polarimeter method of determining the specific rotation of sugar solution.
9. A ray of light is incident on the surface of a glass plate of refractive index 1.55 at the polarising angle. Calculate the angle of refraction.

   [Hint : $\tan i = n = 1.55 \therefore i = 57^\circ 11'$. We know, $i + r = 90^\circ$; $r = 90^\circ - 57^\circ 11' = 32^\circ 49'$]
10. A polariser and an analyser are oriented so that the amount of light transmitted is maximum. To what percentage of its maximum value is the intensity of the transmitted light reduced when the analyser is rotated through (i) 30°, (ii) 45°, (iii) 60° and (iv) 90°. **[Ans:** (i) 7.5%, (ii) 50%, (iii) 25%, (iv) 0]
11. Two polarising sheets have their polarising directions parallel so that the intensity of the transmitted light is maximum. Through what angle must either sheet be turned if the intensity is to drop one half?

    [Hint : $I_m/2 = I_m \cos^2 \theta$.] **[Ans:** ± 45°, ± 135°]
12. Calculate the thickness of a half-wave plate for light of wavelength 589.3 nm. Principal refractive indices are $n_o = 1.544$, and $n_e = 1.553$. **[Ans :** 3.27×10^{-5}m]
13. A plate of thickness 0.020 mm is cut from calcite with optic axis parallel to the face. Given $n_o = 1.648$ and $n_e = 1.481$ (ignoring variations with wavelength), find out those wavelengths in the range 400 nm to 780 nm (visible range) for which the plate behaves as a half-wave plate, and also those for which the plate behaves as a quarter-wave plate.

    [Hint : The given calcite plate will behave as a half wave plate for wavelength λ if

    $$(n_o - n_e)\, t = (2m + 1)\frac{\lambda}{2} \text{ where } m = 0, 1, 2, 3, ...$$

    $$\therefore \lambda = \frac{2(n_0 - n_e)t}{(2m+1)} = \frac{6680 \times 10^{-9}}{(2m+1)}\text{ m} = \frac{6680}{(2m+1)}\text{ nm.}$$

    Substituting $m = 4, 5, 6, 7$, we get λ = 742.2 nm, 607.3 nm, 513.8 nm, 445.3 nm in the visible region.

    Similarly, for a Q.W.P., $\lambda = \dfrac{4(n_o - n_e)t}{(2m+1)} = \dfrac{13360}{(2m+1)}$ nm.

    Substituting $m = 9, 10, 11, 12, 13, 14, 15, 16$ respectively, we get,

    λ = 703.2 nm, 636.2 nm, 580.7 nm, 534.4 nm, 494.8 nm, 460.7 nm, 431 nm, 404.8 nm in the visible region]
14. A beam of linearly polarised light changes to a circularly polarised beam by passing it through a crystalline plate 0.03 mm thick. Assuming this thickness to be minimum, calculate the difference in refractive indices of two rays in the crystal for light of wavelength 600 nm.

    [Hint : The slice is quarter wave plate. **Ans.** 0.005]
15. The values of n_e and n_o for quartz are 1.5508 and 1.5418 respectively. Calculate the phase retardation for λ = 500 nm when the plate thickness is 0.032 mm.

    [Hint : $\left.\begin{matrix}\text{Phase}\\ \text{retardation}\end{matrix}\right\} = \dfrac{2\pi}{\lambda}(n_e - n_o)t = 3.617$ radians.]

16. A 30 cm long tube containing sugar solution rotates the plane of polarization by 12°. If the specific rotation of sugar is 68°, calculate the strength of solution. **[Ans:** 0.06 gm/cc]

17. A glucose solution of unknown concentration is contained in a 12 cm long tube and seen to rotate linearly polarized light by 2.5°. If the specific rotation of glucose is 52°, what is the concentration? **[Ans.** 0.0401 gm/cc].

18. A solution of camphor in alcohol in a tube 20 cm long is found to rotate the plane of vibration of light by 27°. What is the mass of camphor in unit volume of the solution? The specific rotation of camphor is + 45°. **[Ans.** $C = 0.25$ gm/ml.]

19. An impure sample of turpentine gives a rotation of 48° for 22 cm length of solution. The density of pure turpentine is 0.8 gm/cc and it is suspected to be adulterated with kerosene of the same density. Calculate the percentage of purity of the sample if the specific rotatory power of turpentine is 36°. **[Ans:** 75.75%]

20. The specific rotation of quartz for 589.3 nm is 21.7° per mm. What thickness of quartz cut perpendicular to its optic axis and inserted between parallel Nicols, will allow no light of this wavelength to be transmitted?

[Hint : $l = \frac{\theta}{\alpha} = \frac{90}{21.7} = 4.147$ mm].

21. Light waves can be polarised but sound waves cannot be polarised. Why ?

[Ans. Light waves are transverse waves. But sound waves are longitudinal waves. Only transverse waves can be polarised.]

22. Select the only correct alternative:

(*i*) Polarisation of light proves the

(*a*) Corpuscular nature of light.

(*b*) Quantum nature of light.

(*c*) Transverse wave nature of light.

(*d*) Longitudinal wave nature of light.

(*ii*) If θ is the polarising angle, the refractive index *n* of the material is given by

(*a*) sin θ (*b*) tan θ

(*c*) cos θ (*d*) $\cos^{-1}\theta$

(*iii*) Light waves can be polarised because they –

(*a*) have high frequencies (*b*) have short wavelength

(*c*) are transverse (*d*) can be reflected

(*iv*) To analyse a plane polarised light it is allowed to incident upon a rotating analyser. In one complete rotation of the analyser, one finds-

(*a*) one complete extinction.

(*b*) one brightness and two times extinctions.

(*c*) one extinction and two times brightness.

(*d*) two extinctions and two times brightness.

[Ans. (*i*) *c* (*ii*) *b* (*iii*) *c* (*iv*) *d*]

CHAPTER

5

SPECTROSCOPY

5.1. Introduction

Types of spectra. Spectra are broadly classified into two groups:

(i) emission spectra and *(ii)* absorption spectra.

(i) Emission spectra. Emission spectra are of three kinds : (*a*) Continuous spectra, (*b*) band spectra and (*c*) line spectra.

Continuous spectra. A continuous spectrum contains all the wavelengths from one end to the other. When observed through a spectrometer, the spectrum appears as an unbroken continuous band of colours which range from red to violet. For example, incandescent solids like iron or carbon emit continuous spectra when they are heated until they glow. Continuous spectrum is due to the thermal excitation of the molecules of the substance.

Band spectra. The band spectrum consists of a number of bands of different colours separated by dark regions. The bands are sharply defined at one edge called the *head of the band* and shade off gradually at the other edge. When a band is examined through a spectroscope of high resolving power, it is found to consist of a large number of fine lines. Band spectrum is emitted by substances in the molecular state, when the thermal excitement of the substance is not quite sufficient to break the molecules into constituent atoms. *Examples* : Most of the carbon compounds and the blue part of a Bunsen burner show this kind of spectrum. A band spectrum is a discontinuous spectrum.

Line spectra. A line spectrum consists of bright lines in different regions of the visible spectrum against a dark background. All the lines will not have the same intensity. The number of lines, their nature and arrangement depends on the nature of the gas or vapour excited. Line spectra are emitted by incandescent gases and vapours of elements. The line spectrum of sodium contains two bright yellow lines of wavelengths 589 and 589.6 nm called the D-lines. Hydrogen spectrum contains four bright lines in the visible region. The line spectra are characteristic of the elements that produce it. No two elements can ever produce similar line spectra. From an investigation of the bright line spectrum, we can find the nature of the element that has produced it. This process is called *spectrum analysis.*

(ii) Absorption spectra. If a gas, liquid or transparent solid substance is placed between a light source giving a continuous spectrum and a spectrometer, the substance absorbs certain part of the spectrum. Hence in the continuous spectrum of the source appear few dark lines or bands. This spectrum is called the **'absorption spectrum'** of that substance.

The absorption spectrum may be a *line* absorption spectrum or *band* absorption spectrum. A line absorption spectrum is produced when light from a high temperature source, emitting continuous spectrum, is passed through an absorbing gas or vapour, at a comparatively low temperature. For example, when light from the filament of an electric bulb passes through sodium vapour, then two dark lines appear in the yellow part of its continuous spectrum. This is the absorption spectrum of sodium. In this spectrum, the dark lines appear exactly at the same place where two yellow lines appear in the emission spectrum of sodium. This is because the sodium atom absorbs the same wavelengths, which it emits. This is true for all atoms and molecules. If the absorbing substance is a solid or a liquid, a band absorption spectrum is obtained. If we replace sodium vapour by a dilute solution of potassium permanganate, two dark bands are found in the spectrum. This is the absorption band spectrum of potassium permanganate.

The solar spectrum. The solar spectrum was first observed by Newton. He considered it to be a continuous spectrum. Wollaston in 1807, first noticed several dark lines in the solar spectrum. In 1817, Fraunhofer observed as many as 600 dark lines across the solar spectrum. These dark lines are observed in the whole length of the spectrum viz., ultraviolet, visible and infrared. Fraunhofer named the dark lines by *A,B,C,D,E,F,G,*....... in alphabetical order from the red end of the spectrum to the violet end.

Explanation of the Fraunhofer lines. Fraunhofer lines can be explained by applying the 'principle of reversal'. According to Kirchhoff, the sun consists of a solid or liquid core (*photosphere*) at a very high temperature of the order of million degrees. The photosphere is surrounded by a gaseous envelope called the *chromosphere* whose temperature is about 6000 K.

The photosphere gives out white light, which produces the continuous spectrum. When the white light passes through the chromosphere, the vapours absorb those portions of the spectrum which they themselves would emit. Thus, Fraunhofer lines are caused by absorption in Sun's atmosphere. By comparing the wavelengths of these lines with those in the emission spectra of known elements on the earth, the gases and the vapours present in the sun's atmosphere can be known. Nearly forty elements present in earth are found to be in the sun's atmosphere. Helium was thus discovered in the sun's atmosphere before it was known on the earth.

Later investigations have shown that all Fraunhofer lines are not due to absorption in the chromosphere. Some, known as *telluric lines,* are due to absorption by water vapour and oxygen in the earth's atmosphere.

5.2. Infrared Spectroscopy

Discovery. Herschel moved a sensitive thermometer along a spectrum. He found that the temperature reached a maximum at a point some distance beyond the red end of the visible spectrum. Thus he showed that radiations exist beyond the red end of the spectrum. These radiations are called infrared radiations. This portion of the spectrum beyond the visible red is called the infrared region. Their wavelengths extend upto 1 mm. The spectra obtained in this region are called *infrared spectra.* This region is generally studied in absorption *i.e.*, radiations are passed through the absorption tube containing the gas under study and then dispersed by a prism or grating.

Sources: The sun is the natural source of infrared radiations. *Nernst glower* and *Globar* are the most common sources of infrared radiations. The Nernst glower consists of a filament prepared from a mixture of rare earths-particularly the oxides of thorium, zirconium and cesium. The glower is maintained at incandescence electrically. The globar consists of a rod of silicon carbide. It is also electrically heated.

Incandescent solid bodies at temperatures of 1000° to 1500°C emit radiations throughout the infrared. The desired region of wavelengths may be selected by the use of selective filters, selective reflectors or focal isolation.

Detectors. Infrared radiations are detected by *(1) heating effect (2) photography* and *(3) photoelectricity.*

(1) Infrared radiations are detected and also studied by their thermal effects. Thermopiles, bolometers and radio micrometers are the chief detectors used.

Golay cell. It is essentially a sensitive gas thermometer. It is based on the thermal expansion of a small volume of gas as the sensitive element. When radiations fall on the enclosed gas, the change in the pressure of the gas moves a mirror which reflects light on to a photocell.

Photographic method. Infrared radiations do not affect ordinary photographic plates. Photographic plates are made sensitive to infrared radiations by bathing them in solutions of cyanines like neocyanine, kryptocyanine, penta-carbocyanine, etc.

Dispersing systems. The spectrographs used for studying infrared spectra employ mirrors instead of lenses both before and after dispersion. The prism materials depend upon the range of wavelengths to be studied. No single material is transparent to the entire region. The following table shows the wavelength range over which the prism materials are reasonably transparent.

Note: Wavelengths in infrared region are measured in microns represented by μ. $1\mu = 10^{-6}$ m.

Material	*Wavelength Range*
Glass	Visible to 2.5μ
Quartz	Visible to 3.5μ
Fluorite	3μ to 9μ
Rock salt (Sodium chloride)	8μ to 16μ
Sylvine (K Cl)	15μ to 20μ
Potassium Bromide (K Br)	18μ to 28μ
Thallium Bromide	25μ to 40μ

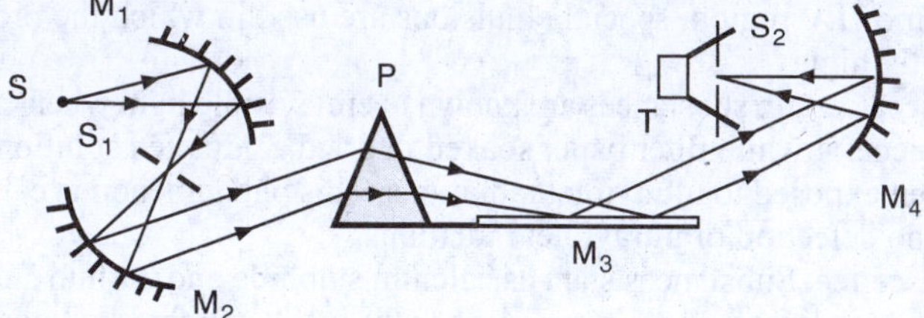

Fig. 5.1.

Wadsworth prism-mirror spectrograph. *S* is the source and the radiations from it are focussed on the slit S_1 by the concave mirror M_1 (Fig.5.1). The light from S_1 is rendered parallel by the concave mirror M_2. The parallel beam passes through the prism-mirror combination and after emergence, falls on the concave mirror M_4 and brought to a focus on the exit slit S_2. A linear thermopile *T* is kept at S_2. The prism is rotated so that spectral lines of different wavelengths pass through *T* in succession. The mirrors and prism are mounted inside a box to avoid stray light. The spectrograph is set focussed for visible monochromatic light and holds good for all other wavelengths.

Infrared photography. Because of its long wavelength, infrared light is scattered by the atmosphere to a lesser extent than visible light ($I \propto 1/\lambda^4$). Hence, if infrared radiations are used for photographing distant scenes, the pictures will be free from haze and will be more sharp. For this, specially designed photographic plates are used with suitable filter, because this transmits the infrared radiations and absorbs the visible light. In the world war II, infrared photography was very useful in detecting objects in the dark through mist, fog and clouds.

Uses. (1) Since infrared radiations are thermal radiations, they are used in treating sprains, dislocations, etc., where heat treatment is required.

(2) Infrared rays have found increasing use in industry due to the fact that heat energy reaches quickly all parts of the object to be heated. Thus food, enamels, automobile, etc., are quickly dried. Dehydration of potatoes during world war II was extensively brought about with the help of infrared rays.

(3) Since infrared rays penetrate more deeply through the human skin than ordinary light, they are used in diagnosing superficial tumors, etc., by infrared photography.

(4) Infrared photography is very helpful in the detection of crime using infrared cameras.

(5) Analysis of chemical compounds, determination of molecular structure and identification of organic molecules can be made by using infrared spectra.

(6) Traces of ammonia and methane have been found in Venus, Saturn and Jupiter using infrared spectroscopy.

(7) Since infrared rays produce photoelectric effect, they are used for illuminating object to be televised.

5.3. Ultraviolet Spectroscopy

Discovery. When silver chloride is exposed to sunlight, it undergoes a change in colour and finally becomes dark brown or black. Ritter found that when silver chloride was placed beyond the violet end of the spectrum, the action considerably increased. This indicated the existence of an invisible spectrum beyond the violet. This part of the spectrum between 400 nm and 10 nm is called the ultraviolet spectrum.

Sources. Sun is the most important and powerful natural source of U.V radiations. An electric arc of carbon, iron or of other materials, mercury vapour lamps and discharge of electricity through hydrogen contained in quartz tubes are used in the laboratory as sources of U.V. light. Mercury arc enclosed in quartz tube is generally used for medical purposes.

Detecting devices. There are four methods for the detection of U.V radiations:

(1) Photography. Photographic plates are very susceptible to U.V. radiations and are quickly blackened even by short exposures. Ordinary photographic plates are used in the near and far U.V region. In the extreme U.V region, special emulsions are used in which the proportion of Ag Br to gelatine is sufficiently high.

(2) Fluorescence. Certain substances are known to emit visible light when exposed to ultraviolet radiations. Thus a piece of white filter paper soaked in dried acidulated solution of quinine sulphate emits blue light when exposed to ultraviolet radiations. This phenomenon is called *fluorescence* and is used widely for the detection of ultraviolet radiations.

(3) Phosphorescence. Substances such as calcium sulphide and barium sulphide, if exposed to U.V light continue to emit light in the dark, even when the incident light is removed. This phenomenon is called *phosphorescence.*

(4) Photoelectric effect. When U.V rays fall on a negatively charged zinc plate, electrons are emitted from the plate. This phenomenon called *photoelectric effect* can be used to detect U.V radiations.

Spectrographs for U.V regions

Quartz spectrograph for near U.V region

Glass absorbs U.V radiations and cannot be used for work in the ultraviolet region. Hence quartz lenses and prisms are used. Fig. 5.2 *(a)* shows the optical components of a quartz spectrograph. It has a collimator with an adjustable slit and a right-handed quartz collimating lens *R*. The camera consists of a left-handed quartz objective *L* and photographic plate *P*. In order to get the whole spectrum in focus on the photographic plate, the plate is inclined at 20° to the axis of the telescope. A Cornu prism (*C*) is mounted in the minimum deviation position on the prism table between the collimator and camera tubes. The Cornu prism [Fig. 5.2 *(b)*] consists of two 30° right angled prisms cemented together, one being made of right handed quartz and the other of left-handed quartz.

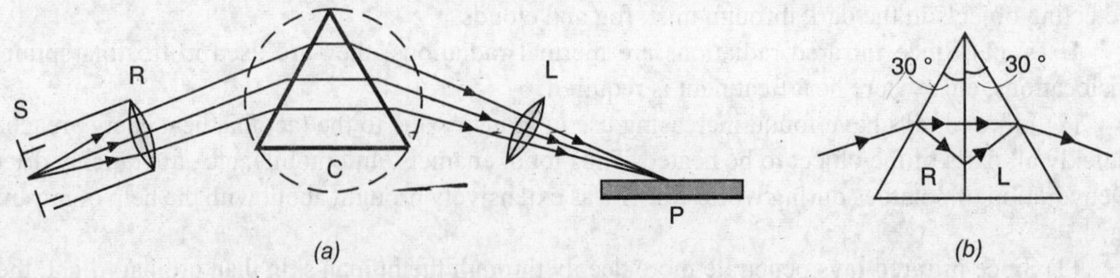

Fig. 5.2.

When used at minimum deviation position, the prism gives a single emergent beam because the phase difference introduced in one half is removed in the other.

Light from the slit (S) is rendered parallel by the lens R. After refraction through the Cornu prism, light emerges out in a direction symmetrical with the incident beam and enters the camera. The camera lens L brings the spectrum to a perfect focus on the photographic plate P. The spectrum is focussed using a fluorescent screen. The screen is then removed and the photographic plate put in position. The U.V. spectrum is thus photographed. Wavelengths may be computed with a comparator.

Vacuum spectrograph for extreme ultraviolet region. The extreme ultraviolet region extends from 200 nm to about 15 nm. Quartz does not transmit ultraviolet beyond 185 nm. Schumann explored the region below 185 nm using fluorite prisms and lenses. Air absorbs ultraviolet light below 170 nm. Hence work has to be carried out in vacuum and hence the region below 170 nm is called the vacuum ultraviolet region. Fig. 5.3 shows a vacuum spectrograph. The light from the slit S falls on the grating G and the diffraction spectrum is received on the photographic plate P.

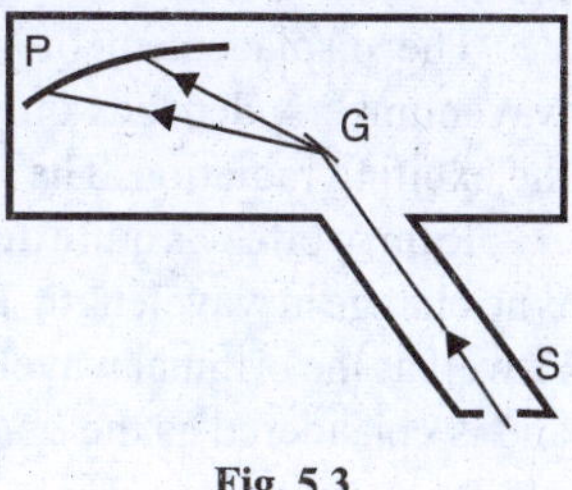

Fig. 5.3.

Application of ultraviolet spectroscopy

(*i*) All elements singly or multiply ionised have spectra in the ultraviolet region below 200 nm. All molecules produce absorption spectra and many organic substances of biochemical interest have both emission and absorption spectra in the region. Their structures have been revealed by the study of vacuum ultraviolet spectroscopy.

(*ii*) Electron transmission in solids are of importance in solid state physics. Studies of ultraviolet spectrum in the region 30 to 10 nm are useful in the theoretical explanation of the phenomenon of photoconductivity, fluorescence and phosphorescence and the transmission and absorption bands of dielectrics.

(*iii*) Observations of emission spectra in the vacuum ultraviolet region have been useful in astrophysics in explaining unusual types of stellar spectra. By such studies, surface temperatures of hot stars has been determined.

5.4. Rayleigh's Scattering

Rayleigh showed theoretically that the intensity of scattered light is inversely proportional to the fourth power of the wavelength *i.e.*, $I \propto \frac{1}{\lambda^4}$. This is known as Rayleigh's law of scattering. I is also proportional to the square of the volume of the particle. The blue colour of the sky is due to the greater scattering of short wavelengths by the dust suspensions in the atmosphere or by the air molecules. This type of scattering simply produces separation of wavelengths originally present in the incident light. No wavelength can be present in the scattered light which is not present in the incident light *i.e*, no new wavelength is created. For this reason, Rayleigh scattering is called coherent scattering.

RAMAN EFFECT

5.5. Discovery

While studying the scattering of light, Raman found that when a beam of monochromatic light was passed through organic liquids such as benzene, toluene, etc., the scattered light contained other frequencies in addition to that of the incident light. This is known as *Raman effect*.

The original arrangement used by Raman was simple in design. A round-bottomed glass flask was filled with pure dust-free benzene. The liquid was strongly illuminated by the mercury blue line of wavelength 435.8 nm. The scattered light was examined through a spectroscope in a direction perpendicular to that of the incident beam. It was observed that the spectrum contained, in addition to the original wavelength (435.8 nm), some lines which had wavelengths longer as well as shorter than the original wavelength. These lines of modified wavelengths are known as *Raman lines*. The lines having wavelengths greater than that of the incident wavelength are called *Stokes lines* and those having shorter wavelengths are called *anti-stokes lines*. The Stokes lines are found to be more intense than the anti-stokes lines. Most of the Raman lines are strongly polarised.

The displacement of the modified spectral lines from the exciting line, when measured in wavenumbers, depends only on the scattering substance and is independent of the wavenumber of the exciting radiation. The Raman spectrum is thus characteristic of the scattering substance.

Raman effect is quite different from Rayleigh scattering. In Rayleigh or coherent scattering, there is no change in wavelength. But in Raman scattering, the scattered light contains modified wavelengths as well as the original wavelengths. For this reason, Raman effect is also called *incoherent scattering* and is considered as the *optical analogue of the Compton effect.*

5.6. Experimental Study of Raman Effect

The apparatus shown in Fig. 5.4 first developed by Wood, is used for studying Raman effect in liquids. It consists of a glass tube *AB* containing the pure experimental liquid free from dust and air bubbles. The tube is closed at one end by an optically plane glass plate *W* and at the other end it is drawn into a horn (*H*) and blackened on the outside.

Light from a mercury arc *S* is passed through a filter *F* which allows only monochromatic radiation of $\lambda = 435.8$ nm to pass through it. The tube is surrounded by a water-jacket (*J*) through which water is circulated to prevent overheating of the liquid. A semi-cylindrical aluminium reflector *R* is used to increase the intensity of illumination. The scattered light coming out of *W* is condensed on the slit of a spectrograph. The spectrograph must have a large light gathering power and the prism must have a large resolving power. A short focus camera is used to photograph the spectrum.

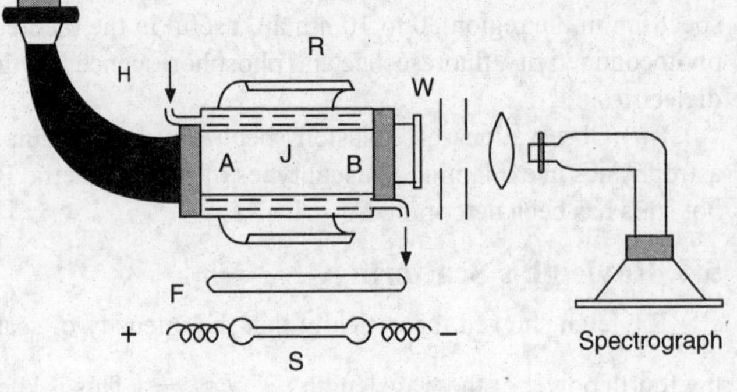

Fig. 5.4.

On developing the photographic plate, it exhibits a number of stokes' lines, a few anti-stokes lines and a strong unmodified line (Fig. 5.5).

Original line ← 435.8 nm
Antistokes' line
Stokes' line
0
λ →

Fig. 5.5.

Characteristics of Raman lines :

(1) The Stokes lines are always more intense than anti-stokes lines.

(2) The Raman lines are symmetrically displaced about the parent line.

(3) The frequency difference between the modified and parent line represents the frequency of the corresponding infrared absorption line.

5.7. Quantum Theory of Raman Effect

Raman effect is due to the interaction between a light photon and a molecule of the scatterer. Suppose a photon of frequency, ν_1 is incident on a molecule and there is a collision between the two.

Let m = mass of the molecule, v_1 and v_2 its velocities before and after impact, E_1 and E_2 the intrinsic energies of the molecule before and after collision. Let ν_2 be the frequency of the scattered photon. Applying the principle of conservation of energy,

$$E_2 + \frac{1}{2}mv_2^2 + h\nu_2 = E_1 + \frac{1}{2}mv_1^2 + h\nu_1 \quad ...(1)$$

We may assume that the K.E. of the molecule is unaltered during the process. Hence,

$$E_2 + h\nu_2 = E_1 + h\nu_1 \text{ or } \nu_2 - \nu_1 = \frac{E_1 - E_2}{h} \text{ or } \nu_2 = \nu_1 + \frac{E_1 - E_2}{h} \quad ...(2)$$

Three cases may arise :

(1) When there is no change in the energy of the molecule, $E_1 = E_2$. Then $\nu_1 = \nu_2$. This represents the unmodified line.

(2) If $E_2 > E_1$, then, $\nu_2 < \nu_1$. This represents the Stokes line. It means that the molecule has absorbed some energy from the incident photon. Consequently the scattered photon has lower energy or longer wavelength.

(3) If $E_2 < E_1$, then, $\nu_2 > \nu_1$. This represents the antistokes line. It means that the molecule was previously in the excited state and it handed over some of its intrinsic energy to the incident photon. The scattered photon thus has greater energy or shorter wavelength.

Since the molecules possess quantised energy levels, we can write,

$$E_1 - E_2 = nh\nu_c \quad ...(3)$$

where n = 1, 2, 3... etc., and ν_c = the characteristic frequency of the molecule.
In the simplest case n = 1, Eq. (2) reduces to

$$\nu_2 = \nu_1 \pm \nu_c \quad ...(4)$$

Eq. (4) shows that the frequency difference $\nu_1 - \nu_2$ between the incident and scattered photon corresponds to the characteristic frequency ν_c of the molecule.

Raman effect and fluorescence. Raman effect and fluorescence resemble each other in that the incident light in both the cases suffers a change of wavelength and lines of new wavelengths appear in the spectrum of either. But the Raman effect is quite different from fluorescence. We compare the two phenomena in the table.

	Raman Spectra	*Fluorescence Spectra*
1.	Spectral lines have frequencies greater and lesser than the incident frequency.	The frequencies of the lines in the fluorescent spectrum are always less than the incident frequency.
2.	The frequency shifts of the Raman lines are determined by the scatterer rather than the frequencies themselves.	Frequencies of the fluorescent lines are determined by the nature of the scatterer.
3.	Raman lines are strongly polarised.	Lines are not polarised.

5.8. Applications

(1) Raman effect and molecular structure. A qualitative identification of Raman spectrum consists in the evaluation of the wavelength of the lines, their intensities, and state of polarisation. Investigation of bond angles, bond stiffness, and other structural confirmation require Raman data in addition to infrared studies.

(*i*) *Diatomic molecules*. From the analysis of Raman spectra of a diatomic molecule, we can have an idea about the nature of the chemical bond existing between the atoms. In a diatomic molecule, the frequency of vibration of the atoms is given by $\nu = \frac{1}{2\pi}\sqrt{\frac{F}{\mu}}$ where F is the restoring force per unit displacement and μ is the reduced mass of the molecule. It is seen that a molecule in which the force binding the atoms is great should have higher characteristic frequency than one in which the force is weak. This force depends upon the nature of interatomic bonds. In covalent molecules, polarisability is considerably changed by the nuclear oscillations due to the nuclei. This appreciable change in polarisability gives rise to intense Raman lines. In electrovalent molecules, the binding electrons definitely change over from one nucleus to the other in the formation of the molecule so that the polarisability of the molecule is little affected by nuclear oscillations and hence no Raman lines will appear.

(*ii*) *Triatomic molecule.* Dealing with triatomic molecules (of type AB_2), the questions to be decided are : Whether each molecule is linear or not and, if linear, whether it is symmetrical (B - A -B) or asymmetrical (B - B - A). From the number and intensity of the observed lines in the Raman effect in conjunction with infrared data, it is possible to draw important conclusions about molecular structure. Theory leads to the following rule, known as the *rule of mutual exclusion.* It states that *for molecules with a centre of symmetry, transitions that are allowed in the infrared are forbidden in Raman spectra and vice versa.* The rule does not imply that all transitions forbidden in one must occur in the other ; *i.e.*, some transitions may be forbidden in both. On the other hand, certain transitions can occur both in the infrared and in the Raman spectra in the case of molecules without a centre of symmetry.

Examples : *(a)* CO_2 has two very strong bands in its infrared absorption spectrum at 66800 and 234900 m^{-1} while only one strong band in its Raman spectrum at 138900 m^{-1}. None of these bands occur both in Raman and infrared spectra. Hence, it follows from the rule of mutual exclusion, that CO_2 molecule has a centre of symmetry. This implies that the molecule is linear and symmetric and hence it should be represented by $O - C - O$.

(*b*) Nitrous oxide (N_2O) has three absorption bands at 222400, 128500 and 58900 m^{-1} of which the first two appear in the Raman spectrum. Thus the molecule cannot have a centre of symmetry though linear. Hence the molecule has the unsymmetrical structure N - N - O .

In a similar manner, the bent symmetric structure of a water molecule represented by Fig. 5.6 is revealed by Raman Effect. Thus the study of Raman spectra of different substances enables one to classify them according to their molecular structure.

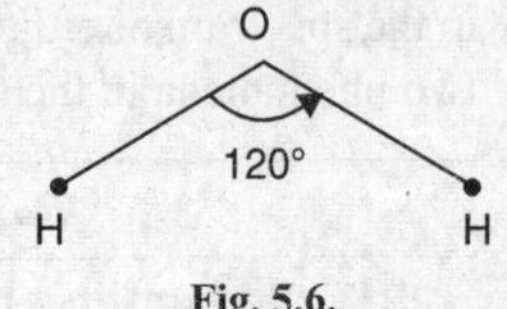

Fig. 5.6.

(2) Raman effect in crystals is complimentary to X-ray crystal study and provides information about the binding forces in crystals.

(3) Various chemical effects like strength of chemical bonds, electrolytic dissociation, hydrolysis, etc., have been understood through Raman effect. Specific heat capacities of solids, brilliance of metals and their molecular structure have been explained by Raman effect.

5.9. Nuclear Magnetic Resonance

Introduction : Spectroscopy may be defined as the interaction between matter and electromagnetic radiation, such that energy is absorbed or emitted according to the relation $\Delta E = h\nu$.

Here, ΔE = energy difference between the initial and final states of matter,

ν = frequency of the electromagnetic radiation.

Nuclear Magnetic Resonance (NMR) is a branch of spectroscopy in which radio frequency waves induce transitions between magnetic energy levels of nuclei of a molecule. The magnetic energy levels are created by keeping the nuclei in a magnetic field.

Without the magnetic field, the spin states of nuclei are degenerate, *i.e.*, possess the same energy, and energy level transition is not possible. When a magnetic field is applied, there is splitting up of nuclear energy levels. Radio frequency radiation can cause transitions between these energy levels.

For any given sample of nuclei, there is a small excess population in the lower energy state. Any net energy absorbed by the sample from the applied radio frequency field depends on the population difference between the energy states. This population difference, though minute, is crucial, since the applied radio frequency field can *induce transitions* corresponding to absorption as well as emission (release) of energy.

I, the spin quantum number of the nucleus, can be zero or can have various integral or half-integral values. *NMR* spectroscopy is often concerned with nuclei with $I = \frac{1}{2}$. Examples of such nuclei are H^1, P^{31}, F^{19}. We shall be mostly concerned with proton, *i.e.*, the nucleus of H^1. Spectra cannot be obtained from nuclei with $I = 0$.

Theory. If a proton is placed in an external magnetic field, its magnetic moment can be aligned in either of the two ways : with or against the external field. Alignment with the field is the more stable situation. Energy is required to "flip" the proton magnet over to the less stable alignment against the field.

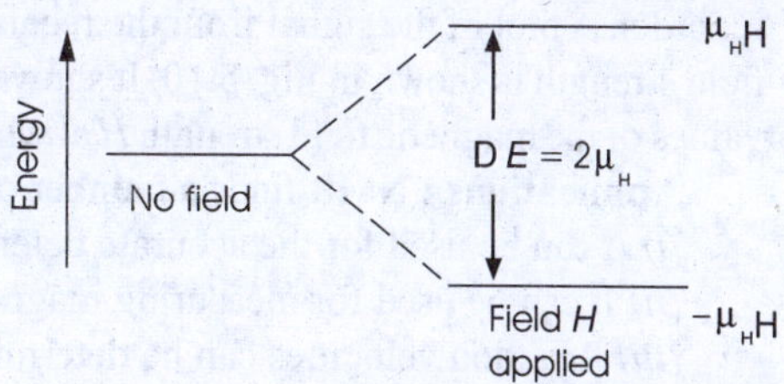

Fig. 5.7.

Let μ_H = magnetic moment of the proton and H = applied external field. Then, energy of the proton = μ_HH (Fig. 5.7).

Difference between the two nuclear energy levels :

$$\Delta E = \mu_H H - (-\mu_H H) = 2\mu_H H$$

The radiation frequency is kept constant and the strength of the magnetic field is varied. At some value of field strength, the energy required to "flip" the proton matches the energy of the radiation. The absorption of radiation occurs when $2\mu_H H = h\nu$.

The occurrence of resonance is marked by a sharp increase in energy absorption. Absorption occurs and a signal is observed. Such a spectrum is called NMR spectrum (Fig. 5.8).

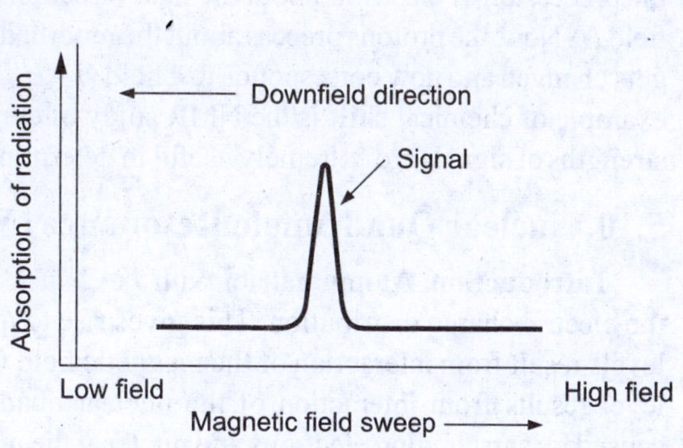

Fig. 5.8.

Experimental Arrangement : Fig. 5.9 shows an N.M.R. spectrometer. A is a radio frequency oscillator. It is adjusted to generate a definite frequency ν (say, 60MHz) which passes through a coil D located between the magnetic poles. Surrounding the sample S, there is a receiver coil R. The receiver coil is used to pick up the broadcast signals. These signals are carried to a receiver and amplifier B, and then to a cathode ray oscillograph recorder C. The strength of the magnetic field can be varied with the help of the device M, called "magnetic field sweep".

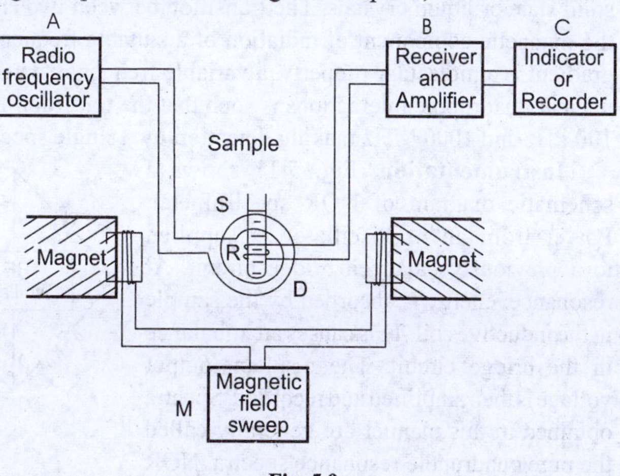

Fig. 5.9.

The external magnetic field acts on the nuclei of the sample to produce energy levels (Fig. 5.7). Transitions between these levels are induced by radiation from the radio frequency transmitter. Radiation will be absorbed by the sample if the frequency of the radiation is such that the quanta of radiation have an energy matching the nuclear level spacing. When such radiation is absorbed, it can be thought of as producing nuclei in the excited state. These nuclei will then tend to re-emit the radiation in order to approach the Boltzmann distribution ratio. It is this emitted radiation that is detected by the receiver coil, which being oriented at 90° to the transmitter, receives no signal unless the sample provides this coupling with the transmitter. The signal from the receiver-coil can be displayed on an oscilloscope or a recorder. A plot of the signal from the receiver versus magnetic field strength is shown in Fig. 5.10. It shows the NMR spectrum of C_2H_5OH. The peaks represent the values of the magnetic field strength H at which the sample is in resonance with the applied frequency.

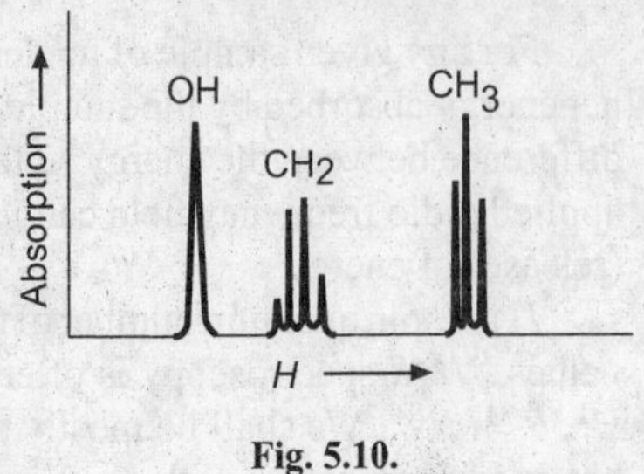

Fig. 5.10.

Applications : NMR finds a number of practical applications :

(i) It can be used for the accurate determination of magnetic moments of Li^7 and F^{19}.

(ii) It can be used for measuring magnetic fields.

(iii) Reaction velocities can be determined by NMR spectroscopy.

(iv) **Application in Organic Chemistry – Chemical Shift :** An important application of NMR is the study of the magnetic properties of materials. A proton in a material is always surrounded by atomic electrons. This electronic nature around each proton is always somewhat diamagnetic. Thus the precession of electrons about the field direction creates a new field H_d which opposes the external field H. Now, the protons precess about the modified field $(H-H_d)$. The observed resonance frequency gets changed and now corresponds to a field $H-H_d$. This effect is called the *chemical shift*. A striking example of chemical shift is the NMR study of ethyl alcohol. Measurements of chemical shifts and strengths of signals are extremely useful in determination of structures of complex organic molecules.

5.10. Nuclear Quadrupole Resonance (NQR)

Introduction. Atomic nuclei with $I \geq 1$ lack the spherical symmetry (along the spin axis) in the electric charge distribution. This gives rise to quadrupole moment eQ to it. In NMR, the energy levels result from interaction of the magnetic field with the magnetic dipole. But in NQR, the energy level results from interaction of the nuclear quadrupole moment with the electric field gradient, caused by surrounding electrons and nuclei at the nucleus. Each energy level corresponds to a certain orientation of the nuclear quadrupole moment w.r.t. the field gradient. In liquids, molecular vibrations as well as tumbling motion is so fast that the field gradient experienced by the nucleus reduces to zero. Hence the effect may be observed only if the field gradient axis is fixed. This is possible in solid state or liquid crystals. The transition between two energy levels takes place by interaction of the magnetic component of radiation of a suitable frequency given by $\nu = (E_2-E_1)/h$. As the field gradient is a molecular property, a variable frequency detection system must be used. The range of nuclear quadrupole interactions is such that the transition frequencies can occur anywhere between 100 kHz and 1000 MHz making detection by a single spectrometer very difficult.

Instrumentation. Fig. 5.11 shows a schematic diagram of NQR spectrometer. Power from an r.f. oscillator is supplied to a previously balanced bridge circuit. At resonance, energy is absorbed by the sample in the inductive coil. This causes an imbalance in the bridge circuit. The resultant output voltage is then amplified and recorded. Spectra obtained in this manner are generally called the pure quadrupole resonance spectra. NQR

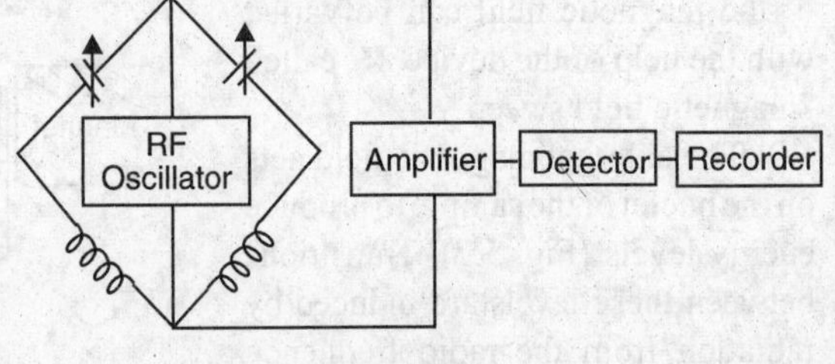

Fig. 5.11.

spectra can be obtained only for solids. Because of this limitation, NQR does not have the same range of application as NMR.

Applications of NQR. (*i*) NQR is mainly used for investigating the electronic structure of molecules. When a comparison of the nuclear quadrupole coupling constant is done in atomic and molecular state for the same nucleus, it provides information regarding the extent of hybridisation and the ionic character of the bond.

(*ii*) NQR technique has been used for studying a number of group III halides of the type MX_3 (*e.g.*, $Al\ Br_3$), chloromethanes and chloroacetyl chlorides.

5.11. Electron Spin Resonance Spectroscopy

Introduction. Electron spin resonance is a branch of absorption spectroscopy in which radiation having frequency in the *microwave region* is absorbed by *paramagnetic substances* to induce transitions between magnetic energy levels of electrons with *unpaired spins*.

Theory of ESR. The presence of an unpaired electron in a molecule or ion allows energy levels to be produced from the interaction of the magnetic moment of the electron with an applied magnetic field. For an electron of spin $s = 1/2$, the spin angular momentum quantum number can have values of $m_s = \pm 1/2$. In the absence of magnetic field, the two values of m_s, *i.e.*, $+ \frac{1}{2}$ and $-\frac{1}{2}$ will give rise to a doubly degenerate spin energy state. If a magnetic field is applied, this degeneracy is removed and this leads to two non-degenerate energy levels. The low energy state has the spin magnetic moment aligned with the field and corresponds to the quantum number, $m_s = -1/2$, while the high energy state $m_s = + 1/2$, has its moment opposed to the field. The two states will have energies that are split from the original state with no applied field by the amounts $+ \mu_e H$ and $-\mu_e H$. Here, μ_e is the magnetic moment of the spinning electron and H the magnetic field acting on the unpaired electron (Fig. 5.12).

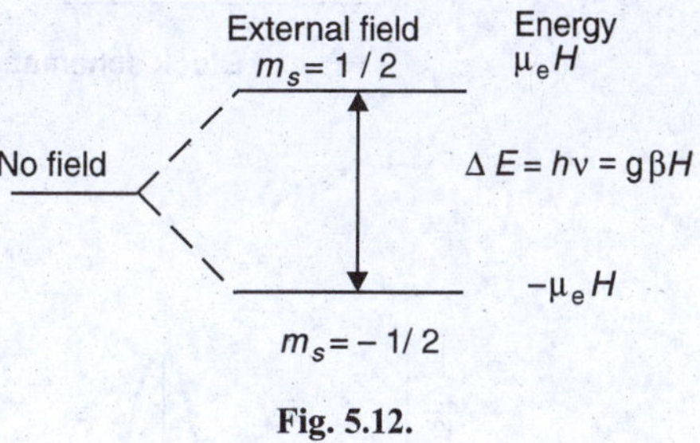

Fig. 5.12.

In ESR, a transition between the two different electron spin energy states occurs upon absorption of a quantum of radiation of frequency ν in microwave region. When the absorption occurs, the following relation holds good :

$$2\mu_e h = h\nu \quad ...(1)$$

Eq. (1) holds good for a free electron.

The energy (ΔE) of transition is more accurately given by:

$$\Delta E = h\nu = g\beta H \quad ...(2)$$

Here, h is Planck's constant, β the Bohr magneton, and g the spectroscopic splitting factor.

Instrumentation. Fig. 5.13 shows the block diagram of a typical ESR spectrometer. The sample is kept in the resonant cavity. The cavity serves as a very long path-length cell in which waves are reflected to and fro thousands of times.

The klystron oscillator is set to produce microwaves which after passing through the isolator, wavemeter and attenuator are received by circulator through arm 1. The microwave power entering arm 1 divides between arms 2 and 3. The arm 2 is generally having a balancing load. If the impedances of arms 2 and 3 are same, microwave power is absorbed completely and no power will be received by the detector through arm 4. If the impedance of arm 3 (the sample cavity) changes because of some ESR resonance absorption by a sample in it, some microwave power will enter into detector through arm 4. The detector acts as a rectifier, converting the microwave power into direct current. If the magnetic field around the resonant cavity having the sample is changed to the value required for resonance, the recorder will show an absorption peak (Fig. 5.14). Further, if the main magnetic field is swept slowly over a period of several minutes, the recorder will show the derivative of the microwave absorption spectrum against magnetic field (Fig. 5.15). Due to instrumental considerations associated with the signal-to-noise ratio, the ESR spectra are generally recorded as first derivative spectra.

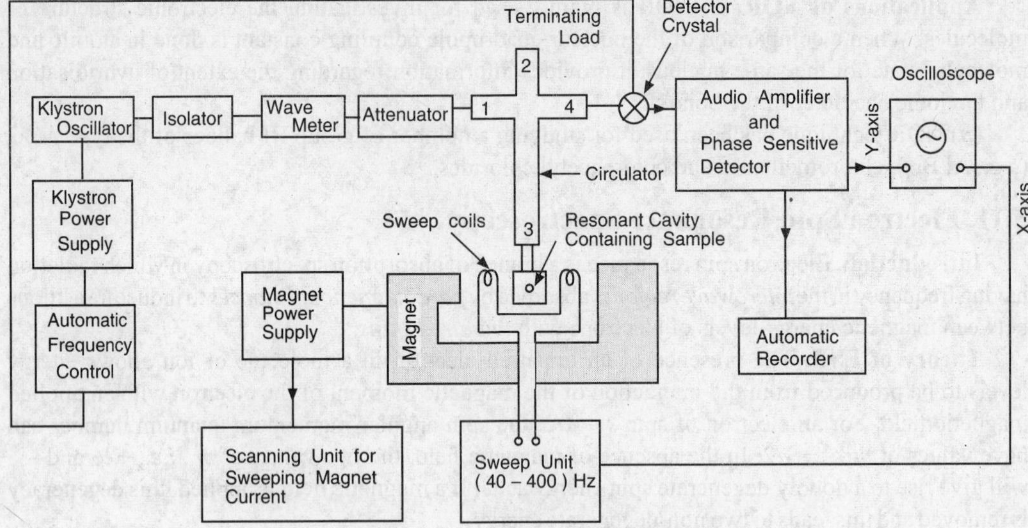

Block schematic of a typical ESR spectrometer.

Fig. 5.13.

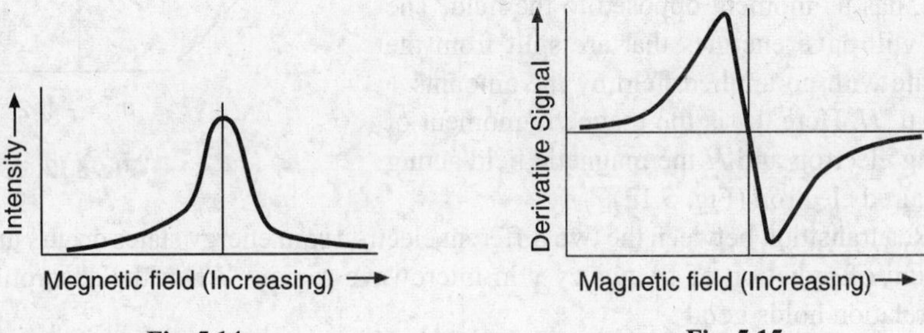

Fig. 5.14. **Fig. 5.15.**

Applications. (*i*) Nuclear spins can be determined by ESR.

(*ii*) Reaction velocities and mechanisms can be studied by ESR.

(*iii*) It is very useful for the study of chemical species with unpaired electrons.

(*iv*) *Study of free radicals.* Free radicals can be readily studied by ESR, even in very low concentrations. For example, a signal for DPPH radical can be detected even if there is 10^{-12} gm of material in the spectrometer.

The ESR spectrum of a quinone-hydroquinone redox system is shown in Fig. 5.16. This proves conclusively that a semiquinone free radical anion exists as an intermediate. The first derivative spectrum of this system contains five-line spectrum which arises due to the magnetic spin interaction between the odd electron and the four hydrogen atoms on the ring. The intensities of five lines are in the ratios 1 : 4 : 6 : 4 : 1.

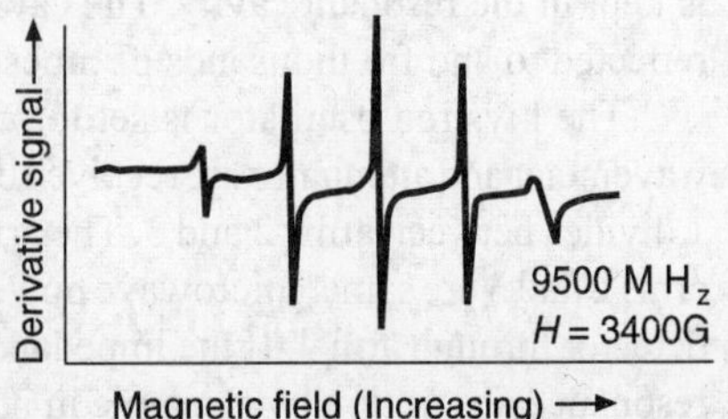

Fig. 5.16.

LASERS

5.12. Three Level Laser System

Consider a three energy level active material (Fig. 5.17).

The atom has three different energy levels. E_1 is ground state energy. E_2 is the energy of metastable state. E_3 is the energy of excited state.

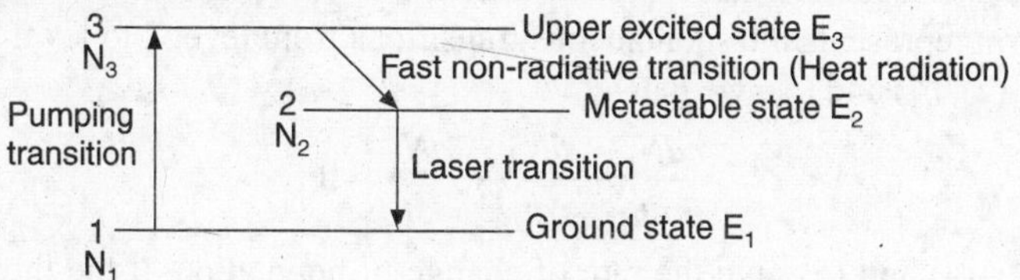

Fig. 5.17.

(*i*) By supplying energy (by means of optical pumping), a large number of atoms are raised from ground state level E_1 to upper excited energy level E_3, where their lifetime is extremely small ($\sim 10^{-8}$ s).

(*ii*) As the lifetime in level 3 is extremely small, the excited atoms decay by non-radiative transition to the metastable state having energy E_2. The decayed atoms remain in the metastable state for a considerably longer time ($\sim 10^{-4}$ to 10^{-2} s). Thus, the population of atoms in level 2 increases and that in the level 1 decreases. So population inversion is achieved.

(*iii*) Now by stimulated emission, the transition $E_2 \to E_1$ is achieved. This is laser-transition between the two levels.

- **Ruby laser** is a very good example of this scheme.

Rate Equation for Three-level Laser System

N_1, N_2 and N_3 are the number of atoms per unit volume in energy levels E_1, E_2 and E_3 respectively. We assume that only these three levels are populated and the transitions take place only between these three levels.

The total number of atoms per unit volume,

$$N = N_1 + N_2 + N_3 = \text{a constant} \quad ...(1)$$

Population of level 3 changes due to induced absorption from level 1 to level 3 (*i.e.*, $1 \to 3$), stimulated emission from level $3 \to 1$ and non-radiative transistion from level $3 \to 2$.

Therefore, the rate of change of population of level 3 is given by

$$\frac{dN_3}{dt} = W_p(N_1 - N_3) - N_3 T_{32} \quad ...(2)$$

Here, W_p = probability of induced absorption for transition $1 \to 3$ and stimulated emission for transition $3 \to 1$;

T_{32} = probability for transition $3 \to 2$.

Further
$$T_{32} = A_{32} + S_{32} \quad ...(3)$$

Here, A_{32} is Einstein's A - coefficient corresponding to a radiative transition rate and S_{32} is the non-radiative transition rate from level 3 to level 2.

Similarly, the rate of change of population of level 2 is given by

$$\frac{dN_2}{dt} = W_i(N_1 - N_2) + N_3 T_{32} - N_2 T_{21} \quad ...(4)$$

Here, first term represents the net stimulated transition between levels 1 and 2; the second term represents the spontaneous transition from level 3 to level 2; the last term represents the spontaneous transition from level 2 to level 1. The quantity W_i is proportional to Einstein's coefficient B_{21} and the energy density associated with lasing transition $2 \rightarrow 1$. If this transition is predominantly radiative, then $T_{21} \propto A_{21}$, where A_{21} is Einstein's coefficient.

Similarly, the rate of change of population of level 1 is given by

$$\frac{dN_1}{dt} = W_p(N_3 - N_1) + W_i(N_2 - N_1) + N_2T_{21} \quad ...(5)$$

Here, the first and second terms represents the stimulated transitions to level 1 from level 3 and level 2; the last term represents the spontaneous transitions from level 2 to level 1.

Adding Eqs. (2), (4) and (5), we get

$$\frac{dN_1}{dt} + \frac{dN_2}{dt} + \frac{dN_3}{dt} = 0 \quad ...(6)$$

Equations (2), (4) and (5) give the rate of change of population of the three levels. They are called the rate equations.

In the steady state, the number of atoms in any level remains constant. *i.e.*,

$$\frac{dN_1}{dt} = 0, \frac{dN_2}{dt} = 0 \text{ and } \frac{dN_3}{dt} = 0.$$

Taking $\frac{dN_3}{dt} = 0$, Eq. (2) gives

$$N_3 = \frac{W_pN_1}{W_P + T_{32}} \quad ...(7)$$

Similarly, taking $\frac{dN_2}{dt} = 0$, Eq. (4) gives

$$N_2 = \frac{W_iN_1 + T_{32}N_3}{W_i + T_{21}}$$

Using Eq. (7), we get

$$N_2 = \left(W_i + \frac{T_{32}W_P}{W_p + T_{32}}\right) \cdot \frac{N_1}{W_i + T_{21}} \quad ...(8)$$

The population difference between levels 1 and 2 is

$$N_2 - N_1 = \left(W_i + \frac{T_{32}W_p}{W_p + T_{32}}\right) \frac{N_1}{(W_i + T_{21})} - N_1$$

$$N_2 - N_1 = N_1 \left[\frac{W_p(T_{32} - T_{21}) - T_{32}T_{21}}{(W_p + T_{32})(W_i + T_{21})}\right] \quad ...(9)$$

The total number of atoms in the three levels is

$$N = N_1 + N_2 + N_3$$

$$= N_1 + \left(W_i + \frac{T_{32}W_p}{W_p + T_{32}}\right) \frac{N_1}{(W_i + T_{21})} + \left(\frac{W_p}{W_p + T_{32}}\right) N_1$$

$$N = N_1 \left[\frac{3W_pW_i + 2W_pT_{21} + 2T_{32}W_i + T_{32}W_p + T_{32}T_{21}}{(W_p + T_{32})(W_i + T_{21})}\right] \quad ...(10)$$

The ratio of the population difference between the levels 2 and 1 and the total number of atoms is

$$\frac{N_2 - N_1}{N} = \frac{W_p(T_{32} - T_{21}) - T_{32}T_{21}}{3W_pW_i + 2W_pT_{21} + 2T_{32}W_i + T_{32}W_p + T_{32}T_{21}} \quad ...(11)$$

From Eq. (11), it is obvious that for population inversion between levels 1 and 2 *i.e.* for $N_2 > N_1$

$$T_{32} > T_{21} \quad ...(12)$$

The relaxation times of atoms in levels 3 and 2 are inversely proportional to the corresponding relaxation rates. Therefore, in accordance with Eq. (12), for achieving population inversion, the life time of level 2 must be greater than the life time of level 3 *i.e.*, energy state 2 must be metastable state.

In addition to this condition, in order to achieve population inversion, a minimum pump power is required. For this, the minimum value of W_p from Eq. (11) is given by

$$(W_p)_{\min}(T_{32} - T_{21}) - T_{32}T_{21} = 0$$

$$\Rightarrow \quad (W_p)_{\min} = \frac{T_{32}T_{21}}{(T_{32} - T_{21})} \quad ...(13)$$

Thus, for population inversion, the following two conditions must be satisfied:

(*i*) The level 2 must be a metastable state.

(*ii*) $W_p > \dfrac{T_{32}T_{21}}{(T_{32} - T_{21})}$.

5.13. Ruby Laser

It is a solid state laser. It uses three energy level scheme of population inversion.

Construction: Fig 5.18 shows the schematic of a ruby laser. It consists of three parts:

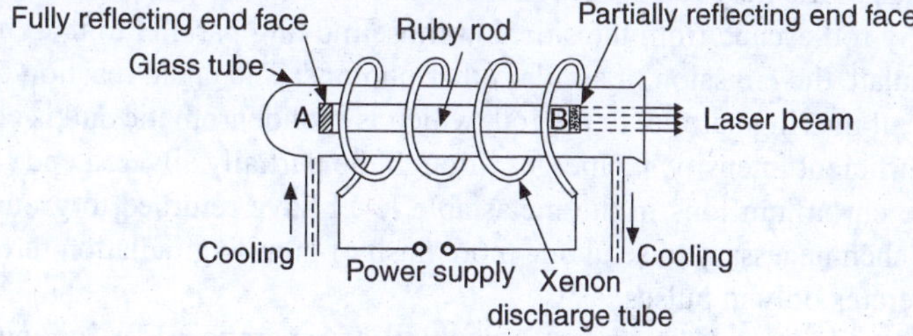

Fig. 5.18.

(i) Active working material: It is a rod of ruby crystal. Ruby is a crystal of aluminium oxide (Al_2O_3) with some Al atoms replaced by chromium (Cr) atoms. The active material in the ruby is Cr^{+++} ions. When ruby crystal contains about 0.5% of chromium, its colour is pink. In a ruby laser, a pink rod of 4 cm length and 0.5 cm in diameter is generally used.

(ii) Resonant cavity: Flat end faces *A* and *B* of the rod are made strictly parallel, plane and silvered. The end face *A* is silvered to achieve 100% reflection. The end face *B* is silvered to give 90% reflection and 10% transmission. The silvered faces constitute the Fabry-Perot resonator.

(iii) The optical pumping system: The ruby rod is arranged along the axis of a helical xenon flash tube. The coils of the tube surround the rod. The flash tube is provided with a power supply. The xenon flash tube provides the pumping light to raise the chromium ions to upper energy level. The tube is made to flash for several milliseconds. Only a small part of energy is used in pumping *i.e.*, in exciting the Cr^{+++} ions while the rest heats up the apparatus. The ruby rod is surrounded by a cylindrical glass tube through which liquid nitrogen or water circulates to keep the rod cool.

Working: Fig. 5.19 shows the energy level diagram of chromium ions.

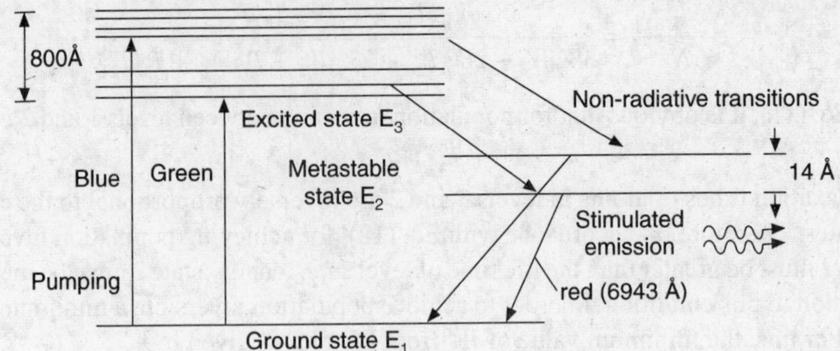

Fig. 5.19.

(*i*) When the ruby crystal is irradiated with white light of Xenon flash, the chromium ions are excited from the ground state E_1 to the wide-band excited state E_3 by absorption of the green and blue components of white light. The energy level E_3 has a band width 800 Å. The energy levels in these bands have a lifetime $\approx 10^{-9}$ s.

(*ii*) The excited ions quickly undergo nonradiative transitions to the level E_2. The E_2 level is a metastable state with a lifetime of about 3×10^{-3} s. The metastable state E_2 is a *doublet*, of separation 14Å. In a non-radiative transition, the energy is transferred to the crystal lattice, increasing its lattice energy. The metastable state E_2 has relatively longer lifetime than the lifetime of E_3 level. Therefore, Cr^{3+} ions accumlate at E_2 level. Now, the population of the E_2 level becomes greater than that of the E_1 level. Thus "population inversion" is achieved.

(*iii*) Some photons are produced by spontaneous transition from E_2 to E_1, and have a wavelength of 6943Å. The ends of the ruby rod act as reflecting mirrors. Therefore, photons that are not moving parallel to the ruby rod escape from the side, but those moving parallel to it are reflected back and forth. These stimulate the emission of similar other photons. The chain reaction quickly develops a beam of photons all moving parallel to the rod, which is monochromatic and is coherent. When the beam develops sufficient intensity, it emerges through the partially silvered end.

Once all the chromium ions in the metastable level have returned to ground level, the laser action stops. It is then necessary to send one more flash of pumping radiation through the rod. Thus the ruby laser operates only in pulses.

• The metastable state E_2 is in fact a doublet, of separation 14Å, giving rise to two lines $\lambda_1 = 6943$ Å and $\lambda_2 = 6929$ Å in the laser output. But the line 6943 Å predominates.

5.14. Four-level Laser System

Figure 5.20 shows a four-level laser system.

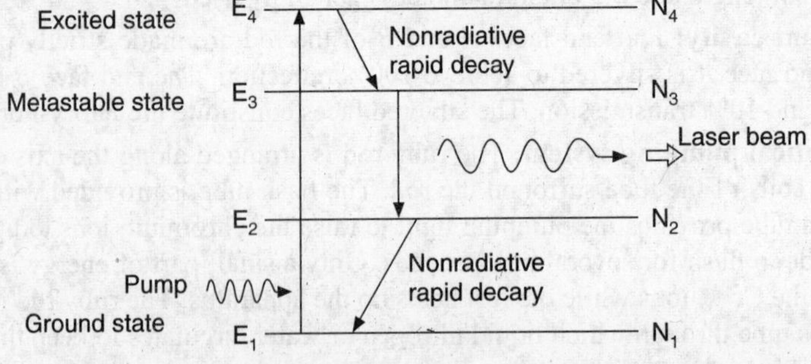

Fig. 5.20.

E_1, E_2, E_3 and E_4 are the energies of levels. E_1 is ground state energy. E_2, E_3 and E_4 are the excited levels. E_3 is meta-stable state.

(*i*) The pumping source carries some of the atoms from ground level E_1 to the higher excited level of energy E_4.

(*ii*) There is a rapid non-radiative decay from level E_4 to level E_3.

(*iii*) The life time of excited stas E_2 and E_4 is about 10^{-8} seconds. The lifetime of level E_3 is about 10^{-3} seconds. So E_3 is metastable state. As E_3 is a metastable state, there is a lasing action between E_3 and E_2.

(*iv*) There is a non-radiative rapid decay from level E_2 to E_1.

- **Population Inversion** is achieved between levels E_2 and E_3. E_2 is not the ground state. Usually energy level E_2 is much above the ground energy level E_1. However it so happens that the transitions $E_4 \rightarrow E_3$ and $E_2 \rightarrow E_1$ are much faster as compared to $E_3 \rightarrow E_2$. Hence it is easy to achieve population inversion with a four level system than with a three level system.
- As the pumping is from E_1 to E_4 level and laser transition is from E_3 to E_2 level, it is convenient to operate 4-level lasers in continuous-wave (C–W) mode.
- Examples of four level lasers are Helium-Neon (He–Ne) and Nd-YAG lasers.

Rate Equation for Four-level Laser System

N_1, N_2, N_3 and N_4 are the number of atoms per unit volume in energy states E_1, E_2, E_3 and E_4 respectively. Then

$$N = N_1 + N_2 + N_3 + N_4 = \text{constant} \qquad \text{...(1)}$$

Let W_p be the probability of stimulated transitions between levels 1 and 4.

The rate of change of population per unit volume of level 4 is given by

$$\frac{dN_4}{dt} = W_p(N_1 - N_4) - T_4 N_4 \qquad \text{...(2)}$$

Here, $W_p(N_1 - N_2)$ = net rate of stimulated transitions occurring between levels 1 and 4 caused by pumping;

$T_4 N_4$ = rate at which atoms decay from level 4.

T_4 = net relaxation rate from level 4 to any lower level (1, 2 and 3) *i.e.*,

$$T_4 = T_{43} + T_{42} + T_{41} \qquad \text{...(3)}$$

Usually T_{43} is much greater than T_{42} and T_{41}. So most of the atoms pumped to level 4 drop down to level 3.

The rate of change of population of level 3 is given by

$$\frac{dN_3}{dt} = T_{43}N_4 + W_i(N_2 - N_3) - T_3 N_3 \qquad \text{...(4)}$$

Here, $T_{43}N_4$ = rate at which the atoms jump from level 4 to level 3;

$W_i(N_2 - N_3)$ = rate of stimulated transition from level 2 to level 3 due to presence of laser radiation;

$T_3 N_3$ = rate of loss of atoms from level 3 to levels 2 and 1 through spontaneous transitions.

$$T_3 = T_{32} + T_{31}$$

Similarly, we can write the rate equations of levels 2 and 1 as

$$\frac{dN_2}{dt} = T_{42}N_4 + W_i(N_3 - N_2) - T_{21}N_2 + T_{32}N_3 \qquad \text{...(5)}$$

and

$$\frac{dN_1}{dt} = W_p(N_4 - N_1) + T_{41}N_4 + T_{31}N_3 + T_{21}N_2 \qquad \text{...(6)}$$

Adding Eqs. (2), (4), (5) and (6), we get

$$\frac{dN_1}{dt} + \frac{dN_2}{dt} + \frac{dN_3}{dt} + \frac{dN_4}{dt} = 0 \qquad ...(7)$$

In the steady state, the number of atoms in any level remains constant.

$$\therefore \qquad \frac{dN_1}{dt} = \frac{dN_2}{dt} = \frac{dN_3}{dt} = \frac{dN_4}{dt} = 0 \qquad ...(8)$$

Using $\frac{dN_4}{dt} = 0$, Eq. (2) gives

$$W_p(N_1 - N_4) - T_4 N_4 = 0$$

$$W_p(N - N_2 - N_3 - N_4 - N_4) - T_4 N_4 = 0 \qquad \text{[using Eq. 1]}$$

$$\therefore \qquad N_4 = \frac{W_p(N - N_2 - N_3)}{(2W_p + T_4)} \qquad ...(9)$$

Now putting $\frac{dN_3}{dt} = 0$, $\frac{dN_2}{dt}$ and $\frac{dN_1}{dt} = 0$, we get

$$T_{43}N_4 + W_i(N_2 - N_3) - T_3 N_3 = 0 \qquad ...(10)$$

$$T_{42}N_4 + W_i(N_3 - N_2) - T_{21}N_2 + T_{32}N_3 = 0 \qquad ...(11)$$

and $$W_p(N_4 - N_1) + T_{41}N_4 + T_{31}N_3 + T_{21}N_2 = 0 \qquad ...(12)$$

Solving these equations using equations (1) and (9), we get

$$\frac{N_3 - N_3}{N} = \frac{T_{43}(T_{21} - T_{32}) - T_{43}T_3}{T_{43}(T_{32} + T_{21}) + T_3 T_{42} + PT_{21}T_3 + W_i[2(T_{43} + T_{42}) + P(T_{21} + T_{31})]} \qquad ...(13)$$

Here, $$P = \left(\frac{2W_p + T_4}{W_p}\right)$$

In most four-level lasers, the atoms from level 4 go primarily to level 3 than to levels 2 and 1 and hence, $T_{42} << T_{43}$ and $T_{41} << T_{43}$.

Also the atoms from level 3 go to level 2 *i.e.*,

$$T_3 \approx T_{32}$$

Under these circumstances, we see that in order to obtain population inversion $N_3 > N_2$, we must have $T_{21} > T_{32}$. This means that the rate of decay of atoms from level 2 to 1 must be greater than the rate of decay from level 3 to 2.

5.15. Helium-Neon Laser

The He-Ne laser is a gas laser. It is a four-level laser in which the population inversion is achieved by electric discharge.

Construction: Fig. 5.21 shows the schematic of a He-Ne laser.

(i) Active Medium: The Helium-Neon laser system consists of a gas discharge tube made up of quartz. The tube is filled with the mixture of Neon under a pressure of 0.1 mm of mercury and Helium under a pressure of 1 mm of mercury. The ratio of the Helium-Neon mixture is 10:1 *i.e.*, the number of Helium atoms is greater than the number of Neon atoms. Neon atom are the active centers and have energy levels suitable for laser transitions. Helium atoms help in exciting neon atoms.

(ii) Resonant Cavity: The resonant cavity is a quartz tube of about 0.8 m length and 5 mm diameter. The end faces of the discharge tube are tilted at the *Brewster angle*. These are called *Brewster* *angle windows*. On the axis of the tube, two concave mirrors are arranged externally which form the Fabry-Perot optical resonator. The distance between the mirrors is equal to an integral number of half-wavelengths of the laser light.

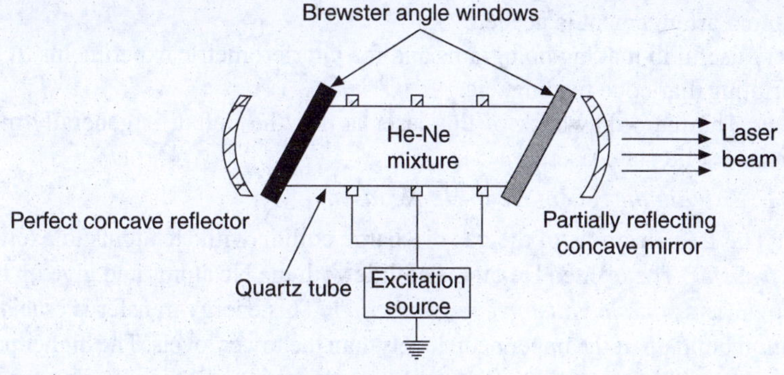

Fig. 5.21.

(iii) Pumping System: An *exciting source* for creating a discharge in the tube, is a radio frequency high voltage source such as a Tesla coil. The radio frequency oscillator produces electrical discharge in the Helium-Neon mixture and pumps the Helium atoms to higher energy states.

Working: Fig. 5.22 shows the energy level diagram.

(*i*) When electromagnetic energy is injected into the tube by means of radio frequency high voltage source, helium atoms get excited to metastable states. One of the excited levels of helium is at 20.61 eV above the ground level.

(*ii*) One of the excited levels of neon is at 20.66 eV. Some of the excited He atoms transfer their energy to ground-state Ne atoms by collisions. Such an energy transfer can take place when the two colliding atoms have approximately identical energy levels. It is known as *resonant energy transfer*.

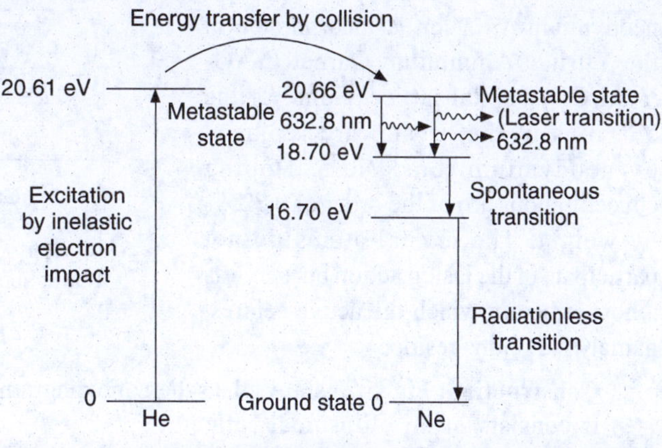

Fig. 5.22.

(*iii*) When an excited Ne atom passes *spontaneously* from the metastable state at 20.66 eV to state at 18.70 eV, it emits a 632.8 nm photon.

This photon stimulates an excited Ne atom and causes it to emit a fresh 632.8 nm photon in phase with the stimulating photon. This stimulated transition from 20.66 eV level to 18.70 eV level is the *laser transition*. This process is continued and a beam of coherent radiation builds up in the tube. When this beam becomes sufficiently intense, a portion of it escapes through the partially-silvered end.

(*iv*) From the 18.70 eV level, the Ne atom passes down spontaneously to a lower metastable state (16.70 eV), emitting incoherent light.

(*v*) Finally, from the 16.70 eV level, the Ne atom passes down to the ground state through collision with the tube walls. The final transition is thus raditionless.

Because the electron impacts that excite the He and Ne atoms occur all the time a He-Ne laser operates continuously.

Merits:

1. The He-Ne laser is more unidirectional and more monochromatic than solid state lasers.
2. It is inexpensive and has high stability of frequency.
3. No cooling arrangement is needed.
4. It is very useful in making holograms and for interferometric experiments. It is also widely used in laser printing, bar code reading, etc.

Drawback: The main drawback of this laser is that the output is generally moderate when compared to that of solid state lasers.

Example 1 : *Explain the role of He in He-Ne laser.*

Solution : The free electrons of the gas discharge collide with He atoms to excite them by the *impact energy transfer*. The excited He-atoms collide with the Ne atoms and give up their energy in the process known as *resonant collision energy transfer*. This energy transfer is crucial for the more intense population build up of the upper neon levels than the lower levels. The higher partial pressure of the He in the gas mixture, contributes favourably to this process, by decreasing the probability of energy transfer in the reverse direction *i.e.*, from Ne $\rightarrow$ He.

5.16. Nd: YAG Laser

This is a four-level solid state laser system. In this system, the trivalent neodymium Nd^{3+} ion is incorporated in the Yttrium Aluminium Garnet (YAG-$Y_3Al_5O_{12}$) crystal lattice. Some of the Y^{3+} ions in the crystal are replaced by neodymium ions, Nd^{3+}. Doping concentrations are of the order of 0.725% by weight. The crystal atoms do not participate in the lasing action but serve as a host lattice in which the active centres, namely Nd^{3+} ions reside.

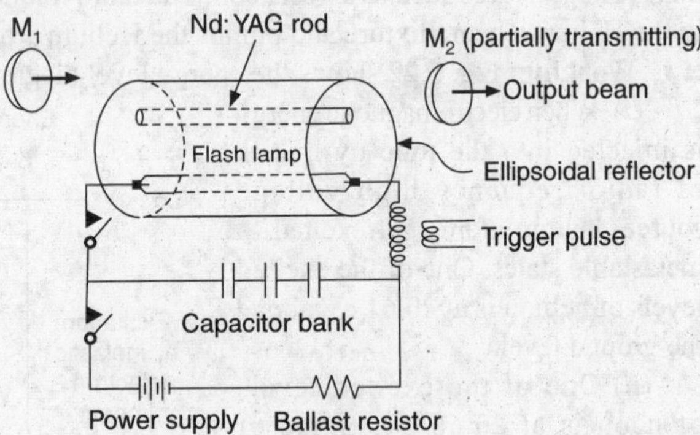

Fig. 5.23.

Construction: Fig. 5.23 shows the schematic diagram of a Nd : YAG laser.

It consists of an ellipsoidal reflector housing the laser rod along one of its focus lines and a flash lamp along the other focus line. The YAG crystal rods are of 10 cm in length and 12 mm in diameter. The light leaving one focus of the ellipse will pass through the other focus after reflection from the silvered surface of the reflector. Thus the entire flash lamp radiation gets focussed on the laser rod (Fig. 5.24).

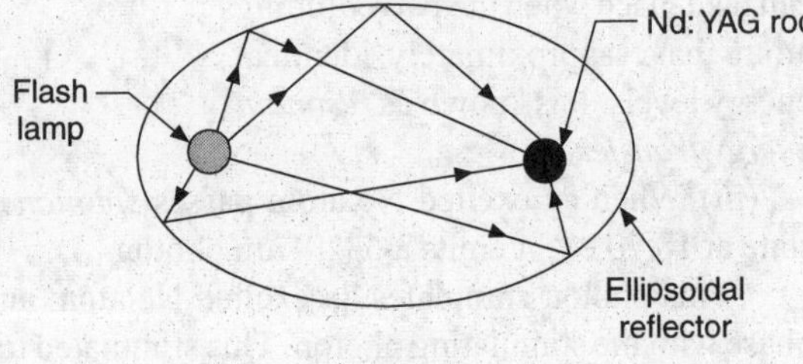

Fig. 5.24.

The optical cavity is formed by using two external reflecting mirrors M_1 and M_2. The system is cooled by either air or water circulation.

Working: Fig. 5.25 shows the energy level diagram for the neodymium ion in YAG crystal.

1. The Neodymium atoms are pumped by a krypton flash lamp, from the ground state to the higher energy levels $4F_{5/2}$ and $4H_{9/2}$ by absorption of radiations of wavelength 0.73 μm and 0.80 μm respectively.

2. Transition from these energy levels to the level $4F_{3/2}$ takes place by non-radiative processes. $4F_{3/2}$ is a metastable state.

Population inversion takes place at $4F_{3/2}$ with respect to $4I_{11/2}$ which is virtually empty at room temperature.

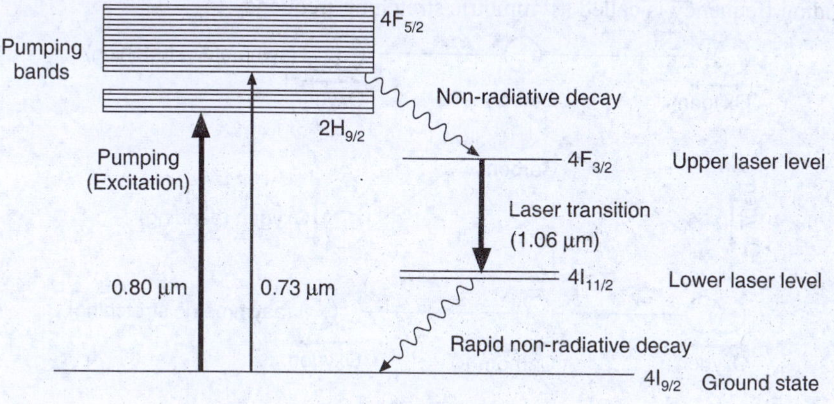

Fig. 5.25.

3. Stimulated emission takes place between $4F_{3/2}$ and $4I_{11/2}$. The laser transition $4F_{3/2} \rightarrow 4I_{11/2}$ has a wavelength 1.064 μm.

4. The transition from $4I_{11/2}$ to ground state $4I_{9/2}$ is non-radiative with a life time of 30 ns. However, this radiation is strongly absorbed by the lattice of the host (YAG crystal) and the resulting energy is converted into heat. As YAG has very good thermal conductivity, the heat energy is removed by conduction. Nd-YAG laser can be operated in CW or pulsed mode.

- The size of Nd-YAG laser is compact and portable. It can be contained in a quarter portion of a brief case.
- The limitation of Nd-YAG laser is the growth of crystals. Only small rods of size 10 cm in length and 1 cm diameter may be grown.

Applications:

1. Nd YAG lasers are widely used in the industries for cutting, welding, drilling and surface hardening of the industrial products.

2. YAG lasers are used in military as range finders and target designators.

3. These lasers are finding increasing uses in the field of medicine. They are used in surgery. 1.064 mm wavelength radiation passes through optical fibre without absorption. So fibre optic endoscopes with YAG lasers are used to treat gastrointestinal bleeding.

5.17. Carbon Dioxide Laser

Principle: It is a four-level molecular laser. The active medium in CO_2 laser is a gas mixture consisting of CO_2, N_2 and He. In a CO_2 laser, one uses the transitions occurring between different vibrational states of the carbon dioxide molecule.

Fig. 5.26 shows the three independent modes of vibration of the carbon dioxide molecule. The state of vibration of the molecule is represented by three quantum numbers ($m\ n\ q$).

(*i*) In the *symmetric stretching* mode, the oxygen atoms oscillate along the axis of the molecule, by simultaneously departing and approaching the carbon atom in between. The corresponding frequency is called symmetric stretching frequency.

(*ii*) In the *bending mode*, all the three atoms of the molecule undergo vibrational oscillations by moving perpendicularly to the molecular axis. The corresponding frequency is called bending frequency.

(*iii*) In the *asymmetrical stretching mode*, the oxygen atoms also move along the axis of the molecule, but both in one direction relative to the carbon atom itself moving to the other side. The corresponding frequency is called asymmetric stretching frequency.

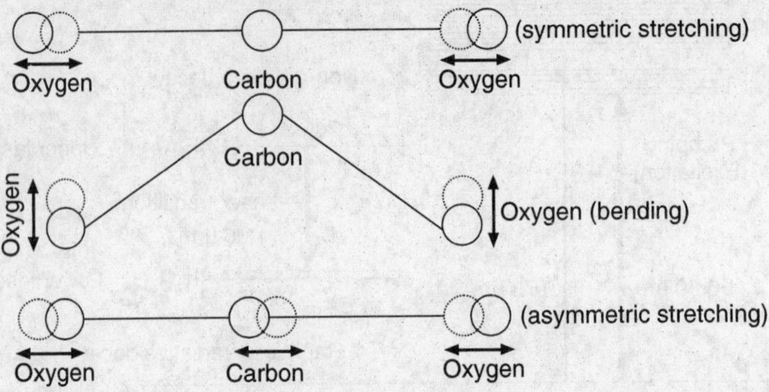

Fig. 5.26.

Construction: Fig. 5.27 shows the schematic diagram of a CO_2 laser. The discharge tube has 2.5 cm diameter and 5 m length. Discharge is produced by d.c. excitation. Sodium chloride Brewster windows are used at the end. The system is cooled by water circulation. The discharge tube is filled with a mixture of carbon dioxide, nitrogen and helium gases in 1 : 2 : 3 proportions respectively. Other additives such as water vapour are also added. The active centres are CO_2 molecules lasing on the transitions between the vibrational levels of the electronic ground state. It is a four level molecular laser.

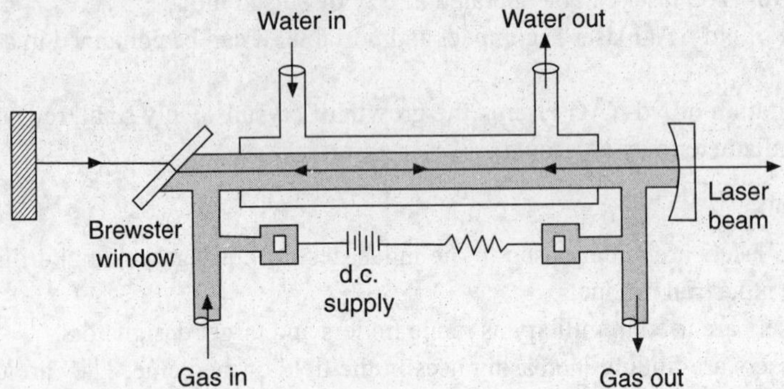

Fig. 5.27.

Working: Fig. 5.28 shows the various vibrational energy levels taking part in the laser transition. E_5 refers to (001) vibrational level of CO_2 molecule. The excited state of an N_2 molecule is metastable and it is identical in energy to E_5.

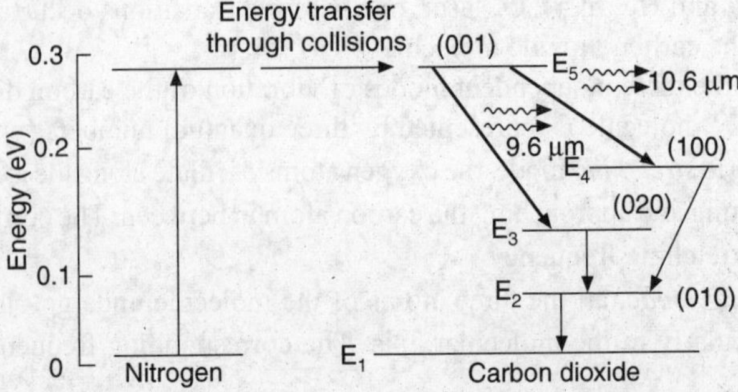

Fig. 5.28.

(*i*) As current passes through the mixture of gases, the N_2 molecules get excited to the metastable state. The excited N_2 molecules return to ground state through inelastic collisions with ground state CO_2 molecules. Thereby the CO_2 molecules are excited to E_5 level.

(*ii*) The E_5 level is the upper laser level. E_3 and E_4 levels act as the lower laser levels. As the population of CO_2 molecules builds up at E_5 level, population inversion is achieved between E_5 level and the levels at E_4 and E_3.

(*iii*) The lasing transition between $E_5 \rightarrow E_4$ levels produces for IR radiation at the wavelength 10.6 μm.

(*iv*) The lasing transition between $E_5 \rightarrow E_3$ levels produces far IR radiation at 9.6 μm. Helium gas is used to depopulate the lower levels.

The CO_2 laser operates in CW mode and is capable of generating high powers of the order of several kilowatts at a relatively high efficiency of about 40%. Therefore, it is the most widely used laser. Its applications include use in communications, weaponry and laser fusion.

Applications:

1. High power CO_2 lasers find applications in materials processing, welding, hole drilling, cutting, etc.

2. Since the atmospheric attenuation is low at 10.6 μm, CO_2 lasers are used in open air communication.

3. It is used in surgery as a direct substitute for a scalpel. In cutting tissues with CO_2 laser beam, the laser light is absorbed in a small volume, causing a little damage to the surrounding tissues. It is a bloodless and precise one. Along with endoscopes, CO_2 lasers are widely used in angioplasty, Gallstone lithotrity, neurosurgery and urology.

4. Laser radar or Lidar (**L**ight **d**etection **a**nd **r**anging) has proved to be a powerful tool for investigation of a variety of atmospheric features. Mostly pulsed laser type is used. Atmospheric pollutants' concentration, ozone concentration, water vapour concentration, etc can be measured. Thus laser remote sensing plays an important role in atmospheric optics.

5.18. Argon Ion Laser

Principle: It is a four level laser. It uses argon ions as an active medium. Argon has 18 electrons with electronic configuration $1s^2 2s^2 2p^6 3s^2 3p^6$. Argon is ionized in the discharge tube by collision with fast moving electrons. Since argon is noble gas atom, it requires large amount of energy for its ionization (≈ 35 eV). So the current density must be high.

The argon atoms are ionized by electron collision in a high current discharge (≈ 15 to 50 A). The ions are excited by further electron collisions up to a group of energy levels (4p) some 35 eV above the atomic ground state. As the electron energies are only a few electron volts, the excitation must be the result of multiple collisions. A population inversion forms between the 4p levels and the 4s level, which is about 33.5 eV (Fig. 5.27) above the ground state so that a series of stimulated lines is emitted. These range in wavelength from 351 nm to 520 nm, although most of the energy is concentrated in the 488 nm and 514.5 nm lines.

Construction: Fig. 5.29 shows the schematic diagram of argon ion laser.

(i) Resonant Cavity. Argon ion laser consists of a resonant cavity (or tube), made of graphite or beryllium oxide having two windows at the ends placed at Brewster's angle. During operation, the positive ions tend to collect at the cathode and may eventually cause the discharge to be extinguished. To prevent this, a gas return path is provided between the cathode and anode to equalize the pressure. To dissipate the large amount of heat generated, the laser tube is water cooled. A series of metal disks act as heat exchangers. Holes in the center of these disks define the active region of the laser.

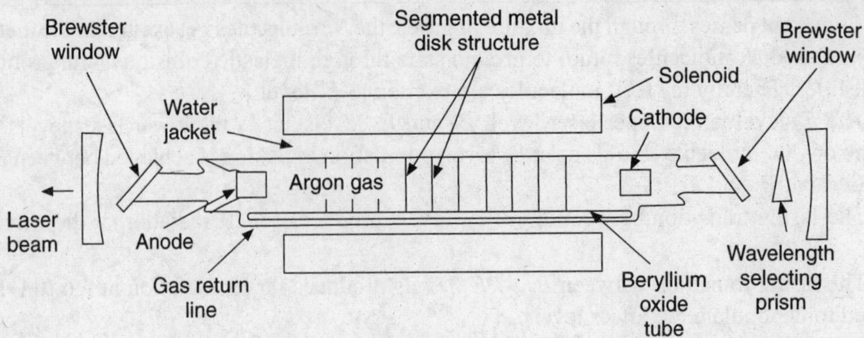

Fig. 5.29.

(ii) Pumping System. The discharge is initiated by a high voltage pulse and then maintained by d.c. voltage of about 200 V. The current density can be increased by concentrating the discharge with a magnetic field applied along the axis of the tube (the ions spiral about the magnetic lines of force). This has the added advantage of reducing the number of ions which collide with and damage the walls of the tube.

(iii) Active Medium. The active medium is argon gas. Gas pressure is 0.1 torr or less in bore region.

Working: Fig. 5.30 shows energy level diagram for the argon ion laser.

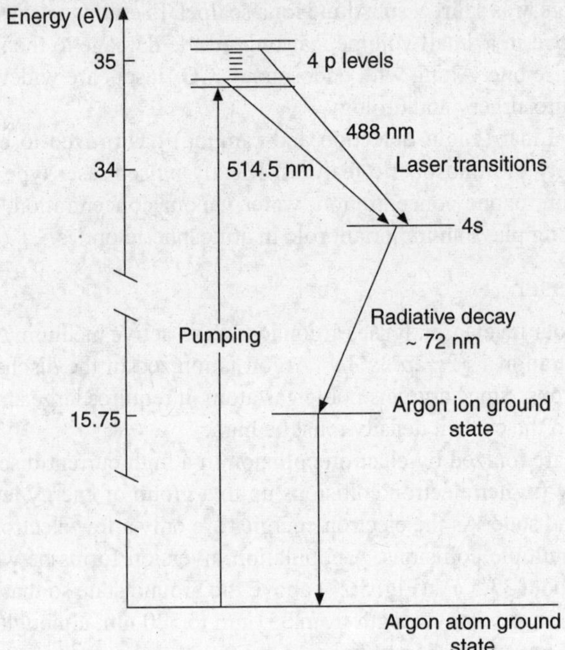

Fig. 5.30.

- When transitions occur from 4p to 4s level, the lasing action emits the two important wavelengths at 488 nm (blue) and 514.5 nm (green).

Selection of wavelength using a prism: To select any desired wavelength, a small prism is inserted into the cavity. The position of the end mirror is changed by rotating it to be normal to the path of the radiation with the desired wavelength. This ensures that radiation of this particular wavelength will be reflected to and fro. The radiation of other wavelengths will be lost from the cavity after only a few round trips.

Applications:

1. Pumping dye lasers. Argon ion lasers are very efficient pumping device for titanium-sapphire lasers and continuous wave dye lasers. The reason is that the wavelength of these lasers lie within the absorption region of these materials. The argon ion lasers have the power required to provide sufficient gain for pumping.

2. Laser printing. Argon ion laser are quite intense. Hence blue and green wavelengths of this laser are suitable for laser printing. Moreover, these lasers have excellent focusability and can be focused on tiny spots. The high intensity and focusing properties provide very fast rate of scanning, which is essential for laser printing.

3. Cell cytometry. Argon ion lasers are used in cell cytometry. This means, these laser beams can count different categories of living cells.

4. Photo-therapy of eye. Ionic lasers are used in phototherapy of eye.

5.19. Liquid Dye Laser

Absorption and Emission (fluorescence) Spectra of a Dye Laser

The *active medium* is an organic dye dissolved in a solvent. When the dye is excited by short wavelength light, it emits radiation at a longer wavelength, *i.e.*, it fluoresces. The energy difference between the absorbed and emitted photons ultimately appears as heat. Figure 5.31 shows absorption and emission spectra.

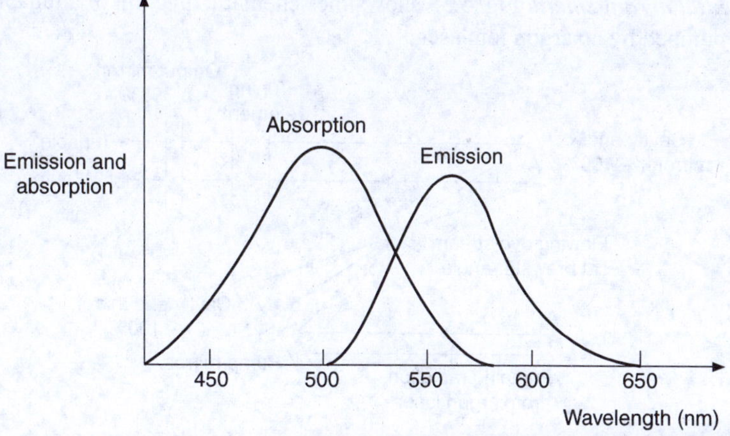

Fig. 5.31.

The broad fluorescence spectrum can be explained by the energy level diagram of a dye molecule.

Energy Level Diagram: Fig 5.32 shows the energy level diagram of a dye molecule.

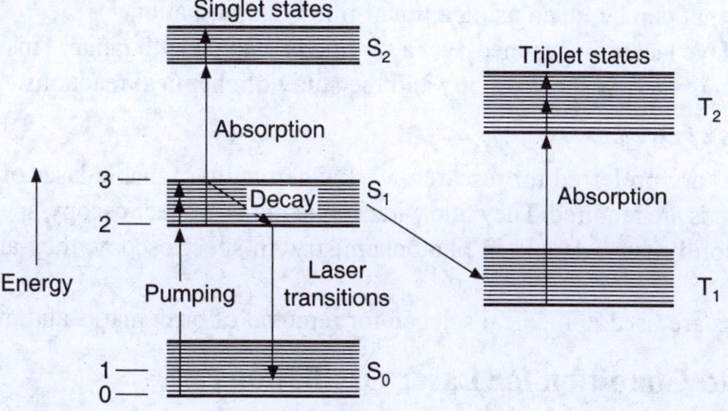

Fig. 5.32.

The molecule has two groups of closely spaced electronic energy levels; the singlet states (S_0, S_1 and S_2) and the triplet states (T_1 and T_2). Each electronic energy level is broadened into a near continuum of levels by the effects of the vibration and rotation of the dye molecule and also by the effects of the solvent molecules.

• Pumping results in the excitation of the molecule from the ground state S_0 to the first excited state S_1. This is followed by very rapid nonradiative decay processes to the lower of the energy levels in S_1. The laser transition is then from these levels to a level in S_0. Since there are many such rotational/vibrational levels within S_0 and S_1, there are many transitions resulting in an emission line which is very broad.

$S_1 \rightarrow T_1$ transitions lead to strong absorption at the laser wavelengths, thereby quenching laser action.

$S_1 \rightarrow S_2$ transitions may also quench laser action in some dyes.

Once a significant number of molecules have made the $S_1 \rightarrow T_1$ transition, $T_1 \rightarrow T_2$ absorption reduces the gain and may stop the lasing action. For this reason, most dye lasers operate in short pulses.

• The dye called rhodamine 6G with methanol as a solvent is the most commonly used laser medium. It has an efficiency of about 20% and a broad tuning range (570 to 660 nm).

• All dye lasers are optically pumped. The pumping source has a wavelength slightly less than that of the laser output. Commercial pumping methods include flashtubes, nitrogen lasers, solid state lasers and ion (A^+ or Kr^+) lasers.

Experimental arrangement. Fig. 5.33 shows the schematic diagram of a tunable CW laminar flow dye laser pumped by an argon ion laser.

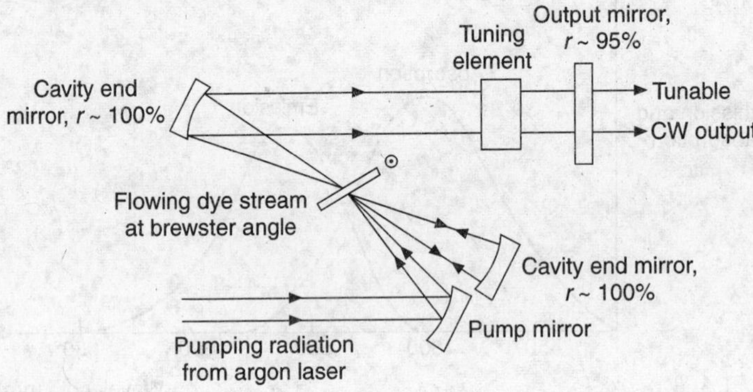

Fig. 5.33.

Here, the cavity has 3 reflectors. Two of these reflectors focus the beam into the jet stream, which is the thin flowing dye region. The jet stream is oriented at Brewster's angle. The dye stream flow is perpendicular to the page.

The laser output can be tuned using a prism-mirror combination.

Advantage. Dye laser can be tuned over a *significant* wavelength range. This is extremely useful in many applications such as spectroscopy and the study of chemical reactions.

Applications of Dye Lasers:

(*i*) Dye lasers are preferred for research work where either tunable laser of ultrafast pulses or tunable narrow bands are required. They find wide applications in spectroscopy, specially in absorption spectroscopy of solid materials and in photochemistry. In spectroscopy, they are used for isotope separation.

(*ii*) Dye lasers are used in medical science for removal of birth marks and treatment of cancer.

5.20. Threshold Condition for Laser Oscillations

Figure 5.34 shows the "*Basic laser structure.*"

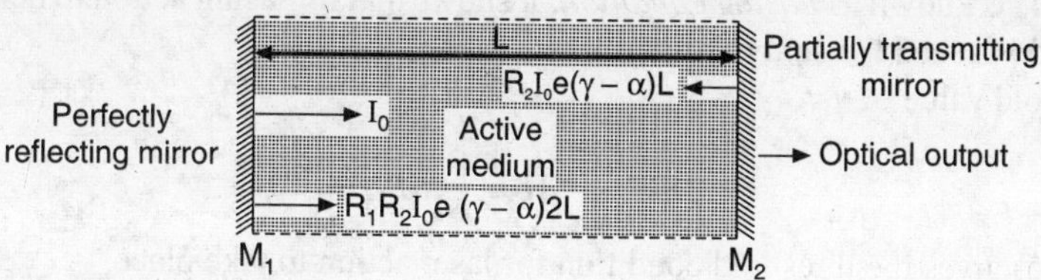

Fig. 5.34.

The laser medium fills the space between the mirrors M_1 and M_2 which have reflectances R_1 and R_2. The separation between the mirrors is L.

As light travels in an optical resonant cavity in the forward and backward direction, it undergoes amplification and also suffers various losses.

The total loss of the system is due to the following processes:

1. Transmission at the mirrors—the transmission from one of the mirrors usually provides the useful output, the other mirror is made as reflective as possible to minimize losses.
2. Absorption and scattering at the mirrors.
3. Absorption in the laser medium due to transitions other than the desired transitions.
4. Scattering at optical inhomogeneities in the laser medium—this applies particularly to solid state lasers.
5. Diffraction losses at the mirrors.

Let the intensity of light starting from the mirror M_1 to begin with be I_0.

Then, as the light reaches the mirror M_2, its intensity will be

$$I_L = I_0\, e^{(\gamma-\alpha)L} \qquad \text{...(1)}$$

Here, γ = amplification factor of the light beam per unit length of the resonator cavity,

α = loss per unit length due to scattering, diffraction and absorption taking place as light travels through the medium in the cavity.

After reflection at mirror M_2, the intensity of light beam will become $R_2\, I_0\, e^{(\gamma-\alpha)L}$.

Again after returning from M_2 and reflection from M_1 *i.e.*, after one complete round trip the intensity of light will become

$$I_{2L} = R_1 R_2\, I_0\, e^{(\gamma-\alpha)2L} \qquad \text{...(2)}$$

$\therefore$ Amplification after one round trip is

$$G = \frac{I_{2L}}{I_0} = R_1 R_2\, e^{(\gamma-\alpha)2L} \qquad \text{...(3)}$$

For balancing all the losses, the amplification $G \geq 1$.

or $$R_1 R_2\, e^{(\gamma-\alpha)2L} \geq 1 \text{ or } e^{(\gamma-\alpha)2L} \geq \frac{1}{(R_1 R_2)}$$

Taking natural logarithm of both sides, we get:

$$2L(\gamma-\alpha) \geq -\ln(R_1 R_2)$$

or $$\gamma - \alpha \geq \frac{1}{2L}\ln(R_1 R_2)$$

or $$\gamma \geq \alpha - \frac{1}{2L}\ln(R_1 R_2)$$

or $$\gamma \geq \alpha + \frac{1}{2L}\ln\left(\frac{1}{R_1 R_2}\right) \qquad \text{...(4)}$$

Equation (4) is known as *Lasing condition*. It shows that for lasing action to be maintained, the initial gain should exceed the sum of all the losses in the cavity.

The threshold value of γ is

$$\gamma_{th} = \alpha + \frac{1}{2L} \ln\left(\frac{1}{R_1 R_2}\right) \qquad ...(5)$$

Equation (5) gives the threshold condition for laser action to take place.

5.21. Laser Modes

Examination of a laser output with a spectrometer of very high resolving power, such as the scanning Fabry-Perot interferometer, reveals that it consists of a number of discrete frequency components (or very narrow spectral lines). To appreciate how these discrete lines arise and how they are related to the laser transition lineshape, we need to examine the effects of the mirrors on the light within the laser cavity.

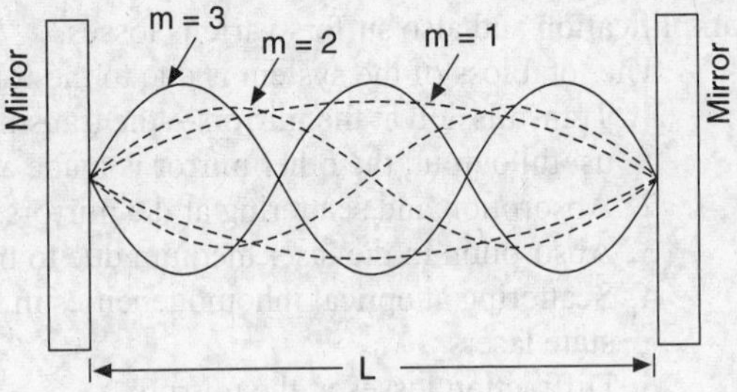

5.21.1. *Axial Modes*

The two mirrors of the laser for a resonant cavity. Standing wave patterns are set up between the mirrors in exactly the same way that standing waves develop on a string (Fig. 5.35).

Fig. 5.35.

The standing waves satisfy the condition

$$m\frac{\lambda}{2} = L \quad (m = 1, 2, 3, ...) \qquad ...(1)$$

The frequencies of the various modes are given by

$$\nu_m = \frac{c}{\lambda_m} = \frac{mc}{2L} \qquad ...(2)$$

The number m is called the *axial mode number*.

Each value of m satisfying Eq. (2) defines an *axial* (or longitudinal) *mode* of the cavity.

Example : If $L = 0.5$ m and $\lambda \simeq 500$ nm, then, $m \simeq 2 \times 10^6$.

From Eq. (2), the frequency separation $\delta\nu$ between adjacent modes is given by

$$\delta\nu = \nu_{m+1} - \nu_m = \frac{c}{2L}(m+1-m) = \frac{c}{2L} \qquad ...(3)$$

For $\quad L = 0.5$ m, $\delta\nu = 300$ MHz.

As Eq. (3) is independent of m, the frequency separation of adjacent modes is the same irrespective of their actual frequencies.

The modes of oscillation of the laser cavity will consist, therefore, of a large number of frequencies, each given by Eq. (2) and separated by $c/2L$. [Fig. 5.36 (b)].

While all the integers m give *possible* axial cavity modes, only those which lie within the gain curve or laser transition line [Fig. 5.36 (a)] will actually oscillate. The gain curve for the 632.8 nm wavelength emitted by neon is about 1.5×10^9 GHz wide. So with the 0.5 m long cavity in the above example, we would expect four or five modes to be present [Fig. 5.36 (c)].

- Axial modes are formed by plane waves traveling axially along the laser cavity on a line joining the centers of the mirrors.
- The length L of the resonant cavity governs the axial modes.

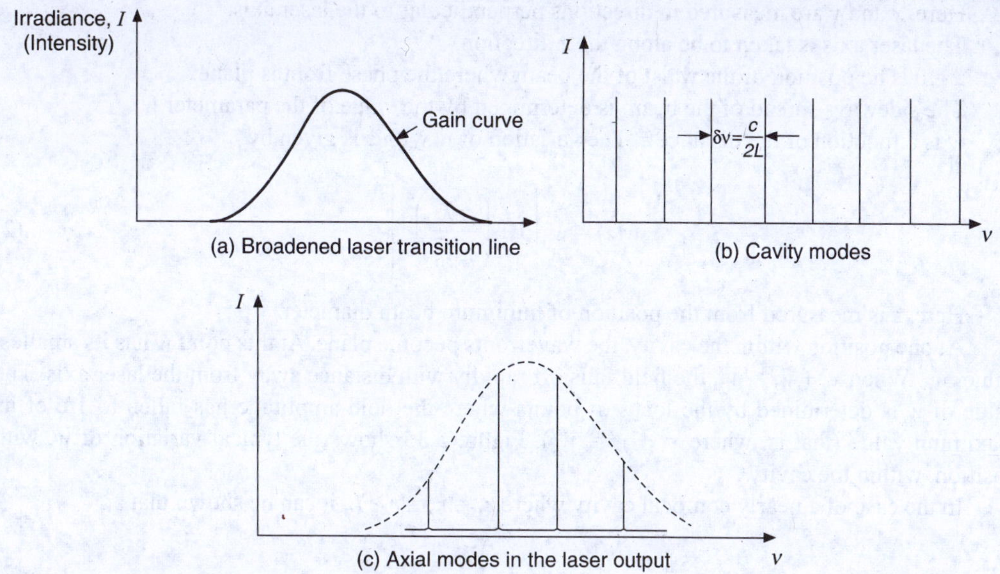

Fig. 5.36.

5.21.2. Transverse Modes

For any real laser cavity, there will be waves traveling just off-axis that are able to replicate themselves after covering a closed path (Fig. 5.37). These will also give rise to resonant modes. These modes have components of their electromagnetic fields which are transverse to the direction of propagation. Therefore, they are termed transverse electromagnetic (or TEM) modes.

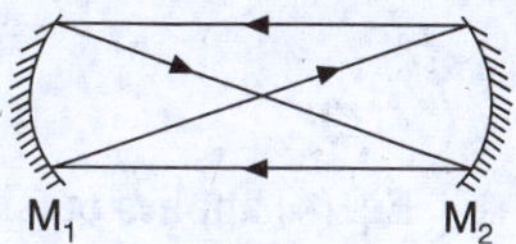

Fig. 5.37.

• The laser modes governed by the cross-sectional dimensions of the optical cavity are called Transverse Electromagnetic Modes (TEM). The transverse modes characterize the intensity distribution across the cross-section of the laser beam. The allowed modes are designated as TEM_{qr}. q and r are integers.

q gives the number of minima as the beam is scanned horizontally.

r gives the number of minima as the beam is scanned vertically.

The lowest order transverse mode is TEM_{00}.

Fig. 5.38 shows TEM_{00} mode within a laser cavity.

The cavity is formed by two concave mirrors of radius of curvature r_1 and r_2 separated by L. Such a cavity is stable when $L \le r_1 + r_2$.

The mode 'adjusts' itself so that the mirror surface are surfaces of constant phase.

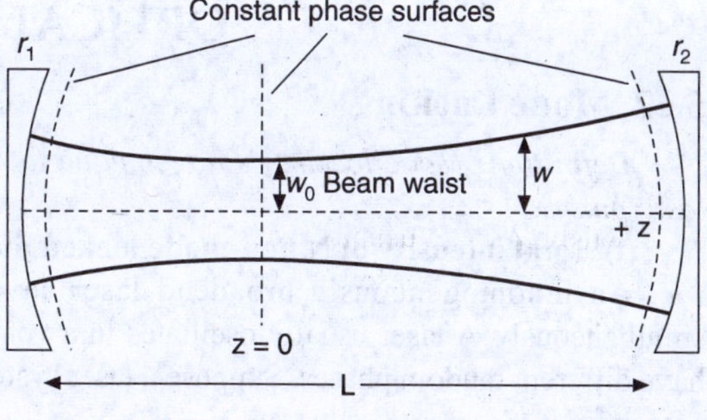

Fig. 5.38.

The irradiance distribution across the beam is Gaussian. So we may write the electric field variation as

$$\mathcal{E}(x, y) = \mathcal{E}_0 \exp\left(-\frac{x^2 + y^2}{w^2}\right) \quad ...(1)$$

Here, x and y are measured in directions perpendicular to the laser axis.
The laser axis is taken to be along the z direction.
$z = 0$ is he position of the waist of the beam where the phase front is plane.
The sideways spread of the beam is determined by the value of the parameter w.
w is a function of the distance z. The variation of w with z is given by

$$w(z) = w_0 \left[1 + \left(\frac{z\lambda}{\pi w_0^2} \right)^2 \right]^{\frac{1}{2}} \quad ...(2)$$

Here, z is measured from the position of minimum beam diameter.

At one position within the cavity, the wavefronts become plane. At this point w has its smallest value, w_0. When $x_2 + y_2 > w_2$, the field falls off rapidly with distance away from the laser axis. The value of w is determined by the locus of points where the field amplitude has fallen to 1/e of its maximum value (that is, where $x^2 + y^2 = w^2$). Figure 5.35 shows the typical variation of w, with position, within the cavity.

In the case of a nearly confocal cavity where $r_1 = r_2 = r \geq L$, it can be shown that

$$w_0^2 = \frac{\lambda r}{2\pi} \quad ...(3)$$

Combining Eq. (2) and (3) yields

$$w(z) = w_0 \left[1 + \left(\frac{2z}{r} \right)^2 \right]^{\frac{1}{2}} \quad ...(4)$$

Eq. (4) applies outside as well as inside the cavity.
At large distances from the laser, that is when $z \gg r$, we have

$$w(z) = \frac{z\lambda}{\pi w_0} \quad ...(5)$$

METHODS OF GENERATING HIGH ENERGY SHORT OPTICAL PULSES

5.22. Mode Locking

Definition: *Mode locking is a technique for producing periodic, high power, short duration laser pulses.*

(i) Total intensity of N non-mode locked modes

An inhomogeneously broadend laser cavity may support oscillations in many modes simultaneously. A laser usually osciliates in a very large number of *spectral modes*. These modes have different random phases. Suppose a laser system has a very large number (N) of spectral modes present in it.

The output of such a laser as a function of time depends on the relative phases, frequencies and amplitudes of the modes.

The total electric field as a function of time is given by

$$\mathcal{E}(t) = \sum_{n=0}^{N-1} (\mathcal{E}_0)_n \exp \{i(\omega_n t + \delta_n)\} \quad ...(1)$$

Here, $(\mathcal{E}_0)_n$, ω_n and δ_n are the amplitude, angular frequency and phase of the *n*th mode. Usually these parameters are all time varying. So the modes are incoherent. The total intensity is simply the sum of the intensities of individual modes *i.e.*,

$$\text{Total intensity } I = N\mathcal{E}_0^2 \quad ...(2)$$

Here, we have assumed that all N modes have the same amplitude $\mathcal{E}_0$.

The total intensity shows random fluctuations [Fig. 5.39 (a)].

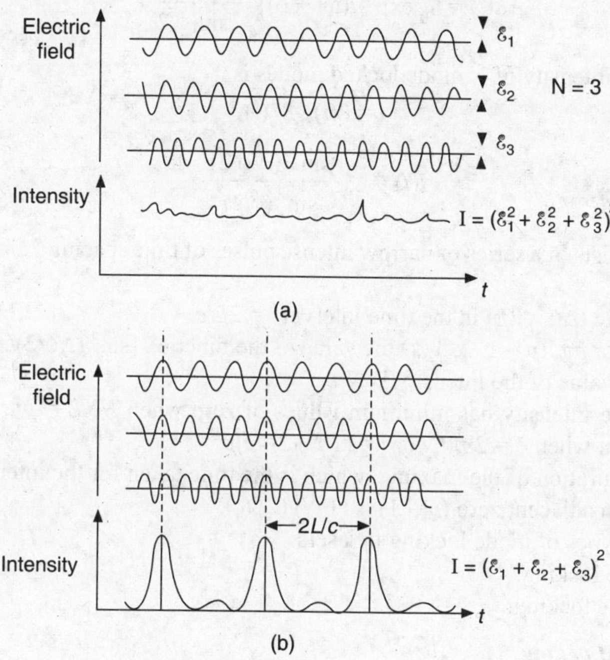

Fig. 5.39.

(ii) Total intensity of N mode-locked laser modes

Suppose that we now force the various modes to maintain the same relative phase δ to one another, *i.e.*, we *mode lock* the laser such that $\delta_n = \delta$.

The total intensity must now be found by adding the individual electric fields rather than the intensities.

The resultant electric field is given by

$$\mathcal{E}(t) = \mathcal{E}_0 \exp(i\delta) \sum_{n=0}^{N-1} \exp(i\omega_n t) \quad ...(3)$$

Let us write the angular frequency ω_n as

$$\omega_n = \omega - n\delta\omega$$

Here, ω = angular frequency of the highest frequency mode and

$\delta\omega$ = angular frequency separation between modes = $\pi \dfrac{c}{L}$

Equation (3) for $\varepsilon(t)$ can then be rewritten as

$$\mathcal{E}(t) = \mathcal{E}_0 \exp(i\delta) \sum_{n=0}^{N-1} \exp(i(\omega - n\delta\omega)t)$$

$$= \mathcal{E}_0 \exp\{i(\omega t + \delta)\} \sum_{n=0}^{N-1} \exp(-\pi inct/L)$$

$$\mathcal{E}(t) = \mathcal{E}_0 \exp\{i(\omega t + \delta)\}\,(1 + \exp(-i\phi) + \exp(-2i\phi) + ... + \exp\{-(N-1)i\phi\})$$

Here, $\phi = \pi ct/L.$

$$\mathcal{E}(t) = \mathcal{E}_0 \exp\{i(\omega t + \delta)\} \frac{\sin(N\phi/2)}{\sin(\phi/2)}$$

Hence the total intensity of N mode locked modes is

$$I = \mathcal{E}(t) \cdot \mathcal{E}^*(t)$$

$$\therefore \qquad I(t) = \mathcal{E}_0^2 \frac{\sin^2(N\phi/2)}{\sin^2(\phi/2)} \qquad ...(4)$$

The output consists of a series of narrow intense pulses of time spacing $2L/c$ and duration $(1/N)(2L/c)$ [Fig. 5.39 (b)].

- $I(t)$ is periodic ($\Delta\phi = 2\pi$) in the time interval $t = 2L/c$.
- When $\phi = 0$ or $p\pi$, ($p = 1, 2, 3$ ...) the value of the function [$\sin^2(N\phi/2)/\sin^2(\phi/2)$ equals N^2.

The maximum value of the intensity is $N^2\varepsilon_0^2$.

- Similarly, the intensity has minimum values of zero when $N\phi/2 = p\pi$, p being an integer which is not zero, *i.e.*, when $\phi = 2p\pi/N$ or $t = (1/N)(2L/c)p$.

Thus the time duration of the maxima, which is the time taken for the intensity to fall from its maximum value to an adjacent zero ($p = 1$) is $(1/N)(2L/c)$.

There are two types of mode locking in lasers:

(*i*) Active mode locking.

(*ii*) Passive mode locking.

5.22.1. Active Mode Locking

Mode locking is achieved by forcing the longitudinal modes to maintain fixed phase relationships. This is accomplished by modulating the loss (or gain) of the laser cavity at a frequency equal to the intermode frequency separation $\delta\nu = c/2L$.

Here, L is the separation between the reflectors of the cavity.

The intermode frequency spacing $\delta\nu = c/2L$ corresponds to a time period of $2L/c$ which is exactly one round trip time inside the cavity.

Figure 5.40 shows schematic arrangement for production of mode-locked pulses. A shutter is placed at one end of the cavity next to one of the mirrors. The loss modulation is provided by the shutter.

- The shutter is closed (corresponding to very high losses) most of the time.
- The shutter is opened for a very short interval of time every $2L/c$ seconds (corresponding to the cavity round trip time of the wave packet).

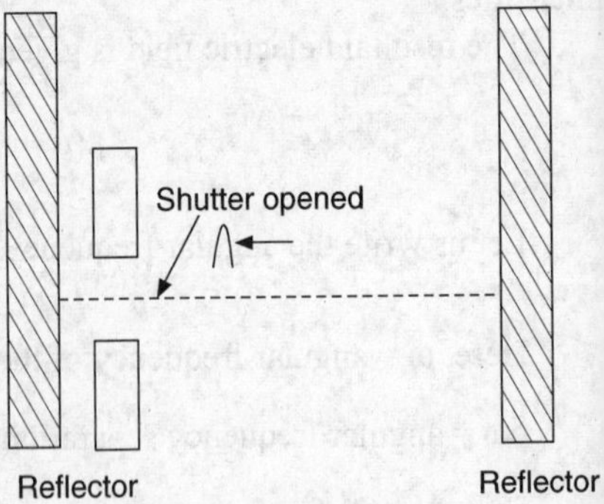

Fig. 5.40.

Now if the wave packet is exactly as long in time as the shutter stays open, and if it arrives exactly when the shutter is open, it will be unaffected by the presence of the shutter. Any parts of the wave packet, however, that arrive before the shutter opens or after it closes will be eliminated. Thus, the phase relationships of the oscillating modes are continuously restored by the periodic operation of the shutter.

The electro-optic or acousto-optic modulators are used as shutters.

5.22.2. Passive Mode Locking

Principle: For certain dyes, absorption decreases with increasing intensity (Fig. 5.41). Materials exhibiting this behaviour are called *saturable absorbers*.

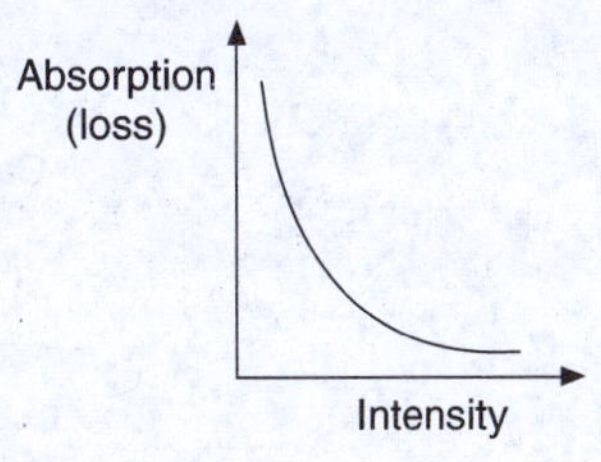

Fig. 5.41.

A dye is chosen which has an absorption band at the lasing transition frequency. Initially, at low light levels, the dye is opaque owing to the large number of unexcited molecules which can absorb the light. As the intensity increases, however, more and more of the excited states are populated until eventually ali of them are filled so that the dye becomes transparent. The dye is now said to be *bleached*.

Experimental Arrangement

Figure 5.42 shows the configuration for mode locking of lasers using a saturable absorber. A dye cell is placed inside a laser cavity. Eastman No. 9740 or Eastman No. 9860 dye solutions and cryptocyanine are used as the saturable absorber for Nd:glass and ruby laser systems respectively. Absorbers with *very small relaxation time* are used so that the absorber *recovers* in a time which is short compared with duration of the pulse.

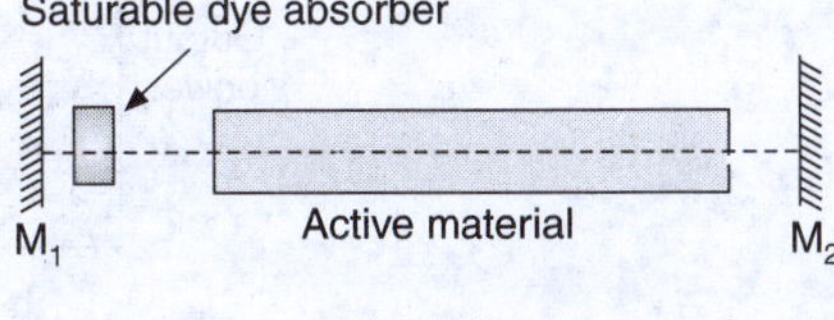

Fig. 5.42.

The growth of the mode locked pulses can be envisaged as follows. Initially, the laser medium emits spontaneous radiation which gives rise to incoherent fluctuations in the energy density within the cavity. Some of these fluctuations, which can be of short duration, may be amplified by the laser medium and grow in intensity to such an extent that the peak part of the fluctuation is transmitted by the saturable absorber with little attenuation. The low power parts of the fluctuation, however, are much more strongly attenuated and thus a high power pulse can grown within the cavity. Because of the non-linear behaviour of the dye, the shortest and most intense fluctuations grow at the expense of the weaker ones. With careful adjustment of the concentration of the dye within the cavity, an initial fluctuation may grow into a narrow pulse 'bouncing' to and fro within the cavity producing a periodic train of mode locked pulses.

- When a saturable absorber is used to mode lock a laser, the laser is simultaneously Q-switched.

5.23. Q-Switching

Definition: *Q*-switching in a laser system means *closing* the optical cavity by a suitable method until a *very strong population inversion*, much higher than the threshold, is built up. When this is accomplished, the optical cavity is suddenly *opened* so that a large fraction of the stored energy is released into a *single short pulse*.

Q-switching is carried out by placing a closed shutter (*i.e.*, the Q-switch) within the cavity (Fig. 5.43). After the laser has been pumped, the shutter is suddenly opened. In this method, the Q factor of the cavity of the laser is switched from a low value to a high value state.

Figure 5.44 shows the variation of the parameters – (a) flashlamp output, (b) cavity *Q*, (c) population inversion and (d) output power – as a function of time during the formation of a *Q*-switched laser pulse.

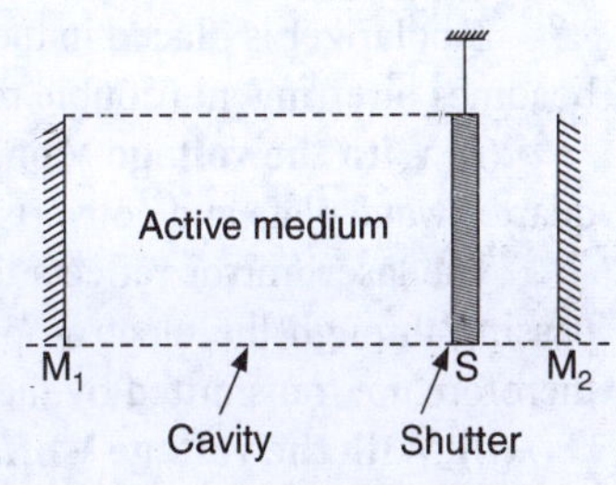

Fig. 5.43.

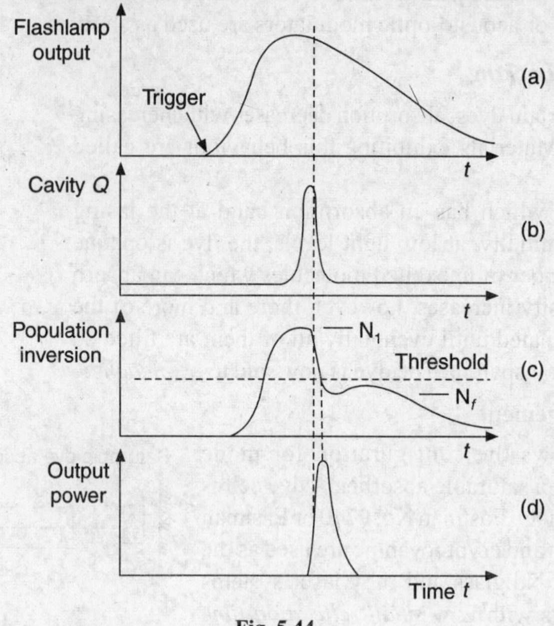

Fig. 5.44.

5.23.1. *Methods of Q-Switching*

1. *Electro-Optic Q-Switching*

Experimental arrangement. Fig. 5.45 shows electro-optic crystal used as a Q-switch.

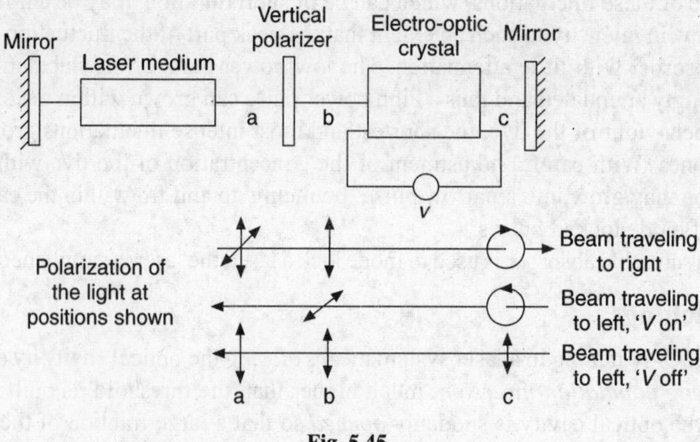

Fig. 5.45.

A polarizer is placed in the cavity along with the electro-optic crystal. The electro-optic crystal becomes birefringent (double refracting) when an electric field is applied across the crystal.

(i) With the voltage V on. A voltage is applied to the crystal. The electro-optic crystal acts as a quarter-wave plate and converts the vertically polarized light at b into circularly polarized light at *c*.

The laser mirror reflects this light and in so doing reverses its direction of rotation. So on re-passing through the electro-optic crystal, it emerges as horizontally polarized light. This light is therefore not transmitted by the polarizer and the cavity is 'switched off'.

(ii) With the voltage V off. When the voltage applied to the electro-optic crystal is switched off, double refraction cannot take place in the crystal. The state of polarisation of light passing through the crystal is unaffected. Light, therefore, travels freely through the crystal in both the directions and returns to the laser cavity where it is further amplified. The cavity, therefore, gets *switched on* and regains its high *Q*-value.

The *Q*-switching process is synchronised with pumping mechanism in such a way that the voltage applied to the electro-optic crystal drops to zero at the time when the population inversion in the laser medium attains peak value.

2. *Q-switching Using Saturable Absorbers*

Figure 5.46 shows the arrangement for achieving *Q*-switching using a saturable absorber.

A saturable absorber (bleachable dye) is placed in the cavity. At the beginning of the excitation flash, the dye is opaque. Laser action is prevented. A larger population inversion is achieved. As the light intensity within the cavity increases, the dye can no longer absorb, *i.e.*, it bleaches and *Q*-switching occurs.

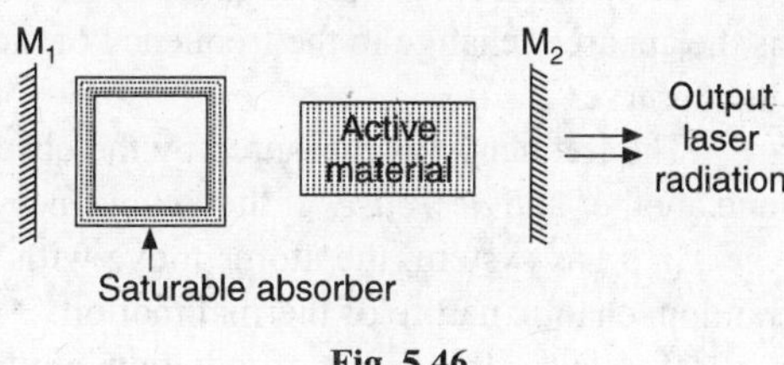

Fig. 5.46.

Suitable dyes include cryptocyanine for ruby lasers and sulfur hexafluoride for CO_2 lasers.

3. Rotating Mirror Method

Figure 5.47 shows the arrangement.

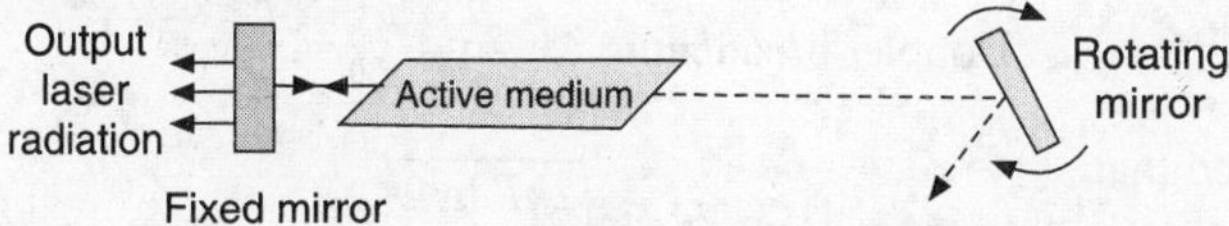

Fig. 5.47.

One of the mirrors is rotated at a very high angular velocity. So the optical losses are large except for the brief interval in each rotation cycle when the mirrors are very nearly parallel. Just before this point is reached, a trigger mechanism initiates the flashlamp discharge to pump the laser. As the mirrors are not yet parallel, population inversion can build up without laser action starting. When the mirrors become parallel, *Q*-switching occurs.

LINE BROADENING MECHANISMS

5.24. Lineshape Function

The radiation coming out of a collection of atoms making transitions between two energy levels is never perfectly monochromatic. This line broadening is described in terms of the *lineshape function* g(ν).

Figure 5.48 shows the emission curve for transitions between energy levels E_1 and E_2. The precise form of these curves (the lineshape) depends on the spectral broadening mechanisms.

We may define $g(\nu)d\nu$ as the probability that a given transition between the two energy levels will result in the emission of a photon whose frequency lies between ν and ν + $d\nu$. g(ν) is normalized such that $\int_{-\infty}^{\infty} g(\nu)d\nu = 1$.

Therefore we see that a photon of energy $h\nu$ may not necessarily stimulate another photon of energy $h\nu$. We then take $g(\nu)d\nu$ as the probability that the stimulated photon will have an energy between $h\nu$ and $h(\nu + d\nu)$.

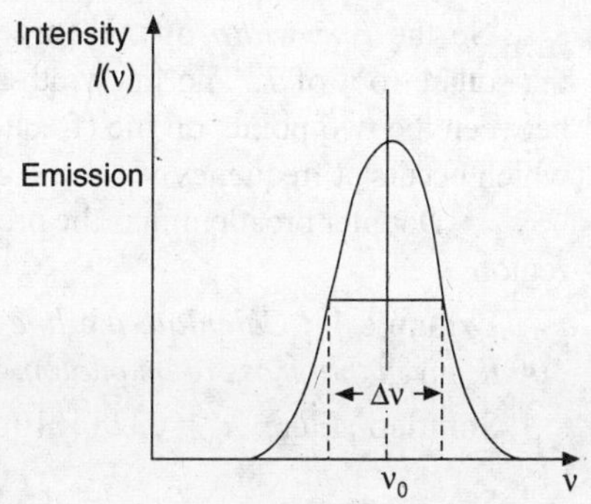

Fig. 5.48.

The form of the lineshape function $g(\nu)$ depends on the particular mechanism responsible for the *spectral broadening* in a given transition. The three most important mechanisms are Doppler broadening, collision (or pressure) broadening and natural broadening.

5.25. Doppler Broadening

Doppler effect occurs because of the relative motion of a source and observer. Doppler effect is the apparent change in the frequency of radiation due to a relative motion between the source and the observer.

The frequency as measured by the observer *increases* if the source and the observer approach one another and *decreases* if they *move away* from each other.

In a gas system, the atoms move with all possible velocities in all possible directions due to random chaotic nature of thermal motion.

If ν_0 is the frequency of radiation emitted by the atom and ν is the observed frequency, then

$$\nu - \nu_0 = \frac{v}{c}\,\nu_0 \qquad ...(1)$$

Here, v is the relative velocity of the atom w.r.t. the observer and c the velocity of radiation.

$$\therefore \qquad \text{Doppler broadening } \Delta\nu = \nu - \nu_0 = \frac{v}{c}\,\nu_0 \qquad ...(2)$$

It can be proved that

$$\Delta\nu = 2\nu_0\sqrt{\frac{2kT\ln 2}{Mc^2}} \qquad ...(3)$$

$\Delta\nu$ = Full width of the curve at half the maximum intensity of emission or FWHM

$\Delta\nu$ = line width (Fig. 5.49)

ν_0 = frequency corresponding to maximum

k = Boltzmann's constant

T = Temperature in Kelvin

M = mass of the atom

c = velocity of light.

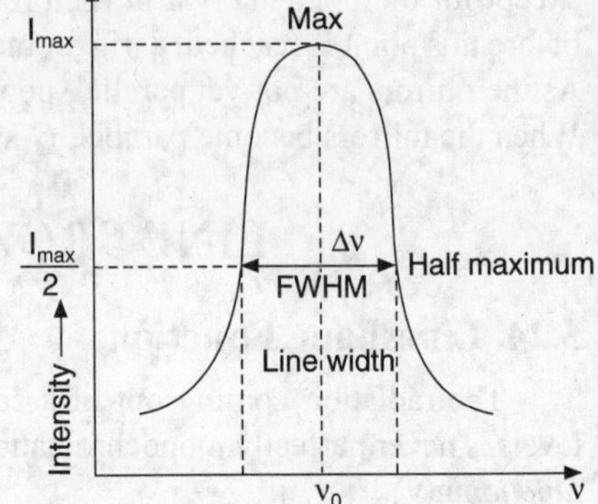

Fig. 5.49.

The mean squared velocity components v depend on the temperature according to

$$\frac{1}{2}Mv^2 = \frac{1}{2}kT \qquad ...(4)$$

So the *line width* of the curve is proportional to the square root of T. The linewidth $\Delta\nu$ is the separation between the two points on the (frequency) curve where the intensity falls to half of its peak value which occurs at frequency ν_0.

- Doppler broadening is the predominant mechanism in most gas lasers emitting in the visible region.

Example 1 : *Calculate the line width for 6328 Å wavelength of a He-Ne laser operating at 300 K, given that mass of neon atom = 20 × 1.66 × 10^{-27} kg.*

Solution : Here, $\lambda_0 = 6328 \times 10^{-10}$ m.

$$\therefore \qquad \nu_0 = \frac{c}{\lambda_0} = \frac{3.0\times10^8}{6328\times10^{-10}} = 4.742\times10^{14}\text{ Hz}$$

$$T = 300\text{ K};\ M = 20\times1.66\times10^{-27}\text{ kg. } \Delta\nu = ?$$

Line width $$\Delta\nu = 2\nu_0\sqrt{\frac{2kT\ \ln 2}{Mc^2}}$$

$$= 2\times(4.742\times10^{14})\left[\frac{2\times(1.38\times10^{-23}\times300\times0.693}{(20\times1.66\times10^{-27})\times(2\times10^{8})^2}\right]^{\frac{1}{2}}$$

$$= 1300\times10^{6}\ \text{Hz} = 1300\ \text{MHz}$$

Example 2 : *Compute the Doppler broadening for the 6328 Å laser transition in the He-Ne Laser, assuming a single isotope of* Ne^{20} *and that the laser operates at a discharge-bore temperature of 373 K.* *(G.N.D.U. 2010)*

Solution : 1500 MHz.

5.26. Collision Broadening

In a gas, atoms are in a state of continuous random motion and collide frequently with each other. If an atom in the process of transition *i.e.*, on its way to emit a photon, undergoes a collision with another atom, the phase of radiated wave changes. Each collision brings about a random phase change (Fig. 5.47). This results in shortening of wave trains and consequent broadening of spectral lines.

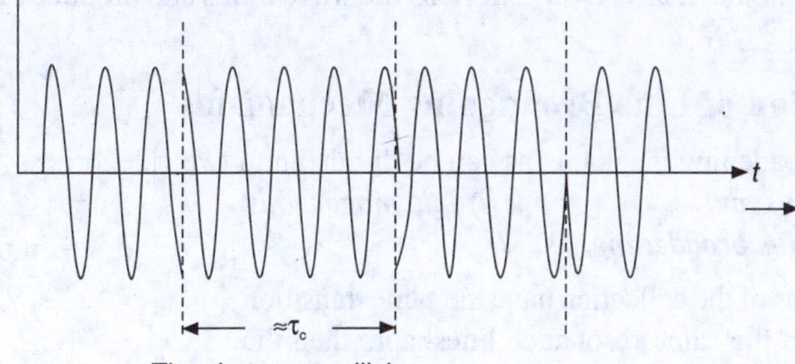

Fig. 5.50.

The wave shown in Fig. 5.47 is no longer monochromatic. This broadening is known as *collision broadening*.

• The higher the pressure (and temperature) of the gas, the more frequently will the atoms suffer collisions and the greater will be the spectral broadening.

• Collision broadening occurs in CO_2 laser.

Collision broadening also occurs in doped insulator lasers. In these lasers, the ions of the active medium may suffer collisions with phonons, *i.e.*, quantized lattice vibrations.

Line Width

The average time of collision of a gas atom is given by

$$t = \frac{\text{Mean free path}}{\text{Average velocity of atom}}$$

Now, mean free path $$\lambda = \frac{kT}{\sqrt{2}\ p\pi d^2} = \frac{kT}{4\sqrt{2}\ p\pi a^2}$$

Here, k = Boltzmann's gas constant; T = Temperature of the gas in Kelvin; p = pressure of the gas and d = diameter of the gas atom = $2a$ where a is the radius of the atom.

The average velocity of the atom, $\bar{v} = \sqrt{\frac{8kT}{\pi M}}$ (M = mass of the atom)

$$\therefore \qquad t = \frac{kT}{4\sqrt{2}\,p\pi a^2}\cdot\frac{1}{\sqrt{\frac{8kT}{\pi M}}} = \frac{kT}{4\sqrt{2}\,p\pi a^2}\cdot\frac{\sqrt{\pi M}}{\sqrt{8kT}} = \frac{1}{16pa^2}\cdot\sqrt{\frac{MkT}{\pi}}$$

Hence frequency spread $\Delta\omega = \frac{1}{t} = \frac{16pa^2}{\sqrt{MkT/\pi}}$

In terms of ν, we have, $\Delta\omega = 2\pi\,\Delta\nu$

$$\therefore \qquad \Delta\nu = \frac{\Delta\omega}{2\pi} = \frac{16pa^2}{2\pi\sqrt{MkT/\pi}} = \frac{8pa^2}{\sqrt{\pi MkT}}$$

5.27. Natural Broadening

Even in the absence of collision and thermal motion, one has the inherent line width because of the finite lifetime of the excited state due to spontaneous emission. This is called natural broadening.

It can be shown that the very act of an atomic electron emitting energy in the form of a photon leads to an exponential damping of the amplitude of the wave train. The effect of this is similar to collision broadening in that it effectively shortens the wave trains and produces a broadened spectral line.

5.28. Classification of Line Broadening Mechanisms

The various broadening mechanisms can be divided into two categories:

(i) Homogeneous and *(ii) Inhomogeneous.*

(*i*) ***Homogeneous broadening.***

If all of the atoms of the collection have the same transition center frequency and the same resonance lineshape, then the broadening is termed homogeneous.

Example : Collision broadening.

Homogeneous broadening mechanisms lead to a Lorentzian lineshape which may be written as:

$$g(\nu)_L = \frac{\Delta\nu}{2\pi}\left[(\nu-\nu_0)^2 + \left(\frac{\Delta\nu}{2}\right)^2\right]^{-1}$$

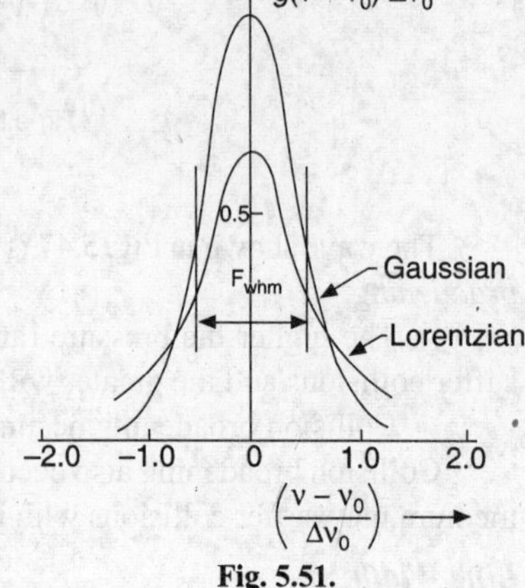

Fig. 5.51.

Here, Δν is the linewidth, that is, the separation between the two points on the (frequency) curve where the function falls to half of its peak value which occurs at frequency ν_0 (Fig. 5.50). Putting $\nu = \nu_0$ gives

$$g(\nu_0)_L = \frac{2}{\pi\Delta\nu} \qquad ...(1)$$

(*ii*) ***Inhomogeneous broadening.***

In the case of inhomogeneous broadening, different groups of atoms are distinguished by different frequency responses. Thus, for example, in Doppler broadening, group of atoms having different velocity components are distinguishable and they have different spectral responses. Similarly, broadening caused by local inhomogeneities of a crystal lattice acts to shift the central frequency of the response of individual atoms by different amounts thereby leading to inhomogeneous broadening.

Inhomogeneous broadening mechanisms lead to a Gaussian frequency distribution, given by:

$$g(\nu)_G = \frac{2}{\Delta\nu}\left(\frac{\ln 2}{\pi}\right)^{\frac{1}{2}} \exp\left[-(\ln 2)\left(\frac{\nu-\nu_0}{\Delta\nu/2}\right)^2\right]$$

Putting $\nu = \nu_0$ gives

$$g(\nu_0)_G = \frac{2}{\Delta\nu}\left(\frac{\ln 2}{\pi}\right)^{\frac{1}{2}} \quad \text{...(2)}$$

Comparison between Gaussian line graph and Lorentzian line graph

Gaussian curve	Lorentzian curve
The Gaussian curve is sharper.	The Lorentzian curve is rather flat.
$g(\nu_0)_G = \frac{2}{\Delta\nu}\left(\frac{\ln 2}{\pi}\right)^{1/2}$	$g(\nu_0)_L = \frac{2}{\pi\Delta\nu}$
The Gaussian line is non-homogeneous in nature.	The Lorentzian line is a homogeneous line.

SOLVED PROBLEMS

1. *If light of wavelength 660 nm has wave trains 13.2×10^{-6} m long, calculate "coherence time".* (*GNDU 2002*)

Solution : We have, $L = c\tau_c$.

Here, coherence length L = Length of one wave train = 13.2×10^{-6} m, coherence time τ_c = ?

$$\therefore \quad \tau_c = \frac{L}{c} = \frac{13.2 \times 10^{-6}}{3 \times 10^8} = 4.4 \times 10^{-14} \text{ s.}$$

2. *Calculate the coherence length of a laser beam for which the bandwidth $\Delta\nu$ = 3000 Hz. Speed of light $c = 3 \times 10^8$ m/s.* (*MDU 2003, 2001*)

Solution : Given : bandwidth $\Delta\nu$ = 30000 Hz.

$$\text{Coherence time } \tau_c = \frac{1}{\delta\nu} = \frac{1}{3000} = \frac{1}{3} \times 10^{-3} \text{s.}$$

$$\text{Coherence length } L = c\,\tau_c = (3 \times 10^8 \text{ m/s}) \times \left(\frac{1}{3} \times 10^{-3} \text{ s}\right) = 10^5 \text{ m.}$$

3. *For an ordinary source, the coherence time $\tau_c = 10^{-10}$ second. Obtain the degree of non-monochromaticity for wavelength λ_0 = 5400 Å.* (*MDU 2002*)

Solution : Given : coherence time $\tau_c = 10^{-10}$ sec.

$$\text{Frequency-spread} \quad \Delta\nu = \frac{1}{\tau_c} = \frac{1}{10^{-10}} \text{ s} = 10^{10} \text{ Hz.}$$

For wavelength λ_0 = 5400 Å = 5400×10^{-10} m,

$$\text{frequency,} \quad \nu_0 = \frac{c}{\lambda_0} = \frac{3 \times 10^8 \text{ m/s}}{5400 \times 10^{-10} \text{ m}} = \frac{1}{18} \times 10^{16} \text{ Hz.}$$

Degree of non-monochromaticity is given by

$$\frac{\Delta\nu}{\nu_0} = \frac{10^{10}}{(1 \times 10^{16}/18)} = \frac{10^{10} \times 18}{1 \times 10^{16}} = 1.8 \times 10^{-5}$$

4. *The Doppler width for orange line of krypton is 550 × 10⁻¹⁵ m. If the wavelength of light is 605.8 nm, calculate the coherence length.* *(GNDU 2007, 2003)*

Solution : We have

$$L = \frac{\lambda^2}{\Delta\lambda} = \frac{(605.8 \times 10^{-9}\text{ m})}{550 \times 10^{-15}\text{ m}} = 667.3 \times 10^{-3}\text{ m.}$$

MULTIPLE CHOICE QUESTIONS: SELECT THE RIGHT CHOICE

1. The relation between Einstein's A_{21} and B_{21} coefficient is

(a) $\frac{A_{21}}{B_{21}} = 1:2$ **(b)** $\frac{A_{21}}{B_{21}} = \frac{8\pi h v^3}{c^3}$

(c) $\frac{A_{21}}{B_{21}} = 1:1$ **(d)** $\frac{A_{21}}{B_{21}} = v^3$

2. The population inversion in helium neon laser is produced by

(a) photon excitation **(b)** chemical excitation

(c) inelastic atomic collisions **(d)** chemical reaction

3. Q-switching is used to get:

(a) a continuous pulse **(b)** a giant laser pulse for a short duration

(c) a giant laser pulse for long duration **(d)** to stop the incoming laser beam

4. Mode-locking in laser means production of pulses:

(a) in same phase **(b)** of same amplitude

(c) both above **(d)** none of the above

5. Second harmonic generation converts laser beam of wavelength 700 nm into a laser beam of wavelength

(a) 1400 nm **(b)** 350 nm

(c) 300 nm **(d)** 175 nm

6. It is possible to change the colour of laser beam through

(a) second harmonic generation **(b)** sum and difference frequency generation

(c) parametric oscillation **(d)** all are correct

ANSWERS

1. **(b)**	2. **(c)**	3. **(b)**
4. **(a)**	5. **(b)**	6. **(d)**

CHAPTER

6 MATRIX METHOD IN PARAXIAL OPTICS

6.1. Introduction

Let us consider a complicated optical system (for example, a system of lenses and surfaces). In order to determine the position of the final image formed by such a system, we have to calculate step by step the position of the image due to each individual surface which is then taken as object for the next surface and so on. This method is lengthy and tedious. Matrix method is a powerful tool to tackle such problems. In this method, we determine the *translation matrix* and *refraction matrix* of the light ray in the optical system. By multiplying the two matrices, we get the *system matrix*. Using the system matrix, we can find out the properties of the system. The matrix method gives the final result without bothering about actual intermediate reflections and refractions. In this way, it can be easily adopted in calculations by computers. The matrix data can be fed to computer to obtain the final result *i.e.,* location, size and nature of the image.

6.2. Designation of a (Para-axial) Ray

Consider a cylindrically symmetric optical system as shown in Fig. 6.1. The axis of symmetry is chosen as the *Z*-axis. Consider a paraxial ray travelling in a homogeneous medium. The light ray can be specified by two parameters : (*i*) its distance (x) from the *Z*-axis and (*ii*) the angle (α) which it makes with the *Z*-axis. Let *P* be a point on the ray *PQ*. This point is at a distance x_1 from the *Z*-axis and makes an angle α_1 with the *Z*-axis. Hence at *P*, the coordinates of the light ray are (x_1, α_1). Define a quantity

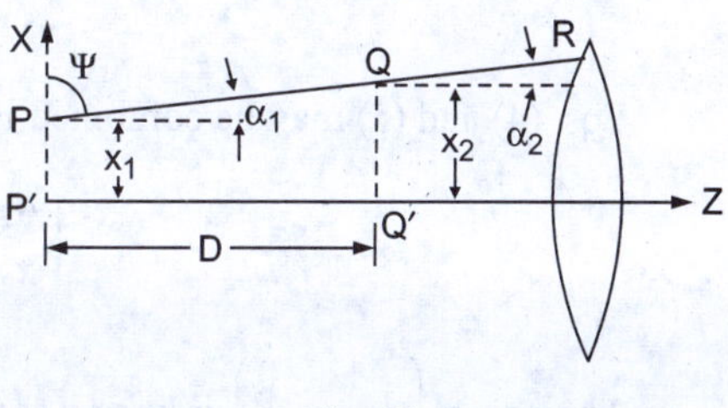

Fig. 6.1.

$$\lambda = n \cos \psi = n \cos (90 - \alpha_1) = n \sin \alpha_1$$

λ is known as *optical direction cosine*. It represents the product of the refractive index n of the medium and the cosine of the angle ψ that the ray makes with the *X*-axis. The ray can be re-designated as (x, λ). $\begin{pmatrix} x \\ \lambda \end{pmatrix}$ matrix denotes a ray vector and determines the ray completely.

When a ray propagates through an optical system, it undergoes the following two operations : (*i*) *Translation* and (*ii*) *Refraction.*

(*i*) **Translation.** The rays undergo translation when they propagate through a homogeneous medium as in the region *PR*.

(*ii*) **Refraction.** When the ray of light strikes an interface of two media, it undergoes refraction. At *R*, the ray suffers refraction.

We will now study the effect of translation and of refraction on the coordinates of the ray.

6.3. Effect of Translation — Translation Matrix

Consider a ray travelling in a homogeneous medium of refractive index n_1. Let (x_1, α_1) be the coordinates of the ray at the point *P* (Fig. 6.2). In travelling from *P* to *Q*, the ray undergoes translation. Let (x_2, α_2) represent the coordinates of the ray at the point *Q*.

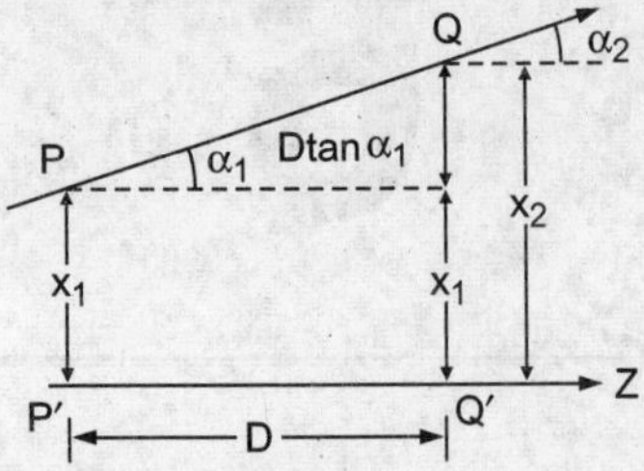

Fig. 6.2.

Since the medium is homogeneous, the ray *PQ* travels in a straight line. Therefore,

$$\alpha_2 = \alpha_1 \qquad \text{...(1)}$$

Let *PP*' and *QQ*' be perpendiculars on the *Z*-axis. Let *P'Q'* = *D*. Then,

$$x_2 = x_1 + D \tan \alpha_1 \qquad \text{...(2)}$$

α_1 is very small for paraxial rays and $\tan \alpha_1 \approx \alpha_1$. Hence

$$x_2 = x_1 + \alpha_1 D \qquad \text{...(3)}$$

Let λ_1 and λ_2 be the optical direction cosines at *P* and *Q* respectively. Then,

$$\lambda_1 = n_1 \sin \alpha_1 \approx n_1\alpha_1 \text{ and } \lambda_2 = n_1 \sin \alpha_2 \approx n_1 \alpha_2$$

Since $\alpha_2 = \alpha_1, \lambda_2 = \lambda_1 = 0x_1 + 1\lambda_1$ ...(4)

Using the equation $\lambda_1 = n_1\alpha_1$, Eq. (3) reduces to

$$x_2 = x_1 + \frac{D}{n_1}\lambda_1 \qquad \text{...(5)}$$

Eqs. (4) and (5) may be combined into the following matrix equation :

$$\begin{pmatrix} \lambda_2 \\ x_2 \end{pmatrix} = \begin{pmatrix} 1 & 0 \\ D/n_1 & 1 \end{pmatrix} \begin{pmatrix} \lambda_1 \\ x_1 \end{pmatrix} \qquad \text{...(6)}$$

The matrix $\begin{pmatrix} 1 & 0 \\ D/n_1 & 1 \end{pmatrix}$ is known as the *translation matrix* denoted by *T*. Thus, if a ray is initially specified by a (2 × 1) matrix with elements λ_1 and x_1, then the effect of translation through a distance *D* in a homogeneous medium of refractive index n_1, is comrletely given by the 2 × 2 matrix,

$$T = \begin{pmatrix} 1 & 0 \\ D/n_1 & 1 \end{pmatrix} \qquad \text{...(7)}$$

Eq. (6) represents the final ray.

$$\det T = \begin{vmatrix} 1 & 0 \\ D/n_1 & 1 \end{vmatrix} = 1 \qquad \text{...(8)}$$

6.4. Effect of Refraction — Refraction Matrix

Consider a spherical surface of radius of curvature R separating two media of refractive indices n_1 and n_2 (Fig. 6.3). Let a ray AP be incident on this spherical surface at a point P (height x from the axis) and get refracted along PB. CN is the normal drawn to the surface at the point of incidence P. M_1M_2 is a line parallel to the Z-axis of the optical system. Let i and r be the angle of incidence and angle of refraction respectively. Suppose the incident and refracted rays make angles α_1 and α_2 with the line M_1M_2.

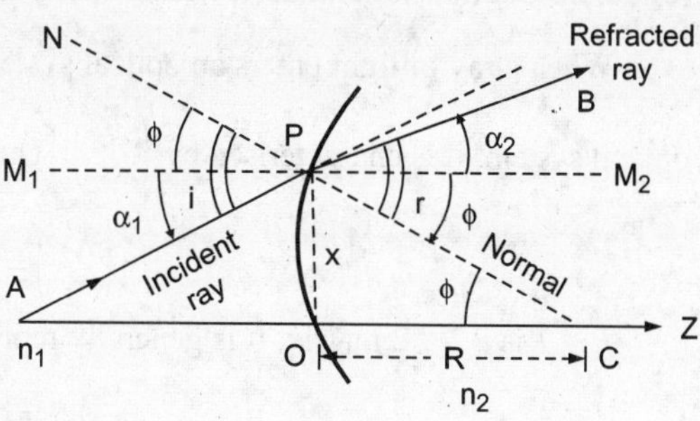

Fig. 6.3.

According to Snell's law

$$n_1 \sin i = n_2 \sin r$$

For paraxial rays, i and r are small $\sin i \approx i$ and $\sin r \approx r$

$$\therefore \qquad n_1 i = n_2 r \qquad \text{...(1)}$$

Let ϕ be the angle between normal CN and the line M_1M_2. Then,

$$i = \alpha_1 + \phi \text{ and } r = \alpha_2 + \phi \qquad \text{...(2)}$$

Substituting these values in Eq. (1), we get

$$n_1(\alpha_1 + \phi) = n_2(\alpha_2 + \phi)$$

or

$$\begin{aligned} n_2\alpha_2 &= n_1\alpha_1 + n_1\phi - n_2\phi \\ &= n_1\alpha_1 - (n_2 - n_1)\phi \\ &= n_1\alpha_1 - (n_2 - n_1)(x/R) \end{aligned} \qquad \text{...(3)}$$

Let λ_1 and λ_2 be the optical direction cosines before and after refraction. Then,

$$\lambda_1 = n_1\alpha_1 \text{ and } \lambda_2 = n_2\alpha_2 \qquad \text{...(4)}$$

Let $\{(n_2 - n_1)/R\} = P =$ Power of the refracting surface. ...(5)

Using Eqs. (4) and (5), Eq. (3) reduces to

$$\lambda_2 = \lambda_1 - Px \qquad \text{...(6)}$$

The height of the ray at P, before and after refraction, is the same (*i.e.*, $x_2 = x_1$). Hence

$$x_2 = x_1 \text{ or } x_2 = 0\lambda_1 + 1x_1 \qquad \text{...(7)}$$

Writing Eqs. (6) and (7) in matrix form, we have

$$\begin{pmatrix} \lambda_2 \\ x_2 \end{pmatrix} = \begin{pmatrix} 1 & -P \\ 0 & 1 \end{pmatrix} \begin{pmatrix} \lambda_1 \\ x_1 \end{pmatrix} \qquad \text{...(8)}$$

The matrix $\begin{pmatrix} 1 & -P \\ 0 & 1 \end{pmatrix}$ is called *refraction matrix* and denoted by R.

$$\det R = \begin{vmatrix} 1 & -P \\ 0 & 1 \end{vmatrix} = 1 \qquad \text{...(9)}$$

6.5. System Matrix

Generally an optical system consists of a number of lenses. This system can be characterised by the refraction and translation matrices. The only two operations a ray undergoes in traversing through an optical system are refraction and translation. Hence the *system matrix* is a product of refraction and translation matrices.

When a ray of light enters an optical system, it is specified by $\begin{pmatrix}\lambda_1\\ x_1\end{pmatrix}$. When the ray leaves the optical system, it is characterised by $\begin{pmatrix}\lambda_2\\ x_2\end{pmatrix}$. Then we can write

$$\begin{pmatrix}\lambda_2\\ x_2\end{pmatrix} = R \times T \begin{pmatrix}\lambda_1\\ x_1\end{pmatrix}$$

$R \times T$ is a 2×2 matrix. It is generally represented by

$$S = \begin{pmatrix} b & -a\\ -d & c\end{pmatrix}$$

S is called the *system matrix* and is determined solely by the optical system. The negative signs in some of the elements of *S* have been chosen for convenience. The determinants of translation and refraction matrices are unity. Hence the determinant of system matrix is also unity. Thus

$$\det S = 1$$

$$\therefore \quad bc - ad = 1$$

Now we may write

$$\begin{pmatrix}\lambda_2\\ x_2\end{pmatrix} = S \begin{pmatrix}\lambda_1\\ x_1\end{pmatrix}$$

6.6. Equations for the Image Plane and Magnification of an Optical System

Let the object plane OO' be at a distance $(-D_1)$ from the first refracting surface of an optical system (Fig. 6.4). The image plane II' is at a distance D_2 from the last refracting surface. Let a ray OP start from the point O', which lies in the object plane, and strike the first refracting surface at P. It emerges from the last refracting surface along QI' from the point Q. I' is a point in the image plane. Point I' is the image of point O'. Let (λ_1, x_1), (λ_1', x_1'), (λ_2', x_2') and (λ_2, x_2) be the coordinates of points O', P, Q and I' respectively.

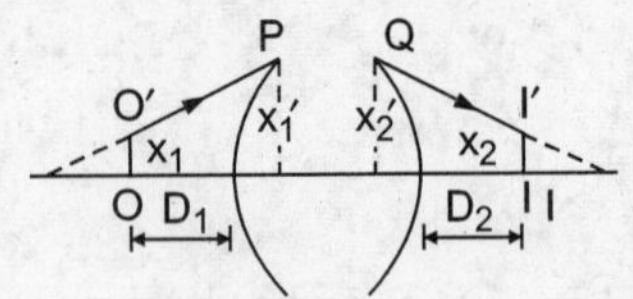

Fig. 6.4.

Between O' and P, there is only translation and medium is the same. Hence

$$\begin{pmatrix}\lambda_1'\\ x_1'\end{pmatrix} = \begin{pmatrix} 1 & 0\\ -D_1 & 1\end{pmatrix}\begin{pmatrix}\lambda_1\\ x_1\end{pmatrix} \qquad \ldots(1)$$

Between P and Q, there will be refraction followed by translation.

$$\therefore \quad \begin{pmatrix}\lambda_2'\\ x_2'\end{pmatrix} = \begin{pmatrix} b & -a\\ -d & c\end{pmatrix}\begin{pmatrix}\lambda_1'\\ x_1'\end{pmatrix} \qquad \ldots(2)$$

Between Q and I', there will be only translation and medium is the same.

$$\therefore \quad \begin{pmatrix}\lambda_2\\ x_2\end{pmatrix} = \begin{pmatrix} 1 & 0\\ D_2 & 1\end{pmatrix}\begin{pmatrix}\lambda_2'\\ x_2'\end{pmatrix} \qquad \ldots(3)$$

$$\begin{pmatrix}\lambda_2\\ x_2\end{pmatrix} = \begin{pmatrix} 1 & 0\\ D_2 & 1\end{pmatrix}\begin{pmatrix} b & -a\\ -d & c\end{pmatrix}\begin{pmatrix} 1 & 0\\ -D_1 & 1\end{pmatrix}\begin{pmatrix}\lambda_1\\ x_1\end{pmatrix} \qquad \ldots(4)$$

The first and third matrices give the effect of translation by distances D_2 and $(-D_1)$ in the medium of refractive index unity. The second matrix corresponds to the system matrix and is determined solely by the optical system.

The product of the 3 matrices on the R.H.S. is

$$\begin{pmatrix} b+aD_1 & -a \\ bD_2+aD_1D_2-d-cD_1 & c-aD_2 \end{pmatrix}$$

$$\therefore \qquad \begin{pmatrix} \lambda_2 \\ x_2 \end{pmatrix} = \begin{pmatrix} b+aD_1 & -a \\ bD_2+aD_1D_2-d-cD_1 & c-aD_2 \end{pmatrix} \begin{pmatrix} \lambda_1 \\ x_1 \end{pmatrix} \qquad \ldots(5)$$

$$\begin{pmatrix} \lambda_2 \\ x_2 \end{pmatrix} = \begin{pmatrix} (b+aD_1)\,\lambda_1 - ax_1 \\ (bD_2+aD_1D_2-d-cD_1)\,\lambda_1 + (c-aD_2)x_1 \end{pmatrix} \qquad \ldots(6)$$

$$\therefore \qquad x_2 = (bD_2+aD_1D_2-d-cD_1)\,\lambda_1 + (c-aD_2)x_1 \qquad \ldots(7)$$

If the object lies at the axis *i.e.*, $x_1 = 0$, then image also lies at the axis *i.e.*, $x_2 = 0$. Thus for the image plane, we have the following relationship between distances D_1 and D_2.

Putting $x_1 = 0$ and $x_2 = 0$ in Eq. (7), we get

$$0 = (bD_2+aD_1D_2-d-cD_1)\,\lambda_1 + (c-aD_2)\,0$$

Now, $\lambda_1 \neq 0$.

$$\therefore \quad bD_2+aD_1D_2-d-cD_1 = 0 \qquad \ldots(8)$$

Now, Eq. (5) for the image plane becomes

$$\begin{pmatrix} \lambda_2 \\ x_2 \end{pmatrix} \begin{pmatrix} b+aD & -a \\ c & aD \end{pmatrix} \begin{pmatrix} \lambda_1 \\ x_1 \end{pmatrix} \qquad \ldots(9)$$

If $x_2 \neq 0$, $x_2 = (c-aD_2)\,x_1$ ...(10)

Magnification of the system

The magnification of the system is given by

$$M = \frac{x_2}{x_1} = c - aD_2 \qquad \ldots(11)$$

Since $\begin{vmatrix} b+aD_1 & -a \\ 0 & c-aD_2 \end{vmatrix} = 1$, $(b+aD_1)(c-aD_2) = 1$

or
$$b+aD_1 = \frac{1}{c-aD_2} \qquad \ldots(12)$$

From Eqs. (11) and (12),

$$b+aD_1 = \frac{1}{M} \qquad \ldots(13)$$

Image plane

Substituting Eq. (11) and Eq. (13) in Eq. (9), we get the equation for the image plane in terms of magnification of the system.

$$\begin{pmatrix} \lambda_2 \\ x_2 \end{pmatrix} = \begin{pmatrix} 1/M & -a \\ 0 & M \end{pmatrix} \begin{pmatrix} \lambda_1 \\ x_1 \end{pmatrix} \qquad \ldots(14)$$

6.7. To Obtain the System Matrix for a Thick Lens and to Derive Thin Lens and Thick Lens Formulae

Consider a lens of thickness t and refractive index n (Fig. 6.5). The lens is immersed in air. Let R_1 and R_2 be the radii of curvature of the two surfaces. Let a ray be incident on first surface of the lens at point P at a height x_1 from the axis and emerge from the point Q at height x_2. Let λ_1 and λ_2 be the optical direction cosines of the ray at P and Q. Then the coordinates of the ray at P and Q are respectively given by

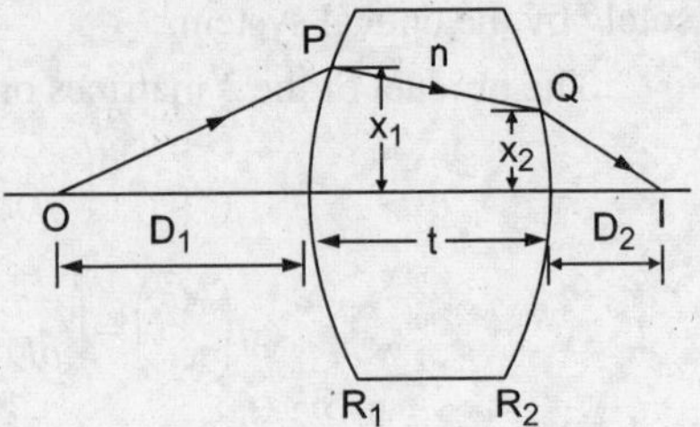

Fig. 6.5.

$$\begin{pmatrix} \lambda_1 \\ x_1 \end{pmatrix} \text{ and } \begin{pmatrix} \lambda_2 \\ x_2 \end{pmatrix} \quad ...(1)$$

The ray in travelling from P to Q suffers two refractions—one at the first surface of radius of curvature R_1 and other at the second surface of radius of curvature R_2 and a translation through a distance t, in a medium of refractive index n. Thus

$$\begin{pmatrix} \lambda_2 \\ x_2 \end{pmatrix} = \begin{pmatrix} 1 & -P_2 \\ 0 & 1 \end{pmatrix} \begin{pmatrix} 1 & 0 \\ t/n & 1 \end{pmatrix} \begin{pmatrix} 1 & -P_1 \\ 0 & 1 \end{pmatrix} \begin{pmatrix} \lambda_1 \\ x_1 \end{pmatrix} \quad ...(2)$$

Here, $P_1 = \dfrac{(n-1)}{R_1}$ and $P_2 = \dfrac{(1-n)}{R_2} = -\dfrac{(n-1)}{R_2}$ ...(3)

represent the powers of the two refracting surfaces.

Now the system matrix S is given by

$$S = \begin{pmatrix} b & -a \\ -d & c \end{pmatrix} = \begin{pmatrix} 1 & -P_2 \\ 0 & 1 \end{pmatrix} \begin{pmatrix} 1 & 0 \\ t/n & 1 \end{pmatrix} \begin{pmatrix} 1 & -P_1 \\ 0 & 1 \end{pmatrix}$$

$$= \begin{pmatrix} 1 - \dfrac{P_2 t}{n} & -P_1 - P_2\left(1 - \dfrac{t}{n} P_1\right) \\ \dfrac{t}{n} & 1 - \dfrac{t}{n} P_1 \end{pmatrix} \quad ...(4)$$

6.8. Thin Lens Formula

For a thin lens, $t \to 0$. The system matrix reduces to the following form :

$$S = \begin{pmatrix} b & -a \\ -d & c \end{pmatrix} = \begin{pmatrix} 1 & -P_1 - P_2 \\ 0 & 1 \end{pmatrix} \quad ...(5)$$

Comparing the values of a, b, c and d, we get

$$a = P_1 + P_2, b = 1, c = 1 \text{ and } d = 0 \quad ...(6)$$

For image plane, $bD_2 + aD_1D_2 - d - cD_1 = 0$

Substituting the values of a, b, c and d from Eq. (6) in it we get

$$1 . D_2 + (P_1 + P_2)\, D_1D_2 - 0 - 1 . D_1 = 0$$

$$\therefore \quad (P_1 + P_2)D_1D_2 = D_1 - D_2$$

Dividing by D_1 D_2, we get

$$P_1 + P_2 = \frac{1}{D_2} - \frac{1}{D_1}$$

or
$$\frac{(n-1)}{R_1}-\frac{(n-1)}{R_2}=\frac{1}{D_2}-\frac{1}{D_1}$$

or
$$(n-1)\left(\frac{1}{R_1}-\frac{1}{R_2}\right)=\frac{1}{f}$$

where
$$\frac{1}{D_2}-\frac{1}{D_1}=\frac{1}{f}$$

$\therefore$
$$\frac{1}{f}=(n-1)\left(\frac{1}{R_1}-\frac{1}{R_2}\right) \qquad \ldots(7)$$

Eq. (7) is the thin lens formula.
The system matrix for a thin lens is

$$S=\begin{pmatrix}1 & -\dfrac{1}{f}\\ 0 & 1\end{pmatrix} \qquad \ldots(8)$$

Example 1. *Obtain the system matrix for a lens placed in air and made of glass of refractive index 1.5 and radii of curvature 50 cm each.*

(*Osmania University, Oct. 89*)

Solution. The system matrix for a thin lens is given by

$$S=\begin{pmatrix}1 & -P_1-P_2\\ 0 & 1\end{pmatrix}$$

The powers of the two refracting surfaces are equal.

$$P_1=P_2=\frac{n-1}{R_1}=\frac{n-1}{|R_2|}=\frac{1.5-1}{50}=\frac{0.5}{50}=0.01\ \text{cm}^{-1}$$

$$S=\begin{pmatrix}1 & -0.01-0.01\\ 0 & 1\end{pmatrix}=\begin{pmatrix}1 & -0.02\\ 0 & 1\end{pmatrix}$$

$\therefore$
$$f=\frac{1}{0.02}=\mathbf{50\ cm.}$$

6.9. Thick Lens Formula

The system matrix for a thick lens is given by

$$S=\begin{pmatrix}b & -a\\ -d & c\end{pmatrix}=\begin{pmatrix}1-\dfrac{P_2 t}{n} & -P_1-P_2\left(1-\dfrac{t}{n}P_1\right)\\ \dfrac{t}{n} & 1-\dfrac{t}{n}P_1\end{pmatrix}$$

Comparing the values of *a, b, c* and *d*, we get

$$a=P_1+P_2\left(1-\frac{P_1 t}{n}\right)=P_1+P_2-\frac{P_1 P_2 t}{n}$$

$$b=1-\frac{P_2 t}{n}$$

$$c=1-\frac{P_1 t}{n} \qquad \ldots(9)$$

$$d = -t/n$$

For image plane, $bD_2 + aD_1 D_2 - d - cD_1 = 0$

Substituting the values of a, b, c and d, we have

$$\left(1 - \frac{P_2 t}{n}\right)D_2 + \left(P_1 + P_2 - \frac{P_1 P_2 t}{n}\right)D_1 D_2 + \frac{t}{n} - \left(1 - \frac{P_1 t}{n}\right)D_1 = 0$$

Simplifying we get

$$P_1 + P_2 - \frac{P_1 P_2 t}{n} = \frac{1}{D_2} - \frac{1}{D_1} + \frac{P_2 t}{nD_1} - \frac{P_1 t}{nD_2} - \frac{t}{nD_1 D_2} \quad \text{...(10)}$$

D_1 or $D_2 >> t$. Hence we can neglect the last three terms on R.H.S. of Eq. (10).

$$\therefore \quad P_1 + P_2 - \frac{P_1 P_2 t}{n} = \frac{1}{D_2} - \frac{1}{D_1} \quad \text{...(11)}$$

But $\frac{1}{D_2} - \frac{1}{D_1} = \frac{1}{f}$, where f is the focal length of thick lens.

Further $P_1 = \frac{(n-1)}{R_1}$ and $P_2 = -\frac{(n-1)}{R_2}$

Substituting these values in Eq. (11) we get the thick lens formula

$$\frac{1}{f} = (n-1)\left(\frac{1}{R_1} - \frac{1}{R_2} + \frac{(n-1)t}{nR_1 R_2}\right) \quad \text{...(12)}$$

Example 2. *A lens of thickness 2 cm and refractive index 1.5 placed in air has radii of curvature 8 cm and –8 cm. Find the system matrix and focal length of this thick lens.*

Solution. Here, $t = 2$ cm, $n = 1.5$, $R_1 = 8$ cm and $R_2 = -8$ cm.

$$\therefore \quad P_1 = \frac{n-1}{R_1} = \frac{1.5-1}{8} = 0.0625$$

and

$$P_2 = \frac{1-n}{-R_2} = \frac{1.5-1}{8} = 0.0625$$

The system matrix of a thick lens is given by

$$S = \begin{pmatrix} 1 - \frac{P_2 t}{n} & -P_1 - P_2\left(1 - \frac{P_1 t}{n}\right) \\ \frac{t}{n} & 1 - \frac{P_1 t}{n} \end{pmatrix}$$

Substituting the above values and simplifying, we get

$$S = \begin{pmatrix} 0.9167 & -0.1198 \\ 1.333 & 0.9167 \end{pmatrix}$$

Comparing it with the system matrix

$$S = \begin{pmatrix} b & -a \\ -d & c \end{pmatrix}, \text{ we get } a = 0.1198$$

$\therefore$ Focal length $f = \frac{1}{a} = \frac{1}{0.1198} = \mathbf{8.347\ cm}$

6.10. System of Two Thin Lenses

Consider a system of two thin lenses of focal lengths f_1 and f_2 placed coaxially at a distance t apart (Fig. 6.6). The matrices of the two lenses L_1 and L_2 are

$$\begin{pmatrix} 1 & -1/f_1 \\ 0 & 1 \end{pmatrix} \text{ and } \begin{pmatrix} 1 & -1/f_2 \\ 0 & 1 \end{pmatrix} \qquad \ldots(1)$$

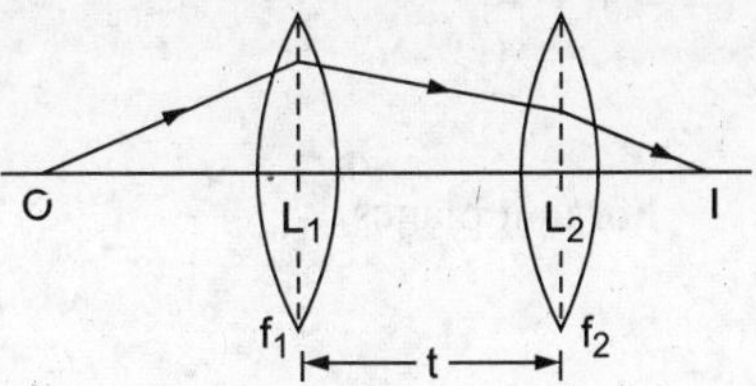

Fig. 6.6.

The matrix for translation through a distance t (in air) is

$$\begin{pmatrix} 1 & 0 \\ t & 1 \end{pmatrix} \qquad \ldots(2)$$

Thus the system matrix S is given by

$$S = \begin{pmatrix} 1 & -\dfrac{1}{f_2} \\ 0 & 1 \end{pmatrix} \begin{pmatrix} 1 & 0 \\ t & 1 \end{pmatrix} \begin{pmatrix} 1 & -\dfrac{1}{f_1} \\ 0 & 1 \end{pmatrix} = \begin{pmatrix} 1-\dfrac{t}{f_2} & -\left(\dfrac{1}{f_1}+\dfrac{1}{f_2}-\dfrac{t}{f_1 f_2}\right) \\ t & \left(1-\dfrac{t}{f_1}\right) \end{pmatrix} \qquad \ldots(3)$$

Comparing it with the general system matrix

$$S = \begin{pmatrix} b & -a \\ -d & c \end{pmatrix},$$

$$\left.\begin{aligned} a &= \frac{1}{f_1}+\frac{1}{f_2}-\frac{t}{f_1 f_2} \qquad & b &= 1-\frac{t}{f_2} \\ c &= 1-\frac{t}{f_1} & d &= -t \end{aligned}\right\} \qquad \ldots(4)$$

The element a in the system matrix represents the inverse of the focal length of the system. Thus, the focal length of the combination is given by

$$\frac{1}{f} = \frac{1}{f_1}+\frac{1}{f_2}-\frac{t}{f_1 f_2} \qquad \ldots(5)$$

6.11. Location of Unit Planes

The unit planes are two planes, one each in the object and the image space, between which the magnification M is unity. Any paraxial ray emanating from the unit plane in the object space will

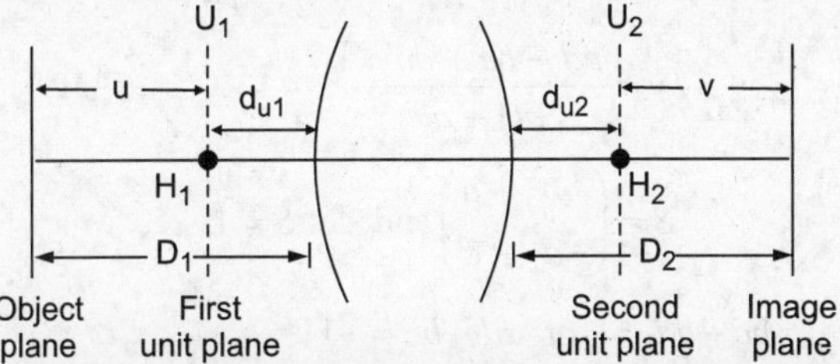

Fig. 6.7.

emerge at the same height from the unit plane in the image space. Let d_{u1} and d_{u2} be the respective distances of the first and second unit planes from the refracting surfaces of a lens (Fig. 6.7).

The magnification of optical system is given by

$$b + aD_1 = \frac{1}{c - aD_2} = \frac{1}{M} \quad \ldots(1)$$

For unit planes, $M = 1, D_1 = d_{u1}$ and $D_2 = d_{u2}$

$\therefore$ $b + a\, d_{u1} = 1$

or $$d_{u1} = \frac{(1-b)}{a} \quad \ldots(2)$$

and $c - a\, d_{u2} = 1$

or $$d_{u2} = \frac{(c-1)}{a} \quad \ldots(3)$$

Hence *the unit planes are completely determined by the elements of the system matrix S.*

It is convenient to measure the distances of object and image from the unit planes. Let u be the distance of the object plane from the first unit plane and v be the distance of the corresponding image plane from the second unit plane. Then

$$D_1 = u + d_{u1} = u + \frac{1-b}{a} \quad \ldots(4)$$

and $$D_2 = v + d_{u2} = v + \frac{c-1}{a} \quad \ldots(5)$$

Now, from Eq. (8) of Article 6.6, we have

$$bD_2 + aD_1D_2 - d - cD_1 = 0 \quad \ldots(6)$$

$\therefore$ $$D_2 = \frac{d + cD_1}{b + aD_1} \quad \ldots(7)$$

Substituting for D_1 and D_2 from Eqs. (4) and (5) in Eq. (7), we get

$$v + \frac{(c-1)}{a} = \frac{d + c\left(u + \dfrac{1-b}{a}\right)}{b + a\left(u + \dfrac{1-b}{a}\right)} = \frac{d + c\left(u + \dfrac{1-b}{a}\right)}{1 + au}$$

$\therefore$ $$v = \frac{d + c\left(u + \dfrac{1-b}{a}\right)}{1 + au} - \frac{(c-1)}{a}$$

Simplifying, $$v = \frac{ad - bc + au + 1}{a(1 + au)} \quad \ldots(8)$$

Now, $$S = \begin{pmatrix} b & -a \\ -d & c \end{pmatrix} \text{ and } \det S = 1$$

$\therefore$ $$bc - ad = 1 \text{ or } ad - bc = -1 \quad \ldots(9)$$

From Eqs. (8) and (9), we get

$$v = \frac{-1 + au + 1}{a(1 + au)} = \frac{u}{1 + au}$$

or $$\frac{1}{v} = \frac{1 + au}{u} = \frac{1}{u} + a$$

$$\therefore \qquad \frac{1}{v}-\frac{1}{u}=a \qquad \text{...(10)}$$

Comparing Eq. (10) with lens equation $\frac{1}{v}-\frac{1}{u}=\frac{1}{f}$, we see that $(1/a)$ represents the tocal length of the system. *Hence, if the distances are measured from the unit planes, then the element 'a' in the system matrix represents the inverse of the focal length of the system.*

6.12. Unit Planes for a Thick Lens

The system matrix for a thick lens is given by

$$S=\begin{pmatrix} b & -a \\ -d & c \end{pmatrix}=\begin{pmatrix} 1-\frac{P_2 t}{n} & -P_1-P_2\left(1-\frac{t}{n}P_1\right) \\ \frac{t}{n} & \left(1-\frac{t}{n}P_1\right) \end{pmatrix} \qquad \text{...(1)}$$

The elements of the system matrix are also called "*Gaussian constants*" of the system.

$$a=P_1+P_2\left(1-\frac{P_1 t}{n}\right),\ b=1-\frac{P_2 t}{n}$$

$$c=1-\frac{P_1 t}{n} \text{ and } d=-\frac{t}{n} \qquad \text{...(2)}$$

The positions of first unit plane and second unit plane are

$$d_{u1}=\frac{1-b}{a} \text{ and } d_{u2}=\frac{c-1}{a}$$

$$\therefore \qquad du_1=\frac{1-\left(1-\frac{P_2 t}{n}\right)}{P_1+P_2\left(1-\frac{P_1 t}{n}\right)}$$

$$=\frac{P_2 t/n}{P_1+P_2\left(1-\frac{P_1 t}{n}\right)} \qquad \text{...(3)}$$

Fig. 6.8.

For a biconvex lens, $|R_1|=|R_2|=R$.

and $$P_1=P_2=\frac{n-1}{R}$$

$$\therefore \qquad d_{u1}=\frac{t}{n}\frac{1}{\left[2-\frac{t}{n}\frac{n-1}{R}\right]}\approx\frac{t}{2n}\ (\because R>>t) \qquad \text{...(4)}$$

$$d_{u2}=\frac{(c-1)}{a}=\frac{1-\frac{P_1 t}{n}-1}{P_1+P_2\left(1-\frac{P_1 t}{n}\right)}$$

$$= \frac{-P_1 t}{n\left\{P_1 + P_2\left(1 - \frac{P_1 t}{n}\right)\right\}}$$

$$\therefore \qquad d_{u2} = -\frac{t}{n}\frac{1}{\left[2 - \frac{t}{n}\frac{n-1}{R}\right]} \approx -\frac{t}{2n} \qquad \ldots(5)$$

The positions of the unit planes of a thick biconvex lens are shown in Fig. 6.8.

Focal length of a thick lens

We know that if the distances are measured from the unit planes, then the element a in the system matrix represents the inverse of the focal length of the system.

$$\frac{1}{f} = a = P_1 + P_2\left(1 - \frac{t}{n}P_1\right) \qquad \ldots(1)$$

But
$$P_1 = \frac{n-1}{R_1} \text{ and } P_2 = -\frac{(n-1)}{R_2}$$

$$\therefore \qquad \frac{1}{f} = (n-1)\left(\frac{1}{R_1} - \frac{1}{R_2}\right) + \frac{(n-1)^2 t}{nR_1R_2} \qquad \ldots(2)$$

6.13. Location of Nodal Planes

Nodal points are two points on the axis which have a relative angular magnification of unity. A ray striking the first point at an angle α emerges from the second point at the same angle (Fig. 6.9). The planes which pass through the nodal points and are normal to the principal axis are called nodal planes.

To locate the positions of nodal points, consider two points N_1 and N_2 on the axis at distances d_{n1} and d_{n2} from the two refracting surfaces of a thick lens (Fig. 6.9). From the definition of nodal points, we consider a ray passing through N_1 at an angle α with the principal axis and emerging from N_2 at the same angle α. Since the medium on either side of the system is the same *i.e.*, air, $\lambda_1 = \lambda_2$. Further, we are considering the object on the axis of the system, $x_1 = 0$.

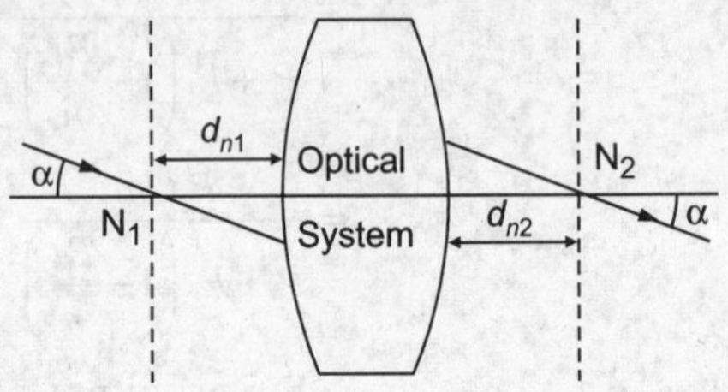

Fig. 6.9.

Now, from Eq. (9) of Art. 6.6, we have

$$\begin{pmatrix}\lambda_2 \\ x_2\end{pmatrix} = \begin{pmatrix} b + aD_1 & -a \\ 0 & c - aD_2\end{pmatrix}\begin{pmatrix}\lambda_1 \\ x_1\end{pmatrix} \qquad \ldots(1)$$

Here, $x_1 = x_2 = 0$, $D_1 = d_{n1}$ and $D_2 = d_{n2}$. Eq. (1) becomes

$$\begin{pmatrix}\lambda_2 \\ 0\end{pmatrix} = \begin{pmatrix} b + ad_{n1} & -a \\ 0 & c - ad_{n2}\end{pmatrix}\begin{pmatrix}\lambda_1 \\ 0\end{pmatrix} \qquad \ldots(2)$$

$$\therefore \qquad \lambda_2 = (b + a\,d_{n1})\,\lambda_1$$

or
$$b + ad_{n1} = 1 \qquad (\because \lambda_1 = \lambda_2)$$

$$\therefore \qquad d_{n1} = \frac{1-b}{a}$$

or $$d_{n1} = d_{u1} \qquad \textbf{...(3)}$$

Now $$b + aD_1 = \frac{1}{c - aD_2} = \frac{1}{M}$$

But $$M = 1$$

$\therefore$ $$c - a\,d_{n2} = 1$$

or $$d_{n2} = \frac{c-1}{a}$$

or $$d_{n2} = d_{u2} \qquad ...(4)$$

Thus when the media on either side of an optical system have the same refractive index, the nodal planes coincide with the unit planes.

Example 3. *A lens of thickness 1 cm, made of a material of refractive index 1.5 is placed in air. The radii of curvatures of the two surfaces are + 4 cm and – 4 cm. Find the system matrix and determine the focal length and the positions of unit points and nodal points.*

Solution. The power P_1 of the first surface is given by

$$P_1 = \frac{(n-1)}{R_1} = \frac{1.5-1}{4} = 0.125 \text{ cm}^{-1}$$

The power of the second surface is given by

$$P_2 = \frac{1-n}{-R_2} = \frac{1-1.5}{-4} = 0.125 \text{ cm}^{-1}$$

The system matrix of a thick lens is given by

$$S = \begin{pmatrix} 1 - \frac{P_2 t}{n} & -P_1 - P_2\left(1 - \frac{P_1 t}{n}\right) \\ \frac{t}{n} & 1 - \frac{P_1 t}{n} \end{pmatrix}$$

$$= \begin{pmatrix} 1 - \frac{0.125 \times 1}{1.5} & -0.125 - 0.125\left(1 - \frac{0.125 \times 1}{1.5}\right) \\ \frac{1}{1.5} & 1 - \frac{0.125 \times 1}{1.5} \end{pmatrix}$$

$$= \begin{pmatrix} 0.9167 & -0.240 \\ 0.6667 & 0.9167 \end{pmatrix}$$

Focal length $= f = \frac{1}{a} = \frac{1}{0.240} \approx 4.2$ cm, $b = 0.9167 = c$, $d = -0.6667$

The position of unit planes are given by

$$d_{u1} = \frac{1-b}{a} = \frac{1-0.9167}{0.240} \approx 0.35 \text{ cm}$$

$$d_{u2} = \frac{c-1}{a} = \frac{0.9167-1}{0.240} \approx -0.35 \text{ cm}$$

The nodal points coincide with the unit points because the lens is immersed in a homogeneous medium.

Example 4. *(a) Consider two coaxially mounted thin lenses, one convex of focal length 20 cm and the other concave of focal length 10 cm, separated by a distance 8 cm. An object 1 cm tall is placed 40 cm from the convex lens. Calculate the position and size of the image assuming lenses to be thin.*

(b) Find cardinal planes of the system and employ them to determine the position and size of the image. Draw the course of rays.

Solution. (*a*) Let the image be formed *v* cm away from the concave lens (Fig. 6.10 *a*).

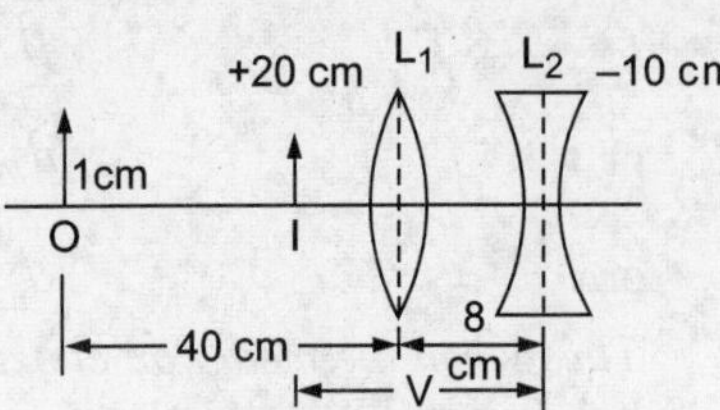

Fig. 6.10(*a*)

Power of convex lens, $P_1 = +\dfrac{1}{20}\ \text{cm}^{-1}$

Power of concave lens, $P_2 = -\dfrac{1}{10}\ \text{cm}^{-1}$

Thus the matrix, which when operated on the object column matrix gives the image column matrix, is given by Σ =

Translation L_2 to I	Refraction at L_2	Translation L_1 to L_2	Refraction at L_1	Translation O to L_1
$\begin{pmatrix}1 & 0\\ v & 1\end{pmatrix}$	$\begin{pmatrix}1 & +1/10\\ 0 & 1\end{pmatrix}$	$\begin{pmatrix}1 & 0\\ 8 & 1\end{pmatrix}$	$\begin{pmatrix}1 & -1/20\\ 0 & 1\end{pmatrix}$	$\begin{pmatrix}1 & 0\\ 40 & 1\end{pmatrix}$

$$= \begin{pmatrix}1 & 0\\ v & 1\end{pmatrix}\begin{pmatrix}2.2 & 0.01\\ +32 & 0.6\end{pmatrix} = \begin{pmatrix}2.2 & 0.01\\ 2.2v+32 & 0.6+0.01\,v\end{pmatrix}$$

The image plane corresponds to $2.2\ v + 32 = 0$

$\therefore\ v \approx -14.5$ cm

The image is formed at a distance of 14.5 cm to the left of the concave lens.

Magnification, $M = 0.6 + 0.01\ v = 0.6 + 0.01 \times (-\ 14.5) = 0.455$.

Magnification is positive. Therefore the image formed is erect and virtual [Fig. 6.10(*a*)].

Size of the image = 1 × 0.455 = 0.455 cm.

(*b*) The system matrix is given by

	Refraction L_2	Translation L_1 to L_2	Refraction L_1
$S =$	$\begin{pmatrix}1 & 1/10\\ 0 & 1\end{pmatrix}$	$\begin{pmatrix}1 & 0\\ 8 & 1\end{pmatrix}$	$\begin{pmatrix}1 & \dfrac{1}{-20}\\ 0 & 1\end{pmatrix}$

$$S = \begin{pmatrix}9/5 & 1/100\\ 8 & 3/5\end{pmatrix} = \begin{pmatrix}b & -a\\ -d & c\end{pmatrix}$$

Here, $a = -\dfrac{1}{100} \Rightarrow f = -100$ cm

$$b = \frac{9}{5},\ c = \frac{3}{5},\ d = -8$$

$$d_{u1} = \frac{1-b}{a} = \frac{1-(9/5)}{-(1/100)} = 80\ \text{cm}$$

Thus the first unit plane is at a distance of 80 cm to the right of the convex lens.

$$d_{u2} = \frac{c-1}{a} = \frac{(3/5)-1}{-(1/100)} = 40\ \text{cm}$$

The second unit plane is at a distance of 40 cm to the right of the concave lens.

Thus the two unit planes are situated on the right side, out of the lens system [Fig. 6.10(*b*)].

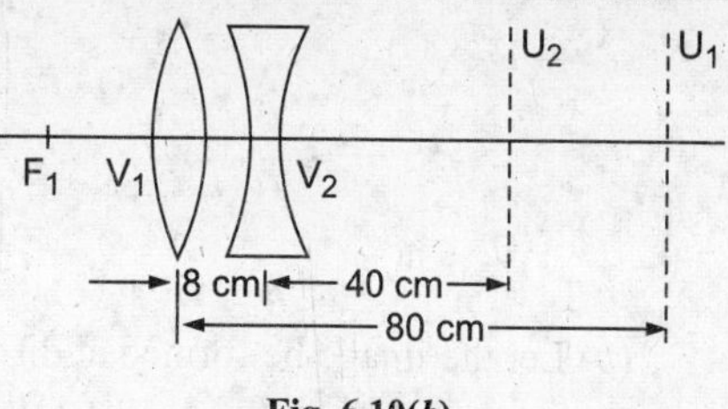

Fig. 6.10(*b*)

Nodal planes coincide with the unit planes.

Distance of the object from the first unit plane is

$$U = -(80 + 40) = -120 \text{ cm}$$

Let the distance of the image from the second unit plane be V. We know that

$$\frac{1}{f} = \frac{1}{V} - \frac{1}{U}$$

$$\therefore \qquad \frac{1}{V} = \frac{1}{f} + \frac{1}{U} = -\frac{1}{100} - \frac{1}{120}$$

$$\therefore \qquad V = -54.54 \text{ cm}$$

Thus the image is at 54.54 cm to the left of the second unit plane or at 14.54 cm to the left of the concave lens.

$$\text{Magnification, } M = \frac{V}{U} = \frac{-54.54}{-120} = 0.455$$

Once again the image formed is virtual and erect of size 0.455 cm. Course of rays is shown in Fig. 6.10(*c*), which has been drawn keeping in view the properties of cardinal points.

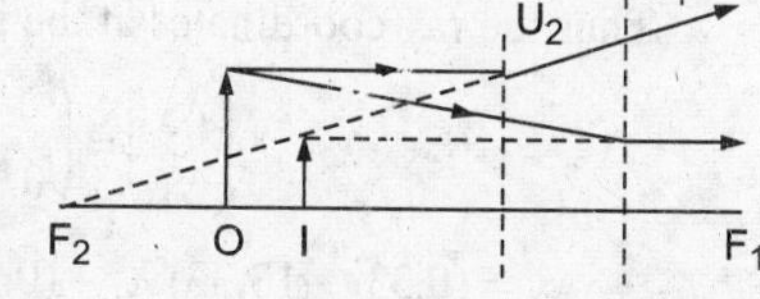

Fig. 6.10(*c*).

Example 5. *For a sphere of radius 10 cm and material of refractive index 1.5, (a) write the refraction matrices at the two surfaces and the translation matrix within the material of the sphere and hence the system matrix, (b) determine the position of the paraxial focal point, (c) find the positions of unit and nodal planes and show them in a diagram.*

Solution. Here, $R_1 = 10$ cm, $R_2 = -10$ cm, $n = 1.5$, $t = 20$ cm.

$$P_1 = \frac{n-1}{R_1} = \frac{1.5-1}{10} = 0.05$$

$$P_2 = \frac{1-n}{R_2} = \frac{1-1.5}{-10} = 0.05$$

(*a*) The refraction matrix at the first surface

$$R_1 = \begin{pmatrix} 1 & -P_1 \\ 0 & 1 \end{pmatrix} = \begin{pmatrix} 1 & -0.05 \\ 0 & 1 \end{pmatrix}$$

The refraction matrix at the second surface

$$R_2 = \begin{pmatrix} 1 & -P_2 \\ 0 & 1 \end{pmatrix} = \begin{pmatrix} 1 & -0.05 \\ 0 & 1 \end{pmatrix}$$

The translation matrix T_{12} between surface S_1 and surface S_2 is

$$T_{12} = \begin{bmatrix} 1 & 0 \\ t/n & 1 \end{bmatrix} = \begin{bmatrix} 1 & 0 \\ 20/1.5 & 1 \end{bmatrix} = \begin{bmatrix} 1 & 0 \\ 13.33 & 1 \end{bmatrix}$$

The system matrix of the sphere is given by

$$S = R_2 T_{12} R_1$$

$$= \begin{pmatrix} 1 & -0.05 \\ 0 & 1 \end{pmatrix} \begin{pmatrix} 1 & 0 \\ 13.33 & 1 \end{pmatrix} \begin{pmatrix} 1 & -0.05 \\ 0 & 1 \end{pmatrix}$$

$$= \begin{pmatrix} 0.33 & -0.066 \\ 13.33 & 0.33 \end{pmatrix}$$

(*b*) Let the image be formed at distance v from second surface.

Translation matrix from second surface to image is

$$\begin{pmatrix} 1 & 0 \\ v & 1 \end{pmatrix}$$

Hence the matrix of operation from the first refracting surface to the image plane would be given by

$$\Sigma = \begin{pmatrix} 1 & 0 \\ v & 1 \end{pmatrix} S = \begin{pmatrix} 1 & 0 \\ v & 1 \end{pmatrix} \begin{pmatrix} 0.33 & -0.066 \\ 13.33 & 0.33 \end{pmatrix}$$

$$= \begin{pmatrix} 0.33 & -0.066 \\ 0.33v + 13.33 & 0.33 - 0.066v \end{pmatrix}$$

Thus the ray coordinates at the image plane would be

$$\begin{pmatrix} \lambda_2 \\ x_2 \end{pmatrix} = \begin{pmatrix} 0.33 & -0.066 \\ 0.33v + 13.33 & 0.33 - 0.066v \end{pmatrix} \begin{pmatrix} \lambda_1 \\ x_1 \end{pmatrix}$$

$$\therefore \quad x_2 = (0.33v + 13.33)\,\lambda_1 + (0.33 - 0.066v)\,x_1$$

In order to determine the focal distance v, we consider a ray incident parallel to axis for which $\lambda_1 = 0$ (Fig. 6.11). The focal plane would be that plane for which x_2 is also zero. Therefore, putting $\lambda_1 = 0$ and $x_2 = 0$ in the above Eq., we get

$$(0.33 - 0.066\,v)\,x_1 = 0$$

Now $x_1 \varpi 0$. Hence $0.33 - 0.066v = 0$

$$\therefore \quad v = \frac{0.33}{0.066} = \mathbf{5\ cm}$$

(*c*) From the system matrix S, we have

$$a = 0.066,\ b = 0.33,\ c = 0.33, \text{ and } d = -13.33$$

Now
$$f_2 = \frac{1}{a} = \frac{1}{0.066} = \mathbf{15\ cm}$$

$$d_{u1} = \frac{1-b}{a} = \frac{1-0.33}{0.066} = \mathbf{10\ cm}$$

$$d_{u2} = \frac{c-1}{a} = \frac{0.33-1}{0.066} = \mathbf{-10\ cm}$$

Thus both the unit planes pass through the centre of the sphere.

$$F_1 = -\frac{b}{a} = \frac{-0.33}{0.066} = \mathbf{-5\ cm}$$

$$F_2 = \frac{c}{a} = \frac{0.33}{0.066} = \mathbf{5\ cm}$$

All the points are shown in Fig. 6.11.

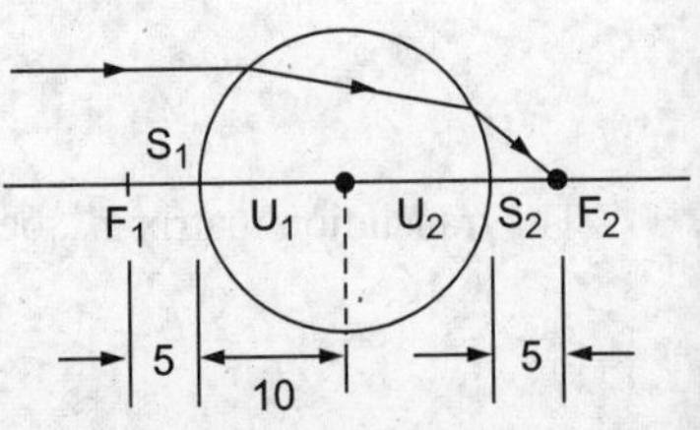

Fig. 6.11.

EXERCISE

1. Explain the translation, refraction and system matrices. Obtain the system matrix for a thin lens and hence obtain the formula for a thin lens. *(Osmania University, 89)*
2. Obtain the system matrix for a thin lens and hence derive a relation for the focal length of a combination of two thin lenses separated by a distance. *(Kakatiya U., 90)*
3. Obtain Snell's law for paraxial rays and hence derive a matrix for the combination of refraction and translation for light rays passing through a thick lens. *(Andhra University, 89)*
4. Explain how translation and refraction matrices are formed and hence build up the system matrix for a thick lens. *(S.V.U., April 90)*
5. Formulate the equations for the image plane and magnification of an optical system using the matrix method. Explain the terms unit planes and nodal planes briefly. *(K.U., 92)*
6. Explain the terms system matrix and nodal planes. Formulate the system matrix for a thick lens and hence derive the thick lens formula. *(K.U., 91)*
7. Calculate the position of the paraxial focal point of a lens in the form of a sphere of radius 20 cm of refractive index 1.6. **[Ans.** 0.67 cm]
8. A convergent thick lens has radii of curvature + 6 cm and –6 cm and thickness 2 cm. If it is made of a material of refractive index 1.6 and is placed in air, obtain the system matrix. Determine its focal-length and the position of unit points and nodal points.

   [Ans. $\begin{pmatrix} 0.875 & -0.1875 \\ 1.25 & 0.875 \end{pmatrix}$, $f = 5.3$ cm, $d_{u1} = 0.67$ cm

   and $d_{u2} = -0.67$ cm
   The nodal points coincide with unit points]

CHAPTER

7

COHERENCE

7.1. Introduction

Coherence—A wave which appears to be a pure sine wave for an infinitely large period of time or in an infinitely extended space is said to be a perfectly coherent wave. In such a wave, there is a definite relationship between the phase of the wave at a given time and at a certain time later, or at a given point and at a certain distance away.

Let E represent the electric field associated with the light wave. We assume that

$$E = A \cos (kx - \omega t + \phi)$$

The above equation predicts that at any value of x, the displacement is sinusoidal for $-\infty < t < \infty$. No actual light source, however, emits a perfectly coherent wave.

There are two different criteria of coherence : the criteria of time and the criteria of space. This gives rise to temporal coherence and spatial coherence.

Temporal Coherence—The oscillating electric field E of a perfectly coherent light wave would have a constant amplitude of vibration at any point, while its phase would vary linearly with time. As a function of time, the field would appear as shown in Fig. 7.1. It is an ideal sinusoidal function of time.

Fig. 7.1.

However, no light emitted by an actual source produces an ideal sinusoidal field for all values of time. The radiation from an ordinary light source consists of finite size wavetrains. Hence the field due to an actual light source will be as shown in Fig. 7.2. Now, in Fig. 7.2, τ_c represents the average duration of the wavetrains, *i.e.,* the electric field remains sinusoidal for time-intervals of the order of τ_c. After this time-interval, the phase changes abruptly. Thus, at a given point, the electric fields at times t and $t + \Delta t$ will, in general, have a definite phase relationship if $\Delta t << \tau_c$ and will (almost) never have any phase relationship if $\Delta t >> \tau_c$. The average time-interval for which the field remains sinusoidal (*i.e.,* definite phase relationship exists) is known as "coherence time" of the source and is denoted by τ_c. The distance L for which the field is sinusoidal is given by

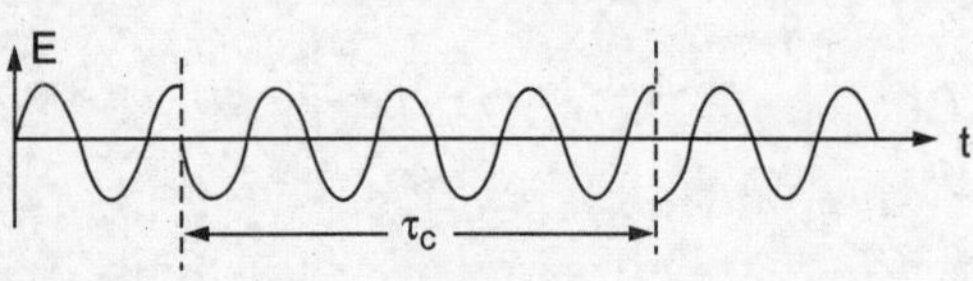

Fig. 7.2.

$$L = c\tau_c.$$

where c is the speed of light in free space. L is called the "coherence length". For example, for the neon 632.8 nm line, $\tau_c \sim 10^{-10}$ s and the corresponding coherence length is 0.03 m.

The finite coherence time could be due to many factors. For example, if a radiating atom undergoes collision with another atom, then the wavetrain undergoes an abrupt phase shift. It may be due to the random motion of atoms or due to the fact that an atom has finite lifetime in the energy level from which it drops to the lower energy level while radiating.

The coherence time (or the coherence length) can be measured by means of Michelson's interferometer (Fig. 7.3). A light beam from the source S falls on a semi-silvered plate P at which it is partly reflected and partly transmitted. The reflected and transmitted beams, 1 and 2, are reflected back from mirrors M_1 and M_2 respectively and enter the telescope T. Interference fringes are observed in T. We know that the two beams can produce a stationary interference pattern only if there is a definite phase relationship between them.

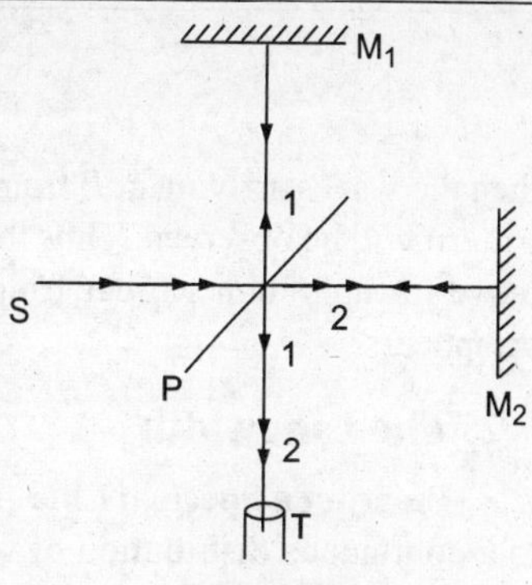

Fig. 7.3.

Let M_2' be the image of M_2 formed by the plate P. Let the distance M_1M_2' be denoted by d. Then the beam which gets reflected by mirror M_2 travels an additional path equal to $2d$. Thus, the beam reflected from M_2 interferes with the beam reflected by M_1 which had originated $2d/c$ seconds earlier.

If the distance d is such that

$$\frac{2d}{c} << \tau_c$$

then there will be a definite phase relationship between the two beams and well defined interference fringes are observed. On the other hand, if

$$\frac{2d}{c} >> \tau_c$$

then, there will be no definite phase relationship between the two beams and the fringes will not be observed. Therefore, starting with equal path-lengths, as the distance d is increased (by moving one mirror), the fringes become gradually poorer in contrast and finally disappear. For the neon 632.8 nm line, the disappearance occurs when the path difference is about a few centimetres giving $\tau_c \sim 10^{-10}$ s

The coherence time for a laser beam is usually much larger in comparison to ordinary light sources. Commercially available helium-neon lasers have $\tau_c \sim 50$ ns implying coherence lengths of about 15 m. Thus using such a laser beam, high contrast interference fringes can be obtained even for a path difference of a few metres.

Spatial Coherence—The spatial coherence is the phase relationship between the radiation fields at different points in space. Two fields at two different points on a wavefront of a given electromagnetic wave are said to be space coherent if they preserve a constant phase difference over any time t, *i.e.,* space coherency requires that the waves not only are of the same frequency, but that they are in phase in space.

Consider Young's double-slit experiment (Fig. 7.4). Light emanating from a *narrow* slit S falls on two slits S_1 and S_2 placed symmetrically with respect to S. Consider a point P on the screen. Let $S_1P = r_1$ and $S_2P = r_2$. The interference pattern observed around the point P at time t is due to the superposition of waves emanating from S_1 and S_2 at times $t - (r_1/c)$ and $t - (r_2/c)$ respectively.

Fig. 7.4.

If

$$\frac{r_2 - r_1}{c} << \tau_c$$

then the waves arriving at P from S_1 and S_2 will have a definite phase relationship and interference fringes of good contrast will be observed. But if the path difference $(r_2 - r_1)$ is large enough such that

$$\frac{r_2 - r_1}{c} >> \tau_c$$

then the waves arriving at P from S_1 and S_2 will have no fixed phase relationship and no interference pattern will be observed. Thus the central fringe (for which $r_1 = r_2$) will have a good contrast. As we move towards higher order fringes, the pattern becomes poorer and poorer in contrast and finally disappears.

7.2. The Linewidth

Purity of a Spectral Line—Every spectral line has a finite width which means that it corresponds to a continuous distribution of wavelengths in some narrow interval between λ and $\lambda + \Delta\lambda$. In the Michelson interferometer experiment, the decrease in the contrast of the fringes is due to the fact that the source is not emitting at a single frequency but over a narrow band of frequencies. When the path difference between the two interfering beams is zero or very small, the different wavelength components produce fringes superimposed on one another. Hence the fringe contrast is good. When the path difference between the interfering beams is gradually increased, different wavelength components produce fringe patterns which are slightly displaced with respect to one another. Hence the fringe contrast becomes poorer. The pattern disappears when the path difference exceeds the coherence length. The poor fringe visibility for a large path difference is due to the non-monochromaticity of the light source. Thus the concept of temporal coherence is directly related to the width (or purity) of the spectral line.

Consider the Michelson interferometer experiment using two closely spaced wavelengths λ_1 and λ_2 (See Sec. 2.12). For two closely spaced wavelengths λ_1 and λ_2, the interference pattern disappears if

$$\frac{2d}{\lambda_2} - \frac{2d}{\lambda_1} = \frac{1}{2}$$

where $2d$ represents the path difference between the two beams. Thus

$$2d = \frac{\lambda_1\lambda_2}{2(\lambda_1 - \lambda_2)} \simeq \frac{\lambda^2}{2(\lambda_1 - \lambda_2)}$$

Let us adopt the criterion that when the path difference between the interfering beams becomes equal to the coherence length L, the rings due to two closely spaced wavelengths λ_1 and λ_2 are completely out of step at the centre. Then we can write

$$L = \frac{\lambda^2}{2(\lambda_1 - \lambda_2)}.$$

Instead of two discrete wavelengths, λ_1 and λ_2, let us assume that the beam consists of all wavelengths lying between λ and $\lambda + \Delta\lambda$ Then the pattern would disappear if

$$L \simeq \frac{\lambda^2}{\Delta\lambda}$$

The spectral width of the source $\Delta\lambda$ is given by

$$\Delta\lambda \sim \frac{\lambda^2}{L} = \frac{\lambda^2}{c\tau_c}.$$

Thus the temporal coherence τ_c of the beam is directly related to the spectral width $\Delta\lambda$.

Further, since $\nu = \frac{c}{\lambda}$, the frequency-spread $\Delta\nu$ of a line would be

$$\Delta\nu = \frac{c}{\lambda^2}\Delta\lambda \simeq \frac{c}{L}$$

We know that $\tau_c = \frac{L}{c}$, where τ_c is coherence time.

$$\therefore \qquad \Delta\nu \simeq \frac{1}{\tau_c}.$$

Thus the frequency-spread of a spectral line is of the order of the inverse of the coherence time. It means that a perfectly sharp monochromatic line ($\Delta\nu = 0$) would correspond to an infinite interval of time (τ_c = ').

The quantity $\Delta\nu / \nu$ represents the monochromaticity (or the spectral purity) of the source.

7.3. The Spatial Coherence

Let us discuss the coherence properties of the field associated with the finite dimension of the source. Consider the case when the Young's double-slit is illuminated by two independent point-sources S and S′ at a distance l apart (Fig. 7.5). We shall find minimum value of l at which pattern on the screen would disappear.

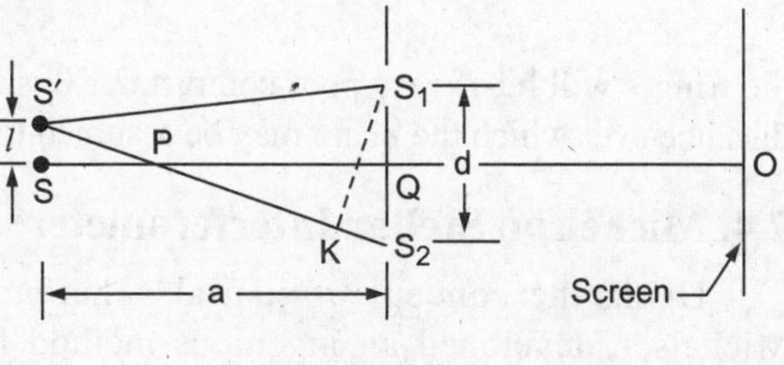

Fig. 7.5.

The waves from S which reach the point O on the screen via S_1 and S_2 have zero path difference. Hence there is a bright fringe at O due to S. Now, the waves from S' reaching the point O via S_1 and S_2 have a path difference $S'S_2 - S'S_1 = KS_2$. We shall obtain a dark fringe at O due to S' when S to t

$$KS_2 = \frac{\lambda}{2} \qquad \text{...(1)}$$

where λ is the wavelength of light. When this is the case the maxima of the interference pattern due to S will fall on the minima due to S' so that the fringes would disappear.

Let a be the distance between S and Q. d = the separation between S_1 and S_2. Now,

$$S'S_2 = \left[a^2 + \left(\frac{d}{2}+l\right)^2\right]^{1/2} \simeq a + \frac{1}{2a}\left(\frac{d}{2}+l\right)^2$$

$$S'S_1 = \left[a^2 + \left(\frac{d}{2}-l\right)^2\right]^{1/2} \simeq a + \frac{1}{2a}\left(\frac{d}{2}-l\right)^2$$

Here, we have assumed $a >> d, l$. Thus

$$KS_2 = S'S_2 - S'S_1 \simeq \frac{ld}{a}.$$

For the fringes to disappear, we must have

$$KS_2 = \frac{\lambda}{2} \simeq \frac{ld}{a}$$

or

$$l \simeq \frac{\lambda a}{2d}$$

From this it is clear that if we illuminate the double-slit with an entended source whose linear dimension exceeds $\frac{\lambda a}{2d}$, then no interference pattern will appear on the screen.

Suppose we have an extended incoherent source whose linear dimension is ~ $\lambda a / d$. Then for every point on the source, there is a point at a distance of $\lambda a / 2d$ which produces fringes which are shifted by half a fringe width. Therefore, the interference pattern will not be observed. Thus for an extended incoherent source, interference fringes of good contrast will be observed only when

$$l << \frac{\lambda a}{d} \qquad ...(2)$$

Let θ be the angle subtended by the extended source SS' at Q. Then, $\theta = l / a$. Eq. (2) becomes

$$d << \frac{\lambda}{\theta} \qquad ...(3)$$

But, if

$$d \sim \frac{\lambda}{\theta} \qquad ...(4)$$

the fringes will be of very poor contrast. λ / θ is termed as the *lateral coherence width*. It gives the distance over which the beam may be assumed to be spatially coherent.

7.4. Michelson Stellar Interferometer

Using the concept of spatial coherence, Michelson developed an ingenious method for determining the angular diameters of stars. In this interferometer, the separation of effective slits for receiving star light may be increased considerably while the separation of slits for producing fringes was small and fixed. He accomplished this by mounting four mirrors M_1, M_2, M_3, M_4 on a long iron girder in front of the telescope objective (Fig. 7.6). S_1 and S_2 are slits in front of the telescope objective L. Light from the star is incident on the mirrors M_1 and M_2 which reflect into M_3 and M_4 and which, in their turn, pass light into telescope through the apertures of double slit. When the mirrors have been adjusted so that M_1 is parallel to M_3 and M_2 to M_4 and the optical paths $M_1M_3S_1$ and $M_2M_4S_2$ are equal, the beams overlap in the focal plane of the objective and the *Airy's disc* is seen crossed by a number of fine fringes. Airy's disc is elongated and much larger than would be formed by the telescope alone because of its restricted aperture in the present case. Starting with M_1 and M_2 close to the fixed mirrors M_3 and M_4, they are moved out symmetrically maintaining optical paths $M_1M_3S_1$ and $M_2M_4S_2$ equal until the fringes disappeared. Let a be the least separation of the mirrors M_1 and M_2 for disappearance of fringes. Then, the angular diameter of the star is

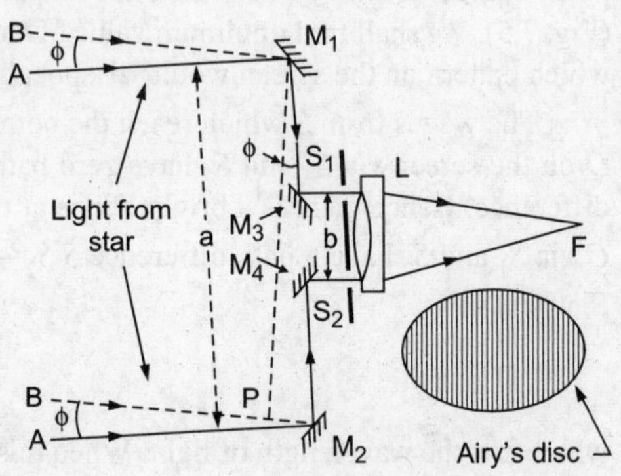

Fig. 7.6.

$$\phi = 1.22\, \lambda / a.$$

λ is the effective wavelength for the star.

The fringe spacing, which depends on the separation b of S_1 and S_2, remains constant as the separation a of the movable apertures M_1 and M_2 is varied.

In a typical experiment, the first disappearance occurred when the distance M_1M_2 was about 7.315 m. He assumed that for the star Arctures $\lambda = 5 \times 10^{-7}$ m.

$$\phi = \frac{1.22 \times (5 \times 10^{-7})}{7.315} \text{ radians} = 8.34 \times 10^{-8} \text{ radians}$$

$$= (8.34 \times 10^{-8}) \times \frac{180}{\pi} \times 3600 \text{ sec. of arc} = 0.0172'' \simeq 0.02''$$

The distance of the star is known from its parallax. Hence we can estimate the linear diameter of the star. It is about 27 times that of the sun.

EXERCISE

1. Explain the concept of coherence. Discuss temporal coherence and spatial coherence, illustrating them with the help of suitable experiments.
2. Explain how the purity of a spectral line is related with the concept of temporal coherence. Show that the frequency-spread of a spectral line is of the order of the inverse of the coherence time.
3. If light of $\lambda = 660$ nm has wavetrains 20 λ long, what is its :

   (*a*) coherence length ;

   (*b*) coherence time? **[Ans.** (*a*) 13.2 μm; (*b*) 4.4×10^{-14} s**]**

CHAPTER 8

FIBRE OPTICS

8.1. Introduction

Optical frequencies are very large (~ 10^{15} Hz) as compared to the conventional radio waves (~ 10^{6} Hz) and microwaves (~ 10^{10} Hz). So a light beam acting as a carrier wave is capable of carrying far more information than radio waves and microwaves. Thus optical communication is superior to radio and microwave communication. *Optical fibre* is a hair structure made up of transparent material which can guide the light beam from one place to another by total internal reflection. The combination of low loss (~ 0.2 dB/km) and wide bandwidth available at optical frequencies make these fibres extremely attractive for the transmission medium in communication systems. Fig. 8.1 compares attenuation versus carrier frequency for a single-mode silica fibre operated at 1.55 μm and three commonly used metallic guiding media. The metallic transmission lines suffer from exponentially increasing conduction losses at higher operating frequency (and band width). But the dielectric fibre exhibits no such losses. For signal frequencies above several kilohertz, the losses of the fibre line are far less, thereby allowing repeaters (amplifiers) to be spaced farther apart. *Optical fibres* are replacing the *copper cables* previously used in telecommunications. Currently, throughout the world fibre optics is being used to transmit voice, television and digital data signals by light waves over flexible hair-thin threads of glass or plastic. Optical fibres are also being extensively used for local area networks (LANS).

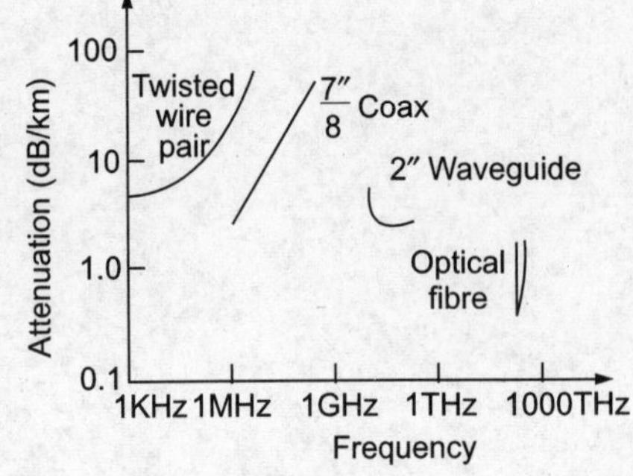

Fig. 8.1.

8.2. Fibre Construction

An optical fibre consists of a central cylindrical core with refractive index n_1 surrounded by a layer of material, called the *cladding*, with a lower refractive index n_2 (Fig. 8.2). The core transmits the light waves, and the cladding keeps the light waves within the core and strengthens the fibre. The core and the cladding are made of either glass or plastic. Core diameters range from 5 to 600 μm, and cladding diameters vary from 125 to 750 μm. To keep the light within the core, the cladding must have a minimum thickness of one or two wavelengths of the light transmitted.

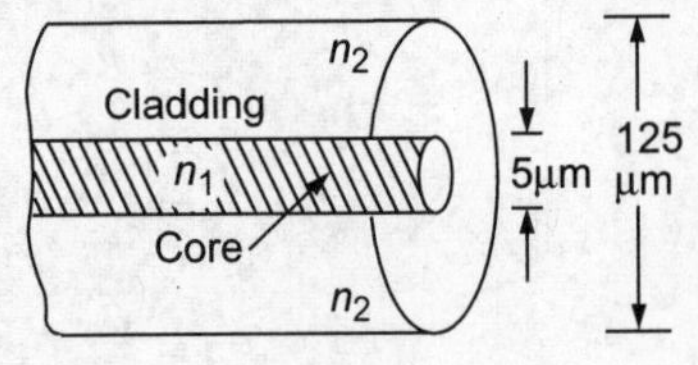

Fig. 8.2.

8.3. Light Propagation in Fibres

Let a = core radius and b = cladding radius.

The refractive index distribution (in the transverse direction) is given by

$$n(r) = n_1 \quad 0 < r < a \qquad \text{CORE}$$
$$= n_2 \quad r > a \qquad \text{CLADDING} \quad ...(1)$$

The variation in refractive index along the radial direction of the fibre is shown in Fig. 8.3.

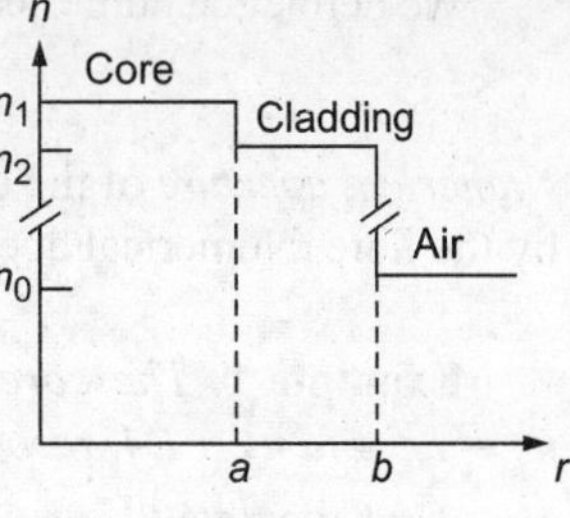

Fig. 8.3.

Consider a ray which is incident on the entrance aperture of the fibre making an angle *i* with the axis (Fig. 8.4). Let the refracted ray make an angle θ with the axis. Let ϕ be the angle incidence of the ray at the core-cladding interface. Let ϕ_c be the critical angle of the core-cladding interface. Then,

$$\sin \phi_c = n_2 / n_1.$$

If ϕ is greater than the critical angle $\phi_c = (\sin^{-1} n_2 / n_1)$, then the ray will undergo total internal reflection at that interface. Further, because of the cylindrical symmetry in the fibre structure, this ray will suffer total internal reflection at the lower interface also and therefore get guided through the core by repeated total internal reflections. In this way, by total internal reflection, a ray of light entering one end of a fibre can travel along the fibre by *multiple reflections* with a fairly high intensity. Even for a bent fibre, light guidance can occur through multiple total internal reflections.

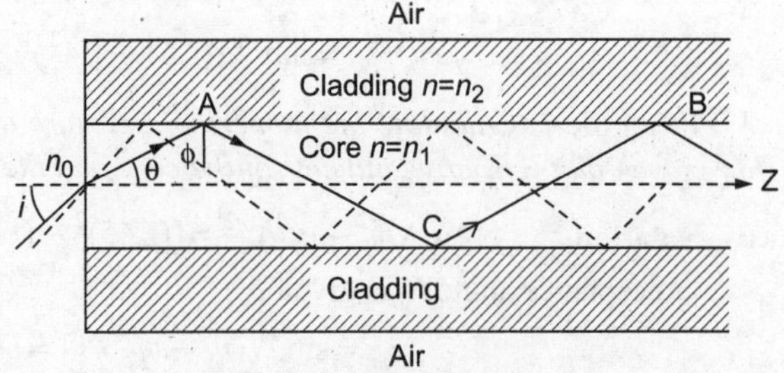

Fig. 8.4.

The Numerical Aperture. Let n_0 be the refractive index of the outside medium.

Then, using Snell's law,

$$\frac{\sin i}{\sin \theta} = \frac{n_1}{n_0} \quad ...(1)$$

If this ray has to suffer total internal reflection at the core-cladding interface,

$$\sin \phi = (\cos \theta) > n_2 / n_1 \quad ...(2)$$

$$\therefore \quad \sin \theta < [1 - (n_2 / n_1)^2]^{1/2}$$

But $$\sin i = \frac{n_1}{n_0} \sin \theta$$

$$\therefore \quad \sin i < \frac{n_1}{n_0}\left[1 - \left\{\frac{n_2}{n_1}\right\}^2\right]^{1/2} = \left\{\frac{n_1^2 - n_2^2}{n_0^2}\right\}^{1/2}$$

If $(n_1^2 - n_2^2) \geq n_0^2$, then for all values of *i*, total internal reflection will occur.

Assuming $n_0 = 1$. the maximum value of sin *i* for a ray to be guided is given by

$$\sin i_m = \begin{cases} (n_1^2 - n_2^2)^{1/2} & \text{when } n_1^2 - n_2^2 < 1 \\ 1 & \text{when } n_1^2 - n_2^2 > 1 \end{cases} \quad ...(3)$$

Thus, if a cone of light is incident on one end of the fibre, it will be guided through it provided the semi-angle of the cone is less than i_m. *Angle of acceptance* is the maximum angle i_m with which a ray can enter the end of the fibre and still be totally internally reflected. This angle is a measure of the *light gathering power* of the fibre.

We define the numerical aperture (NA) of the fibre by the following equation :

$$NA = (n_1^2 - n_2^2)^{1/2} \quad \ldots(4)$$

Numerical aperture of the optical fibre is a measure of the amount of light rays that can be accepted by the fibre. Numerical aperture is the sine of angle of acceptance.

$$NA = \sin i_m \quad \ldots(5)$$

Example 1. *The core and the cladding of a silica optical fibre have refractive indices of $n_1 = 1.5$ and $n_2 = 1.4$, respectively.*

(a) Calculate the critical angle of reflection for the core-cladding boundary.

(b) Calculate the acceptance angle of the fibre.

Solution. (*a*) $\phi_c = \sin^{-1}(n_2/n_1) = \sin^{-1}(1.4/1.5) = 68.96°$

(*b*) $i_m = \sin^{-1}[(n_1^2 - n_2^2)^{1/2}]$

$= \sin^{-1}[(1.5^2 - 1.4^2)^{1/2}] = 32.58°$

Example 2. *Calculate the numerical aperture and hence the acceptance angle for an optical fibre given that refractive indices of the core and the cladding are 1.45 and 1.40 respectively.*

Solution. $NA = (n_1^2 - n_2^2)^{1/2} = [(1.45)^2 - (1.40)^2]^{1/2} = 0.3775$

Acceptance angle i_m is given by

$$i_m = \sin^{-1}(n_1^2 - n_2^2)^{1/2} = \sin^{-1} 0.3775 = 22.18°$$

8.4. The Coherent Bundle

If a large number of fibres are put together, it forms a bundle. A bundle may consist of thousands of individual fibres with diameter in the range 0.0002 cm to 0.001 cm. If the fibres are aligned properly, *i.e.*, if the relative positions of the fibres in the input and output ends are the same, the bundle is said to form a *coherent bundle*. A bright (or dark) spot at the input end of the fibre produces a bright (or dark) spot at the output end. Thus an image will be transmitted (in the form of bright and dark spots) through a bundle of aligned fibres. The bundle has the advantage of flexibility.

The lasers in combination with fibre optic probes permit the visualization of internal portions of human body. A typical fibre optic endoscope is shown in Fig. 8.5. It can be put inside a human body and the inner parts of the human body can be viewed from outside.

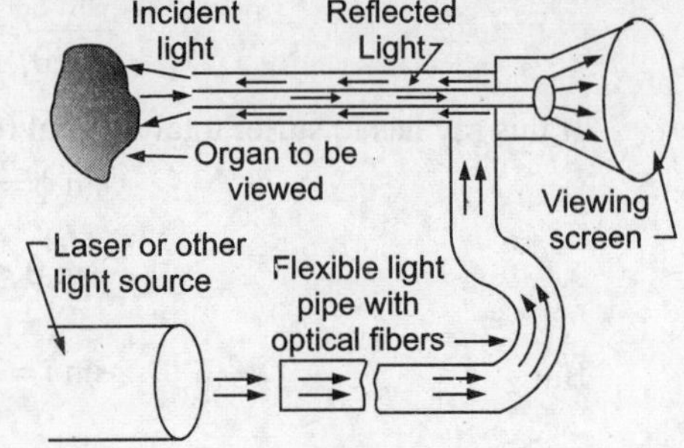

Fig. 8.5.

8.5. The Fibre-Optic Communication System

Fig. 8.6 is a schematic diagram of a fibre optic communication system. The major components of an optical communication system are the optical transmitter, the optical fibre and the optical receiver. The transmitter usually consists of a laser diode or a light emitting diode (LED). The optical fibre essentially serves the purpose of transmitting the light waves. The detector can be either an avalanche photo diode (APD) or a positive intrinsic negative (PIN) diode. Basically, a fibre optic system simply converts an electrical signal to an infrared light signal. This signal is transmitted through an optical fibre. At the end of the optical fibre it is reconverted into an electric signal.

Two types of light wave modulation are possible. (*a*) Analog and (*b*) Digital.

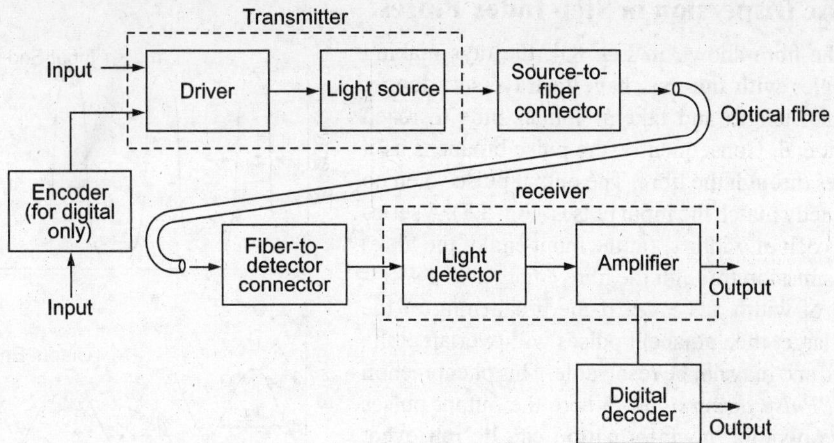

Fig. 8.6.

In the analog modulation the intensity of the light from the laser or LED is varied continuously *i.e.*, the light source emits a continuous beam of varying intensity. In digital modulation, the intensity is changed in an on/off fashion *i.e.*, the light flashes on and off at an extremely fast rate.

In the fibre optic system, the encoder and decoder circuits are necessary only for digital signals. Encoder and decoder are electrical circuits in which information is encoded and decoded respectively. The information is encoded and decoded into binary sequence of zeros and ones. In the light wave transmitter each 'one', corresponds to an electrical pulse and 'zero' corresponds to an absence of a pulse. These electrical pulses are used to turn a light source on and off very rapidly. The driver converts the incoming electrical signal, whether digital or analog into a form that will operate the light source. Even though light sources and light detectors are minute in size, special connectors must be used to couple the light from the source to the fibre and from the fibre to the detectors. In order to overcome the problem of attenuation of light waves in fibres, repeaters are employed. These amplify and regenerate the light signals at regular distances.

Using the system of *time division multiplexing*, many thousands of telephone calls can be transmitted through a single fibre with the use of digital pulses.

8.6. Advantages of Fibre-Optic Communication Systems

(*i*) *Extremely wide band width* : Optical frequencies are very large (~ 10^{15} Hz) as compared to radio frequencies (~ 10^6 Hz) and microwave frequencies (~ 10^{10} Hz). The rate at which information can be transmitted is directly related to signal frequency. Therefore a transmission system that operates at the frequency of light can theoretically transmit information at a higher rate than systems that operate at radio frequencies or microwave frequencies.

(*ii*) *Lack of cross talk between parallel fibres*: There is virtually no signal leakage from fibres and hence cross-talks between neighbouring fibres are almost absent. This is quite frequent in conventional metallic systems.

(*iii*) *Immunity to inductive interference*: Since optical fibres are not metallic, they do not pick up electromagnetic waves. The result is noise free transmission *i.e.*, fibre optic cables are immune to interference caused by lightning or other electromagnetic equipment.

(*iv*) *Smaller-diameter, lighter-weight cables :* Optical fibres because of their light weight and flexibility can be handled more easily than copper cables.

(*v*) Since signals do not leak away from the fibres, they are resistant to intrusion and are accordingly highly suited to secure communications in defence communication networks.

8.7. Pulse Dispersion in Step Index Fibres

In the fibre shown in Fig. 8.4, the rays making larger angles with the axis have to traverse a longer optical path length and take a longer time to reach the output end. Consequently, the pulse broadens as it propagates through the fibre. The output pulses then no longer exactly match the input pulses (Fig. 8.7). A series of pulses each of width τ_1 (at the input end of the fibre) after transmission through the fibre emerges as a series of pulses of width τ_2 ($> \tau_1$). If the broadening of the pulses is large, then adjacent pulses will overlap at the output end and may not be resolvable. This phenomenon is called "*Pulse dispersion*". Where the output pulses are not resolvable, no information can be retrieved. *Thus the smaller the pulse dispersion, the greater will be the information carrying capacity of the system.* Therefore, for a very high information carrying system, it is necessary to reduce the pulse dispersion.

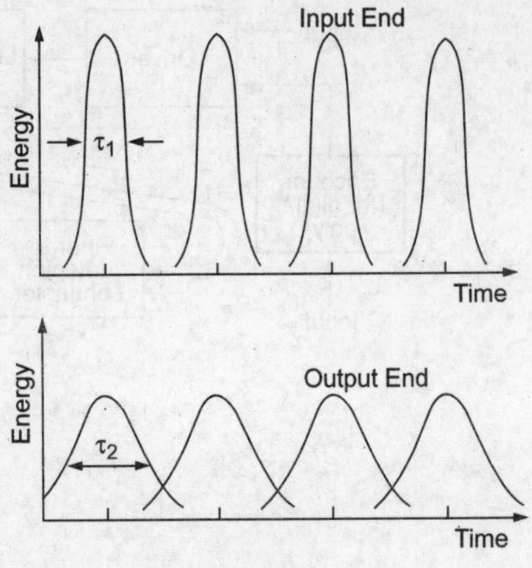

Fig. 8.7.

Let us calculate the amount of dispersion in a step index fibre. Step-index fibre consists of a transparent core of glass of uniform refractive index n_1 surrounded by a cladding layer of uniform lower refractive index n_2 (Fig. 8.4). Consider a ray making an angle θ with the Z-axis. The speed of light in a medium of refractive index n_1 is c/n_1. The distance AB is traversed in time

$$\tau = \frac{AC + CB}{c / n_1} = \frac{n_1 (AB)}{c \cos \theta}.$$

The ray path repeats itself. Hence the time taken by a ray to traverse a length L of the fibre is

$$\tau = \frac{n_1 L}{c \cos \theta} \qquad \text{...(1)}$$

We assume that all rays lying between 0 and θ_c are present.

Time taken by rays corresponding to $\theta = 0$ is

$$\tau_{min} = \frac{n_1 L}{c} \qquad \text{...(2)}$$

Time taken by rays corresponding to $\theta = \theta_c = \cos^{-1} (n_2 / n_1)$ is

$$\tau_{max} = \frac{n_1^2 L}{n_2 c} \qquad \text{...(3)}$$

Let all the input rays be excited simultaneously. Then the rays at the output end occupy a time interval of duration

$$\Delta\tau = \tau_{max} - \tau_{min} = \frac{n_1 L}{c} \left(\frac{n_1}{n_2} - 1 \right) \qquad \text{...(4)}$$

8.8. Graded Index Fibre

GRIN fibre is ~~one~~ in which refractive index varies radially, decreasing continuously in a parabolic manner, from a maximum value of n_1 at the centre of the core to a constant value of n_2 at the core-cladding interface. (Fig. 8.8).

In a graded-index fibre, light rays travel at different speeds in different parts of the fibre because the refractive index varies throughout the fibre. Near the outer edge the refractive index is lower. As a result, rays near the outer edge travel faster than rays in the centre of the core. Because of this all the rays arrive at the end of the fibre at approximately the same time. In effect light rays in these fibres are continuously refocussed as they travel down the fibre (Fig. 8.9). All rays (in the paraxial approximation) take the same amount of time in traversing the fibre. This leads to small pulse dispersions.

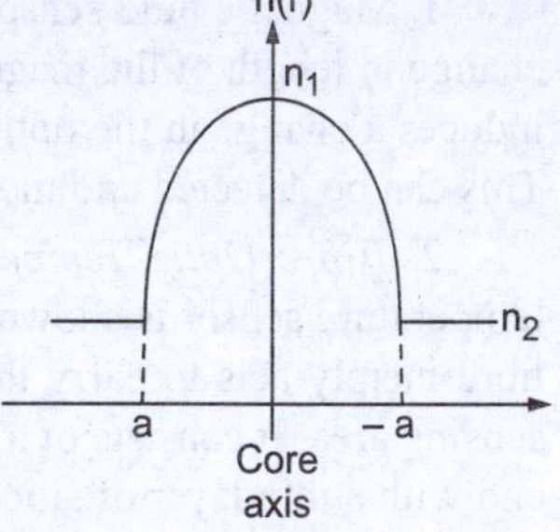

Fig. 8.8.

The pulse dispersion is given by $\Delta\tau = \tau_{max} - \tau_{min} = \dfrac{n_2 L}{2c}\Delta^2$

where $\Delta \equiv \dfrac{n_1 - n_2}{n_2}$.

For a parabolic index fibre, the pulse dispersion is reduced by a factor of about 200 in comparison to the step index fibre. It is because of this reason that first and second generation optical communication systems used near parabolic index fibres.

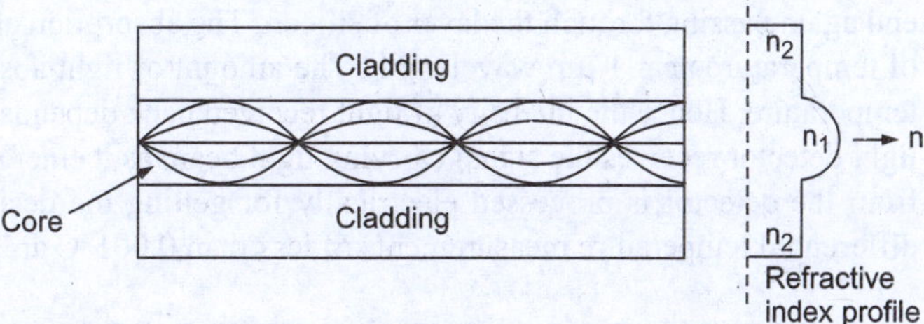

Fig. 8.9.

8.9. Single Mode Fibres

In general, optical fibres can be divided into two broad categories, *single mode* and *multimode*. In single-mode fibres, only one mode is allowed to propagate in the fibre. Mode means simply the various paths light can take in a fibre. Multimode means that several paths are available. According to wave theory, if the diameter of the fibre (or if the refractive index difference between the core and the cladding) becomes very small, in a fibre, rays making only certain discrete angles θ are possible. If the core radius a or $n_1 - n_2$ decreases, then the possible value of θ also decreases. If the combinations of these two fibre parameters are such that the following quantity (often called the V-number of the fibre) defined as

$$V = \frac{2\pi}{\lambda_0} a \sqrt{n_1^2 - n_2^2}$$

is less than 2.4045, then only one guided mode (as if there is only one discrete value of θ) is possible and the fibre is known as a single mode fibre.

Only a single ray path is possible in a single mode fibre. Hence, the dispersion caused due to the differences in the transit times of different rays in a multimode fibre would be completely absent.

The dispersion in a single-mode fibre is due primarily to absorption by the material of the core. Thus, a single-mode fibre is the most suitable for long-distance communication and is capable of handling very broad-band signals. For example, figures of 10 Gbits/sec over 10 km have been reported. Since the core of a single-mode fibre is very small, launching light into the fibre is a difficult task, but keeping the insertion of light constant over a long period of time is even more difficult. Nevertheless, these basic problems have been solved, and single-mode fibres are expected to be in common use for high-capacity communication links in the near future.

8.10. Fibre Optic Sensors

In fibre optic sensors we try to convert various input variables into light signals. Fibre optic sensors are used for sensing and measuring of acoustic fields, magnetic fields, current, rotation, acceleration, strain, pressure, temperature and so on.

1. Magnetic field sensors are based on attachment of fibre to a magnetostrictive material. The change in length of the magnetostrictive material along its length in the presence of magnetic field induces a change in the optical path in an optical fibre by means of longitudinal strain in the fibre. This can be detected and measured using an interferometer.

2. *Fibre Optic Temperature Sensor*: A fibre optic temperature sensor is shown in Fig. 8.10. Here the optical fibre merely acts to carry the light beam to and from the sensing area. It consists of a multimode fibre coated at one end with a thin layer of silicon which is then covered with a reflective coating. Light from a light source is coupled into the fibre. Light from one end passes first through the fibre, then through the silicon and after reflection (from the reflection coating) back to the same fibre end again passing through the layer of silicon. The absorption characteristics of silicon is a function of temperature near 1 μm wavelength. The amount of light absorbed by silicon varies widely with temperature. Hence the intensity of light received back depends sensitively on the temperature. The light detector receives the signal carrying light beam as it emerges from the optical fibre. The signal from the detector is processed electrically for getting the desired output. With such a technique, differential temperature measurements of less than 0.001°C are possible.

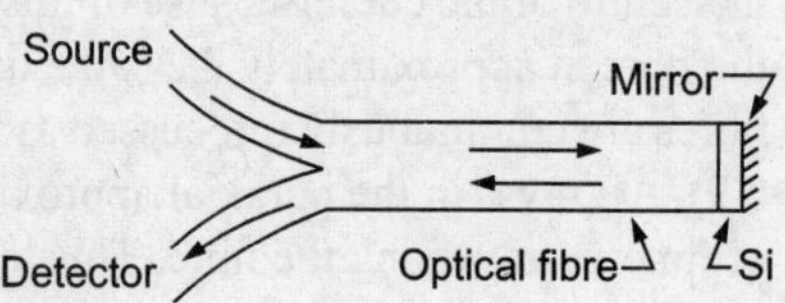

Fig. 8.10.

EXERCISE

1. How is optical communication superior to radio and microwave communication?
2. Explain the two types of optical fibres. What is angle of acceptance? Define numerical aperture.
3. What is the advantage of GRIN fibre over step-index fibre?
4. Explain the principle used in the working of fibre optics materials. Explain their uses in communication and sensing applications.
5. What are the advantages of optical fibres over the coaxial cables? Describe the communication system based on optical fibres.
6. Given the numerical aperture to be 0.2441 and the refractive index of the core to be 1.50, calculate the refractive index of the cladding as well as the acceptance angle.

   [**Ans.** 1.48, Acceptance angle 14.13°]
7. For a step-index optical fibre, using Snell's law of refraction

   (*a*) Calculate the numerical aperture of the fibre.

   (b) Calculate the acceptance angle of the fibre.

CHAPTER

9

HOLOGRAPHY

9.1. Introduction

When an object is photographed by a camera, the photograph records only the *intensity distribution* in a particular plane. The details of the field nearer and farther than the focussed plane are not recorded. Also the phase distribution which prevailed at the plane of the photograph is lost. Thus the three dimensional character of the object scene is lost and we get only a two-dimensional recording of a three-dimensional scene.

The *method of recording the amplitude as well as the phase of the light waves* was evolved by Gabor in 1947 by using interferometric techniques. This photographic technique is known as *holography*. The resulting photograph is called a *hologram.*

The holograms so obtained contain details of both amplitudes and phases of light received from different parts of the three dimensional object. In 1963, E. Leith and J. Upatneiks prepared laser holograms for the first time.

Principle. The principle of holography can best be explained in two steps:

(*i*) recording of the hologram and (*ii*) reconstructing the image.

Recording of a Hologram

Fig. 9.1 shows the arrangement for recording a hologram.

First of all the laser beam is divided into two parts (1 and 2). The second beam illuminates the object. The reflected or scattered beam falls on the photograph plate *P*. The first beam (reference beam) is reflected onto photographic plate by means of plane mirror *M*. In this way, the film is exposed simultaneously to reference beam and reflected beam from the object. Since both beams belong to the same laser wavefront, the beams interfere on the plate. Thus we obtain a complicated interference pattern on the film. The film is called a hologram. The hologram contains information not only about the amplitude but also about the phase of the object wave. Unlike a photograph, the hologram has little resemblance with the object; in fact, information about the object is coded into the hologram.

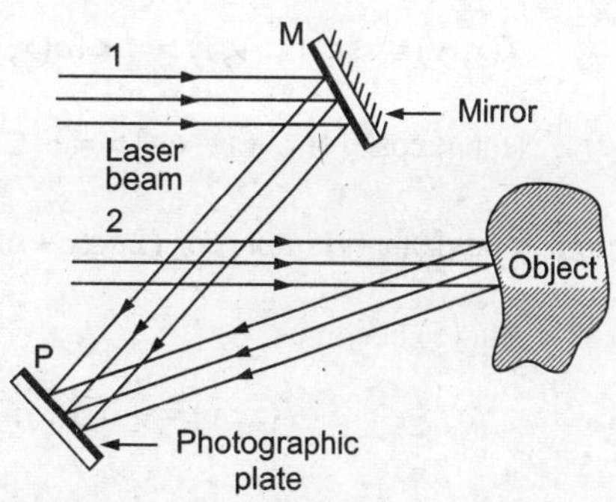

Fig. 9.1.

Viewing a Hologram

Fig. 9.2 shows how the hologram is used to produce the real and virtual image of the object. The hologram is illuminated by a single beam from laser, called the reconstruction wave. This reconstruction wave is identical in wavelength to the reference wave used for recording the hologram. When the hologram is illuminated by the reconstruction wave, two waves are produced. One wave appears to diverge from the object and provides the virtual image of the object. The second wave converges to form a second image which is real and thus can be recorded on a screen or photographed.

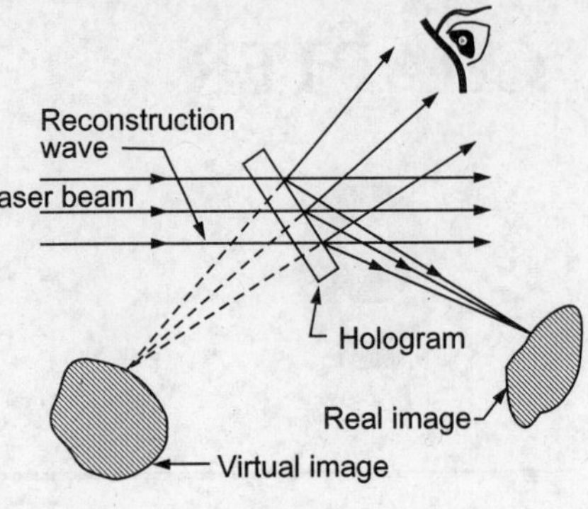

Fig. 9.2.

9.2. Theory

Let us consider the recording process. Let

$$O(x, y) = a(x, y) \cos[\phi(x, y) - \omega t] \quad \ldots(1)$$

represent the object wave in the plane of the photographic plate which is assumed to be $z = 0$.

Consider a plane reference wave propagating in the x–z-plane inclined at an angle θ with the z-axis. The field associated with this plane wave is given by

$$r(x, y, z) = A \cos[\mathbf{k} \cdot \mathbf{r} - \omega t]$$

$$= A \cos(kx \sin\theta + kz \cos\theta - \omega t) \quad \ldots(2)$$

Let $r(x, y)$ represent the field at the plane $z = 0$ due to this reference wave. Then,

$$r(x, y) = A \cos[kx \sin\theta - \omega t]$$

$$= A \cos[2\pi\alpha x - \omega t] \quad \ldots(3)$$

Here $\alpha = \dfrac{\sin\theta}{\lambda} = \dfrac{k \sin\theta}{2\pi}$ is the spatial frequency.

The total field at the photographic plate (which is coincident with the plane $z = 0$) is given by

$$u(x, y, t) = a(x, y) \cos[\phi(x, y) - \omega t] + A \cos(2\pi\alpha x - \omega t) \quad \ldots(4)$$

The photographic plate responds only to the intensity which is proportional to the time average of $[u(x, y, t)]^2$. Thus, the intensity pattern recorded by the photographic plate is

$$I(x, y) = \langle u^2(x, y, t) \rangle = \langle [a(x, y) \cos\{\phi(x, y) - \omega t\} + A \cos(2\pi\alpha x - \omega t)]^2 \rangle \quad \ldots(5)$$

But $\langle \cos^2[\phi(x, y) - \omega t] \rangle = 1/2 = \langle \cos^2(2\pi\alpha x - \omega t) \rangle$

and $\langle \cos[\phi(x, y) - \omega t] \cos(2\pi\alpha x - \omega t) \rangle = \dfrac{1}{2} \cos[\phi(x, y) - 2\pi\alpha x]$

Eq. (5) becomes

$$I(x, y) = \frac{1}{2} a^2(x, y) + \frac{1}{2} A^2 + Aa(x, y) \cos[\phi(x, y) - 2\pi\alpha x] \quad \ldots(6)$$

Eq. (6) shows that the phase information of the object wave, which is contained in $\phi(x, y)$, is recorded in the *intensity pattern*. The photographic plate containing the above intensity pattern when developed represents the hologram. Under suitable conditions of developing, the amplitude transmittance may be linearly related to $I(x, y)$. Under this situation, let $R(x, y)$ represent the field of the reconstruction wave at the hologram plane.

Then the transmitted field is given by

$$v(x, y) = KR(x, y)\, I(x, y) = KR(x, y) \left[\frac{1}{2} a^2(x, y) + \frac{1}{2} A^2 + Aa(x, y) \cos\{\phi(x, y) - 2\pi\alpha x\} \right] \ldots(7)$$

where K is a constant.

If reconstructed wave is identical with reference wave, $K = 1$, and $r(x, y) = R(x, y)$ so that Eq. (7) takes the form

$$v(x,y) = A\cos(2\pi\alpha x - \omega t)\left[\frac{1}{2}a^2(x,y) + \frac{1}{2}A^2 + A\,a(x,y)\cos\{\phi(x,y) - 2\pi\alpha x\}\right]$$

$$= \left[\frac{1}{2}a^2(x,y) + \frac{1}{2}A^2\right]A\cos(2\pi\alpha x - \omega t) + A^2 a(x,y)\cos(2\pi\alpha x - \omega t)\cos\{\phi(x,y) - 2\pi\alpha x\}$$

$$= \left[\frac{1}{2}a^2(x,y) + \frac{1}{2}A^2\right]A\cos(2\pi\alpha x - \omega t) + \frac{1}{2}A^2 a(x,y)\cos[\phi(x,y) - \omega t]$$

$$+\frac{1}{2}A^2 a(x,y)\cos\{4\pi\alpha x - \phi(x,y) - \omega t\} \quad \text{...(8)}$$

This equation represents the transmitted field in the plane $z = 0$. In this equation first term represents the reconstruction wave whose amplitude is modulated due to the presence of the term $a^2(x, y)$. The second term merely represents the original object wave and gives rise to a virtual image. The third term represents the *real image of the object.*

9.3. Applications of Holography

1. Holographic Interferometry. In the technique called double exposure holographic interferometry, one first partially exposes an object to the photographic plate with a reference wave. Now the object is stressed and one makes another exposure along with the same reference wave. If the resulting hologram is developed and illuminated by a reconstruction wave, then two object waves emerge from the hologram. One of them corresponds to the unstressed object and the other to the stressed object. These two object waves would interfere to produce interference fringes. These interference fringes are characteristic of the strain suffered by the body. A quantitative study of the fringe pattern produced in the body gives the distribution of strain in the object.

2. Holographic Microscopy. A conventional microscope has a small depth of field (the range of depth over which an object is in focus at any microscope setting). Biological specimen are generally suspended in a fluid and move about, becoming in and out of focus of the microscope. This motion can be freezed in a hologram taken through the microscope. The reconstructed three dimensional image can then be studied at leisure.

3. Acoustic Holography. The principle of holography can be used to study the image formed by sound. Let us apply the principle to view the image of an object in water. Two ultrasound generators are used. One emits the reference signal while the signal from the second is directed towards an under-water object. Now ripples are produced on the calm surface of water which is the hologram and can be photographed.

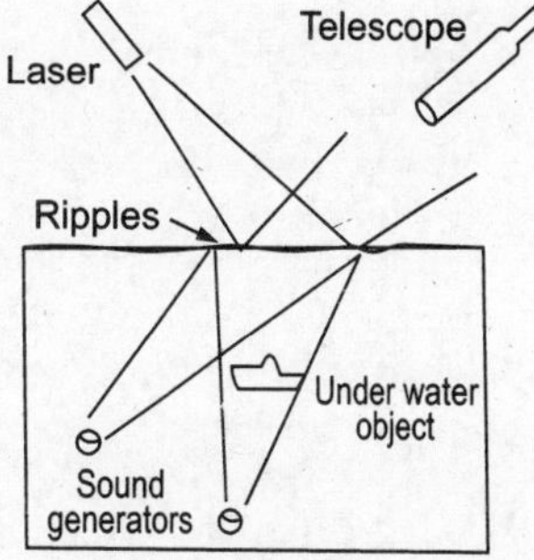

Fig. 9.3.

4. In Cinematography. Holography can be used in the formation of three dimensional images in cinematography and television.

5. Other Applications. Spatial filtration and character recognition, long distance holography using microwaves, etc.

EXERCISE

1. Explain the basic technique of holography. *(S.V.U., 89)*
2. Explain how a hologram is prepared and viewed. *(S.K.U., 95)*

3. Outline the theory of wavefront reconstruction in holography. Mention some important applications of holography.
4. Consider a three-dimensional object and image and discuss the essential difference between *stereoscopic* and *holographic* image formation.
5. Distinguish between photography and holography. (*Anna University, May 2002*)

S. No.	Photography	Holography
1.	Photography is a two dimensional recording process.	Holography is a three dimensional recording process.
2.	Ordinary light can be used for recording.	Only laser beam is used for recording or constructing a hologram.
3.	Lens is used.	No lens is used.
4.	Amplitude alone can be recorded.	Both amplitude and phase can be recorded.
5.	Image is recorded totally.	Image is recorded bit by bit.
6.	Image has poor resolution.	Image has high resolution.

CHAPTER 10 ELECTROMAGNETIC WAVES

10.1. Maxwell's Field Equations

The differential form of Maxwell's four field equations is given below:

$$\nabla \cdot \mathbf{D} = \rho \qquad \text{...(1)}$$

$$\nabla \cdot \mathbf{B} = 0 \qquad \text{...(2)}$$

$$\nabla \times \mathbf{E} = -\frac{\partial \mathbf{B}}{\partial t} \qquad \text{...(3)}$$

$$\nabla \times \mathbf{H} = \mathbf{J} + \frac{\partial \mathbf{D}}{\partial t} \qquad \text{...(4)}$$

D = electric displacement vector
ρ = volume density of charge
B = magnetic induction
E = electric field intensity
H = magnetic field intensity
J = current density.

For a linear, isotropic and homogeneous medium the *constitutive relations* are given by the following equations :

$$\mathbf{D} = \varepsilon \mathbf{E} \qquad \text{...(5)}$$

$$\mathbf{B} = \mu \mathbf{H} \qquad \text{...(6)}$$

and

$$\mathbf{J} = \sigma \mathbf{E} \qquad \text{...(7)}$$

where ε, μ and σ denote respectively the dielectric permittivity, magnetic permeability and conductivity of the medium respectively.

10.2. Derivation of Maxwell's Equations

1. Let us consider a surface *S* bounded by a volume *V* in a dielectric medium.

In a dielectric medium total charge consists of *free charge* plus *polarisation charge.*

Let ρ and ρ_p be the charge densities of free charge and polarisation charge at a point in small volume element *dV*. Then Gauss law can be expressed as

$$\int_s \mathbf{E} \cdot d\mathbf{S} = \frac{1}{\varepsilon_0} \int (\rho + \rho_p)\, dV. \text{ But } \rho_p = -\operatorname{div} \mathbf{P}$$

∴

$$\int_s \varepsilon_0 \mathbf{E} \cdot d\mathbf{S} = \int_V \rho\, dV - \int_V \operatorname{div} \mathbf{P}\, dV \qquad \text{...}(i)$$

But from Gauss divergence theorem

$$\int_s \varepsilon_0 \mathbf{E} \cdot d\mathbf{S} = \int_V \text{div } (\varepsilon_0 \mathbf{E})\, dV$$

Therefore Eq. (*i*) gives

$$\int_V \text{div } (\varepsilon_0 \mathbf{E})\, dV = \int_V \rho dV - \int_V \text{div } \mathbf{P}\, dV$$

i.e.,
$$\int_\text{v} \text{div } (\varepsilon_0 \mathbf{E} + \mathbf{P})\, dV = \int_\text{v} \rho\, dV$$

i.e.,
$$\int_V \text{div } \mathbf{D}\, dV = \int_V \rho\, dV \qquad (\text{since } \mathbf{D} = \varepsilon_0 \mathbf{E} + \mathbf{P})$$

i.e.,
$$\int_V (\text{div } \mathbf{D} - \rho)\, dV = 0$$

As volume is arbitrary, we must have

$$\text{div } \mathbf{D} - \rho = 0$$

i.e.,
$$\text{div } \mathbf{D} = \rho \qquad \ldots(1)$$

2. We know that isolated magnetic poles do not exist. This fact is also expressed by saying that magnetic lines of force are, in general, closed curves. Because of this, the magnetic flux leaving and entering any closed surface is always same. In other words, the total magnetic induction flux over any closed surface is always zero, *i.e.,*

$$\int_s \mathbf{B} \cdot \mathbf{dS} = 0$$

Using Gauss's theorem we change surface integral into a volume integral.

$\therefore$
$$\int_V (\text{div } \mathbf{B})\, \text{dV} = 0$$

Since the above statement is true for any arbitrary volume, *V*, it is only possible if the integrand itself is zero, *i.e.,*

$$\text{div } \mathbf{B} = 0$$

or
$$\nabla \cdot \mathbf{B} = 0$$

3. By Faraday's law of electromagnetic induction, the induced e.m.f., in a closed loop is

$$e = -\frac{\partial \phi}{\partial t} \qquad \ldots(1)$$

By definition, the magnetic flux ϕ over any arbitrary surface *S* is,

$$\phi = \int_\text{s} \mathbf{B} \cdot d\mathbf{S}$$

$$\text{Therefore } e = -\frac{\partial}{\partial t}\left[\int_\text{s} \mathbf{B} \cdot \mathbf{dS}\right] = -\int_s \frac{\partial \mathbf{B}}{\partial t} . \text{dS} \qquad \ldots(2)$$

($\because$ surface *S* is fixed and hence only **B** is a function of time).

Again e.m.f., by definition, is the work done in carrying a unit charge round a closed loop. If *C* is the loop constituting the surface **S**, then

$$e = \int_\text{c} \mathbf{E} \cdot \mathbf{d}\boldsymbol{l} \qquad \ldots(3)$$

where **E** is the electric field intensity at a point where elementary loop $\mathbf{d}\boldsymbol{l}$ is located.

Comparing Eqs. (2) and (3), we get

$$\int_\text{c} \mathbf{E} \cdot \text{d}l = -\int_s \frac{\partial \mathbf{B}}{\partial t} . \mathbf{dS} \qquad \ldots(4)$$

From Stokes' theorem, $\int_\text{c} \mathbf{E} \cdot \mathbf{d}\boldsymbol{l} = \int_s \text{curl } \mathbf{E} \cdot \text{dS}$

Eq. (4) becomes, $\int_s \text{curl}\,\mathbf{E}\cdot d\mathbf{S} = -\int_s \frac{\partial \mathbf{B}}{\partial t}\cdot d\mathbf{S}$

Since surface is arbitrary, curl $\mathbf{E} = -\frac{\partial \mathbf{B}}{\partial t}$

or $$\nabla\times\mathbf{E} = -\frac{\partial \mathbf{B}}{\partial t} \qquad \text{...(5)}$$

4. The work done in carrying a unit magnetic pole once around a closed arbitrary path linked with the current is expressed by

$$\oint \mathbf{H}\cdot \mathbf{d}l = \mathbf{I} = \int \mathbf{J}\cdot \mathbf{d}s \qquad \text{...(1)}$$

However, the use of Stokes' theorem gives

$$\int_s (\nabla\times\mathbf{H})\cdot d\mathbf{S} = \int \mathbf{J}\cdot \mathbf{d}s$$

or $$\nabla\times\mathbf{H} = \mathbf{J} \qquad \text{...(2)}$$

This equation is incomplete and accounts for steady currents only. But for varying electric fields, the current density should be modified. The difficulty with above equation is that if we take divergence of above equation, then,

$$\nabla\times(\nabla\times\mathbf{H}) = \nabla\times\mathbf{J} = 0 \qquad \text{...(3)}$$

However this fact is disallowed by *Continuity Equation* according to which

$$\nabla\cdot\mathbf{J} = -\partial\rho/\partial t$$

On this basis Maxwell added another current density $\mathbf{J}'$ to $\mathbf{J}$ to make Eq. (2) as

$$\nabla\times\mathbf{H} = \mathbf{J}+\mathbf{J}' \qquad \text{...(4)}$$

so that continuity equation too is satisfied. Now divergence of Eq. (4) gives

$$\nabla\cdot(\nabla\times\mathbf{H}) = \nabla\cdot\mathbf{J}+\nabla\cdot\mathbf{J}' = 0$$

or $$\nabla\cdot\mathbf{J}' = -\nabla\cdot\mathbf{J} = \frac{\partial\rho}{\partial t} \qquad \text{...(5)}$$

We know that $$\nabla\times\mathbf{D} = \rho$$

Eq. (5) becomes, $$\nabla\cdot\mathbf{J}' = \frac{\partial}{\partial t}(\nabla\cdot\mathbf{D})$$

or $$\nabla\cdot\mathbf{J}' = \nabla\cdot\frac{\partial \mathbf{D}}{\partial t}$$

or $$\mathbf{J}' = \frac{\partial \mathbf{D}}{\partial t} \qquad \text{...(6)}$$

In this way we observe that the additional current density $\mathbf{J}'$ is due to time variations of electric displacement $\mathbf{D}$. It is termed as *Displacement current density*. According to Maxwell, it is just as effective as $\mathbf{J}$ in producing magnetic field.

Thus Eq. (4) is written as

$$\nabla\times\mathbf{H} = \mathbf{J}+\frac{\partial \mathbf{D}}{\partial t} \qquad \text{...(7)}$$

10.3. Propagation of Electromagnetic Wave through Free Space

Maxwell's equations are

$$\nabla\cdot\mathbf{D} = \rho \qquad \text{...(1)}$$
$$\nabla\cdot\mathbf{B} = 0 \qquad \text{...(2)}$$

$$\nabla \times \mathbf{E} = -\frac{\partial \mathbf{B}}{\partial t} \quad \text{...(3)}$$

$$\nabla \times \mathbf{H} = \mathbf{J} + \frac{\partial \mathbf{D}}{\partial t} \quad \text{...(4)}$$

For free space,
$$\left.\begin{array}{l} \mathbf{J} = 0, \rho = 0, \sigma = 0 \\ \mathbf{D} = \varepsilon_0 \, \mathbf{E} \text{ and } \mathbf{B} = \mu_0 \mathrm{H} \end{array}\right] \quad \text{...(5)}$$

Therefore, Maxwell's equations in free space are :

$$\nabla \cdot \mathbf{E} = 0 \quad \text{...(6)}$$

$$\nabla \cdot \mathbf{B} = 0 \quad \text{...(7)}$$

$$\nabla \times \mathbf{E} = -\mu_0 \frac{\partial \mathbf{H}}{\partial t} \quad \text{...(8)}$$

and
$$\nabla \times \mathbf{H} = \varepsilon_0 \frac{\partial \mathbf{E}}{\partial t} \quad \text{...(9)}$$

Taking curl of Eq. (8), we get

$$\nabla \times (\nabla \times \mathbf{E}) = -\mu_0 \frac{\partial}{\partial t} (\nabla \times \mathbf{H})$$

$$= -\mu_0 \frac{\partial}{\partial t} \left(\varepsilon_0 \frac{\partial \mathbf{E}}{\partial t} \right) \quad \text{from Eq. (9)}$$

$$= -\mu_0 \, \varepsilon_0 \frac{\partial^2 \mathbf{E}}{\partial t^2} \quad \text{...(10)}$$

Further,
$$\nabla \times (\nabla \times \mathbf{E}) = \nabla(\nabla \cdot \mathbf{E}) - \nabla^2 \mathbf{E}$$

$$= -\nabla^2 \mathbf{E} \quad [\because \nabla \cdot \mathbf{E} = 0 \text{ from Eq. (6)}]$$

Eq. (10) becomes
$$-\nabla^2 \mathbf{E} = -\mu_0 \varepsilon_0 \frac{\partial^2 \mathbf{E}}{\partial t^2}$$

or
$$\nabla^2 \mathbf{E} = \mu_0 \varepsilon_0 \frac{\partial^2 \mathbf{E}}{\partial t^2} \quad \text{...(11)}$$

Similarly starting with Eq. (9), we get

$$\nabla^2 \mathbf{H} = \mu_0 \varepsilon_0 \frac{\partial^2 \mathbf{H}}{\partial t^2} \quad \text{...(12)}$$

Eqs. (11) and (12) are the wave equations for **E** and **H**.
These equations can be compared with the general wave equation

$$\nabla^2 u = \frac{1}{v^2} \cdot \frac{\partial^2 u}{\partial t^2}, \ v \text{ being velocity of wave} \quad \text{...(13)}$$

Comparing Eqs. (11), (12) with Eq. (13), we see that the *field vectors* **E** *and* **H** *are propagated in free space as waves* at a speed equal to

$$v = \frac{1}{\sqrt{\mu_0 \varepsilon_0}} = \frac{1}{\sqrt{4\pi \times 10^{-7} \times 8.8542 \times 10^{-12}}} \approx 3 \times 10^8 \text{ ms}^{-1} = c = \text{the speed of light}$$

10.4. Poynting Vector

As em waves propagate through matter from their source to distant receiving point, there is a transfer of energy from source to receiver. *The rate at which energy is transmitted through unit area perpendicular to the direction of propagation of energy is called Poynting vector and is represented by* **P**. Let us derive an expression for Poynting vector.

Two of the Maxwell's equations are

$$\nabla \times \mathbf{H} = \mathbf{J} + \frac{\partial \mathbf{D}}{\partial t} \qquad ...(1)$$

$$\nabla \times \mathbf{E} = -\frac{\partial \mathbf{B}}{\partial t} \qquad ...(2)$$

Taking dot product of both sides of Eqs. (1) and (2) with **E** and **H** respectively,

$$\mathbf{E} \cdot (\nabla \times \mathbf{H}) = \mathbf{E} \cdot \mathbf{J} + \mathbf{E} \cdot \frac{\partial \mathbf{D}}{\partial t} = \mathbf{E} \cdot \mathbf{J} + \frac{\partial}{\partial t}\left(\frac{1}{2}\varepsilon E^2\right) \qquad ...(3)$$

$$\mathbf{H} \cdot (\nabla \times \mathbf{E}) = \mathbf{H} \cdot \left(\frac{-\partial \mathbf{B}}{\partial t}\right) = -\frac{\partial}{\partial t}\left(\frac{1}{2}\mu \mathrm{H}^2\right) \qquad ...(4)$$

Subtracting Eq. (4) from Eq. (3), we get

$$\mathbf{E} \cdot (\nabla \times \mathbf{H}) - \mathbf{H} \cdot (\nabla \times \mathbf{E}) = \mathbf{E} \cdot \mathbf{J} + \frac{\partial}{\partial t}\left(\frac{1}{2}\mu \mathrm{H}^2 + \frac{1}{2}\varepsilon \mathrm{E}^2\right) \qquad ...(5)$$

But

$$\nabla \cdot (\mathbf{E} \times \mathbf{H}) = \mathbf{H} \cdot (\nabla \times \mathbf{E}) - \mathbf{E} \cdot (\nabla \times \mathbf{H})$$

$$-\nabla \cdot (\mathbf{E} \times \mathbf{H}) = \mathbf{E} \cdot \mathbf{J} + \frac{\partial}{\partial t}\left(\frac{1}{2}\mu \mathrm{H}^2 + \frac{1}{2}\varepsilon \mathrm{E}^2\right) \qquad ...(6)$$

Let us now consider a volume *V* enclosed by a surface *S*.

Integrating the above relation over the volume *V*, we have

$$\frac{\partial}{\partial t}\int_V \left(\frac{\mu H^2}{2} + \frac{\varepsilon E^2}{2}\right) dV + \int_V (\mathbf{E} \cdot \mathbf{J})\, dV = -\int_V \nabla \cdot (\mathbf{E} \times \mathbf{H})\, dV$$

Using Gauss divergence theorem,

$$\int_v \nabla \cdot (\mathbf{E} \times \mathbf{H})\, dV = \oint_s (\mathbf{E} \times \mathbf{H}) \cdot d\mathbf{S}$$

$$-\frac{\partial}{\partial t}\int_V \left(\frac{\mu H^2}{2} + \frac{\varepsilon E^2}{2}\right) dV - \int_V (\mathbf{E} \cdot \mathbf{J})\, dV = \oint_s (\mathbf{E} \times \mathbf{H}) \cdot d\mathbf{S} \qquad ...(7)$$

The first term on L.H.S. of Eq. (7) represents the rate of decrease of energy stored in volume *V* due to electric and magnetic fields. The second term on L.H.S. represents the rate at which electromagnetic energy is lost through Joule heating. Hence the R.H.S. represents the rate of flow of energy over the surface *S* enclosing the volume *V*.

Therefore, $(\mathbf{E} \times \mathbf{H})$ gives the rate of flow of energy through unit area enclosing the volume *V*. This is denoted by **P**, called Poynting vector.

$$\mathbf{P} = \mathbf{E} \times \mathbf{H} \qquad ...(8)$$

The direction of **P** is perpendicular to both **E** and **H**. Eq. (8) shows that the flow of energy takes place at right angles to a plane containing vectors **E** and **H**. However, it will be shown ahead that vectors **E** and **H** are also at right angles to each other and in this way the pure transverse character of electromagnetic waves is to be established, *i.e.*, vectors **P, E** and **H** are mutually at right angles.

10.5. Transverse Nature of Electromagnetic Waves

Consider an electromagnetic wave propagating in free space along *z*-direction.

E and **H** will vary only in the direction of *z*-axis, *i.e.*,

$$\frac{\partial}{\partial x} = \frac{\partial}{\partial y} = 0 \text{ and } \frac{\partial}{\partial z} \neq 0.$$

Such a wave is said to be *plane wave*. Since its vectors are functions of (z, t) only, we write

$$\left.\begin{array}{l}\mathbf{E} = \mathbf{E}(z, t)\\ \mathbf{H} = \mathbf{H}(z, t)\end{array}\right\} \quad \ldots(1)$$

Maxwell's first equation gives :

$$\nabla \cdot \mathbf{D} = 0, \text{ or } \varepsilon_0 (\nabla \cdot \mathbf{E}) = 0$$

or $$\nabla \cdot \mathbf{E} = 0, \text{ or } \partial E_z / \partial z = 0$$

$\therefore$ $$E_z = \text{constant in space}$$

Similarly second equation gives

$$\nabla \cdot \mathbf{B} = 0, \text{ or } \mu_0 (\nabla \cdot \mathbf{H}) = 0$$

or $$\nabla \cdot \mathbf{H} = 0, \text{ or } \partial H_z / \partial z = 0$$

$\therefore$ $$H_z = \text{constant in space}$$

Using third equation,

$$\nabla \times \mathbf{E} = -\partial \mathbf{B} / \partial t$$

or $$(\nabla \times E)_z = -\mu_0 \, \partial H_z / \partial t$$

or $$\mathbf{k}\left(\frac{\partial E_y}{\partial x} - \frac{\partial E_x}{\partial y}\right) = -\mu_0 \frac{\delta Hz}{\partial t}$$

or $$\partial H_z / \partial t = 0$$

$\therefore$ $$H_z = \text{constant in time}$$

Similarly, by using fourth equation it can be shown that

$$E_z = \text{constant in time.}$$

Thus we arrive at the conclusion that E_z and H_z are constant as regards to time and space. Thus they represent static components and consequently, no part of wave motion. We can , therefore, write

$$E_z = H_z = 0$$

so that

$$\left.\begin{array}{l}\mathbf{E} = \mathbf{i}\, E_x + \mathbf{j}\, E_y\\ \mathbf{H} = \mathbf{i}\, H_x + \mathbf{j}\, H_y\end{array}\right\} \quad \ldots(2)$$

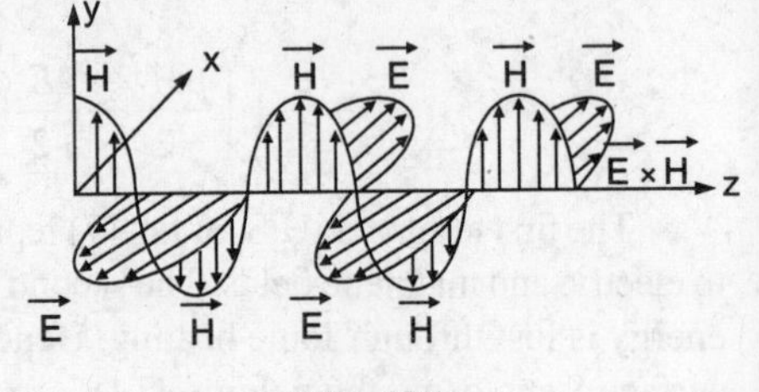

Fig. 10.1.

Since the electric vector **E** and the magnetic vector **H** do not have any z component, the z-direction being the direction of propagation, *both these vectors are perpendicular to the direction of propagation. Maxwell's electromagnetic waves are, therefore, purely transverse.* (Fig. 10.1).

10.6. Propagation of Electromagnetic Wave through a Homogeneous, Isotropic Dielectric Medium (Non-conducting Isotropic Medium)

Maxwell's equations are

$$\nabla \cdot \mathbf{D} = \rho$$

$$\nabla \cdot \mathbf{B} = 0$$

$$\nabla \times \mathbf{E} = -\frac{\partial \mathbf{B}}{dt}$$

and $$\nabla \times \mathbf{H} = \mathbf{J} + \frac{\partial \mathbf{D}}{dt}$$

In an *isotropic dielectric* (or non-conducting isotropic medium)

$\mathbf{D} = \varepsilon\mathbf{E}$, $\mathbf{B} = \mu\mathbf{H}$, $\mathbf{J} = \sigma\mathbf{E} = 0$ and $\rho = 0$.

Therefore, Maxwell's equations in this case take the form

$$\nabla \cdot \mathbf{E} = 0 \quad \ldots(1)$$

$$\nabla \cdot \mathbf{H} = 0 \quad \ldots(2)$$

$$\nabla \times \mathbf{E} = -\mu \frac{\partial \mathbf{H}}{\partial t} \quad \ldots(3)$$

and

$$\nabla \times \mathbf{H} = \varepsilon \frac{\partial \mathbf{E}}{\partial t} \quad \ldots(4)$$

Equation of propagation of magnetic vector, H

Taking curl of Eq. (4), we get

$$\nabla \times (\nabla \times \mathbf{H}) = \varepsilon \nabla \times \left(\frac{\partial \mathbf{E}}{\partial t}\right)$$

or

$$\nabla(\nabla \cdot \mathbf{H}) - \nabla^2 \mathbf{H} = \varepsilon \frac{\partial}{\partial t} (\nabla \times \mathbf{E}) \quad \ldots(5)$$

Putting values from Eqs. (2) and (3), we get

$$\nabla^2 \mathbf{H} = \mu\varepsilon \frac{\partial^2 H}{\partial t^2} \quad \ldots(6)$$

Equation of propagation of electric vector, E

Taking curl of Eq. (3), we get

$$\nabla \times (\nabla \times \mathbf{E}) = \nabla \times \left(-\mu \frac{\partial \mathbf{H}}{\partial t}\right)$$

$$\nabla(\nabla . \mathbf{E}) - \nabla^2 \mathbf{E} = -\mu \frac{\partial}{\partial t} (\nabla \times \mathbf{H})$$

Putting values from Eqs. (1) and (4), we get

$$\nabla^2 \mathbf{E} = \mu\varepsilon \frac{\partial^2 \mathbf{E}}{\partial t^2} \quad \ldots(7)$$

Eqs. (6) and (7) can be compared with the general wave equation,

$$\nabla^2 u = \frac{1}{v^2} \frac{\partial^2 u}{\partial t^2} \text{ where } v \text{ is the speed of wave.}$$

$$\therefore \quad v = 1/\sqrt{\mu\varepsilon}$$

This means *that the field vectors* **E** and **H** are propagated in isotropic dielectric as waves with speed v given by

$$v = \frac{1}{\sqrt{(\mu\varepsilon)}} \quad \ldots(8)$$

Now, $\dfrac{1}{\sqrt{\mu_0 \varepsilon_0}} = c$, speed of electromagnetic waves in free space.

Refractive index is

$$n = \frac{c}{v} = \sqrt{\frac{\mu\varepsilon}{\mu_0 \varepsilon_0}} = \sqrt{\mu_r \, \varepsilon_r}$$

In a non-magnetic medium $\mu_r = 1$

$$\therefore \qquad n = \sqrt{\varepsilon_r} \quad \text{or} \quad n^2 = \varepsilon_r \qquad \ldots(9)$$

Eq. (9) is termed as Maxwell's relation and is experimentally verified in case of air, hydrogen, benzene and carbon etc , having non-polar molecules. But for substances containing polar molecules *e.g.*, water, glass etc., ε_r is dependent on frequency and decreases with increasing frequency. Thus refractive index exhibits variation with frequency—a phenomenon termed as *dispersion*.

10.7. Propagation of Plane Electromagnetic Waves through Conducting Media

Maxwell's equations are

$$\nabla \cdot \mathbf{D} = \rho$$

$$\nabla \cdot \mathbf{B} = 0$$

$$\nabla \times \mathbf{E} = -\frac{\partial \mathbf{B}}{\partial t}$$

$$\nabla \times \mathbf{H} = \mathbf{J} + \frac{\partial \mathbf{D}}{\partial t}$$

Consider a conducting medium with permeability = μ, permitivity = ε, and conductivity = σ. Charge density ρ = 0. Further, **D = εE, B = μH, J = σE.**

The Maxwell's equations become,

$$\nabla \cdot \mathbf{E} = 0 \qquad \ldots(1)$$

$$\nabla \cdot \mathbf{H} = 0 \qquad \ldots(2)$$

$$\nabla \times \mathbf{E} = -\mu \frac{\partial \mathbf{H}}{\partial t} \qquad \ldots(3)$$

$$\nabla \times \mathbf{H} = \sigma \mathbf{E} + \varepsilon \frac{\partial \mathbf{E}}{\partial t} \qquad \ldots(4)$$

Taking the curl of Eq. (3), we get

$$\nabla \times (\nabla \times \mathbf{E}) = -\mu \frac{\partial}{\partial t} (\nabla \times \mathbf{H})$$

$$\nabla (\nabla \cdot \mathbf{E}) - \nabla^2 \mathbf{E} = -\mu \frac{\partial}{\partial t}\left[\sigma \mathbf{E} + \varepsilon \frac{\partial \mathbf{E}}{\partial t}\right] \qquad \text{from Eq. (4)}$$

$$-\nabla^2 \mathbf{E} = -\mu\sigma \frac{\partial \mathbf{E}}{\partial t} - \mu\varepsilon \frac{\partial^2 \mathbf{E}}{\partial t^2} \qquad \text{from Eq. (1)}$$

$$\therefore \qquad \nabla^2 \mathbf{E} - \mu\sigma \frac{\partial \mathbf{E}}{\partial t} - \mu\varepsilon \frac{\partial^2 \mathbf{E}}{\partial t^2} = 0 \qquad \ldots(5)$$

Similarly we can get

$$\nabla^2 \mathbf{H} - \mu\sigma \frac{\partial \mathbf{H}}{\partial t} - \mu\varepsilon \frac{\partial^2 \mathbf{H}}{\partial t^2} = 0 \qquad \ldots(6)$$

Eqs. (5) and (6) are modified wave equations for **E** and **H**.

Eq. (5) admits plane wave solution

$$\mathbf{E} = \mathbf{E}_0 \exp\left[i\,(k_z - \omega t)\right] \qquad \ldots(7)$$

This can be easily checked by putting Eq. (7) into Eq. (5).

For a plane wave of this type we obtain

$$-k^2 + i\,\omega\mu\sigma + \omega^2 \varepsilon\mu = 0 \qquad \ldots(8)$$

which shows that the wave number k must be a complex number. Let

$$k = \alpha + i\beta \quad \ldots(9)$$

Then, $$-(\alpha^2 + 2i\alpha\beta - \beta^2) + i\omega\mu\sigma + \omega^2\varepsilon\mu = 0$$

Equating real and imaginary parts, we get

$$\alpha^2 - \beta^2 = \omega^2\varepsilon\mu \quad \ldots(10)$$

and $$\beta = \frac{\omega\mu\sigma}{2\alpha} \quad \ldots(11)$$

Substituting for β in Eq. (10) and solving for α, we get

$$\alpha = \omega\sqrt{\varepsilon\mu}\left[\frac{1}{2} \pm \frac{1}{2}\left(1 + \frac{\sigma^2}{\omega^2\varepsilon^2}\right)^{1/2}\right]^{1/2} \quad \ldots(12)$$

The negative sign would make α complex. So, we must choose the positive sign.

$$\therefore \quad \alpha = \omega\sqrt{\varepsilon\mu}\left[\frac{1}{2} + \frac{1}{2}\left(1 + \frac{\sigma^2}{\omega^2\varepsilon^2}\right)^{1/2}\right]^{1/2} \quad \ldots(13)$$

and $$\beta = \frac{\omega\mu\sigma}{2\alpha}.$$

Now, when k is complex, Eq. (7) becomes

$$\mathbf{E} = \mathbf{E}_0 \exp(-\beta z) \exp[i(\alpha z - \omega t)] \quad \ldots(14)$$

which represents an attenuated wave. The attenuation is due to the Joule-loss. Eq. (14) represents a plane wave travelling along z-direction.

The imaginary part of k results in an *attenuation* of the wave.

The distance it takes to reduce the amplitude by a factor of $1/e$ is called the *skin depth* :

$$\delta = \frac{1}{\beta} \quad \ldots(15)$$

It is a measure of the depth to which an electromagnetic wave can penetrate the conductor.

Meanwhile, the real part of k determines the wavelength, the propagation speed, and the index of refraction in the usual way :

$$\lambda = \frac{2\pi}{\alpha}, \; v = \frac{\omega}{\alpha}, \; n = \frac{c\alpha}{\omega} \quad \ldots(16)$$

The quantity ck/ω is called the *complex index of refraction.*

We call the material a "poor" conductor if $\frac{\sigma}{\omega\varepsilon} << 1$.

In this case, $$\alpha \approx \omega\sqrt{\varepsilon\mu}, \; \beta \approx \frac{\sigma}{2}\sqrt{\mu/\varepsilon} \quad \ldots(17)$$

and the skin depth is independent of frequency.

For a "good" conductor, $\frac{\sigma}{\omega\varepsilon} >> 1$. We get

$$\alpha \approx \beta \approx \left(\frac{\omega\mu\sigma}{2}\right)^{1/2} \quad \ldots(18)$$

The skin depth decreases with increasing frequency.

For a typical metal, σ is on the order of 10^7 mhos/m, μ is 10^{-6} N/A^2, and ε is around 10^{-11} C^{2}N^{-1}m^{-2}. Therefore, $(\sigma/\varepsilon) \approx 10^{18}$ sec^{-1}, and the material is a good conductor well into the ultraviolet region. The skin depth at optical frequencies ($\omega \approx 10^{15}$ sec^{-1}) is roughly 10^{-8}m, which explains why metals are opaque.

10.8. Reflection and Refraction of Electromagnetic Waves

Let us consider a plane interface at Z = 0 (*X-Y* plane) separating two homogeneous, charge free and non-conducting isotropic media, characterised by permittivities ε_1 and ε_2, permeabilities μ_1 and μ_2 respectively (Fig. 10.2). Let a plane wave with wave vector $\mathbf{k}_1$ and frequency ω_1 be incident from medium '1' at point *O* on the interface. Part of this wave will reflect into region 1 and part will refract and transmit into region 2. Let the reflected and refracted waves have wave vectors $\mathbf{k}_1'$ and $\mathbf{k}_2'$, frequencies ω_1' and ω_2' respectively. Also let $\mathbf{n}$ be the unit vector normal to the interface and directed from medium '1' into medium '2'. The field vectors for incident, reflected and refracted waves may be expressed as

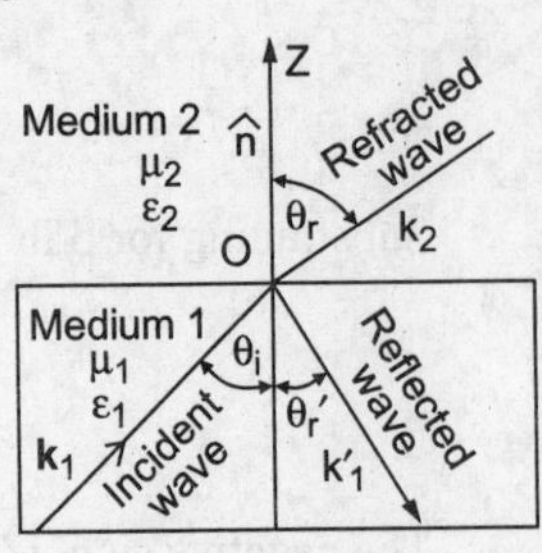

Fig. 10.2.

For incident wave
$$\left.\begin{aligned} \mathbf{E}_1 &= \mathbf{E}_{01} e^{i\mathbf{k}_1 \cdot \mathbf{r} - i\omega_1 t} \\ \mathbf{B}_1 &= \frac{\mathbf{k}_1 \times \mathbf{E}_1}{\omega_1} \text{ or } \mathbf{H}_1 = \frac{\mathbf{k}_1 \times \mathbf{E}_1}{\mu_1 \omega_1} \end{aligned}\right\} \quad \ldots(1)$$

For reflected wave
$$\left.\begin{aligned} \mathbf{E}_1' &= \mathbf{E}_{01}' e^{i\mathbf{k}_1' \cdot \mathbf{r} - i\omega_1' t} \\ \mathbf{B}_1' &= \frac{\mathbf{k}_1' \times \mathbf{E}_1'}{\omega_1'} \text{ or } \mathbf{H}_1' = \frac{\mathbf{k}_1' \times \mathbf{E}_1'}{\mu_1 \omega_1'} \end{aligned}\right\} \quad \ldots(2)$$

For refracted wave
$$\left.\begin{aligned} \mathbf{E}_2 &= \mathbf{E}_{02} e^{i\mathbf{k}_2 \cdot \mathbf{r} - i\omega_2 t} \\ \mathbf{B}_2 &= \frac{\mathbf{k}_2 \times \mathbf{E}_2}{\omega_2'} \text{ or } \mathbf{H}_2 = \frac{\mathbf{k}_2 \times \mathbf{E}_2}{\mu_2 \omega_2'} \end{aligned}\right\} \quad \ldots(3)$$

Let us use the boundary condition : *the tangential components of the electric field vector remain continuous at the boundary of two media.*

$$(E_1)_{\text{tangential}} + (E_1')_{\text{tangential}} = (E_2)_{\text{tangential}} \quad \ldots(4)$$

or $$(E_{01})_{\text{tangential}}\, e^{i\mathbf{k}_1 \cdot \mathbf{r} - i\omega t} + (E_{01}')_{\text{tangential}}\, e^{i\mathbf{k}_1' \cdot \mathbf{r} - i\omega_1' t} = (E_{02})_{\text{tangential}}\, e^{i\mathbf{k}_2 \cdot \mathbf{r} - i\omega_2 t}$$

or $$(E_{01})_{\text{tangential}}\, e^{i\mathbf{k}_1 \cdot \mathbf{r}}\, e^{-i\omega t} + (E_{01}')_{\text{tangential}}\, e^{i\mathbf{k}_1' \cdot \mathbf{r}} e^{-i\omega_1' t} = (E_{02})_{\text{tangential}}\, e^{i\mathbf{k}_2 \cdot \mathbf{r}}\, e^{-i\omega_2 t} \quad \ldots(5)$$

Eq. (5) is valid for all values of *x*, *y* and *t* (as *z* component is normal to the boundary). This can be satisfied if the time and space varying components of the phases are equal.

Equating the time varying components, we get

$$\omega_1 = \omega_1' = \omega_2 = \omega \text{ (say)} \quad \ldots(6)$$

That is the incident, reflected and refracted waves all have the same frequency.

Equating the space varying components in Eq. (5), we have

$$(\mathbf{k_1} \cdot \mathbf{r})_{z=0} = (\mathbf{k_1}' \cdot \mathbf{r})_{z=0} = (\mathbf{k_2} \cdot \mathbf{r})_{z=0}$$

or $$k_{1x}x + k_{1y}y = k_{1x}'x + k_{1y}'y = k_{2x}x + k_{2y}y \qquad \ldots(8)$$

or $$\left.\begin{aligned} k_{1x} &= k_{1x}' = k_{2x} \quad \ldots(a) \\ k_{1y} &= k_{1y}' = k_{2y} \quad \ldots(b) \end{aligned}\right\} \qquad \ldots(9)$$

Since incident beam is in *XZ* plane, $k_{1y} = 0$. Therefore, $k_{1y}' = k_{2y} = 0$.

This shows that both $\mathbf{k_1}'$ and $\mathbf{k_2}$ also lie in *XZ* plane. As normal **n** is along *Z*-axis, we conclude that all the *three wave vectors and normal to the interface* **n** *all lie in the same plane.* In other words, *the incident, reflected, refracted waves and the normal to the interface all lie in the same plane.*

Furthermore, from Fig. 10.2, we have

$$\left.\begin{aligned} \mathbf{k_1} \cdot \mathbf{r} &= k_1(x \sin\theta_i + z\cos\theta_i) \quad \ldots(a) \\ \mathbf{k_1}' \cdot \mathbf{r} &= k_1'(x \sin\theta_r' - z\cos\theta_r') \quad \ldots(b) \\ \mathbf{k_2} \cdot \mathbf{r} &= k_2(x \sin\theta_r + z\cos\theta_r) \quad \ldots(c) \end{aligned}\right\} \qquad \ldots(10)$$

Substituting values from (10 *a*) and (10 *b*) in Eq. (7) *viz.*,

$$(\mathbf{k_1} \cdot \mathbf{r})_{z=0} = (\mathbf{k_1}' \cdot \mathbf{r})_{z=0}$$

we get $$k_1 x \sin\theta_i = k_1' x \sin\theta_r'$$

or $$k_1 \sin\theta_i = k_1' \sin\theta_r' \qquad \ldots(11)$$

Since wave vectors $\mathbf{k_1}$ and $\mathbf{k_1}'$ lie in the same medium, $k_1 = k_1' = \omega\sqrt{\mu_1\varepsilon_1} = (\omega / v_1)$, v_1 being phase velocity of electromagnetic wave in medium 1. Therefore, Eq. (11) gives

$$\sin\theta_i = \sin\theta_r' \text{ or } \theta_i = \theta_r' \qquad \ldots(12)$$

i.e., the angle of incidence (θ_i) *is equal to the angle of reflection* $(\theta_r)'$

Now substituting values of $\mathbf{k_1}.\mathbf{r}$ and $\mathbf{k_2}.\mathbf{r}$ from Eq. (10) in Eq. (7) *viz.*,

$$(\mathbf{k_1} \cdot \mathbf{r})_{z=0} = (\mathbf{k_2} \cdot \mathbf{r})_{z=0}$$

we get $$k_1 x \sin\theta_i = k_2 x \sin\theta_r$$

i.e., $$k_1 \sin\theta_i = k_2 \sin\theta_r$$

or $$\frac{\sin\theta_i}{\sin\theta_r} = \frac{k_2}{k_1} = \frac{\omega_2\sqrt{(\mu_2\varepsilon_2)}}{\omega_1\sqrt{(\mu_1\varepsilon_1)}} = \frac{\sqrt{(\mu_2\varepsilon_2)}}{\sqrt{(\mu_1\varepsilon_1)}} \qquad (\text{since } \omega_1 = \omega_2 = \omega)$$

This gives $$\frac{\sin\theta_i}{\sin\theta_r} = \frac{n_2}{n_1} \qquad \ldots(13)$$

(since refractive index, $n = \sqrt{(\mu\varepsilon / \mu_0\varepsilon_0)} = (c / v)$)

This is known as *Snell's law of refraction.*

EXERCISE

1. Derive Maxwell's equations from basic laws of electricity and magnetism.
2. Write down Maxwell's equations in differential form. Derive expression for Maxwell's wave equation and evaluate velocity of light in free space.
3. Define Poynting vector and derive the expression **P = E × H.**
4. Discuss the propagation of E. M. waves through free space. Establish the transverse character of such waves.
5. Discuss the propagation of E.M. waves through isotropic homogeneous dielectric media.
6. Derive wave equation for electromagnetic wave propagation through a non-conductor. Show that it exhibits dispersive property.
7. Discuss the propagation of plane monochromatic electromagnetic waves in conducting media. Derive the dispersion equation and thus obtain

   (*a*) phase velocity, (*b*) refractive index, (*c*) skin depth.
8. Deduce the laws of reflection and refraction from electromagnetic theory of light.

CHAPTER

11

DISPERSION AND SCATTERING

11.1. Introduction

When E.M. radiation passes through a medium it is dispersed. This implies that the medium exhibits different refractive indices for different wavelengths and this *variation of refractive index with wavelength is termed as dispersion.* Usually it is observed that refractive index (n) decreases with increase in wavelength and such a dispersion is known as *normal dispersion*. But in absorption region, the refractive index increases with increase in wavelength and the phenomenon is known as *anomalous dispersion.*

11.2. Dispersion in Gases (Lorentz Theory)

An adequate theory of dispersion, based on electromagnetic theory, was given by Lorentz. The theory is based on the following assumptions :

(1) There is no appreciable interaction between the atoms (in case of atomic gases) or between the molecules (in case of molecular gases).

(2) The electric field of the electromagnetic wave induces a dipole moment in the gas molecules.

(3) Electrons are bound to the nucleus in an atom by linear restoring force.

(4) Over an atom or a molecule **E** is constant in space *i.e.,*

$$\mathbf{E} = \mathbf{E_0} \exp[-i(\omega t - \mathbf{k}\cdot\mathbf{r})]$$

$$\approx \mathbf{E}_0 \exp(-i\omega t) \qquad \ldots(1)$$

Let m and e represent the mass and charge of the electron, **E** the external electric field. He assumed that the atomic dipoles in a medium execute forced oscillations under the influence of the electric field due to the incident electromagnetic radiation. The electric force on the electron is $e\mathbf{E}$. The differential equation of motion of a bound electron under the action of impressed field given by Eq. (1) is

$$m\frac{d^2\mathbf{r}}{dt^2} = e\mathbf{E} - \alpha\frac{d\mathbf{r}}{dt} - k\mathbf{r}.$$

Here, $-k\,\mathbf{r}$ is elastic restoring force;

$-\alpha\dfrac{d\mathbf{r}}{dt}$ is damping frictional force against the motion of the electron.

$$\therefore \qquad \frac{d^2\mathbf{r}}{dt^2} + \frac{\alpha}{m}\frac{d\mathbf{r}}{dt} + \frac{k}{m}\mathbf{r} = \frac{e}{m}\mathbf{E}$$

or
$$\frac{d^2\mathbf{r}}{dt^2}+\gamma\frac{d\mathbf{r}}{dt}+\omega_0{}^2\mathbf{r}=\frac{e}{m}\mathbf{E} \quad ...(2)$$

where $\gamma=\alpha/m$ and $\omega_0{}^2=k/m$.

The solution of Eq. (1) is
$$\mathbf{r}=\frac{(e/m)\,\mathbf{E_0}\,e^{-i\omega t}}{[(\omega_0{}^2-\omega^2)-i\,\gamma\,\omega]} \quad ...(3)$$

Eq. (3) gives the electron displacement at time t. Natural frequency of electronic dipole oscillator is ω_0; impressed frequency of the field is ω and damping coefficient is γ. These three parameters control the dispersion.

The polarization P of the medium is equal to the dipole moment per unit volume.

If the number of dispersion electrons per unit volume is N, then
$$\mathbf{P}=N\,e\mathbf{r} \quad ...(4)$$

Substituting the value of **r** given by (3) in (4), we get
$$\mathbf{P}=\frac{Ne\,(e/m)\,\mathbf{E_0}\,e^{-i\omega t}}{[(\omega_0{}^2-\omega^2)-i\gamma\omega]}=\frac{N(e^2/m)\,\mathbf{E}}{[(\omega_0{}^2-\omega^2)-i\gamma\omega]} \quad ...(5)$$

In general, differently situated electrons within a given molecule exhibit different natural frequencies and damping coefficients. Let us say there are f_k electrons with frequency ω_k and damping γ_k in each molecule. Then,
$$\mathbf{P}=Ne\,\Sigma f_k\mathbf{r}_k$$
$$=\frac{N\,e^2\mathbf{E}}{m}\sum_k\frac{f_k}{(\omega_k^2-\omega^2-i\gamma_k\omega)} \quad ...(6)$$
$$\frac{\mathbf{P}}{\mathbf{E}}=\frac{Ne^2}{m}\sum_k\frac{f_k}{(\omega_k^2-\omega^2-i\gamma_k\omega)}$$

Now dielectric constant is
$$\varepsilon=1+\frac{1}{\varepsilon_0}\frac{\mathbf{P}}{\mathbf{E}}$$
$$=1+\frac{1}{\varepsilon_0}\frac{Ne^2}{m}\sum_k\frac{f_k}{(\omega_k^2-\omega^2-i\gamma_k\omega)}$$

Thus the dielectric constant is complex. Dielectric constant is equal to the square of the refractive index. We thus conclude that refractive index is also complex.
$$\therefore \quad n^{*2}=1+\frac{1}{4\pi\varepsilon_0}\cdot\frac{4\pi Ne^2}{m}\sum_k\frac{f_k}{(\omega_k^2-\omega^2-i\gamma_k\omega)} \quad ...(7)$$

This is known as dispersion formula. The significance of the complex refractive index is that there is an absorption of energy in the medium. The refractive index has become frequency dependent. Now we consider the following situations :

(*a*) Normal dispersion :

In the region remote from the natural frequencies of oscillators *i.e.,* the absorption frequencies of the medium, $(\omega_k{}^2-\omega^2)>>i\gamma_k\omega$. Eq. (7) reduces to

$$n^2 = 1 + \frac{1}{4\pi\varepsilon_0} \cdot \frac{4\pi N e^2}{m} \sum_k \frac{f_k}{(\omega_k^2 - \omega^2)} \qquad \ldots(8)$$

This expression shows that the refractive index is real and increases with frequency of the incident wave. This is normal dispersion.

Substituting $\omega_k = 2\pi\nu_k = 2\pi c/\lambda_k$ and $\omega = 2\pi\nu = 2\pi c/\lambda$, we get

$$n^2 = 1 + \frac{Ne^2}{\varepsilon_0 m} \sum_k \frac{f_k}{\left(\frac{4\pi^2 c^2}{\lambda_k^2} - \frac{4\pi^2 c^2}{\lambda^2}\right)}$$

or
$$n^2 = 1 + \sum_k \frac{A_k \lambda^2}{(\lambda^2 - \lambda_k^2)} \text{ where } A_k = \frac{Ne^2 \lambda_k^2 f_k}{\varepsilon_0 m \times 4\pi^2 c^2} \qquad \ldots(9)$$

Eq. (9) represents Sellmeier's dispersion formula.

If $\lambda >> \lambda_k$, Eq. (9) becomes

$$n^2 = 1 + \sum_k A_k \left(1 - \frac{\lambda_k{}^2}{\lambda^2}\right)^{-1}$$

$$= 1 + \sum_k A_k + \sum_k A_k \frac{\lambda_k{}^2}{\lambda^2} + \sum_k A_k \frac{\lambda_k{}^4}{\lambda^4} + \ldots$$

$$= A + \frac{B}{\lambda^2} + \frac{C}{\lambda^4} \qquad \ldots(10)$$

This is known as **Cauchy's dispersion** relation in which term A corresponds to refraction and B corresponds to dispersion.

$(dn/d\lambda)$ gives the dispersive power of the medium.

By differentiating Eq. (10), it can be shown that

$$\frac{dn}{d\lambda} = -\frac{L}{\lambda^3}$$

where L is a constant.

(*b*) Anomalous Dispersion

So long as the impressed frequencies do not include a natural frequency of the electrons, the dispersion is always normal. But when the impressed frequencies contain the natural frequency from the so many natural frequencies of the electron, the normal course of dispersion is disturbed. Here, for simplicity we assume that there is one natural frequency *i.e.*, $\omega_k = \omega_0$. Then Eq. (7) becomes

$$n^2 = 1 + \frac{1}{4\pi\varepsilon_0} \cdot \frac{4\pi N(e^2/m)}{(\omega_0{}^2 - \omega^2 - i\gamma\omega)} \qquad \ldots(11)$$

$\therefore$
$$n = \left[1 + \frac{1}{4\pi\varepsilon_0} \cdot \frac{4\pi N(e^2/m)}{(\omega_0^2 - \omega^2 - i\gamma\omega)}\right]^{1/2}$$

or
$$n = 1 + \frac{1}{4\pi\varepsilon_0} \cdot \frac{2\pi N(e^2/m)}{(\omega_0^2 - \omega^2 - i\gamma\omega)} \qquad \ldots(12)$$

Multiplying numerator and denominator of R.H.S. by $(\omega_0^2 - \omega^2 + i\gamma\omega)$, we get

$$n = 1 + \frac{2\pi N(e^2/m)}{4\pi\varepsilon_0}\left[\frac{(\omega_0{}^2 - \omega^2) + i\gamma\omega}{(\omega_0^2 - \omega^2)^2 + \gamma^2\omega^2)}\right]$$

or

$$n = 1 + \frac{2\pi Ne^2}{4\pi\varepsilon_0 m}\left\{\frac{(\omega_0^2 - \omega^2)}{(\omega_0^2 - \omega^2)^2 + \gamma^2\omega^2}\right\}$$

$$+ i\,.\frac{2\pi Ne^2}{4\pi\varepsilon_0 m}\left\{\frac{\gamma\omega}{(\omega_0^2 - \omega^2)^2 + \gamma^2\omega^2)}\right\}$$

$\therefore$

$$n = \eta + i\kappa \qquad \text{...(13)}$$

where

$$\eta = 1 + \frac{2\pi Ne^2}{4\pi\varepsilon_0 m}\left\{\frac{(\omega_0^2 - \omega^2)}{(\omega_0^2 - \omega^2)^2 + \gamma^2\omega^2)}\right\} \qquad \text{...(14)}$$

and

$$\kappa = \frac{2\pi Ne^2}{4\pi\varepsilon_0 m}\left\{\frac{(\gamma\omega)}{(\omega_0^2 - \omega^2)^2 + \gamma^2\omega^2)}\right\} \qquad \text{...(15)}$$

Fig. 11.1 shows the real and imaginary part of n plotted as a function of frequency ω. It is clear from the figure that at low frequencies η is slightly greater than unity and it increases as ω increases. It reaches a maximum for $\omega_1 \approx \omega_0 - \gamma/2$ and then decreases rapidly until $\omega = \omega_2 = \left(\omega_0 + \frac{\gamma}{2}\right)$. When $\omega = \omega_0$, $\eta = 1$. After ω_2, the curve again increases and approaches unity asymptotically.

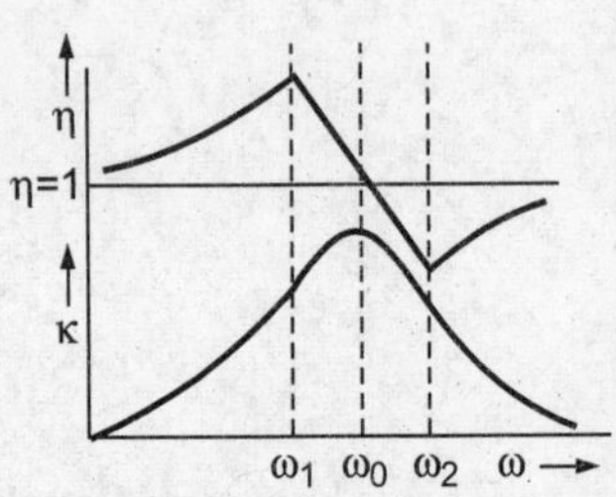

Fig. 11.1.

The imaginary part corresponds to absorption of electromagnetic waves propagating through the gas. It is also evident from the figure that κ is maximum when $\omega = \omega_0$ where η is unity. It has a width at half maximum approximately equal to γ. Thus in the region where η changes rapidly, the gas is relatively highly absorbing.

11.3. Experimental Demonstration of Anomalous Dispersion

The anomalous dispersion of sodium vapour was demonstrated by Wood. The apparatus is shown in Fig. 11.2. Pieces of sodium are placed along the bottom of a steel tube T which is fitted with water cooled glass window at each end. The tube is partially evacuated. When sodium is heated, its vapour diffuses upwards across the tube and thus the density of vapour gradually decreases upwards. In this way the vapour is optically equivalent to a prism whose thickness increases downwards.

White light from the horizontal slit S_1 is collimated by lens L_1, and then refracted by sodium vapour prism and is brought to a focus in the plane of vertical slit S_2 of the spectrometer by means of lens L_2.

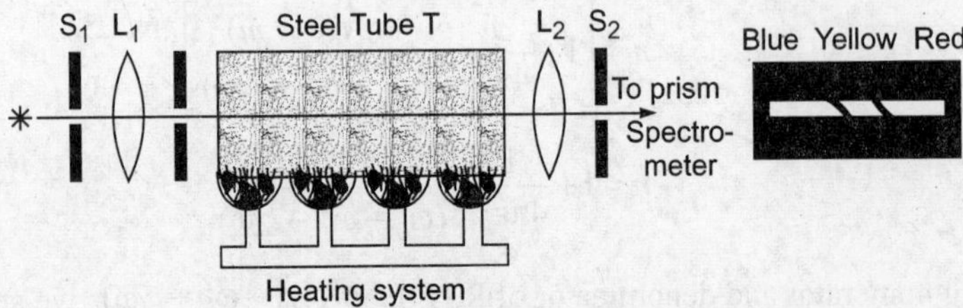

Fig. 11.2.

When the tube is cold, the gas is homogeneous and a white image of S_1 is produced across the slit S_2 giving a continuous spectrum along the horizontal strip of the focal plane of the spectrometer. When sodium is vaporised, due to its prismatic action, the images of S_1 corresponding to different wavelengths are dispersed across the length of the slit S_2. For wavelengths remote from D lines, $n = 1$ and the horizontal continuous spectrum is unaffected. But as the D lines are approached from the short wavelength side, the index of refraction of the vapour becomes less than one and the light is deviated upwards. Since the spectrometer lens inverts the image of the slit, the spectrum on the green side of the D lines is bent down in the spectroscope. Similarly, the spectrum on the long wavelength side of the D lines is bent up. The spectrum (Fig. 11.2) illustrates the anomalous dispersion for two neighbouring wavelengths 589 nm and 589.6 nm.

11.4. Rayleigh Scattering

When E.M. energy is incident on any matter, the electric vector associated with the energy sets the electron system of the matter to vibrate. Scattering may be treated as a case of absorption of E.M. energy and its re-emission due to vibrations of the electrons. We can regard scattering as the case of forced vibrations of an electron under the action of an electric vector of type

$$\mathbf{E} = \hat{\mathbf{x}}\, E_0 \cos(kz - \omega t) \qquad \ldots(1)$$

i.e., the field is in the x-direction having an amplitude E_0 and propagating in the $+z$-direction; $\hat{\mathbf{x}}$ represents the unit vector in the x-direction.

ω is the frequency of the wave with propagation constant $k = 2\pi/\lambda$.

Consider an electron of mass m and charge e. The equation of motion for the electron, in the presence of an external electric field $\mathbf{E}$, would be

$$m\frac{d^2\mathbf{x}}{dt^2} + k_0\mathbf{x} = e\mathbf{E}$$

or

$$\frac{d^2\mathbf{x}}{dt^2} + \omega_0{}^2\mathbf{x} = \frac{e}{m}\hat{\mathbf{x}}\, E_0 \cos(kz - \omega t)$$

k_0 is the force constant and $\omega_0 (= \sqrt{k_0/m})$ represents the frequency of the oscillator.

The displacement and the electric field are in the same direction. So we can replace the vectors by the corresponding scalar quantities. Thus

$$\frac{d^2x}{dt^2} + \omega_0{}^2 x = \frac{eE_0}{m}\cos(kz - \omega t) \qquad \ldots(2)$$

We have neglected the effect of damping in Eq. (2).

The solution corresponding to the forced vibrations is given by

$$x = \frac{eE_0}{m(\omega_0{}^2 - \omega^2)}\cos(kz - \omega t) \qquad \ldots(3)$$

Eq. (3) gives the electron displacement x at time t.

The dipole moment of the atom is

$$\mathbf{P} = e\,\mathbf{x} = \frac{e^2}{m(\omega_0{}^2 - \omega^2)}\mathbf{E} \qquad \ldots(4)$$

where

$$|\,p_0\,| = \frac{e^2 E_0}{m(\omega_0^2 - \omega^2)}$$

This oscillating dipole radiates energy. Now, an oscillating dipole given by

$$\mathbf{p} = \mathbf{p_0}\, e^{-i\omega t} \qquad \ldots(5)$$

radiates energy at a rate

$$\bar{P} = \frac{\omega^4 p_0^2}{12\pi\varepsilon_0 c^3} \qquad \ldots(6)$$

or

$$\bar{P} = \frac{\omega^4}{12\pi\varepsilon_0 c^3} \frac{e^4}{m^2(\omega_0^2 - \omega^2)^2} E_0^2 \qquad \ldots(7)$$

Let N be the number of atoms per unit volume.

Then

the total energy radiated away per unit volume = $N\bar{P}$

The intensity of the incident wave is given by

$$I = \frac{1}{2} \varepsilon_0 c E_0^2 \qquad \ldots(8)$$

Thus the change in the intensity of the electromagnetic wave as it propagates through a distance dx is given by

$$dI = -N\bar{P}\, dx$$

or

$$\frac{dI}{I} = -\gamma dx \qquad \ldots(9)$$

where

$$\gamma = \frac{N\omega^4}{6\pi\varepsilon_0^2 c^4} \frac{e^4}{m^2(\omega_0^2 - \omega^2)^2} \qquad \ldots(10)$$

$\therefore$

$$I = I_0\, e^{-\gamma x} \qquad \ldots(11)$$

implying that γ represents the attenuation coefficient.

Thus if we assume $\omega << \omega_0$, then γ becomes proportional to ω^4

or

$$\gamma \propto \frac{1}{\lambda^4} \qquad \ldots(12)$$

Eq. (12) represents Rayleigh scattering law.

(*i*) **Fluorescence.** When ultra-violet light is incident on certain substances, they emit visible light. For example, when a solution of quinine sulphate in dilute sulphuric acid is exposed to the ultra-violet portion of the spectrum, it is found to glow brightly emitting blue light. *This phenomenon of absorbing light of one wavelength and then giving out light of another wavelength is called* **fluorescence.** According to *Stoke's law, the wavelength of the emitted (fluorescent) radiation is always greater than that of exciting light.* Fluorescence stops as soon as the incident radiation is cut off.

(*ii*) **Phosphorescence.** There are certain substances *which continue to emit visible radiations even after the exciting light has been turned off.* This phenomenon is called **phosphorescence.** When calcium sulphide or zinc sulphide is exposed to sunlight and then taken to a darkened room, they glow with bluish and greenish light respectively.

Explanation. The origin of the phenomenon of fluorescence and phosphorescence can be explained on the basis of the quantum theory of radiations. When the light of some suitable frequency is allowed to fall upon a **fluorescent** substance, the incident light photons are able to raise many electrons from their normal state E_0 to excited state E_1 (Fig. 11.3). $E_a, E_b, \ldots E_n$ are the energies of the intermediate levels. While returning to the normal state, the excited electrons can do so either in a single step (E_1 to E_0) with the

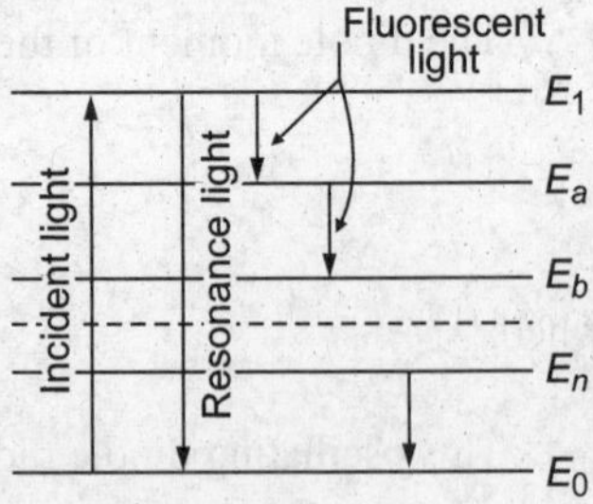

Fluorescent light
E_1
E_a
E_b
E_n
E_0
Incident light
Resonance light

Fig. 11.3.

emission of light of the same frequency or in stages from E_1 to E_a, E_a to E_b, ... E_n to E_0. This results in the emission of radiations of several frequencies. When a substance re-emits radiation of the same frequency, the emitted radiation is called resonance radiation. We thus find that the fluorescent light consists of a range of frequencies mainly lower than the frequency of the incident light.

Applications. 1. Gems such as diamonds are often recognized by the characteristic glow given off by them in the presence of ultraviolet radiations.

2. Fluorescent materials have been widely used in the manufacture of fluorescent lamps used for general illumination.

11.5. Mie Scattering

When the particles are larger than about one-tenth of a wavelength, the light scattered from one point on the particle may well be out of phase with light scattered from another point. The two contributions then interfere and the scattered-light distribution is no longer symmetrical as in the Rayleigh case. With increasing particle size, the scattered light becomes more concentrated in the forward direction (Fig. 11.4).

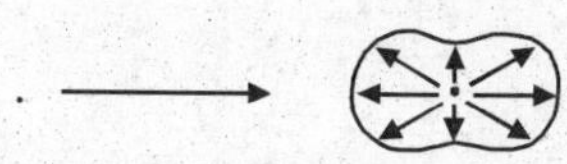

Small particles : Rayleigh scattering

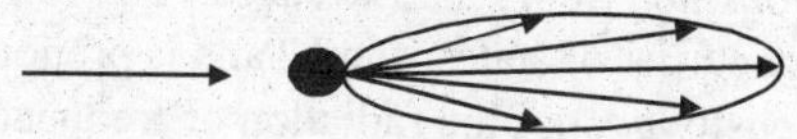

Large particles : Mie scattering

Fig. 11.4.

Fig. 11.4 shows angular distribution in Rayleigh and Mie scattering. The difference between forward and back scatter is called *dissymmetry*, which is defined as the ratio $i_{45°}/i_{135°}$. Dissymmetry can be used to determine the average size of particles, provided that their shape is known. For larger particles, however, the angular dependence of the scattering becomes quite complicated, showing a number of maxima and minima, and cannot be adequately described by a single parameter such as the dissymmetry.

EXERCISE

1. Give the theory of normal and anomalous dispersion and describe how the latter has been studied in the case of sodium vapour.
2. Explain Rayleigh scattering. Differentiate between Rayleigh and Raman scattering.
3. At a wavelength of 546 nm the scattering intensity of pure carbon tetrachloride is found to be 5.9 arbitrary units. What is the intensity at 436 nm ? **[Ans.** 14.51]

CHAPTER 12

LASER

12.1. The Einstein Coefficients

(1) The probable rate of occurrence of the *absorption transition* from state 1 to state 2 (Fig. 12.1) depends on the properties of states 1 and 2 and is proportional to energy density $u(\nu)$ of the radiation of frequency ν incident on the atom. Thus

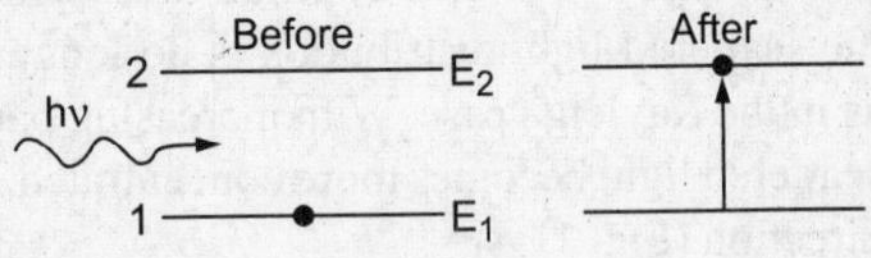

Fig. 12.1.

$$P_{12} \propto u(\nu) \text{ or } P_{12} = B_{12}\, u(\nu) \quad \ldots(1)$$

The proportionality constant B_{12} is called *Einstein's coefficient of absorption of radiation.*

(2) The probability of spontaneous emission from state 2 to state 1 (Fig. 12.2) depends only on the properties of states 1 and 2. This is independent of energy density $u(\nu)$ of incident radiation. Einstein denoted this probability per unit time by A_{21}.

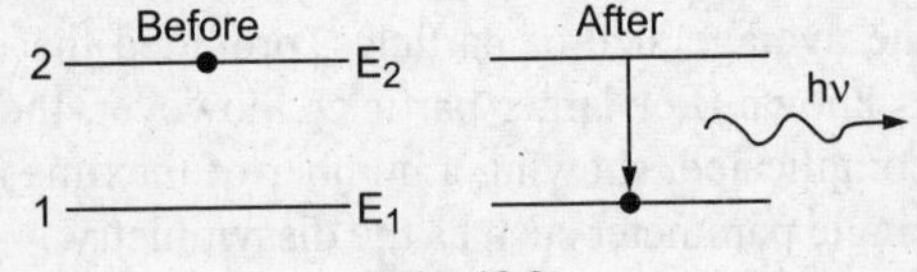

Fig. 12.2.

$$(P_{21})_{spontaneous} = A_{21} \quad \ldots(2)$$

A_{21} is called *Einstein's coefficient of spontaneous emission of radiation.*

(3) The probability of stimulated emission transition from state 2 to state 1 (Fig. 12.3) is proportional to the energy density $u(\nu)$ of the stimulating radiation as well as on the properties of two states involved *i.e.,*

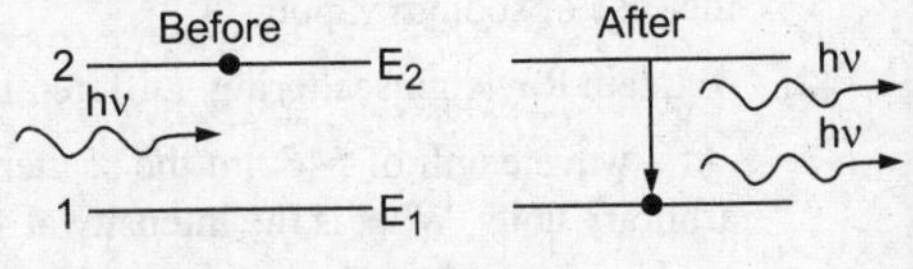

Fig. 12.3.

$$(P_{21})_{stimulated} = B_{21}\, u\,(\nu) \quad \ldots(3)$$

B_{21} is *Einstein's coefficient of stimulated emission of radiation.*

The total probability for an atom in state 2 to drop to the lower state 1 is, therefore

$$P_{21} = A_{21} + B_{21}\, u\,(\nu) \quad \ldots(4)$$

Relation between Einstein's *A* and *B* coefficients

Consider an assembly of atoms in thermal equilibrium at temperature T with radiation of frequency ν and energy density $u(\nu)$. Let N_1 and N_2 be the number of atoms in energy states 1 and 2 respectively at any instant (Fig. 12.4).

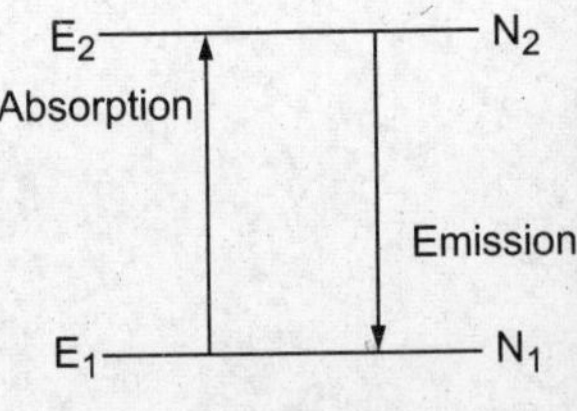

Fig. 12.4.

The number of atoms in state 1 that absorb a photon and rise to state 2 per unit time is given by

$$N_1P_{12} = N_1B_{12}\,u(\nu) \qquad \ldots(5)$$

The number of photons in state 2 that can cause emission process [spontaneous + stimulated] per unit time is given by

$$N_2P_{21} = N_2[A_{21} + B_{21}\,u(\nu)] \qquad \ldots(6)$$

For equilibrium, the absorption and emission rates must be equal *i.e.*,

$$N_1P_{12} = N_2P_{21}$$

$$N_1B_{12}\,u(\nu) = N_2[A_{21} + B_{21}\,u(\nu)]$$

or
$$N_1B_{12}\,u(\nu) = N_2A_{21} + N_2B_{21}\,u(\nu)$$

or
$$u(\nu)[N_1B_{12} - N_2B_{21}] = N_2A_{21}$$

$$\therefore \qquad u(\nu) = \frac{N_2A_{21}}{N_1B_{12} - N_2B_{21}} = \frac{A_{21}}{B_{21}}\frac{1}{\left[\frac{N_1}{N_2}\left(\frac{B_{12}}{B_{21}}\right) - 1\right]} \qquad \ldots(7)$$

According to Boltzmann distribution law, the number of atoms N_1 and N_2 in energy states E_1 and E_2 in thermal equilibrium at temperature T is given by

$$N_1 = N_0e^{-E_1/kT} \text{ and } N_2 = N_0e^{-E_2/kT}$$

where N_0 = total number of atoms present and k = Boltzmann's constant.

$$\therefore \qquad \frac{N_2}{N_1} = \frac{e^{-E_2/kT}}{e^{-E_1/kT}} = e^{-(E_2-E_1)/kT} = e^{-h\nu/kT}$$

$$[\because E_2 - E_1 = h\nu \text{ (energy of photon emitted or absorbed)}]$$

or
$$\frac{N_1}{N_2} = e^{h\nu/kT} \qquad \ldots(8)$$

Substituting this value of N_1/N_2 in Eq. (7), we get

$$u(\nu) = \frac{A_{21}}{B_{21}}\frac{1}{[e^{h\nu/kT}\left(\frac{B_{12}}{B_{21}}\right) - 1]} \qquad \ldots(9)$$

Comparing it with Planck's radiation formula

$$u(\nu) = \frac{8\pi h\nu^3}{c^3}\frac{1}{[e^{h\nu/kt} - 1]} \qquad \ldots(10)$$

we get
$$\frac{A_{21}}{B_{21}} = \frac{8\pi h\nu^3}{c^3} \text{ and } \frac{B_{12}}{B_{21}} = 1 \qquad \ldots(11)$$

(*i*) $B_{21} = B_{12}$, *i.e.*, the probability of stimulated emission is the same as that of (induced) absorption.

(*ii*) $\frac{A_{21}}{B_{21}} \propto \nu^3$.

That is, the ratio of *spontaneous emission and stimulated emission* is proportional to ν^3. It means that the probability of spontaneous emission dominates over induced emission more and more as the energy difference between the two states increases.

12.2. Population Inversion

An atom can interact with radiation over a range of frequencies and the strength of the interaction is a function of the frequency. This function is called the line shape function. Let $g(\omega)$ represent the normalized line shape function corresponding to the transition between levels 1 and 2. Then, $N_1 g(\omega)d\omega$ and $N_2 g(\omega)d\omega$ would represent the number of atoms per unit volume in levels 1 and 2 respectively which are capable of interacting with radiation of frequency between ω and $\omega + d\omega$.

Now,

$$\int_0^\infty g(\omega)\, d\omega = 1$$

so that the total number of atoms in levels 1 and 2 is N_1 and N_2 respectively.

Einstein coefficients are related by

$$B_{21} = \frac{\pi^2 c^3}{\hbar\, \omega^3 n_0^3} A_{21} = \frac{\pi^2 c^3}{\hbar\, \omega^3 n_0^3 t_{sp}} \quad \text{...(1)}$$

where $t_{sp}(=1/A_{21})$ represents the spontaneous lifetime of the upper level, $\hbar = \frac{h}{2\pi}$, h being the Planck's constant, n_0 = refractive index of the medium, $\omega = 2\pi\nu$.

Let $u(\omega)$ be the energy per unit volume per unit frequency range.

The total number of stimulated emissions per unit time per unit volume is given by

$$W_{21} = N_2 \int_0^\infty B_{21}\, u(\omega)\, g(\omega)\, d\omega$$

$$= N_2 \frac{\pi^2 c^3}{\hbar\, t_{sp} n_0^3} \int_0^\infty \frac{u(\omega)}{\omega^3} g(\omega)\, d\omega \quad \text{...(2)}$$

For a near monochromatic radiation field, $u(\omega)$ is very sharply peaked at a particular value of ω (say ω'). So $g(\omega)/\omega^3$ is constant. Integrating Eq. (2),

$$W_{21} \simeq N_2 \frac{\pi^2 c^3}{\hbar\, t_{sp} n_0^3} \frac{u_{\omega'}}{\omega'^3} g(\omega') \quad \text{...(3)}$$

Here, $u_{\omega'}$ = energy density associated with the radiation field and $g(\omega')$ = value of the lineshape function evaluated at the radiation frequency ω'. Similarly, the number of stimulated absorptions per unit time per unit volume is given by

$$W_{12} = N_1 \frac{\pi^2 c^3}{\hbar\, t_{sp} n_0^3} \frac{u_{\omega'}}{\omega'^3} g(\omega') \quad \text{...(4)}$$

Now, consider a monochromatic beam of frequency ω propagating through a collection of atoms along the z-direction (Fig. 12.5).

Consider two planes of area S perpendicular to the z-direction at z and $z + dz$.

The volume of the medium between planes P_1 and $P_2 = S\,dz$

Number of stimulated absorptions per unit time $= W_{12}\, S\, dz$

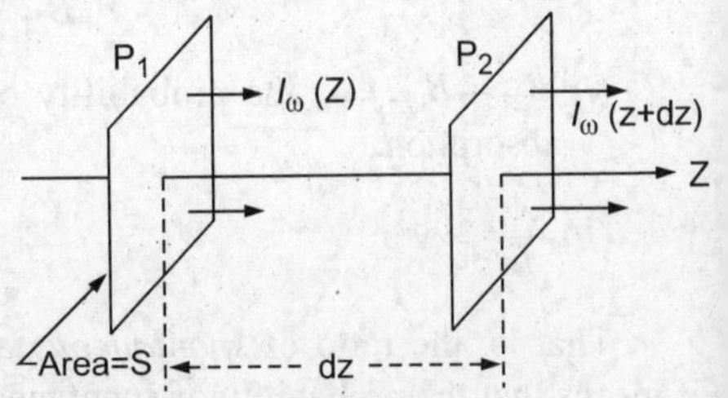

Fig. 12.5.

Energy of each photon = $\hbar\omega$

∴ the energy absorbed per unit time in the volume element $S\,dz$ is

$$W_{12}\,\hbar\,\omega S\,dz.$$

Similarly, the corresponding energy gain (because of stimulated emissions) is

$$W_{21}\,\hbar\,\omega S\,dz$$

where we have neglected the radiation arising out of spontaneous emissions, which is randomly radiated.

Thus, the net amount of energy absorbed per unit time in the volume element $S\,dz$ is

$$(W_{12} - W_{21})\,\hbar\,\omega\,S\,dz.$$

Let $I_\omega(z)$ be the intensity of the beam in the plane P_1.

Then,

the total energy entering the volume element $S\,dz$ per unit time $= I_\omega(z)\,S$.

Similarly, if $I_\omega(z + dz)$ represents the intensity in the plane P_2, then the total energy leaving the volume element per unit time is

$$I_\omega(z + dz)\,S = I_\omega(z)\,S + \frac{\partial I_\omega}{\partial z}\,dz\,S$$

Hence, the net amount of energy leaving the volume element per unit time is

$$\frac{\partial I_\omega}{\partial z}\,dz\,S.$$

This must be equal to the negative of the net energy absorbed by the medium between z and $z + dz$. Thus,

$$\frac{\partial I_\omega}{\partial z}\,S\,dz = -\,(W_{12} - W_{21})\,\hbar\,\omega\,S\,dz$$

$$= -\,\frac{\pi^2 c^3}{\hbar\,t_{sp}\,\omega^3 n_0^3}\,g(\omega)\,u_\omega\,\hbar\,\omega S\,dz\,(N_1 - N_2)$$

or
$$\frac{\partial I_\omega}{\partial z} = -\,\frac{\pi^2 c^3}{\omega^2 t_{sp} n_0^3}\,g(\omega)\,u_\omega\,(N_1 - N_2) \qquad \text{...(5)}$$

Now the energy density u_ω and the intensity I_ω are related by

$$I_\omega = v u_\omega = \frac{c}{n_0}\,u_\omega \qquad \text{...(6)}$$

Here, $v = c/n_0$ = velocity of the radiation field in the medium.

Thus Eq. (5) reduces to $\dfrac{1}{I_\omega}\dfrac{dI_\omega}{dz} = -\,\alpha_\omega$ ...(7)

where
$$\alpha_\omega = \frac{\pi^2 c^2}{\omega^2 t_{sp} n_0^2}\,(N_1 - N_2)\,g(\omega) \qquad \text{...(8)}$$

Integrating Eq. (7),
$$I_\omega(z) = I_\omega(0)\exp(-\alpha_\omega z) \qquad \text{...(9)}$$

Thus, if α_ω is positive, *i.e.*, if $N_1 > N_2$, then the intensity of the beam decreases exponentially. But if there are more atoms in the excited level than in the lower level, (*i.e.*, there is a population inversion) then $\alpha_\omega < 0$ and there would be an exponential increase in intensity of the beam.

This is called light amplification.

In an actual laser system, the active medium is placed between a pair of mirrors forming a resonant cavity (Fig. 12.6).

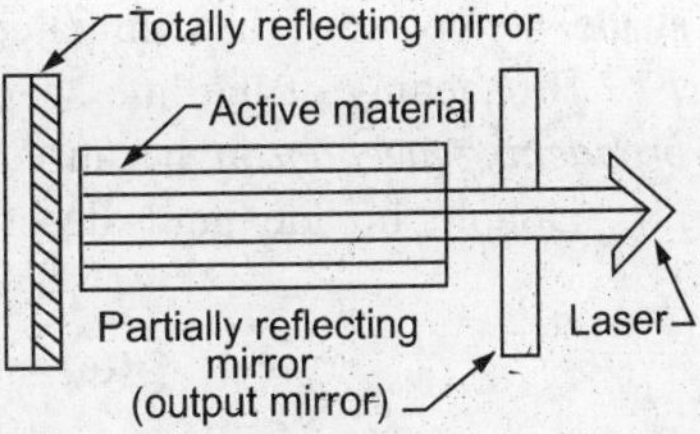

Fig. 12.6.

In order that oscillations be sustained in the cavity, it is essential that the net losses suffered by the beam be compensated for by the gain of the medium. At threshold and under steady state operation, the two are exactly compensated.

Let d be the length of the active medium. Let R_1 and R_2 be the reflectivities of the mirrors. In one round trip the beam traverses a distance $2d$ through the active medium. Therefore, the beam gets amplified by a factor exp $[-2\alpha_\omega d]$ and diminished by a factor R_1R_2 exp $[-2\alpha_c d]$. Here, exp $[-2\alpha_c d]$ represents losses caused by absorption, scattering, diffraction, etc. In an amplifying medium $\alpha_\omega < 0$ and $e^{-\alpha_\omega d} > 1$.

Thus, for the laser oscillation to begin, we must have

$$R_1 R_2 \exp[2d(-\alpha_\omega - \alpha_c)] \geq 1 \qquad \ldots(10)$$

or

$$-2\alpha_\omega d \geq 2\alpha_c d - \ln R_1 R_2 \qquad \ldots(11)$$

Now, in the absence of the active medium, due to finite reflectivities of the mirror and the losses due to diffraction and scattering, the energy stored in a mode of the resonator decays exponentially as $W(t) = W(t=0)\, e^{-t/t_c}$ where t_c is known as the passive cavity lifetime.

Since the time taken for one round trip is $2n_0 d/c$, we have

$$R_1 R_2 e^{-2\alpha_c d} = e^{-2n_0 d/ct_c} \qquad \ldots(12)$$

or

$$\frac{2n_0 d}{ct_c} = 2\alpha_c d - \ln R_1 R_2 \qquad \ldots(13)$$

Eq. (11) becomes

$$-2\alpha_\omega d \geq \frac{2n_0 d}{ct_c} \qquad \ldots(14)$$

Substituting for α_ω from Eq. (8), we get

$$N_2 - N_1 \geq \frac{4\nu^2}{c^3} \frac{n_0^3}{g(\omega)} \frac{t_{sp}}{t_c} \qquad \ldots(15)$$

Eq. (15) gives the threshold population inversion required for the oscillation of the laser.

12.3. The Line Shape Function

Atoms in gaseous media move randomly at different velocities and in different directions. The frequencies of the radiation emitted by these moving atoms are Doppler-shifted. The observed frequency of the radiation emitted by an atom moving at a velocity v along the observed direction of propagation would become $\omega = \omega_0 + (v/c)\omega_0$, where ω_0 is the frequency of the field radiated by stationary atoms. Thus, even though all the atoms of the laser medium possess the same internal resonance, the observed frequencies of the radiation emitted are different because of the Doppler effect. The frequency spectrum of the radiation emitted by all the atoms has a certain characteristic line shape and width. Emission can be stimulated at any frequency within this line shape. The power of the stimulated emission would be determined by the line shape. Thus, if the laser medium is stimulated by radiation at a frequency corresponding to the centre of the Doppler-broadened line shape, the power of the stimulated emission would be the maximum obtained from the medium.

Broadening of the line shape through the Doppler effect is classified as an *inhomogeneous broadening mechanism*, because different atoms "see" different Doppler effects.

Doppler broadening is described by a Gaussian function

$$g(\omega) = \frac{2}{\Delta\omega}\left(\frac{\ln 2}{\pi}\right)^{1/2} \exp\left[-4\ln 2\,\frac{(\omega - \omega_{21})^2}{(\Delta\omega)^2}\right] \qquad \ldots(1)$$

where $$\Delta\omega = 2\omega_{21}\left[\frac{2k_B T}{Mc^2}\ln 2\right]^{1/2} \qquad ...(2)$$

with M representing the mass of the atom, c the speed of light in free space and T the absolute temperature. The quantity $\Delta\omega$ is the full width at half maximum (Fig. 12.7) and is referred to as the line width.

Certain line broadening mechanisms, such as collision broadening or natural broadening, act to broaden the response of each atom in an identical fashion. This effect is classified as a *homogeneous broadening mechanism.*

The line shape function corresponding to the collisions that occur in a collection of atoms is a Lorentzian curve with line width $\Delta\omega$ given by

$$\Delta\omega = \frac{2}{\tau_0} \qquad ...(3)$$

τ_0 representing the mean time between two collisions. The frequency distribution takes the following form :

$$g(\omega) = \frac{2}{\pi\Delta\omega}\,\frac{1}{1+4(\omega-\omega_0)^2/(\Delta\omega)^2} \qquad ...(4)$$

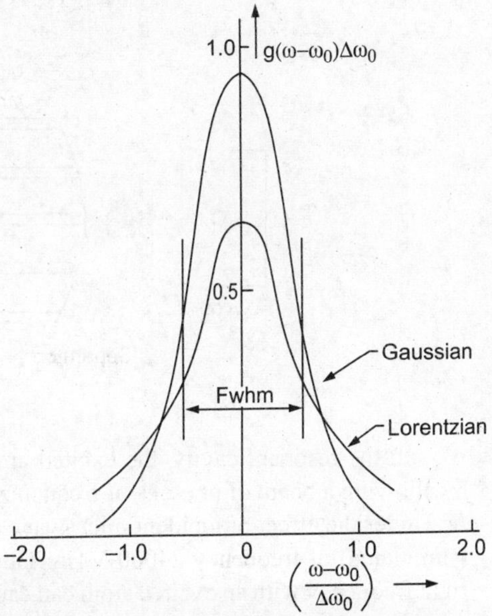

Even in the absence of collision and thermal motion, one has the inherent line width because of the finite lifetime of the excited state due to spontaneous emission. This is called natural broadening. The line shape is Lorentzain (Eq. 4) with

$$\Delta\omega = \frac{1}{t_{sp}} = A \qquad ...(5)$$

where $t_{sp}(=1/A)$ is the spontaneous emission lifetime and A is the Einstein A coefficient.

Fig. 12.7.

12.4. Maser

The term Maser stands for "*Molecular* (formerly microwave) *amplification by stimulated emission of radiation*". The working of maser is similar to that of laser. Actually the principle of stimulated emission on which laser is based was first employed in microwave region (in frequency range ~10^{10} Hz).

The *principle of maser* can be stated as follows :

The maser action is based on the principle of '*population inversion*' *followed by* '*stimulated emission*'. For stimulated emission to predominate, most of the atoms should be in the excited state. So population of atoms in higher energy level is increased by pumping the matter by photons of appropriate size. If during the excited state of the atom, it is struck by a matching photon having precisely the energy as that emitted by the atom in spontaneous emission, then the excited atom emits photons that fall precisely in phase with the stimulating photons. So the excited atom emits waves in step with incoming wave and increases its intensity. Finally resulting emergent wave is extremely intense and diverges the least.

12.5. The Ammonia Gas Maser

Principle. The NH_3 molecule has a pyramidal structure with N at the apex. Ammonia maser is based on the fact that ammonia molecules can exist in two energy states which differ in energy by about 10^{-4} eV. It corresponds to a frequency

$$\nu = \frac{\Delta E}{h} = \frac{10^{-4} \times 1.6 \times 10^{-19}}{6.6 \times 10^{-34}} \text{s}^{-1} \approx 24000 \text{ MHz}$$

which is in the microwave range ($\lambda \approx 1.25$ cm). Thus the transition of a molecule from the higher to the lower level will emit a photon of frequency 24000 MHz. The transition process may be set by an input signal of correct frequency or by the spontaneous fall of one molecule which stimulates another and so on till the process becomes self sustaining. To sustain the production of photons, more excited molecules are fed till the chamber becomes a *resonant cavity.*

Construction and Working. The different parts are shown in Fig. 12.8.

Ammonia molecules are excited in the oven O. The excited molecules stream out through the outlet H and then pass through the separator S. The separator consists of charged rods and is cylindrical in shape. Excited ammonia molecules are repelled by the electric field of charged rods of separator while unexcited molecules are attracted. But since the force of repulsion from all sides of the cylindrical separator will balance, the excited molecules will continue to move along the axis of the separator and reach the resonant cavity R. The unexcited molecules are removed from the molecular beam by attraction of the electric field.

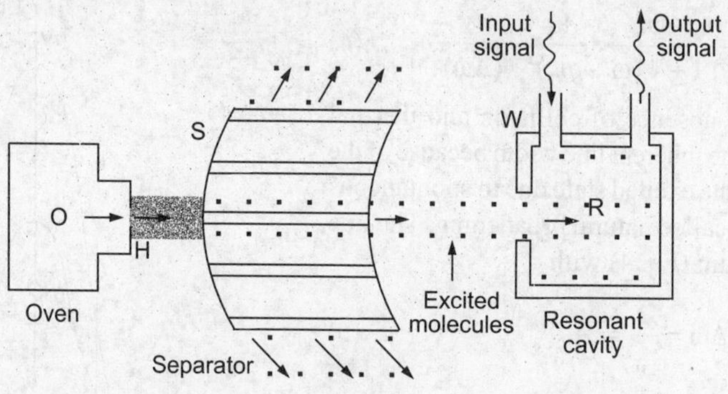

Fig. 12.8.

In the resonant cavity, the excited ammonia molecules are subjected to stimulated emission by allowing a beam of photons of frequency 24000 MHz to enter the cavity through the waveguide W. Under the effect of incident microwave beam, the excited ammonia molecules give out photons (stimulated) of frequency 24000 MHz. This takes place because the interaction of a photon of the right frequency with an excited atom can cause a transition from the higher to the lower state with the emission of an additional photon of the same energy. The stimulated photons suffer reflection from the wall of the cavity, collide with excited ammonia molecules and produce additional photons. The cavity is designed so as to build up the stimulated photons of 24000 MHz.

The resonant cavity of ammonia maser thus acts as an oscillator, the frequency of which is determined by the ammonia molecules itself. The ammonia maser is superb as a stable oscillator and has a narrow band-width as an amplifier. The frequency deviation from its central frequency in only 3 to 5 kHz.

Applications of Maser

(*i*) The ammonia maser provides a low frequency oscillator which ticks like a clock. Such a clock is called an *atomic clock* and has got such a high degree of precision that it can keep correct time to within 1 second in 1000 years.

(*ii*) Maser clocks have been employed to reconfirm the results of Michelson-Morley experiment *to prove* once again *that the speed of light is constant.*

(*iii*) Maser (atomic) clocks have also been used to *check the time-dilation effects of relativity*, thus providing a firm confirmation to relativistic predictions.

(*iv*) Maser amplifiers are being used to *detect* extremely weak *signals from outer space*.

(*v*) A super-sensitive magnetometer *for measuring earth's magnetic field* has been built using maser principles.

EXERCISE

1. Establish the relation between Einstein's Coefficients.

2. What is threshold population inversion ?

Ans. The radiation which bounces back and forth between the reflecting mirrors of the resonator is amplified by the active medium. This radiation also suffers losses due to finite reflectivity of the mirrors, scattering and diffraction. If the oscillations have to be sustained in the cavity, then the losses must be exactly compensated by the gain. For this compensation a minimum population inversion density is required. This is called the *threshold population inversion*.

3. State Schawlow-Towne's condition (threshold condition) for laser oscillation.

Ans. $N_2 - N_1 = \dfrac{4\nu^2 n_0^3 \, t_{sp}}{c^3 t_c} \dfrac{1}{g(\omega)}$

where N_1 and N_2 are the number of atoms per unit volume present in the energy levels E_1 and E_2,
$n_0 \rightarrow$ refractive index of the medium
$c \rightarrow$ velocity of light in the free space
$\nu \rightarrow$ frequency of the light
$t_{sp} \rightarrow$ spontaneous life time
$t_c \rightarrow$ passive cavity life time
$g(\omega) \rightarrow$ the frequency distribution in the line.

4. A two-level pumping scheme cannot be used for lasing. Explain.

Ans. In a two-level pumping scheme, pumping radiation would, on the one hand, move the ground state atoms to the upper level and, on the other hand, initiate transitions from the upper level to the ground level. In other words, pumping would simultaneously populate and deplete the upper level. So population inversion cannot be achieved.

5. Explain ***spiking*** in ruby laser.

Ans. The optical pumping of ruby laser is effected by a gas discharge xenon lamp. The flash operation of the pump leads to a pulsed output of the laser. Even in the short period of a few tens of microseconds in which ruby is lasing, one finds that the emission is made up of spikes of high intensity emissions as shown in Fig. 12.9.

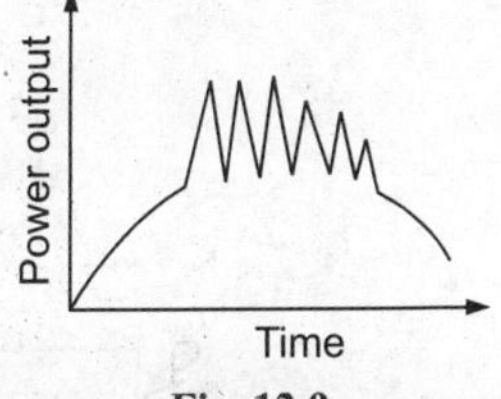

Fig. 12.9.

6. Describe the laser action in CO_2 laser with suitable energy level diagram.

7. Distinguish between laser and maser. Discuss the principle and working of ammonia maser.

8. A relative population, or Boltzmann ratio, of $1/e$ is often considered representative of the ratio of populations in two energy states at room temperature, $T = 300$ K. As an example, determine the wavelength of the radiation emitted at that temperature.

[**Hint.** $\dfrac{N_2}{N_1} = e^{-(E_2-E_1)/kT}$; Here, $\dfrac{N_2}{N_1} = \dfrac{1}{e} = e^{-1}$. $\therefore \dfrac{E_2 - E_1}{kT} = 1$

or $E_2 - E_1 = kT = h\nu = hc/\lambda$

$$\therefore \lambda = \frac{ch}{kT} = \frac{(3\times10^8)(6.63\times10^{-34})}{(1.38\times10^{-23})(300)} = 48\times10^{-6} \text{ m}]$$

CHAPTER

13 INTERFERENCE-II

13.1. Standardisation of the Metre

Principle. If the movable mirror of Michelson i nterferometer is moved through a distance x and correspondingly n fringes of wavelength λ are shifted, then $2x = n\lambda$. Thus by counting n, x can be determined in terms of λ. On this principle Michelson and Benoit used the Michelson interferometer to standardise the metre scale in terms of λ of red cadmium light (λ = 643.84722 nm). It is not possible to count the number of fringes which cross the field of view when the movable mirror moves through the whole length of one metre. Hence, nine sub-standards called 'etalons' were used for the determination. The length of each etalon was twice the length of the preceding etalon. The length of longest etalon was 10 cm while that of shortest was $10/2^8 = 0.039$ cm.

Etalon. An etalon consists of two front silvered optically plane mirrors m_1 and m_2 (Fig. 13.1*a*). The distance between their surfaces is *d*. The mirrors can be made perfectly parallel by means of screws attached to them. The mirrors are at different levels.

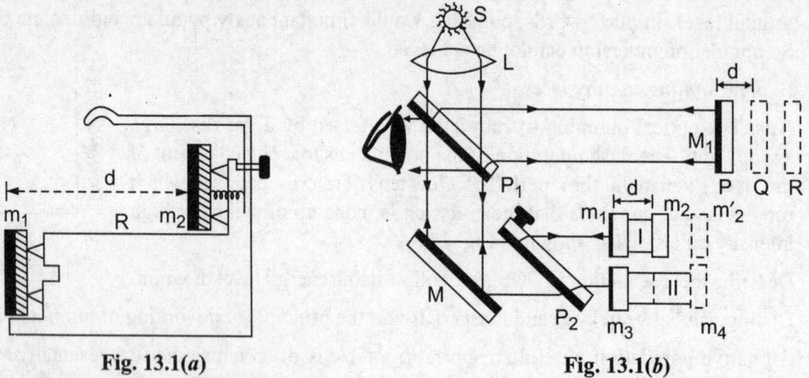

Fig. 13.1(*a*)   Fig. 13.1(*b*)

The standardisation is performed in three steps.

(*i*) **Determination of number of wavelengths corresponding to shortest elaton** — Michelson interferometer, slightly modified in form as shown in Fig. 13.1(*b*) is used. The fixed mirror M_2 is replaced by two shortest etalons placed side by side. An extra mirror M and a compensating plate P_2 are inserted as shown. P_2 can be titled by a screw fitted at its back. It enables us to count the fraction of the fringe.

Now first the distances P_1M_1, P_1Mm_1 or P_1Mm_3 are adjusted equal. This is ascertained by using white light and by moving M_1 till we get the straight line fringes with center coinciding with cross-wires using mirror m_1 or m_3. Now white light is replaced by monochromatic red cadmium light and tilt of m_1 and m_3 is adjusted to observe the circular fringes. Then M_1 is moved back slowly and number of fringes crossing the crosswires is counted very carefully. This is done until M_1 reaches Q by shifting it exactly equal to the length of etalon I. This is judged by switching back to white light and observing the central white fringe now using the mirror m_2. The fraction of last fringe (if required) is measured by rotating the compensating plate P_2 (and noting this rotation) till central white fringe comes exactly at centre of crosswires. The angle of tilt for shift of one complete fringe for cadmium light is previously determined.

Thus the length m_1m_2 of shortest etalon is determined in terms of λ of this light.

(*ii*) **Comparison of successive etalons**—The etalon I is moved backward through its own length till the mirror m_1 occupies the position previously occupied by the mirror m_2. Now the mirror m_1 will be coplanar with the mirror M_1 in the position Q, as indicated by the appearance of white fringe in the centre of the field of view of m_1.

The mirror M_1 is moved from the position Q to the position R till it becomes coplanar with the mirror m_2 in the position m_2'. This is indicated by the appearance of white fringe in the mirror m_2 (now in the position m_2'). At this stage the mirror M_1 has been moved exactly through twice the length of the etalon I or $2n\frac{\lambda r}{2}$ in terms of wavelength of cadmium red line. But still the mirror M_1 may not be coplanar with the mirror m_4 of the etalon II. Hence, the mirror M_1 is shifted further beyond the position R, till it is coplanar with m_4 as indicated by the appearance of white fringes in the centre of the field of view of m_4. This excess to make M_1 coplanar with m_4 is determined in terms of the number of fringes (x) shifted. The total length through which the mirror M_1 has been moved is $(2n + x)\ \frac{\lambda r}{2}$ and this is exactly the length of the second etalon.

Then by same method, the second etalon is compared with the third and so on until the length of 10 cm long etalon is determined in terms of number of cadmium fringes or λ of cadmium light.

(*iii*) **Comparison of 10 cm etalon with the standard metre**—Now the metre to be standardised, is placed along side the 10 cm etalon and its I mirror is made coplanar with the first fiducary mark of the metre scale by help of a good travelling microscope. Then M_1 is moved till white light fringes appear centrally using II mirror of etalon. Now keeping M_1 fixed, the etalon is moved by its own length such that its I mirror occupies the position previously occupied by II mirror and hence now white light fringes appear centrally using I mirror. Now again M_1 is moved until white light fringes appear using II mirror and then again the etalon is moved as above till white light fringes appear using I mirror. This process is repeated 10 times. This makes M_1 nearly coplanar with the last fiducary mark of the metre. The difference (if any) is determined by counting the red cadmium fringes until M_1 becomes exactly coplanar with last mark of metre (to be judged by switching on to white light fringes).

Thus the given metre is evaluated in terms of λ of cadmium light.

13.2. Non-reflecting Films (Anti-reflecting Coatings)

One of the important applications of the thin film interference phenomenon lies in reducing the reflectivity of lens surfaces. Let us consider a ray of light travelling in a medium of refractive index n_1 (say air) is incident normally on a substance of refractive index n_2 (say glass). Let I be the intensity of the incident light and I_r, the intensity of reflected light. Then according to Fresnel's equations,

$$I_r = \left(\frac{n_2 - n_1}{n_2 + n_1}\right)^2 I$$

or

$$\frac{I_r}{I} = \left(\frac{n_2 - n_1}{n_2 + n_1}\right)^2 \qquad ...(1)$$

For air-glass surface [$n_1 = 1$ and $n_2 = 1.5$], $I_r/I = 4\%$. This shows that only 96% of the incident light is transmitted. This results in a loss of intensity of the image formed by refraction through the lens-system. In order to reduce the reflection loss from the surface, a transparent film of proper thickness is deposited on the surface. This film is known as *non-reflecting film.* The coating material should not only be transparent but also should be insoluble and scratch resistant. The best material known is magnesium fluoride (MgF_2); its refractive index is 1.38. In general, the thickness of the coating material should be $\lambda/4n$, where λ is the wavelength of light used and n is the refractive index of the coated material.

In the absence of reflection, most of the light is transmitted. This improves the brightness of the image formed by the device (lens, prism, etc.). Almost all the optical parts of higher quality instruments are coated with such films.

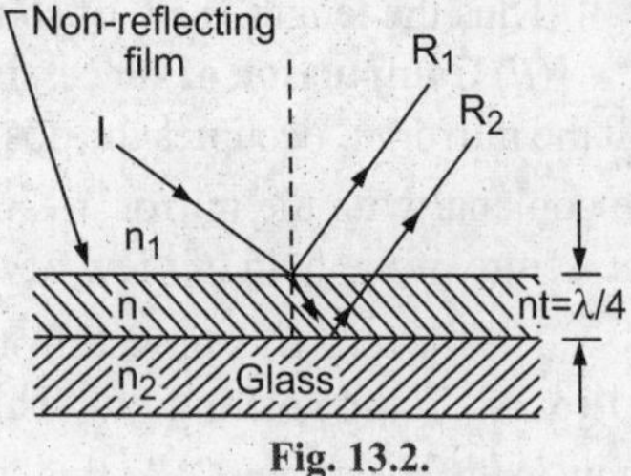

Fig. 13.2.

Let us consider that a normal ray of light is incident on the film of MgF_2 coated on glass (Fig. 13.2). This ray is reflected from the upper surface as well as from the lower surface of MgF_2. These two reflected rays R_1 and R_2 interfere with each other. Both the reflections take place at denser medium. Therefore, there will be no additional phase difference due to reflections at two surfaces. The path difference is 2 nt where t is thickness of the film.

So, if path difference is equal to $(2m + 1)\ \lambda/2$, where $m = 0, 1, 2$ ...., destructive interference will take place.

For $m = 0$, we have $$2nt = \frac{\lambda}{2}$$

$$\therefore \qquad t = \frac{\lambda}{4n} \qquad \ldots(2)$$

So the thickness of the film must be $\lambda/4\ n$.

For complete destructive interference, the intensities of two reflected beams should be equal. Thus

$$\left(\frac{n - n_1}{n + n_1}\right)^2 = \left(\frac{n_2 - n}{n_2 + n}\right)^2 \qquad \ldots(3)$$

If outside medium is air *i.e.*, $n_1 = 1$, then

$$\left(\frac{n - 1}{n + 1}\right)^2 = \left(\frac{n_2 - n}{n_2 + n}\right)^2$$

or $$\frac{n - 1}{n + 1} = \frac{n_2 - n}{n_2 + n} \qquad \ldots(4)$$

Adding 1 on both sides, we get

$$\frac{n}{n + 1} = \frac{n_2}{n_2 + n} \qquad \ldots(5)$$

Similarly, subtracting 1 on both sides of Eq. (4),

$$\frac{1}{n + 1} = \frac{n}{n_2 + n} \qquad \ldots(6)$$

Dividing Eq. (5) by Eq. (6), we get

$$n^2 = n_2 \text{ or } n = \sqrt{(n_2)} \qquad \ldots(7)$$

Thus the refractive index of coating material should be approximately equal to the square root of the refractive index of glass.

Example 1. *What percentage of light is reflected at normal incidence on a surface between air and glass (n = 5/3) ?*

Solution. $\frac{I_r}{I} = \frac{(n_2 - n_1)^2}{(n_2 + n_1)^2} = \frac{(5/3 - 1)^2}{(5/3 + 1)^2} = 0.0625 = 6.25\%$

Thus 6.25% of the light energy is reflected.

Example 2. *Magnetisum fluoride has a refractive index of 1.38 and is frequently used to coat lenses. How thick this coating should be for maximum transmittance at a wavelength of 530 nm?*

(A.U. 89)

Solution. The thickness t of the coating for maximum transmittance is given by

$$t = (\lambda / 4n)$$

where λ is the wavelength of incident light.

$$n = 1.38 \text{ and } \lambda = 530 \text{ nm} = 530 \times 10^{-9} \text{ m.}$$

$$\therefore \qquad t = \frac{530 \times 10^{-9}}{4 \times 1.38} = 9.601 \times 10^{-8} \text{ m}$$

13.3. Interference Filters

Interference filters work on the principle of Fabry-Perot interferometer. When a parallel beam of white light is incident normally on a pair of plane parallel plates silvered on the inner surfaces (just like in Febry-Perot etalon), multiple reflections take place and interference occurs for all the monochromatic components of incident light. Maxima of different orders are formed in the transmitted beam corresponding to wavelengths given by

$$2nt = m\lambda$$

where n is the refractive index of the medium, t is the plate separation and $m = 1, 2, 3, ...$

If t is large, a large number of maxima will be observed in the visible region. But when t is reduced considerably, only one or two maxima are observed in the visible region. For example, if $t = 500 \times 10^{-9}$ m and $n = 1$ (for air), the transmitted wavelengths for $m = 1, 2, 3, ...$ are 1000 nm, 500 nm, 333.3 nm, ... Out of these only 500 nm lies in the visible region. Thus a particular wavelength can be filtered out of a white light beam. Such an arrangement is known as interference filter.

Interference filters using this principle can be prepared by modern vacuum deposition techniques. The interference filter is shown in Fig. 13.3. A reflecting metal film is first evaporated on a glass plate. Now a thin film of dielectric material (like quartz or MgF_2) is evaporated on the top of reflecting metal film. Further, the dielectric layer is coated with another similar film of reflecting material. Finally, a glass plate is placed over the thin film for protection. Thus a Fabry-Perot structure is formed between the two glass plates. By varying the thickness of the dielectric film, one can filter out any particular wavelength. However, the filtered light will have a finite width, that is, it will have a narrow spectrum sharply peaked about one wavelength. The sharpness of the transmitted spectrum is determined by the reflectivity of the metallic surfaces. The larger the reflectivity, the narrower is the transmitted spectrum. But it is not possible to increase the thickness of metallic films indefinitely as absorption will reduce the intensity of the transmitted light. To overcome this difficulty, metallic films are replaced by all dielectric structures.

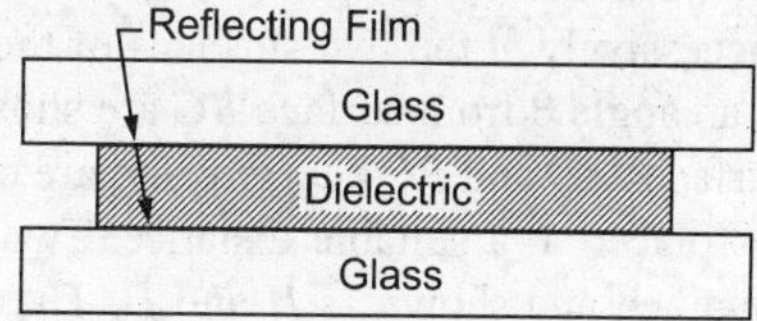

Fig. 13.3.

In an all-dielectric structure, a $\lambda/4$ thick film of titanium oxide ($n = 2.8$) is deposited on a glass substrate. Then a thin layer of dielectric material with lower refractive index (such as cryolite or magnesium fluoride) is deposited. On this is again deposited a λ/4 thick layer of a material of higher refractive index. To increase the reflectivity, multilayer structures of alternate higher and lower refractive index materials are used. In this way, it is possible to achieve a reflectivity of more than 90% for any particular wavelength. Such filters are capable of transmitting over a bandwidth as small as 1.1 nm or even less with peak at any wavelength within the visible region.

Interference filters are used in spectroscopic work for studying the spectra in a narrow range of wavelengths. Furthermore, such filters absorb practically no energy and so they are free from the trouble of overheating.

Example 3. *(a) Calculate the minimum thickness required for a layer of cryolite of refractive index 1.35 in an interference filter which is required to isolate light of wavelength 589.3 nm.*

(b) Also determine the wavelength of peak transmittance if the filter is tilted by 10°.

Solution. (*a*) The condition for the maxima in transmitted light is

$$2nt \cos \theta = m\lambda.$$

$$\therefore \quad t = \frac{m\lambda}{2n \cos \theta} = \frac{1 \times 589.3 \text{ nm}}{2 \times 1.35 \times 1} = 218.3 \text{ nm}.$$

(*b*) The angle of tilt is the angle of incidence $i = 10°$. The angle of refraction θ is given by

$$\sin \theta = \frac{\sin i}{n} = \frac{\sin 10°}{1.35} = 0.1286$$

$$\therefore \quad \cos \theta = \sqrt{[(1 - \sin^2 \theta)]} = \sqrt{[1 - (0.1286)^2]} = 0.9917$$

$$\text{Now} \quad \lambda = \frac{2nt \cos \theta}{m} = \frac{2 \times 1.35 \times 218.3 \text{ nm} \times 0.9917}{1} = 584.5 \text{ nm}$$

13.4. The Lummer-Gehrcke Plate

It consists of an accurately plane-parallel plate of glass or quartz about 10–15 cm long, 1–2 cm wide, and a few mm thick (Fig. 13.4). A small totally-reflecting prism *P* is cemented at one end of the plate. A beam of light from an extended sources *S* enters into the plate through the prism *P* and meets the lower surface of the plate at an angle slightly smaller than the critical angle. Therefore, a small fraction of the light emerges out, while the rest is internally reflected. This happens many times successively at the two surfaces of the plate. The incidence at an angle *i* in *AD* and an emergence at an angle θ from surface *BC* are shown in the figure 13.4. Multiple beams emerging out from the surfaces *AD* and *BC* interfere and are brought to a focus by an achromatic convex lens *L* on a screen *XY* placed at a suitable distance. Two sets of interference fringes, separated by a dark space, are obtained and shown as I_1 and I_2. The two sets of fringes are identical and of Haidinger type. The large value of path difference in this case enables one to use it for following specific measurements.

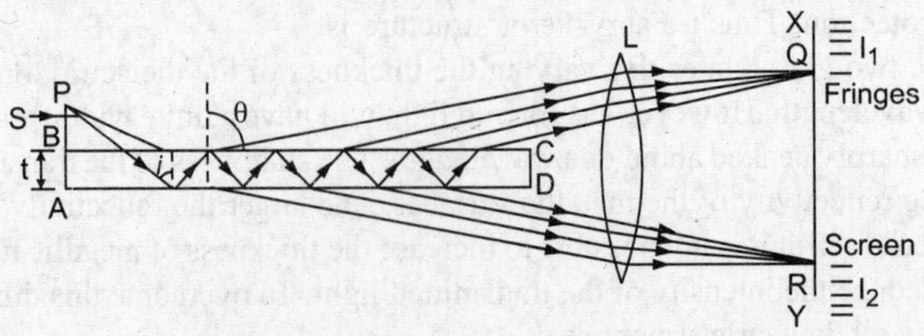

Fig. 13.4.

1. To determine the wavelength of monochromatic light.
2. To determine the hyperfine structure of spectral lines.
3. To study Zeeman effect.

Theory of Lummer Plate. Let t be the thickness of plate and n its refractive index for wavelength λ. Then for maxima,

$$2nt \cos i = m\lambda \qquad \text{...(1)}$$

where m is the order of fringe.

According to Snell's law, $\sin\theta = n \sin i$

$$\therefore \qquad \cos^2 i = 1 - \sin^2 i = \frac{n^2 - \sin^2\theta}{n^2}$$

Therefore, Eq. (1) becomes

$$2t\sqrt{(n^2 - \sin^2\theta)} = m\lambda \qquad \text{...(2)}$$

or

$$4t^2(n^2 - \sin^2\theta) = m^2\lambda^2 \qquad \text{...(3)}$$

This is the basic equation to explain interference in case of Lummer plate and it can be used to study the optical dispersion produced by this plate.

In case of monochromatic light n is constant and the value of m changes with angle of emergence θ. Hence differentiating in terms of m and θ we get

$$4t^2(0 - 2\sin\theta\cos\theta\,\delta\theta) = (2m\,\delta m)\,\lambda^2$$

or

$$\delta\theta = -\frac{m\lambda^2}{2t^2\sin 2\theta}\,\delta m$$

For successive orders $\delta m = 1$

$$\therefore \qquad \delta\theta = -\frac{m\lambda^2}{2t^2\sin 2\theta} \qquad \text{...(4)}$$

This equation gives the angular separation of successive orders and shows its dependence on λ, t and θ.

Substituting the value of m from Eq. (2), we get

$$\delta\theta = -\frac{\sqrt{(n^2 - \sin^2\theta)}}{t\sin 2\theta}\,\lambda \qquad \text{...(5)}$$

Thus by measuring the separation $\delta\theta$ and knowing the values of m, θ, t, n, the wavelength λ may be determined.

Dispersion of Spectral Lines by Lummer Plate. In case we use composite wavelength, the change of angle θ with wavelength λ is known as dispersion.

As n depends on wavelength, differentiation of Eq. (3) with respect to λ gives

$$4t^2\left[2n\frac{dn}{d\lambda} - 2\sin\theta\cos\theta\,.\frac{d\theta}{d\lambda}\right] = m^2\,.\,2\lambda$$

or

$$\frac{d\theta}{d\lambda} = \frac{4t^2\,.\,n(dn/d\lambda) - m^2\lambda}{2t^2\sin 2\theta}$$

$$= \frac{4t^2\,.\,n\,(dn/d\lambda)\,\lambda - m^2\lambda^2}{2t^2\lambda\sin 2\theta}$$

Substituting for $m^2\lambda^2$ from Eq. (3), we get

$$\frac{d\theta}{d\lambda} = \frac{2n\lambda\,(dn/d\lambda) - 2(n^2 - \sin^2\theta)}{\lambda \sin 2\theta} \qquad \ldots(6)$$

This shows that dispersion due to Lummer plate depends upon n and $(dn/d\lambda)$ of the glass plate, the wavelength λ and the angle of emergence θ.

EXERCISE

1. Describe how you will use Michelson's interferometer to calibrate metre in terms of wavelength of light.
2. Describe the principle and applications of non-reflecting films. **(A.U. 89)**
3. What are interference filters ? Explain. **(K.U. 92)**
4. Give a short account of theory, construction and working of Lummer-Gehrcke plate.
5. The thickness of an interference filter is chosen to transmit a narrow band and the wavelength $\lambda = 400$ nm. The optical path difference in the filter is 4 λ. Neglecting dispersion, find the wavelengths of the other bands transmitted in the visible region.

   [**Ans.** 700 nm, 560 nm and 466.7 nm for $m = 4, 5, 6$]
6. Explain the terms: fringes of equal thickness (Newton's rings) and fringes of equal inclination (Haidinger's fringes). **(Allahabad 87, Rajasthan 84, Kanpur, 80)**

Ans. (*i*) **Fringes of equal thickness.** The basic formula for the path difference between two interfering rays obtained due to division of amplitude by a film of thickness t and refractive index n is $2nt \cos r$, where r is the inclination of ray inside the film. If the thickness t of the film is rapidly varying, then the path difference $2nt \cos r$ changes mainly due to change in t. Therefore, the occurrence of alternate maxima and minima is due to variation in the thickness of the film. Each fringe will be the locus of all the points of same thickness. These are called 'fringes of equal thickness'. Such fringes are localised on the film itself and are observed by a microscope focussed on the film. Newton's rings belong to this class of fringes.

(*ii*) **Fringes of equal inclination.** On the other hand, if the thickness of the film is uniform, then the path difference $2nt \cos r$ between coherent rays can change only with inclination r. In this case we include wide cones of light in our observations and each fringe corresponds to a particular value of r. Such fringes are called "fringes of equal inclination" and are formed at infinity. These were first observed by Haidinger by a telescope focussed at infinity and are called Haidinger's fringes. The circular fringes in case of Michelson interferometer fall in this group of fringes in which each circular fringe corresponds to one value of r.

CHAPTER

14

MAGNIFICATION OF A LENS SYSTEM

14.1. Lateral, Longitudinal and Angular Magnifications of a Lens System

AB is the principal axis of a convergent coaxial lens-system (Fig. 14.1).

F_1 and F_2 are the first and second focal points.

H_1 and H_2 are the first and second principal points.

OO' is a small object of size y_1 placed perpendicular to the axis.

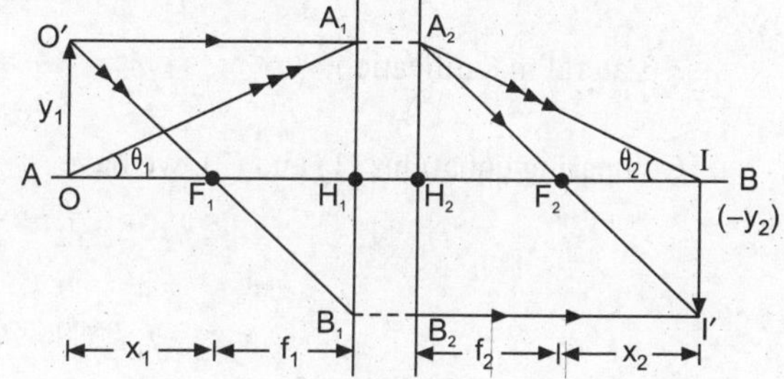

Fig. 14.1.

(*i*) An incident ray $O'A_1$ parallel to the axis emerges through A_2 where $H_1A_1 = H_2A_2$ and passes through F_2.

(*ii*) Another ray $O'B_1$ passing through F_1 emerges through B_2 where $H_1B_1 = H_2B_2$ and becomes parallel to the axis.

• The two emergent rays meet at I'.

I' is the image of O'.

• The perpendicular II' is the complete image of OO'.

The size of the image is y_2.

x_1 and x_2 are the distances of the object OO' and the image II' respectively from F_1 and F_2.

f_1 and f_2 are the first and second focal lengths of the lens system.

According to sign convention,

$$F_1H_1 = -f_1,\ F_2H_2 = +f_2$$

$$OO' = H_1A_1 = H_2A_2 = +y_1,$$

$$II' = H_2B_2 = H_1B_1 = -y_2,$$

$$OF_1 = -x_1 \text{ and } F_2I = x_2$$

(*i*) Lateral (or transverse) Magnification:

The lateral magnification (m_y) of a lens system is defined as the ratio of the length of the image (y_2) to the length of the object (y_1), both being measured perpendicular to the principal axis.

$$m_y = \frac{y_2}{y_1}$$

From similar $\Delta^s OO'F_1$ and $H_1B_1F_1$, we have

$$\frac{H_1B_1}{OO'} = \frac{H_1F_1}{OF_1}$$

or

$$\frac{-y_2}{y_1} = \frac{-f_1}{-x_1};$$

$\therefore$ Lateral magnification,

$$m_y = \frac{y_2}{y_1} = \frac{-f_1}{x_1} \quad \text{... (1)}$$

From similar $\Delta^s A_2H_2F_2$ and $I'I F_2$, we have

$$\frac{II'}{H_2A_2} = \frac{IF_2}{H_2F_2}$$

or

$$-\frac{y_2}{y_1} = \frac{x_2}{f_2}.$$

$\therefore$ Lateral magnification,

$$m_y = \frac{y_2}{y_1} = -\frac{x_2}{f_2} \quad \text{... (2)}$$

Comparing equations (1) and (2), we have

$$-\frac{f_1}{x_1} = \frac{-x_2}{f_2}$$

$\therefore$

$$x_1x_2 = f_1f_2 \text{ (Newton's formula)} \quad \text{... (3)}$$

***(ii)* Longitudinal Magnification**

The ratio of the extension of image (dx_2) to the extension of the object (dx_1) along the principal axis is called the longitudinal magnification (m_x) of the lens system.

Differentiating Eq. (3), we have

$$x_1(dx_2) + x_2(dx_1) = 0$$

$\therefore$ Longitudinal magnification,

$$m_x = \frac{dx_2}{dx_1} = -\frac{x_2}{x_1}. \quad \text{... (4)}$$

***(iii)* Angular Magnification**

The ratio of the slopes of emergent ray and conjugate incident ray with principal axis is called the angular magnification.

If the incident ray and the conjugate emergent ray make angles θ_1 and θ_2 with the principal axis, then the angular magnification of the lens system,

$$m_\theta = \frac{\tan\theta_2}{\tan\theta_1}$$

Consider the incident ray OA_1. The corresponding emergent ray is A_2I. θ_1 and θ_2 are the angles which these rays make with the principal axis. According to sign convention, θ_1 is (+)ve and θ_2 is (–)ve.

Now, we have from the figure

$$\tan\theta_1 = \frac{A_1H_1}{H_1O} = \frac{y_1}{-x_1 - f_1} = \frac{-y_1}{x_1 + f_1} \quad \text{... (5)}$$

and

$$\tan(-\theta_2) = \frac{A_2H_2}{H_2I} = \frac{y_1}{x_2 + f_2}$$

or $$\tan \theta_2 = -\frac{y_1}{x_2 + f_2} \qquad \text{... (6)}$$

From Eqs. (5) and (6), we have angular magnification,

$$m_\theta = \frac{\tan \theta_2}{\tan \theta_1} = \frac{-y_1/(x_2+f_2)}{-y_1/(x_1+f_1)} = \frac{x_1+f_1}{x_2+f_2}$$

$$= \frac{x_1+f_1}{\frac{f_1 f_2}{x_1} + f_2} \qquad [\because x_1 x_2 = f_1 f_2]$$

$$= \frac{x_1+f_1}{\left(\frac{f_2}{x_1}\right)(f_1+x_1)} = \frac{x_1}{f_2}$$

$$\therefore \qquad m_\theta = \frac{x_1}{f_2} = \frac{f_1}{x_2} \quad (\text{Since } x_1 x_2 = f_1 f_2) \qquad \text{... (7)}$$

Relation between various magnifications of a lens system

From Eqs. (4), (7) and (2), we get

$$m_x\, m_\theta = \left(\frac{-x_2}{x_1}\right)\left(\frac{x_1}{f_2}\right) = -\frac{x_2}{f_2} = m_y. \qquad \text{... (8)}$$

i.e., longitudinal magnification × angular magnification = lateral magnification.

EXERCISE

1. Define linear, lateral and angular magnifications of a lens system. Establish a relation between them. (Allahabad 2008; Agra 1981, 80)

CHAPTER

15

STIMULATED RAMAN SCATTERING

15.1. Introduction

Stimulated Raman scattering is a process where an optical medium absorbs a higher energy photon and emits a lower energy one with simultaneous absorption of the balance of the original photon energy for transition to an excited level.

Figure 15.1 illustrates this process.

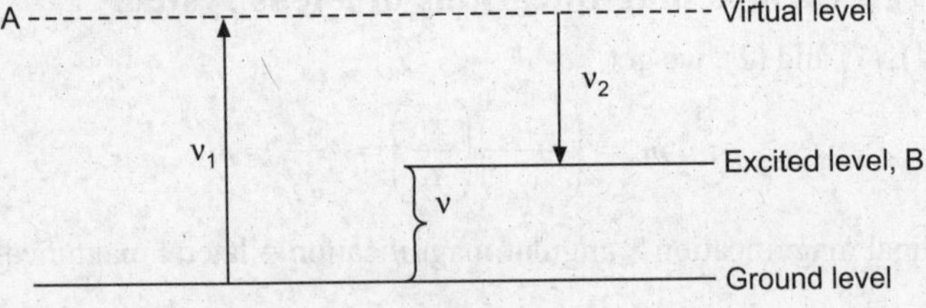

Fig. 15.1.

- A photon of frequency ν_1 (and hence of energy $h\nu_1$) is incident on the material. A molecule in the material absorbs the photon and goes to a virtual energy level *A*.
- Then the molecule emits a photon of lower frequency ν_2 (and hence of energy $h\nu_2$). The molecule *simultaneously* absorbs the balance of energy $h(\nu_1 - \nu_2) = h\nu$. As a result, the molecule is left at the excited level *B*.

In Stimulated Raman Scattering (SRS), the radiation is coherent.

15.2. Experimental Arrangement for Observing SRS

Figure 15.2 shows the experimental arrangement.

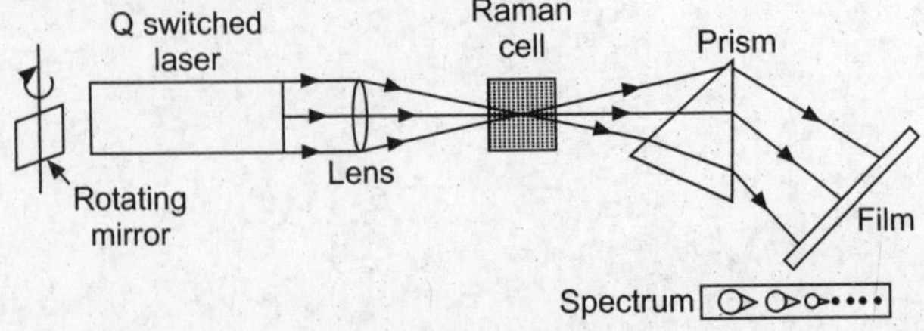

Fig. 15.2.

The beam from a *Q*-switched laser is focussed through a benzene filled Raman cell by a lens. The emerging radiation when passed through a prism of a spectrograph to a photographic film shows a spectrum. The spectrum shows ruby radiation at 6940 Å and several orders of Stokes radiation as bright spots. It also shows several orders of anit-Stokes radiation in the form of cones whose radius increases with order number. These cones of anti-Stokes radiation can be photographed as concentric rings when no prism is placed between the Raman cell and the photographic film as shown in Fig. 15.3.

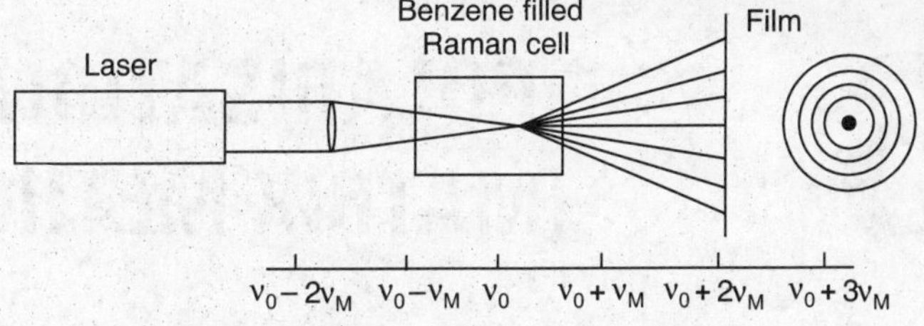

Fig. 15.3.

If the scattered radiation is photographed on a colour film, a distinct pattern of coloured concentric rings is obtained. The additional wavelengths are as coherent and well collimated as the main beam.

(*i*) *The Stokes bands.* The Stokes bands, which lie at lower wave numbers, are essentially emitted along the direction of the laser beam. Parametric amplification is responsible for the enhancement of the stimulated Raman-Stokes' lines. The power transferred to the first Stokes line at $(\nu_0 - \nu_M)$ increases with the power in the laser at ν_0. The first Stokes line rapidly becomes intense enough and acts as a powerful source at wave number $(\nu_0 - \nu_M)$ and another Stokes line at

$$(\nu_0 - \nu_M) - \nu_M = (\nu_0 - 2\,\nu_M)$$

is generated. As this line gains in intensity, it acts as another source giving rise to a third line and so on.

(*ii*) *Anti-Stokes lines.* The coloured rings correspond to the successive anti-Stokes lines at higher wave numbers, which are emitted only along directions that make specific small angles with the direction of the laser beam.

Although strong anti-Stokes lines are observed in SRS, the Stokes lines are much more intense than the anti-Stokes lines.

- The *main importance* of SRS is that it is now a mechanism for generating coherent radiation in infrared region.
- There is an essential difference between SRS and the stimulated emission in lasers. There is no need for any population inversion of states in producing the coherent radiation in SRS.

CHAPTER

16

POLARISATION-II (MATRIX METHOD)

16.1. Matrix Representation of Polarisation

We shall discuss the method of R. Clark Jones to study the phenomenon of polarisation by matrix method.

Consider a light disturbance propagating along X-axis (Fig. 16.1). Mathematically, it can be represented as

$$\mathbf{y} = \hat{\mathbf{j}}\ a_y e^{i(kx-\omega t)} \quad \ldots(1)$$

Here, $\hat{j}$ is unit vector along Y-axis and a_y is the amplitude of oscillation.

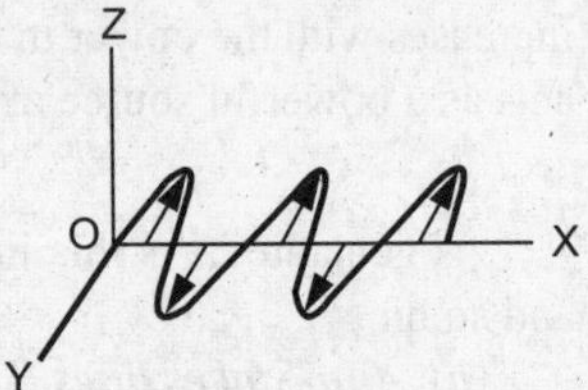

Fig. 16.1.

The horizontally polarised light disturbance can be expressed as

$$y = a_y e^{i\phi_x} \quad \ldots(2)$$

Here, $\phi_x = (kx - \omega t)$ and $z = 0$.

According to Jones, the polarised light can be represented by a column vector whose first element represents the amplitude and phase of Y-component and the second element represents the Z-component. Hence

$$J_1 = \begin{bmatrix} y \\ z \end{bmatrix} = \begin{bmatrix} a_y e^{i\phi_x} \\ 0 \end{bmatrix} \quad \ldots(3)$$

Let us consider that intensity is normalised and phase indicating factor is omitted. Then,

$$J_1 = \begin{bmatrix} y \\ z \end{bmatrix} = \begin{bmatrix} 1 \\ 0 \end{bmatrix}$$

Thus the horizontally polarised light $(\leftrightarrow)$ can be represented by Jones vector J_1 as

$$J_1 = \begin{bmatrix} 1 \\ 0 \end{bmatrix} \leftrightarrow \text{horizontally polarised light} \quad \ldots(4)$$

$\begin{bmatrix} 1 \\ 0 \end{bmatrix}$ represents a wave linearly polarized in the x direction.

Further, consider the disturbance in vertical plane (Y-Z plane). This can be represented as

$$\mathbf{z} = \hat{\mathbf{k}}\, a_z\ e^{i(kx-\omega t+\phi)} \quad \ldots(5)$$

Here ϕ is the phase difference with respect to horizontal disturbance.

The vertically polarised light disturbance can be expressed as

$$y = 0 \text{ and } z = a_z\ e^{i\phi_x} \,.\, e^{i\phi}$$

$$\therefore \qquad \begin{bmatrix} y \\ z \end{bmatrix} = \begin{bmatrix} 0 \\ a_z\, e^{i\phi_x}\, e^{i\phi} \end{bmatrix} \qquad \ldots(6)$$

Omitting the phase factor and considering that intensity is normalised, the Jones vector J_2 for vertically polarised light ($\updownarrow$) can be written as

$$J_2 = \begin{bmatrix} 0 \\ 1 \end{bmatrix} \text{ for vertically polarised light } (\updownarrow) \qquad \ldots(7)$$

$\begin{bmatrix} 0 \\ 1 \end{bmatrix}$ represents a wave linearly polarised in the y-direction.

Consider a disturbance which is polarised at 45° with respect to horizontal or vertical direction. Jones vector for such polarised light can be written as

$$J_3 = \begin{bmatrix} y \\ z \end{bmatrix} = \begin{bmatrix} ae^{i\phi_x} \cos \pi/4 \\ ae^{i\phi_x} \sin \pi/4 \end{bmatrix} = \begin{bmatrix} 1/\sqrt{2} \\ 1/\sqrt{2} \end{bmatrix}$$

$$= \frac{1}{\sqrt{2}} \begin{bmatrix} 1 \\ 1 \end{bmatrix} \qquad \ldots(8)$$

One important application of the Jones notation is finding the result of adding two or more waves in given states of polarization. The corresponding Jones vectors are added.

(*i*) The linearly polarised light may be obtained by adding a horizontally polarized light and a vertically polarised light. Thus

$$\begin{bmatrix} 1 \\ 0 \end{bmatrix} + \begin{bmatrix} 0 \\ 1 \end{bmatrix} = \begin{bmatrix} 1 \\ 1 \end{bmatrix}$$

So, a normalised linearly polarised light at 45° is written as

$$J_3 = \frac{1}{\sqrt{2}} \begin{bmatrix} 1 \\ 1 \end{bmatrix}$$

(*ii*) If right- and left-handed circularly polarised light (of equal amplitude) is combined,

$$\begin{bmatrix} 1 \\ -i \end{bmatrix} + \begin{bmatrix} 1 \\ i \end{bmatrix} = \begin{bmatrix} 2 \\ 0 \end{bmatrix} = 2 \begin{bmatrix} 1 \\ 0 \end{bmatrix}$$

This means that the result is linearly polarized light of twice the amplitude of either of the initial contributions.

The following table lists the Jones vectors that represent some states of polarization.

State of polarization of a light beam	***Jones vector***
Linearly polarized with the vibration direction along the x-axis (horizontal).	$\begin{bmatrix} 1 \\ 0 \end{bmatrix}$
Linearly polarized with the vibration direction along the y-axis (vertical).	$\begin{bmatrix} 0 \\ 1 \end{bmatrix}$
Linearly polarized with the vibration direction at 45° relative to the x-axis.	$\begin{bmatrix} 1 \\ 1 \end{bmatrix}$
Right circular polarization.	$\begin{bmatrix} 1 \\ -i \end{bmatrix}$
Left circular polarization.	$\begin{bmatrix} 1 \\ i \end{bmatrix}$

There is no Jones representation of unpolarized light.

16.2. Jones Matrices for Linear Optical Elements such as Polarizers and Retardation Plates

Jones vectors are useful in computing the effect of inserting linear optical elements into the paths of beams of light of given polarization.

A polarizer (such as a Nicol prism), a quarter-wave plate and a half-wave plate are examples of 'linear optical elements'. We can associate with each one of them a 2×2 matrix called a Jones matrix.

We have seen that the state of polarization of a beam can be represented by a column vector $\begin{bmatrix} A \\ B \end{bmatrix}$. Let $\begin{bmatrix} a & b \\ c & d \end{bmatrix}$ reprsesent the Jones matrix of a linear optical element. When the beam passes through the element, the emergent beam will, in general, have a different state of polarization represented by, say $\begin{bmatrix} A' \\ B' \end{bmatrix}$. Jones matrices can be used to find the state of polarisation of the emergent beam.

The Jones matrix is a transformation matrix which transforms the vector $\begin{bmatrix} A \\ B \end{bmatrix}$ into the vector $\begin{bmatrix} A' \\ B' \end{bmatrix}$ as follows :

$$\begin{bmatrix} A' \\ B' \end{bmatrix} = T \begin{bmatrix} A \\ B \end{bmatrix}$$

Here, $T = \begin{bmatrix} a & b \\ c & d \end{bmatrix}$ is the transformation matrix.

Let us express the general transformation by the following matrix operation:

$\begin{bmatrix} a & b \\ c & d \end{bmatrix}$	$\begin{bmatrix} A \\ B \end{bmatrix}$	=	$\begin{bmatrix} A' \\ B' \end{bmatrix}$
Transformation matrix	Incident light matrix		Emergent light matrix

Let us denote the incident light by the following matrix.

$$\begin{bmatrix} ae^{i\phi_1} \\ be^{i\phi_2} \end{bmatrix}$$

Consider the following transformation matrix.

$$\begin{bmatrix} 1 & 0 \\ 0 & 0 \end{bmatrix}$$

By the operation of transformation matrix on incident light matrix, we get the emergent light as horizontally linearly polarised light.

$$\begin{bmatrix} 1 & 0 \\ 0 & 0 \end{bmatrix} \begin{bmatrix} ae^{i\phi_1} \\ be^{i\phi_2} \end{bmatrix} = \begin{bmatrix} ae^{i\phi_1} \\ 0 \end{bmatrix} \rightarrow \begin{bmatrix} 1 \\ 0 \end{bmatrix} \quad ...(1)$$

The matrix $\begin{bmatrix} 1 & 0 \\ 0 & 0 \end{bmatrix}$ represents the linear polariser with axis horizontal.

Similarly, the matrix $\begin{bmatrix} 0 & 0 \\ 0 & 1 \end{bmatrix}$ represents the linear polariser with axis vertical as shown below :

$$\begin{bmatrix} 0 & 0 \\ 0 & 1 \end{bmatrix}\begin{bmatrix} ae^{i\phi_1} \\ be^{i\phi_2} \end{bmatrix} = \begin{bmatrix} 0 \\ be^{i\phi_2} \end{bmatrix} \rightarrow \begin{bmatrix} 0 \\ 1 \end{bmatrix} \quad ...(2)$$

If $\phi_1 = \phi_2 = 0$ and $a = b = 1$, then

$$\frac{1}{2}\begin{bmatrix} 1 & 1 \\ 1 & 1 \end{bmatrix}\begin{bmatrix} 1 \\ 1 \end{bmatrix} = \frac{1}{2}\begin{bmatrix} 2 \\ 2 \end{bmatrix} = \begin{bmatrix} 1 \\ 1 \end{bmatrix} \quad ...(3)$$

Here matrix $\frac{1}{2}\begin{bmatrix} 1 & 1 \\ 1 & 1 \end{bmatrix}$ represents a linear polariser with axis at + 45°.

Similarly, matrix $\frac{1}{2}\begin{bmatrix} 1 & -1 \\ -1 & 1 \end{bmatrix}$ represents a linear polariser with axis at –45°.

$$\frac{1}{2}\begin{bmatrix} 1 & -1 \\ -1 & 1 \end{bmatrix}\begin{bmatrix} 1 \\ 1 \end{bmatrix} = \frac{1}{2}\begin{bmatrix} 0 \\ 0 \end{bmatrix} \quad ...(4)$$

The emergent light has zero intensity *i.e.*, cut off. So the matrix represents a polariser of –45°.

There is no Jones vector representation for unpolarized light. Jones matrices are useful only when we commence with light that is initially polarized in some manner.

If light in some state of polarization given by $\begin{bmatrix} A \\ B \end{bmatrix}$ is sent through a train of n optical elements having Jones matrices $T_1, T_2, \ldots T_n$, the final state of polarization is given by

$$\begin{bmatrix} A' \\ B' \end{bmatrix} = T_n \ldots T_2 T_1 \begin{bmatrix} A \\ B \end{bmatrix}.$$

The following table lists Jones matrices for some linear polarisers.

Linear Polariser	***Jones Matrix***
Linear polarizer with the transmission axis horizontal. (↔)	$\begin{bmatrix} 1 & 0 \\ 0 & 0 \end{bmatrix}$
Linear polarizer with the transmission axis vertical. (↕)	$\begin{bmatrix} 0 & 0 \\ 0 & 1 \end{bmatrix}$
Linear polarizer with the transmission axis at + 45°.	$\frac{1}{2}\begin{bmatrix} 1 & 1 \\ 1 & 1 \end{bmatrix}$
Linear polarizer with the transmission axis at – 45°.	$\frac{1}{2}\begin{bmatrix} 1 & -1 \\ -1 & 1 \end{bmatrix}$

Jones' Matrices for Quarter Wave Plate and Half Wave Plate

In Jones' matrix formalism, a linear retarder, either right handed or left handed can be represented by a matrix

$$[M] = \begin{bmatrix} 1 & 0 \\ 0 & e^{i\delta} \end{bmatrix}.$$

A $\lambda/4$ plate introduces a phase change of $\pi/2$. Hence Jones' matrix for $\lambda/4$ plate can be represented as

$$[J]_{\lambda/4} = \begin{bmatrix} 1 & 0 \\ 0 & e^{\pm i\pi/2} \end{bmatrix}$$

The positive sign is used when a $\lambda/4$ plate matrix produces a right circularly polarised light. The negative sign is used when $\lambda/4$ plate matrix produces left circularly polarised light. Thus

$$[J]_{\lambda/4} = \begin{bmatrix} 1 & 0 \\ 0 & i \end{bmatrix} \lambda/4 \qquad \lambda/4 \text{ plate matrix producing } RCP.$$

$$[J]_{\lambda/4} = \begin{bmatrix} 1 & 0 \\ 0 & -i \end{bmatrix} \lambda/4 \qquad \lambda/4 \text{ plate matrix producing } LCP.$$

A $\lambda/2$ plate introduces a phase change of π. Hence the Jones' matrix for half wave plate is

$$\begin{bmatrix} 1 & 0 \\ 0 & -1 \end{bmatrix}$$

Some simple results for retardation plates are shown in the following Table.

Optical Element	*Jones Matrix*
Quarter-wave plate with the fast axis horizontal.	$\begin{bmatrix} 1 & 0 \\ 0 & -i \end{bmatrix}$
Quarter-wave plate with the fast axis vertical.	$\begin{bmatrix} 1 & 0 \\ 0 & i \end{bmatrix}$
Quarter-wave plate with the fast axis at ± 45°.	$\frac{1}{\sqrt{2}}\begin{bmatrix} 1 & \pm i \\ \mp i & 1 \end{bmatrix}$
Half-wave plate with the fast axis either vertical or horizontal.	$\begin{bmatrix} 1 & 0 \\ 0 & -1 \end{bmatrix}$
Half-wave plate with the fast axis at + 45°.	$\begin{bmatrix} 0 & 1 \\ 1 & 0 \end{bmatrix}$

Example 1. *Consider a linear polariser with its transmission axis horizontal. Describe what would happen when*

(i) Linearly polarised light of horizontal orientation is incident on it.

(ii) The same light but now with vertical orientation is incident on it.

Solution. $\begin{bmatrix} A' \\ B' \end{bmatrix} = T \begin{bmatrix} A \\ B \end{bmatrix}$

$$= \begin{bmatrix} 1 & 0 \\ 0 & 0 \end{bmatrix} \begin{bmatrix} 1 \\ 0 \end{bmatrix} = \begin{bmatrix} 1 \\ 0 \end{bmatrix}.$$

The emergent light is linearly horizontally polarised.

(*ii*) Let us denote the emergent light by the matrix $\begin{bmatrix} A' \\ B' \end{bmatrix}$

Then, $\begin{bmatrix} A' \\ B' \end{bmatrix} = T \begin{bmatrix} A \\ B \end{bmatrix}$

$$= \begin{bmatrix} 1 & 0 \\ 0 & 0 \end{bmatrix} \begin{bmatrix} 0 \\ 1 \end{bmatrix} = \begin{bmatrix} 0 \\ 0 \end{bmatrix}$$

There is no emergent light.

Example 2. *What is the polarisation of emergent light when the light first be horizontally polarised, then passed, through a + 45° polariser and finally through a – 45° polariser ?*

Solution. The matrix operation is shown below.

$$\frac{1}{2} \begin{bmatrix} 1 & -1 \\ -1 & 1 \end{bmatrix} \frac{1}{2} \begin{bmatrix} 1 & 1 \\ 1 & 1 \end{bmatrix} \begin{bmatrix} 1 \\ 0 \end{bmatrix} = \frac{1}{4} \begin{bmatrix} 0 \\ 0 \end{bmatrix} = \begin{bmatrix} 0 \\ 0 \end{bmatrix}$$

Hence no light is transmitted.

Example 3. *(i) Determine the polarization of the emerging light when a quarter-wave plate is inserted into a beam of linearly polarized light with the vibration direction making an angle of 45° with the x-axis. Assume the fast axis as horizontal.*

(ii) Work out the same example, for a half-wave plate (instead of the quarter-wave plate) with its fast axis at 45°.

Solution. (*i*) $\begin{bmatrix} A \\ B \end{bmatrix} = \begin{bmatrix} 1 \\ 1 \end{bmatrix}; T = \begin{bmatrix} 1 & 0 \\ 0 & -i \end{bmatrix}$

$$\therefore \qquad \begin{bmatrix} A' \\ B' \end{bmatrix} = T \begin{bmatrix} A \\ B \end{bmatrix} = \begin{bmatrix} 1 & 0 \\ 0 & -i \end{bmatrix} \begin{bmatrix} 1 \\ 1 \end{bmatrix} = \begin{bmatrix} 1 \\ -i \end{bmatrix}$$

The emerging light beam is right circularly polarized.

(*ii*) Here $T = \begin{bmatrix} 0 & 1 \\ 1 & 0 \end{bmatrix}$

$$\therefore \qquad \begin{bmatrix} A' \\ B' \end{bmatrix} = T \begin{bmatrix} A \\ B \end{bmatrix} = \begin{bmatrix} 0 & 1 \\ 1 & 0 \end{bmatrix} \begin{bmatrix} 1 \\ 1 \end{bmatrix} = \begin{bmatrix} 1 \\ 1 \end{bmatrix}$$

This is a case where the direction of vibration of the incident light is along a privileged direction of the crystal. The polarized beam is transmitted as it is.

Example 4. *Linearly polarized light oscillating in the horizontal direction is incident on a half-wave plate oriented at + 45°. What is the state of polarization of the emerging light ?*

Solution. $\begin{bmatrix} 0 & 1 \\ 1 & 0 \end{bmatrix} \begin{bmatrix} 1 \\ 0 \end{bmatrix} = \begin{bmatrix} 0 \\ 1 \end{bmatrix}$

The emerging light, therefore, is linearly polarized in the vertical direction.

16.3. Circularly-Polarised Light—Jones Matrices

Consider a vibration in Y-direction represented as

$$\mathbf{y} = \hat{\mathbf{j}}\, a_1\, e^{i(kx-\omega t)} = \hat{\mathbf{j}}\, a_1\, e^{i\phi_x} \qquad \text{...(1)}$$

Here, $\phi_x = (kx - \omega t)$ and $k =$ propagation constant $= 2\pi/\lambda$.

Further consider an orthogonal vibration propagating along X-axis and having displacement along Y-axis. This can be represented as

$$\mathbf{z} = \hat{\mathbf{k}}\, a_2\, e^{i(\phi_x + \phi)} \qquad \text{...(2)}$$

Here, ϕ is the phase difference between two linear vibrations.

When $a_1 = a_2$ and $\phi = \pi/2$, then the resultant of two vibrations is $a(\mathbf{j} + \hat{\mathbf{k}}\, e^{i\pi/2})e^{i\phi_x}$. Thus the two components fluctuate in different phases. The two components can be represented as

$$y = \text{real part of } ae^{i\phi_x} = a\cos\phi_x \qquad \text{...(3)}$$

and

$$z = \text{real part of } ae^{i(\phi_x + \pi/2)}$$

$$= a\cos(\phi_x + \pi/2) = -a\sin\phi_x \qquad \text{...(4)}$$

From Eqs. (3) and (4) we get

$$\frac{y^2}{a^2} + \frac{z^2}{a^2} = 1 \qquad \text{...(5)}$$

Thus the light is circularly polarised light.

Special Cases

(*a*) **When Z-component leads in phase by π/2**

(*i*) Let $\phi_x = 0$ at time $t = t_1$.

So $\omega t_1 = kx$ or $t_1 = kx/\omega$.

From Eqs. (3) and (4), we get $y = a$ and $z = 0$.

(*ii*) Let $\phi_x = \pi/2$ at time $t = t_2$.

Then $t_2 = \dfrac{kx}{\omega} - \dfrac{\pi}{2\omega}$ $(t_2 << t_1)$.

From Eqs. (3) and (4), we get $y = 0$ and $z = -a$.

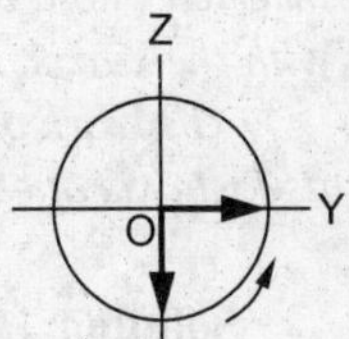

Fig. 16.2.

Thus, as seen from Fig. 16.2, as the time passes the resultant displacement vector passes from a configuration [$y = 0$, $z = -a$] to a configuration [$y = a$, $z = 0$]. Thus the resultant displacement vector rotates in a circle in *anti-clockwise sense*. In general, if the Z vibration leads Y vibration by $(2n + 1/2)\pi$, where $n = 0, 1, 2 \ldots$ etc., then the resultant vibration would be rotating anticlockwise in a circle. The light in this case is called right circularly polarised (RCP) light. In Jones' matrix formalism

$$[J]_{RCP} = \begin{bmatrix} a \\ ae^{i\pi/2} \end{bmatrix} e^{i\phi_x} \rightarrow \frac{1}{\sqrt{2}}\begin{bmatrix} 1 \\ i \end{bmatrix} \qquad \text{...(6)}$$

with intensity normalised.

(*b*) **When Z-component lags in phase by π/2**

Let $y = a\cos\phi_x$ and $z = a\cos\left[\phi_x - \dfrac{\pi}{2}\right] = a\sin\phi_x$

(*i*) When $\phi_x = 0, t_1 = \frac{kx}{\omega}$.

Now $y = a$ and $z = 0$.

(*ii*) When $\phi_x = \frac{\pi}{2}, t_2 = \frac{kx}{\omega} - \frac{\pi}{2\omega}$.

Now $y = 0$ and $z = a$.

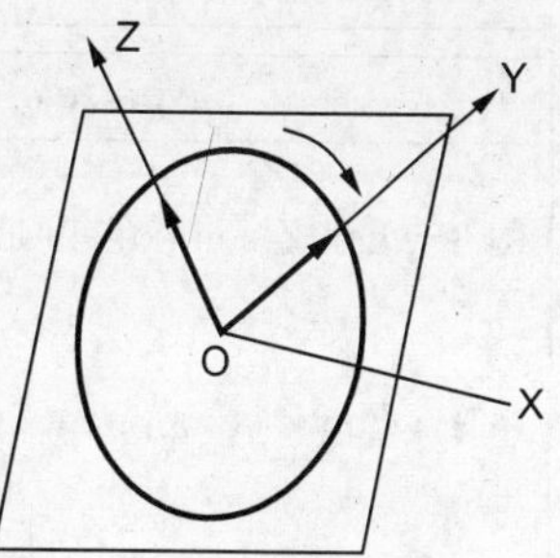

Fig. 16.3.

As the time passes, the resultant vector passes from the configuration $[y = 0, z = a]$ to configuration $y = a, z = 0$ as shown in Fig. 16.3. The light is called left circularly polarised (LCP) light.

In Jones' formalism

$$[J]_{LCP} = \begin{bmatrix} a \\ ae^{-i\pi/2} \end{bmatrix} e^{i\phi_x} \rightarrow \frac{a}{\sqrt{2}}\begin{bmatrix} 1 \\ -i \end{bmatrix} \rightarrow \frac{1}{\sqrt{2}}\begin{bmatrix} 1 \\ -i \end{bmatrix}$$

(*c*) **When Y-component leads by π/2**

$$\text{Let } y = a\cos\left[\frac{\pi}{2} + \phi_x\right] = -a\sin\phi_x \text{ and } z = a\cos\phi_x.$$

(*i*) When $\phi_x = 0, t_1 = \frac{kx}{\omega}$. Now $y = 0$ and $z = a$.

(*ii*) When $\phi_x = \frac{\pi}{2}, t_2 = \frac{kx}{\omega} - \frac{\pi}{2\omega}$. Now $y = -a$ and $z = 0$.

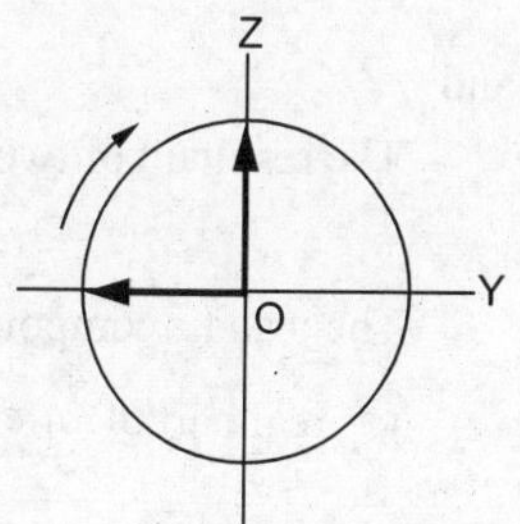

Fig. 16.4.

Thus as the time passes, the resultant vector passes from configuration
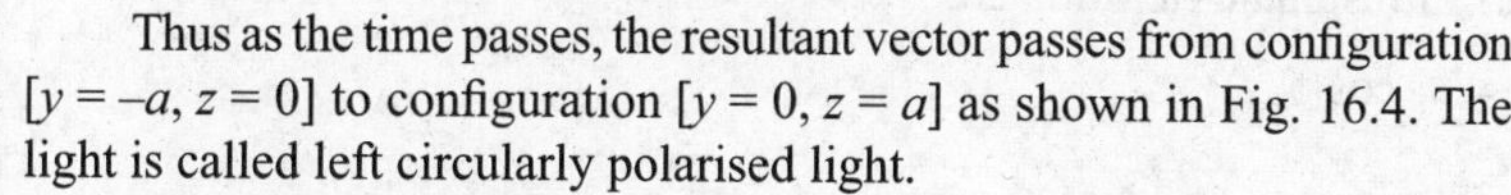
$[y = -a, z = 0]$ to configuration $[y = 0, z = a]$ as shown in Fig. 16.4. The light is called left circularly polarised light.

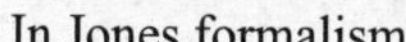
In Jones formalism

$$[J]_{LCP} = \frac{1}{\sqrt{2}}\begin{bmatrix} i \\ 1 \end{bmatrix}$$

(*d*) **When Y component lags by π/2**

$$\text{Let } y = a\cos\left[\phi_x - \frac{\pi}{2}\right] = a\sin\phi_x \text{ and } z = a\cos\phi_x.$$

(*i*) When $\phi_x = 0, t_1 = \frac{kx}{\omega}$.

Now $y = 0$ and $z = a$.

(*ii*) When $\phi_x = \frac{\pi}{2}, t_2 = \frac{kx}{\omega} - \frac{\pi}{2\omega}$.

Now $y = a$ and $z = 0$.

Now we have right circularly polarised light.

In Jones' formalism $[J]_{RCP} = \frac{1}{\sqrt{2}}\begin{bmatrix} -i \\ 1 \end{bmatrix}$

The following Table lists Jones matrices for circularly polarised light.

Nature of light	*Jones matrix*
RCP when Z-component leads.	$\frac{1}{\sqrt{2}}\begin{bmatrix} 1 \\ i \end{bmatrix}$
LCP when Z-component lags.	$\frac{1}{\sqrt{2}}\begin{bmatrix} 1 \\ -i \end{bmatrix}$
LCP when Y-component leads.	$\frac{1}{\sqrt{2}}\begin{bmatrix} i \\ 1 \end{bmatrix}$
RCP when Y-component lags.	$\frac{1}{\sqrt{2}}\begin{bmatrix} -i \\ 1 \end{bmatrix}$

16.4. Elliptically-Polarised Light—Jones Matrices

Elliptically polarised light is the superposition of horizontally polarised light and vertically polarised light. The two lights are represented as

$$\mathbf{y} = \hat{\mathbf{j}}\, a_1\, e^{i(kx-\omega t)} = \hat{\mathbf{j}}\, a_1\, e^{i\phi_x}$$

and

$$\mathbf{z} = \hat{\mathbf{k}}\, a_2\, e^{i(kx-\omega t+\phi)} = \hat{\mathbf{k}} a_2\, e^{i\phi_x}\, e^{i\phi}$$

The resultant of two vibrations is

$$(\hat{\mathbf{j}}\, a_1 + \hat{\mathbf{k}} a_2\, e^{i\phi})\, e^{i\phi_x}$$

The y and z components of resultant vibration are

y = real part of $a_1\, e^{i\phi_x} = a_1 \cos \phi_x$

z = real part of $a_2\, e^{i(\phi_x+\phi)} = a_2 \cos(\phi_x + \phi)$.

From these equations, we get

$$\frac{y^2}{a_1^2} + \frac{z^2}{a_2^2} - \frac{2yz}{a_1 a_2} \cos\phi = \sin^2 \phi$$

When $\phi = (2n+1)\,\pi/2$, $n = 0, 1, 2, 3 \ldots.$, the equation takes the form

$$\frac{y^2}{a_1^2} + \frac{z^2}{a_2^2} = 1.$$

This represents an ellipse with its axes along y and z directions.

Thus we get elliptically polarised light.

In Jones' formalism, elliptically polarised light is represented as

$$[J]_{REP} = \begin{bmatrix} a_1 \\ a_2 e^{i\phi} \end{bmatrix} \rightarrow \begin{bmatrix} a_1 \\ i a_2 \end{bmatrix}$$

This is right elliptically polarised light.

Similarly, the left elliptically polarised light is represented as

$$[J]_{LEP} = \begin{bmatrix} a_1 \\ -i a_2 \end{bmatrix}$$

The following Table lists Jones matrices for elliptically polarised light :

Nature of light	*Jones matrix*
Right handed elliptically polarised light [REP]	$\begin{bmatrix} a_1 \\ ia_2 \end{bmatrix}$
Left handed elliptically polarised light [LEP]	$\begin{bmatrix} a_1 \\ -ia_2 \end{bmatrix}$

16.5. Production of Circularly-Polarised Light

Ordinary monochromatic light is passed through a Nicol prism. The light emerging from Nicol prism is plane polarised. This plane polarised light is then allowed to fall *normally* on a quarter wave plate such that the vibrations in the incident plane polarised light make an angle of 45° with the optic axis of the plate. The light emerging from the quarter wave plate is circularly polarised (Fig. 16.5).

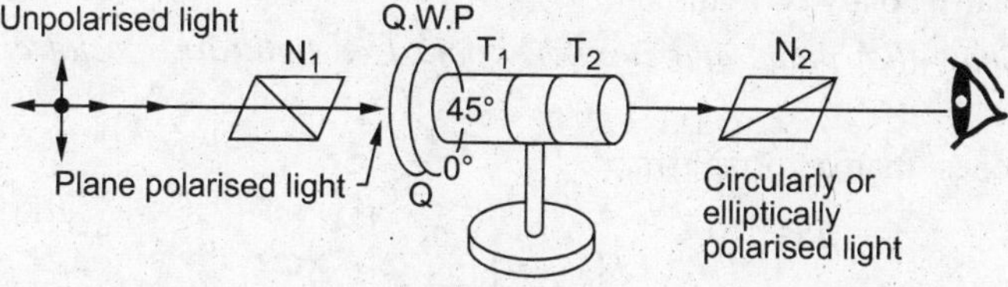

Fig. 16.5.

Consider a plane polarised light at 45°. This is represented by Jones' vector $\frac{1}{\sqrt{2}}\begin{bmatrix} 1 \\ 1 \end{bmatrix}$.

A λ/4 plate with fast axis vertical is represented by Jones' matrix as

$$\begin{bmatrix} 1 & 0 \\ 0 & i \end{bmatrix}$$

$$\therefore \qquad \begin{bmatrix} 1 & 0 \\ 0 & i \end{bmatrix} \frac{1}{\sqrt{2}}\begin{bmatrix} 1 \\ 1 \end{bmatrix} = \frac{1}{\sqrt{2}}\begin{bmatrix} 1 \\ +i \end{bmatrix} \qquad \text{Right circularly polarised light.}$$

For a λ/4 plate with fast axis horizontal, we have

$$\begin{bmatrix} 1 & 0 \\ 0 & -i \end{bmatrix} \frac{1}{\sqrt{2}}\begin{bmatrix} 1 \\ 1 \end{bmatrix} = \frac{1}{\sqrt{2}}\begin{bmatrix} 1 \\ -i \end{bmatrix} \qquad \text{Left circularly polarised light.}$$

16.6. Production of Elliptically Polarised Light

A parallel beam of monochromatic light is allowed to fall on two Nicols N_1 and N_2 in crossed position (Fig. 16.5). In this case, the field of view is dark. A λ/4 plate is now introduced between the two Nicols so that the field of view may be bright. The quarter wave plate is rotated in such a way that the field of view is again dark. Again the quarter wave plate is rotated such that vibration of light incident on it makes any angle other than 45°. This makes the amplitudes of ordinary and extra-ordinary rays unequal. Hence the resulting light from quarter-wave plate is elliptically polarised.

In Jones' matrix formulation, the elliptical polariser can be represented as

$$[J]_{EP} = \begin{bmatrix} a_1 & 0 \\ 0 & a_2 e^{i\phi} \end{bmatrix}$$

When a circularly polarised light is incident on an elliptical polariser, the emergent light is elliptically right handed polarised light as shown below:

$$\begin{bmatrix} a_1 & 0 \\ 0 & a_2 e^{i\phi} \end{bmatrix} \begin{bmatrix} 1 \\ i \end{bmatrix} = \begin{bmatrix} a_1 \\ i a_2\, e^{i\phi} \end{bmatrix}$$

Eliptical polariser CPL EPL

A quarter wave plate introduces a phase difference of $\pi/2$ between the orthogonal components. Hence quarter wave plate matrix producing elliptically polarised light is represented as

$$\begin{bmatrix} a_1 & 0 \\ 0 & a_2\, e^{i\pi/2} \end{bmatrix} \rightarrow \begin{bmatrix} a_1 & 0 \\ 0 & i a_2 \end{bmatrix} \quad \text{(Right handed)}$$

Similarly,

$$\begin{bmatrix} a_1 & 0 \\ 0 & i a_2 \end{bmatrix} \begin{bmatrix} A \\ b \end{bmatrix} = \begin{bmatrix} A a_1 \\ i B a_2 \end{bmatrix} = \begin{bmatrix} b_1 \\ i b_2 \end{bmatrix}$$

This is elliptically polarised light.

Example 1. *Show that plane and circularly polarised light are special cases of elliptically polarised light.*

Solution. In Jones' matrix formalism,

$$[J]_{REP} = \begin{bmatrix} a_1 \\ a_2\, e^{i\phi} \end{bmatrix}$$

and

$$[J]_{LEP} = \begin{bmatrix} a_1 \\ a_2\, e^{-i\phi} \end{bmatrix}$$

(*i*) If $a_1 = a_2$ and $\phi = 0$, then

$$[J] = \begin{bmatrix} a \\ a \end{bmatrix} \rightarrow \begin{bmatrix} 1 \\ 1 \end{bmatrix} \text{ Linearly polarised light}$$

(*ii*) If $a_1 = a_2$ and $\phi = \pi/4$

$$[J]_{RCP} = \begin{bmatrix} 1 \\ +i \end{bmatrix} \text{ Right circularly polarised light.}$$

16.7. Analysis of Polarised Light

A Nicol prism is used in conjunction with a quarter-wave plate to distinguish between various kinds of light by applying the following tests:

The given light is passed through the Nicol prism and the changes in intensity are noted as the Nicol is rotated. Three possibilities may arise:

(*i*) *The intensity becomes maximum and zero alternatively.*

If so, the light incident will be **plane polarised** since this has only one direction of vibration and rotating Nicol transmits the light when this direction is parallel to principal section of Nicol and does not transmit when two are perpendicular.

In Jones' formalism, the interpretation is as follows:

Consider the nature of emergent light when a linearly polarised light with orientation horizontal is passed through analyser whose fast axis is vertical. We have the following matrix operation :

$$\begin{bmatrix} 0 & 0 \\ 0 & 1 \end{bmatrix} \begin{bmatrix} 1 \\ 0 \end{bmatrix} = \begin{bmatrix} 0 \\ 0 \end{bmatrix}$$

So, the intensity of emergent light is zero.

Further consider the nature of emergent light when a linear polarised light with orientation horizontal is passed through the analyser whose fast axis is horizontal. We have the following matrix operation.

$$\begin{bmatrix} 1 & 0 \\ 0 & 0 \end{bmatrix}\begin{bmatrix} 1 \\ 0 \end{bmatrix} = \begin{bmatrix} 1 \\ 0 \end{bmatrix}$$

So, the intensity of the emergent light is maximum.

(ii) The intensity becomes maximum and minimum (not zero) alternatively.

If so, the light incident may be either **elliptically polarised or partially polarised.**

To distinguish between partially plane-polarised light and elliptically polarised light, a quarter-wave plate is interposed between the source of light and analyser, with its optic axis parallel to transmission axis of analyser (Fig. 16.6). The analyser is again rotated about the direction of propagation of light. If the intensity varies with zero minimum, the given light is elliptically polarised. If on the other hand, emergent light shows variation in intensity with a non-zero minimum, the given beam of light is partially plane polarised.

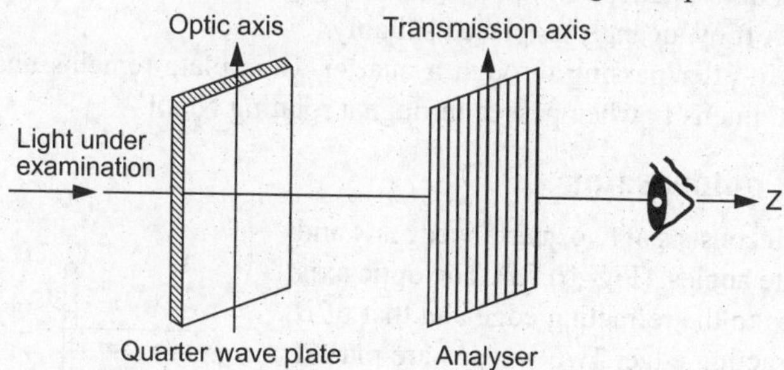

Fig. 16.6.

Elliptically polarised light after passing through a quarter wave plate becomes plane polarised. Hence it *will give two maxima and two complete extinctions* when observed through a rotating Nicol.

Consider the action of λ/4 plate on elliptically polarised light. The matrix operation is as follows:

$$\begin{bmatrix} 1 & 0 \\ 0 & i \end{bmatrix}\begin{bmatrix} a_1 \\ ia_2 \end{bmatrix} = \begin{bmatrix} a_1 \\ -a_2 \end{bmatrix}$$

Polariser matrix   REP   Linearly polarised

So the emergent light is linearly polarised. This light makes an angle θ with the slow axis, where

$$\tan\theta = -(a_2 / a_1).$$

So, the intensity would be minimum (zero) when the transmission axis of the analyser placed after λ/4 plate is at right angle to the emergent linear vibration.

Partially polarised light after passing through a quarter-wave plate remains practically unaffected. Hence *the light transmitted after rotating Nicol, will now show variations having maximum and non-zero minimum.*

(iii) The intensity shows no variation.

In this case, the incident light is either unpolarised or circularly polarised.

To settle between circularly polarised and ordinary unpolarised light, the given beam of light is first passed through a quarter wave plate and then examined by a rotating Nicol. If the intensity varies with zero minimum, the given light is circularly polarised. If, however, there is no variation in intensity, the given beam of light is unpolarised.

Consider the action of $\lambda/4$-plate on right circularly polarised light. The matrix operation is as follows :

$$\begin{bmatrix} 1 & 0 \\ 0 & i \end{bmatrix} \frac{1}{\sqrt{2}} \begin{bmatrix} 1 \\ i \end{bmatrix} = \frac{1}{\sqrt{2}} \begin{bmatrix} 1 \\ -1 \end{bmatrix}$$

Polariser matrix RCP Emergent light

This shows that emergent light is plane polarised at 45° with the slow axis of $\lambda/4$ plate.

Similarly, the action of a $\lambda/4$ plate on left circularly polarised light can be considered as follows :

$$\begin{bmatrix} 1 & 0 \\ 0 & i \end{bmatrix} \frac{1}{\sqrt{2}} \begin{bmatrix} 1 \\ -i \end{bmatrix} = \frac{1}{\sqrt{2}} \begin{bmatrix} 1 \\ 1 \end{bmatrix}$$

Polariser matrix LCP matrix Emergent light

The emergent light is plane polarised light at 45° with the slow axis of $\lambda/4$ plate.

So by the action of $\lambda/4$ plate, a circularly polarised light is converted into plane polarised light. The light may be cut off by an analyser placed suitably.

Unpolarised light after passing through a quarter-wave plate remains unaffected and hence shows no variation in intensity when passed through a rotating Nicol.

16.8. Babinet's Compensator

Construction. It consists of two quartz-wedges A and B of equal small acute angles. (Fig. 16.7*a*). The optic axis of A is at right angles to the refracting edge and that of B is parallel to the refracting edge. Two wedges are placed with their hypotenuses in contact to form a rectangular block. One of the wedges is fixed, while the other can be moved in its own plane by means of a micrometer screw S. The angle of rotation of screw may be read on a graduated circular scale.

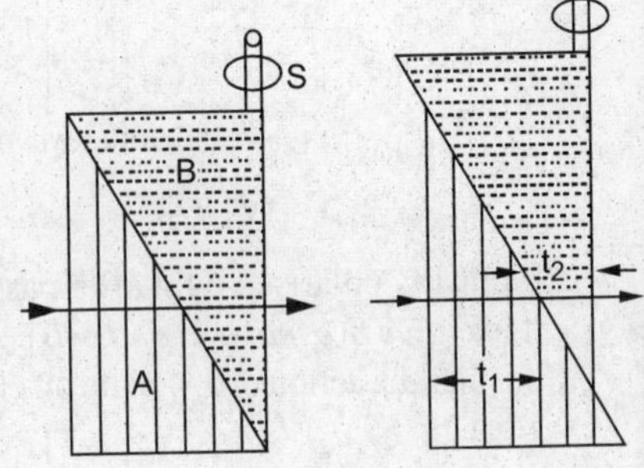

Fig. 16.7.

Theory. When plane-polarised light falls normally on the first wedge with its plane of vibration making an angle θ with the optic axis, it is broken up into E and O components. As quartz is a positive crystal, the ordinary component travels faster than extra-ordinary component. On entering the second wedge, the E-component becomes the O-component and vice-versa. In other words, the two components exchange velocities in passing from one wedge to the other. Thus the two wedges tend to cancel each other's effect.

Let t_1 and t_2 be the thicknesses of the two wedges traversed by a particular ray (Fig. 16.7*b*), and n_e and n_o the refractive indices of quartz for the E and O components respectively. Then the path difference introduced between the two components by the first wedge is $t_1 (n_e - n_o)$ and that introduced by the second wedge is $-t_2 (n_e - n_o)$. Hence the resultant path difference

$$\Delta = (t_1 - t_2)(n_e - n_0).$$

The value of $\Delta = 0$ for $t_1 = t_2$ at the centre of the compensator and emergent light in this region is plane polarised. On either side of this point, the path difference gradually increases and the emergent light is polarised in various ways depending on the corresponding values of $(t_1 - t_2)$.

By moving the second wedge relative to the first, any value of $(t_1 - t_2)$ can be arranged at the centre of the compensator. Then this portion can be used as a quarter-wave plate, half-wave plate or a wave plate of any other thickness. Thus the compensator has an advantage over a quarter-wave plate.

The quarter-wave plate produces a path difference of $\lambda/4$ for one particular wavelength, whereas the compensator can be adjusted for any wavelength.

Calibration of compensator. The experimental arrangement is shown in Fig. 16.8(*a*). The compensator *C* is placed between two crossed Nicols N_1 and N_2 (polariser and analyser). The polariser N_1 is so oriented that the plane polarised light emerging from it and falling normally on the compensator makes an angle θ with the optic axis of the first edge of compensator. The plane-polarised light traversing those portions of the compensator for which the path difference is $0, \lambda, 2\lambda, \ldots m\lambda$, the emergent light from *C* is also plane polarised with its plane of vibration parallel to that of incident light. Hence dark bands appear at all such places with equal spacing (Fig. 16.8*b*). For places where $\Delta = \lambda/2, 3\lambda/2, 5\lambda/2$...., the emergent light is plane polarised with vibrations at an angle 2θ with that of the incident light. However, if $\theta = 45°$ then $2\theta = 90°$ and N_2 does not stop this light at all and at such places bright equispaced bands appear. The clear cut appearance of bright and dark bands is indication of such a correct setting. In the intermediate positions, the light transmitted by compensator shall be

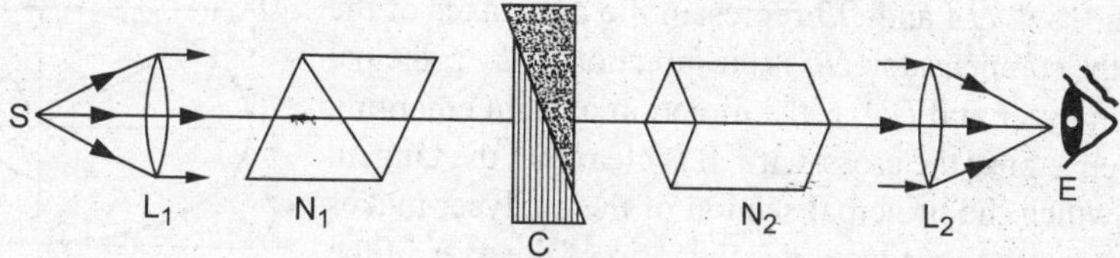

Fig. 16.8(*a*)

elliptically polarised having different orientations. However with white light, the central band shall be dark while others shall be coloured.

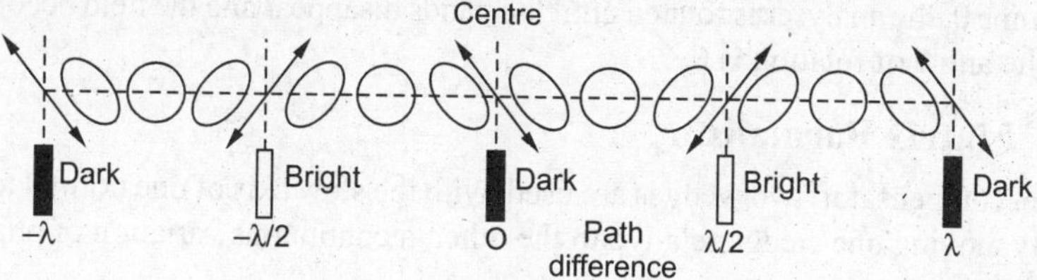

Fig. 16.8(*b*)

To calibrate the compensator, the movable wedge is displaced with the help of micrometer screw. Now the dark bands move laterally across the field of view. The movable wedge is adjusted in such a way that a dark band appears on the cross wire. The reading of micrometer screw is noted. The screw is again turned to bring the next dark band on the cross-wire and its reading is again noted. The difference of two readings gives the angle of rotation α of the screw which corresponds to a path difference of λ or a phase difference of 2π. The procedure is repeated for number of bands and mean α is obtained. This calibration is then used to determine following constants of elliptic vibrations.

Analysis of Elliptically Polarised Light

(*a*) **Phase difference**. Treating elliptic vibration to be made up of two rectangular linear vibrations with phase difference ϕ, we proceed as given below to find the value of phase difference between two such vibrations.

Illuminate the compensator with white plane polarised light and work the micrometer to set cross-wire on the central dark band. Now replace white light by elliptic light (use sodium lamp and $\lambda/4$ plate) and examine through N_2. The central band shifts at a point where the original phase difference ϕ is exactly compensated by additional phase difference introduced by the compensator. Work the micrometer screw to bring the central band under crosswire. If this rotation is β, then

$$\frac{\phi}{2\pi} = \frac{\beta}{\alpha}$$

$$\therefore \qquad \phi = \frac{2\pi\beta}{\alpha}.$$

(*ii*) **Position of the axes**. The position of the major and minor axes of the given elliptical vibration can be found as follows :

Using a white light source, plane polarised light is obtained by nicol N_1 (Fig. 16.8). The compensator is illuminated by this light and micrometer screw is adjusted to bring the central dark band on the cross-wire. The micrometer screw is then turned through an angle $\alpha/4$ so that the compensator at the cross-wire introduces a phase difference of $\pi/2$. The central dark band is now no more on the cross-wire. The given elliptically-polarised light is made to fall once more on the compensator. Now the compensator is *rotated* in its own plane until the central dark band is on the cross-wire. The analyser N_2 is also rotated to make the bands as distinct as possible. *The axes of the incident elliptically-polarised light are now parallel to the optic axes of the wedges of the compensator* (Fig. 16.9).

(*iii*) **Ratio of the axes**. The situation in the above adjustment is shown in Fig. 16.9. *OA* and *OB* represent the optic axes of the two wedges of the compensator. *OC* is the direction of the principal section of the analyser and *DD′* is the direction of light emerging from the compensator at the cross-wire. It is clear that the tangent of the angle θ which the principal section of the analyser makes with the optic axis of the wedge gives the ratio of the axes of the incident elliptical vibration.

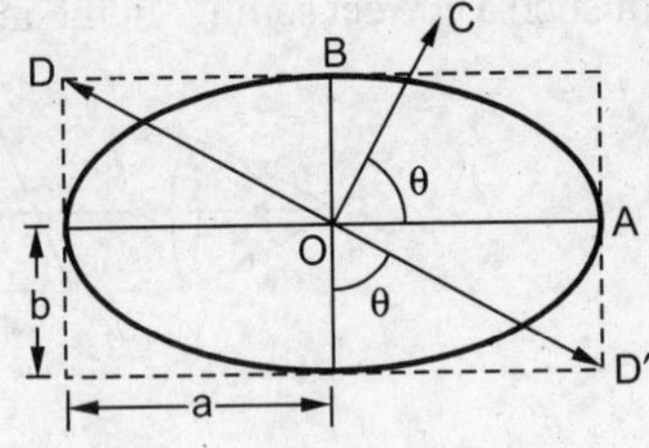

Fig. 16.9.

$$\tan\theta = \frac{a}{b}.$$

To determine θ, the analyser is rotated until the bands disappear and the field becomes uniformly illuminated. The angle of rotation is θ.

16.9. Jones' Matrix Formalism

In Babinet compensator, two wedges are used, with the slow axis of one normal to the slow axis of the other. By moving one wedge relative to the other, a continuous variation of phase retardation can be produced.

The first piece of the Babinet compensator may be regarded as the retarder plate with fast axis horizontal. The Jones' matrix is represented as

$$[J_1] = \begin{bmatrix} 1 & 0 \\ 0 & e^{-i\delta_1} \end{bmatrix} \qquad \ldots(1)$$

Similarly, the second piece of Babinet compensator may be regarded as a retarder plate with fast axis vertical. The Jones' matrix is represented as

$$[J_2] = \begin{bmatrix} 1 & 0 \\ 0 & e^{i\delta_2} \end{bmatrix} \qquad \ldots(2)$$

Consider a plane polarised light whose vibration is oriented at an angle θ with the horizontal. This is represented as

$$\begin{bmatrix} A \\ B \end{bmatrix} \text{ where } \frac{B}{A} = \tan\theta \qquad \ldots(3)$$

This light is incident on Babinet compensator. Consider the nature of emergent light.

$$\begin{bmatrix} 1 & 0 \\ 0 & e^{i\delta_2} \end{bmatrix}\begin{bmatrix} 1 & 0 \\ 0 & e^{-i\delta_1} \end{bmatrix}\begin{bmatrix} A \\ B \end{bmatrix} = \begin{bmatrix} 1 & 0 \\ 0 & e^{i\delta_2} \end{bmatrix}\begin{bmatrix} A \\ Be^{-i\delta_1} \end{bmatrix}$$

$$= \begin{bmatrix} A \\ Be^{i(\delta_2 - \delta_1)} \end{bmatrix} \qquad ...(4)$$

Here, $A = a \cos \theta$ and $B = a \sin \theta$.

Special cases. Let us consider four cases.

(*i*) *Symmetric passage of the incident light through compensator.* (Fig. 16.10*a*).

In this case, $\delta = \delta_2 - \delta_1 = 0$.

The emergent light matrix $[J_e]$ is given by

$$J_e = \begin{bmatrix} A \\ B \end{bmatrix} = \begin{bmatrix} a \cos \theta \\ a \sin \theta \end{bmatrix} \qquad ...(5)$$

So the emergent light is linearly polarised light with orientation parallel to incident orientation.

(*ii*) Let the incident light be departed from the symmetric line such that $\delta = \delta_2 - \delta_1 = \pi$. In this case, the compensator acts as a $\lambda/2$ plate. Hence

$$J_{\lambda/2} = \begin{bmatrix} 1 & 0 \\ 0 & -1 \end{bmatrix} \qquad ...(6)$$

The nature of emergent light is

$$\begin{bmatrix} 1 & 0 \\ 0 & -1 \end{bmatrix} \begin{bmatrix} a\cos\ \theta \\ a\sin\ \theta \end{bmatrix} = \begin{bmatrix} a\cos\ \theta \\ -a\sin\ \theta \end{bmatrix} \qquad ...(7)$$

The emergent light is a plane polarised light. The orientation is given by (Fig. 16.10*b*),

$$\tan \phi = \frac{-a \sin \theta}{a \cos \theta} = -\tan \theta$$

or

$$\phi = (\pi - \theta) \qquad ...(8)$$

(*iii*) When $\delta = \delta_2 - \delta_1 = \pi/2$ and $\theta = 45°$.

In this case, the compensator acts as a $\lambda/4$ plate. Hence

$$J_{\lambda/4} = \begin{bmatrix} 1 & 0 \\ 0 & e^{\pm i\pi/2} \end{bmatrix} = \begin{bmatrix} 1 & 0 \\ 0 & \pm i \end{bmatrix}$$

The nature of emergent light is

$$\begin{bmatrix} 1 & 0 \\ 0 & \pm i \end{bmatrix} \frac{1}{\sqrt{2}} \begin{bmatrix} 1 \\ 1 \end{bmatrix} = \frac{1}{\sqrt{2}} \begin{bmatrix} 1 \\ \pm i \end{bmatrix}$$

So, the emergent light is circularly polarised light (either *RCP* or *LCP*).

(*iv*) $\delta = \delta_2 - \delta_1 = \pi/2$ but $\theta \neq 45°$.

In this case, the Babinet compensator acts as a retarder.

The matrix is given as

$$J = \begin{bmatrix} A_1 & 0 \\ 0 & A_2 e^{\pm i\pi/2} \end{bmatrix}$$

The nature of emergent light is given by

$$\begin{bmatrix} A_1 & 0 \\ 0 & A_2 e^{\pm i\pi/2} \end{bmatrix} \begin{bmatrix} a \\ b \end{bmatrix} = \begin{bmatrix} aA_1 \\ bA_2 e^{\pm i\pi/2} \end{bmatrix} = \begin{bmatrix} A \\ \pm iB \end{bmatrix}$$

The emergent light is elliptically polarised light.

(Right handed for + sign and left handed for – sign).

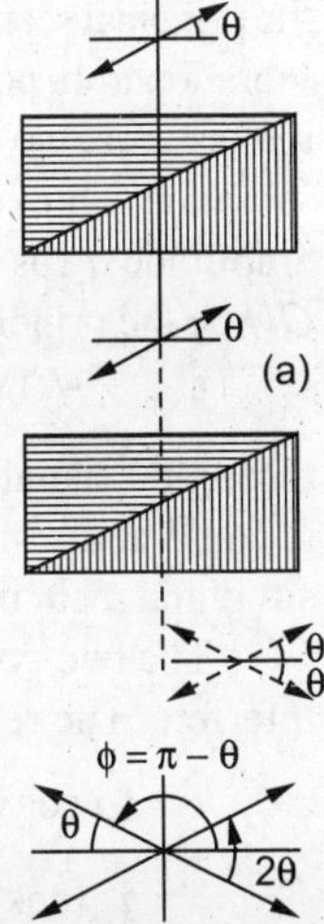

Fig. 16.10.

Thus Babinet compensator under specified conditions produces linearly polarised light, circularly polarised light or elliptically polarised light.

16.10. Interference of Polarised Light

The interference of polarised light was studied by Fresnel and Arago. They gave the following laws governing the interference of polarised light.

(1) Two rays polarised at right angles to each other do not interfere.

(2) Two rays polarised in the same plane interfere in the same manner as ordinary light.

(3) The rays polarised at right angles will interfere only when they have been brought in the same plane provided they originally belonged to the same beam of plane polarised light.

Light from a monochromatic source *S* rendered parallel by a lens is allowed to fall on polariser *P* (Fig. 16.11). The emergent plane polarised light is then allowed to fall normally on a calcite plate *C* cut with its faces parallel to optic axis. On emerging from the crystal plate the light is, in general, elliptically polarised. The emergent light from calcite plate is then allowed to pass through analyser Nicol *A* which is placed in a crossed position without inserting the calcite plate. Due to the presence of calcite plate *C*, some of the light is transmitted by the analysing Nicol. The transmitted light consists of part of each (*O* and *E* components) which has vibrations in the principal plane of the analyser. As component vibrations of elliptically polarised light have been derived from the same plane polarised light and the analyser brings some part of these into same plane of polarisation, the transmitted light satisfies the laws of interference of polarised light given by Fresnel and Arago.

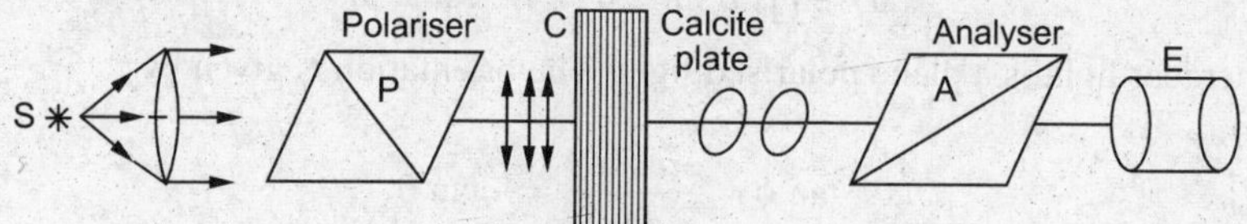

Fig. 16.11.

Theory. Let *ABCD* be the crystal plate of thickness *t* (Fig. 16.12). *OX* and *OY* are the directions parallel and perpendicular to the principal section of crystal plate respectively. The plane polarised vibration from polariser *P* incident on calcite plate is along *OE* which makes an angle α with *OX*. Let its amplitude be *a*. The incident vibration is broken up into extra-ordinary and ordinary vibration of amplitude *a* cos α and *a* sin α along *OX* and *OY* respectively. Let *OH* be the principal section of analyser *A*.

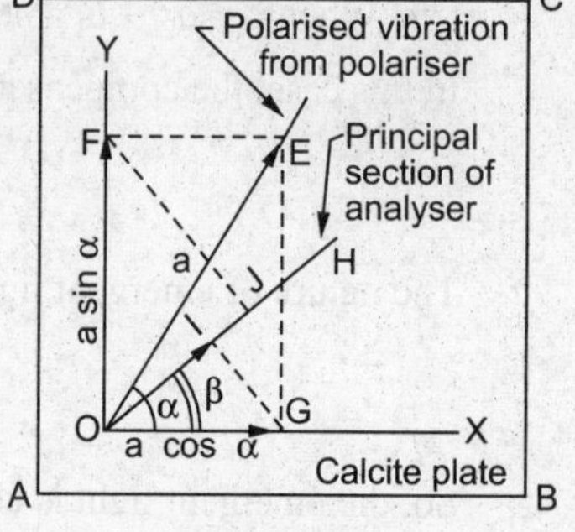

Fig. 16.12.

Let $\angle HOX = \beta$. Now components of vibration transmitted through the analyser *A* are *OI* and *OJ*. Their amplitudes are *a* cos α cos β and a sin α sin β respectively. The components *a* cos α sin β and *a* sin α cos β being normal to *OH* will not be transmitted.

The two components *a* cos α cos β and *a* sin α sin β which retain their phase difference δ now interfere. The resultant intensity is given by

$$I = (a \cos \alpha \cos \beta)^2 + (a \sin \alpha \sin \beta)^2 + 2\,(a \cos \alpha \cos \beta)\,(a \sin \alpha \sin \beta) \cos \delta \qquad \ldots(1)$$

$$= a^2 \left[\cos^2 \alpha \cos^2 \beta + \sin^2 \alpha \sin^2 \beta + 2 \cos \alpha \cos \beta \sin \alpha \sin \beta \,(1 - 2 \sin^2 \delta/2)\right]$$

$$= a^2 \left[(\cos \alpha \cos \beta + \sin \alpha \sin \beta)^2 - 4 \cos \alpha \cos \beta \sin \alpha \sin \beta \sin^2 \delta/2\right]$$

$$\therefore \qquad I = a^2 \left[\cos^2 (\alpha - \beta) - \sin 2\alpha \sin 2\beta \sin^2 \delta/2\right] \qquad \ldots(2)$$

Eq. (2) shows that the intensity varies with variation in α, β and δ.

(*i*) *When polariser and analyser are crossed, i.e.*, $\alpha - \beta = 90°$, then the resultant intensity is given by

$$I_c = a^2 \sin^2 2\alpha \sin^2 \delta/2 \qquad \ldots(3)$$

If the plate is rotated, then in four positions of the plate defined by

$$\alpha = \frac{\pi}{4}, \frac{3\pi}{4}, \frac{5\pi}{4}, \frac{7\pi}{4}$$

the intensity becomes maximum, while in positions defined by

$$\alpha = 0, \frac{\pi}{2}, \pi, \frac{3\pi}{2}$$

the intensity becomes minimum.

(*ii*) *When polariser and analyser are parallel i.e*, $\alpha - \beta = 0$.

In this case, the intensity is given by

$$I_p = a^2 [1 - \sin^2 2\alpha \sin^2 \delta/2] \qquad \ldots(4)$$

If the plate is rotated, then in positions defined by

$$\alpha = 0, \pi/2, \pi, 3\pi/2$$

the intensity becomes minimum while in positions defined by

$$\alpha = \frac{\pi}{4}, \frac{3\pi}{4}, \frac{5\pi}{4}, \frac{7\pi}{4}$$

the intensity becomes maximum.

The following patterns will be seen for different positions of analyser and polariser (Fig. 16.13).

When the two Nicols are crossed, the pattern consists of concentric rings of alternate brightness and darkness with a dark spot at the centre. There are dark brushes at right angles to each other superimposed on the ring pattern as shown in Fig. 16.13*a*. When the Nicols are set parallel, the pattern consists of rings of alternate darkness and brightness with a bright spot at the centre. There are brushes at right angles to each other superimposed on ring system. (Fig. 16.13*b*).

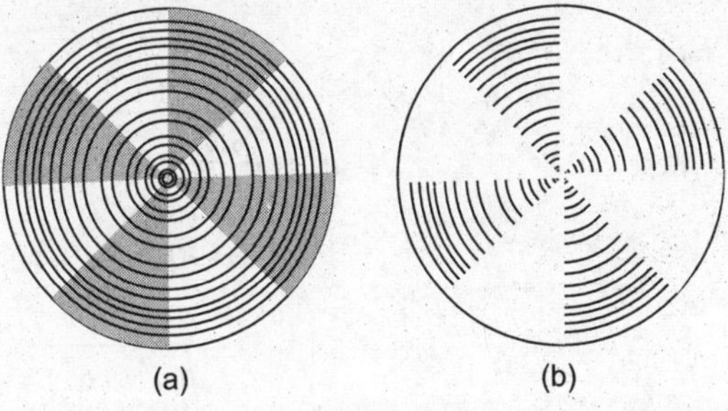

Fig. 16.13.

16.11. Fresnel's Explanation of Optical Rotation

A linearly polarised light is a combination of two circularly polarised light, one left circularly polarised (LCP) and other right circularly polarised (RCP). In Jones' formalism

$$\underbrace{\frac{1}{\sqrt{2}}\begin{bmatrix}1\\-i\end{bmatrix}}_{\text{LCP}} + \underbrace{\frac{1}{\sqrt{2}}\begin{bmatrix}1\\i\end{bmatrix}}_{\text{RCP}} = \frac{1}{\sqrt{2}}\begin{bmatrix}2\\0\end{bmatrix} = \sqrt{2}\begin{bmatrix}1\\0\end{bmatrix}$$

or

$$\frac{1}{2}\begin{bmatrix}1\\-i\end{bmatrix} + \frac{1}{2}\begin{bmatrix}1\\i\end{bmatrix} = \begin{bmatrix}1\\0\end{bmatrix}$$

The incident plane-polarised light on entering an optically-active substance is broken up into two circularly-polarised waves, one clockwise and the other anti-clockwise. Let l be the length of active medium. We assume that the right-circularly polarized wave sees an index of refraction n_R and the left-circularly polarized wave sees n_L.

When the medium is dextro-rotatory, a phase difference is introduced between two circularly polarised light, such that RCP advances in phase over LCP. The case is reversed when the medium is laevo-rotatory. At emergence, the RCP and LCP light are represented respectively as

$$\frac{1}{\sqrt{2}}\begin{bmatrix} e^{i\phi_1} \\ ie^{i\phi_1} \end{bmatrix} \text{ where } \phi_1 = k_R\, l$$

and

$$\frac{1}{\sqrt{2}}\begin{bmatrix} e^{i\phi_2} \\ -ie^{i\phi_2} \end{bmatrix} \text{ where } \phi_2 = k_L\, l.$$

At emergence, the light is represented as

$$\frac{1}{\sqrt{2}}\begin{bmatrix} e^{i\phi_1} \\ ie^{i\phi_1} \end{bmatrix} + \frac{1}{\sqrt{2}}\begin{bmatrix} e^{i\phi_2} \\ -ie^{i\phi_2} \end{bmatrix} = \frac{1}{\sqrt{2}}\begin{bmatrix} e^{i\phi_1} \\ e^{i(\phi_1+\pi/2)} \end{bmatrix} + \frac{1}{\sqrt{2}}\begin{bmatrix} e^{i\phi_2} \\ e^{i(\phi_2-\pi/2)} \end{bmatrix}$$

$$= \frac{1}{\sqrt{2}}\begin{bmatrix} e^{i\phi_1} + e^{i\phi_2} \\ e^{i(\phi_1+\pi/2)} + e^{i(\phi_2-\pi/2)} \end{bmatrix}$$

$$= \frac{1}{\sqrt{2}}\begin{bmatrix} 2\cos\left(\dfrac{\phi_1-\phi_2}{2}\right) e^{i\theta_1} \\ 2\sin\left(\dfrac{\phi_1-\phi_2}{2}\right) e^{i\theta_2} \end{bmatrix}$$

$\therefore$

$$(e^{i\phi_1} + e^{i\phi_2}) = (\cos\phi_1 + i\sin\phi_1) + (\cos\phi_2 + i\sin\phi_2)$$

$$= (\cos\phi_1 + \cos\phi_2) + i\,(\sin\phi_1 + \sin\phi_2)$$

$$= a + b\,i = \sqrt{(a^2+b^2)}\; e^{i\theta_1}$$

where

$$a^2 + b^2 = (\cos\phi_1 + \cos\phi_2)^2 + (\sin\phi_1 + \sin\phi_2)^2$$

$$= 2 + 2\cos(\phi_1 - \phi_2) = 4\sin^2\left(\frac{\phi_1-\phi_2}{2}\right)$$

$$\sqrt{(a^2+b^2)} = 2\sin\left(\frac{\phi_1-\phi_2}{2}\right).$$

Similarly, the value of $e^{i(\phi_1+\pi/2)} + e^{i(\phi_2-\pi/2)}$ can be obtained.

So, the emergent light can be represented by Jones' vector as

$$\frac{1}{\sqrt{2}}\begin{bmatrix} 2\cos\left(\dfrac{\phi_1-\phi_2}{2}\right) e^{i\theta_1} \\ 2\sin\left(\dfrac{\phi_1-\phi_2}{2}\right) e^{i\theta_2} \end{bmatrix} = \sqrt{2}\begin{bmatrix} A_1 e^{i\theta_1} \\ A_2 e^{i\theta_2} \end{bmatrix}$$

This represents a linear vibration. The angle α with y direction is given by

$$\tan\alpha = \frac{A_2}{A_1} = \frac{\sin\left(\dfrac{\phi_1-\phi_2}{2}\right)}{\cos\left(\dfrac{\phi_1-\phi_2}{2}\right)} = \tan\left(\frac{\phi_1-\phi_2}{2}\right)$$

$$\therefore \qquad \alpha = \frac{\phi_1 - \phi_2}{2}.$$

Thus the light rotates through an angle α after traversing a length l of the medium.

The specific rotation S is defined as

$$S = \frac{\alpha}{l} = \left(\frac{\phi_1 - \phi_2}{2l}\right)$$

Now $$k_R = \frac{2\pi}{\lambda_R} \text{ and } k_L = \frac{2\pi}{\lambda_L}$$

Further $$n_R = \frac{c}{v} = \frac{v\lambda}{v\lambda_R} \text{ or } \lambda_R = \frac{\lambda}{n_R}$$

$$k_R = \frac{2\pi}{\lambda} n_R \text{ and } k_L = \frac{2\pi}{\lambda} n_L.$$

$$\therefore \qquad S = \frac{(k_R - k_L)l}{2l} = \frac{(2\pi n_R - 2\pi n_L)}{2\lambda} = \frac{\pi}{\lambda}(n_R - n_L).$$

The specific rotation of any material has an explicit wavelength dependence of $1/\lambda$.

The specific rotation is proportional to the difference between the two indices of refraction.

16.12. Biquartz Polarimeter

It is a simple and accurate instrument, much more sensitive than the half shade one, for finding the angle of rotation produced by an optically active substance.

Apparatus. The experimental arrangement is shown in Fig. 16.14. S is a source of *white light* placed at the focus of convex lens L. The beam, rendered parallel by lens L, falls on a polarising

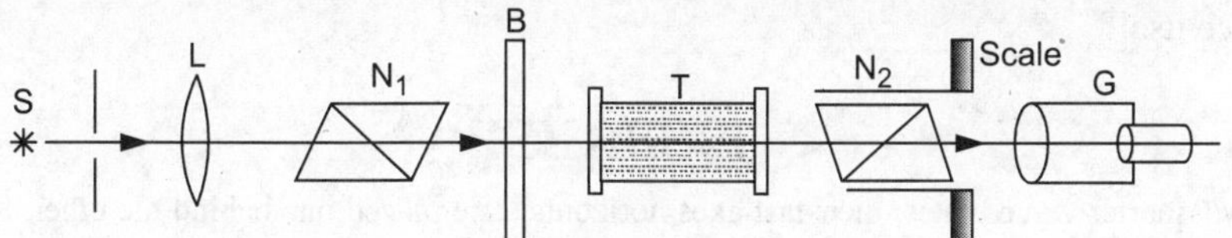

Fig. 16.14.

Nicol N_1. After passing through N_1, the light becomes plane polarized with its vibrations in the principal plane of the Nicol. This plane polarized light passes through a *biquartz plate B* and then through the tube T containing the active substance, say sugar solution. The transmitted light then falls on the analysing Nicol N_2. The emergent light is viewed through a telescope G. The analysing Nicol N_2 can be rotated about the light as axis, and its rotation can be measured on a circular degree-scale by means of a vernier.

Action of the Biquartz Plate. A biquartz plate consists of two semicircular plates one of left-handed quartz and the other of right-handed quartz, both cut perpendicular to optic axis and joined together along the diameter PP' so as to form a complete circular plate (Fig. 16.15). The thickness of each plate is such that it rotates the plane of polarisation for yellow light by 90°. This thickness is 3.75 mm for quartz.

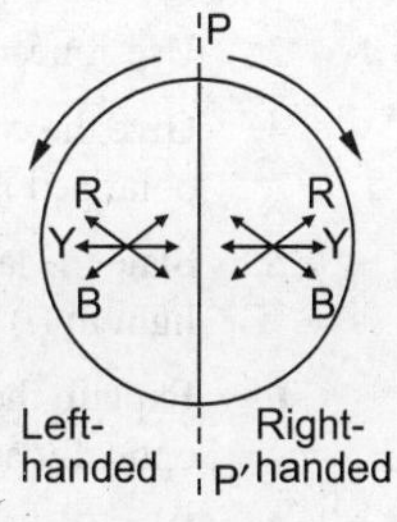

Fig. 16.15.

When *white* light, rendered plane-polarised by the polarising Nicol, travels through the biquartz normally, it is travelling along the optic axis. Therefore, rotatory dispersion occurs in each half plate of the biquartz. If the vibrations in the incident light are along PP' then, after passing through the biquartz, the vibrations of yellow light are along a direction perpendicular to PP'. The vibrations of red and blue lights are along different directions as shown in the Fig. 16.15. If the principal section of the analysing Nicol be parallel to PP', the yellow light will be quenched in both halves of the field, while the red and blue lights will be present in the same proportion in each half. Therefore, the two halves of the field will be equally-illuminated with a reddish-violet tint. This colour is called the *sensitive tint or tint of passage*. The position of sensitive tint is very sensitive because even a slight rotation of the Nicol from this position makes one half blue and the other half red.

Determination of specific rotation. The tube is first filled with water and the analyser is adjusted to obtain the position of *tint of passage*. This position of analyser, is read on the scale. Now the tube is filled with optically active solution of known concentration, and analyser is rotated to obtain the position of *tint of passage*. This position of analyser is also read on the scale. The difference between these two positions of the analyser gives the optical rotation θ produced by the optically active solution. The specific rotation S is then obtained from the formula

$$S = \frac{\theta}{l \times C},$$

where l is the length of the solution in decimetres and C its concentration.

Relative Merits of Half-shade and Biquartz—The half shade device is fairly sensitive and is very suitable for colourblind persons. It is, however, suitable only for one particular wavelength for which the path difference between E and O components is $\lambda/2$ after transmission through it.

The biquartz, on the other hand, is more sensitive than the half-shade device and works well with a white light source. It, however, suffers from two disadvantages : (*i*) Colour-blind persons cannot work with biquartz. Even different persons with normal vision will get different readings because the colour sensitivity of the eyes varies from person to person. (*ii*) Usually (and specially when the optically-active substance is coloured) it is difficult to obtain the accurate setting of the analyser for which the tint in both halves of the field is same. This is due to the rotatory dispersion produced by the substance itself.

EXERCISE

1. Two quarter-wave plates, their fast axes horizontal, are placed one behind the other. Determine the system matrix of the combination.

   Ans. $\begin{bmatrix} 1 & 0 \\ 0 & -1 \end{bmatrix}$

2. In terms of the Jones' calculus, describe a linear polarizer oriented with its transmission axis horizontal and:

   (*a*) Linearly polarized light of horizontal orientation incident on it.

   (*b*) The same light but now with vertical orientation incident on it.

   What is the result ? [**Ans.** (*a*) Linear horizontal ; (*b*) no transmission]

3. Use Jones calculus to describe the effect of a right-circular polarizer on vertically polarized light.

4. Give the construction and working theory of Babinet compensator to produce and analyse elliptically polarised light. How is it superior to $\lambda/4$ plate and why it is known so ?

5. State the laws of interference of polarised light. Give detailed explanation of interference of polarised light in (*i*) crystal plate with parallel faces, (*ii*) quartz wedge.

6. Explain the working of biquartz polarimeter. How would you use it to find the specific rotation of an optically active substance ?

7. Discuss the relative merits of biquartz and half shade polarimeters.

CHAPTER

17 FERMAT'S PRINCIPLE

17.1. Fermat's Principle of Least Time

"Fermat's principle of stationary time" or "Fermat's principle of extremum path" may be stated as follows:

A ray of light in passing from one point to another through a set of media by any number of reflections and refractions selects a path for which the time taken is either a minimum or a maximum.

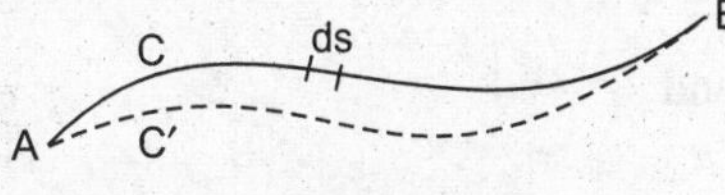

Fig. 17.1.

Let ds be the small distance travelled by light between two points A and B in a medium of refractive index n (Fig. 17.1). Let v be the velocity of light in that medium. Then,

$$\int_A^B \frac{ds}{v} = \text{maximum or minimum}$$

or

$$\int_A^B \frac{n\,ds}{c} = \text{maximum or minimum}.$$

Here $n = c/v$, c being the velocity of light in vacuum.

Since c is a constant,

$$\int_A^B n\,ds = \text{maximum or minimum, or } \delta \int_A^B n\,ds = 0$$

where $n\,ds$ is the optical path in a medium of refractive index n.

If the path ACB represents the actual ray path, then the time taken in traversing the path ACB will be an extremum in comparison to any nearby path $AC'B$. Since the shortest distance between two points is along a straight line, light rays in a homogeneous medium are straight lines.

One can deduce the laws of reflection and refraction from it.

17.2. Law of Reflection

Let a light ray from P be reflected at point O of mirror MM' to another point Q (Fig. 17.2).

Let the angles of incidence and reflection be i and r respectively. Let a and b be the lengths of the perpendiculars drawn from points P and Q on the mirror.

Let $MM' = c$ and $MO = x$, $\therefore OM' = (c - x)$.

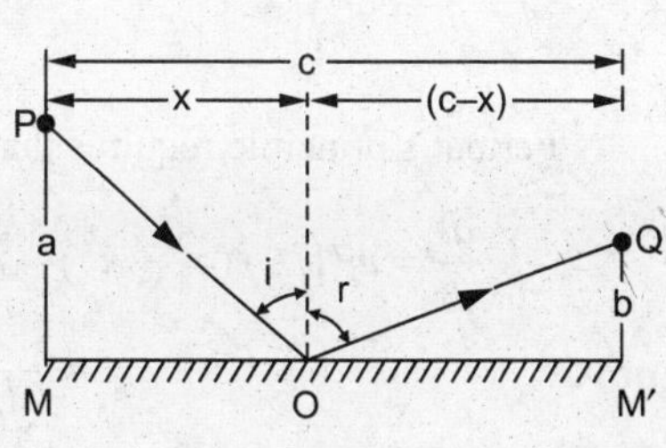

Fig. 17.2.

As the entire path *POQ* is travelled in air, the *optical path* between *P* and *Q* is given by

$$l = POQ = PO + OQ$$

$$= \sqrt{(a^2 + x^2)} + \sqrt{[b^2 + (c - x)^2]} \quad ...(1)$$

According to Fermat's principle, *O* will have a position such that the optical path *l* (or the time of travel) is a minimum or a maximum.

$$\therefore \quad \frac{dl}{dx} = 0$$

$$\frac{dl}{dx} = \frac{1}{2}(a^2 + x^2)^{-1/2}\,(2x) + \frac{1}{2}[b^2 + (c - x)^2]^{-1/2}(2)\,(c - x)\,(-1) = 0$$

or

$$\frac{x}{\sqrt{(a^2 + x^2)}} = \frac{c - x}{\sqrt{[b^2 + (c - x)^2]}} \quad ...(2)$$

From Fig. 17.2, we have

$$\frac{x}{\sqrt{(a^2 + x^2)}} = \frac{MO}{PO} = \sin i$$

and

$$\frac{c - x}{\sqrt{[b^2 + (c - x)^2]}} = \frac{OM'}{OQ} = \sin r$$

Hence from Eq. (2), we get

$$\sin i = \sin r \text{ or } i = r$$

This is law of reflection.

In addition, the incident ray, the surface normal, and the reflected ray all lie in the same plane, called the plane of incidence.

17.3. Law of Refraction

XY is a plane surface separating two media of refractive indices n_1 and n_2 (Fig. 17.3). Let a ray starting from the point *P*, intersect the interface at *O* and proceed to *Q* along *OQ*. Let *i* be the angle of incidence and *r* the angle of refraction.

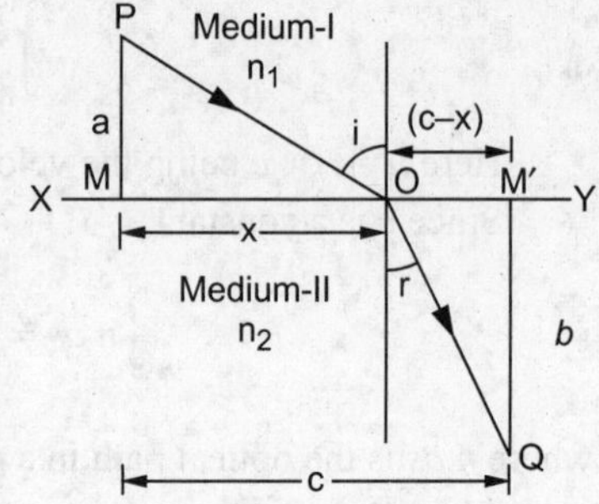

Fig. 17.3.

Clearly, for minimum optical path length, the incident ray, the refracted ray and the normal to the interface must all lie in the same plane.

Let *a* and *b* be the lengths of the perpendiculars drawn from *P* and *Q* to the refracting surface.

Let $MM' = c$ and $MO = x$; $\therefore OM' = (c - x)$.

The *optical path* between *P* and *Q* is given by

$$l = n_1\,(PO) + n_2\,(OQ)$$

$$= n_1\sqrt{(a^2 + x^2)} + n_2\sqrt{[b^2 + (c - x)^2]} \quad ...(1)$$

Fermat's principle requires that *l* be a minimum or a maximum. So, $\frac{dl}{dx} = 0.$

$$\frac{dl}{dx} = n_1(1/2)\,(a^2 + x^2)^{-1/2}(2x) + n_2(1/2)\,[b^2 + (c - x)^2]^{-1/2}\,(2)\,(c - x)\,(-1) = 0$$

or

$$n_1\frac{x}{\sqrt{(a^2 + x^2)}} = n_2\frac{c - x}{\sqrt{[b^2 + (c - x)^2]}} \quad ...(2)$$

But from Fig. 17.3, we have

$$\frac{x}{\sqrt{(a^2+x^2)}} = \frac{MO}{PO} = \sin i$$

and
$$\frac{c-x}{\sqrt{[b^2+(c-x)^2]}} = \frac{OM'}{OQ} = \sin r$$

Hence, from Eq. (2), we get

$$n_1 \sin i = n_2 \sin r.$$

This in *Snell's law of refraction.*

17.4. Refraction of Rays at the Interface Between an Isotropic Medium and an Anisotropic Medium (Optic Axis Normal to the Surface)

Fermat's principle can be used to determine the direction of the refracted ray for a ray incident at the interface of an isotropic and an anisotropic medium.

In a uniaxial medium, the refractive index variation for the extraordinary ray is given by

$$n^2(\theta) = n_o^2 \cos^2\theta + n_e^2 \sin^2\theta.$$

Here, θ represents the angle that the ray makes with the optic axis.

When the extraordinary ray propagates parallel to the optic axis (*i.e.*, when θ = 0), its speed is c/n_0 and when it propagates perpendicular to the optic axis (θ = π/2) its speed is c/n_e.

Fig. 17.4 shows the direction of the refracted extraordinary ray when the optic axis (of the uniaxial crystal) is normal to the surface. n_1 is the refractive index of medium *I*.

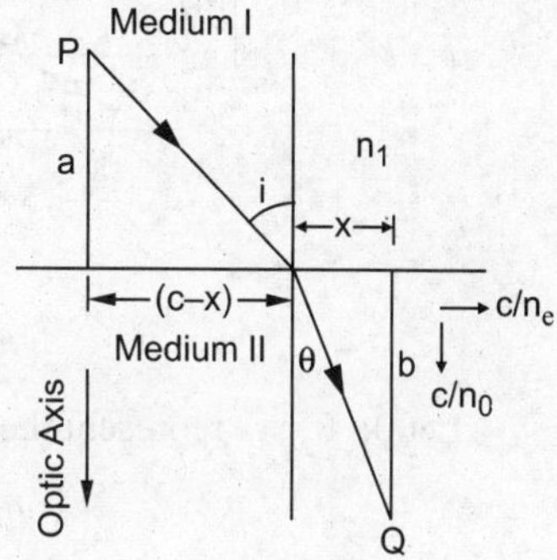

Fig. 17.4.

The optical path length from *P* to *Q* is given by

$$l = n_1[a^2+(c-x)^2]^{1/2} + n(\theta)[b^2+x^2]^{1/2} \quad \text{...(1)}$$

Now,
$$\cos\theta = \frac{b}{(b^2+x^2)^{1/2}} \text{ and } \sin\theta = \frac{x}{(b^2+x^2)^{1/2}}$$

$$\therefore \quad l = n_1[a^2+(c-x)^2]^{1/2} + [n_0^2 b^2 + n_e^2 x^2]^{1/2}$$

According to Fermat's principle, $dl/dx = 0$.

$$\frac{dl}{dx} = n_1\left(\frac{1}{2}\right)[a^2+(c-x)^2]^{-1/2}(2)(c-x)(-1) + \frac{1}{2}[n_o^2 b^2 + n_e^2 x^2]^{-1/2} n_e^2(2x) = 0$$

or
$$\frac{n_1(c-x)}{[a^2+(c-x)^2]^{1/2}} = \frac{n_e^2 x}{[n_o^2 b^2 + n_e^2 x^2]^{1/2}}$$

From Fig. 17.4, we have

$$\frac{(c-x)}{[a^2+(c-x)^2]^{1/2}} = \sin i; \ \frac{x}{b} = \tan\theta = \tan r.$$

Here, *r* is the angle of refraction.

$$\therefore \quad n_1 \sin i = \frac{n_e^2 \tan r}{[n_o^2 + n_e^2 \tan^2 r]^{1/2}}$$

or
$$\tan r = \frac{n_o n_1 \sin i}{n_e\sqrt{n_e^2 - n_1^2 \sin^2 i}} \quad \text{...(2)}$$

Using Eq. (2) we can calculate the angle of refraction for a given angle of incidence.

17.5. Ray Paths in an Inhomogeneous Medium

In an inhomogeneous medium, the refractive index varies in a continuous manner. Fermat's principle can be used to determine the ray paths in an inhomogeneous medium. Consider the case when the refractive index changes continuously along the *x*-axis only. This medium can be thought of as a limiting case of a medium consisting of a continuous set of thin slices of media of different refractive indices (Fig. 17.5*a*). When the thickness of each slice goes to zero and at the same time the number of slices increases indefinitely, then we get a medium in which the refractive index changes continuously along the *x*-direction. At each interface, the light ray satisfies Snell's law, *i.e.*,

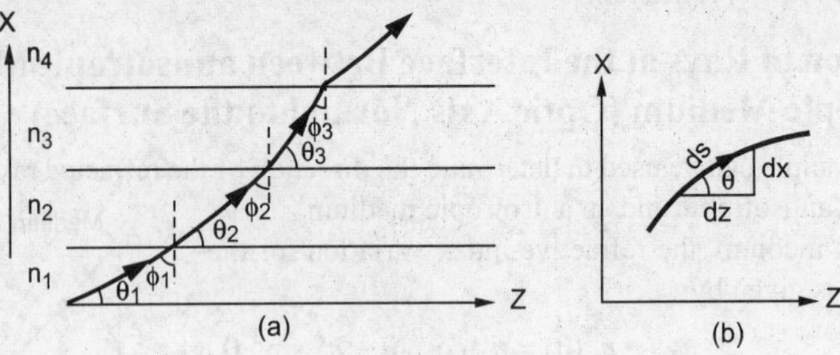

Fig. 17.5.

$$n_1 \sin \phi_1 = n_2 \sin \phi_2 = n_3 \sin \phi_3 = ...$$

Let $\theta_1, \theta_2 \ldots$ represent the corresponding angles that the ray makes with the *z*-axis. Then,

$$n_1 \cos \theta_1 = n_2 \cos \theta_2 = ... = \text{a constant}$$

or $$n(x) \sin \phi(x) = n(x) \cos \theta(x) = \text{constant} \qquad ...(1)$$

Let the ray initially make an angle θ_1 (with the *z*-axis) at a point where the refractive index is n_1. Then the constant is $n_1 \cos \theta_1$. Thus, in the limiting case of the continuous variation of refractive index, the piecewise straight lines shown in Fig. 17.5(*a*) form a continuous curve. The curve is determined from the equation

$$n(x) \cos \theta(x) = n_1 \cos \theta_1 \qquad ...(2)$$

Thus as the refractive index changes, the ray path bends in such a way that the product $n(x) \cos \theta(x)$ remains constant [Fig. 17.5(*b*)]. We denote the constant by $\bar{\beta}$.

$$n(x) \cos \theta(x) = \bar{\beta} \qquad ...(3)$$

The Ray Equation and its Solutions. Eq. (3) can be used to derive the ray equation. Let *ds* represent the infinitesimal arc length along the curve. Then,

$$(ds)^2 = (dx)^2 + (dz)^2$$

or $$\left(\frac{ds}{dz}\right)^2 = \left(\frac{dx}{dz}\right)^2 + 1 \qquad ...(4)$$

From Fig. 17.5(*b*), we have,

$$\frac{dz}{ds} = \cos \theta = \frac{\bar{\beta}}{n(x)} \qquad ...(5)$$

Thus Eq. (4) becomes

$$\left(\frac{dx}{dz}\right)^2 = \frac{n^2(x)}{\bar{\beta}^2} - 1 \qquad ...(6)$$

Differentiating Eq. (6) with respect to z, we get

$$2\frac{dx}{dz}\frac{d^2x}{dz^2} = \frac{1}{\beta^2}\frac{dn^2}{dx}\frac{dx}{dz}$$

or
$$\frac{d^2x}{dz^2} = \frac{1}{2\beta^2}\frac{dn^2}{dx} \qquad ...(7)$$

Eq. (7) represents the ray equation when the refractive index depends only on the x-coordinate. The solution of Eq. (7) will give the precise ray paths in an inhomogeneous medium.

Optical Path Extremum or Maximum. In the case of reflection and refraction, if we evaluate $\partial^2 l/\partial x^2$, we find that it comes out positive. This shows that the actual path is a minimum, *i.e.*, path of least time. However, this is not always the case.

Consider two points P and Q and a curved mirror MON (Fig. 17.6.)

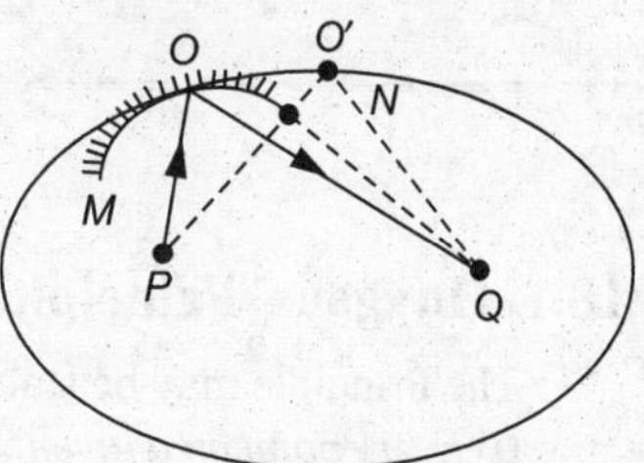

Fig. 17.6.

Let POQ be the actual path or optical path of the ray travelling from P to Q. Draw an ellipse having P and Q as its focii and passing through the point O. From the property of the ellipse, we have

$PO + OQ$ = constant (for all positions of O on the ellipse).

Thus $PO + OQ = PO' + O'Q$

From Fig. 17.6 it is evident that an alternative neighbouring path for the ray of light by reflection at the mirror is PNQ. The difference between the actual path POQ and the neighbouring path PNQ is

$$\begin{aligned}\Delta = POQ - PNQ &= (PO + OQ) - (PN + NQ)\\ &= (PO + OQ) - (PO' - NO' + NQ)\\ &= (PO' + O'Q) - (PO' - NO' + NQ)\\ &= O'Q + NO' - NQ = +\text{ve quantity}\end{aligned}$$

because $O'Q$ and NO' are two sides of a triangle of which NQ is the third side. Thus the neighbouring path PNQ is smaller than the actual path POQ, *i.e.*, the actual or optical path is maximum as compared to all neighbouring paths.

EXERCISE

1. State Fermat's principle and use it to establish the laws of reflection and refraction of light.
2. Determine the ray paths inside a homogeneous medium for which $n(x)$ is a constant.

   [**Hint.** $\frac{d^2x}{dz^2} = 0$. Integrating twice with respect to z, $x = Az + B$.

   The ray path is a straight line.]
3. State Fermat's principle of stationary time.
   Derive the laws of reflection using this principle.s
   Give an example where the path of light is maximum rather than minimum.

   (Rohilkhand 1974, Agra 75)

CHAPTER

18 HUYGENS' PRINCIPLE AND ITS APPLICATIONS

18.1. Huygens' Principle

The Principle may be stated in three parts :

(*i*) *Every point on a given wavefront may be regarded as the source of a new disturbance, called secondary wavelets.*

(*ii*) *The secondary spherical wavelets from each point spread out in all directions with the velocity of light.*

(*iii*) *The envelope of these wavelets in the forward direction at any instant gives the new wavefront at that instant.*

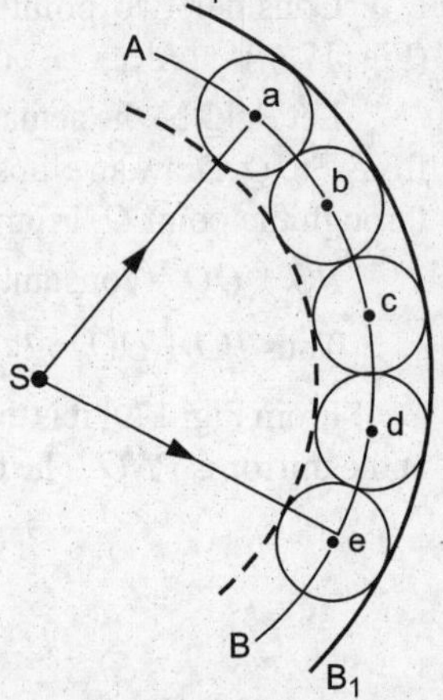

Fig. 18.1.

Consider a number of points say *a, b, c, d, e* ... on the spherical wavefront *AB* (Fig. 18.1). With each point in turn as a centre and radius *ct* draw spheres. These spheres represent the secondary waves originating from these points respectively. The envelope or tangential surface A_1B_1 touching all these spheres in the forward direction is the new *spherical wave front*. At a large distance from the source *S*, the wavefront *AB* becomes *plane*.

The Huygens' construction gives a backward wavefront also (shown by dotted lines) which is contrary to observation. This result is avoided by assuming that the intensity of the secondary wavelets is not uniform, but varies continuously from a maximum in the forward direction to a minimum of zero in the backward direction.

18.2. Huygens' Principle and the Laws of Reflection

Reflection of a Plane Wave at a Plane Surface—Let *XY* be the section of a plane reflecting surface, and *AB* that of a plane wavefront striking it at *A* (Fig. 18.2). Let *v* be the velocity of light. Let *t* be the time for the edge *B* of the wavefront to reach the surface at *A′*.

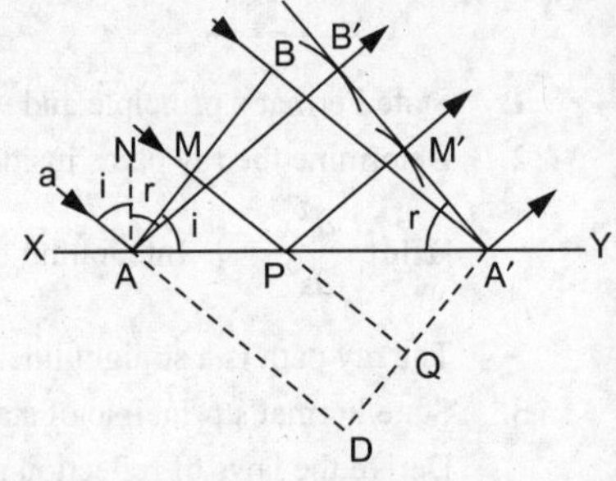

Fig. 18.2.

According to Huygens' principle, each point on the wavefront acts as a source of secondary wavelets. In the absence of reflecting surface *XY*, the wavefront *AB* would have advanced to the position *A′D* after a time *t*. *A′D* is parallel to *AB*. But due to the presence of the reflecting surface, as the wavefront advances, the points on *XY* successively struck by the wavefront become the source of secondary spherical wavelets. Thus, after a time *t* when the wavefront strikes the point *A′*, the secondary wavelet from *A* has acquired a radius $AB' = BA'$

= $AD = vt$, whereas the secondary wavelet from A' just starts. Let us draw a tangent $A'B'$ from A' to the secondary wavelet from A. If $A'B'$ is a tangent *common* to all the secondary wavelets starting from the different points between A and A', then $A'B'$ represents the reflected wavefront.

Let M be any point on AB. Draw MPQ parallel to BA', and PM' perpendicular to $A'B'$.

In Δ^s $AA'B'$ and $AA'D$, we have

$$AB' = AD,$$

$$\angle AB'A' = \angle ADA' \qquad \text{(right angles)}$$

and AA' is common.

Hence Δs $AA'B'$ and $AA'D$ are congruent.

$$\angle AA'B' = \angle AA'D.$$

Again in Δ^s $PA'M'$ and $PA'Q$, we have

$$\angle PA'M' = \angle PA'Q \qquad \text{(just proved)}$$

$$\angle PM'A' = \angle PQA' \qquad \text{(right angles)}$$

and PA' is common.

Hence Δs $PA'M'$ and $PA'Q$ are congruent.

$\therefore$ $PM' = PQ$ = the radius of the secondary wavelet from P.

Hence the wavelet from P is touched by $A'B'$ at M'. Since P is *any* point between A and A', $A'B'$ is the common tangent of the wavelets from all points on AA'. Hence $A'B'$ is the reflected wavefront.

Deduction of the Laws. Let AN be a normal to the reflecting surface XY, i and r be the angles of incidence and reflection respectively.

Since the angle between two straight lines is equal to the angle between their normals, we have

$$i = \angle aAN = \angle BAA'$$

and

$$r = \angle NAB' = \angle B'A'A$$

Now, in Δ^s ABA' and $AB'A'$, we have

$$BA' = AB' = vt$$

$$\angle ABA' = \angle AB'A' \qquad \text{(right angles)}$$

and AA' is common.

Hence Δs ABA' and $AB'A'$ are congruent.

$$\therefore \qquad \angle BAA' = \angle B'A'A$$

or

$$i = r$$

i.e., angle of incidence is equal to the angle of reflection.

Also the incident ray, reflected ray and normal all lie in the same plane.

18.3. Huygens' Principle and the Laws of Refraction

Refraction of a Plane Wave at a Plane Surface. Let XY be the plane refracting surface separating the two media '1' and '2' of refractive indices n_1 and n_2 respectively. (Fig. 18.3). Let a plane wavefront AB be incident obliquely on the surface XY striking it at A. Let v_1 and v_2 be the velocities of light in the two media where $v_1 > v_2$. Let t be the time for edge B of the wavefront to reach A' so that $BA' = v_1 t$.

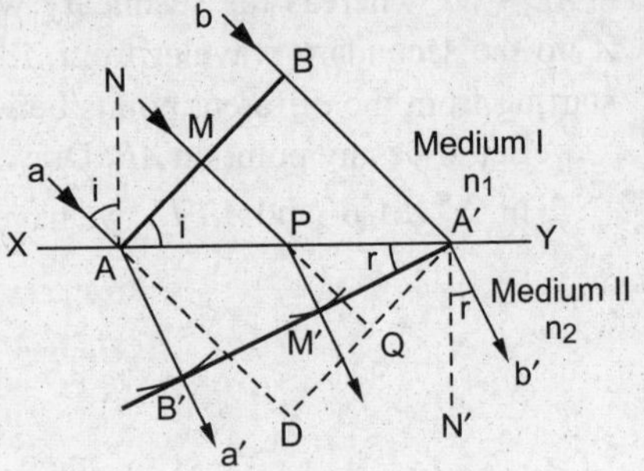

Fig. 18.3.

In the absence of XY, the wavefront AB would have advanced to the position $A'D$ after a time t. $A'D$ is parallel to AB and $AD = BA' = v_1 t$. But in the presence of XY, as the wavefront advances, the points between A and A' successively struck by the wavefront become the sources of secondary spherical wavelets. Thus, after the time t, the secondary wavelet from A in medium 2 has acquired a radius $AB' = v_2 t$, whereas the secondary wavelet from A' has just started.

Now,
$$AB' = v_2 t = v_2 \frac{BA'}{v_1} = \frac{v_2}{v_1} AD$$

or
$$\frac{v_2}{v_1} = \frac{AB'}{AD} = \frac{\text{radius in medium 2}}{\text{radius if medium 2 were absent}} \quad \ldots(1)$$

Draw a tangent $A'B'$ from A' to the secondary wavelet from A. If $A'B'$ is a tangent *common* to all the secondary wavelets starting from different points between A and A′, then $A'B'$ represents the refracted wavefront.

Let M be any point on AB. Draw MPQ parallel to BA', and PM' perpendicular to $A'B'$.

Now, in similar $\Delta^s\, PA'M'$ and $AA'B'$, we have
$$\frac{PA'}{AA'} = \frac{PM'}{AB'} \quad \ldots(2)$$

In similar $\Delta^s\, PA'Q$ and $AA'D$, we have
$$\frac{PA'}{AA'} = \frac{PQ}{AD} \quad \ldots(3)$$

Comparing (2) and (3), we have
$$\frac{PM'}{AB'} = \frac{PQ}{AD}$$

or
$$\frac{PM'}{PQ} = \frac{AB'}{AD}.$$

Substituting for $\frac{AB'}{AD}$ from Eq. (1), we get
$$\frac{PM'}{PQ} = \frac{\text{radius in medium 2}}{\text{radius if medium 2 were absent}}.$$

PQ is the radius of wavelet from P if medium 2 were absent. Therefore, PM' must be the radius of the wavelet from P in medium 2. As PM' is drawn perpendicular to $A'B'$, the wavelet from P must touch A'B′. Since P is any point between A and A', $A'B'$ is the common tangent to the wavelets from all points between A and A'. Hence $A'B'$ is the refracted wavefront.

Deduction of the law. Let i and r be the angles of incidence and refraction respectively, AN and $A'N'$ the normals at surface XY.

As the angle between two straight lines is equal to the angle between their normals, we have

$$i = \angle aAN = \angle BAA'$$

and

$$r = \angle N'A'b' = \angle AA'B'$$

$$\therefore \quad \frac{\sin i}{\sin r} = \frac{\sin BAA'}{\sin AA'B'} = \frac{BA'/AA'}{AB'/AA'}$$

$$= \frac{BA'}{AB'} = \frac{v_1 t}{v_2 t} = \frac{v_1}{v_2} = \text{constant.}$$

This is Snell's law.

The ratio $\frac{\sin i}{\sin r}$ is called the 'refractive index of the second medium with respect to the first and is denoted by ${}_1n_2$. Thus

$${}_1n_2 = \frac{v_1}{v_2} = \frac{\text{velocity of light in medium 1}}{\text{velocity of light in medium 2}}.$$

If the second medium is optically denser, ${}_1n_2 > 1$, then $v_2 < v_1$, *i.e.*, the velocity of light in a denser medium is less than that in a rarer medium.

18.4. Total Internal Reflection

The essential conditions for total internal reflection are that the ray must pass from denser (n_1) to rarer (n_2) medium and the angle of incidence *i* in denser medium should be greater than the critical angle. The angle of incidence *i* in denser medium for which angle of refraction in the rarer medium is 90° is known as critical angle, *i.e.*, $i = i_c$. For all angles of incidence greater than i_c, we will have total internal reflection.

Let *XY* be a plane surface dividing two media of refractive indices n_1 and n_2 (Fig. 18.4*i*). $n_1 > n_2$. Let v_1 and v_2 be the velocities of light in the two media respectively. Then, $v_2 > v_1$.

$$\frac{n_1}{n_2} = \frac{v_2}{v_1} \qquad \text{...(1)}$$

Consider a plane wavefront *AB*. According to Huygens concept the point *A* becomes the source of spherical wave, which propagates in 2nd medium with velocity v_2. In time *t* in which *B* reaches *C*, the secondary wavelet at *A* describes a sphere of radius *AP*. A tangent from *C* on this surface gives *CT* as the refracted wavefront. Now,

$$BC = v_1 t, \text{ and } AT = AP = v_2 t.$$

$$\therefore \quad AT = \frac{v_2}{v_1} BC = \frac{n_1}{n_2} BC \qquad \text{...(2)}$$

In ΔABC, $BC = AC \sin i$ ...(3)

$$\therefore \quad AT = AP = \frac{n_1}{n_2} \times AC \sin i \qquad \text{...(4)}$$

Thus there are three cases :

(*i*) If value of *i* is such that $(n_1/n_2)\sin i < 1$, then $AP < AC$.

The wavefront *AB* is refracted as *CT* in the 2nd medium (Fig. 18.4*i*).

(*ii*) When *i* is such that $(n_1/n_2)\sin i = 1$, we get $AP = AC$.

The refracted wavefront CT becomes normal to XY. This angle of $i = i_c$ is called critical angle and for this we have, (Fig. 18.4*ii*),

$$\frac{n_1}{n_2} \sin i_c = 1. \quad \therefore \sin i_c = \frac{n_2}{n_1}$$

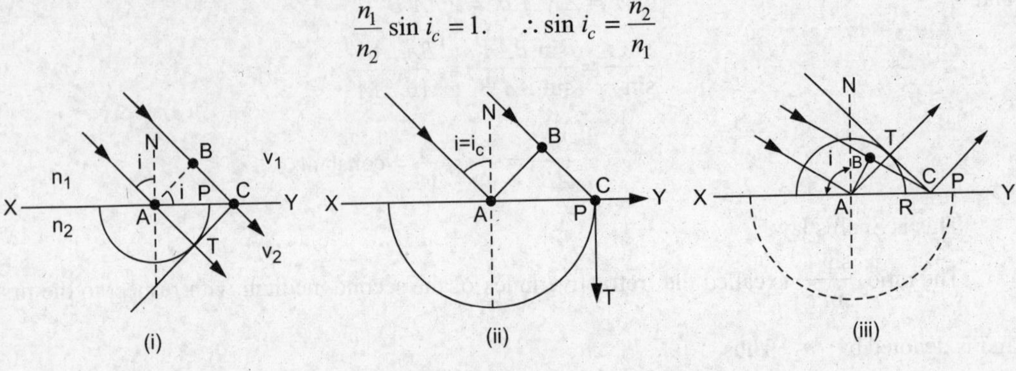

Fig. 18.4.

Thus $\sin i_c = n_{12}$ ...(5)

where i_c denotes the critical angle and n_{12} represents the refractive index of the second medium with respect to the first.

(*iii*) If the value of $i > i_c$ such that $(n_1/n_2) \sin i > 1$, then $AP > AC$ and we cannot draw tangent on wave surface in 2nd medium from point C. Then we cannot have refracted wavefront in 2nd medium. Therefore in such a case *we consider the secondary wavefront in the 1st medium* where its radius AR in time t will be equal to BC as discussed in reflection at plane surface. *Consequently we get CT as the reflected wavefront*. (Fig. 18.4*iii*). This is known as total internal reflection.

18.5. Refraction Through a Lens

Let O be a point object on the principal axis of a convex lens LM of refractive index n (> 1) placed in air (Fig. 18.5). A and B are the poles of two surfaces LAM and LBM having radii of curvature R_1 and R_2 respectively. Let v_1 and v_2 be the velocities of the light in air and medium respectively.

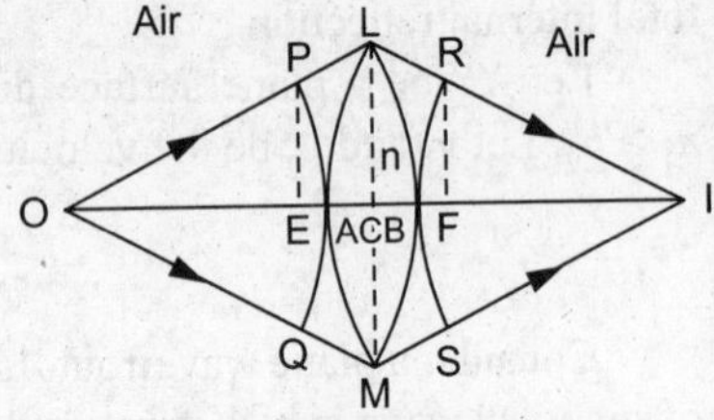

Fig. 18.5.

Consider a spherical wavefront diverging from O. When it reaches the position PAQ, touching the lens at A, the point A begins to send out secondary wavelets in the medium of the lens with velocity v_2, while the wavelets from P and Q are still travelling in air with velocity v_1. Since $v_2 < v_1$, during the time the wavelet from A reaches B, the wavelets from P and Q reach R and S respectively. Thus RBS is the emergent wavefront. If the aperture of the lens be small, the wavefront RBS will be spherical, and convergent if O is situated at a distance greater than the focal length of the lens. Therefore, as it travels further, its radius decreases uniformly and becomes zero at the point I. Thus I is the real image of O.

Draw PE, LC and RF perpendiculars to the principal axis. If the aperture of the lens is small, P, L, R are then approximately at the same height y above the axis, *i.e.*,

$$\left.\begin{array}{c} PE = LC = RF = y \text{ (say)} \\ PL + LR = EF \end{array}\right\} \quad ...(1)$$

and

Since the light travels from A to B, and from P to R (through the point L) during the same time, we have

$$\frac{PL + LR}{v_1} = \frac{AB}{v_2}$$

or $$PL + LR = \frac{v_1}{v_2} AB = n\ AB \qquad ...(2)$$

Now, $$PL + LR = EF = AE + AB + BF \qquad ...(3)$$

Comparing Eqs. (2) and (3), we get

$$AE + AB + BF = n\ AB$$

or $$AE + BF = (n-1)\ AB = (n-1)\ (AC + BC) \qquad ...(4)$$

Let $AO = -u$ and $BI = +v$. R_1 is positive and R_2 is negative.

Now, *PAQ*, *LAM*, *LBM* and *RBS* are the arcs of circles of radii $-u$, $+R_1$, $-R_2$ and $+v$ respectively. Hence by Sagitta theorem, we have

$$AE = \frac{PE^2}{2\ AO} = \frac{y^2}{-2u}$$

$$BF = \frac{RF^2}{2\ BI} = \frac{y^2}{2v}$$

$$AC = \frac{LC^2}{2R_1} = \frac{y^2}{2R_1}$$

and $$BC = \frac{LC^2}{-2R_2} = \frac{y^2}{-2R_2}.$$

Substituting these values in Eq. (4), we get

$$-\frac{y^2}{2u} + \frac{y^2}{2v} = (n-1)\left(\frac{y^2}{2R_1} - \frac{y^2}{2R_2}\right)$$

or $$\frac{1}{v} - \frac{1}{u} = (n-1)\left(\frac{1}{R_1} - \frac{1}{R_2}\right).$$

This is the required relation.

When the incident wavefront is plane *i.e.*, $u = \infty$, then $v = f$.

$\therefore$ $$\frac{1}{f} = (n-1)\left(\frac{1}{R_1} - \frac{1}{R_2}\right) \qquad ...(5)$$

This is the desired lens formula.

Consider a thin lens made of a material of refractive index n_2 placed in a medium of refractive index n_1. Then,

$$\frac{1}{f} = \frac{n_2 - n_1}{n_1}\left(\frac{1}{R_1} - \frac{1}{R_2}\right) \qquad ...(6)$$

EXERCISE

1. State Huygens' principle. How are the laws of reflection and refraction of light explained on the basis of Huygens' principle ?
2. Account for the formation of images by refraction through lenses on the wave theory and prove the relation

$$\frac{1}{v} - \frac{1}{u} = (n-1)\left(\frac{1}{R_1} - \frac{1}{R_2}\right)$$

where the symbols have their usual significance.

CHAPTER

19

CLASSICAL AND MODERN GEOMETRICAL OPTICS

19.1. Fermat's Principle Applied to Refraction at a Spherical Refracting Surface

We can use Fermat's principle to deduce the relation governing the refraction at a spherical refracting surface.

Let *POQ* be a spherical refracting surface of radius r and centre of curvature C (Fig.19.1). It separates two media of refractive indices n_1 and n_2. A point object at A on the principal axis OC is at

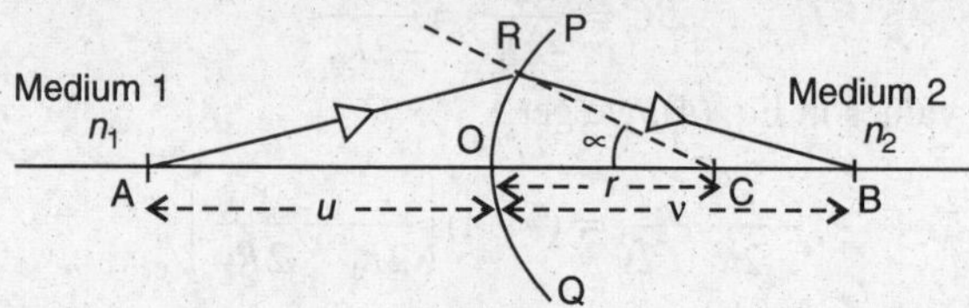

Fig. 19.1.

a distance u from O. Its image B lies at a distance $OB = v$ from O. AR is the incident ray. RB is the refracted ray. In the triangle ARC, let $\angle ACR = \alpha$.

$$AR^2 = AC^2 + CR^2 - 2AC.CR \cos \alpha$$

or

$$AR^2 = (u + r)^2 + r^2 - 2(u + r)\, r \cos \alpha \qquad ...(1)$$

Similarly in the triangle CRB,

$$RB^2 = (v - r)^2 + r^2 + 2(v - r)\, r \cos \alpha \qquad ...(2)$$

Total optical path,
$$l = n_1.AR + n_2.RB \qquad ...(3)$$

Eq. (3) when combined with Eqs. (1) and (2) yields

$$l = [n_1 \{(u + r)^2 + r^2 - 2\,(u + r)\, r \cos \alpha\}^{1/2} + n_2\{(v - r)^2 + r^2 + 2\,(v - r)\, r \cos \alpha\}^{1/2}] \qquad ...(4)$$

The angle α varies if the path is changed.

According to Fermat's principle of extremum path, $\dfrac{dl}{d\alpha} = 0$ ...(5)

Differentiating partially Eq. (4) with respect to α,

$$\frac{n_1(u + r)\, r \sin \alpha}{\left[(u + r)^2 + r^2 - 2\,(u + r)\, r \cos \alpha\right]^{1/2}} + \frac{n_2\,[-(v - r)\, r \sin \alpha]}{\left[(v - r)^2 + r^2 + 2\,(v - r)\, r \cos \alpha\right]^{1/2}} = 0 \qquad ...(6)$$

For small values of α, $\cos \alpha = 1$ and Eq.(6) reduces to

$$\frac{n_1(u + r)}{u} = \frac{n_2(v - r)}{v}$$

or
$$\frac{n_2}{v}+\frac{n_1}{u}=\frac{n_2-n_1}{r}$$

According to sign convention, v and r are positive and u is negative. Therefore,
$$\frac{n_2}{v}-\frac{n_1}{u}=\frac{n_2-n_1}{r} \qquad ...(7)$$

19.2. Fermat's Principle Applied to Refraction at a Lens

LAB is a convex lens whose refractive index is n (Fig. 19.2). P is a point object placed on the axis of the lens at a distance u from the optical centre *i.e.*, $OP = u$. Any ray PL striking the lens at L will after refraction converge at the point Q. Similarly another ray PA directed along the axis of the lens will go undeviated after refraction and cut the first refracted ray LQ at Q such that $OQ = v$. Then Q is the image of P.

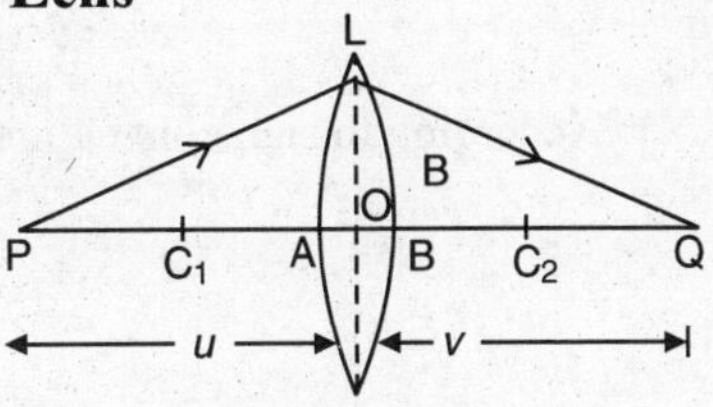

Fig. 19.2.

Since the rays starting from P intersect at Q, the optical paths along the two directions, according to the law of extreme paths, must be the same.

$$\therefore \quad PL + LQ = PA + n.\,AB + BQ = \text{max., min. or constant.} \qquad ...(1)$$

In the triangle PLO, $PL^2 = PO^2 + OL^2 = PO^2\left(1+\frac{OL^2}{OP^2}\right)$

or
$$PL \approx PO + \frac{OL^2}{2OP} \; (\because OL \ll OP) \qquad ...(2)$$

Similarly the triangle OLQ gives
$$QL \approx QO + \frac{OL^2}{2OQ} \qquad ...(3)$$

Substituting these values of PL and QL in Eq. (1),
$$PO + QO + \frac{OL^2}{2}\left(\frac{1}{OP}+\frac{1}{OQ}\right) = PA + n.\,AB + BQ$$

or
$$PO + QO + \frac{1}{2}OL^2\left(\frac{1}{OP}+\frac{1}{OQ}\right) = PO - OA + QO - OB + n\,(AO + OB)$$

or
$$\frac{1}{2}OL^2\left(\frac{1}{OP}+\frac{1}{OQ}\right) = (n-1)\,(AO + OB)$$

$$\therefore \quad \frac{1}{v}+\frac{1}{u} = 2(n-1)\,\frac{AO+BO}{OL^2} \qquad ...(4)$$

For the surface whose radius of curvature is r_1 and centre is C_1,
$$OL^2 = OB(2r_1 - OB)$$
If OB is small as compared to r_1 *i.e.*, if the lens is thin,
$$OL^2 = OB.2r_1 \qquad ...(5)$$
Similarly, for the surface of radius of curvature r_2 and centre C_2,
$$OL^2 = OA.2r_2 \qquad ...(6)$$

$$\therefore \qquad \frac{OB}{OL^2} = \frac{1}{2r_1} \text{ and } \frac{OA}{OL^2} = \frac{1}{2r_2}$$

or

$$\frac{AO + BO}{OL^2} = \frac{1}{2r_1} + \frac{1}{2r_2}$$

Eq. (4) becomes,

$$\frac{1}{v} + \frac{1}{u} = (n-1)\left(\frac{1}{r_1} + \frac{1}{r_2}\right)$$

According to sign convention, v and r_1 are positive and u and r_2 are negative.

$$\therefore \qquad \frac{1}{v} - \frac{1}{u} = (n-1)\left(\frac{1}{r_1} - \frac{1}{r_2}\right)$$

19.3. Ellipsoidal Mirror

Fig. 19.3 depicts a segment of a hollow three-dimensional ellipsoidal mirror. If the source S and the observer, P are at the foci of the ellipsoid, then by definition the length SQP will be constant, regardless of where on the perimeter Q happens to be.

i.e., $\qquad SQ + QP = \text{constant.}$

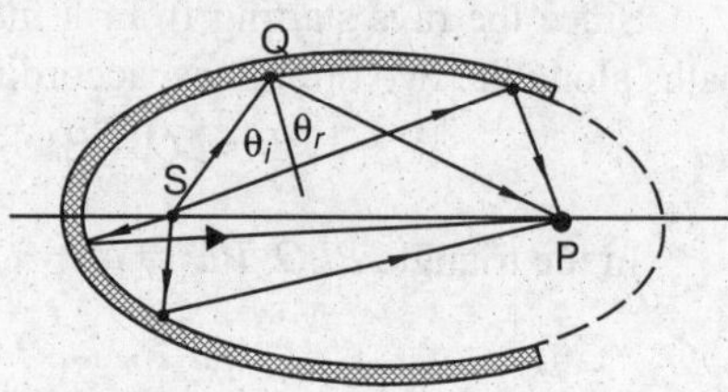

Fig. 19.3.

It is also a geometrical property of the ellipse that $\theta_i = \theta_r$ for any location of Q. All optical paths from S to P via a reflection are therefore precisely equal. None is a minimum, and the *optical path length* is clearly stationary with respect to variations. Rays leaving S and striking the mirror will arrive at the focus P. From another viewpoint we can say that radiant energy emitted by S will be scattered by electrons in the mirrored surface such that the wavelets will substantially reinforce each other only at P, where they have travelled the same distance and have the same phase. The two conjugate points S and P are called aplanatic foci.

Notice that here we have an example where the time taken by the ray is stationary, *i.e.*, it is neither a maximum nor a minimum but has a constant value for all points lying on the mirror. Because of the abovementioned property of elliptical reflectors, they are often used in laser systems. For example, in a ruby laser we have a scheme in which the laser rod and the flash lamp coincide with the focal lines of a cylindrical reflector of elliptical cross section. Such a configuration leads to an efficient transfer of energy from the lamp to the ruby rod.

19.4. Paraboloidal Mirror

As the distance between the foci of the ellipse increases, it approximates more and more to a parabola. If, therefore, it is desired to converge all the rays coming from infinity to one single point, the mirror to be used must has a principal section which is a parabola. The mirror itself is called a paraboloidal mirror (Fig. 19.4). A paraboloid is obtained by rotating a parabola about its axis.

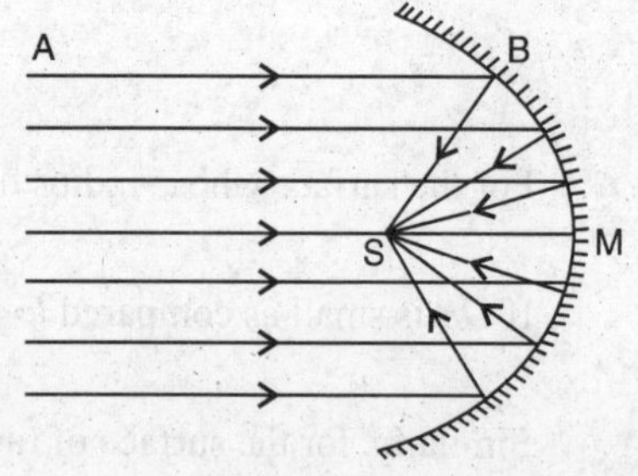

Fig. 19.4.

In a paraboloidal mirror incident rays, parallel to the axis after reflection, are all concentrated at a single point as a consequence of the geometrical property of the parabola *viz.*, the incident ray AB, parallel to the axis and the reflected ray BS are equally inclined to the normal at *B*, wherever the point *B* may be. There is, thus, a complete absence of spherical aberration, no matter how large the aperture of the mirror. This property has commended its use in the construction of high resolving power concave reflectors of the modern telescopes. The Mount Palomar telescope has a concave paraboloidal reflector with 5.08 m (200 inch) aperture.

Antennas (for collecting electromagnetic waves in radio astronomy) and solar collectors are often paraboloidal in shape.

Conversely, a point source of light at the focus of a paraboloidal mirror will send out parallel rays after reflection from the surface of the mirror. It is for this reason that such mirrors are used in automobile headlights and searchlights.

Example 1: *Consider a set of rays, parallel to the axis, and incident on a paraboloidal reflector. Show, by using Fermat's principle, that all the rays will pass through the focus of the paraboloid.*

Solution. Let us find out the ray connecting the focus *S* and an arbitrary point *P* (Fig. 19.5). Let the ray path be *PQ′S*. According to Fermat's principle, the ray path will correspond to a minimum value of $PQ' + Q'S$.

From the point Q' draw a perpendicular, $Q'L'$, on the directrix *ACB*.

From the definition of the parabola,

$$Q'L' = Q'S$$

Fig. 19.5.

$$\therefore \qquad PQ' + Q'S = PQ' + Q'L'$$

Let *L* be the foot of the perpendicular drawn from the point *P* on *AB*. Then, for $PQ' + Q'L'$ to be a minimum, the point *Q* should lie on the straight line *PL*. Thus the actual ray which connects the points *P* and *S* will be $PQ + QS$ where *PQ* is parallel to the axis. Therefore, all rays parallel to the axis of a paraboloidal reflector will pass through the focus *S* after reflection.

19.5. Nodal Slide Method for Determination of Cardinal Points of a Coaxial Lens System

Principle : Properties of nodal and focal points

An incident ray directed towards the first nodal point N_1, after refraction through the system, emerges through the second nodal point N_2 in a direction parallel to the incident ray.

The method of locating the nodal points with the help of the nodal slide involves the following principle.

If a parallel beam of light is incident on a coverging lens system, an image of the object is formed on a screen in its second focal plane. When the system is rotated through a small angle about a vertical axis passing through its second nodal point, the image situated in the second focal plane does not shift laterally and remains stationary.

Let N_1 and N_2 be the nodal points of a co-axial lens system (Fig. 19.6 *a*). Let a beam of light parallel to the principal axis be incident on the system. After refraction through the system, the emergent rays converge to a point

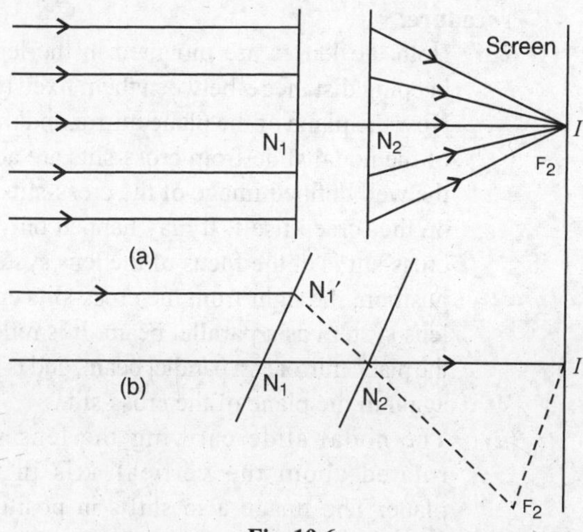

Fig. 19.6.

I which is also the second focal point of the lens system.

Consider now that the system is rotated through a small angle about a vertical axis passing through N_2 in clock-wise direction. N_1' is the new position of N_1 and N_2 remains fixed (Fig. 19.6 *b*). The position of the parallel emergent ray N_2I is unchanged and hence the image remains stationary at I. The point I still lies in the second focal plane though F_2 is now shifted to one side.

Thus it is obvious that if a beam parallel to the principal axis is incident on a lens system, the position of the axis of rotation, for which there is no displacement of the image can be easily located. The inter-section of the principal axis with this position of the axis locates the second nodal point. Since the medium on both sides of the system is air, the second nodal point coincides with second principal point. Hence the distance between the axis of rotation and the screen gives the second focal length.

Experimental arrangement

A source of light, a screen having a cross slit, the nodal slide assembly and a plane mirror are mounted in uprights of a long optical bench (Fig. 19.7). The heights of these uprights are so adjusted that the line joining the centres of each is parallel to the bed of the bench. Light from the lamp passes through the slit and is incident on the lens system. It is rendered parallel and on passing through the lens system it is reflected back by the vertical plane mirror. The reflected light once again passes through the lens system and is brought to a focus in the plane of the slit.

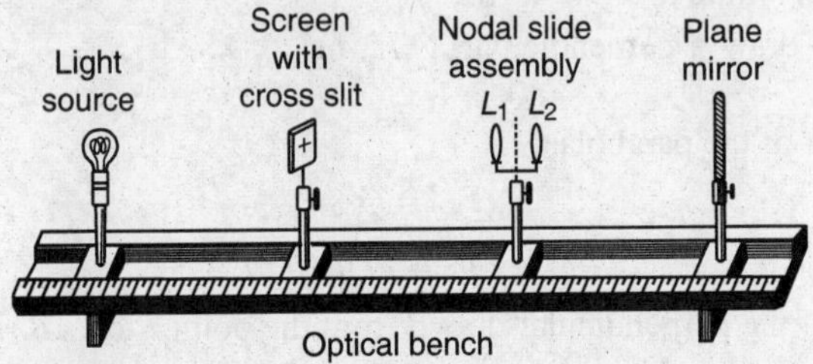

Fig. 19.7.

The nodal slide assembly has an upright which carries a platform with two lens holders (Fig. 19.8). The platform can be made to move forward or backward and rotated in a horizontal plane. The positions of two lenses are noted on a scale on engraved platform. The whole assembly is made to move on an optical bench. The position of the axis of rotation of the assembly is noted on the optical bench.

Procedure:

(*i*) Both the lenses are mounted in the lens holders keeping distance d between them fixed (Fig. 19.9). Now the plane of the plane mirror and the distance of the nodal slide from cross-slits are adjusted till the well defined image of the cross-slits is formed on the screen itself. It may happen only when the cross-slit is at the focus of the lens system. In this position, the light from the cross-slits emerges from the lens system as a parallel beam. It is reflected back from the plane mirror as a parallel beam, and is brought to focus again in the plane of the cross-slit.

(*ii*) The nodal slide carrying the lens system is now rotated about the vertical axis in the horizontal plane. The image also shifts in position. It indicates that the axis of rotation does not pass through the

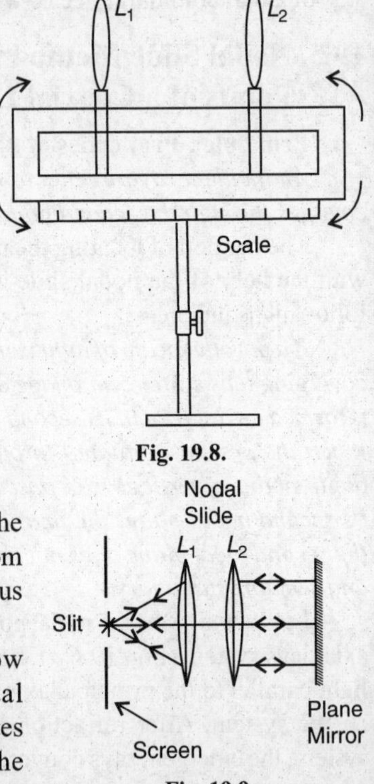

Fig. 19.8.

Fig. 19.9.

second nodal point. Now the bar carrying lens system is shifted back and forth along the bench in the upright till a well-defined image of the cross-slits with no shift is obtained. The axis of rotation, then, passes through the second nodal point and the image of the cross-slits lies in the second focal plane. Thus the positions of both the second nodal plane and the second focal plane can be read on the optical bench. As the medium on either side of the lens system is air, the second nodal point is also the second principal point. The distance between the screen and the axis of rotation for the stationary image is an accurate measure of the first focal length of the lens system.

The nodal slide is rotated by 180° so that L_2 now becomes towards cross-slit. By repeating the experiment, the other focal point and the other nodal (or principal) point are determined. The second value of focal length is found out.

Thus the cardinal points of combination are experimentally located and mean value of focal length of combination is found out.

19.6. Astronomical Telescope

Principle. A telescope is an instrument for viewing distant objects. The angular size of distant object at an ordinary eye is very small and the objective lens of a telescope is used to increase the angular size. The objective forms a real, inverted and diminished image of a distant object. This image is seen by an eyepiece and generally the final image is formed at infinity.

Construction. It consists of two lenses both of which are convex (Fig. 19.10.). The objective is a convex lens of large focal length and large surface area. The eyepiece is a convex lens of small aperture and small focal length. They are mounted on the same axis at the opposite ends of a metallic tube. The eyepiece can be slided in or out as desired.

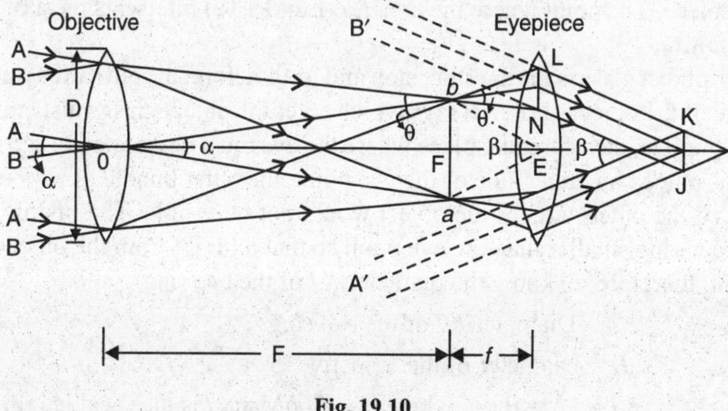

Fig. 19.10.

Normal adjustment. In the *normal adjustment of the telescope,* the final image is formed at infinity and can be seen without strain by the unaccommodated eye. In this adjustment, the inverted image of the object formed by the object glass must lie at the focus of the eyepiece. Thus the distance between the eyepiece and objective is equal to the sum of their focal lengths.

Working. Fig. 19.10 shows the optical path of two bundles of rays through the telescope. The rays *AAA* coming from one part and *BBB* from the other part of a distant object form the inverted real image *ba* in the second focal plane of the **objective.** If this is also the first focal plane of the eye lens, these rays spread out from *ab* and eventually on emergence from the eye lens come out as two bundles of parallel rays. When these emergent bundles fall on the eye, the final inverted and magnified image is seen at *A′B′*.

Magnifying Power of a Telescope:

The magnifying power of the telescope is defined as the *ratio of the angle subtended by the final virtual image formed at infinity at the eye to the angle subtended by the distant object at the eye.*

$$\text{Magnifying power} = \frac{\text{angle subtended at the eye by the final image}}{\text{angle subtended at the eye by the object}}$$

Let the object subtend an angle α at the objective. Then it would subtend approximately the same angle at the eye, as the distance between the eye and the objective is negligible compared to the distance of the object from the telescope.

Let the angle subtended at the eye by the final image be β.

By definition, magnifying power $(M) = \beta/\alpha$....(1)

Now, $\beta = \angle bEa = \dfrac{ab}{f}$, where f is the focal length of eyepiece.

$\alpha = \angle AOB = \angle bOa = \dfrac{ab}{F}$, where F is the focal length of the objective.

$$\therefore \quad M = \frac{\beta}{\alpha} = \frac{ab}{f} \div \frac{ab}{F} = \frac{F}{f} = \frac{\text{Focal length of the objective}}{\text{Focal lenght of the eyepiece}} \quad ...(2)$$

Hence, the *magnifying power of a telescope in normal adjustment is the ratio between the focal length of the objective and that of the eyepiece.*

The exit pupil and the entrance pupil

All rays which pass through both the lenses pass through the area KJ which is in fact the image of the aperture of the object glass in the eye lens. This area is called the **exit pupil** or the **eye ring** or the **Ramsden's circle**. This should be the position of the eye and it is the real image of the objective formed by the eyepiece. The point where the area KJ (Fig. 19.10) intersects the axis of the telescope is called the **eye point.**

In telescopes, object glass is the aperture stop and is, therefore, the **entrance pupil**. The eye ring is in fact the image of the object glass. All rays which enter the object glass must emerge through the eye ring. The exit pupil is, therefore, the place where the eye must be situated to see the largest range of objects. If eye is placed on either side of the eye point, the outer bundle of rays would not fall on the eye at all and so the outer parts of the object would not be visible. The instrument is generally designed with its eye ring smaller than the eye pupil so that all light from the object glass enters the eye. It is important, therefore, to know the diameter KJ of the eye ring.

Let d = Diameter KJ of the *eye ring;*

D = Diameter of the objective

$KJ = LN = \theta \times f$, where $\theta = \angle LbN$ and f is the focal length of the eyepiece.

$\theta = D/F$, where F is the focal length of the objective.

$$\therefore \quad KJ = d = \frac{D}{F} \times f = \frac{D}{F/f} = \frac{D}{M}. \quad ...(1)$$

That is, the diameter of the eye ring varies *directly* as the diameter of the objective and *inversely* as the magnifying power.

The diameter of pupil of the eye is 2.54 mm. The size of the exit pupil should be less than 2.54 mm but it should not be less than 0.85 mm, for below it the image will not be bright enough.

In Yerkes' refracting telescope, the diameter D of the aperture of the objective is 1016 mm.

If $\dfrac{Df}{F} \not> 2.54$ mm, then, $\dfrac{F}{f} \not< 400$.

If $\frac{Df}{F} \nless 0.85$ mm, then, $\frac{F}{f} \ngtr 1200$.

Therefore, the minimum and maximum magnifying powers are 400 and 1200 respectively.

Example 1: *A spectrometer telescope objective of diameter 3.5 cm and focal length 35 cm is fitted with an eyepiece of focal length 4 cm. Find the magnifying power of the telescope and the diameter of Ramsden's circle.*

Solution. Here, $F = 35$ cm, $f = 4$ cm, $D = 3.5$ cm. $M = ?$ $d = ?$

The magnifying power, $M = \frac{F}{f} = \frac{35}{4} = 8.75.$

The diameter of Ramsden's circle, $d = \frac{D}{F/f} = \frac{D}{M} = \frac{3.5}{8.75} = \mathbf{0.4\ cm.}$

Example 2: *A refracting telescope has an objective of diameter 100 cm and a focal length of 20 m. What magnifying power should be used to give an exit pupil of 2.0 mm matching the entrance pupil of the eye? (a) What is the focal length of the eyepiece required for this magnification?*

Solution. Here, $D = 100$ cm, $F = 20$ m, diameter of exit pupil $= d = 2$ mm.

$$\text{Magnification} = \frac{D}{d} = \frac{1000 \text{ mm}}{2 \text{ mm}} = 500$$

$$M = \frac{F}{f}$$

$\therefore$ Focal length of the eyepiece $= f = \frac{F}{M} = \frac{20 \text{ m}}{500} = 0.04$ m

19.7. Astronomical Telescope with the Image at any Desired Position

When a telescope is employed to see near objects as in our laboratory experiments, then it is not in normal adjustment . The final image may be formed at any desired position convenient to the observer.

Fig. 19.11 shows the course of rays through an astronomical telescope with the image in any desired position.

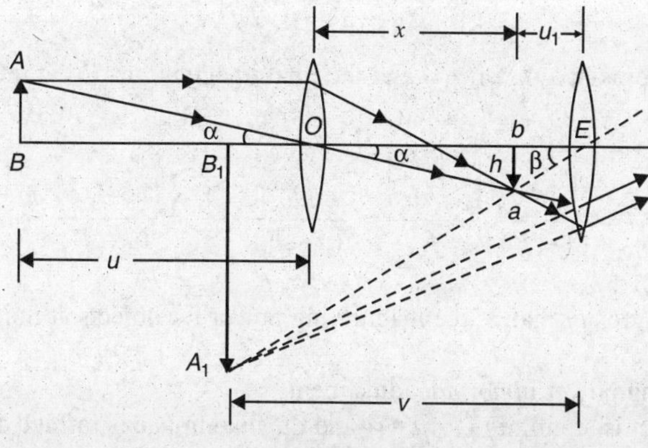

Fig. 19.11.

O and E represent the objective and the eye lens respectively, AB is the object at a distance u from the objective, ba its image in the objective. B_1A_1 is the final image as seen through the system at a distance v, say, from the eye lens.

The magnifying power is given by

$$m = \frac{\beta}{\alpha} = \frac{\text{angle subtended by the image at the eye}}{\text{angle subtended by the object seen directly by the eye}}$$

Angle subtended by the image at the eye $(\beta) = \frac{h}{u_1}$.

Angle subtended by the object $(\alpha) = \frac{h}{x}$.

$$\therefore \qquad M = \frac{\beta}{\alpha} = \frac{h}{u_1} \div \frac{h}{x} = \frac{x}{u_1} \qquad ...(1)$$

Here, x is obtained by the relation (using numerical values),

$$\frac{1}{x} + \frac{1}{u} = \frac{1}{F} \text{ for the objective.}$$

$$\therefore \qquad x = \frac{uF}{(u - F)} \qquad ...(2)$$

u_1 is obtained by the relation, $-\frac{1}{v} + \frac{1}{u_1} = \frac{1}{f}$ for the eye lens.

$$\therefore \qquad u_1 = \frac{vf}{(v + f)} \qquad ...(3)$$

Substituting these values of x and u_1 in Eq. (1), we get

$$M = \frac{uF}{(u - F)} \times \frac{(v + f)}{vf} \qquad ...(4)$$

Special cases:

1. When the object is at a distance u and the image is formed at the least distance of distinct vision *i.e.*, $v = D$, then Eq. (4) gives

$$M = \frac{uF}{(u - F)} \times \frac{(D + f)}{Df} \qquad ...(5)$$

2. When the object is at a distance u and the final image is formed at infinity *i.e.*, $v = \infty$, then Eq. (4) gives

$$M = \frac{uF}{(u - F)} \cdot \frac{\left(1 + \frac{f}{v}\right)}{f} = \frac{uF}{(u - F)} \times \frac{1}{f} = \frac{u}{(u - F)} \cdot \frac{F}{f} \qquad ...(6)$$

Since $\frac{u}{(u - F)}$ is greater than 1, the magnifying power for objects at finite distances is greater than F/f, the magnifying power in *normal* adjustment.

3. When the object is at infinity *i.e.*, $u = \infty$ and the final image is formed at the least distance of distinct vision *i.e.*, $v = D$, then Eq. (4) gives

$$M = \frac{F}{\left(1 - \frac{F}{u}\right)} \times \frac{(D + f)}{Df} = \frac{F}{f} \cdot \frac{(D + f)}{D} \qquad ...(7)$$

4. When both u and v are infinite, the telescope is in normal adjustment. Then Eq. (4) gives

$$M = \frac{F}{\left(1 - \frac{F}{u}\right)} \times \frac{\left(1 + \frac{f}{v}\right)}{f} = \frac{F}{f} \qquad \text{...(8)}$$

19.8. To Derive the Relation $\frac{n_2}{v} - \frac{n_1}{u} = \frac{n_2 - n_1}{R}$ for Refraction at a Convex Spherical Surface by Matrix Method

Consider a convex spherical surface of radius of curvature R separating two media of refractive indices n_1 and n_2 (Fig. 19.12). $n_2 > n_1$. Consider a point object O placed on the principal axis. A ray starting from O is incident on the surface at A making an angle i with the normal CAN. It is refracted along AI making an angle of refraction r with the normal.

Fig. 19.12.

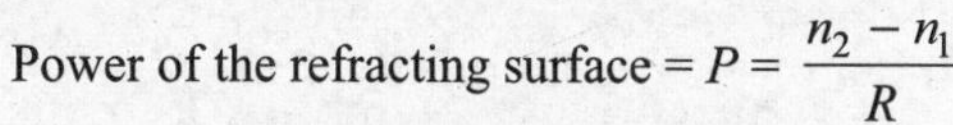

Power of the refracting surface $= P = \frac{n_2 - n_1}{R}$

Distance of the object from the pole $= u$

Distance of the image from the pole $= v$

The system matrix of the system can be written as

$$S = \begin{pmatrix} 1 & 0 \\ v/n_2 & 1 \end{pmatrix}\begin{pmatrix} 1 & -P \\ 0 & 1 \end{pmatrix}\begin{pmatrix} 1 & 0 \\ u/n_1 & 1 \end{pmatrix} = \begin{pmatrix} 1 - \frac{u}{n_1}P & -P \\ \frac{v}{n_2}\left(1 - \frac{u}{n_1}P\right) + \frac{u}{n_1} & -\frac{v}{n_2}P + 1 \end{pmatrix} \qquad \text{...(1)}$$

$$\therefore \quad \begin{pmatrix} \lambda_2 \\ x_2 \end{pmatrix} = \begin{pmatrix} 1 - \frac{u}{n_1}P & -P \\ \frac{v}{n_2}\left(1 - \frac{u}{n_1}P\right) + \frac{u}{n_1} & -\frac{v}{n_2}P + 1 \end{pmatrix}\begin{pmatrix} \lambda_1 \\ x_1 \end{pmatrix}$$

or

$$x_2 = \left(\frac{v}{n_2}\left(1 - \frac{u}{n_1}P\right) + \frac{u}{n_1}\right)\lambda_1 + \left(1 - \frac{v}{n_2}P\right)x_1 \qquad \text{...(2)}$$

For paraxial rays emanating from the object O, $x_1 = 0$.

The image plane is determined for the condition $x_2 = 0$.

$$\therefore \quad \frac{v}{n_2}\left(1 - \frac{u}{n_1}P\right) + \frac{u}{n_1} = 0$$

or

$$\frac{v}{n_2} + \frac{u}{n_1} = \frac{uv}{n_1\, n_2}P = \frac{uv}{n_1 n_2}\left(\frac{n_2 - n_1}{R}\right)$$

or

$$\frac{n_2}{v} + \frac{n_1}{u} = \frac{n_2 - n_1}{R}$$

According to sign convention, u is negative and both v and R are positive.

$$\therefore \qquad \frac{n_2}{v} - \frac{n_1}{u} = \frac{n_2 - n_1}{R} \qquad ...(3)$$

19.9. To Derive an Expression for the Focal Length of a Thin Convex Lens by Matrix Method

Consider a thin convex lens L of refractive index n placed in air (Fig. 19.13). R_1 and R_2 are the radii of curvature of the first and the second surfaces respectively. O is a point object placed on the axis of the lens at a distance D_1 from the optical centre.

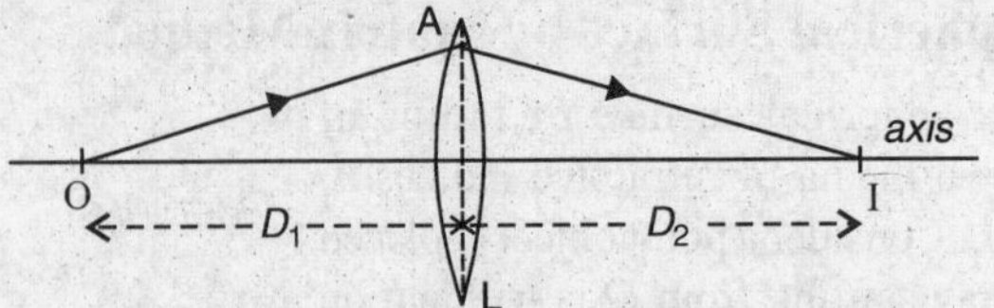

Fig. 19.13.

A ray of light starting from O strikes the lens at A and after refraction moves along AI. I is the image of O.

Distance of the image I from the lens = D_2

According to sign convention, R_1 is positive, R_2 is negative, D_1 is negative and D_2 is positive.

Power of the first surface = $P_1 = \dfrac{(n-1)}{R_1}$

Power of the second surface = $P_2 = \dfrac{(n-1)}{-R_2}$

$\therefore$ System matrix for the thin lens is

$$S = \begin{pmatrix} b & -a \\ -d & c \end{pmatrix} = \begin{pmatrix} 1 & -P_2 \\ 0 & 1 \end{pmatrix} \begin{pmatrix} 1 & -P_1 \\ 0 & 1 \end{pmatrix}$$

or

$$\begin{pmatrix} b & -a \\ -d & c \end{pmatrix} = \begin{pmatrix} 1 & -P_1 - P_2 \\ 0 & 1 \end{pmatrix}$$

$$\therefore \qquad a = P_1 + P_2,\ b = 1,\ c = 1 \text{ and } d = 0 \qquad ...(1)$$

For image plane, $bD_2 + aD_1D_2 - cD_1 - d = 0$

Substituting the values of a, b, c and d from Eq. (1) in it, we get

$$D_2 + (P_1 + P_2)\, D_1\, D_2 - D_1 = 0$$

Substituting the values of P_1 and P_2, we get

$$D_2 + \left[\frac{(n-1)}{R_1} - \frac{(n-1)}{R_2}\right] D_1 D_2 - D_1 = 0$$

Dividing by $D_1\, D_2$,

$$\frac{1}{D_1} - \frac{1}{D_2} = -(n-1)\left(\frac{1}{R_1} - \frac{1}{R_2}\right)$$

or

$$\frac{1}{D_2} - \frac{1}{D_1} = (n-1)\left(\frac{1}{R_1} - \frac{1}{R_2}\right) = \frac{1}{f} \qquad ...(2)$$

19.10. Huygens EyePiece

Construction: It consists of two plano-convex lenses of focal lengths $3f$ (field lens) and f (eye lens), placed a distance $2f$ apart (Fig. 19.14). The lenses are made up of same kind of glass with their curved faces facing incoming light and are coaxially mounted. The eyepiece satisfies the conditions of minimum spherical and chromatic aberrations.

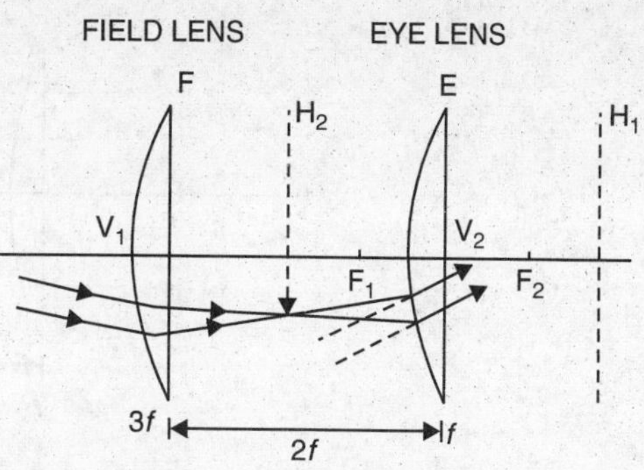

Fig. 19.14.

Determination of position of cardinal points of the eye piece

Let S be system matrix between V_1 and V_2 planes. It can be written as

$$S = \underset{\text{Refraction at } V_2}{\begin{pmatrix} 1 & -1/f \\ 0 & 1 \end{pmatrix}} \underset{\text{Translation } V_1 \text{ to } V_2}{\begin{pmatrix} 1 & 0 \\ 2f & 1 \end{pmatrix}} \underset{\text{Refraction at } V_1}{\begin{pmatrix} 1 & -1/3f \\ 0 & 1 \end{pmatrix}}$$

$$\begin{pmatrix} b & -a \\ -d & c \end{pmatrix} = \begin{pmatrix} -1 & -2/3f \\ 2f & \frac{1}{3} \end{pmatrix}$$

Thus, $-a = -\frac{2}{3f}$ or $a = \frac{2}{3f}$; $b = -1$; $c = \frac{1}{3}$; $-d = 2f$ or $d = -2f$

Focal length of the equivalent lens $= F = \frac{1}{a} = \frac{3f}{2}$

Position of first principal plane H_1 is given by

$$d_{H1} = \frac{1-b}{a} = \frac{1-(-1)}{(2/3f)} = 3f$$

Thus the first principal plane is at a distance $3f$ from the field lens towards right.

Position of second principal plane H_2 is given by

$$d_{H2} = \frac{c-1}{a} = \frac{\frac{1}{3}-1}{2/3f} = -f$$

The second principal plane is at a distance f from the eye lens towards left.

The first focal point F_1 is at a distance $3f/2$ towards left of H_1.

The second principal focal point F_2 is at a distance $3f/2$ away from the plane H_2 and is towards right.

Since the system is in air, the nodal points coincide with the principal points.

The position of the principal points H_1 and H_2 and the principal foci F_1 and F_2 are shown in Fig. 19.15.

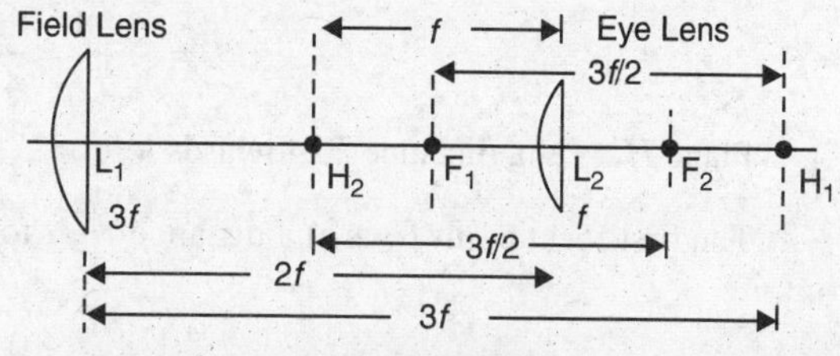

Fig. 19.15

19.11. Ramsden's Eyepiece

Construction: It consists of two plano convex lenses each of focal length f (Fig. 19.16.). The distance between them is $(2/3)f$. The two lenses are made up of same kind of glass. They are co-axially mounted with their curved faces facing each other and plane face of field lens towards incident light.

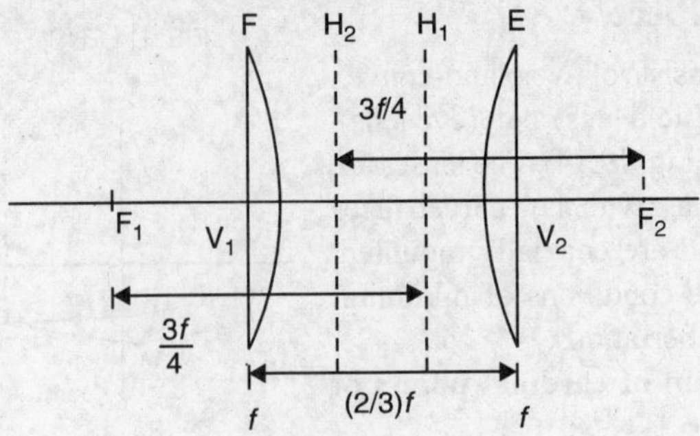

Fig. 19.16.

To determine cardinal points of the eyepiece

The system matrix S between V_1 and V_2 planes can be written as

$$S = \begin{pmatrix} b & -a \\ -d & c \end{pmatrix} = \overset{\text{Refraction at } V_2}{\begin{pmatrix} 1 & -1/f \\ 0 & 1 \end{pmatrix}} \overset{\text{Translation } V_1 \text{ to } V_2}{\begin{pmatrix} 1 & 0 \\ \frac{2}{3}f & 1 \end{pmatrix}} \overset{\text{Refraction at } V_1}{\begin{pmatrix} 1 & -1/f \\ 0 & 1 \end{pmatrix}}$$

$$\begin{pmatrix} b & -a \\ -d & c \end{pmatrix} = \begin{pmatrix} \frac{1}{3} & -4/3f \\ \frac{2}{3}f & \frac{1}{3} \end{pmatrix}$$

Thus, $a = \dfrac{4}{3f}$; $b = \dfrac{1}{3}$; $c = \dfrac{1}{3}$; $d = -\left(\dfrac{2}{3}\right)f$.

Focal length of equivalent lens $F = \dfrac{1}{a} = \dfrac{3}{4}f$

Position of first principal plane H_1 is given by

$$d_{H1} = \frac{1-b}{a} = \frac{1-\frac{1}{3}}{4/3f} = \frac{f}{2}$$

Thus plane H_1 is at a distance $f/2$ towards right of V_1.
Position of second principal plane H_2 is given by

$$d_{H2} = \frac{c-1}{a} = \frac{\frac{1}{3}-1}{4/3f} = -\frac{f}{2}$$

Plane H_2 is at a distance $f/2$ towards left of V_2.

The first focal point F_1 is at a distance $\dfrac{3}{4}f$ towards left of H_1.

The second focal point F_2 is at a distance $\dfrac{3}{4}f$ towards right of H_2.

Since the system is in air, the nodal points coincide with principal points.

Position of crosswire and course of rays

The objective forms the image in front of field lens which then is projected to infinity by the eye-piece (Fig. 19.17). Here crosswire can be used where the image is formed. Both image and cross-wire are equally magnified by the eyepiece. Hence no error in the measurement shall be introduced. Such eyepieces are employed in telescopes and travelling microscopes.

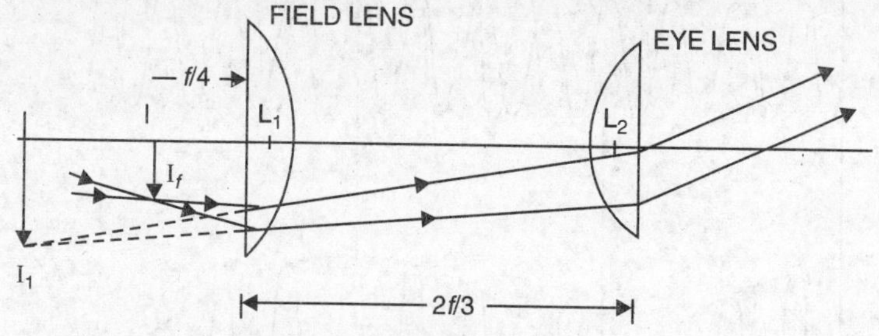

Fig. 19.17.

19.12. Wave Velocity and Group Velocity

The phase velocity or wave velocity v_p of a monochromatic wave is the velocity with which a definite phase of the wave, such as its crest or trough, is propagated in a medium.

An individual sine wave propagates through a medium with a definite velocity. Hence the phases (crests and troughs) of the wave advance with the same velocity. The velocity of propagation of the phases is called **wave** or **phase velocity.**

But when a large number of waves start from any source of light, then in general, the successive waves are found to have different λ and to travel with different velocities. *These waves may be supposed to travel in the form of a wave group, made by superposition of the individual waves. In such a case, the rate of travel of resultant disturbance (or rate of flow of energy) is called group velocity v_g.* Thus, *the group velocity is the velocity with which the energy in the group is transmitted.*

Consider a wavetrain having two wavelengths [Fig. 19.18 (*a*)]

The dotted curve represents the longer wave travelling faster than the shorter. At the instant considered, the two are in phase at the point *A* giving rise to a maximum amplitude shown in Fig. 19.18. (*b*).

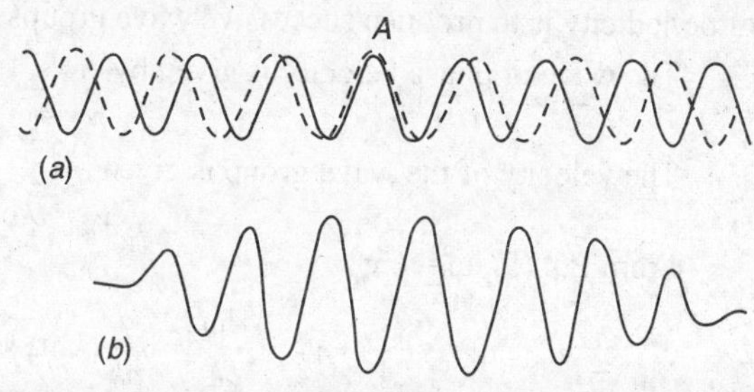

Fig. 19.18.

Some time later the amplitude will build up a little to the left of the point *A*. In other words, the group will advance less rapidly than the separate trains. The energy belongs to the group rather than to the waves and the observed velocity will be that of the group.

Relation between the Group Velocity (v_g) and the Phase Velocity (v_p)

Consider the superposition of two sinusoidal waves of equal amplitude *a* (Fig. 19.19 *a*)

Their separate displacements are given by

$$y_1 = a \sin(\omega t + kx) \qquad ...(1)$$

and

$$y_2 = a \sin[(\omega + d\omega)t + (k + dk)x] \qquad ...(2)$$

The two waves have slightly different angular frequencies ω and $\omega + d\omega$ and propagation constants k and $k + dk$.

The resultant displacement at any time t and at any position x is given by

$$y = y_1 + y_2 = a \sin(\omega t + kx) + a \sin[(\omega + d\omega)t + (k + dk)x]$$

$$y = 2a \sin \frac{1}{2}[(2\omega + d\omega)t + (2k + dk)x] \times \cos \frac{1}{2}(d\omega\, t - dk\, x) \qquad ...(3)$$

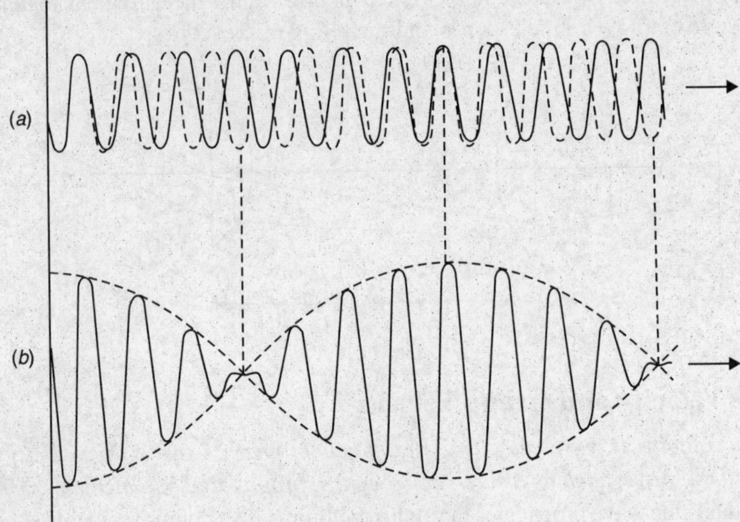

Fig. 19. 19.

Since $d\omega$ and dk are small as compared to ω and k respectively, we can write

$$2\omega + d\omega = 2\omega \text{ and } 2k + dk = 2k$$

$$\therefore \qquad y = 2a\left[\sin(\omega t + kx)\cos\left(\frac{d\omega\, t}{2} - \frac{dk\, x}{2}\right)\right] \qquad ...(4)$$

Thus the resultant consists of a vibration of angular frequency ω and propagation constant k superimposed by a periodicity of angular frequency $d\omega/2$ and propagation constant $dk/2$. The effect of periodicity is to produce successive wave groups as shown in Fig. 19.19(b).

The wave or phase velocity is given by

$$v_p = \omega/k \qquad ...(5)$$

The velocity of the wave group is given by

$$v_g = d\omega/dk \qquad ...(6)$$

From Eq. (5), $\omega = k\, v_p$

$$\therefore \qquad v_g = \frac{d\omega}{dk} = \frac{d}{dk}\left(k\, v_p\right) = v_p + k \times \frac{d\, v_p}{dk}$$

Now, $k = \dfrac{2\pi}{\lambda}$. $\dfrac{dk}{d\lambda} = -\dfrac{2\pi}{\lambda^2}$. $\quad \therefore\ dk = -\dfrac{2\pi}{\lambda^2}d\lambda$

$$v_g = v_p + \frac{2\pi}{\lambda} \times \frac{d\, v_p}{(-2\pi\, d\lambda/\lambda^2)}$$

$$\therefore \qquad v_g = v_p - \lambda\frac{d\, v_p}{d\lambda} \qquad ...(7)$$

This is the relation between group velocity (v_g) and phase velocity (v_p) in a dispersive medium. In free space, phase velocity and group velocity are the same. In matter they are different.

19.13. Kerr Cell Method for Determining the Velocity of Light

Principle. This method is based on the *Kerr electro-optical effect, viz.,* that *certain transparent substances become doubly refracting when placed in a strong electric field.* Making use of this effect, an *electro-optical shutter* is devised in the form of a Kerr cell.

Action of Kerr cell as an electro-optic shutter: A Kerr cell is a glass vessel containing a suitable insulating liquid like nitrobenzene, with two metal electrodes immersed in it. When a high voltage is applied across the electrodes, the cell acts as a uni-axial crystal.

Fig. 19.20.

This cell is placed between two crossed nicols N_1 and N_2. N_1 and N_2 act as a polariser and an analyser respectively [Fig. 19.20]. When the electric field is off, no light is transmitted by the analyser N_2. But when the electric field is switched on, the liquid becomes doubly refracting and some light is transmitted by the analyser N_2. If a high frequency alternating voltage is applied to the cell, it acts as an *electro-optic shutter. It allows the light to pass through when the field is maximum, and stops it when the field is zero.* Thus it is possible to interrupt the light at the rate of several million times per second.

Experimental Arrangement

The experimental arrangement is shown in Fig. 19.21.

S is a source of light, K_1 and K_2 are two Kerr cells; N_1, N_2 and N_3 are three nicol prisms, *M* is a plane mirror. Light from a source *S* is converged by lens L_1 towards the middle of the Kerr cell K_1. Light after coming out of N_2 is again rendered parallel by lens L_2. After reflection at mirror *M*, the beam is converged by lens L_3 at the centre of Kerr cell K_2. Finally the beam passes through the nicol prism N_3 and can be viewed by the eye. Here it should be remembered that the light coming from N_1 will be transmitted by N_2 only when a potential is applied to the plates of Kerr cell K_1 because the two nicols N_1 and N_2 are in crossed position. Similarly, light coming from N_2 will be transmitted through N_3 only when a potential is applied to the plates of K_2 because the two nicols N_2 and N_3 are in crossed position.

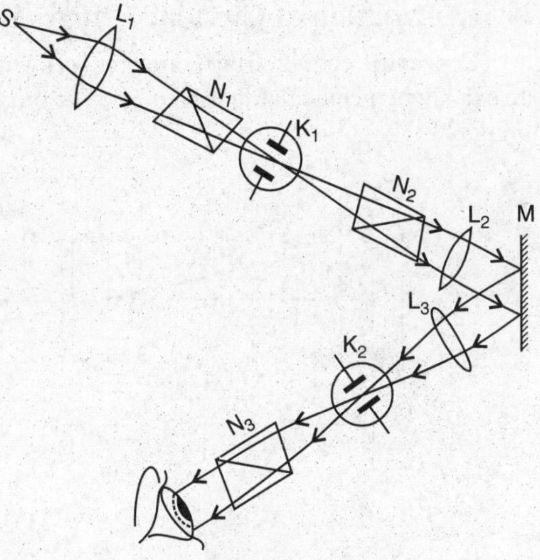

Fig. 19.21.

Procedure. An alternating high frequency voltage is applied to both Kerr cells K_1 and K_2 so that it attains its maximum and minimum values in both simultaneously. Now the two cells will act as two shutters which open and close simultaneously. The distance between K_1 and K_2 is so adjusted that when light from N_1 falls on K_1, the voltage across the electrodes in both K_1 and K_2 is the maximum; So a flash of maximum intensity passes through N_2. But by the time it arrives at K_2, the voltage falls to the minimum. Hence no light can pass through N_3 and an *eclipse or extinction* occurs. This will happen when the time taken by light to travel from K_1 to K_2 is $T/4$. Here, T is the time-period of the alternating voltage applied.

Let d = distance between the two cells and
n = frequency of the high frequency voltage. Then,

$$c = \frac{d}{T/4} = \frac{d}{1/4n} = 4\,nd.$$

The mean value of the velocity of light is found to be 2,99778 × $10^3 \pm 20 \times 10^3$ m/s.

Advantages: (*i*) This experiment can be performed in a laboratory.

(*ii*) The frequency of the high frequency oscillator is known accurately.

CHAPTER

20

INTERFERENCE AND DIFFRACTION OF LIGHT

20.1. Formation of Circular Fringes in Michelson's Interferometer

Concentric circular fringes are produced with monochromatic light when the mirrors M_1 and M_2 are exactly perpendicular to each other. The origin of these fringes may be explained with reference to Fig. 20.1.

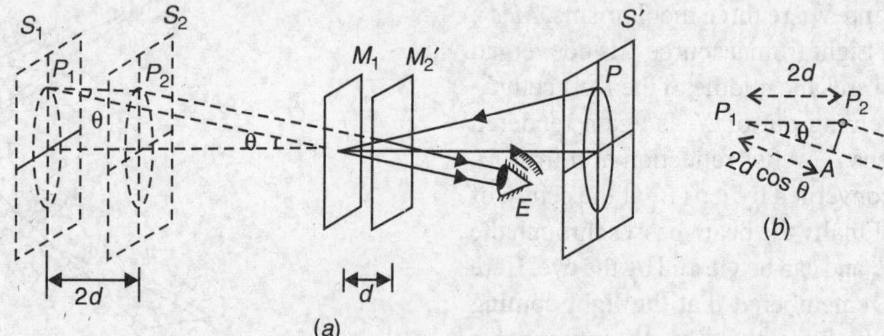

Fig. 20.1.

The real source is replaced by its image S' formed in the plate G_1. The image M_2' of M_2 in G_1 is exactly parallel to M_1. S_1 and S_2 are the images of S' in M_1 and M_2' respectively. If the separation between M_1 and M_2' is d, the separation between S_1 and S_2 is $2d$. The interference pattern is due to the interference between rays reflected from the two surfaces of the air film contained in between M_1 and M_2'.

Light starting from a point P in S' appears to come from P_1 and P_2 in S_1 and S_2. Draw P_2A perpendicular to P_1E [Fig.20.1(*b*)].

Path difference Δ between the rays coming from P_1 and P_2 is P_1A.

$$\therefore \quad \Delta = P_1A = P_1P_2 \cos\theta = 2d\cos\theta \quad ...(1)$$

For the angle θ satisfying the relation $2d\cos\theta = n\lambda$, the rays would reinforce each other to produce a maximum. Therefore for a bright fringe

$$2d\cos\theta = n\lambda \quad ...(2)$$

Here n is an integer. For a particular fringe n is constant. Thus for a constant n, λ and d, cos θ has a constant value. Therefore, the fringe will have a circular shape having its centre on the foot of perpendicular drawn from E upon the mirrors.

When viewed along the normal (θ = 0) to the mirrors, Eq.(2) can be written as

$$2d = m\lambda \quad ...(3)$$

Here m is an integer.

Subtracting Eq.(2) from Eq.(3), we get

$$2d - 2d\cos\theta = (m - n)\lambda$$

or $$2d(1 - \cos\theta) = k\lambda$$

Here $k = m - n$ is also an integer.

or $$(1 - \cos\theta) = \frac{k\lambda}{2d} \qquad ...(4)$$

Now $$\cos\theta = 1 - \frac{\theta^2}{2!} + \frac{\theta^4}{4!} - ....$$

$$(1 - \cos\theta) = 1 - \left(1 - \frac{\theta^2}{2!} + \frac{\theta^4}{4!} - .....\right) = \frac{\theta^2}{2}$$

(Neglecting $\frac{\theta^4}{4!}$ and higher powers of θ.)

Eq. (4) becomes

$$\frac{\theta^2}{2} = \frac{k\lambda}{2d}$$

or $$\theta^2 = \frac{k\lambda}{d}$$

or $$\theta \propto \sqrt{k} \qquad ...(5)$$

If $\theta_1, \theta_2, \theta_3$... are the angles and r_1, r_2, r_3... the radii of the first, second, third... rings respectively, then

$$\theta_1 = \frac{r_1}{x}, \theta_2 = \frac{r_2}{x}, \theta_3 = \frac{r_3}{x} ...$$

Here x is the distance of M_2' from S'. As x is constant $r \propto \theta$.

Thus the radii of the rings are proportional to θ.

Since by Eq. (5) θ is proportional to the square root of natural numbers we conclude that the radii of circular fringes are proportional to the square roots of the natural numbers.

Keeping M_2 fixed, if M_1 is moved backwards or forwards by half a wavelength, the change in the optical path will be λ and a circular fringe would appear or disappear at the centre of the field of view.

However, if M_1 and M_2' are exactly coincident, the whole field of view will be dark because under these conditions the path difference will be zero for all angles of incidence.

Fringes of the type described in this section, where parallel beams produce interference with a phase difference determined by the angle of inclination θ, are referred to as **fringes of equal inclination.**

Example 1. *If the path difference between the mirrors of a Michelson's interferometer is 5 mm, what will be the angular radius of the 8th bright fringe in the circular pattern of the fringes obtained with green light of wavelength 5×10^{-7}m?*

Solution. Here, $k = 8$, $d = 5$ mm $= 5 \times 10^{-3}$ m and $\lambda = 5 \times 10^{-7}$ m.

$$1 - \cos\theta = \frac{k\lambda}{2d} \text{ or } \frac{\theta^2}{2} = \frac{k\lambda}{2d}$$

$$\theta = \sqrt{\frac{k\lambda}{d}} = \sqrt{\frac{8 \times (5 \times 10^{-7})}{(5 \times 10^{-3})}} = 0.02828 \text{ radian} = \mathbf{16.9°}.$$

20.2. Adjustment of Michelson Interferometer for Circular Fringes

Michelson interferometer is said to be in normal adjustment when the lightly silvered surface of glass plate G_1 accurately bisects the angle between the highly silvered front surfaces of mirrors M_1 and M_2, and normals to their surfaces are exactly co-planar. Under this condition, the image M_2' of M_2 in G_1 would be exactly parallel to the real mirror M_1. When this is achieved, concentric circular fringes can be seen in the field of view.

To obtain circular fringes, the Michelson interferometer is adjusted as follows:

(*i*) The distances of the silvered surfaces M_1 and M_2 from the silvered plane of G_1 are adjusted nearly the same within a few millimetres.

(*ii*) Place a metallic screen with a fine pin hole between G_1 and the monochromatic light source. On looking along EG_1, two pairs of images are seen (Fig. 20.2).

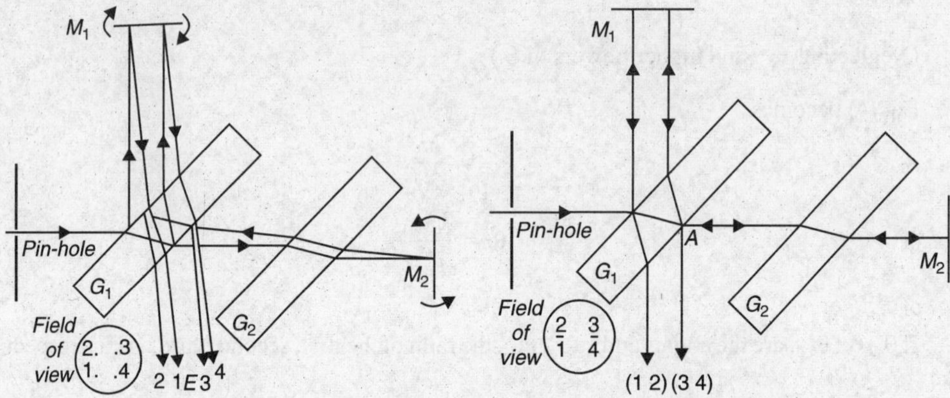

Fig. 20.2. Fig. 20.3.

One pair, 1 and 2, is formed due to partial reflections at the unsilvered surface of G_1 and reflections at M_1 and M_2 respectively. The second pair of images, 3 and 4, is formed due to partial reflections at the lightly silvered surface of G_1 and reflections at M_1 and M_2 respectively.

(*iii*) Adjust the tilt screws of Mirrors M_1 and M_2, till 2 and 3 coincide with 1 and 4 respectively (Fig.20.3). On removing the screen, fringes should appear. By a further finer tilt of mirrors so as to continously increase the width of fringes, concentric circular fringes can be brought into view. But the centre of fringes may be outside the field of view. The angular setting of M_2 is now slightly altered with the help of special screws until the centre of circular fringes is in the centre of the field of view. Now M_1 is slightly displaced so as to obtain large number of concentric fringes in the field of view.

(*iv*) The perfectness of this adjusment may be tested by observing the fringes while the eye is moved vertically or sideways. If the circular fringes do not expand or contract as the eye is thus moved, we conclude that the fringe system is situated at infinity. If otherwise, finer tilt should be given to M_2 until the adjusment is perfect.

20.3. Fraunhofer Diffraction at a Single Slit

Method I

A monochromatic light of wavelength λ is incident normally on the slit of width b (Fig. 20.4). We wish to calculate the intensity distribution on the focal plane of the lens L.

Assume that the slit consists of a large number of equally spaced point sources.

Let the point sources be at $A_1, A_2, A_3, \ldots A_n$. The number of point sources is n.

Distance between two consecutive points = Δ

$$\therefore \qquad b = (n-1)\Delta \qquad \ldots(1)$$

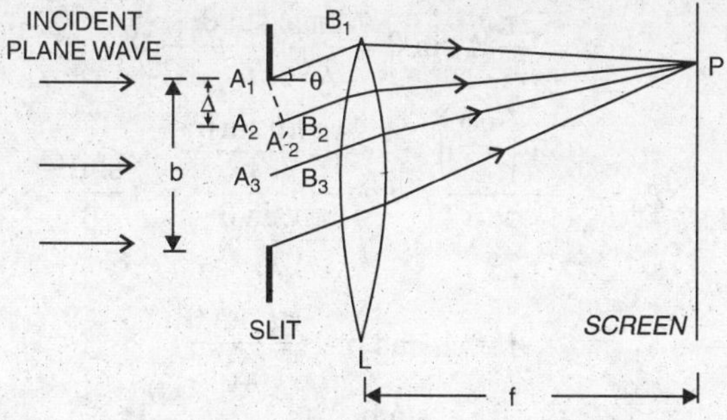

Fig. 20.4.

Each point on the slit is a source of Huygens' secondary wavelets which interfere with the wavelets emanating from other points. P is an arbitrary point on the focal plane of the lens. It receives parallel rays making an angle θ with the normal to the slit.

We will now calculate the resultant field produced by these n sources at the point P.

Now, at the point P, the amplitudes of the disturbances reaching from $A_1, A_2, ...$ will be nearly the same because the point P is at a distance which is very large in comparison to b.

Because of different path lengths to the point P, the field produced by A_1 will differ in phase from the field produced by A_2.

A_2' is the foot of the perpendicular drawn from A_1 on $A_2 B_2$.

For an incident plane wave, the points $A_1, A_2, ...$ are in phase.

The optical paths A_1B_1P and $A_2' B_2P$ are the same.

Therefore, the additional path traversed by the disturbance emanating from the point A_2 is A_2A_2'.

The path difference between the wavelets diffracted from A_1 and A_2 is

$$A_2A_2' = \Delta \sin \theta$$

The corresponding phase differnce, ϕ, is

$$\phi = \frac{2\pi}{\lambda} \Delta \sin \theta \qquad ...(2)$$

Thus, if the field at the point P due to the disturbance emanating from the point A_1 is $a \cos \omega t$, then the field due to the disturbance emanating from A_2 is a $\cos (\omega t - \phi)$.

The difference in the phases of the disturbance reaching from the points A_2 and A_3 is also ϕ.

The resultant field at the point P is given by

$$E = a[\cos \omega t + \cos (\omega t - \phi) + ... + \cos[\omega t - (n - 1)\phi)] \qquad ...(3)$$

Now, $\cos \omega t + \cos (\omega t - \phi) + ... + \cos [\omega t - (n - 1)\phi] = \dfrac{\sin n\phi / 2}{\sin \phi / 2} \cos\left[\omega t - \dfrac{1}{2}(n-1)\phi\right]$ ...(4)

$$\therefore \qquad E = E_\theta \cos\left[\omega t - \frac{1}{2}(n-1)\phi\right] \qquad ...(5)$$

Here, the amplitude E_θ of the resultant field is given by

$$E_\theta = a\frac{\sin(n\phi / 2)}{\sin \phi / 2} \qquad ...(6)$$

The slit actually consists of a continuous distribution of sources.

So $n \to \infty$ and $\Delta \to 0$ in such a way that $n\Delta \to b$.

$$\frac{n\phi}{2} = \frac{\pi}{\lambda} n\Delta \sin \theta \to \frac{\pi}{\lambda} b \sin \theta$$

$$\phi = \frac{2\pi}{\lambda}\Delta \sin\theta = \frac{2\pi}{\lambda}\frac{b\sin\theta}{n} \to 0 .$$

$$\therefore \qquad E_\theta \approx \frac{a\sin\left(\frac{n\phi}{2}\right)}{\frac{\phi}{2}} = na\frac{\sin\frac{\pi b\sin\theta}{\lambda}}{\frac{\pi b\sin\theta}{\lambda}} = A\frac{\sin\beta}{\beta} \qquad ...(7)$$

Here,

$$A = na \text{ and } \beta = \frac{\pi b\sin\theta}{\lambda} \qquad ...(8)$$

Thus

$$E = A\frac{\sin\beta}{\beta}\cos(\omega t - \beta) \qquad ...(9)$$

The corresponding intensity distribution is given by

$$I = I_0\frac{\sin^2\beta}{\beta^2} \qquad ...(10)$$

Here, I_0 represents the intensity at $\theta = 0$.

The variation of the intensity with β is shown in Fig.20.5.

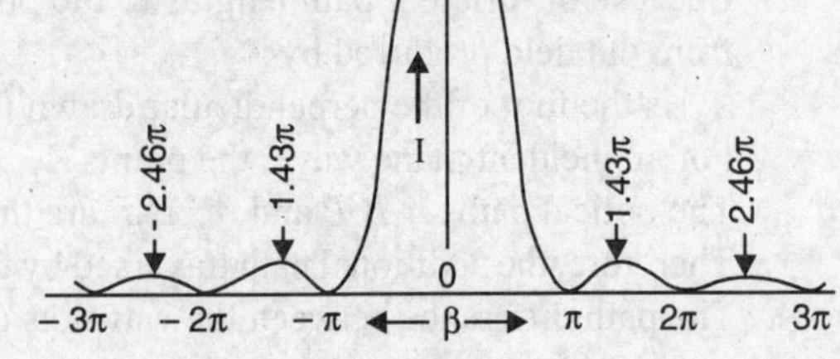

Fig. 20.5.

Positions of Minima

The intensity is zero when

$\beta = \pm m\pi$, where m is an integer ($m \neq 0$, *i.e.*, $m = 1, 2, 3...$) ...(11)

Substituting the value of β, we get

$$\frac{\pi b\sin\theta}{\lambda} = \pm m\pi$$

or

$$b\sin\theta = m\lambda;\ m = \pm1, \pm2, \pm3,....\ \text{(minima)} \qquad ...(12)$$

Eq. (12) is the condition for minimum intensity.

By putting $m = 1, 2, 3,...$ in Eq. (12), we obtain positions of different minima.

The first minimum occurs at $\theta = \pm\sin^{-1}\left(\frac{\lambda}{b}\right)$; the second minimum at $\theta = \pm\sin^{-1}\left(\frac{2\lambda}{b}\right)$, etc.

Positions of Maxima

(*i*) When $\beta = 0$, $\left(\frac{\sin\beta}{\beta}\right) = 1$ and $I = I_0$ is the intensity of the principal maximum.

(*ii*) In addition to the principal maximum at $\theta = 0$, there are weak secondary maxima between the equally spaced minima. The exact positions of these maxima is obtained by differentiating the expression for I, Eq.(10), with respect to β and equating it to zero.

$$\frac{dI}{d\beta} = I_0\left[\frac{2\sin\beta\cos\beta}{\beta^2} - \frac{2\sin^2\beta}{\beta^3}\right] = 0$$

or

$$\frac{2\sin\beta}{\beta}\left[\frac{\beta\cos\beta - \sin\beta}{\beta^2}\right] = 0$$

or

$$\beta = \tan\beta \qquad ...(13)$$

The points of intersection of the two curves $y = \beta$ and $y = \tan\beta$ give the values of β which satisfy Eq. (13). [Fig. 20.6].

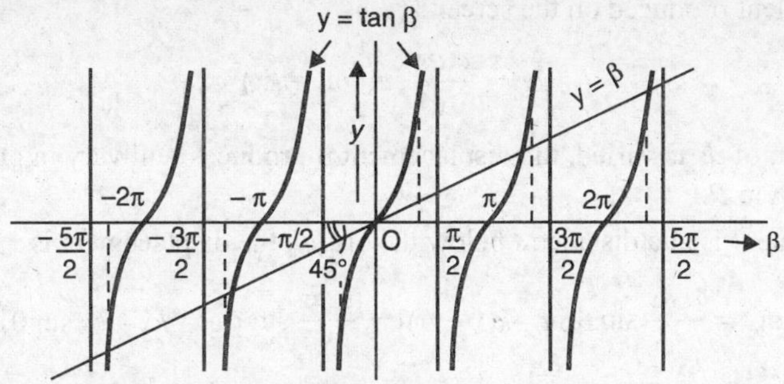

Fig. 20.6.

The intersections occur at $\beta = 1.43\pi$, $\beta = 2.46\pi$, etc.

Substituting the values of β in Eq.(10), we get the intensities for various maxima.

The intensity of the first maximum is

$$I_1 = I_0 \left[\frac{\sin(1.43\pi)}{1.43\pi}\right]^2 = 0.0496\ I_0 \qquad ...(14)$$

The intensity of the second maximum is

$$I_2 = I_0 \left[\frac{\sin(2.46\pi)}{2.46\pi}\right]^2 = 0.0168\ I_0 \qquad ...(15)$$

Secondary maxima are very feeble and their intensities diminish very rapidly.

Method II

Parallel light is incident on a slit of width *b* (Fig. 20.7).

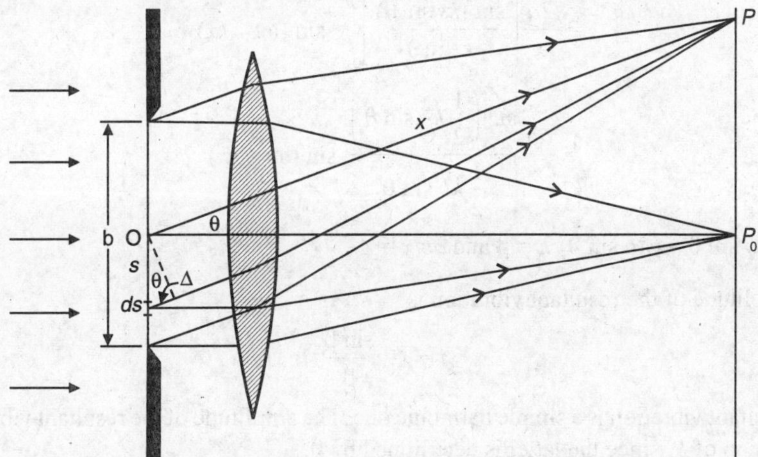

Fig. 20.7.

The centre of the slit *O* is the origin of coordinates. Consider an element *ds* at a distance *s* from the centre *O*.

The parts of each secondary wave which travel at an angle θ with the normal to the plane of the slit reach *P*.

Consider first the wavelet emitted by the element *ds* situated at the origin, *O*. Its amplitude is directly proportional to the length *ds* and inversely proportional to the distance *x*.

The displacement produced on the screen at P is

$$dy_0 = \frac{a\,ds}{x} \sin(\omega t - kx)$$

As the position of ds is varied, the displacement it produces will vary in phase because of the different path length to P.

When the element is at a distance s below the origin, the displacement is

$$dy_s = \frac{a\,ds}{x} \sin\,[\omega t - k(x + \Delta)] = \frac{a\,ds}{x} \sin(\omega t - kx - ks \sin\theta) \qquad ...(1)$$

The total displacement from the pair of elements symmetrically placed at s and $-s$ is

$$dy = dy_{-s} + dy_s$$

$$= \frac{a\,ds}{x} [\sin(\omega t - kx - ks \sin\theta) + \sin(\omega t - kx + ks \sin\theta)]$$

$$dy = \frac{a\,ds}{x} [2 \cos(ks \sin\theta) \sin(\omega t - kx)] \qquad ...(2)$$

Here we have used the identity $\sin\alpha + \sin\beta = 2\cos\frac{1}{2}(\alpha - \beta)\sin\frac{1}{2}(\alpha + \beta)$.

To sum the effects of all elements from one edge of the slit to the other, we integrate Eq. (2) from $s = 0$ to $b/2$.

x is regarded as constant. Thus

$$y = \frac{2a}{x} \sin(\omega t - kx) \int_0^{b/2} \cos(ks \sin\theta)\, ds$$

$$= \frac{2a}{x} \left[\frac{\sin(ks \sin\theta)}{k \sin\theta}\right]_0^{b/2} \sin(\omega t - kx)$$

$$= \frac{ab}{x} \frac{\sin\left(\frac{1}{2} kb \sin\theta\right)}{\frac{1}{2} kb \sin\theta} \sin(\omega t - kx) \qquad ...(3)$$

Put $\frac{1}{2} kb \sin\theta = (\pi b \sin\theta)/\lambda = \beta$ and $ab/x = A_0$.

The amplitude of the resultant vibration is

$$A = A_0 \frac{\sin\beta}{\beta} \qquad ...(4)$$

The resultant vibration is a simple harmonic one. The amplitude of the resultant vibration varies with the position of P, since the latter is determined by θ.

The intensity on the screen is

$$I \approx A^2 = A_0^2 \frac{\sin^2\beta}{\beta^2} \qquad ...(5)$$

Intensity Distribution:

Fig. 20.8(a) shows the intensity distribution curve.

(i) At the point P_0 of Fig. 20.7, all the secondary wavelets will arrive in phase because the path difference $\Delta = 0$.

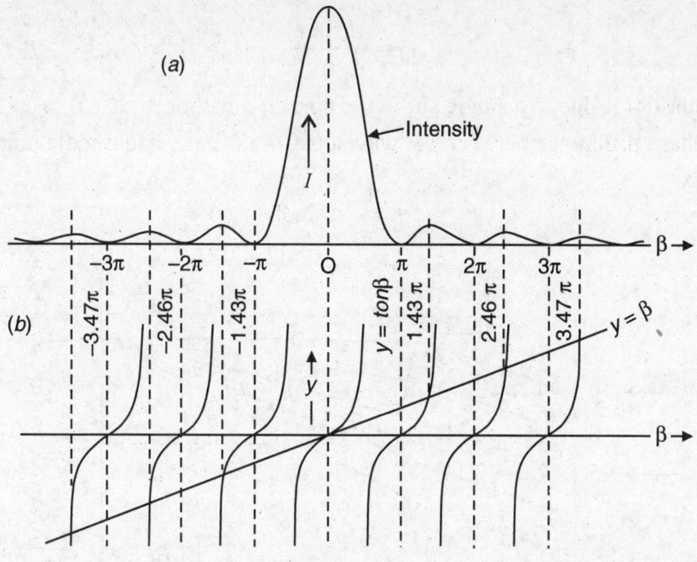

Fig. 20.8.

For $\beta = 0$, $(\sin \beta)/\beta = 1$. $\therefore I_0 = A_0^2$ = Intensity of the *principal maximum*.

$A_0{}^2$ is then the value of the maximum intensity, at the centre of the pattern.

(*ii*) The intensity falls to zero at $\beta = \pm \pi, \pm 2\pi, \pm 3\pi,...$, or in general $\beta = m\pi$.

(*iii*) **Secondary Maxima.** The exact values of β for secondary maxima can be found by differentiating Eq. (5) with respect to β and equating to zero.

$$I = A_0^2 (\sin^2 \beta)/\beta^2 = A_0^2 \left(\frac{\sin \beta}{\beta}\right)^2$$

$$\frac{dI}{d\beta} = 2 A_0^2 \left(\frac{\sin \beta}{\beta}\right)\left[\frac{\beta \cos \beta - \sin \beta}{\beta^2}\right] = 0 \text{ or } \beta \cos \beta - \sin \beta = 0$$

$$\therefore \qquad \tan \beta = \beta \qquad ...(6)$$

The values of β satisfying this relation are found graphically as the intersections of the curve $y = \tan \beta$ and the straight line $y = \beta$ [Fig. 20.8(*b*)].

The intersections occur at $\beta = 1.43\ \pi$, $2.46\ \pi$, $3.47\ \pi$, ...etc.

The intensity of the first maximum is $I_1 = I_0 \left[\frac{\sin (1.43\ \pi)}{1.43\ \pi}\right]^2 = 0.0496\ I_0$ ...(7)

The first secondary maximum is only 4.96 per cent the intensity of the central maximum.

The second and third secondary maxima are only 1.68% and 0.83% respectively the intensity of the central maximum.

20.4. Theory of Diffraction Grating (Diffraction at N slits)

A parallel beam of monochromatic light of wavelength λ is incident normally on *N* parallel slits (Fig. 20.9). Width of each slit is *a*. Width of each opaque space is *b*. The grating element is $(a + b)$.

According to Huygens principle, as soon as the plane wave front is incident on the slits, all points in each slit become the sources of secondary disturbances in all directions. By the theory of Fraunhofer diffraction at a single slit, the secondary waves from all points in a slit diffracted in a direction θ are equivalent to a single wave of amplitude

$$A_0 = \frac{A \sin \alpha}{\alpha}$$

Here, A = amplitude due to a single slit in the normal direction and

α = phase difference between the wavelets from the end and middle points of the slit

$$\alpha = \frac{\pi}{\lambda} .a \sin \theta$$

Fig. 20.9.

$S_1, S_2, S_3, \ldots S_n$ are the middle points of the slits. The resultant disturbance in each slit acts at its middle point.

Let us now find the resultant effect of these N vibrations.

Draw S_1K_1 perpendicular to S_2K_1.

Path difference between the rays from the slits S_1 and $S_2 = S_2K_1 = S_1S_2 \sin \theta = (a + b) \sin\theta$

Phase difference between successive waves $= 2\beta = \frac{2\pi}{\lambda} (a + b) \sin \theta$

If we consider the phase of the wave from S_1 to be zero, then the phases of waves from S_2, S_3, S_4, etc., should be 2β, 4β, 6β, 8β, etc. Thus the phases increase in arithmetical progression.

Hence the problem of determining the intensity along a given direction θ reduces to finding the resultant amplitude of N vibrations in a direction θ, each of amplitude $A_0 = \frac{A \sin \alpha}{\alpha}$ having a common phase difference 2β.

Vector Addition of Amplitudes

Draw equal lengths $MP_1, P_1P_2, P_2P_3 \ldots P_{N-1}P_N$ representing equal amplitudes A_0 (Fig.20.10). The closing side MP_N of the polygon represents the resultant amplitude R. O is the centre of the polygon.

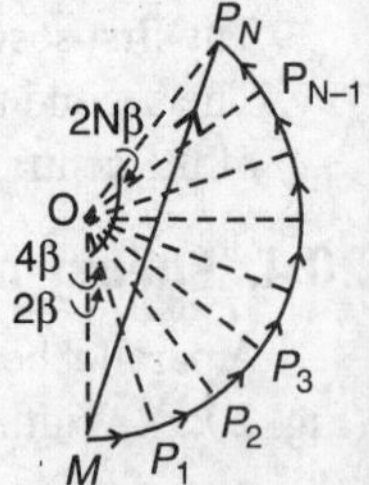

Fig. 20.10.

From geometry, we have

$$MP_1 = 2.\ OM \sin \beta$$

and
$$MP_N = 2.\ OM \sin N\beta$$

Eliminating OM, we get

$$MP_N = 2.\frac{MP_1}{2 \sin \beta}.\sin N\beta = A_0 \frac{\sin N\beta}{\sin \beta} = \left(\frac{A \sin \alpha}{\alpha}\right)\frac{\sin N\beta}{\sin \beta}$$

Hence the resultant amplitude is given by

$$R = \frac{A \sin \alpha}{\alpha} \cdot \frac{\sin N\beta}{\sin \beta} \qquad \text{... (1)}$$

The resultant intensity is

$$I = A^2 \frac{\sin^2 \alpha}{\alpha^2} \cdot \frac{\sin^2 N\beta}{\sin^2 \beta} \qquad \text{... (2)}$$

(*i*) The first factor $A^2 \dfrac{\sin^2 \alpha}{\alpha^2}$ gives the intensity distribution due to diffraction at a single slit.

(*ii*) The second factor $\sin^2 N\beta/\sin^2\beta$ arises from interference between the N diffracted waves from the N slits.

Let $$Z = \frac{\sin^2 N\beta}{\sin^2 \beta}$$

Differentiating with respect to β, we get

$$\frac{dZ}{d\beta} = \frac{\sin\beta \times 2N \sin N\beta \cos N\beta - 2\sin^2 N\beta \sin\beta \cos\beta}{\sin^4 \beta}$$

$$= \frac{2N \sin N\beta \cos N\beta}{\sin^2 \beta} - 2\sin^2 N\beta \frac{1}{\sin^2 \beta} \frac{\cos\beta}{\sin\beta}$$

$$= \frac{2N \sin^2 N\beta \cos N\beta}{\sin^2 \beta \sin N\beta} - \frac{2\sin^2 N\beta}{\sin^2\beta} \cdot \frac{\cos\beta}{\sin\beta} = \frac{2\sin^2 N\beta}{\sin^2\beta}(N \cot N\beta - \cot\beta)$$

For a maximum or a minimum, $\dfrac{dZ}{d\beta} = 0$

$$\therefore \qquad \frac{2\sin^2 N\beta}{\sin^2\beta}(N\cot N\beta - \cot\beta) = 0$$

Hence either $\dfrac{\sin N\beta}{\sin\beta} = 0$ or $N \cot N\beta - \cot\beta = 0$ ...(3)

(*i*) Principal Maxima:

Considering Eq.(3), when $\sin\beta = 0$, $\beta = \pm n\pi$, $[n = 1, 2, 3, ...]$,

$$\frac{\sin N\beta}{\sin\beta} = \frac{0}{0} \text{ becomes indeterminate.}$$

But it may be evaluated by finding its limit as $\beta \to \pm n\pi$

$$\lim_{\beta\to\pm n\pi} \frac{\sin N\beta}{\sin\beta} = \lim_{\beta\to\pm n\pi} \frac{\frac{d}{d\beta}\sin N\beta}{\frac{d}{d\beta}\sin\beta} = \lim_{\beta\to\pm n\pi} \frac{N\cos N\beta}{\cos\beta} = \frac{N\cos N(\pm n\pi)}{\cos(\pm n\pi)} = \pm N$$

Substituting this result in Eq.(2), we get

$$I = A^2 \frac{\sin^2\alpha}{\alpha^2} N^2 \qquad \text{...(4)}$$

These maxima are most intense and hence they are called *principal maxima*.
They are obtained for $\beta = \pm n\pi$

or $$\frac{\pi}{\lambda}(a+b)\sin\theta = \pm n\pi \text{ or } (a+b)\sin\theta = \pm n\lambda \text{ where } n = 0, 1, 2, \ldots \quad \ldots(5)$$

For $n = 0$, we get the *central* or *zero order maximum*. For $n = \pm1, \pm2, \pm3$, etc., we get the first, second and third order principal maxima respectively on either side of the central maximum.

***(ii)* Minima of Intensity:**

When $\sin N\beta = 0$ but $\sin\beta \neq 0$, the factor $\dfrac{\sin N\beta}{\sin\beta}$ becomes zero.

From Eq.(2), we have $I = 0$. Hence intensity is minimum.
Thus for minima, we have

$$\sin N\beta = 0 \text{ or } N\beta = \pm m\pi \text{ or } N\frac{\pi}{\lambda}(a+b)\sin\theta = \pm m\pi \text{ or } N(a+b)\sin\theta = \pm m\lambda$$

Here m has all the integral values except $O, N, 2N, \ldots nN$, because these values of m make $\sin\beta = 0$ which corresponds to the positions of principal maxima. Thus $m = 0$ gives a principal maximam, $m = 1, 2, 3, \ldots (N-1)$ give minima and then $m = N$ gives again a principal maximum. Hence there are $(N-1)$ equally spaced minima between two adjacent maxima.

***(iii)* Secondary Maxima**

There are $(N-1)$ minima between two adjacent principal maxima. So there must be $(N-2)$ other maxima, known as secondary maxima, between two principal maxima.

Intensity of secondary maxima

To find their intensity, the fraction $\dfrac{\sin^2 N\beta}{\sin^2\beta}$ has to be evaluated subject to the condition, $N\cot N\beta - \cot\beta = 0$.

$$\frac{\sin^2 N\beta}{\sin^2\beta} = \frac{1}{\operatorname{cosec}^2 N\beta}\cdot\frac{1}{\sin^2\beta} = \frac{1}{1+\cot^2 N\beta}\cdot\frac{1}{\sin^2\beta}$$

$$= \frac{N^2}{N^2 + N^2\cot^2 N\beta}\times\frac{1}{\sin^2\beta} = \frac{N^2}{N^2+\cot^2\beta}\times\frac{1}{\sin^2\beta} (\because N\cot N\beta = \cot\beta)$$

$$= \frac{N^2}{N^2 + \dfrac{\cos^2\beta}{\sin^2\beta}}\times\frac{1}{\sin^2\beta} = \frac{N^2}{N^2\sin^2\beta + \cos^2\beta} = \frac{N^2}{N^2\sin^2\beta + 1 - \sin^2\beta}$$

or $$\frac{\sin^2 N\beta}{\sin^2\beta} = \frac{N^2}{1+(N^2-1)\sin^2\beta}$$

The intensity of secondary maxima varies as $\dfrac{N^2}{1+(N^2-1)\sin^2\beta}$

The intensity of principal maxima varies as N^2.

$$\therefore \quad \frac{\text{Intensity of secondary maxima}}{\text{Intensity of principal maxima}} = \frac{N^2}{1+(N^2-1)\sin^2\beta}\times\frac{1}{N^2}$$

$$\frac{I_{\text{sec. maxima}}}{I_{pri.\text{ maxima}}} = \frac{1}{1+(N^2-1)\sin^2\beta}$$

When N is large, these secondary maxima are weak and invisible. But when N is small they may be observed.

The intensity distribution as observed on the screen is shown in Fig. 20.11.

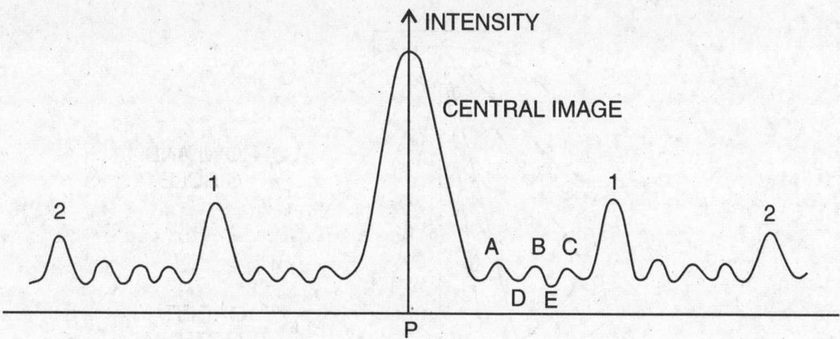

Fig. 20.11.

P corresponds to the position of the central maximum and 1, 2, etc., on both sides of P, show the 1st, 2nd etc., principal maxima. The positions A, B, C, etc., represent the secondary maxima and D, E etc., are the secondary minima.

20.5. Methods of Mounting the Concave Reflection Grating

Rowland Mounting

SA and SB are two strong iron rails lying at right angles to each other in a horizontal plane (Fig. 20.12).

The grating G and photographic plate-holder P are mounted on two carriages C_1 and C_2 which can move along the rails and are joined by a rigid beam of length R, the radius of curvature of the grating. A slit S is placed at the junction of the two rails.

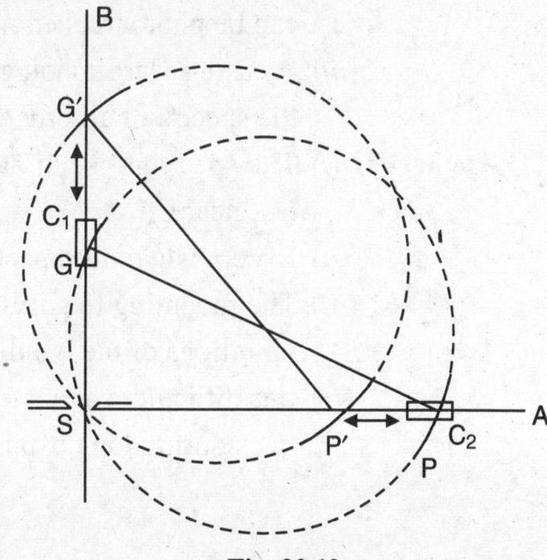

Fig. 20.12.

Since $\angle BSA$ is a right angle and the length of the beam C_1C_2 is equal to the diameter of the Rowland circle, G, S and P all lie on this circle. Hence when white light is passed through the slit and falls on the grating, it is diffracted and the spectra of various orders are focussed on the circle. However, only that portion of the spectrum is used which falls on the photograhic plate P which is always at the normal to the grating. When the beam C_1C_2 is moved, then G and P are moved to positions such as G' and P' (but the two always remain diametrically opposite on the Rowland circle) and spectra of different orders are obtained.

Merit: The spectrum is normal. Hence the rail SP can be graduated to read wavelengths directly.

Demerits: (*i*) It needs large space.

(*ii*) It is expensive and involves a considerable amount of mechanism.

This mounting is now rarely used.

Runge-Paschen Mounting

The grating G is mounted at *a fixed* point on the Rowland circle on a rigid support (Fig. 20.13) The slit S is *fixed* on the same circle at a proper position so as to give a desired angle of incidence.

The source is placed behind the slit. The photographic plate is clamped to a circular steel frame (plate-holder) along the Rowland circle. The spectra of different orders are focussed simultaneously on the plate.

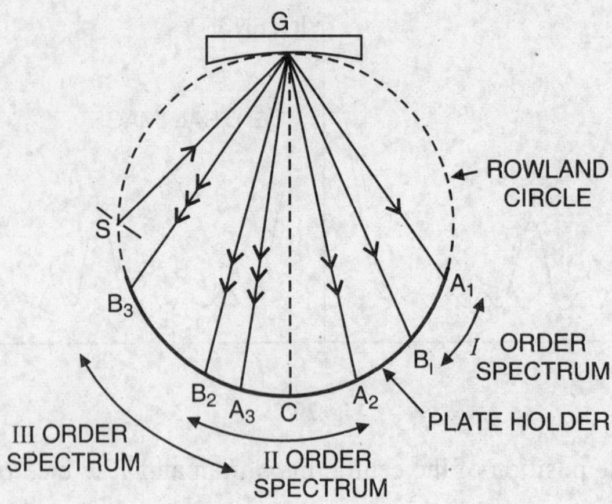

Fig. 20.13.

Merits: (*i*) The arrangement is very rigid.

(*ii*) The plate may be placed anywhere along the frame and any portion of the spectrum can be photographed *without altering the adjustment.*

(*iii*) By using a large photographic film, curved to the curvature of the Rowland circle, the spectra of all orders may be simultaneously photographed.

Demerits: (*i*) The spectrum is normal only in the region which falls near the grating normal *i.e.,* near *C*.

(*ii*) Except where the spectrum is normal, there is a strong astigmatism.

(*iii*) The mounting requires a large space and a solid steady floor insulated from the vibrations of the building.

(*iv*) It is difficult to avoid temperature-changes which may cause shift in the spectral lines during long exposures.

CHAPTER

21

DOPPLER EFFECT IN LIGHT AND ITS APPLICATIONS

21.1. Doppler Effect in Light

The apparent change in frequency of a wave due to relative motion between the source and the observer is known as Doppler effect. For example, when a railway engine approaches an observer blowing its whistle, the pitch of the whistle appears to rise and when it moves away from the observer to fall. A similar phenomenon is observed in the case of light. Whenever there is relative motion between a source and an observer, the frequency and hence the wavelength received by the observer are found to be different from those actually emitted by the source. This is known as Doppler effect in light. If the source moves towards the observer or the observer moves towards the source, the apparent wavelength becomes less and the spectral line is shifted towards the violet end of the spectrum. If the observer moves away from the source or the source moves away from the observer, there is an increase in the wavelength and the spectral line is shifted towards the red end of the spectrum. The apparent change in wavelength known as the Doppler shift is given by $\lambda - \lambda' = d\lambda = \pm \frac{v}{c}\lambda$ according as the source is moving towards or away from the observer. Here, v = velocity of motion of the source. c = velocity of light, λ = the real wavelength of the source and λ' = apparent wavelength for the observer.

Note that v is the component velocity of the source in the line of sight. No Doppler shift can happen if the motion of the source is perpendicular to the line of sight of the observer. By measuring the shift of the spectral lines, the speed with which a star is moving can be calculated.

Example. *Consider the H_α line of hydrogen which has a wavelength of 656.1 nm on the earth. The value obtained from the absorption spectrum of hydrogen from a star is 662.7 nm. Hence, the velocity of recession of the star*

$$= v = c\frac{d\lambda}{\lambda} = 3 \times 10^8 \times \frac{6.6}{656.1} = 3.017 \times 10^6 \text{ ms}^{-1}.$$

21.2. Applications of Doppler Effect

(*a*) The expanding universe : Certain elements like hydrogen present on the earth are also present in the stars. In the spectrum of the light received from distant stars and galaxies, the characteristic spectral lines are shifted towards the red end of the spectrum. By Doppler's principle, this is due to the recession of the stars and galaxies from the earth with large velocities. The more distant galaxies suffer the greater shift and hence their velocity of recession is greater. This phenomenon is called *recessional red shift* and lends support to the theory *of expanding universe.*

(*b*) Nature of Saturn's rings : The main body of the planet Saturn is surrounded by three concentric rings. Two theories were put forward regarding the nature of the rings. According to one theory, the rings were solid discs rotating as a whole round Saturn with a constant angular velocity. According to the second theory, the rings consisted of swarms of particles each moving under the gravitational attraction of Saturn.

(*i*) Let r_1 and r_2 be the radii of the inner and outer edges of a ring respectively (Fig. 21.1). If the rings are solid discs, the angular velocity (ω) is the same for both the edges. Therefore, the velocities of a particle on the inner and outer edges are $v_1 = r_1\omega$ and $v_2 = r_2\omega$. Since $r_2 > r_1$, $v_2 > v_1$. Hence, the outer edge of the ring should have greater velocity than the inner edge.

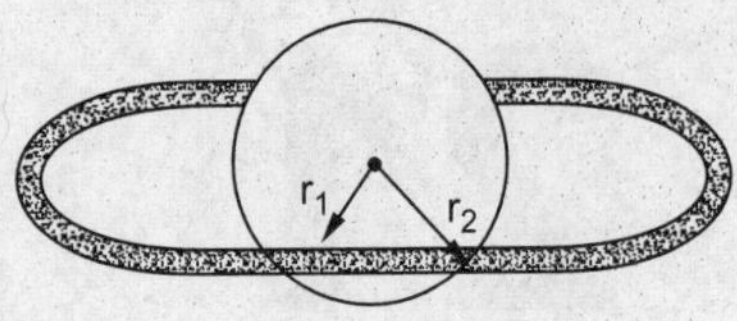

Fig. 21.1.

(*ii*) Let M be the mass of Saturn and m the mass of a particle, r = distance of the particle from the centre of Saturn and v = velocity of the particle. If the disc consists of a swarm of particles, then, centrifugal force on the particle = gravitational force of attraction of Saturn. Or

$$\frac{mv^2}{r} = \frac{GMm}{r^2} \text{ or } v^2 = \frac{GM}{r}$$

This shows that as r increases, v decreases or $v_1 > v_2$ *i.e.*, the inner edge of the ring should have greater velocity than the outer edge.

Doppler shift in the spectrum of light, coming from the innermost and outermost edges of a ring shows that the inner edge is moving with a greater velocity than the outer edge. *This reveals the fact that Saturn's rings are a swarm of particles.*

(*c*) Broadening of Spectral Lines: In a line spectrum, each spectral line should be a geometric line without any breadth. But in practice, it is found that even if the slit of the spectrometer is very narrow, the spectral lines have a definite width. This is explained by the Doppler principle.

The atoms of the incandescent gas emitting light are not at rest but in random motion with different velocities. The motion of the atoms affects the frequency of radiation emitted by them. Let ν be the frequency of light emitted by atoms at rest. Then the frequency of light emitted by atoms moving towards the collimator is greater than ν. The frequency of light emitted by atoms moving away from the collimator is less than ν. There are many atoms moving with velocities ranging from 0 to ∞, some moving towards the slit, some away from it and others in other directions. This leads to different shift in frequencies and broaden out the original line. This is an evidence in favour of the kinetic theory of gases. Further it has been proved that the band width varies exactly in accordance with the change in velocity of atoms due to alterations in the temperature of the gas.

(*d*) Spectroscopic Binaries (Double stars) : Some of the stars that appear to be single are actually double stars and are known as spectroscopic binaries. The two stars revolve about a common centre. Some of the stars are so close that a telescope of even high resolving power fails to resolve them. In such cases, a spectroscope is able to detect them from the Doppler effect.

In Fig.21.2, A and B are two bright stars and they are seen by the spectroscope S. When they are in positions A and B, then A is receding and B is approaching the spectroscope (S). The spectral lines of A are shifted towards the red and the spectral lines of B are shifted towards the violet. A double line would thus appear in the spectrum. But when the stars are in the position A' and B', the motion of both the stars is at right angles to the line of sight. Hence only a single line would appear.

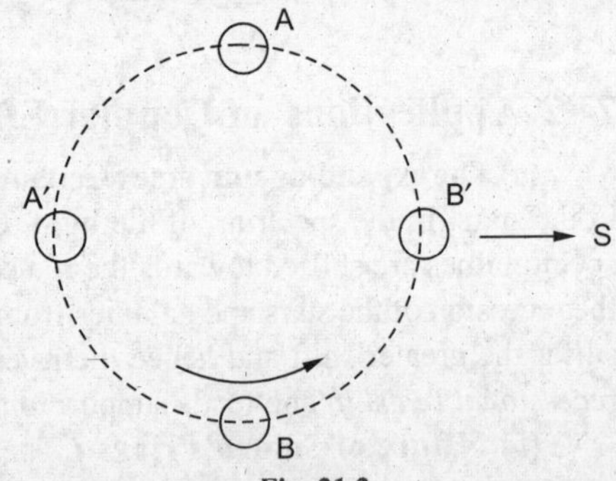

Fig. 21.2.

During the next quarter of a revolution, these two lines appear again and so on. Thus the spectral lines given by this type of binary star double and undouble periodically.

In another type of binary, one is a bright star and the other, a dark star. The spectral lines formed by the bright star exhibit a *periodic shift* about their mean positions. When the star is receding, the line shifts towards the red side. After a time, equal to half the period of its revolution, there will be an equal shift from the mean position towards the violet side.

Hundreds of such spectroscopic binaries have been discovered by applying Doppler effect.

Example: *In the spectrum of a certain star, the* H_α *line is shifted towards the red end by 0.01 nm. Calculate the velocity of the star in the line of sight. Wavelength of* H_α *line = 656.3 nm. Velocity of light =* 3×10^8 ms^{-1}.

Solution. $v = c\dfrac{d\lambda}{\lambda}$.

Here, $d\lambda = \lambda - \lambda' = (656.3 - 656.31)$ nm $= -0.01$ nm; $\lambda = 656.3$ nm.

$$\therefore \qquad v = (3 \times 10^8) \times \frac{(-0.01)}{656.3} = -4.571 \times 10^3 \text{ ms}^{-1}.$$

The negative sign indicates that the star is moving away from the earth.

CHAPTER

22 CORNU'S SPIRAL

22.1. Fresnel's Integrals

Fresnel's integrals are used to find the intensity at a point in the diffraction pattern. Let S be a narrow slit perpendicular to the plane of the paper and illuminated by monochromatic light of wavelength λ (Fig. 22.1). Let W_1W_2 be the section of a cylindrical wavefront, perpendicular to the plane of the paper. Let P be a point on the screen (also $\perp$ to the paper) at which the intensity due to the wavefront is required. O is the pole of the wavefront with respect to P. Consider a small element ds of the wavefront at A. The distance of the element measured along the curve is s, *i.e.*, $OA = s$. The secondary wavelets from the various elements ds of the wavefront OA travel different distances to P and make their contributions.

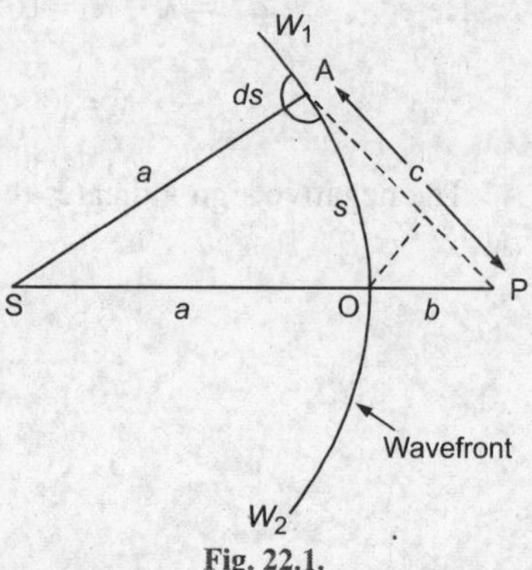

Fig. 22.1.

Let $PA = c$ be the distance from ds to point P. Let us calculate c.

From ΔSAP, $PA^2 = SP^2 + SA^2 - 2\, SP.SA \cos (s/a)$

or $$PA^2 = (a+b)^2 + a^2 - 2(a+b)a\left(1 - \frac{s^2}{2a^2}\right)$$ (neglecting higher powers of s/a)

$$= b^2 + \frac{(a+b)\,s^2}{a}$$

$$= b^2\left[1 + \frac{(a+b)\,s^2}{ab^2}\right]$$

$$c = b\left[1 + \frac{(a+b)\,s^2}{ab^2}\right]^{1/2} = b\left[1 + \frac{1}{2}\frac{(a+b)\,s^2}{ab^2}\right]$$ (neglecting higher powers of s^2)

$\therefore$ $$c = b + \frac{(a+b)\,s^2}{2ab} \qquad \ldots(1)$$

Path difference δ between the vibrations from O and A is

$$\delta = AP - OP = b + \frac{(a+b)\,s^2}{2ab} - b = \frac{(a+b)\,s^2}{2ab} \qquad \ldots(2)$$

Fresnel assumed the wavefront to be divided into infinitesimal elements of width ds, each contributing its effect at P. Fresnel further assumed that

(*i*) the amplitude of vibration contributed by each element ds is proportional to its width ds.

(*ii*) only the portion of the wavefront in the neighbourhood of O, as far as its effect at P is concerned, is of importance. Therefore, the change of amplitude with distance and obliquity of the wavefront may be neglected.

Let the displacement of vibration at O (say, of unit amplitude) be $\frac{2\pi t}{T}$.

Then the displacement at P due to element ds at O

$$= \sin 2\pi \left(\frac{t}{T} - \frac{b}{\lambda} \right) ds.$$

$\therefore$ displacement at P due to element ds at A is

$$dy = \sin 2\pi \left(\frac{t}{T} - \frac{b + \delta}{\lambda} \right) ds.$$

Resultant displacement due to all the wavelets each of width ds,

$$y = \int \sin 2\pi \left(\frac{t}{T} - \frac{b+\delta}{\lambda} \right) ds$$

$$= \int \left[\sin 2\pi \left(\frac{t}{T} - \frac{b}{\lambda} \right) \cos \frac{2\pi\delta}{\lambda} - \cos 2\pi \left(\frac{t}{T} - \frac{b}{\lambda} \right) \sin \frac{2\pi\delta}{\lambda} \right] ds$$

$$= \sin 2\pi \left(\frac{t}{T} - \frac{b}{\lambda} \right) \int \cos \frac{2\pi\delta}{\lambda} ds - \cos 2\pi \left(\frac{t}{T} - \frac{b}{\pi} \right) \int \sin \frac{2\pi\delta}{\lambda} ds \qquad \text{...(3)}$$

Put $R \cos \phi = \int \cos \frac{2\pi\delta}{\lambda} ds$ and $R \sin \phi = \int \sin \frac{2\pi\delta}{\lambda} ds$ ...(4)

Substituting these values in Eq. (3),

$$y = R \sin 2\pi \left(\frac{t}{T} - \frac{b}{\lambda} - \frac{\phi}{2\pi} \right) \qquad \text{...(5)}$$

Thus the resultant displacement at P due to portion OA of the wavefront has the same period as the incoming wave, but a different amplitude R and different phase constant.

The intensity I_p at P is proportional to R^2.

$$\therefore \qquad I_p = R^2 = \left[\int \cos \frac{2\pi\delta}{\lambda} ds \right]^2 + \left[\int \sin \frac{2\pi\delta}{\lambda} ds \right]^2 \qquad \text{...(6)}$$

But from Eq. (2), $\delta = \frac{(a+b)\, s^2}{2ab}$

$$\therefore \qquad I_p = \left[\int \cos \frac{2\pi}{\lambda} \cdot \frac{(a+b)s^2}{2ab} ds \right]^2 + \left[\int \sin \frac{2\pi}{\lambda} \cdot \frac{(a+b)\, s^2}{2ab} \cdot ds \right]^2 \qquad \text{...(7)}$$

Put
$$\frac{\pi\,(a+b)\, s^2}{\lambda\, ab} = \frac{\pi}{2} v^2 \qquad \text{...(8)}$$

$$\therefore \qquad s = \sqrt{\frac{ab\,\lambda}{2(a+b)}}\, v \text{ and } ds = \sqrt{\frac{ab\,\lambda}{2(a+b)}}\, dv \qquad \text{...(9)}$$

$$\therefore \qquad I_p = \frac{ab\,\lambda}{2(a+b)} \left[\left\{ \int_0^v \cos \frac{\pi v^2}{2} dv \right\}^2 + \left\{ \int_0^v \sin \frac{\pi v^2}{2} dv \right\}^2 \right] \qquad \text{...(10)}$$

s is finite and λ is very small. Hence from Eq. (8) v is very large compared to s. Hence we may take ∞ as the upper limit of v in the two integrals.

$$\therefore \qquad I_p = \text{constant} \left[\left\{ \int_0^\infty \cos \frac{\pi v^2}{2} dv \right\}^2 + \left\{ \int_0^\infty \sin \frac{\pi v^2}{2} dv \right\}^2 \right]$$

$\therefore \qquad I_p = k[X^2 + Y^2].$

Here k is constant and

$$X = \int_0^\infty \cos \frac{\pi v^2}{2}\, dv \text{ and } Y = \int_0^\infty \sin \frac{\pi v^2}{2}\, dv \qquad \ldots(12)$$

These integrals are called Fresnel's integrals. When these are evaluated between the specified values of v, they give the resultant disturbance from the corresponding portion of the wavefront.

22.2. Cornu's Spiral

Fresnel's integrals define a curve called *Cornu's spiral.*

The coordinates of any point (x,y) on Cornu's spiral are

$$x = X(v) = \int_0^v \cos \frac{\pi v^2}{2}\, dv \qquad \ldots(1)$$

and

$$y = Y(v) = \int_0^v \sin \frac{\pi v^2}{2}\, dv \qquad \ldots(2)$$

The values of these integrals can be calculated for different values of v. The graph obtained by plotting one of the integrals as the abscissa and the other as ordinate is called Cornu's spiral. The graph is shown in Fig. 22.2. It has two branches and has asymptotic points J_1 and J_2 at $X = Y = \pm 0.5$. The two integrals represent the horizontal and vertical components of the resultant amplitude. The radius vector drawn from the origin to any point (x, y) on the curve gives r and $r^2 = x^2+y^2$.

$$I_P = k\,(x^2+y^2) \qquad \ldots(3)$$

Eq. (3) represents the intensity of illumination at a point and the angle the radius vector makes with x-axis gives the phase of the vibration at the point.

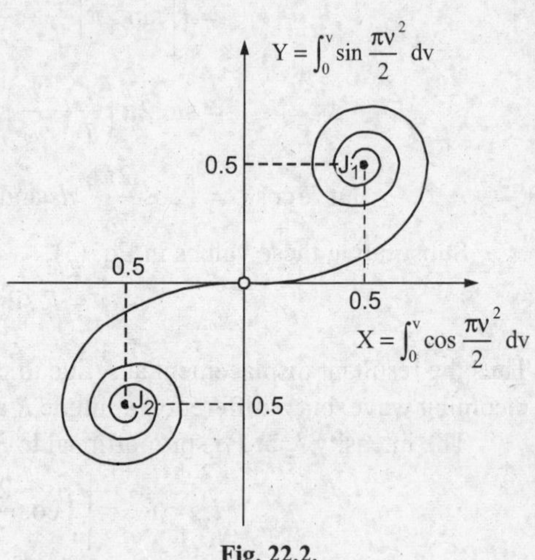

Fig. 22.2.

22.3. Properties of Cornu's Spiral

(1) When $v = 0$, $x = y = 0$. Therefore, the curve passes through the origin. This origin is taken as the pole of the wavefront while considering the diffraction pattern.

(2) The curve is symmetrical about the origin.

(3) Now, $dx = \cos \frac{\pi v^2}{2}\, dv$ and $dy = \sin \frac{\pi v^2}{2}\, dv$.

$$ds = \sqrt{(dx)^2 + (dy)^2} = dv.$$

$\therefore$ Length s of the arc of the curve measured from the origin to a given point (x, y) is

$$\int_0^s ds = \int_0^v dv$$

$$\therefore \qquad s = v$$

Thus, the distance measured along the curve from the origin is equal to v.

(4) If ψ is the angle which the tangent drawn at any point (x, y) of the curve makes with x-axis, then

$$\tan \psi = \frac{dy}{dx} = \frac{dy/dv}{dx/dv} = \frac{\sin \frac{\pi v^2}{2}}{\cos \frac{\pi v^2}{2}} = \tan \frac{\pi v^2}{2}$$

i.e., $$\psi = \frac{\pi v^2}{2}.$$

Obviously, ψ increases monotonically with $|v|$. At the origin $v = 0$.

$\therefore\ \psi = 0$ *i.e.,* x-axis is tangent to the curve at the origin.

(5) We have, $v = s$. $\therefore \psi = \frac{\pi s^2}{2}$.

Radius of curvature of the curve $= \rho = \frac{ds}{d\psi} = \frac{1}{\pi s}$

$\therefore$ curvature $= \frac{1}{\rho} = \pi s$.

When $v = s = 0$, $\rho \to \infty$. Hence there is a point of inflexion at the origin. As we move along the curve from the origin both ways, curvature increases continuously (since curvature $\propto s$). When the arc length $v = s = \infty$, the curvature is infinite.

(6) When the whole of the wavefront is exposed to the point, $v \to \infty$. Then

$$x = \int_0^\infty \cos\left(\frac{\pi v^2}{2}\right) dv = \frac{1}{2},\ y = \int_0^\infty \sin\left(\frac{\pi v^2}{2}\right) dv = \frac{1}{2}.$$

Thus for the point J_1 in Fig.22.2, the x and y coordinates are $\left(\frac{1}{2}, \frac{1}{2}\right)$.

Similarly, for the point J_2, the coordinates are $\left(-\frac{1}{2}, -\frac{1}{2}\right)$.

22.4. Explanation of Fresnel Diffraction at a Straight Edge by the use of Cornu's Spiral

Let O be a sharp straight edge of an opaque obstacle (Fig. 22.3). S is a narrow source slit. XY is a screen. Consider any point P on the screen. One branch of the Cornu's spiral corresponds to elements of the wavefront above SP and other branch to those below SP.

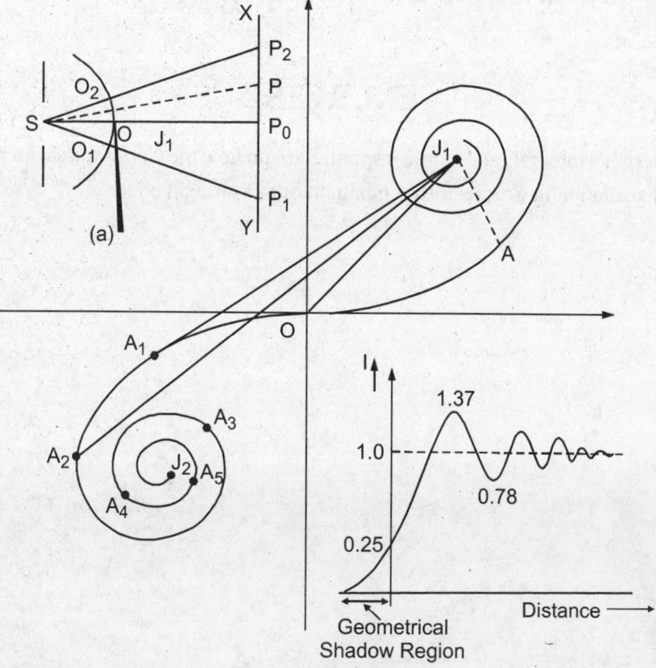

Fig. 22.3.

Intensity at the point P_0. P_0 is the edge of geometrical shadow. The disturbance at P_0 is due to exactly upper half wavefront. Thus the amplitude at P_0 is half, or the intensity one quarter, of that due to unobstructed wavefront. One complete branch of the Cornu's spiral is effective. The intensity at P_0 is $(OJ_1)^2 = 1/4\ (J_1J_2)^2$ = 1/4th of intensity due to complete unobstructed wavefront [= 1/4 if it is assumed that $(J_1J_2)^2 = 1$].

Intensity at the point P_1. When P moves down to P_1 inside the geometrical shadow, the half wavefront below the pole O_1 is completely obstructed. So the lower branch of the spiral is again ineffective. In addition, the elements of wavefront immediately above O_1 are obstructed so that the portion near O of the upper branch of the spiral is ineffective also. The effective part of the spiral commence at A_1, a distance l_1 along the spiral from O where l_1 is determined by $s_1 : l_1 = \sqrt{\dfrac{2(a+b)}{\lambda ab}}\, s_1$. Thus the effective part of the spiral is that between A_1 and J_1. The intensity at P_1 is proportional to $(A_1J_1)^2$ [*i.e.*, $1/2\ (A_1J_1)^2$]. As P_1 moves away from P_0, s_1 and hence l_1 increases and point A_1 moves round the spiral so that $(A_1J_1)^2$ gradually decreases. This means that the intensity at P_1 gradually decrease to zero as P_1 recedes from P_0.

Intensity at the point P_2. Consider a point P_2 above P_0. The whole of the upper part of the wavefront (that above SO_2P_2) is unobstructed. So the whole of the upper branch of the spiral is effective. The corresponding part of the spiral is that between O and A_2. Therefore, the total effective part of the spiral is that between A_2 and J_1. So the intensity at P_2 is $1/2\ (A_2J_1)^2$. As P_2 moves away form P_0, A_2 moves round the lower branch of the spiral. It will be seen that the intensity at P_2 increases and reaches a maximum when A_2 reaches A_3. The intensity is then $1/2\ (A_3J_1)^2$ and is greater than that due to unobstructed wavefront. It is approximately 1.37. As P_2 moves farther above P_0, A_2 moves farther round the spiral. The intensity passes through a minimum (approximately 0.78) when P_2 reaches the point for which A_2 reaches A_4. It passes through successive maxima and minima $(A_5J_1)^2$, $(A_6J_1)^2$, etc., as P_2 recedes from P_0. As P moves farther above P_0, the terminal point of the effective part of the spiral approaches J_2 and the intensity approaches that due to the unobstructed wavefront, *i.e.*, $1/2(J_1J_2)^2$. Fig.22.3 shows how the intensity passes through a series of maxima and minima and approaches unity.

EXERCISE

1. Discuss Fresnel's integrals and Cornu's spiral. Explain how the Cornu spiral can be used to obtain the intensity distribution in Fresnel diffraction pattern of a straight edge.

CHAPTER

23

LASERS II

23.1. Nd: YAG Laser

This is a four-level laser system. In this system, the trivalent neodymium Nd^{3+} ion is incorporated in the $Y_3Al_5O_{12}$ crystal lattice. This crystal is represented by the symbol YAG–acronym for *Y*ttrium *A*luminium *G*arnet. A simplified energy level scheme for Nd: YAG is shown in Fig. 23.1.

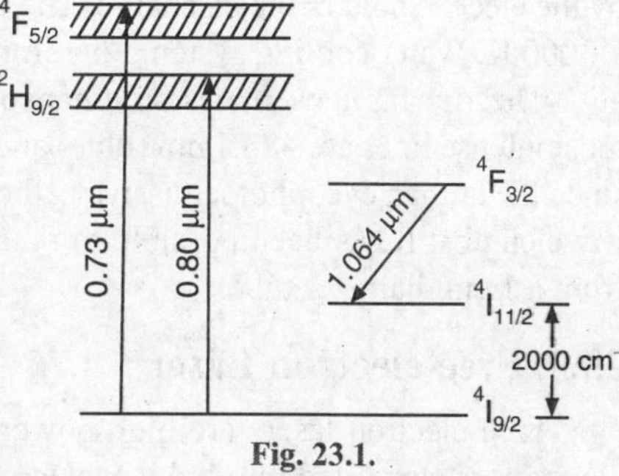

Fig. 23.1.

The levels shown in the figure are all associated with the inner 4*f* shell. The 4*f* shell is screened from external fields by the 5*d* and 6*s* outer shells. So the lower levels are very narrow. The Nd^{3+} ion surrounds itself with several oxygen atoms that largely shield from its surroundings. Some of the upper levels, however, are broadened due to the interaction with 5*d* configuration.

F levels are metestable levels. The probability of transition to the level $^4I_{11/2}$ is an order of magnitude higher than that to the other members of the multiplet. This level is about 2000 cm^{-1} ($v = 6 \times 10^{13}$ Hz) above the ground state. Hence at room temperature $\hbar\omega >> kT$. The $^4I_{11/2}$ level, therefore, is virtually empty in thermal equilibrium. Any population in the level $^4F_{3/2}$ gives rise to an inversion. Transition from the level $^4I_{11/2}$ to the lowest level takes place by non-radiative processes and is quite fast. The laser transition $^4F_{3/2} \rightarrow {}^4I_{11/2}$, has a wavelength 1.064 μm. This laser operates in the *continuous wave* (c.w) mode.

23.2. Argon Ion Laser

Fig. 23.2 shows Energy level scheme of Ar^+. 4*p* band corresponds to $3p^4 4p$ configuration and 4*s* to $3p^4\,4s$ configuration. Each of these configurations consists of several levels. Laser oscillations occur in the 4*p*-4*s* transition. The argon ion ground state is ~ 15.6 eV above the ground state of the neutral atom. The levels involved in the laser transition lie ~ 20 eV above the argon ion ground state *i.e.,* in the neighbourhood of 35 eV. There are three possible ways by which the 4*p* level can be populated: (*a*) electron collisions with Ar^+ ions in their ground state, (*b*) collision with ions in metastable states, and (*c*) radiative cascade from higher states. An interesting feature of this system is that the rate of excitation of the upper level is less than that of the lower level. In fact the former is

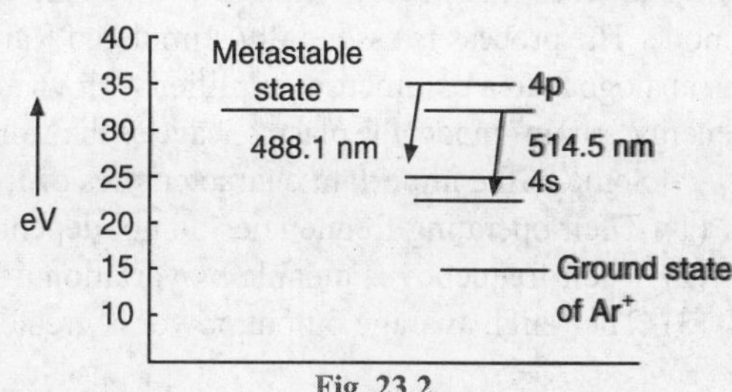

Fig. 23.2.

about one-half of the latter. This condition is not at all satisfactory for creating population inversion. However, the lower laser level $4s$ is radiatively connected with the ground state and its lifetime is very short $\sim 10^{-9}$ sec; while that of the $4p$ level is 10^{-8} sec. The inversion is achieved because the lower level is depopulated considerably faster than the upper level.

Fig. 23.3 shows the schematic diagram of an Argon ion laser. The Ar^+ laser tube has two windows at the end, inclined at Brewster angle. Because of the high current density there is a continuous migration of ions towards the cathode and electrons towards the anode. Because of the low mobility, the ions tend to pile up at the cathode where they are neutralized and diffuse slowly back into the discharge. To equalize their distribution, the discharge tube is provided with a return tube. The ions are accelerated by the electric field in the discharge. The Doppler width is ~ 3500 MHz which implies a temperature ~ 3000 K. Water cooling system, therefore is necessary.

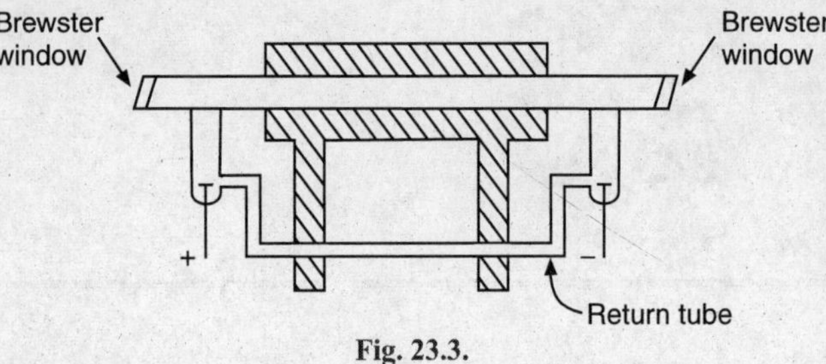

Fig. 23.3.

The Ar^+ laser oscillates simultaneously on various transitions. The most important and the most intense lines are 488.1 nm (blue) and 514.5 nm (green). These two wavelengths are very useful since the human eye, photoemissive cathode and the photographic films have peak response at these wavelengths. It has been possible to obtain very high output powers from Ar^+ lasers up to 100 W from a 1 cm diameter tube.

23.3. Free-electron Laser

Free electron lasers are high power tunable lasers in which the kinetic energy of a beam of electrons is converted into light. McDermolt has demonstrated the feasibility of *free-electron laser* (FEL). An electron beam emerging from an accelerator, passes through a region in which a strong longitudinal field has a spatially periodic ripple imposed on it. The ripple is produced by inserting within the solenoid, a set of aluminium rings that carry currents in alternate directions. Light emitted, while the electrons are in this region, is reflected by annular mirrors, which define the optical cavity.

The successful operation of FEL requires the maintenance of a very precise relationship between the phase of the radiation field and the position of the electron relative to the periodic magnetic field of the wiggler. The wavelength of peak gain is determined by the electron beam's energy, the strength of the magnetic field and the period over which the magnetic field reverses. These parameters are design variables and, hence, the laser can be operated at any chosen wavelength. The FEL, thus, has a broad range of application.

Accelerators and storage rings generally used for high energy physics, have provided electron beams of sufficient energy to yield laser output in the near infrared at 1.6 μm and even in the visible at 650 nm.

The non-linear interaction of the signal wave in the electron plasma and the rippled field produces a disturbance that propagates as a wave along the beam and reinforces the scattering into the laser mode. The process is essentially stimulated Raman scattering from the electron beam and the laser is analogous to a parametric amplifier with virtual photons of the magnetic field serving as the high energy "pump" mode; the plasma wave as the "idler" mode and the scattered light as the "signal" mode.

Some of the important characteristics of free electron lasers are:

(1) Their operating frequencies do not depend upon bound energy levels of atoms or molecules;
(2) Their frequency is tunable by variation of electron energy;
(3) Their high average output power is greater than 10 kW;

(4) Their high efficiency is greater than 10%; and

(5) The radiation generated by them spreads over the electromagnetic spectrum from 100 nm to ~ 1 mm.

23.4. Resonators

The condition $N_2 >> N_1$ in itself is not enough for laser oscillations to occur. A large number of excited atoms decay spontaneously and each spontaneous photon can initiate in the active medium many other stimulated transitions. Photon states are characterized by a definite energy, momentum and polarization of photons. The photons occupying the same state are indiscernible. They are discernible only when they are in different states. Since the spontaneous photons initiating the transitions are generally in different photon states, they are useless from the point of view of coherent amplification. In order to obtain coherent radiation, it is necessary to restrict the number of photon states. The selectivity of photon states and the positive feedback that is necessary to offset the various types of losses, can be obtained by enclosing the radiation in a resonator tuned to the frequency concerned.

The Confocal Resonator: In this system, spherical mirrors are employed in a confocal or near confocal configuration (Fig. 23.4). In a confocal resonator, the radius of curvature of each mirror is equal to the cavity length. The region marked by the dotted lines is covered by the bouncing light beams. The surface that includes this volume is a hyperboloid of revolution. The diffraction losses of such a configuration are low and the alignment is much less critical. The divergence, however, is large, but can be reduced by placing a circular aperture in front of the exit mirror.

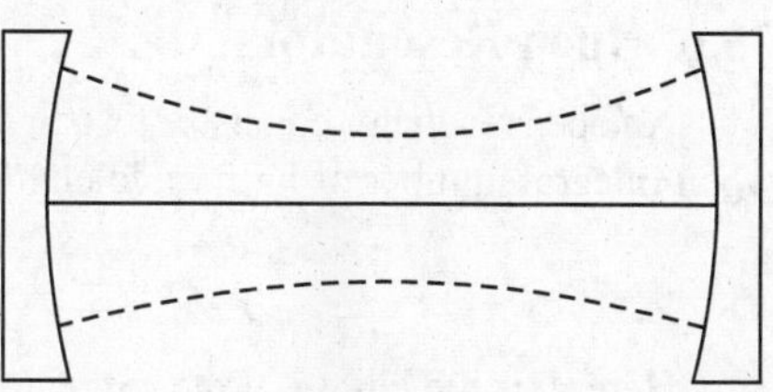

Fig. 23.4.

23.5. Vibrational Modes of a Resonator

The electric field of an electromagnetic wave satisfies the equation

$$\nabla^2 E = \frac{1}{c^2}\frac{\partial^2 E}{\partial t^2}$$

where c is the velocity of light.

Suppose the waves are confined to a region inside a rectangular cavity, the walls of which are assumed to be perfectly conducting. This is the simplest cavity whose resonance frequencies can be easily calculated. The waves in the cavity are obviously standing waves. Note that we are considering here a cavity which does not contain any amplifying active medium. Such a resonator is called a *passive resonator* as distinguished from an *active resonator* which contains an active medium.

For simplicity let us assume the cavity to be a cube of length L (Fig. 23.5). Since $E_x(t)$ must vanish on the planes $y = 0$; $y = L$ and $z = 0$; $z = L$; $k_y = \frac{n\pi}{L}$; $k_z = \frac{q\pi}{L}$. Similarly, $E_y(t)$ must vanish on the planes $x = 0$, $x = L$ and $z = 0$, $z = L$,

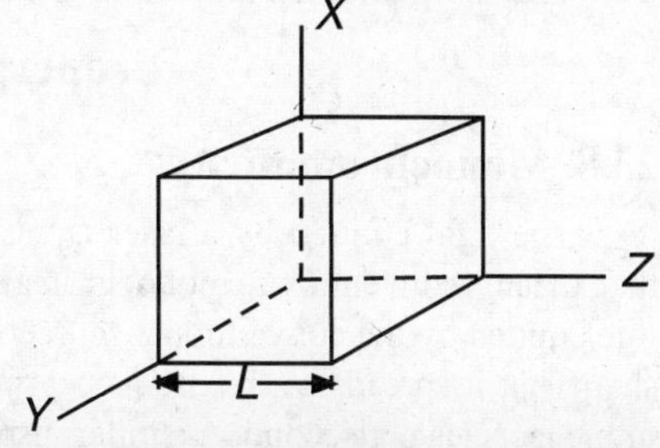

Fig. 23.5.

$$k_x = \frac{m\pi}{L}$$

where $m, n, q, = 0, 1, 2$.

Thus, the boundary conditions force the wave vector components to take discrete values specified by m, n, q. Each set of integers (m, n, q) defines a mode of vibration. These modes are known as Transverse Electromagnetic Modes and are designated by TEM_{mnq} where m, n are the transverse indices and q the longitudinal index. Each mode has a characteristic distribution of amplitude and phase. This distribution in a plane perpendicular to the resonator axis is described by transverse indices m, n. A given combination of m, n characterizes a *transverse mode* regardless of the value of q. This is designated as a TEM_{mn} mode. For each combination of m, n, there correspond a number of modes with different values of q. These are called *longitudinal* or *axial modes*. There is, thus, an infinite number of resonant frequencies of the cavity corresponding to the different values of m, n, q. Any excitation of the electromagnetic field observed in the resonator can be expressed as a linear combination of these modes.

23.6. Open Resonators

An open resonator consists of a pair of reflecting mirrors set on an optical axis. Resonance occurs when integral number of half-wavelengths cover the length between the mirrors.

$$L = \frac{\lambda q}{2}.$$

Here, L is the length of the resonator, *i.e.*, the separation of the two mirrors. $q = 0, 1, 2....$

23.7. The Quality Factor Q

For a definite field configuration, the cavities have some discrete resonant frequencies. This means, if we try to excite a particular mode of oscillation in the cavity, the right sort of field will not be built up unless the exciting frequency is equal to the resonant frequency.

The resonant frequency in a resonator, however, is not quite sharp. Appreciable excitation occurs over a narrow band of frequencies around the resonant frequency. This smearing out of the frequency is due to the various losses associated with the modes in the cavity (scattering, absorption, diffraction etc.). The measure of these losses is often expressed in terms of the *quality factor-Q* of the cavity which is defined as

$$Q = \frac{\omega \times \text{energy stored in the mode}}{\text{energy dissipated per second in the mode}}.$$

That is, smaller the amount of energy dissipated, higher is the Q of the cavity.

Properties of a Laser Beam

23.8. Monochromaticity

The light emitted by a laser is vastly more monochromatic than that of any conventional monochromatic source. An inspection of a line emitted by the conventional source shows that it is never perfectly sharp, but it spreads over a frequency range of the order of thousands of megacycles per second. A similar inspection of the light from a laser would reveal virtually very little spreading (Fig. 23.6).

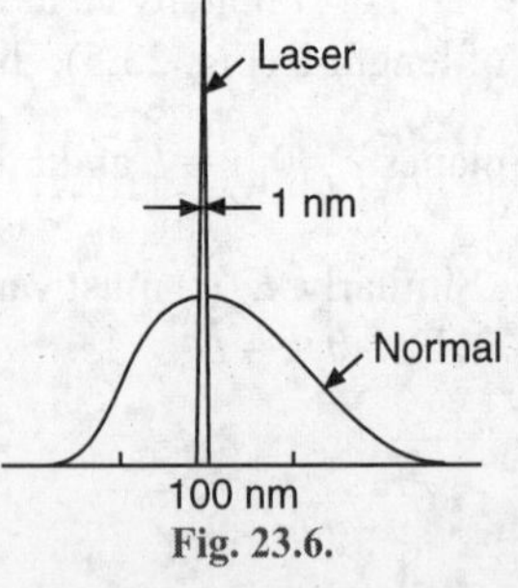

Fig. 23.6.

No light source is capable of producing absolute monochromatic light. The spread in frequency of a line is characterized by the line width $\Delta\nu$. The degree of non-monochromaticity of a wave may be defined as its relative bandwidth given by

$$\xi = \frac{\Delta \nu}{\nu_0}.$$

where ν_0 is the central frequency of the light beam. Absolute monochromaticity for which $\Delta\nu = 0$ is an unattainable goal. We can say that the laser light has a higher degree of monochromaticity. The output from a laser is very nearly a perfectly monochromatic sine wave with a very small bandwidth about 1 kHz.

The output from a high quality stable gas laser, locked to the centre of the absorption line has a bandwidth $\Delta\nu \sim 500$ Hz *i.e.*, $\Delta\lambda = 10^{-9}$ nm at $\lambda = 600$ nm. Even a poor quality solid state laser can have a bandwidth of $\Delta\nu \sim 10^9$ Hz *i.e.*, $\Delta\lambda \sim 10^{-3}$ nm. For example, the linewidth of a line emitted by ruby, ordinarily is 0.3 nm, while, the linewidth of the laser line with the same material is 5×10^{-5} nm.

23.9. Directionality

An ordinary source of light (like a sodium lamp or an incandescent source) radiates in all possible directions. Lasers, on the other hand, emit only in one direction. The directionality of laser beams is usually expressed in terms of the *full angle beam divergence,* which is twice the angle that the outer edge of the beam makes with the axis of the beam. The outer edge of a beam is defined as a point at which the strength of the beam has dropped to $1/e$ times its value at the centre. When a light beam of wavelength λ with planar wavefront radiates from an aperture of diameter d, the beam propagates as a parallel beam for a distance of about d^2/λ. Then the beam begins to spread linearly with distance because of the unavoidable effects of diffraction (Fig. 23.7).

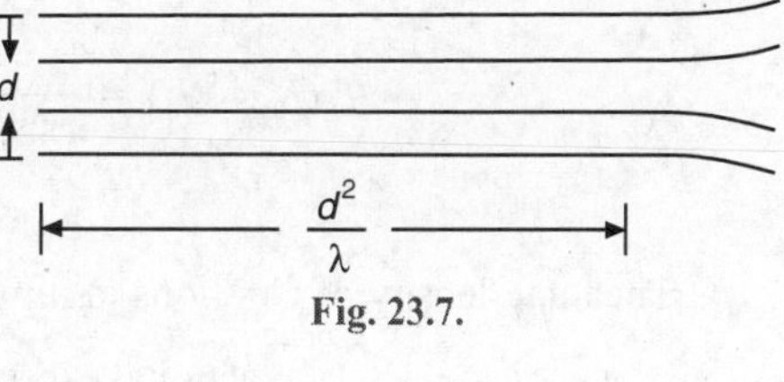

Fig. 23.7.

The angular spread $\Delta\theta$ of the far field beam is related to the aperture diameter d by

$$\Delta\theta = \frac{\lambda}{d}.$$

For a typical laser, the beam divergence is less than 0.01 milliradian. That is, the beam spreads less than 0.01 mm for every metre. Output in a laser beam is many millions of times more concentrated than the best search-light available. Thus laser beam is highly parallel and directional.

The beam from the search light spreads to about a kilometre in diameter for every kilometre that the light traverses. One can imagine its spread on the surface of the moon, which is at a distance of 384,400 km, if it ever reaches the moon. As against this, the spread of the laser beam on the surface of the moon would be just a few kilometres.

Intensity. As the laser beam has the ability to focus over as small an area as 10^{-6} cm^2, it is highly intense beam. For example, a 1 watt laser when focussed over an area 10^{-6} cm^2, has intensity 10^6 watt/cm^2 whereas ordinary 100 watt lamp will not have intensity even 10,000th of this. It is because the light from the lamp cannot be focussed over an area less than a cm^2.

23.10. Threshold Power Calculation for the Ruby Laser

We consider a ruby laser. We first calculate the magnitude of population inversion required for laser oscillation. We consider the laser to be oscillating at the frequency corresponding to the peak of the emission line. We assume a concentration of 0.05% of Cr^{3+} ions in the crystal. This corresponds to a population of $N = 1.6 \times 10^{25}$ Cr^{3+} ions/m^3. The line is homogeneously broadened and the value of $g(\omega)$ at the peak of the line is $2/(\pi\Delta\omega)$. Hence the threshold population inversion density is

$$(N_2 - N_1)_t = \frac{4\pi^2}{\lambda^3} \frac{\Delta\nu}{\nu} \frac{t_{sp}}{t_c} \qquad \text{... (1)}$$

Here, N_1 and N_2 are the number of atoms in energy states 1 and 2 respectively.

$t_{sp} \rightarrow$ spontaneous relaxation time of the upper laser level,

$t_c \rightarrow$ cavity lifetime.

For ruby laser transition, we have

$$\nu \simeq 4.3 \times 10^{14}\ \text{s}^{-1},\ \Delta\nu \simeq 1.5 \times 10^{11}\ \text{s}^{-1},\ t_{sp} \simeq 3 \times 10^{-3}\ \text{s};$$

$$\lambda \simeq \frac{0.6943}{1.76} \times 10^{-6} \simeq 4 \times 10^{-7}\ \text{m} \quad (\because \text{the refractive index of ruby is 1.76})$$

t_c is calculated using the formula

$$t_c = \frac{2n_0 d}{c \ln\left(\dfrac{1}{1-x}\right)} \qquad \text{... (2)}$$

$d \rightarrow$ length of the optical cavity,

$n_0 \rightarrow$ refractive index of the medium filling the cavity,

$x \rightarrow$ fractional loss per round trip.

Let us assume $d = 0.05$ m and $x = 10\% = 0.1$. Then, we get

$$t_c \simeq 6 \times 10^{-9}\ \text{s}.$$

Substituting these values in Eq. (1), we get

$$(N_2 - N_1)_t = \frac{4\pi^2}{(4 \times 10^{-7})^3}\left(\frac{1.5 \times 10^{11}}{4.3 \times 10^{14}}\right)\left(\frac{3 \times 10^{-3}}{6 \times 10^{-9}}\right)$$

$$\simeq 1.1 \times 10^{23}\ \text{Cr}^{3+}\ \text{ions/m}^3$$

The total density of Cr^{3+} ions in ruby = 1.6×10^{25} m^{-3}.

$$\left.\begin{array}{l}\text{Fractional excess}\\ \text{population required}\end{array}\right\} = \frac{1.1 \times 10^{23}}{1.6 \times 10^{25}} \simeq 0.7 \times 10^{-2}.$$

Hence, the fractional degree of inversion required is very small, about 0.7 in 100.

Let us calculate the minimum power required to maintain population inversion.

No. of atoms decaying per unit time from the upper laser level = N_2/t_{sp}.

Let ν_p represent the average pump frequency.

For each atom lifted to level 2, we have to supply at least an amount of energy $h\nu_p$.

Hence in order to maintain N_2 atoms in the level 2, the minimum power P to be spent is $N_2 h\nu_p/t_{sp}$ per unit volume of the active material. Thus,

$$P = \frac{N_2\, h\nu_p}{t_{sp}} \qquad (3)$$

For a three-level laser system, $N_2 \simeq \dfrac{N}{2}$ (4)

Thus, the minimum pumping power per unit volume required to maintain population inversion in a three level laser system is

$$P_t \simeq \frac{N}{2}\frac{h\nu_p}{t_{sp}} \qquad (5)$$

Let us take the average pumping frequency as $\nu_p = 6.25 \times 10^{14}$ Hz.

$$P_t \simeq \frac{(1.6 \times 10^{25})}{2} \times \frac{6.6 \times 10^{-34} \times 6.25 \times 10^{14}}{3 \times 10^{-3}}$$

$$\simeq 1100 \times 10^6 \text{ Wm}^{-3} \simeq 1100 \text{ W/cm}^3.$$

Let us assume that the efficiency of the pumping source is 25% and also that only 25% is absorbed in passage through the ruby rod. Then the electrical threshold power comes out to be about 18 kW/cm^3 of the active material. This is consistent with the threshold powers determined experimentally.

EXERCISE

1. A laser beam emerging from a laser tube operating at 800 nm has a cross sectional diameter of 2 mm. The diameter of the beam at a distance of 1 km is approximately given by

(A) 10 mm (B) 8 cm (C) 80 cm (D) 10 m **(Gate-1999)** [**Ans.** (C)]

Explanation. Angle of emergence,

$$\theta = \frac{\lambda}{D} = \frac{800 \times 10^{-9}}{2 \times 10^{-3}} = 400 \times 10^{-6} \text{ radian}$$

$\therefore$ Diameter of the beam 1000 m away = $2 \times (400 \times 10^{-6})10^3 = 0.8$ m = **80 cm**

2. A laser beam of wavelength 600 nm with a circular cross section having a radius of 10 mm falls normally on a lens of radius 20 mm and focal length 10 cm. The radius of the focussed spot is appxorimately

(A) 0.3 μm (B) 0.6 μm (C) 3.0 μm (D) 6.0 μm

(Gate-2002) [**Ans.** (C)]

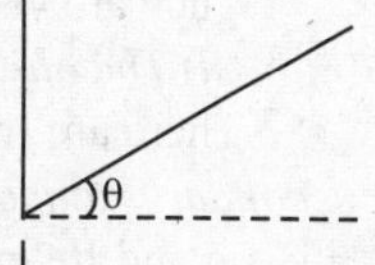

Fig. 23.8.

Explanation. $\theta \simeq \dfrac{\lambda}{D} = \dfrac{600 \times 10^{-9}}{20 \times 10^{-3}} = 3 \times 10^{-5}$ radian

$$R = \theta \times d = (3 \times 10^{-5})(10 \times 10^{-2}) = 3 \times 10^{-6} \text{ m}$$

CHAPTER 24

RAINBOW

24.1. Rainbow

Rainbows are formed by dispersion and internal reflection of sun-rays by rain drops.

They are visible only when (*i*) *The sun is on the back of the observer*

(*ii*) *The altitude of the sun rays is less than* 42°.

There are two types of rainbows. One is called a *primary rainbow* and the other is known as *secondary rainbow.* In primary rainbow, the outer arc is red, the inner arc is violet. The other colours mix up and lie in between two extreme colours. The primary rainbow is the result of one internal reflection. Secondary rainbow is produced by two internal reflections. Hence the order of colours is reversed, *i.e.,* the outer arc is violet and the inner arc is red. The secondary rainbow is quite faint as most of the energy is lost in two reflections.

24.2. Theory of Primary Rainbow

A ray of light *SA* from the sun is incident on a spherical raindrop with centre *O* (Fig. 24.1). It is refracted along *AB* and suffers total internal reflection at *B*. It undergoes refraction at *C*. Finally the ray emerges as *CE*. Since the drop behaves as a prism, white light incident at *A* is also dispersed into its constituent colours and the emergent light appears in the form of coloured arcs.

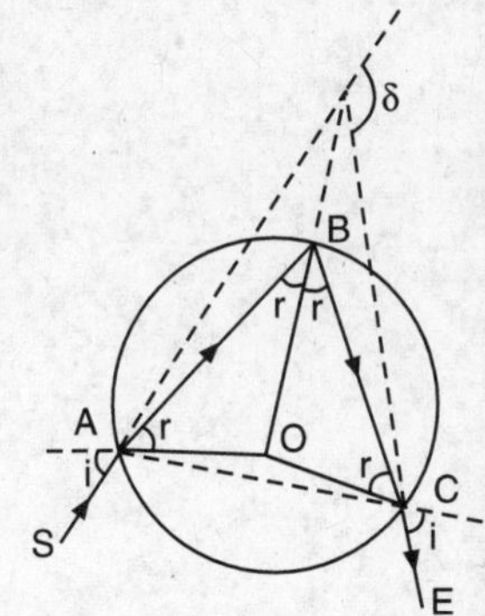

Fig. 24.1.

At point *A*, the angles of incidence and refraction are i and r respectively.

The ray is deviated at points *A, B* and *C.*

The deviation produced at $A = (i - r)$

The deviation produced at $B = (180 - 2r)$

The deviation produced at $C = (i - r)$

Let δ be the total angle of deviation produced.

$$\therefore \quad \delta = (i - r) + (180 - 2r) + (i - r)$$

i.e.,
$$\delta = 2i - 4r + 180 \qquad ...(1)$$

The deviation will be minimum if $\dfrac{d\delta}{di} = 0$

$$\therefore \quad \frac{d\delta}{di} = 2 - 4\frac{dr}{di} = 0$$

or
$$\frac{dr}{di} = \frac{1}{2} \qquad ...(2)$$

Let n be the refractive index of water.

Using Snell's law for refraction, we get

$$n = \frac{\sin i}{\sin r}$$

$$\sin i = n \sin r \qquad ...(3)$$

Differentiating with respect to i, we get

$$\cos i = n \cos r \frac{dr}{di} = n \cdot \frac{1}{2} \cdot \cos r$$

or $$\cos^2 i = \frac{1}{4} n^2 \cos^2 r$$

or $$4 \cos^2 i = n^2 (1 - \sin^2 r) = n^2 - n^2 \sin^2 r$$

or $$4 \cos^2 i = n^2 - \sin^2 i, \qquad [\because \sin i = n \sin r, \text{ by Eq. (3)}]$$

or $$4 \cos^2 i = n^2 - 1 + \cos^2 i,$$

or $$3 \cos^2 i = n^2 - 1,$$

$\therefore$ $$\cos i = \sqrt{\left(\frac{n^2 - 1}{3}\right)} \qquad ...(4)$$

The refractive indices of water for violet and red colours are 1.343 and 1.329 respectively. Using Eq. (4), we get the values of the angle of incidence i as 58.8° and 59.6° respectively. Next using relation given by Eq. (3), we get the respective values of r as 39.6° and 40.5°. Knowing i and r, the respective values of δ and $(180 - \delta)$ can be found by using relation (1). For violet and red, the values of $(180 - \delta)$ come out to be 40.8° and 42.8° respectively. This shows that the emergent rays from the water drop are confined within right circular cones of semi-vertical angles of 40.8° and 42.8° for violet and red respectively (Fig. 24.2). The angle of inclination of red light is more on the eye than the violet. Therefore, the outside of the rainbow appears red and the inner violet. The other spectral colours lie in between violet and red in their order.

Fig. 24.2.

24.2. Theory of Secondary Rainbow

The secondary rainbow is formed by two internal reflections inside the drop. A ray of light SA from the sun is incident at A on the drop (Fig. 24.3). It undergoes refraction at A, and total internal reflections at B and C. Finally after refraction at D, it emerges along DE. O is the centre of the drop.

Let i be the angle of incidence and r the angle of refraction at A.

The deviation produced at $A = (i - r)$

The deviation produced at $B = (180 - 2r)$

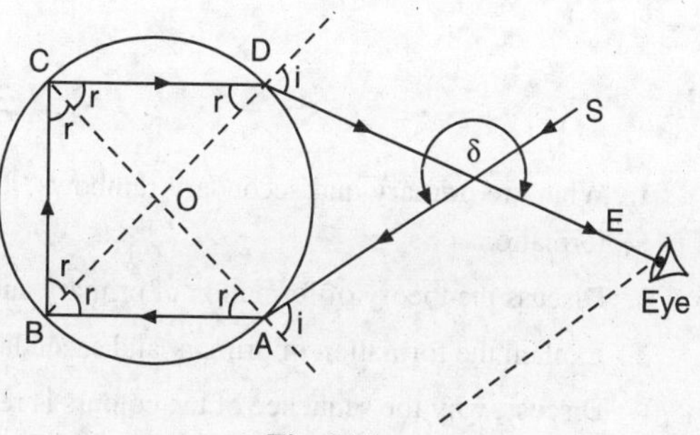

Fig. 24.3.

The deviation produced at $C = (180 - 2r)$

The deviation produced at $D = (i - r)$.

Let δ be the total angle of deviation produced.

$$\delta = (i - r) + (180 - 2r) + (180 - 2r) + (i - r)$$

$$\delta = 2i - 6r + 360 \qquad \text{...(1)}$$

For the angle of deviation to be minimum, $\frac{d\delta}{di} = 0$

$$\frac{d\delta}{di} = 0 = 2 - 6\frac{dr}{di}$$

$$\therefore \qquad \frac{dr}{di} = \frac{1}{3} \qquad \text{...(2)}$$

Let n be the refractive index of water.

Using Snell's law, $\sin i = n \sin r$ (3)

Differentiating with respect to i,

$$\cos i = n \cos r \times \frac{dr}{di} \qquad \text{...(4)}$$

$$\cos i = n \cos r \times \frac{1}{3}$$

$$\cos^2 i = \frac{n^2 \cos^2 r}{9}$$

$$9 \cos^2 i = n^2 [1 - \sin^2 r] = n^2 - n^2 \sin^2 r = n^2 - \sin^2 i = n^2 - [1 - \cos^2 i]$$

$$8 \cos^2 i = n^2 - 1$$

$$\therefore \qquad \cos i = \sqrt{\frac{n^2 - 1}{8}} \qquad \text{...(5)}$$

n of water for violet and red are 1.343 and 1.329 respectively. Hence the values of i, r and δ can be calculated using equations (5), (3) and (1) respectively. Angle δ for violet and red comes out to be 234° and 231° respectively. Hence the angles at which these colours are seen are given by $(\delta - 180)$. Thus violet is seen at the semivertical angle of 54° and red is seen at an angle of 51° (Fig. 24.4).

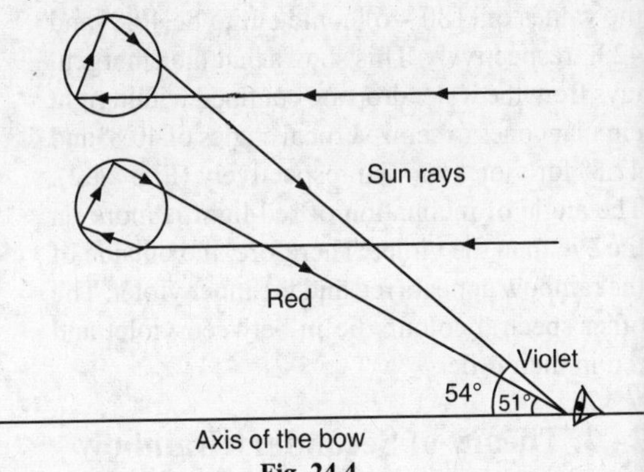

Fig. 24.4.

EXERCISE

1. What are primary and secondary rainbows? Derive and compare the conditions of incidence for their formation. *(Kurukshetra 75, Punjab 76)*
2. Discuss the theory of formation of primary and secondary rainbows.
3. Explain the formation of primary and secondary rainbows.

   Discuss, why the sequence of the colours is reverse in these two. *(Punjab 76, Delhi 78, Lucknow 74, Bihar 73)*

4. A ray of light is incident on the surface of a transparent sphere of refractive index 1.32. After refraction, it is internally reflected (not total) and then refracted out of the sphere. Find the deviation suffered by the ray and the angle of incidence to the nearest degree, so that the deviation may be a minimum.

(*Culcutta University, 1949*)

Solution. For one internal reflection,

$$\cos i = \sqrt{\left(\frac{n^2 - 1}{3}\right)} = \sqrt{\frac{(1.32)^2 - 1}{3}} \quad i.e., i = 60° \text{ (approx.)}$$

$$\sin r = \frac{\sin i}{1.32} = \frac{\sin 60°}{1.32} = \frac{0.8660}{1.32} = 0.6560 \qquad \therefore r = 41°$$

$$\delta = 2(i - r) + (180° - 2r) = 2(60° - 41°) + (180° - 2 \times 41°) = 136°.$$

5. A ray of light is incident at an angle of 60° on a sphere of material whose index of refraction is $\sqrt{3}$ and emerges after one reflection. Show that its direction on emergence is parallel to the original direction. (The problem is solved if total deviation is proved to be equal 180°)

6. In what part of the sky is a rainbow seen in the early morning and how would its appearance change to a man ascending vertically from the ground in a balloon? (*Culcutta University, 1949*)

CHAPTER 25

MAGNETO-OPTICS AND ELECTRO-OPTICS

25.1. Introduction

In support of the electromagnetic character of light, there are two classes of optical experiments which demonstrate the interaction between light and matter. Experiments which depend for their action on an applied magnetic field are classed under *magneto-optics.* Experiments which depend for their action on an electric field are classed under *electro-optics.*

Magneto-optics	*Electro-optics*
Zeeman effect	Stark effect
Inverse Zeeman effect	Inverse Stark effect
Voigt effect	Electric double refraction
Cotton-Mouton effect	Kerr electro-optic effect
Faraday effect	
Kerr magneto-optic effect	

Certain materials, under the influence of an electric field, change their polarization characteristics; in particular, they change in birefringence. Such materials are called *electro-optic materials.* Other materials show similar changes when exposed to a magnetic field; these are called *magneto-optic materials.*

25.2. Zeeman Effect

When a source of light, giving line spectrum is placed in a magnetic field, the spectral lines are split up into two or more components. Doublets, triplets and even more complex systems are observed. This is called *Zeeman effect.* If the magnetic field is very strong, each spectral line is split up into two components in the longitudinal view and into three components in transverse view. This is called *Normal Zeeman effect.* When the magnetic field is comparatively weak, each line splits into more than three components. This is called *Anomalous Zeeman Effect.*

Experimental Arrangement: The source of light L producing line spectrum (*e.g.*, a sodium vapour lamp) is placed symmetrically between the pole-pieces PP of a very strong electromagnet (Fig. 25.1). The pole-pieces have longitudinal holes drilled through them. The spectral lines are observed with the help of a high resolving power spectrograph (S).

The Zeeman effect may be observed in two ways:

1. Transverse View. When the spectral line is viewed perpendicular to the direction of magnetic field, it is observed that the spectral line of frequency ν is split up symmetrically into three components of frequencies ν, $\nu - \Delta\nu$ and $\nu + \Delta\nu$. The central line of frequency ν is linearly polarised with vibrations parallel to the magnetic field. The component lines on either side are linearly polarised at right angles to the magnetic field. This effect is called the *normal transverse Zeeman effect.*

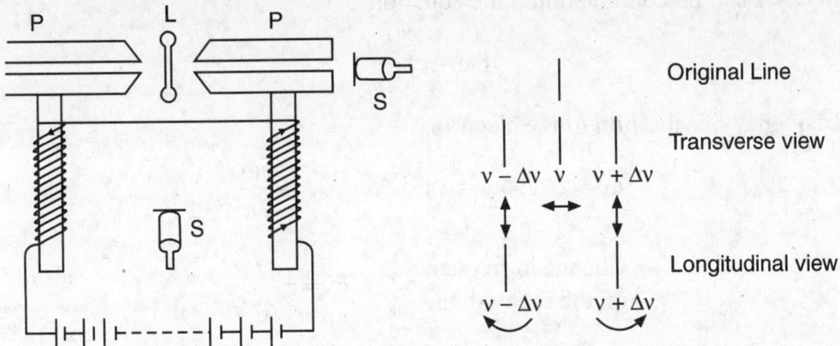

Fig. 25.1.

2. Longitudinal view: When the light is viewed along the direction of magnetic field, the same two additional lines with frequencies ν – Δν and ν + Δν are observed. The central line having frequency ν is missing. These two lines are circularly polarised in opposite directions. This effect is called the *normal longitudinal Zeeman effect.*

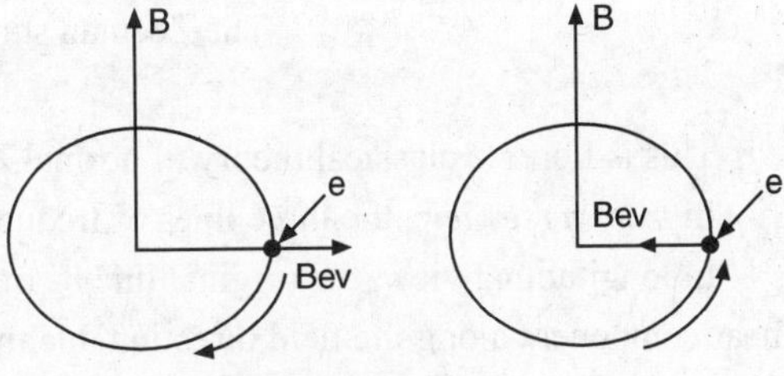

Fig. 25.2.

Expression for the Zeeman shift

Consider an electron in the atom moving in a circular orbit of radius r with a linear velocity v and angular velocity ω. Let e be the charge of the electron and m its mass (Fig. 25.2).

$$\left.\begin{array}{l}\text{The centripetal force on}\\ \text{the electron towards the}\\ \text{centre in the absence of}\\ \text{the magnetic field}\end{array}\right\} = F = \frac{mv^2}{r} = m\omega^2 r \quad (1)$$

Now let an external magnetic field of induction B be applied in a direction perpendicular to the plane of the orbits of the two circular components and directed towards the reader. Then an additional radial force of magnitude Bev acts on the electron. The direction of this force will be outwards from the centre for clockwise motion but inwards towards the centre for anticlockwise motion. The resulting complex motion of the electron subjected to an additional radial force is called *Larmor precession.* This produces a change in the angular velocity without any change in the form of the orbit.

Let $\delta\omega$ be the change in angular velocity caused by the field. For the circular motion in the *clockwise* direction, the additional radial force is directed away from the centre.

$$\therefore \quad F - Bev = m(\omega + \delta\omega)^2 r \quad ...(2)$$

$$m\omega^2 r - m(\omega + \delta\omega)^2 r = Be\omega r$$

or $$-2mr\omega\,\delta\omega = Be\omega r \quad \text{neglecting } (\delta\omega)^2$$

or $$\delta\omega = -Be/2m \quad ...(3)$$

For the circular motion in the anticlockwise direction, the additional radial force is directed towards the centre.

$$\therefore \quad F + Bev = m(\omega + \delta\omega)^2 r$$

or $$\delta\omega = +Be/2m \quad ...(4)$$

The two cases can be combined into the equation

$$\delta\omega = \pm\frac{Be}{2m} \quad ...(5)$$

If ν = frequency of vibration of the electron,

$$\omega = 2\pi\nu;\ \delta\omega = 2\pi\ \delta\nu \text{ or } \delta\nu = \frac{\delta\omega}{2\pi}.$$

$$\therefore \quad \left.\begin{array}{l}\text{Change in frequency}\\ \text{of the spectral line}\end{array}\right\} = \delta\nu = \pm\frac{Be}{4\pi m} \quad ...(6)$$

If ν and λ are the frequency and wavelength of the original line,

$$\nu = \frac{c}{\lambda} \text{ or } \delta\nu = \frac{-c}{\lambda^2}\,\delta\lambda.$$

$$\therefore \quad \text{The Zeeman shift} = \delta\lambda = \pm\frac{Be\lambda^2}{4\pi mc}$$

This is Lorentz classical theory of normal Zeeman effect.

In *transverse view*, the three lines of frequencies $\nu - \Delta\nu$, ν and $\nu + \Delta\nu$ are observed.

In longitudinal view, the original line of frequency ν is not observed because of the unaltered linear component along the field direction due to transverse nature of light waves.

25.3. Inverse Zeeman Effect

The Zeeman effect obtained in absorption is called the *inverse Zeeman effect.* The phenomenon is observed by sending white light through an absorbing vapor when the vapor is subjected to a uniform magnetic field. Let us consider the longitudinal Zeeman effect. Fig. 25.3 shows Zeeman patterns for a normal triplet, showing the polarization of the light. Unpolarized light of any particular frequency may be imagined as consisting of right and left circularly polarized components with all possible phase relations. Let ν_0 represents a natural resonance frequency of the vapor in the absence of a field. Then, the plus circular components of frequency $\nu_0 + \Delta\nu$ will be strongly absorbed in the presence of a field. The corresponding minus circular components of frequency $\nu_0 + \Delta\nu$ pass on through with little decrease in intensity, since to be absorbed these must have the frequency $\nu_0 - \Delta\nu$. Hence at frequency $\nu_0 + \Delta\nu$, looking against the field direction as in Fig. 25.3, right circularly polarized light is transmitted. For a thick absorbing layer this is one-half as intense as the background of continuous light (Fig. 25.4).

A similar argument can be given for $\nu_0 - \Delta\nu$.

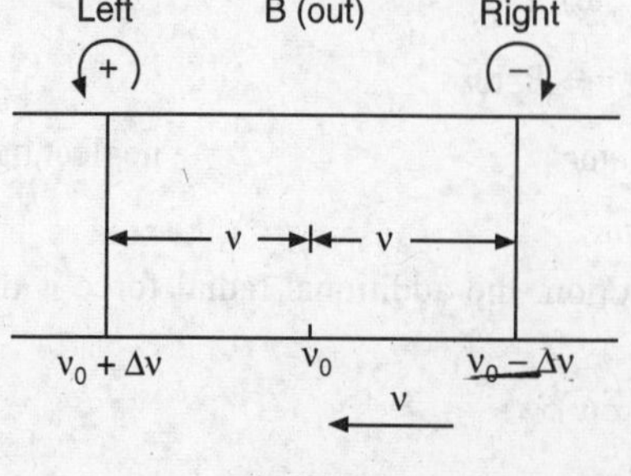

Fig. 25.3.

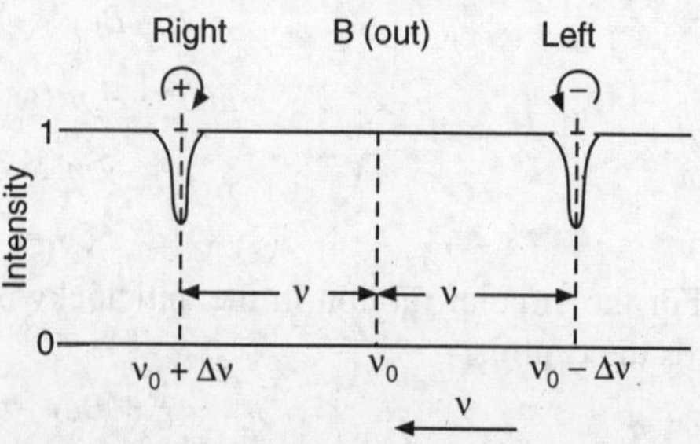

Fig. 25.4.

The Zeeman components of any spectrum line obtained in absorption along the field direction are therefore not completely absorbed, and the light that does get through is found to be circularly polarized in directions opposite to those of the corresponding components obtained in emission. This is verified by experiment even in anomalous patterns of many components.

25.4. The Faraday Effect

In 1845, Michael Faraday discovered that when a block of glass is subjected to a strong magnetic field, it becomes optically active. When plane-polarized light is sent through glass in a direction parallel to the applied magnetic field, the plane of vibration is rotated. This was one of the earliest indications of a connection between light and electromagnetism. This phenomenon is called the *Faraday effect.*

Fig. 25.5 shows Faraday's experiment. Light is passed through slab of glass mounted on top of electromagnet. The magnetic field changes the index of refraction. But it is the index for right-circularly polarized light that changes with respect to the index for left-circularly polarized light. The net result is a rotation of the plane of oscillation.

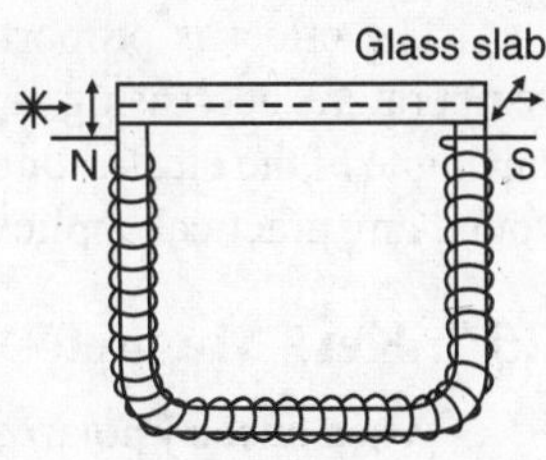

Fig. 25.5.

The effect has been observed in many solids, liquids, and gases. The rotation of the plane of polarization for light of a given wavelength is given by

$$\theta = VBl.$$

Here, B is the magnetic induction, l is the length of the path in the field, θ is the angle of rotation and V is a constant called the *Verdet constant.* Verdet's constant is defined as the rotation per unit path per unit field strength. V is different for different substances. A positive value indicates that the rotation is in the clockwise direction *when looking along the direction of the field.*

Magnetic rotation can be explained by assuming that in the presence of a magnetic field the velocity of circularly polarized light depends on the direction of rotation. This variation of velocity is associated with the fact that the absorption frequencies, and therefore the dispersion of the medium, depend on the polarization of the light. Away from an absorption band, Verdet's constant is approximately proportional to $(1/\lambda^2)$

Very strong effects are observed with thin transparent films of ferromagnetic materials. For example, for iron the effect is many thousand times larger than for water.

25.5. The Voigt Effect or Magnetic Double Refraction

In 1902, Voigt discovered that when a strong magnetic field is applied to a vapor through which light is passing perpendicular to the field, double refraction takes place. This phenomenon is called the *Voigt effect* or *magnetic double refraction.* It is attributed to the effect of the field on the absorption frequencies which become dependent on the polarization of the light. The phenomenon is readily explained from absorption and dispersion curves.

25.6. Cotton-Mouton effect

When a liquid such as nitrobenzene or carbon disulphide is exposed to a transverse magnetic field, the liquid becomes birefringent. This phenomenon is called *Cotton-Mouton effect.* This effect is 10^3 times greater than the Voigt effect. In pure liquids like nitrobenzene, very strong double refraction is observed. This double refraction is attributed to a lining up of the magnetically and optically anisotropic molecules in the applied field direction. This lining up would result whether the magnetic dipole moments of the molecules were permanent or induced by the field.

The birefringence is interpreted as due to the difference in the propagation velocities of two waves. One wave, with **D** parallel to the external magnetic field **H**, has an index of refraction represented by $n_{\parallel}$. The second wave, with **D** perpendicular to the magnetic field, has an index of refraction represented by $n_{\perp}$.

The phase shift that is produced when the two orthogonally polarized waves propagate, perpendicular to the applied magnetic field, through a distance d in the material, is given by

$$\delta = \frac{n_{\parallel} - n_{\perp}}{\lambda} d = CdH^2.$$

C is called the Cotton-Mouton coefficient.

The effect is proportional to the square of the field strength. The effect is dependent upon temperature, decreasing rapidly with a rise in temperature. The Cotton-Mouton effect is the magnetic analogue of the electro-optic Kerr effect. The Cotton-Mouton and Voigt effects are too small to have found any practical application.

25.7. Kerr Magneto-optic Effect

When linearly polarized light is reflected from the polished pole of an electromagnet, it becomes elliptically polarized, with the major axis of the ellipse rotated with respect to the incident vibrations. This is the Kerr magneto-optic effect. The *Kerr magneto-optic effect* amounts to a rotation through only a few arc min. It can be enhanced by applying a multilayer coating to the pole surface, thereby causing multiple reflections.

25.8. Stark Effect

The Stark effect is the electrical analogue of Zeeman Effect. *The Stark effect is the splitting of spectral lines due to the action of an external electric field on the radiating substance.*

Experimental study. Here the hydrogen atoms emitting spectral lines are subjected to a powerful electric field. The arrangement is shown in Fig. 25.6. The canal rays are produced in an ordinary glass discharge tube provided with a perforated cathode C. Discharge takes place between the anode A and cathode C maintained at a suitable *P.D.* A stream of canal rays is obtained behind the cathode. An auxiliary electrode F is placed parallel and close to C. A very strong electric field of several thousand volts per metre is maintained between F and C. The effect produced can be studied both transversely [Fig. 25.6 (*a*)] and longitudinally [Fig. 25.6 (*b*)].

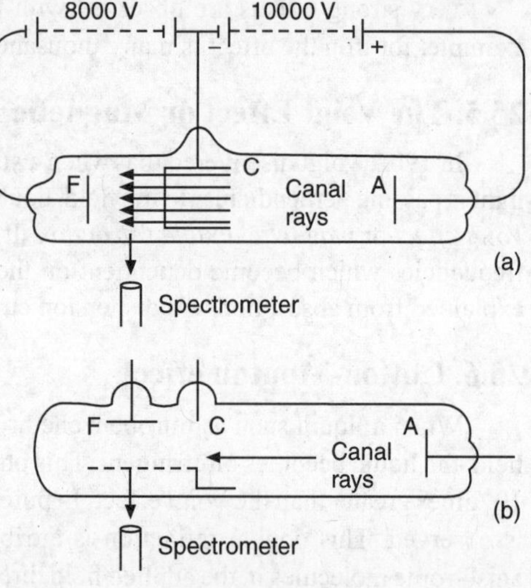

Fig. 25.6.

When viewed perpendicular to the electric field, some of the components of each line pattern are observed to be plane-polarized with the electric vector parallel to the field (p components) and the others plane-polarized with the electric vector normal to the field (s components). This is the transverse Stark effect. When viewed parallel to the field, only the s components appear, but as ordinary unpolarized light. This is the longitudinal Stark effect.

The widest Stark patterns are observed in the hydrogen and helium spectra. In the case of all other spectra, we observe a slight shift of the line, usually toward longer wavelengths. This effect is called the *quadratic Stark effect.* In hydrogen and helium, we have *linear* Stark effect. In quadratic Stark effect, the shifts are proportional to the square of the electric field strength, E^2. In linear Stark effect, the splittings depend on the first power of this field (E).

25.9. Inverse Stark Effect

The Stark effect with the lines appearing in absorption is called the inverse Stark effect. Grotrian and Ramsauer investigated the phenomenon, using a long tube containing potassium vapor at low pressure and two long parallel metal plates 1.5 mm apart. With a potential of 14 kV on the plates, the absorption lines λ 4044, λ 4047, and λ 3447 were found to be shifted from the field-free position toward longer wavelength. This shift is proportional to the square of the field strength. This is therefore a case of a quadratic Stark effect.

25.10. Electric Double Refraction

Electric double refraction is observed in the region of absorption lines. It is found that in the presence of an electric field the absorption frequencies are dependent upon the polarization of the light. Thus for frequencies near an absorption line the velocity of light will depend on the state of polarization, *i.e.*, the medium is birefringent.

Electric double refraction is related to the transverse Stark effect. Ladenberg observed the absorption of the sodium resonance lines when produced with and without a strong transverse electric field applied to the vapor. Although the shift of the lines predicted by the quadratic Stark effect was too small to observe even with very high resolving power, double refraction was observed at frequencies close to the absorption lines. This double refraction is attributed to the very small difference in the frequency of the absorption line for light polarized parallel and perpendicular to the electric lines of force.

25.11. The Kerr Electro-optic Effect

Kerr discovered that when a plate of glass is subjected to a strong electric field, it becomes doubly refracting. The effect is called the *Kerr electro-optic effect.* It is observed in both solids (glass) and liquids. This effect is attributed to the alignment of the molecules in presence of the electric field. The substance then behaves optically as if it were a uniaxial crystal in which the electric field defines the optic axis. The magnitude of the effect is found to be proportional to the square of the electric field strength. The Kerr constant K is defined by the relation

$$n_{\parallel} - n_{\perp} = KE^2\lambda .$$

Here $n_{\parallel}$ and $n_{\perp}$ are the indices of refraction in the direction of the applied field **E** and perpendicular to **E** respectively. λ is the wavelength of the light in vacuum.

The Kerr electro-optic effect is utilized to produce a high speed light modulator known as a *Kerr cell. A Kerr cell* (Fig. 25.7) *consists of a glass container filled with nitrobenzene and fitted with two electrodes.* If the polariser and analyser are crossed and are oriented at ± 45° w.r.t. the electric axis of the Kerr cell, then no light is transmitted except when the electric field is turned on. If a high frequency alternating voltage is applied to the cell, it acts as an *electro-optic shutter. It allows the light to pass through when the field is maximum and stops it when the field is zero.* Thus it is possible to interrupt the light at the rate of several million times per second.

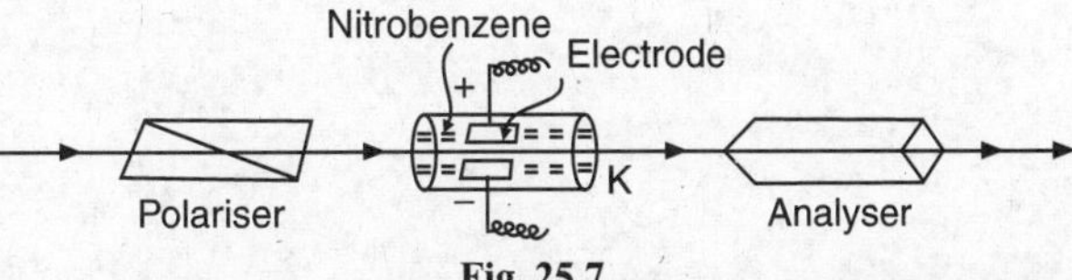

Fig. 25.7.

25.12. Pockels Electro-optic Effect

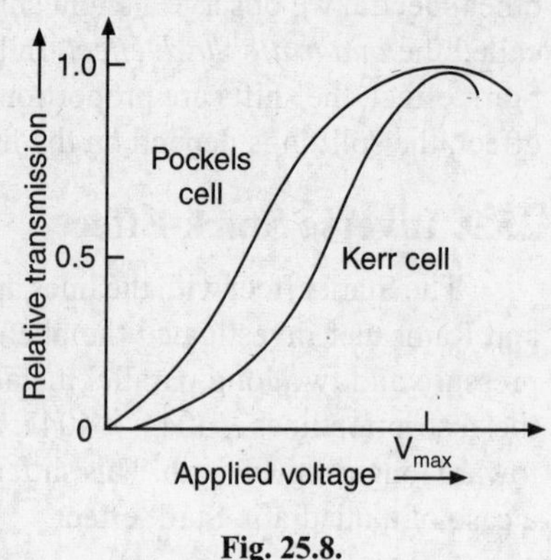

Fig. 25.8.

When certain kind of birefringent crystals are placed in an electric field, their indices of refraction are altered by the presence of the field. This effect is known as the Pockels effect. In the *Pockels effect,* the field is *parallel* to the direction of the light. Often used are crystals of potassium dihydrogen phosphate, KH_2PO_4 (KDP), mounted between two crossed polarizers. Other electro-optic materials include lithium niobate, $LiNbO_3$, lithium tantalate, $LiTaO_3$, and barium sodium niobate, $Ba_2NaNb_5O_{15}$. When a high enough dc field, usually several hundred V mm^{-1}, is applied to the crystal, the refractive index (in one axis of the crystal) changes and linearly polarized light becomes elliptically polarized. Such electro-optic crystals make compact and reliable high-speed shutters. The transmission as a function of applied voltage is shown in Fig. 25.8. Such electro-optic crystals are used as light switches and modulators.

CHAPTER

26

NONLINEAR OPTICS

26.1. Introduction

When a light wave propagates through an optical medium, the oscillating electromagnetic field exerts a polarizing force on all of the electrons comprising the medium. Since the inner electrons of the atoms are tightly bound to the nuclei, the major polarizing effect is exerted on the outer or valence electrons. With ordinary light sources, the radiation fields are much smaller than the fields that bind the electrons to the atoms. Hence the radiation acts as a small perturbation. This produces a polarization **P** that is proportional to the electric field **E** of the light wave. The ratio P/E remains independent of intensity E, and we have what is called *linear optics.* The various properties of the medium such as polarizability (χ), refractive index (n), etc., are independent of the field intensity (**E**). If we have simultaneous fields E_1, E_2, etc., present, then the total result is the linear sum of the result due to individual fields. The 'principle of superposition' holds, *i.e.,*

$$P_1 = \varepsilon_0 \chi E_1, P_2 = \varepsilon_0 \chi E_2 \text{ implies } P_1 + P_2 = \varepsilon_0 \chi (E_1 + E_2)\,.$$

If the radiation field is comparable with the atomic fields ($\sim 10^8$ Vm^{-1}), then the relation between the polarization and radiation field is no longer a linear one.

The requisite light fields needed to exhibit this non-linearity are obtainable with laser sources. Some lasers have electric field strengths as high as 10^9 Vm^{-1}. Non-linear optical effects that have been observed include optical harmonic generation, production of combination frequencies, optical rectification, and many others.

With intense laser radiation, an electric field E would create in a molecule an electric polarization P according to the general relation:

$$P(E) \approx \varepsilon_0 \chi_1 E + \varepsilon_0 \chi_2 E^2 + \varepsilon_0 \chi_3 E^3 + \ldots\ldots \qquad \ldots(1)$$

Here, χ_n is the nth-order non-linear electric susceptibility.

In the case of lasers, we have such high energy densities that the corresponding E fields are able to produce substantial contributions to P from the higher order terms in Eq. (1).

In such a situation, P/E is not independent of intensity.

Then we have *nonlinear optics.*

We shall consider three applications of Nonlinear optics.

26.2. Harmonic Generation

The relation between electric field E and electric polarization P is

$$P = \varepsilon_0 \chi_1 E + \varepsilon_0 \chi_2 E^2 + \varepsilon_0 \chi_3 E^3 + \ldots\ldots \qquad \ldots(1)$$

Suppose that the field incident on a medium has the form

$$E(t) = E_0 \cos \omega t \qquad \text{...(2)}$$

Substituting (2) in (1), we get

$$P = \varepsilon_0 \chi_1 E_0 \cos \omega t + \varepsilon_0 \chi_2 E_0^2 \cos^2 \omega t + \varepsilon_0 \chi_3 E_0^3 \cos^3 \omega t + ... \qquad \text{...(3)}$$

Now,

$$\cos^2 \omega t = \frac{1 + \cos 2\omega t}{2}; \cos^3 \omega t = \frac{\cos 3\omega t + 3 \cos \omega t}{4} \qquad \text{...(4)}$$

Eq. (3) becomes,

$$P = \frac{1}{2} \varepsilon_0 \chi_2 E_0^2 + \left(\varepsilon_0 \chi_1 + \frac{3}{4} \varepsilon_0 \chi_3 E_0^2 \right) E_0 \cos \omega t$$

$$+ \frac{1}{2} \varepsilon_0 \chi_2 E_0^2 \cos 2\omega t + \frac{1}{4} \varepsilon_0 \chi_3 E_0^3 \cos 3\omega t + ... \qquad \text{...(5)}$$

The first term in Eq. (5) is a constant term and gives rise to a *d.c.* field across the medium. The effect of this term is of comparatively little practical importance. The second term follows the external polarization and is called *the first* or *fundamental harmonic* of polarization. The third term builds a field component with a frequency 2ω, *i.e.*, oscillates at frequency 2ω and is called the *second harmonic of polarization.* This is also called *frequency doubling.* The fourth term is called the *third harmonic of polarization.* In general, higher harmonics with frequencies 4ω, etc., also may be present.

Frequency doubling is the generation of second harmonics. If red light from a ruby laser (694.3 nm) is passed through a KDP crystal, part of the light is converted into 347.15 nm UV, exactly one-half the wavelength of the incident light.

26.3. Optical Mixing

The equation for non-linear polarization is

$$P = \varepsilon_0 \chi_1 E + \varepsilon_0 \chi_2 E^2 + \varepsilon_0 \chi_3 E^3 + ...$$

We have assumed that the factor E^2 in the second term is the dot product of E with E. However, such a non-polarization term may also result from the interaction of two fields with different frequencies, ω_1 and ω_2, at least one of them being a laser beam. When these are incident on the medium, the non-linear term will have frequencies $2\omega_1$, $2\omega_2$, $\omega_1 + \omega_2$ and $\omega_1 + \omega_2$. Obviously, the emerging beam will contain components with these frequencies. This reveals that the effect of a low frequency beam, *e.g.*, ω_2 in the infra-red region, may be observed in the optical region by choosing ω_1 in the optical region.

26.4. Self-focussing

It has been observed that the refractive index of many substances increases as the intensity of radiation increases. This means, the effective refractive index of the material is larger near the centre of propagating laser beam so that the rays bend towards the beam axis. This is called as self-focussing. This property is used in *fibre-optics communication.*

Self focussing is one of the threshold phenomenon in non-linear optics. It occurs when intensity of light beam reaches a certain limiting value called threshold intensity. Since the laser beam has high intensity and high frequency, we can expect the self-focussing effect in optical fibers when we use laser as light source. Thus the spreading of the modulated light energy in optical fibers is avoided.

26.5. Momentum Mismatch

In producing the harmonic 2ω out of a laser beam ω, the medium acts as an agency to absorb two photons of momentum $\hbar k$ (ω) each and emitting one photon of momentum $\hbar k$ (2ω). Now, due to colour dispersion k (2ω) is *not* equal to $2k$ (ω). The difference is called *momentum mismatch.*

$$|\Delta k| = k(2\omega) - 2k(\omega)$$

($\hbar$ is conventionally omitted).

So the second harmonic generation in the medium remains coherent only over a length of the order $(\Delta k)^{-1}$. To obtain a harmonic generation which is directed like a laser beam, we must achieve the condition $\Delta k \rightarrow 0$. In general, crystals which show second harmonic generation are double refracting. In an anisotropic crystal, this difference between ordinary and extraordinary components varies with direction. We may choose the direction such that this difference just balances the effect of colour dispersion.

Significance of Momentum Matching

Algebra may indicate components of frequencies 0, ω, 2ω, 3ω, etc., being present simultaneously according to the coefficients. The physics of momentum conservation decides which of the components will dominantly show up in the experimental set up.

Example 1. *Harmonic 2ω is to be generated out of a laser beam ω = 4 × 10^14 Hz. If the medium has refractive index 1.528 at ω and 1.542 at 2ω, deduce the momentum mismatch due to colour dispersion. Also estimate the length of the medium for which the 2ω beam will remain coherent if the mismatch is not compensated.*

Solution. $k = \frac{2\pi}{\lambda} = \frac{2\pi\nu}{c'} = \frac{\omega}{c}n$

Here, c' is speed of light in the medium and n is the appropriate refractive index.

$$\Delta k = \frac{8\times10^{14}\times1\cdot542}{3\times10^{8}} - 2\frac{4\times10^{14}\times1\cdot528}{3\times10^{8}} \approx 4\times10^{4}\,\omega^{-1}$$

$$\therefore \qquad L \approx (\Delta k)^{-1} \approx 2\times10^{-5}\text{ m}$$

26.6. Second Harmonic Generation

Experimental arrangement for observing second harmonic generation

PeterA. Franken was the first to observe SHG experimentally. The experimental arrangement is shown in Fig. 26.1.

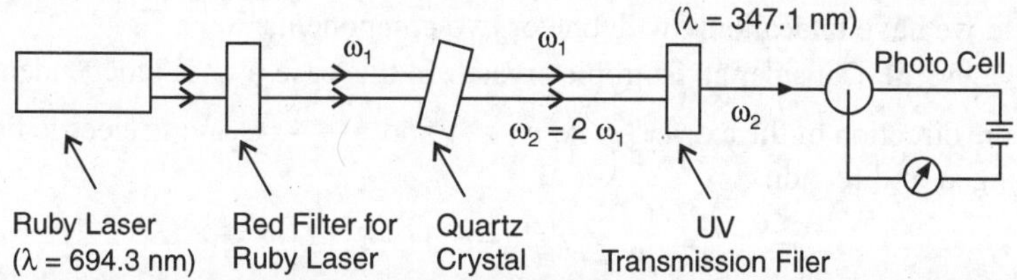

Fig. 26.1.

Light from a ruby laser is passed through a red filter to remove light from the flash lamp used to excite the ruby rod. The light is then passed through a quartz crystal where the second harmonic radiation at frequency $2\omega_1$ is generated. The transmitted light with frequency ω_1 and $2\omega_1$ is passed through an ultraviolet transmitting filter. The ultraviolet transmitting filter absorbs the ruby red light of frequency ω_1 and allows ultraviolet second harmonic radiation at frequency $2\omega_1$ to pass through. The emerging light is incident on a photocell. Therefore the response of the photomultiplier cell corresponds to the intensity of the second harmonic radiation.

In terms of the photon representation, we can envision two identical photons of energy $\hbar\omega$ coalescing within the medium to form a single photon of energy $\hbar 2\omega$ (Fig. 26.2).

The relationship between the applied electric field E and the polarization P is

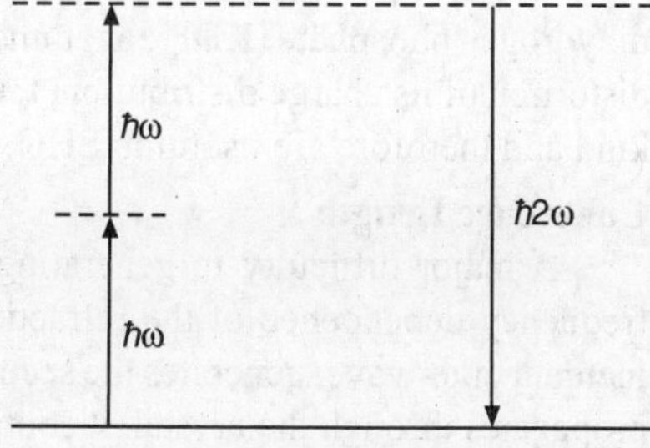

Fig. 26.2.

$$P = \epsilon_0 (\chi^{(1)} E + \chi^{(2)} E^2) \quad \text{... (1)}$$

Suppose that the lightwave incident on the medium has the form

$$E = E_0 \sin \omega t \quad \text{... (2)}$$

The resulting electric polarization

$$P = \epsilon_0 \chi^{(1)} E_0 \sin \omega t + \epsilon_0 \chi^{(2)} E_0^2 \sin^2 \omega t$$

$$P = \epsilon_0 \chi^{(1)} E_0 \sin \omega t + \frac{\epsilon_0 \chi^{(2)}}{2} E_0^2 (1 - \cos 2\omega t) \quad \text{... (3)}$$

The cos $2\omega t$ term in Eq. (3) corresponds to a variation in electric polarization at twice the fundamental frequency (*i.e.* at twice that of the incident wave). The reradiated light that arises from the driven oscillators also has a component at this same frequency, 2ω, and the process is spoken of as SHG (Second harmonic generation). Franken focused a 3-kW pulse of red (694.3 nm) ruby laserlight onto a quartz crystal. Just about one part in 10^8 of this incident wave was converted to the 347.15–nm ultraviolet second harmonic.

SHG is important because it is one of the principal methods of effective conversion of infrared radiation into visible and visible into ultraviolet.

Crystal Symmetry

In most crystalline materials, the electric polarization $\left(\vec{P}\right)$ and the applied field $\left(\vec{E}\right)$ are not necessarily parallel. Hence $\vec{P}$ and $\vec{E}$ are related not by a single scalar but by a group of quantities arranged in the particular form of a tensor, namely, the susceptibility tensor. We can express the second order polarization as

$$P_i^{(2)} = \epsilon_0 \sum_{j.k.} \chi_{ijk}^{(2)} E_j E_k \quad \text{...(1)}$$

Here, i, j and k represent the coordinates x, y, z. Most of the coefficients χ_{ijk}, however, are usually zero and we have to deal only with one or two components.

Isotropic Crystal. Consider an isotropic crystal. In this case χ_{ijk} is independent of direction. Let us reverse the direction of the axis ($x \to -x$, $y \to -y$ and $z \to -z$) leaving electric field and dipole moment unchanged in direction.

$$-P_i^{(2)} = \epsilon_0 \sum_{j.k.} \chi_{ijk}^{(2)} (-E_j)(-E_k) = P_i^{(2)} \quad \text{...(2)}$$

Obviously, $P_i^{(2)} = 0$, *i.e.*, $\chi_{ijk}^{(2)} = 0$. This clearly reveals that SHG cannot occur in an isotropic medium such as liquids or gases nor in centro-symmetric crystals (*i.e.*, crystals symmetrical about a point). Thus, SHG cannot be produced by crystal like calcite, which have *a center of symmetry* or an *inversion center.* The requirement for SHG is that a crystal should not have inversion symmetry.

Anisotropic Crystals. Under pressure, a piezoelectric crystal [such as quartz, potassium dihydrogen phosphate (KDP), or ammonium dihydrogen phosphate (ADP)] undergoes an asymmetric distortion of its charge distribution, thus producing a voltage. Of the 32 crystal classes, 20 are of this kind and therefore are useful in SHG.

Coherence Length

A major difficulty in generating copious amounts of second-harmonic light arises from the frequency dependence of the refractive index, that is, dispersion. At some initial point where the incident or ω-wave, generates the second-harmonic or 2ω-wave, the two are coherent. As the ω-wave propagates through the crystal, it continues to generate additional contributions of second-harmonic light, which all combine totally constructively only if they maintain a proper phase relationship. Yet the ω-wave travels at a phase velocity v_ω, which is ordinarily different from the phase velocity, $v_{2\omega}$, of the 2ω-wave. Thus the newly emitted second harmonic periodically falls out-of-phase with some of the previously generated 2ω-waves.

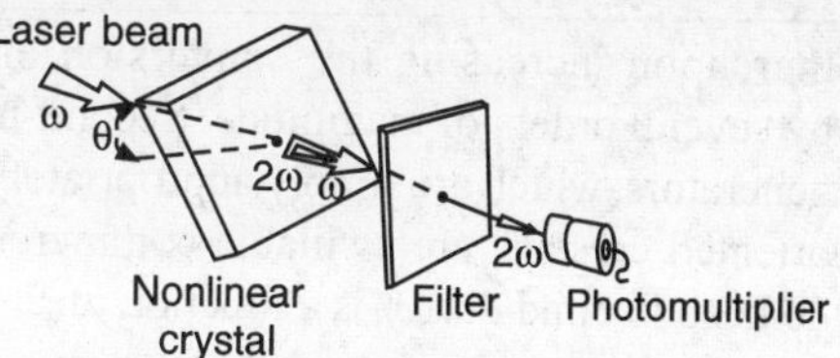

Fig. 26.3.

Consider a nonlinear crystal plate of thickness ℓ. (Fig. 26.3).

The intensity of the second harmonic light, $I_{2\omega}$, emerging from the plate is given by

$$I_{2\omega} \propto \frac{\sin^2 [2\pi (n_\omega - n_{2\omega}) \ell / \lambda_0]}{(n_\omega - n_{2\omega})^2} \quad ...(1)$$

Here, λ_0 is the vacuum wavelength of the fundamental frequency.

$I_{2\omega}$ has its maximum value when $\ell = \ell_c$, where

$$\ell_c = \frac{1}{4} \frac{\lambda_0}{|n_\omega - n_{2\omega}|} \quad ...(2)$$

The thickness ℓ_c is commonly known as the *coherence length.* It is usually of the order of only about $20\lambda_0$. Coherence length equals the length of a nonlinear material effective in generating second harmonic radiation.

Fig. 26.4 shows SHG as a function of θ for a 0.78-mm thick quartz plate. Peaks occur when the effective thickness is an even multiple of ℓ_c.

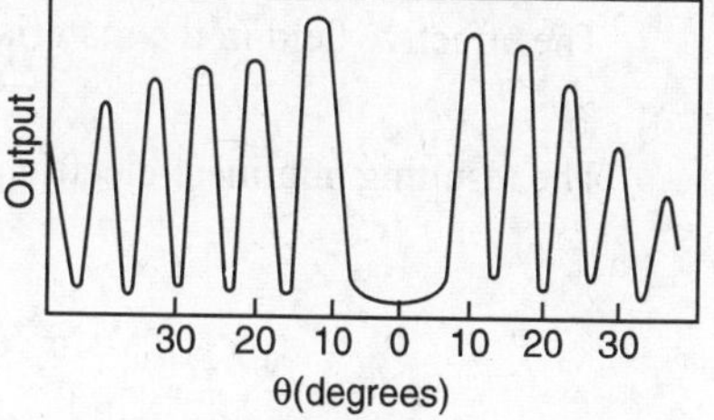

Fig. 26.4.

The coherence lengths for SHG are quite small. For KDP crystal, λ_0 = 0.694 μm, ℓ_c exp. = 18.5 μm, ℓ_c calc. = 18.8 μm.

For SHG to be useful, it is necessary to obtain larger coherence lengths.

The short coherence lengths demonstrate that it is important to keep $\Delta k \approx 0$ if the nonlinear properties of materials are to be used to generate new wavelengths. In nonlinear optics, the process of reducing the phase difference Δk is called *phase matching.* If we view harmonic generation as a scattering process, phase matching is equivalent to insuring that momentum is conserved.

Index Matcing Condition

Efficient SHG can be accomplished by a procedure known as *index matching,* which negates the undesirable effects of dispersion. We arrange things so that $n_\omega = n_{2\omega}$. A commonly used SHG material is KDP. It is piezoelectric, transparent, and also negatively uniaxially birefringent. Furthermore, it has the interesting property that if the fundamental light is a linear polarized *ordinary wave,* the resulting second harmonic will be an *extraordinary wave.* Refractive index surface for KDP is shown in Fig. 26.5. If light propagates within a KDP crystal at the specific angle θ_0 with respect to the optic axis, the index, $n_{0\omega}$, of the ordinary fundamental wave will precisely equal the index of the extraordinary second harmonic $n_{e2\omega}$. The second-harmonic wavelets will then interfere constructively,

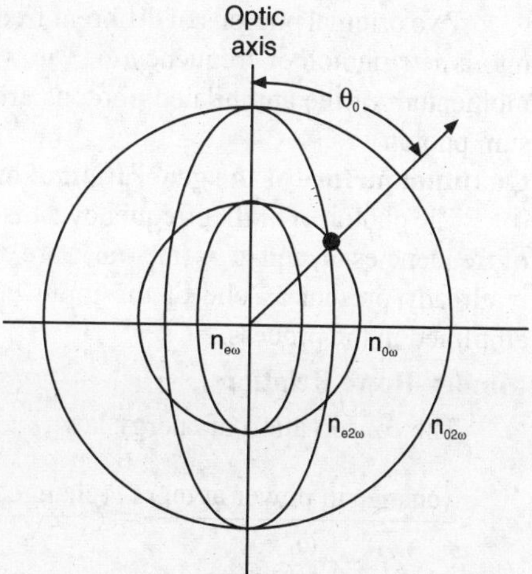

Fig. 26.5.

thereupon increasing the conversion efficiency by several orders of magnitude. Second-harmonic generators, which are simply appropriately cut and oriented crystals, are available commercially, but do keep in mind that θ_0 is a function of λ, and each such device performs at one frequency.

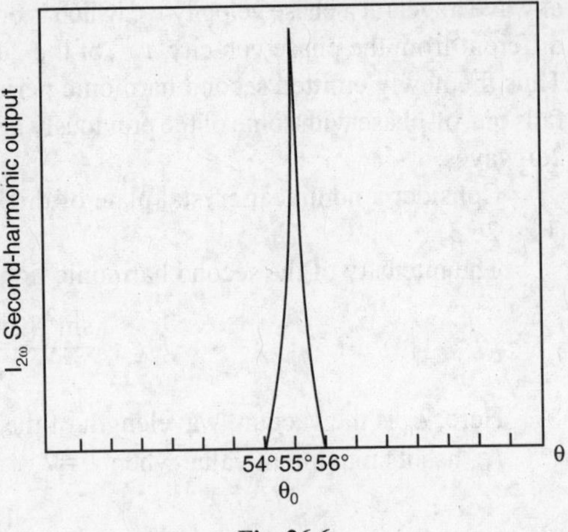

Fig. 26.6.

Fig. 26.6 shows $I_{2\omega}$ versus crystal orientation in KDP.

SHG was first realized successfully in quartz and subsequently generated in many other crystals, *e.g.*, potassium dihydrophosphate (KDP), ammonium dihydrophosphate (ADP), barium titanate, lithium iodate etc. SHG was also successfully realized in gases and in semiconductors.

26.7. Frequency Mixing [Sum and Difference Frequency Generation]

Suppose two coherent light wave trains of unequal frequencies ω_1 and ω_2 are propagating simultaneously in a nonlinear dielectric.

The effective field in the material due to the superposition of the two waves is

$$E = E_{01} \sin \omega_1 t + E_{02} \sin \omega_2 t \qquad \text{...(1)}$$

The resulting nonlinear electric polarization is

$$P_{nl} = \varepsilon_0 \chi_2 E^2$$

$$P_{nl} = \varepsilon_0 \chi_2 (E_{01}^2 \sin^2 \omega_1 t + E_{02}^2 \sin^2 \omega_2 t + 2E_{01} E_{02} \sin \omega_1 t \sin \omega_2 t) \qquad \text{...(2)}$$

$$= \varepsilon_0 \chi_2 E_{01}^2 \left[\frac{1 - \cos 2\omega_1 t}{2}\right] + \varepsilon_0 \chi_2 E_{02}^2 \left[\frac{1 - \cos 2\omega_2 t}{2}\right]$$

$$+ \varepsilon_0 \chi_2 E_{01} E_{02} [\cos (\omega_1 - \omega_2) t - \cos (\omega_1 + \omega_2) t] \qquad \text{...(3)}$$

The last term in Eq. (3) shows that the non-linear polarization and, therefore, emitted radiation contains frequencies $\omega_1 + \omega_2$ and $\omega_1 - \omega_2$.

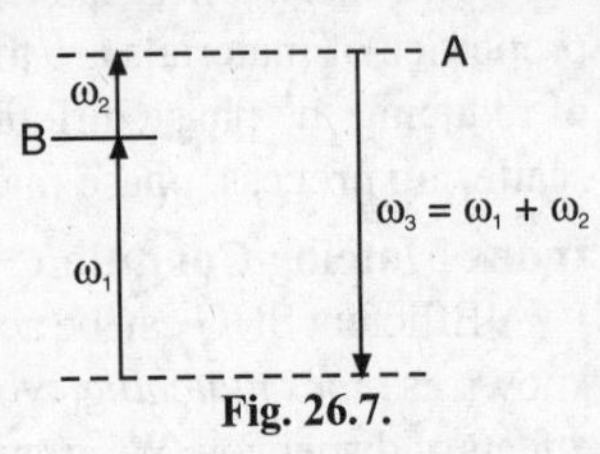

Fig. 26.7.

Quantum Picture of Sum Frequency Generation

Two original photons of different frequencies ω_1 and ω_2 coalesce into a new photon of frequency $\omega_1 + \omega_2$ (Fig. 26.7). The energy and momentum of the annihilated photons are carried off by the created sum photon.

Quantum picture of the generation of an $\omega_1 - \omega_2$ difference photon

The photon of higher frequency ω_1 is destroyed. Two new photons of frequencies ω_2 and $\omega_3 = (\omega_1 - \omega_2)$ are created (Fig. 26.8). ω_2-photon is already present as one of the input beams. It indicates that ω_2 is amplified in the process.

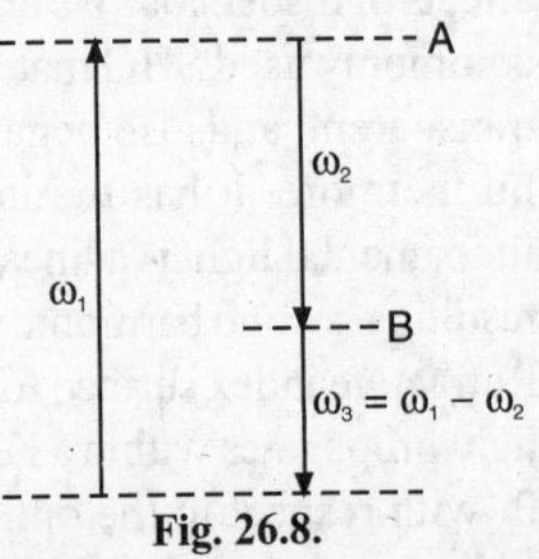

Fig. 26.8.

Manley-Rowe Relation

The conservation of energy law

$$\frac{(\text{change in power at } \omega_1)}{\omega_1} = \frac{(\text{change in power at } \omega_2)}{\omega_2} = \frac{(\text{change in power at } \omega_3)}{\omega_3}$$

is known as the *Manley-Rowe relation* and is valid for all sum- or difference-frequency generation (mixing) processes. In other words, it states that if waves at frequencies ω_1 and ω_2 are used to generate a wave at frequency ω_3, then the waves at frequencies ω_1 and ω_2 will lose the energy gained by the wave at frequency ω_3.

A quantum view of the Manley-Rowe relation can be obtained by noticing that EE^*/ω is a measure of the photon density

$$\frac{EE^*}{\omega} = \frac{I}{\omega} = \frac{N\hbar\omega}{\omega} = N\hbar$$

Two photons with energies of $\hbar\omega_1$ and $\hbar\omega_2$ combine to produce a single photon with energy $\hbar\omega_3$, the number of photons is not conserved but the energy is.

The mixing of optical frequencies in crystals, thus, uncovers additional possibilities for optical frequency conversion. It has provided a source of narrow band coherent radiation in various regions including those in which there are no primary lasers available. Primary laser radiation is available from the *IR* through the visible and *UV* down to ~116 *nm.* Optical frequency mixing has enabled us to extend the range of wavelength to the *XUV*.

Phase matching condition is also important in frequency mixing. In the case of sum or difference frequencies, three waves must be matched. If

$$\omega_3 = \omega_1 \pm \omega_2 \quad \text{... (1)}$$

the condition to be satisfied is

$$k_3 = k_1 \pm k_2 \quad \text{... (2)}$$

26.8. Parametric Amplification of Light

If three waves satisfying the condition

$$\vec{k} = \vec{k_1} + \vec{k_2}$$

propagate in a medium, an exchange of energy takes place between them. If one of the waves (say, with wave vector *k*) is much more powerful than the other two waves, the energy is transferred from this wave to the weaker waves which are consequently amplified. This phenomenon is called parametric amplification of light, since it can be considered as the modulation of the optical parameters of the medium by the wave $\boldsymbol{k}$, leading to the amplification of waves $\boldsymbol{k}_1$ and $\boldsymbol{k}_2$.

PARAMETRIC GENERATORS OF LIGHT.

These generators are based on the parametric amplification of light.

Fig. 26.9 shows the schematic diagram of a parametric generator of light. A parametric generator is a laser whose cavity contains a nonlinear crystal cut in such a way that one of the vector conditions of phase matching, viz.,

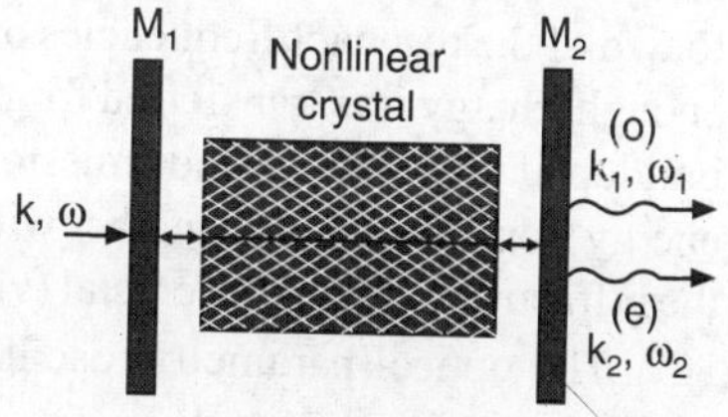

$$\boldsymbol{k}_1^{(0)} + \boldsymbol{k}_2^{(0)} = \boldsymbol{k}^{(e)} \text{ or } k_1^{(0)} + k_2^{(e)} = k^{(e)}$$

M$_1$ Nonlinear crystal M$_2$ (o) k$_1$, ω$_1$ k, ω (e) k$_2$, ω$_2$

Fig. 26.9.

is satisfied for the axial rays. The lasing frequencies of the generator are ω_1 and ω_2, and the reflectivities of the mirrors M_1 and M_2 are chosen in accordance with these frequencies. The generator is pumped by a beam of light with frequency ω, and the mirror M_1 should be quite transparent for this frequency.

If the orientation of the crystal or its parameters are changed (say, by varying the temperature, external electric field, etc.) the conditions of spatial synchronism change for axial rays, resulting in a change in the frequency of the laser radiation. This affords a possibility of a continuous variation of the generation frequency.

Suppose that a strong wave of frequency ω_p, called the *pump light,* is applied to a nonlinear crystal and a weak *signal wave* of lower frequency ω_s. (called as signal frequency) is introduced at one end. The fields at the original frequencies are regarded as fixed parameters. 'Mixing' of the signal and the pump frequency may result into a secondary wave at frequency ω_i, *i.e.*,

$$\omega_i = \omega_p - \omega_s \quad \text{...(1)}$$

ω_i is called idler frequency.

The corresponding field strength is

$$E_p E_s = E_i \qquad ...(2)$$

The field generated by the polarization component oscillation at ω_i, (idler frequency) and the original pump field, when mixed would make a contribution to the signal field. Thus

$$\omega_p - \omega_i = \omega_p - (\omega_p - \omega_s) = \omega_s \qquad ...(3)$$

The corresponding field strength is

$$E_p E_i = E_p^{\,2} E_s \qquad ...(4)$$

Eq. (4) shows that field strength is proportional to E_s, in accordance with the usual requirement for parametric amplification. In this way both the signal and idler waves are amplified. This reveals that the secondary light waves at frequencies ω_s and ω_i can be excited parametrically at the expense of the part of energy of pumping wave. Initial signals required for triggering the process of parametric generation are always available in any crystal in the form of spontaneous photons.

Optical parametric oscillator

Fig. 26.10 shows the construction and working of an optical-parametric oscillator.

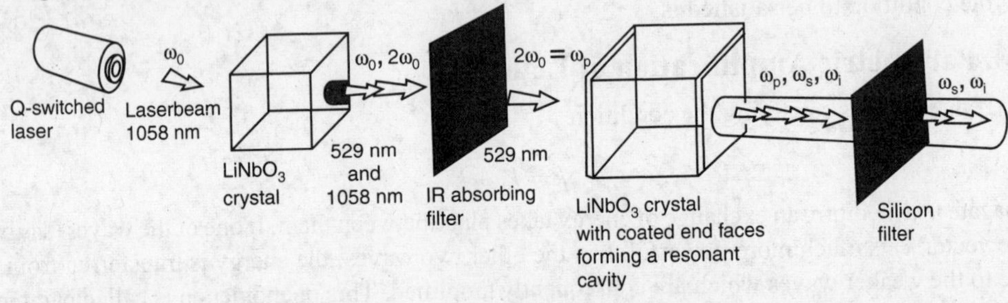

Fig. 26.10.

The flat parallel end faces of a nonlinear crystal $LiNbO_3$ (lithium niobate), were coated to form an optical Fabry-Perot cavity. The signal and idler frequencies (both about 1000 nm) corresponded to two of the resonant frequencies of the cavity. When the flux density of the pumping light was high enough, energy was transferred from it into the signal and idler oscillatory modes, with the consequent buildup of those modes and emission of coherent radiant energy at those frequencies. This transfer of energy from one wave to another within a lossless medium typifies parametric processes. By changing the refractive index of the crystal (via temperature, electric field, etc.), the oscillator becomes tunable.

The optical parametric oscillator is a laser-like, broadly tunable source of coherent radiant energy in the IR to the UV.

CHAPTER 27

FIBRE OPTICS II

27.1. Material Dispersion

Most light sources produce light output that contains energy at a number of wavelengths around the designed "center" wavelength. Fig. 27.1 shows the spectral output of a typical laser light source. Although the output is centered at 1.3 μm, it contains some lower level energy at wavelengths both above and below that value. The device is said to have a finite *spectral width.* Laser diodes have spectral widths typically on the order of 4 nm. LED sources have a significantly greater spectral width, on the order of 40 to 100 nm.

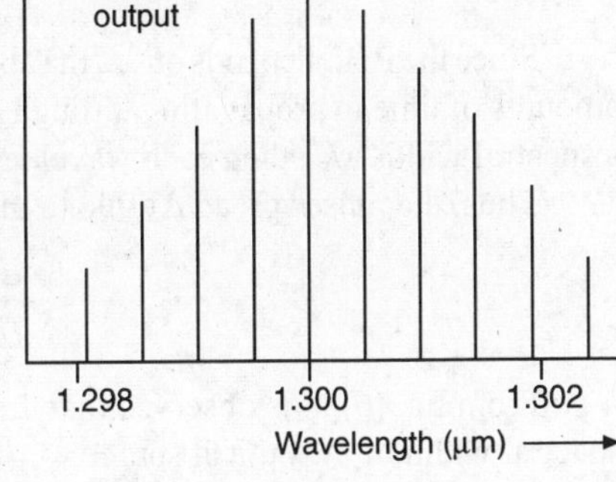

Fig. 27.1.

The energy launched into the fiber includes the several wavelengths shown in the figure. The refractive index of optical materials is a function of wavelength.

These different wavelengths transit the fibre at different speeds. Thus, a light pulse at the far end of the fibre will be dispersed in time, or broadened. This phenomenon is referred to as *material dispersion.*

Fig. 27.2 shows a typical curve of material dispersion as a function of wavelength for a single mode fibre made of pure silica. Note that the pulse broadening goes to zero for $\lambda_0 \cong 1.3$ μm. Material dispersion will be dramatically reduced (and transmission bit rates will increase) when single-frequency light sources are developed.

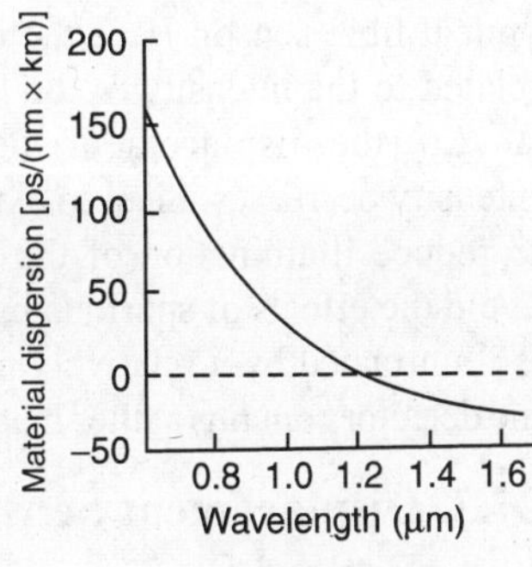

Fig. 27.2.

Let us consider a plane wave propagating through a uniform medium having refractive index $n(\lambda)$. The propagation constant k for this wave is given by

$$k = n\frac{\omega}{c} \qquad ...(1)$$

ω represents the angular frequency of the wave and c the speed of light in free space.

Taking the derivative of k with respect to ω, we get

$$\frac{dk}{d\omega} = \frac{1}{c}\left(n + \omega\frac{dn}{d\omega}\right) \qquad ...(2)$$

Let λ_0 be the free space wavelength defined by

$$\frac{\omega}{2\pi}\lambda_0 = c \qquad \text{...(3)}$$

Now

$$\frac{dn}{d\omega} = \frac{dn}{d\lambda_0}\frac{d\lambda_0}{d\omega} = \frac{dn}{d\lambda_0}\left(\frac{-2\pi c}{\omega^2}\right) \qquad \text{...(4)}$$

$$\therefore \quad \frac{dk}{d\omega} = \frac{1}{c}\left(n - \lambda_0\frac{dn}{d\lambda_0}\right).$$

The group velocity v_g is defined by the equation

$$\frac{1}{v_g} = \frac{dk}{d\omega}$$

The delay time τ required for a monochromatic signal to travel a distance L along the fiber is L/v_g.

$$\tau = \frac{L}{c}\left[n - \lambda_0\frac{dn}{d\lambda_0}\right] \qquad \text{...(5)}$$

Since the RHS depends on λ_0, the above equation implies that different wavelengths take different amounts of time in propagating through a certain length of the fibre. If the source is characterized by a spectral width $\Delta\lambda_0$, then each wavelength component will take different amounts of time.

Thus, the pulse spread $\Delta\tau$ due to material dispersion is

$$\Delta\tau = \frac{d\tau}{d\lambda_0}\Delta\lambda_0 = -\frac{\lambda_0 L}{c}\frac{d^2 n}{d\lambda_0^2}\Delta\lambda_0 \qquad \text{...(6)}$$

From Eq. (6), it is observed that the amount of pulse broadening is directly proportional to the spectral width of the optical source.

27.2. Measurement of Displacement using Fibre Optic Sensor

Fig. 27.3 shows 'Displacement measurement' using an optical fibre sensor. Here the displacement of an object is related to the intensity of the light arriving at the detector. As ΔL (the displacement) increases, the detected light intensity decreases. Sleeves extend beyond the end of fibres to reduce illumination of the fibers by spurious light. To avoid the effects of spurious light at the detector, the object is illuminated by a relatively narrow line light source, and the detector contains a filter matching the source wavelength.

Light source
Detector
Optical filter
ΔL
Reference position
Displaced by ΔL

Fig. 27.3.

27.3. Displacement Sensing Using Diffraction Patterns

Fig. 27.4 shows the use of diffraction gratings for sensing displacement. Two diffraction gratings are butted against each other, forming the equivalent of a new grating. As one of the gratings is displaced, the effective grating distance is changed.

Thus the diffraction angle (shown by dashed lines) is changed. The detector senses this as an intensity change. Here, we digitally count the high- and low-intensity cycles as the gratings move. This yields a direct digital measurement of displacement.

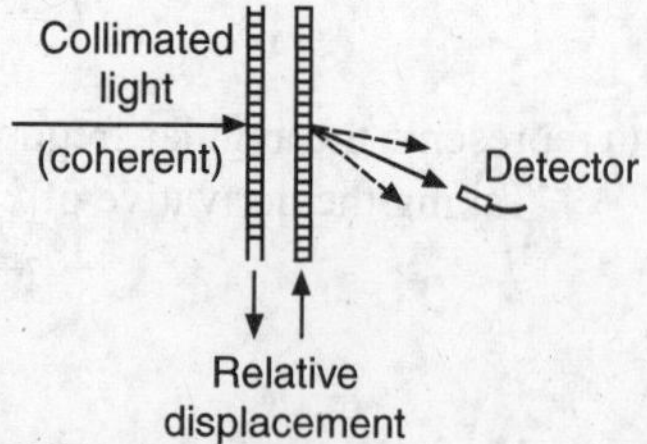

Fig. 27.4.

27.4. Measurement of Displacement — Moire-Fringe Modulation Sensor

In digital measurement technique, the measurement of displacement is done using a Moire-Fringe modulator. The arrangement is shown in Fig. 27.5.

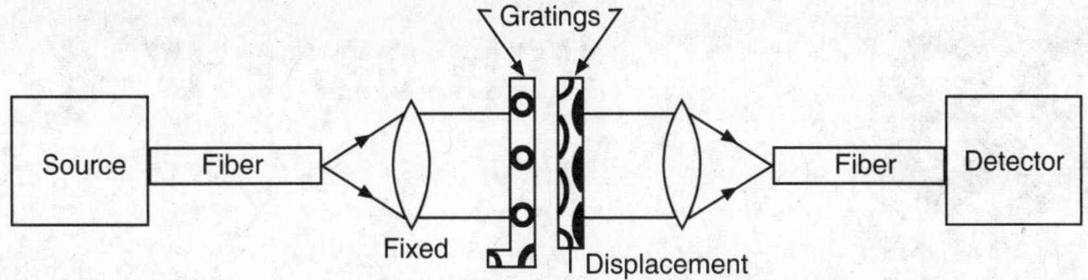

Fig. 27.5.

In this case, the opaque lined gratings produce dark Moire fringes. Transverse movement of one grating with respect to the other one causes the fringes to move up or down. Thus, a count of fringes as the gratings are moved gives a measurement of displacement. The counting of fringes is independent of instabilities with the system components which affect the optical intensity. However, mechanical vibrations may severely affect the accuracy of the measurement and prove difficult to eradicate. There are also problems involved with loss of counts if the optical power to the sensor is interrupted.

27.5. Microbending Pressure Sensor

Principle: This sensor is based on measurement of the light losses caused by controlled microbending.

Working : A fibre is mounted in between a pair of plates containing parallel grooves (Fig. 27.6).

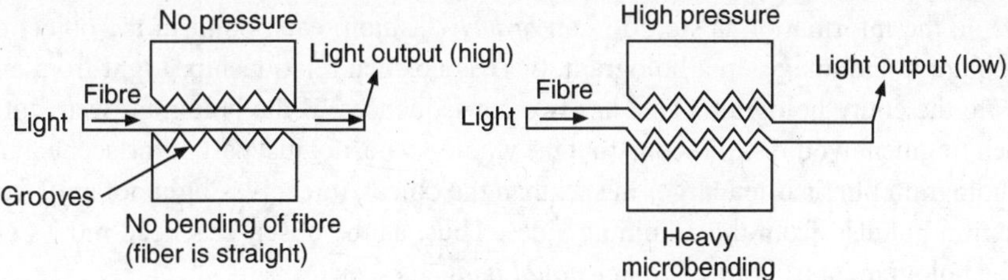

Fig. 27.6.

The force of pressure on the top plate will create bends in the fibre, which will lead to loss of light (known as *microbend losses*). Hence, intensity of light coming out of the fibre at the other end changes with load (pressure) applied. The pressure variation can be measured by measuring light intensity.

EXERCISE

1. Explain "Material dispersion". Derive a formula for the pulse spread $\Delta\tau$ due to material dispersion.
2. What is an example of a sensor that uses light intensity variations?
3. How are the diffraction gratings used to measure displacement?

CHAPTER 28

HOLOGRAPHY–II

28.1. Characteristics of Holograms

(*i*) The hologram acts somewhat like a simple periodic diffraction grating. It produces a beam direct and two first-order diffracted beams on either side of the direct beam. The zero order of the diffracted light is the direct beam. The two first orders on either side comprise the virtual and the real images.

(*ii*) The virtual image produced by a hologram appears in complete three-dimensional form. Each section of a hologram is capable of reconstructing an image of the entire object. As the pieces become smaller, the quality of the reconstructed image will become poorer, small details will vanish, and a characteristic speckled structure will appear.

There is a fundamental difference between a hologram and a conventional photograph. In a photograph, the information is stored in an *orderly* fashion: each point in the object relates to a *conjugate* point in the image. In a hologram, there is no such relationship. Light from every object point goes to the entire hologram. This has two consequences. If the hologram were cut into small pieces, each fragment would still reconstruct the whole scene, not just part of the scene. Furthermore, since the hologram film is often larger in size than the object, it receives light not only from the side facing the film but also from the adjoining sides. Thus, as the observer's head moves sideways in viewing the hologram, a truly *three-dimensional image* is seen.

(*iii*) In holography, the viewer always sees a positive image whether a positive or a negative photographic transparency is used for the hologram. The reason for this is that a negative hologram merely produces a wave field that is shifted 180° in phase w.r.t that of a positive hologram. The eye is insensitive to this phase difference. So the view seen by the observer is identical in the two cases.

(*iv*) Holograms have extremely high information capacity. Superposition of several images on a photographic plate is pointless. A hologram, on the other hand, may contain a number of consecutively recorded scenes that can be recorded independently. Different scenes can be recorded on a hologram, each time changing the angle at which the reference wave is incident on the holographic plate, which can be done by rotating the plate. Reconstruction of a specific scene only requires that the hologram be properly oriented with respect to the read out wave.

(*v*) The curious property of the wavefront reconstruction process is that it does not produce negatives. One may consider a hologram itself as a negative. But the image it produces is a positive. A hologram copied from another by contact printing would be reversed in the sense that opaque areas now become transparent and *vice versa.* The image reconstructed from it, would be identical in all respects to that produced by the original. This is because the information is recorded on the plate in the form of a modulated spatial carrier.

28.2 Holographic Computer Memories

A holographic memory records and reads out a large number of bits simultaneously. Figure 28.1 shows the schematic diagram of a holographic memory system.

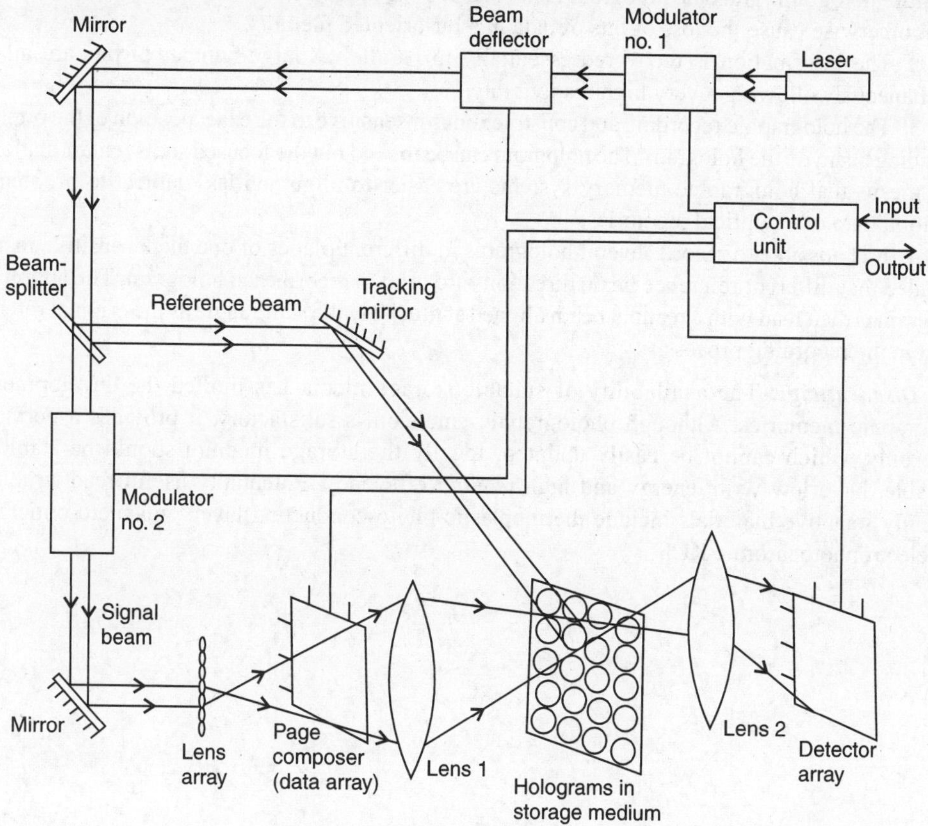

Fig. 28.1.

Storage of Data. The information to be stored is formed into a two-dimensional array of bits by a device called a *page composer*. The page composer may be thought of as an array of light valves which may be open or closed corresponding to 'ones' and 'zeros' respectively. This array of up to perhaps 10^4 bits is then stored at one time in a particular location of the holographic memory. During recording, the light modulators allow maximum irradiance in both the signal and reference beams. To store a different array as another hologram on the storage medium, the *beam deflector* moves the beams to the appropriate position.

Readout of Data. Readout of data occurs when the hologram is addressed only by the *reference beam*. The *first modulator* is partly closed to reduce the irradiance of the light reaching the holographic storage medium. The *second modulator* is closed to cut off the signal beam. The deflector directs the beam to the hologram to be read via the tracking mirror. An image representing the arrays of 'ones' and 'zeros' is produced. This image is focused by a lens and projected onto the *detector array*. Each bit originally stored in the page composer is incident on one photodetector in the array. The stored information is thus converted into electrical signals.

• *Holographic computer memories* have a very high storage capacity theoretically up to 10^{10} mm^{-3} with rapid access.

Advantages

The storage of data in this way offers a number of advantages compared to a bit-oriented memory.

1. The information about a particular array is distributed in a holographic fringe pattern across the entire hologram. Thus the hologram is not sensitive to small scratches or particles of dust which might otherwise cause the loss of bits of data in a bit-oriented memory.

2. The information is recovered essentially in parallel. A large number of bits are all read simultaneously allowing a very high readout rate.

3. The holographic recording and construction is insensitive to the exact position of the reference or reading beam on the hologram. The hologram can be moved but the focused spots remain stationary. This means that holographic memory systems are easier to align and less subject to problems of vibration than other optical memories.

4. It is possible to record several holograms in different planes of one thick sensitive material provided that different reference beam directions are used to record each hologram. The appropriate hologram is then read with a readout beam aligned at precisely the same angle as the original reference beam in the recording process.

Disadvantage. The availability of suitable storage media has limited the development of holographic memories. Although photographic emulsion is satisfactory, it provides a permanent store only, which cannot be easily updated. Ideally the storage medium should be stable and alterable, have low write energy and high readout efficiency. Potentially useful read/write/erase optically sensitive materials include thermoplastic-photo-conducting layers, magneto-optic MnBi and electrophotochromic KCl.

CHAPTER 29

INTERFERENCE–III

29.1. Principle of Superposition

Phenomenon of interference is based upon the principle of superposition.

It states that a number of waves travelling, simultaneously, in a medium behave independent of each other and the net displacement of the particle, at any instant, is equal to the vector sum of the individual displacements due to all the waves.

29.2. Analytical Treatment of Interference

Consider a monochromatic source of light S emitting waves of wavelength λ (Fig. 29.1). S_1 and S_2 are two narrow pin holes equidistant from the source S. S_1 and S_2 act as two virtual coherent sources. Let us find the resultant intensity of light at point P on a screen placed parallel to S_1 and S_2.

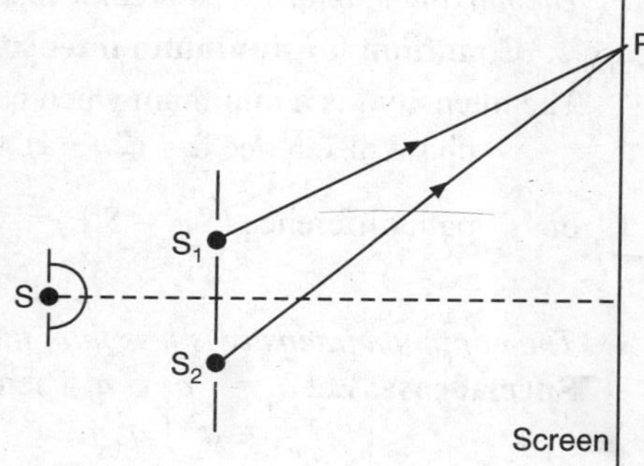

Fig. 29.1.

Let a_1 and a_2 be the amplitudes of the waves from S_1 and S_2. Let δ be the phase difference between the two waves reaching at point P.

Let y_1 and y_2 be the displacements of the waves. Then,

$$y_1 = a_1 \sin \omega t \quad ...(1)$$

and

$$y_2 = a_2 \sin (\omega t + \delta) \quad ...(2)$$

Now according to the principle of superposition, when a particle is disturbed simultaneously by two or more waves, the resultant displacement is equal to the sum of the displacements due to each wave separately. Hence the resultant displacement is

$$y = y_1 + y_2 = a_1 \sin \omega t + a_2 \sin (\omega t + \delta)$$

$$= a_1 \sin \omega t + a_2 \sin \omega t \cos \delta + a_2 \cos \omega t \sin \delta$$

$$= \sin \omega t (a_1 + a_2 \cos \delta) + \cos \omega t (a_2 \sin \delta) \quad ...(3)$$

Let

$$a_1 + a_2 \cos \delta = R \cos \theta \quad ...(4)$$

and

$$a_2 \sin \delta = R \sin \theta \quad ...(5)$$

Here R and θ are new constants.

Substituting Eqs. (4) and (5) in Eq. (3), we get

$$y = \sin \omega t . R \cos \theta + \cos \omega t . R \sin \theta$$

or
$$y = R \sin(\omega t + \theta) \qquad ...(6)$$

Hence the resultant displacement at P is simple harmonic and of amplitude R.
Squaring Eqs. (4) and (5) and adding, we get

$$R^2 \cos^2\theta + R^2 \sin^2\theta = (a_1 + a_2 \cos\delta)^2 + (a_2 \sin\delta)^2$$

or
$$R^2 = a_1^2 + a_2^2 + 2a_1a_2 \cos\delta. \qquad ...(7)$$

The resultant intensity at P is given by the square of the amplitude R. Thus

$$I = R^2 = a_1^2 + a_2^2 + 2a_1a_2 \cos\delta \qquad ...(8)$$

The phase difference δ between the two waves arriving at P is

$$\delta = \frac{2\pi}{\lambda} \text{ (Path difference)} = \frac{2\pi}{\lambda}(S_2P - S_1P) \qquad ...(9)$$

Case 1. Condition for maximum intensity:

The intensity I is a maximum when $\cos\delta = +1$. That is, when
phase difference $\delta = 2n\pi$, $n = 0, 1, 2...$ etc. ...(10)
or path difference $(S_2P - S_1P) = n\lambda$. ...(11)

$$I_{max} = a_1^2 + a_2^2 + 2a_1a_2 = (a_1 + a_2)^2 \qquad ...(12)$$

The maximum intensity is greater than the sum of two separate intensities $(a_1^2 + a_2^2)$.

Case 2. Condition for minimum intensity:

The intensity I is a minimum when $\cos\delta = -1$. That is, when
phase difference $\delta = (2n+1)\pi$ $n = 0, 1, 2, 3...$ etc. ...(13)

or path difference $(S_2P - S_1P) = (2n+1)\lambda/2$. ...(14)

$$I_{min} = a_1^2 + a_2^2 - 2a_1a_2 = (a_1 - a_2)^2 \qquad ...(15)$$

The minimum intensity is less than the sum of two separate intensities $(a_1^2 + a_2^2)$.

Special case: Let $a_1 = a_2 = a$. Then,

$$I_{max} = a^2 + a^2 + 2a^2 = 4a^2 \qquad ...(16)$$

and
$$I_{min} = a^2 + a^2 - 2a^2 = 0 \qquad ...(17)$$

Energy distribution. The intensity at bright points is $4a^2$. The intensity at dark points is zero. Fig. 29.2 shows the variation of intensity I as a function of δ.

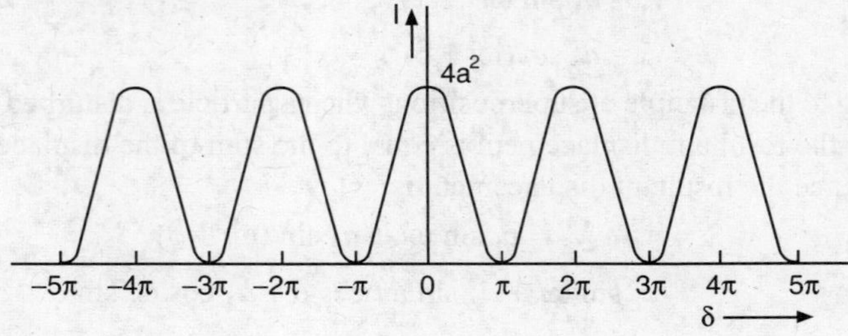

Fig. 29.2.

Interference and Conservation of Energy. The average intensity between the range $\delta = 0$ to $\delta = 2\pi$ is

$$I_{average} = \frac{\int_0^{2\pi} I\, d\delta}{\int_0^{2\pi} d\delta} = \frac{\int_0^{2\pi} (a_1^2 + a_2^2 + 2a_1a_2 \cos\delta)\, d\delta}{\int_0^{2\pi} d\delta}$$

$$= \frac{[a_1^2\,\delta + a_2^2\,\delta + 2a_1a_2\,\sin\delta]_0^{2\pi}}{[\delta]_0^{2\pi}} = \frac{2\pi(a_1^2 + a_2^2)}{2\pi}$$

$$= a_1{}^2 + a_2{}^2.$$

Thus the average intensity is equal to the sum of the separate intensities.

That is, whatever energy apparently disappears at the minima is actually present at the maxima. Thus there is no violation of the law of conservation of energy in the phenomenon of interference. *The formation of interference fringes is in accordance with the law of conservation of energy.*

Example 1. *What are nodal lines? Show that the equation of a nodal line is a hyperbola.*

Solution. The locus of all points which correspond to interference minima is called a nodal line.

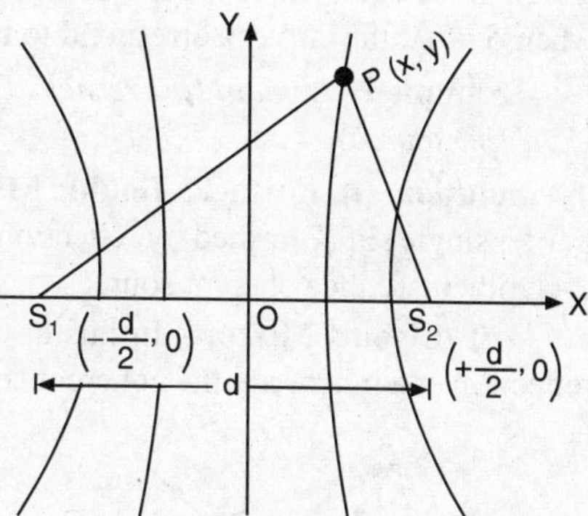

Fig. 29.3.

S_1 and S_2 are two sources separated by a distance d (Fig. 29.3). We choose the midpoint of S_1S_2 as the origin, with the x-axis along S_1S_2 and the y-axis perpendicular to it. The co-ordinates of the points S_1 and S_2 are $\left(-\frac{d}{2},0\right)$, $\left(\frac{d}{2},0\right)$. The co-ordinates of the point P are (x, y). Then,

$$S_1P = \left[\left(x+\frac{d}{2}\right)^2 + y^2\right]^{1/2} \quad \text{and } S_2P = \left[\left(x-\frac{d}{2}\right)^2 + y^2\right]^{1/2}$$

$$S_1P - S_2P = \left[\left(x+\frac{d}{2}\right)^2 + y^2\right]^{1/2} - \left[\left(x-\frac{d}{2}\right)^2 + y^2\right]^{1/2} = \delta$$

$$\left[\left\{x+\frac{d}{2}\right\}^2 + y^2\right]^{1/2} = \delta + \left[\left(x-\frac{d}{2}\right)^2 + y^2\right]^{1/2}$$

Squaring both sides

$$\left(x+\frac{d}{2}\right)^2 + y^2 = \delta^2 + \left(x-\frac{d}{2}\right)^2 + y^2 + 2\delta\left[\left(x-\frac{d}{2}\right)^2 + y^2\right]^{1/2}$$

$$\left(x+\frac{d}{2}\right)^2 - \left(x-\frac{d}{2}\right)^2 = \delta^2 + 2\delta\left[\left(x-\frac{d}{2}\right)^2 + y^2\right]^{1/2}$$

$$2xd - \delta^2 = 2\delta\left[\left(x-\frac{d}{2}\right)^2 + y^2\right]^{1/2}$$

Squaring both sides

$$4x^2d^2 - 4xd\delta^2 + \delta^4 = 4\delta^2\left[x^2 - xd + \frac{d^2}{4} + y^2\right]$$

$$\therefore \qquad \frac{x^2}{\delta^2/4} - \frac{y^2}{(d^2 - \delta^2)/4} = 1$$

which is the equation of a hyperbola. When $\delta = (2n + 1)\ \lambda/2$, the curves correspond to minima and when $\delta = n\lambda$, the curves correspond to maxima.

Example 2. *Explain (i) Fresnel's Double Mirror and (ii) Llyod's Mirror methods of producing Coherent sources.*

Solution. (*i*) Fresnel's Double Mirror—In this device (Fig. 29.4), two virtual images S_1 and S_2 of a single slit S, formed by reflection at two plane mirrors M_1 and M_2 inclined at a small angle to each other, are the coherent sources

(*ii*) Llyod's Mirror—In this device (Fig. 29.5), a slit S and its virtual image S' formed by reflection at a mirror are the coherent sources.

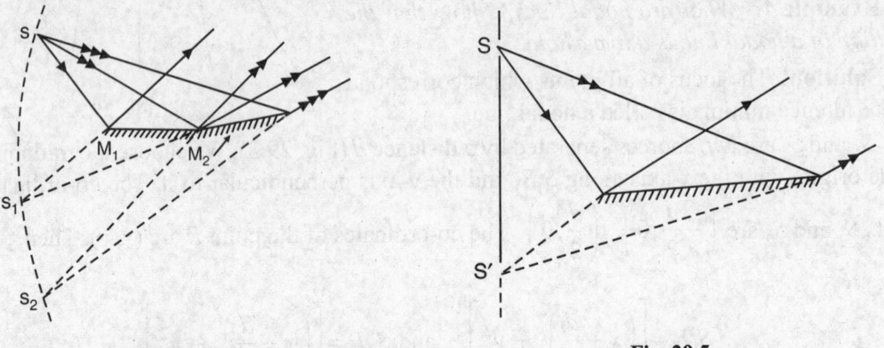

Fig. 29.4. **Fig. 29.5**

Example 3. *Explain what happens when white light is used to illuminate the slit in Fresnel's biprism experiment.*

Solution. In this case, the interference pattern consists of a central white fringe having few coloured fringes on both sides and then a general illumination.

Explanation—White light is a mixture of many wavelengths varying from 400 nm (violet) to 750 nm (red). So the coherent sources of monochromatic light are now equivalent to an infinite number of monochromatic sources of different wavelengths. According to the formula $\beta = \lambda\ D/d$, each pair of monochromatic sources of white light produces its own system of fringes with different fringe widths.

At the centre of the pattern, the path difference, and hence the phase difference, between the interfering waves is zero *for all wavelengths.* Therefore, all different coloured interfering waves produce a bright fringe at the centre. The superposition of the different colours makes the central fringe 'white'. This is the 'zero order fringe'.

The fringe width for violet is minimum and maximum for red. Therefore, the first dark band of violet is obtained first and that of red the last on either side of the zero order. The inner-edge of the first minimum of violet receives sufficient intensity from red because the maximum of red falls in its vicinity and the edge is reddish. The first maximum of violet falls close to the inner edge of the minimum of red and the edge appears violet. For points at large distances from the centre, the maxima and minima due to large number of wavelengths overlap and it results in uniform illumination.

Example 4. *Explain the difference between interference and diffraction of light.*

Solution. (*i*) In the phenomenon of interference, the interaction occurs between two separate wavefronts originating from the two coherent sources; while in case of diffraction, the interaction occurs between the secondary wavelets originating from different points of the exposed parts of the same wavefront.

(*ii*) In an interference pattern, all the regions of minimum intensity are usually almost perfectly dark while it is not so in a diffraction pattern.

(*iii*) In an interference pattern, all the maxima are of the same intensity but in a diffraction pattern they are of varying intensity.

(*iv*) The interference fringes are usually of the same width while the diffraction fringes are never of the same width.

29.3. Newton's Rings Formed by Two Curved Surfaces

Case 1. When a concave surface is in contact with the convex surface.

In Fig. 29.6, a convex surface of radius of curvature R_1 is in contact with a concave surface of radius of curvature R_2 at point O $(R_2 > R_1)$. An air film is enclosed between the two surfaces. Now a system of dark and bright rings is formed which can be viewed by a travelling microscope. Let t be the thickness of air film at point P. Let m^{th} dark ring pass through P.

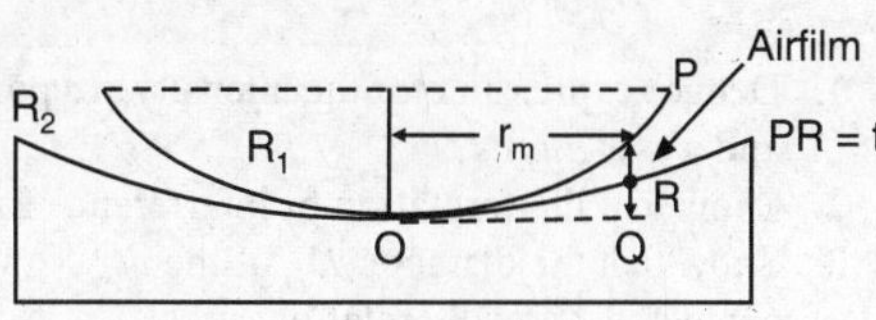

Fig. 29.6.

Its radius is r_m.

$$t = PQ - QR = \left[\frac{r_m^2}{2R_1} - \frac{r_m^2}{2R_2}\right]$$

For normal incidence and reflected light, we have

$$2nt \cos\theta = m\lambda \text{ for } m^{\text{th}} \text{ dark ring.}$$

Here $\quad n = 1$ for air, and $\cos\theta = 1$.

$$\therefore \quad 2t = m\lambda$$

or
$$2\left[\frac{r_m^2}{2R_1} - \frac{r_m^2}{2R_2}\right] = m\lambda$$

or
$$r_m^2\left[\frac{1}{R_1} - \frac{1}{R_2}\right] = m\lambda$$

$$(m = 0, 1, 2, 3\text{..... etc.}) \qquad \text{...(1)}$$

For bright rings

$$r_m^2\left[\frac{1}{R_1} - \frac{1}{R_2}\right] = (2m-1)\frac{\lambda}{2} \qquad \text{...(2)}$$

Case II. When two convex surfaces are in contact with each other.

In this case (Fig. 29.7),

$$t = PQ = PR + RQ$$

$$= \frac{r_m^2}{2R_1} + \frac{r_m^2}{2R_2}$$

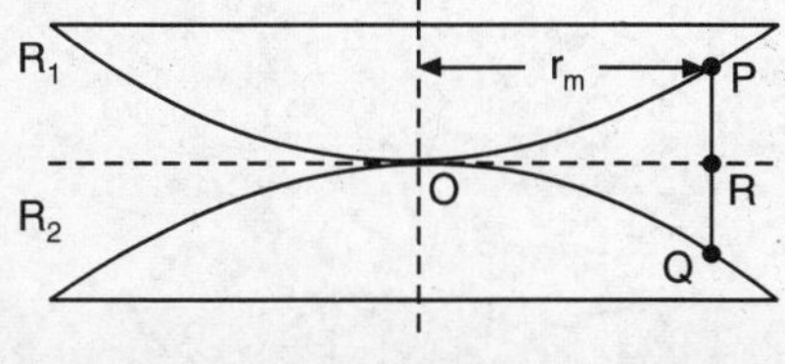

Fig. 29.7.

For normal incidence, we have

$$2t = m\lambda \text{ for dark rings.}$$

or
$$2\left[\frac{r_m^2}{2R_1} + \frac{r_m^2}{2R_2}\right] = m\lambda$$

or
$$r_m^2\left[\frac{1}{R_1}+\frac{1}{R_2}\right]=m\lambda \qquad ...(3)$$

For bright rings
$$r_m^2\left[\frac{1}{R_1}+\frac{1}{R_2}\right]=\frac{(2m-1)}{2}\lambda \qquad ...(4)$$

EXERCISE

1. Deduce expression for the intensity at a point in the region of superposition of two waves of same periods and wavelengths.
2. Show that the formation of interference fringes is in accordance with the law of conservation of energy.
3. Show that the diameter D_m of the mth Newton's ring, when two surfaces of radii R_1 and R_2 are placed in contact, is given by relation
$$\frac{1}{R_1}\pm\frac{1}{R_2}=\frac{4m\lambda}{D_m^2}$$
4. Explain why two independent sources of light of the same wavelength cannot produce observable interference pattern?
[**Hint.** The two independent sources of light cannot produce sustained interference because they are not coherent (vibrating in the same phase) or the atoms emitting light in the two sources are independent of one another.
5. What happens to the interference pattern when entire arrangement of double slit experiment is dipped in water?
[**Hint.** The fringes will become narrower because in water the wavelength λ decreases and hence fringe width
$$\left(\beta=\frac{\lambda D}{2d}\right) \text{ decreases.]}$$
6. In Young's double slit experiment, the monochromatic source of light is replaced by white light source. Would the central fringe be white or dark?
[**Hint.** The central fringe will be white because the central fringe corresponds to zero path difference for two wave trains. This is true for all wevelengths.]
7. What will be the effect on interference fringes obtained in Young's double slit experiment if
(*i*) one slit is covered, (*ii*) a source of light of higher wavelength is used, (*iii*) distance between two slits be increased to one cm, (*iv*) distance between screen and double slit is increased.
[**Hint.** We know that $\beta = \lambda D/2d$. (*i*) The interference pattern will be replaced by diffraction due to a single slit. (*ii*) The fringe width β will increase. (*iii*) The fringes will disappear and there will be uniform illumination. (*iv*) The fringe width β will increase.]

CHAPTER

30

DIFFRACTION OF LIGHT–(IV)

30.1. Phase Reversal Zone Plate

Suppose the alternate zones of a zone plate, instead of being blackened, are covered by some transparent substance which introduces an additional optical path of $\lambda/2$. Then, the amplitudes from successive zones will help each other. The intensity of the image approaches that for an equivalent lens. Such zone plates are called phase reversal zone plates. They are prepared in the following way:

A chemically cleaned glass plate is coated with a thin layer of gelatine solution and dried. It is then sensitised by immersing it in a weak solution of potassium dichromate for a few seconds. It is now again dried in the dark and after placing in contact with an ordinary zone plate, it is exposed to sunlight. Light passing through transparent zones acts on gelatine and makes it insoluble in water while the gelatine in contact with opaque zones remains soluble in water. By immersing the glass plate in water, the gelatine of the unexposed parts is dissolved to such a depth that the optical paths from successive zones on the glass plate have an additional optical path of $\lambda/2$.

30.2. Distinction Between Interference and Diffraction Bands

1. Interference bands are alternate bright and dark bands. Diffraction bands are bands of alternate maximum and minimum intensity.

2. All interference bright bands are equally bright. For diffraction maxima, the intensity goes on diminishing as order increases and for minima bands the intensity goes on increasing. After a few bands, the maxima and minima coincide.

3. Interference bands are bands due to superposition of wavelets from two coherent sources. Diffraction bands are due to superposition of wavelets from successive half period zones.

4. The diffraction bands are less in number than interference bands.

30.3. Fresnel Diffraction at an Opaque Circular Disc

An opaque circular disc AB (Fig. 30.1) is placed in the path of light from a point-source S. The diffraction pattern is obtained on a screen. The pattern consists of a *bright spot* at the centre C of the geometrical shadow $A'B'$. Surrounding the bright spot, there are a number of *faint alternate dark and bright rings within the geometrical shadow* $A'B'$. Outside the shadow, brighter and broader rings are obtained.

Explanation—Let WW' be the spherical wavefront starting from S and incident on the disc. O is the pole of the wavefront with respect to C. To find the intensity at C, let us divide the wavefront into half-period zones. Some of the central zones are blocked by the disc and the intensity at C is due to the exposed zones on the wavefront.

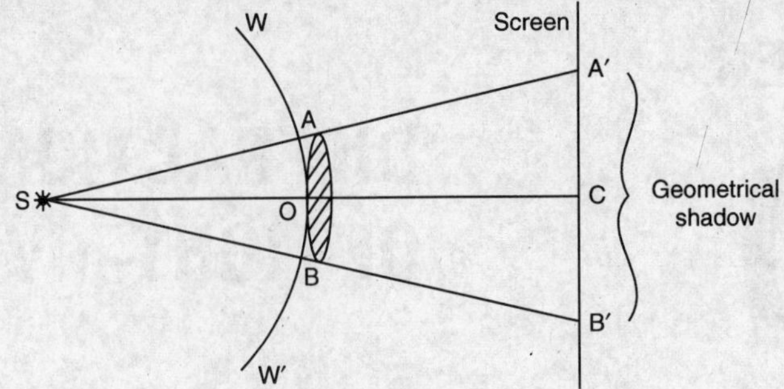

Fig. 30.1.

Let d_1, d_2, d_3, ...etc., be the amplitudes due to the first, second, third.... zones respectively. Let the disc be so small that it obstructs only the first zone.

Then the resultant amplitude at C due to the remaining exposed zones is

$$A = -d_2 + d_3 - d_4 + d_5 - ..... = -\frac{d_2}{2}.$$

The intensity is proportional to $d_2^2/4$. Hence C is bright.

If AB contains respectively 2 *hpzs,* 3 *hpzs*, ...

$$A = \frac{d_3}{2}, -\frac{d_4}{2}, ...$$

The intensity is proportional to $d_3^2/4$, $d_4^2/4$, ...

Thus the intensity at C is proportional to $d_2^2/4$, $d_3^2/4$, $d_4^2/4$, etc., according as the disc covers one, two, three, zones etc. Hence, C is always bright. But the intensity of illumination gradually decreases as the size of the disc increases.

The concentric bright and dark rings inside the geometrical shadow are formed due to interference between the exposed parts of the wavefront above A and below B.

The rings outside the shadow are the unequally-spaced 'diffraction' rings formed in the same way as the unequally-spaced bands in the illuminated region of a straight edge.

30.4. Fraunhofer Diffraction at a Circular Aperture

Consider a plane wavefront of monochromatic light of wavelength λ incident normally on a circular aperture AB of diameter d (Fig. 30.2.).

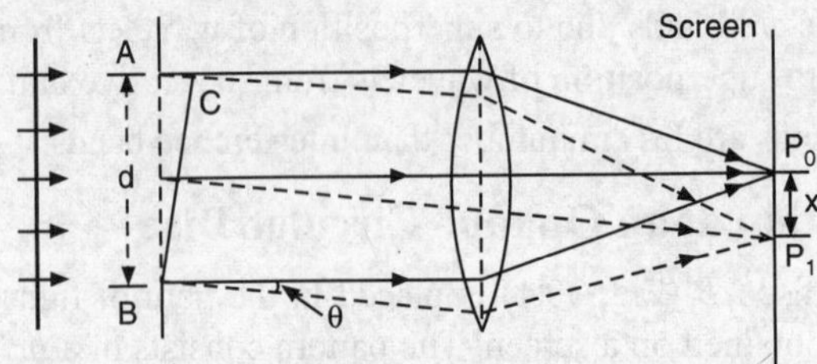

Fig. 30.2.

Every point on the aperture becomes a secondary source of light. The wavelets travelling along the normal to the circular aperture come to focus at P_0. Here all the wavelets travel the same distance before reaching P_0. Hence they all reinforce one another. Therefore, P_0 corresponds to the position of the central maximum.

Consider the wavelets travelling in a direction inclined at an angle θ with the normal to the aperture. They meet at P_1. Let $P_0P_1 = x$.

The path difference between the waves emanating from the points A and B is

$$AC = AB \sin\theta = d \sin\theta.$$

The point P_1 will be of minimum intensity if this path difference is an integral multiple of λ and of maximum intensity if the path difference is odd multiple of $\lambda/2$.

$$d \sin\theta = m\lambda \text{ minima} \qquad ...(1)$$

$$d \sin\theta = (2m+1)\lambda/2 \text{ maxima} \qquad ...(2)$$

Here $m = 1, 2, 3...$ etc. The term $m = 0$ corresponds to central maximum at P_0.

If the point P_1 is of minimum intensity, then all the points which are at the same distance from P_0 as P_1 *i.e.*, all points lying on a circle of radius x, will be of minimum intensity. Thus the diffraction pattern consists of a central bright disc, called the *Airy's disc*, surrounded by alternate dark and bright concentric rings. The intensities of the bright rings decrease as the order increases.

If the lens is very near the circular aperture, then the angle subtended by the radius of the first dark ring at the centre of the lens (or the aperture) is

$$\sin\theta = \theta = x/f \qquad ...(3)$$

Here f is the focal length of the lens.

Further, for the first secondary minimum

$$d \sin\theta = 1.\lambda$$

or

$$\sin\theta = \theta = \lambda/d. \qquad ...(4)$$

From Eqs. (3) and (4),

$$\frac{x}{f} = \frac{\lambda}{d}$$

or

$$x = \frac{f\lambda}{d} \qquad ...(5)$$

Airy modified the equation as follows.

$$x = \frac{1.22 f\lambda}{d} \qquad ...(6)$$

This concept is often used to study the resolving power of optical instruments.

30.5. Fraunhofer Diffraction at N Slits (Integral Calculus Method)

Let a parallel beam of monochromatic light of wavelength λ be incident normally on N parallel slits (Fig. 30.3). Let a be the width of each slit and b the width of each opaque space between the slits. Then the grating element is $d = (a + b)$. Take the centre of the first slit as the origin O.

Let us consider Fraunhofer diffraction at a single slit (Fig. 30.4). Consider a small element dz of the wavefront with co-ordinates (o, z). The distance of the element from the point P' is ρ. Let $OP' = r$.

The displacement at the point P' due to the element dz at any instant

$$= K\, dz \sin 2\pi\left(\frac{t}{T} - \frac{\rho}{\lambda}\right) = K \sin\left[2\pi\left(\frac{t}{T} - \frac{r}{\lambda} + \frac{z \sin\theta}{\lambda}\right)\right] dz$$

We sum the effects of all elements from one edge of the slit to the other.

Therefore, for a single slit,

$$dy = K \int_{-\frac{a}{2}}^{+\frac{a}{2}} \sin\left[2\pi\left(\frac{t}{T} - \frac{r}{\lambda} + \frac{z \sin\theta}{\lambda}\right)\right] dz \qquad ...(1)$$

Let
$$\sin 2\pi\left(\frac{t}{T}-\frac{r}{\lambda}+\frac{z\sin\theta}{\lambda}\right)=\phi(z).$$

$$dy = K\int_{-\frac{a}{2}}^{+\frac{a}{2}}\phi(z)\,dz$$

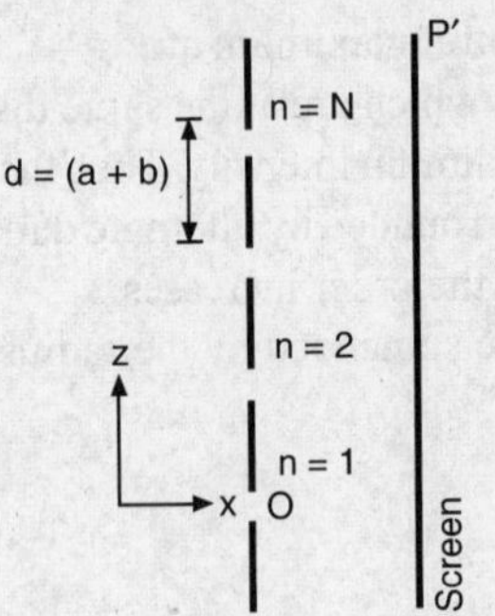

Fig. 30.3.

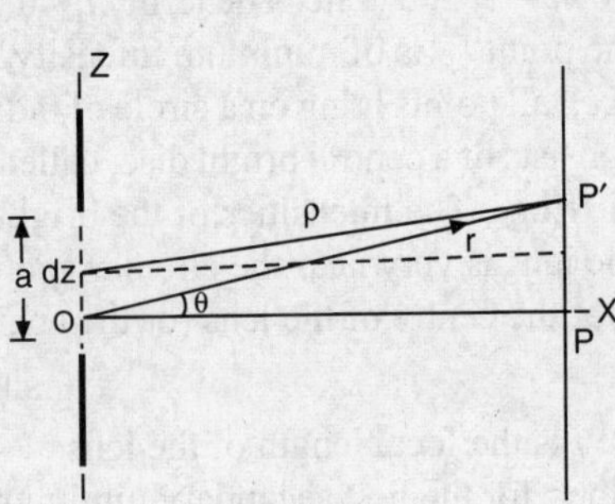

Fig. 30.4.

To calculate the total displacement at P', we have to integrate over all the N slits.

For first slit, the integration limit is from $-(a/2)$ to $(+a/2)$.

For second slit, from $(d-a/2)$ to $(d+a/2)$.

For third slit, from $(2d-a/2)$ to $(2d+a/2)$.

For N^{th} slit, from $\left\{(N-1)d-\frac{a}{2}\right\}$ to $\left\{(N-1)\,d+\frac{a}{2}\right\}$.

$$y=K\left[\int_{-\frac{a}{2}}^{+\frac{a}{2}}\phi(z)dz+\int_{d-\frac{a}{2}}^{d+\frac{a}{2}}\phi(z)dz+\int_{2d-\frac{a}{2}}^{2d+\frac{a}{2}}\phi(z)dz+\ldots+\int_{(N-1)d-\frac{a}{2}}^{(N-1)d+\frac{a}{2}}\phi(z)dz\right]$$

On simplification

$$y=Ka\frac{\sin\alpha}{\alpha}\left[\sin 2\pi\left(\frac{t}{T}-\frac{r}{\lambda}\right)+\sin 2\pi\left(\frac{t}{T}-\frac{r}{\lambda}+\frac{d\sin\theta}{\lambda}\right)\right.$$
$$+\sin 2\pi\left(\frac{t}{T}-\frac{r}{\lambda}+\frac{2d\sin\theta}{\lambda}\right)+\ldots$$
$$\left.+\sin 2\pi\left(\frac{t}{T}-\frac{r}{\lambda}+\frac{(N-1)\,d\sin\theta}{\lambda}\right)+\ldots\right]$$

Here
$$\alpha=\frac{\pi\,a\sin\theta}{\lambda}.$$

We know that

$$\sum_{p=0}^{p=n}\sin(x+pm)=\frac{\sin\left(x+\frac{nm}{2}\right)\sin\left[\left(\frac{n+1}{2}\right)m\right]}{\sin\left(\frac{m}{2}\right)}$$

Here, $$x = 2\pi\left(\frac{t}{T} - \frac{r}{\lambda}\right), m = \frac{2\pi\, d \sin\theta}{\lambda} = 2\beta$$

where $$\beta = \frac{\pi\, d \sin\theta}{\lambda}$$

and $$n = (N - 1).$$

$$\therefore \qquad y = \frac{Ka\left(\dfrac{\sin\alpha}{\alpha}\right)\left[\sin\left(x + \dfrac{(N-1)m}{2}\right)\sin\left(\dfrac{Nm}{2}\right)\right]}{\sin\left(\dfrac{m}{2}\right)}$$

$$= Ka\left(\frac{\sin\alpha}{\alpha}\right)\frac{\sin\left(\dfrac{Nm}{2}\right)}{\sin\left(\dfrac{m}{2}\right)}\left[\sin\left(x + \frac{(N-1)m}{2}\right)\right]$$

$$y = Ka\left(\frac{\sin\alpha}{\alpha}\right)\frac{(\sin N\beta)}{\sin\beta}\left[\sin 2\pi\left(\frac{t}{T} - \frac{r}{\lambda} + \frac{(N-1)\, d\sin\theta}{2\lambda}\right)\right] \qquad ...(1)$$

The intensity at a point P' is

$$I = K^2a^2\left(\frac{\sin^2\alpha}{\alpha^2}\right)\left(\frac{\sin^2 N\beta}{\sin^2\beta}\right) \qquad ...(2)$$

The maximum intensity, when $\alpha = 0$ and $\beta = 0$ is

$$I_0 = K^2a^2$$

$$\therefore \qquad I = I_0\left(\frac{\sin^2\alpha}{\alpha^2}\right)\left(\frac{\sin^2 N\beta}{\sin^2\beta}\right) \qquad ...(3)$$

The first factor $\dfrac{\sin^2\alpha}{\alpha^2}$ expresses intensity distribution due to diffraction at a single slit.

The second factor $\dfrac{\sin^2 N\beta}{\sin^2\beta}$ arises from interference between the N diffracted waves from the N slits.

Intensity of Principal Maxima

If $$\sin\beta = 0$$

i.e., $$\beta = \pm\, n\pi \qquad (n = 0, 1, 2, 3 \text{ etc.})$$

then $$\sin N\beta = 0$$

Thus $\dfrac{\sin N\beta}{\sin\beta} = \dfrac{0}{0}$ becomes indeterminate. Its value can be evaluated as follows.

$$\lim_{\beta \to \pm n\pi} \frac{\sin N\beta}{\sin\beta} = \lim_{\beta \to \pm n\pi} \frac{\dfrac{d}{d\beta}\sin N\beta}{\dfrac{d}{d\beta}\sin\beta}$$

$$= \lim_{\beta \to \pm n\pi} \frac{N\cos N\beta}{\cos\beta} = \frac{N\cos N(\pm\, n\pi)}{\cos(\pm\, n\pi)} = \pm\, N.$$

The intensity is then [from Eq. (3)],

$$I = N^2 I_0 \left(\frac{\sin^2 \alpha}{\alpha^2}\right) \quad ...(4)$$

These maxima are most intense and are called 'principal maxima'.

Therefore the condition for a principal maxima is,

$$\beta = \pm n\pi$$

or $$\frac{\pi d \sin\theta}{\lambda} = \pm n\pi \quad \text{or} \quad \frac{\pi}{\lambda}(a+b)\sin\theta = \pm n\pi$$

or $$(a+b)\sin\theta = \pm n\lambda \quad ...(5)$$

Here $n = 0, 1, 2, 3$ etc., represents the order of the interference maximum. For $n = 0$, we get the *zero order maximum.* For $n = 1, 2, 3$ etc., we obtain the first, second, third etc., principal maxima. The ± sign indicates that there are two principal maxima for each order lying on either side of the zero order maximum.

Minima—When $\sin N\beta = 0$, but $\sin\beta \neq 0$, then

$$\frac{\sin N\beta}{\sin\beta} = 0.$$

From Eq. (3), we have $I = 0$ which is a minimum.

Thus for minima, we have $\sin N\beta = 0$

or $$N\beta = \pm m\pi$$

or $$N.\frac{\pi}{\lambda}(a+b)\sin\theta = \pm m\pi$$

or $$N(a+b)\sin\theta = \pm m\lambda \quad ...(6)$$

Here m takes all integral values except 0, N, $2N$,... nN, because these values of m make $\sin\beta = 0$, which gives principal maxima.

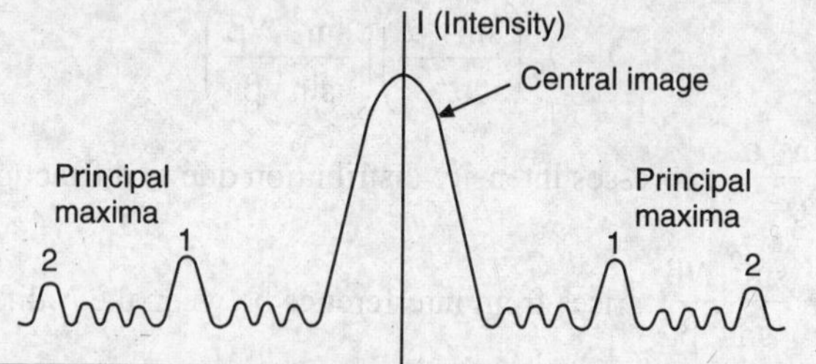

Fig. 30.5.

It is clear from above that $m = 0$ gives a principal maximum, $m = 1, 2, 3., (N-1)$ give minima and then $m = N$ gives again a principal maximum. Thus *there are $(N-1)$ minima between two consecutive principal maxima.* The intensity distribution due to diffraction at N slits is shown in Fig. 30.5.

30.6. Width of Principal Maxima

It is the angular separation between the first two minima lying on either side of a principal maximum.

The direction of the nth principal maximum is given by

$$(a+b)\sin\theta_n = n\lambda \quad ...(1)$$

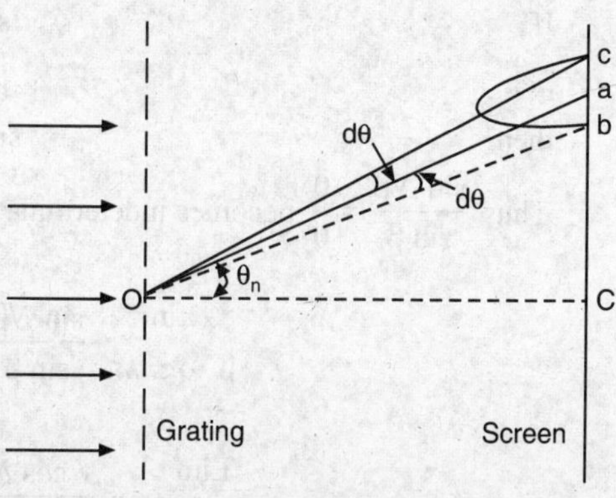

Fig. 30.6.

Let $(\theta_n + d\theta)$ and $(\theta_n - d\theta)$ represent the directions of the first secondary minima on the two sides of the nth primary maxima (Fig. 30.6).

Then $$(a+b)\sin(\theta_n \pm d\theta) = n\lambda \pm \frac{\lambda}{N} \qquad ...(2)$$

Here, N is the total number of lines on the grating surface.

$d\theta_n$ is the 'angular half width' of the nth maximum.

Dividing (2) by (1),

$$\frac{(a+b)\sin(\theta_n \pm d\theta)}{(a+b)\sin\theta_n} = \frac{n\lambda \pm \lambda/N}{n\lambda}$$

or $$\frac{\sin\theta_n \cos d\theta \pm \cos\theta_n \sin d\theta}{\sin\theta_n} = 1 \pm \frac{1}{Nn} \qquad ...(3)$$

For small values of $d\theta$, $\cos d\theta \simeq 1$ and $\sin d\theta \simeq d\theta$.

$$1 \pm \cot\theta_n \, d\theta = 1 \pm \frac{1}{Nn}$$

$$\therefore \qquad d\theta = \frac{1}{Nn\cot\theta_n} \qquad ...(4)$$

Eq. (4) gives half the width of principal maxima. The half width $d\theta$ is
(*i*) inversely proportional to N, the total number of lines and (*ii*) inversely proportional to $n \cot\theta_n$.

30.7. Phase Contrast Microscope

Principle. Suppose that the object is completely transparent but has an optical thickness which varies from point to point. Such an object is called a *phase object.* It introduces phase differences between disturbances which pass through different parts of it. Consequently, the disturbances in the conjugate image plane have the same amplitude at all points but will show variations in phase from point to point. The eye can distinguish only changes in intensity but not changes in phase. To see a small *transparent object,* it is necessary to magnify it and also to convert differences in phase into differences in intensity. *Zernike* in 1935 introduced the concept of phase contrast.

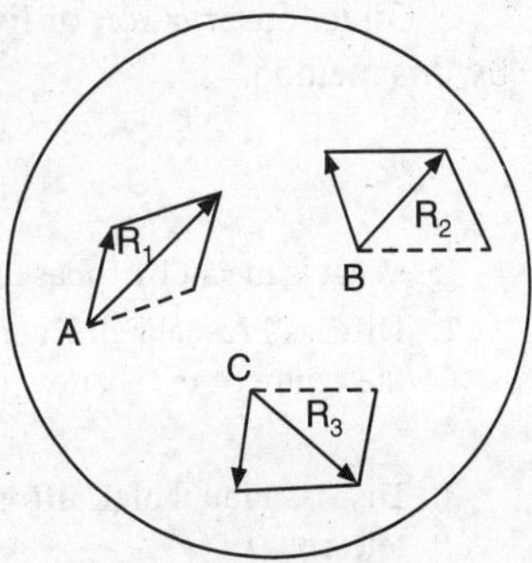

Fig. 30.7.

Consider a beam of light passing through a transparent plate of varying thickness. The amplitude vector at the points *A, B, C* has the same magnitude but is in different directions (Fig. 30.7). The intensity is the same at all points but there is difference in phase between the vectors. If a constant phase difference is introduced with the help of a phase plate (represented by the dotted vectors), the resultant amplitudes at the points *A, B* and *C* are R_1, R_2 and R_3 respectively. Their magnitudes are different. Hence the intensities are different and can be seen by the eye. The variations in optical thickness in the object cause variations in intensity in the image. This method of converting differences in phase into differences in intensity is employed in phase-contrast microscope.

Construction and Working

A phase contrast microscope consists of a transparent object slide placed in between the condenser and the objective (Fig. 30.8).

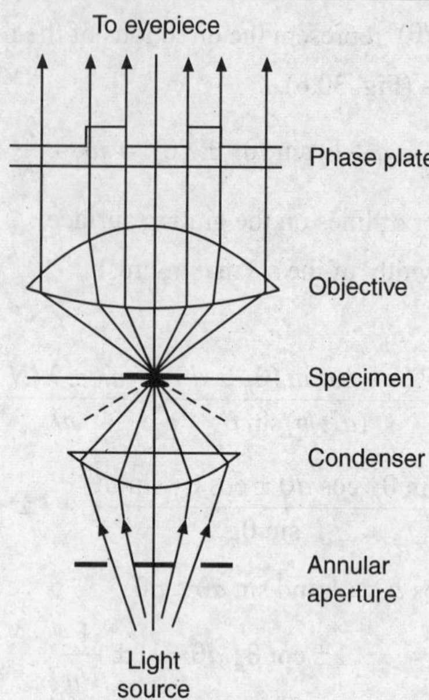

Fig. 30.8.

An annular aperture is placed at the first focal plane of the condenser. The phase plate is kept at the second focal plane of the objective. The phase plate has an annular depression exactly conjugate to the annular aperture. The image formed on the image plane is seen by the eyepiece. The final image is formed at the first focal plane of the eyepiece.

Direct observation of living material such as bacteria, yeast and fungi has been made possible by this method.

EXERCISE

1. What is meant by 'phase reversal zone plate'?
2. Discuss Fresnel's diffraction at a small opaque disc illuminated by monochromatic light. Explain why the centre of the shadow of the disc is bright. What happens if the size of the disc is increased? *(Kanpur 87, 85, Gauhati 86)*
3. Discuss Fraunhofer diffraction at a circular aperture. How are these results helpful in designing a good telescope? *(Meerut 87S, Rohilkhand 88)*
4. Discuss the Fraunhofer diffraction pattern due to *N* slits.
5. Deduce an expression for the half angular width of a principal maximum in the diffraction pattern of a grating having *N* slits. *(Kanpur 88, Agra 89, Gorakhpur 86)*
6. Describe the construction, theory and special application of Zernike's phase contrast microscope.

CHAPTER

31

POLARISATION OF LIGHT–(III)

31.1. Experimental Determination of Principal Refractive Indices

The principal refractive index for the *O*-ray

$$n_0 = \frac{\text{velocity of light in vacuum}}{\text{velocity of O-ray in the crystal}}$$

The principal refractive index for the *E*-ray

$$n_e = \frac{\text{velocity of light in vacuum}}{\text{velocity of E-ray normal to the optic axis}}.$$

The crystal whose principal refractive indices are to be determined is taken in the form of a prism cut with optic axis parallel to the refracting edge (Fig. 31.1). The optic axis is indicated by dots.

The prism is mounted on the prism table of an adjusted spectrometer. The slit is illuminated with sodium light. Light inside the prism is travelling at right angles to the optic axis. So the ordinary and extraordinary refractive indices are different and are equal to the principal refractive indices. Hence two refracted images of the slit are formed. Each image is adjusted in succession for minimum deviation position by rotating the prism table. The position of minimum deviation of each ray is measured. The prism is then removed and the direct reading of the slit is taken. The angle of minimum deviation for the ordinary ray δ_0 and that for the extraordinary ray δ_e are then calculated.

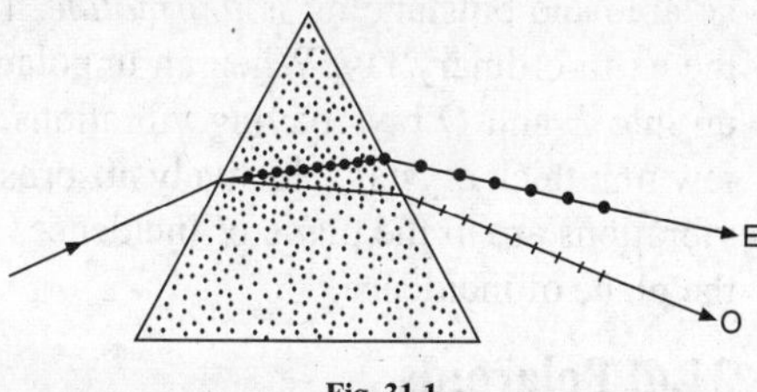

Fig. 31.1.

The refracting angle of the prism *A* is found by putting the edge on the centre of the table and taking the reading of the reflected images on both sides of the direct image. Half of the difference between the two readings gives *A*.

The refractive indices are calculated from the formulae:

$$n_0 = \frac{\sin \frac{1}{2}(A + \delta_0)}{\sin \frac{1}{2} A}$$

and

$$n_e = \frac{\sin \frac{1}{2}(A + \delta_e)}{\sin \frac{1}{2} A}$$

31.2. Brewster Window

When an ordinary beam of light is incident normally on a glass window, about 8% of light is lost by reflection on its two surfaces. Approximately 92% intensity is transmitted. Some laser designs use windows set at the Brewster's angle to reduce the reflective losses in the laser cavity for one polarization (Fig. 31.2). By tilting these plates to the polarizing angle i_p, the windows will have a 100 percent transmission for light whose electric vector is parallel to the plane of incidence. Brewster's windows are used in gas lasers.

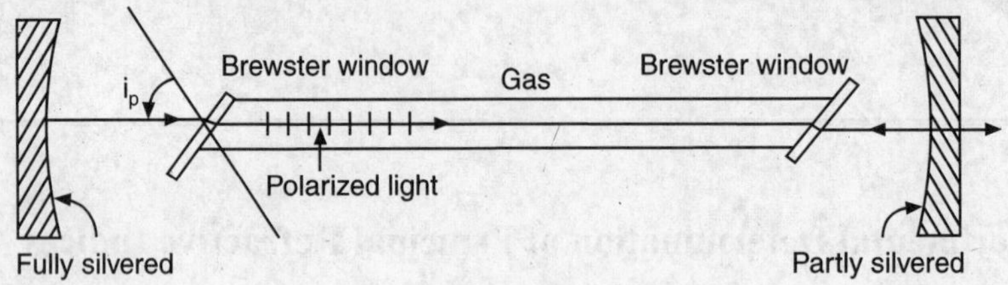

Fig. 31.2.

The polarizing angle is given by

$$\tan i_p = n$$

where n is the refractive index of the medium.

31.3. Dichroism

There are certain crystals and minerals of doubly refracting class which have the property of absorbing one of the doubly refracted beams to a greater extent than the other. The crystals showing this property are called *dichroic* crystals and the phenomenon is known as *dichroism.*

This phenomenon of selective absorption is shown by a number of substances but the most notable and outstanding is *tourmaline.* Tourmaline absorbs ordinary ray much more strongly than the extra-ordinary ray. When an unpolarised beam of light falls on a tourmaline crystal, it is split up into E and O rays having vibrations in the plane of incidence and perpendicular to it. Even a few mm thick crystal selectively absorbs the O-ray completely and transmits only the E-ray whose vibrations are in the plane of incidence. In this way we get plane polarised light with vibrations in the plane of incidence.

31.4. Polaroids

The phenomenon of selective absorption is utilized in the construction of polaroids. They are large sized polarizing films mounted between two thin glass sheets and are used to obtain plane polarised light for commercial purposes. The film consists of a thin sheet of nitrocellulose. Ultra-microscopic crystals of herpathite (an organic compound of iodine and quinine sulphate) are embedded in this film in such a manner that optic axes of all of them are parallel so that they function together as a single crystal of large dimensions. These crystals exhibit dichroism to such a marked degree that a very small thickness of the order of 0.005 inch can absorb one of the doubly refracted beams completely.

Recently H-polaroids have been prepared by making a sheet of polyvinyl alcohol which is stretched from 3 to 8 times of its length. The molecules are oriented in the direction of the stress and the material under stress becomes doubly refracting. This acts as a dichroic crystal when impregnated with iodine. If instead of iodine impregnation, the stretched film of polyvinyl is heated in the presence of dehydrating agent, then it becomes strongly dichroic and it is very stable. This is called the *K-Polaroids*.

When the two pieces of polaroids are uncrossed, the emergent beam is plane polarised [Fig. 31.3(*a*)]. When the two polaroids are crossed, there is perfect extinction of light [Fig. 31.3(*b*)].

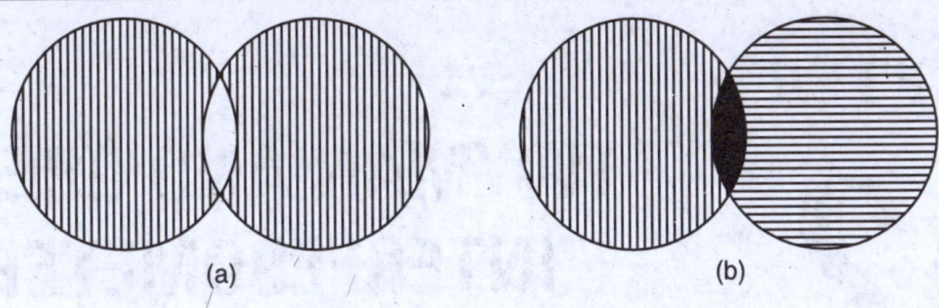

Fig. 31.3.

Uses of polaroids

(*i*) As polaroids have got a relatively much lower cost than Nicols, they are used in modern polarizing instruments and as such utilized to produce and analyse the beams of plane polarized light.

(*ii*) They are used in sun-glasses to cut off the glare produced by light reflected from polished and semi-smooth surface just as in the road and car hoods etc. The light reflected from such surface is partially polarized by reflection with horizontal vibrations and so can be easily avoided by looking at them through polaroids with a vertical transmission plane.

(*iii*) They are used to eliminate the head light glare in motor cars. The head lights and the wind-screen through which the driver sees, are fitted with polaroids with their planes of transmission inclined at 45° to vertical. When two cars approach each other from opposite directions, the transmission planes of the wind-screen of each are at right angles to the incident light. Thus the glare of the opposite head lights is completely eliminated by the wind-screen of the driver's car while the road will be visible to the driver by light sent out from his own car.

(*iv*) Polaroids are used to control the intensity of light in trains and aeroplanes.

(*v*) Two crossed polaroids subjected to pre-determined stress are used in optical stress analysis.

(*vi*) Polaroids are used to project stereoscopic pictures and also used to observe three-dimensional moving pictures.

EXERCISE

1. Explain what you understand by the principal refractive indices of a crystal. How can they be experimentally determined? [*Madras 1950; B.U. 1959*]

2. Unpolarized light of intensity I_0 falls on a combination of two linear polarizers where the second polarizer is having its pass axis rotated clockwise by 30° with respect to the first. The transmitted intensity will be

(A) $3I_0/4$ (B) $3I_0/8$ (C) $\sqrt{3}\,I_0/4$ (D) $I_0/4$ [*Gate-1999*] [**Ans.** (B)]

Explanation. Intensity of light coming out of first polarizer $= \dfrac{I_0}{2}$

(Because unpolarized light is incident on the first polariser)

Now, $$\text{transmitted intensity} = \frac{I_0}{2}\cos^2 30^\circ = \frac{3I_0}{8}$$

3. What do you understand by induced optical activity? [*Calcutta University, 1961*]

Ans. There are certain substances which can acquire optical activity when they are strained in some way or other.

(i) When glass is strained by an electric field, it acquires the property of double refraction and optical rotation. This phenomenon is *Kerr effect*.

(ii) When a block of glass is subjected to a strong magnetic field, it becomes optically active. When plane - polarized light is sent through glass in a direction parallel to the applied magnetic field, the plane of vibration is rotated.

This phenomenon is known as *Faraday effect*.

CHAPTER

32

TWYMAN AND GREEN INTERFEROMETER

32.1. Twyman and Green Interferometer

The essential optical parts of Twyman and Green interferometer are shown in Fig. 32.1.

Light from the monochromatic point source S is rendered parallel by the lens L_1 and falls on the glass plate G_1. The plate G_1 is semi-silvered on the back side. G_1 is a beam splitter. The single plane wavefront W, emerging from the lens L_1, after partial reflection at the lightly silvered surface of the plate G_1, gives rise to two plane wavefronts W_1 and W_2.

The wavefronts W_1 and W_2 are respectively reflected at the plane mirrors M_1 and M_2 at normal incidence. Finally, the reflected plane wavefronts, by partial reflection and refraction at G_1, give rise to superposed plane wavefronts. The interfering wave-fronts are focussed at E by another lens L_2.

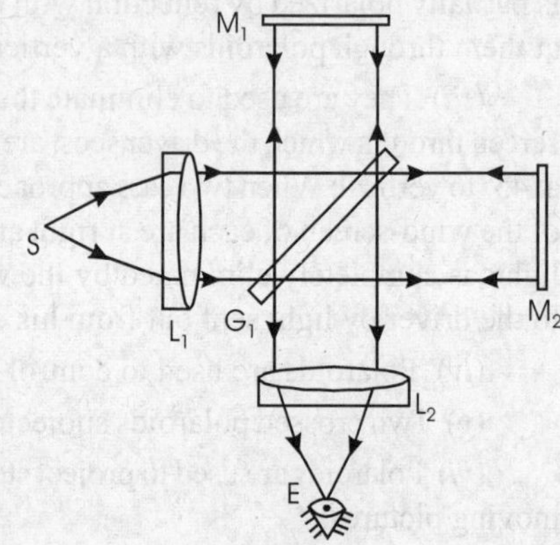

Fig. 32.1.

When M_1 and M_2 are exactly perpendicular so that the lightly silvered side of G_1 exactly bisects the angle between them, the superposed wavefronts are exactly parallel. Therefore, the phase difference between the superposed disturbances is the same at every point in the field of view. The field is, therefore, of uniform colour or intensity, the value depending on the extent to which wavefronts reinforce each other. The intensity is obviously a maximum when $2\,(G_1 M_1 - G_1 M_2) = \lambda$.

Here, $n = 0, 1, 2, 3$ etc.

In order to adjust the interferometer, a pin-hole in a screen illuminated with opalescent lamp from behind is placed at S in the focal plane of L_1 and $G_1 M_1$ is made equal to $G_1 M_2$. Two images of the pin-hole, one from each mirror are reflected back on the screen along side the pin-hole itself. Similarly, two images are seen on a screen at E in the focal plane of L_2. The mirrors are adjusted by tilting screws until the two images at S coincide with the pin hole *simultaneously* with the coincidence of two images at E. On removing the screen and using a point monochromatic source at the focus of L_1 no fringes should be seen. If, otherwise, the mirrors should be adjusted further until all traces of fringes disappear. The interferometer is now in perfect adjustment.

Uses. The interferometer is used for testing the optical homogeneity of prisms, lenses and glass plates. It is also used for testing the rulings of plane reflection grating for the absence of *ghosts*. Thus it assists in the manufacture of perfect prism, lenses and gratings.

32.2. Testing a Prism with Twyman Green Interferometer

Principle. The piece to be tested is placed in one of the light beams. The mirror behind it is so chosen that the reflected waves, after traversing the test piece a second time, again become plane. These waves are then brought to interference with the plane waves from the other arm of the interferometer by another lens, at the focus of which the eye is placed. If the prism or lens is optically perfect, so that the returning waves are strictly plane, the field will appear uniformly illuminated. Any local variation of the optical path will, however, produce fringes in the corresponding part of the field. These fringes are essentially the contour lines of the distorted wave front.

Experimental Arrangement

The experimental arrangement is shown in Fig. 32.2.

The plane mirror M_2 is replaced by a prism set at minimum deviation and a plane mirror. The plane mirror is adjusted to reflect the emerging beam from the prism normally along its own path. Therefore, if the prism is perfect, the returning wavefront W_4 will be plane and parallel to the incident wavefront W_2. The field of view will be of uniform intensity. If however, the prism is not optically perfect, the returning wavefront W_4 will be no longer plane but it will be deformed from the plane formation as shown in Fig. 32.2. Therefore, the phase difference between the superposed disturbances in the returning wavefronts from M_1 and M_2 will vary across the field of view and fringes will be seen of irregular shape depending upon the deformation of the wavefront W_4.

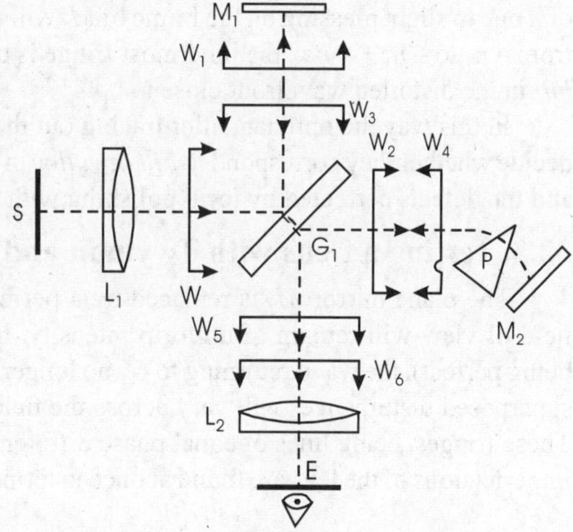

Fig. 32.2.

Since a fringe is a locus of equal phase difference, the observed fringes are contours of the deformed wavefront, the fringes corresponding to different levels, one wavelength apart, of the deformed wavefront. These fringes are fringes of equal optical thickness and may be regarded as a contour map of the imperfection in the prism. The eye is focussed on the prism face as seen through the lens L_2 and fringes are observed coincident with one of the prism faces. These fringes are marked out on the prism face with a grease pencil and serve as a guide to subsequent local polishing. The fringe system shown in Fig. 32.3 correspond to either increased retardation (*Valley* in the distorted wavefront) or decreased retardation (*Hill* in the distorted wavefront) of light in passing the inhomogeneous glass portion as compared to the passage through the homogeneous portion.

This method was originally used in the manufacture of prisms. The errors caused by inhomogeneity of the prism is corrected by local *figuring* of the surface of the prism until the field of view in the interferometer is of uniform intensity.

Fig. 32.4 shows the distorted wavefront and interference pattern obtained with it. The undistorted wavefront W_5 from M_1 is shown by dotted vertical line. The distorted wavefront has one *valley A* and one *hill B*.

The optical path difference between distorted and undistorted wavefronts decreases as we consider disturbances from A towards C or C'. Therefore, in the corresponding fringe system the innermost fringe is of the highest order say n and the next is of order $(n - 1)$ and so on. By gently pressing the mounting of the mirror M_2, the optical path AM_2 is slightly increased so that more retardation is introduced in the optical disturbance at A, G, E etc. If the increased retardation is exactly λ, then the innermost fringe in the fringe system P will be of $(n + 1)$th order and the next of nth order and so on *i.e.*, the pattern at P opens out.

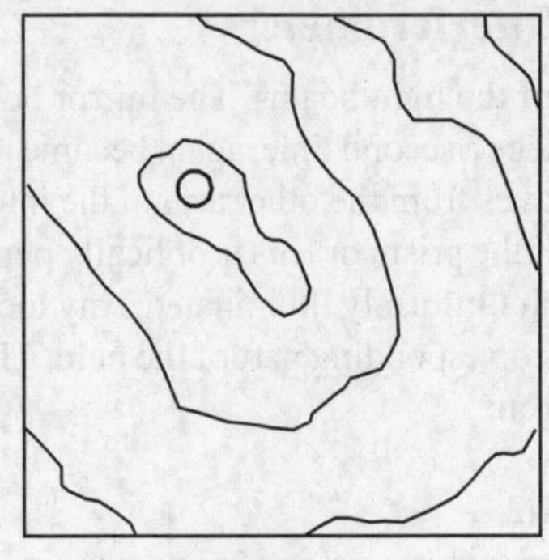
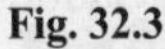

Fig. 32.3.

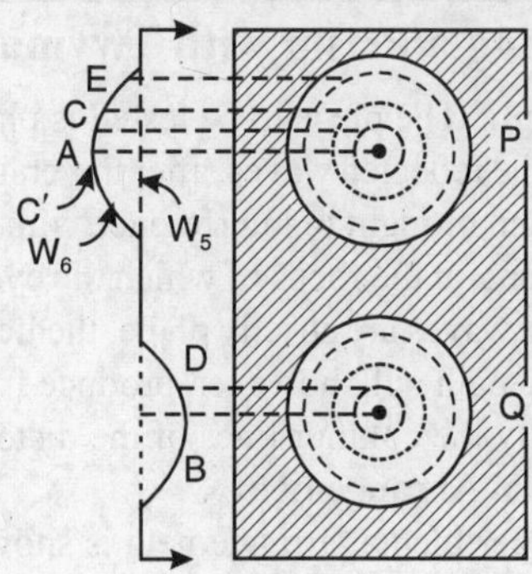

Fig. 32.4.

In the fringe system, Q corresponding to *hill* in distorted wavefront, the innermost ring is of order $-n_1$ (say), the next fringe is of order $-(n_1 - 1)$ and so on. In this case, the increased retardation of λ due to slight pressing on the frame of M_2 will result in the change of the order of innermost fringe from $-n_1$ to $-n_1 + 1$ *i.e.*, the innermost fringe is of order $-(n_1 - 1)$ *i.e.*, the fringes corresponding to *hill* in the distorted wavefront close in.

In this way the optician, after tracing out the fringes on the prism face with grease pencil, can decide whether they correspond to *hill* or *valley* in the distorted wavefront. The prism can be removed and the defects corrected by local polishing with pencil polisher.

32.3. Testing a Lens with Twyman and Green Interferometer

The plane mirror M_2 is replaced by a perfect lens and spherical mirror (Fig. 32.5). Now the field of view will remain of uniform intensity. If, however, the lens is not perfect (the mirror M_2 being perfect), the wave returning to G_1 no longer will be plane and the phase difference between the superposed disturbances will vary across the field of view and fringes will be seen (see Fig. 32.6). These fringes, being lines of equal phase difference, are contours of the distorted wavefront. So the imperfections of the lens are found at once in terms of the wavefront aberration which they introduce.

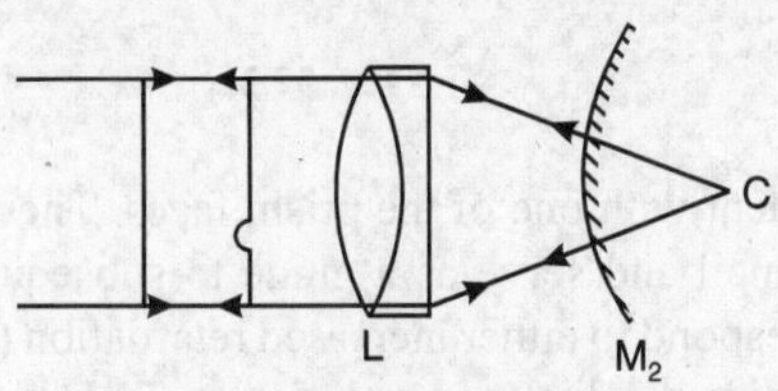

Fig. 32.5.

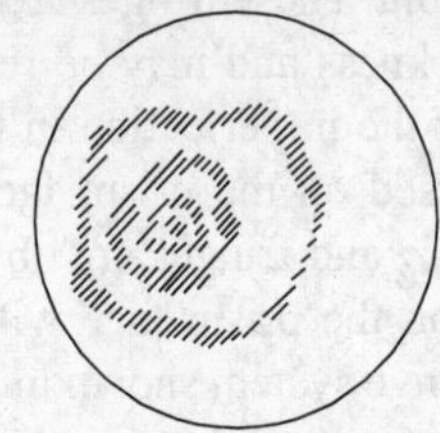

Fig. 32.6.

In order to see whether a fringe system such as that shown in Fig. 32.6 corresponds to an additional optical path or the reverse, it is only necessary to press on the mounting of the mirror M_2 so as to increase the path G_1M_2 slightly. The fringes will then move to points corresponding to relatively smaller optical paths so that if the ring system corresponds to additional retardation the rings will open out. By mounting a lens under test on a nodal slide which automatically maintains the centre of curvature of M_2 at the Gaussian focus of the lens, it is possible to find the aberrations present at any point of the field of the lens.

EXERCISE

1. Outline the construction and theory of Twyman and Green interferometer. Show how it may be used for testing the optical homogeneity of prisms and lenses.

CHAPTER

33

CRYSTAL OPTICS

33.1. Introduction

Many crystals have symmetry sufficiently low that their optical properties are anisotropic. The result is double refraction. We have already introduced this subject in section 4.5. Here we wish to elaborate on the phenomenon with a presentation based on Maxwell's equations.

33.2. Plane Wave Propagation in an Anisotropic Medium

In an anisotropic medium, the relation between D and E is

$$\begin{aligned} D_x &= \varepsilon_{xx}E_x + \varepsilon_{xy}E_y + \varepsilon_{xz}E_z \\ D_y &= \varepsilon_{yx}E_x + \varepsilon_{yy}E_y + \varepsilon_{yz}E_z \\ D_z &= \varepsilon_{zx}E_x + \varepsilon_{zy}E_y + \varepsilon_{zz}E_z \end{aligned} \quad ...(1)$$

The nine quantities $\varepsilon_{xx}, \varepsilon_{xy}$ ...are constants of the medium, and constitute the dielectric tensor.

$$\varepsilon_{xy} = \varepsilon_{yx}, \ \varepsilon_{xz} = \varepsilon_{zx} \text{ and } \varepsilon_{yz} = \varepsilon_{zy} \quad ...(2)$$

We choose a coordinate system such that

$$D_x = \varepsilon_x E_x$$

$$D_y = \varepsilon_y E_y \quad ...(3)$$

and

$$D_z = \varepsilon_z E_z \,.$$

This coordinate system is known as the principal axis system.

The quantities $\varepsilon_x, \varepsilon_y$ and ε_z are called the principal dielectric permittivities of the medium.

For a uniaxial crystal,

$$\varepsilon_x = \varepsilon_y \neq \varepsilon_z \quad ...(4)$$

Here the z-axis represents the optic axis of the medium.

The quantities

$$n_o = \sqrt{\varepsilon_x / \varepsilon_0} = \sqrt{\varepsilon_y / \varepsilon_0} \text{ and } n_e = n_z = \sqrt{\varepsilon_z / \varepsilon_0} \quad ...(5)$$

are called ordinary and extraordinary refractive indices.

For a uniaxial medium, since $\varepsilon_x = \varepsilon_y$, the x and y directions can be arbitrarily chosen as long as they are perpendicular to the optic axis.

Consider the propagation of a plane electromagnetic wave. For such a wave,

$$\left.\begin{aligned}\mathbf{E} = \mathbf{E}_0 e^{i(\mathbf{k}\cdot\mathbf{r}-\omega t)}, \ \mathbf{H} = \mathbf{H}_0 e^{i(\mathbf{k}\cdot\mathbf{r}-\omega t)} \\ \mathbf{D} = \mathbf{D}_o e^{i(\mathbf{k}\cdot\mathbf{r}-\omega t)}, \ \mathbf{B} = \mathbf{B}_0 e^{i(\mathbf{k}\cdot\mathbf{r}-\omega t)}\end{aligned}\right\} \quad ...(6)$$

Here, the vectors $\mathbf{E}_0$, $\mathbf{H}_0$, $\mathbf{D}_0$ and $\mathbf{B}_0$ are independent of space and time. **k** represents the propagation vector of the wave and ω the angular frequency. The wave velocity v_w and the wave refractive index n_w are defined through the following equation:

$$v_w = \frac{\omega}{k} = \frac{c}{n_w} \quad ...(7)$$

Thus $|\mathbf{k}| = k = \frac{\omega}{c} n_w$.

The medium is nonmagnetic so that $\mathbf{B} = \mu_0 \mathbf{H}$.

Maxwell's curl equations become

$$\nabla \times \mathbf{E} = -\frac{\partial \mathbf{B}}{\partial t} = i\omega\mathbf{B} = i\omega\mu_0\mathbf{H} \quad ...(8)$$

and $$\nabla \times \mathbf{H} = \frac{\partial \mathbf{D}}{\partial t} = -i\omega\mathbf{D} \quad ...(9)$$

We have $$\mathbf{E} = \mathbf{E}_0 \exp[i(\mathbf{k}\cdot\mathbf{r} - \omega t)]$$

$$(\nabla \times \mathbf{E})_x = \frac{\partial E_z}{\partial y} - \frac{\partial E_y}{\partial z} = (ik_y E_{0z} - ik_z E_{0y}) e^{i(\mathbf{k}\cdot\mathbf{r}-\omega t)}$$

$$= i(\mathbf{k}_y \mathbf{E}_z - k_z \mathbf{E}_y) = i(\mathbf{k} \times \mathbf{E})_x$$

Thus $$\nabla \times \mathbf{E} = i(\mathbf{k} \times \mathbf{E}) = i\omega\mu_0\mathbf{H} \Rightarrow \mathbf{H} = \frac{1}{\omega\mu_0}(\mathbf{k} \times \mathbf{E}) \quad ...(10)$$

and $$\nabla \times \mathbf{H} = i(\mathbf{k} \times \mathbf{H}) = -i\omega\mathbf{D} \Rightarrow \mathbf{D} = \frac{1}{\omega}(\mathbf{H} \times \mathbf{k}) \quad ...(11)$$

Equations (10) and (11) show that **H** is at right angles to **k**, **E** and **D**.

Thus **k**, **E** and **D** must lie in the same plane.

Further **D** is at right angles to **k** (Fig. 33.1).

Substituting for **H** in Eq. (11), we get

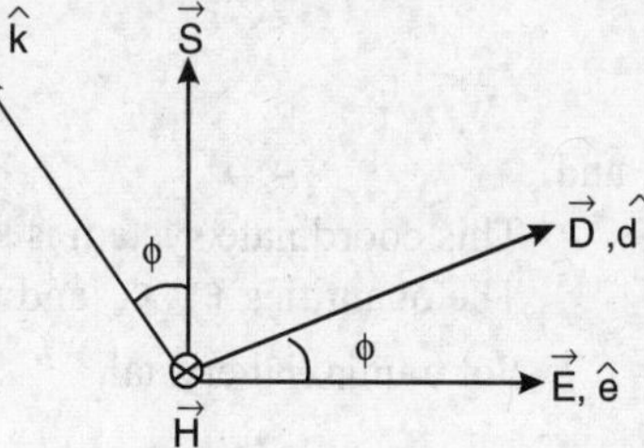

Fig. 33.1.

$$\mathbf{D} = \frac{1}{\omega^2\mu_0}[(\mathbf{k} \times \mathbf{E}) \times \mathbf{k}]$$

$$\mathbf{D} = \frac{1}{\omega^2\mu_0}[(\mathbf{k}\cdot\mathbf{k})\mathbf{E} - (\mathbf{k}\cdot\mathbf{E})\mathbf{k}] \quad ...(12)$$

$$\therefore \quad \mathbf{D} = \frac{k^2}{\omega^2\mu_0}\left[\mathbf{E} - (\hat{k}\cdot\mathbf{E})\hat{k}\right]$$

$$\mathbf{D} = \frac{n_\omega^2}{c^2\mu_0}\left[\mathbf{E} - (\hat{k}\cdot\mathbf{E})\hat{k}\right] \quad ...(13)$$

Here, $\hat{k} = \dfrac{\mathbf{k}}{k}$ represents the unit vector along **k**.

Now $D_x = \varepsilon_x E_x = \varepsilon_0 n_x^2 E_x$.

The x-component of Eq. (13) is

$$\frac{\varepsilon_0 \mu_0 c^2 n_x^2}{n_w^2} E_x = E_x - k_x (k_x E_x + k_y E_y + k_z E_z)$$

Since $c^2 = 1/(\varepsilon_0 \mu_0)$, we have

$$\left(\frac{n_x^2}{n_w^2} - k_y^2 - k_z^2 \right) E_x + k_x k_y E_y + k_x k_z E_z = 0 \qquad \text{...(14)}$$

Here we have used the relation $k_x^2 + k_y^2 + k_z^2 = 1$ (since $\hat{k}$ is a unit vector).
Similarly,

$$k_x k_y E_x + \left(\frac{n_y^2}{n_w^2} - k_x^2 - k_z^2 \right) E_y + k_y k_z E_z = 0 \qquad \text{...(15)}$$

$$k_x k_z E_x + k_y k_z E_y + \left(\frac{n_z^2}{n_w^2} - k_x^2 - k_y^2 \right) E_z = 0 \qquad \text{...(16)}$$

Eqs. (14), (15) and (16) form a set of three homogeneous equations.
For non-trivial solutions, we must have

$$\begin{vmatrix} \dfrac{n_x^2}{n_w^2} - k_y^2 - k_z^2 & k_x k_y & k_x k_z \\ k_x k_y & \dfrac{n_y^2}{n_w^2} - k_x^2 - k_z^2 & k_y k_z \\ k_x k_z & k_y k_z & \dfrac{n_z^2}{n_w^2} - k_x^2 - k_y^2 \end{vmatrix} = 0 \qquad \text{...(17)}$$

Eq. (17) appears to be cubic in n_w^2.

But when Eq. (17) is written in terms of a polynomial in n_w^2, the coefficient of n_w^6 will always be zero, leaving a quadratic equation in n_w^2.

The quadratic equation in n_w^2 can be shown to have two positive roots, n_{1w}^2 and n_{2w}^2.

Thus for a given propagation direction **k**, there are in general two values of the refractive index, n_{1w} and n_{2w}.

33.3. Propagation Along a Principal Axis

Consider the propagation of the wave along one of the principal axes–say along the z-axis. Thus $k_x = 0 = k_y$ and $k_z = 1$.

Eqs. (14), (15), (16) and (17) become

$$\left.\begin{aligned}\left(\frac{n_x^2}{n_w^2}-1\right)E_x=0\\ \left(\frac{n_y^2}{n_w^2}-1\right)E_y=0\\ \frac{n_z^2}{n_w^2}E_z=0\end{aligned}\right\} \quad ...(18)$$

and
$$\begin{vmatrix}\left(\frac{n_x}{n_w}\right)^2-1 & 0 & 0\\ 0 & \left(\frac{n_y}{n_w}\right)^2-1 & 0\\ 0 & 0 & \left(\frac{n_z}{n_w}\right)^2\end{vmatrix}=0 \quad ...(19)$$

$\therefore$
$$\left(\frac{n_x^2}{n_w^2}-1\right)\left(\frac{n_y^2}{n_w^2}-1\right)\frac{n_z^2}{n_w^2}=0 \quad ...(20)$$

Eq. (20) gives us only two roots for n_w :

$$n_w = n_{w_1} = n_x \text{ with } E_y = 0 = E_z \text{ (}x\text{-polarized wave)}$$

$$n_w = n_{w2} = n_y \text{ with } E_x = 0 = E_z \text{ (}y\text{-polarized wave)} \quad ...(21)$$

These are the two modes of the system.
Corresponding to the solution $n_w = n_{w_1} = n_x$, the **E** and **D** vectors are along the *x*-axis.

For the solution $n_w = n_{w_2} = n_y$, the **E** and **D** vectors are along the *y*-axis.

If the crystal is uniaxial, then, $n_x = n_y = n_o$.

$\therefore$
$$n_{w_1} = n_{w_2} = n_o \quad ...(22)$$

Both waves propagate with the same wave velocity

$$v_w = v_{w_1} = v_{w_2} = \frac{c}{n_o} \quad ..(23)$$

Similarly, for a wave propagating along the *x*-axis, $k_x = 1$, $k_y = 0 = k_z$.
We have the following two solutions :

$$\left.\begin{aligned} n_w = n_{w1} = n_y \quad (y\text{-polarized})\\ n_w = n_{w2} = n_z \quad (z\text{-polarized})\end{aligned}\right\} \quad ...(24)$$

For a uniaxial medium, $n_y = n_o$ and $n_z = n_e$.
The two waves propagate with different wave velocities.
The *y*-polarized wave is the ordinary wave.
The *z*-polarized wave is the extraordinary wave.

33.4. Propagation in Uniaxial Crystals

For uniaxial crystals,

$$n_x = n_y = n_o \text{ and } n_z = n_e \qquad \text{...(25)}$$

Here, n_o and n_e are the ordinary and the extraordinary indices of refraction respectively.

For a uniaxial crystal, the x and y directions can be arbitrarily chosen as long as they are perpendicular to the optic axis.

The optic axis is assumed to lie along the z-axis (Fig. 33.2).

For a wave propagating along *any* direction **k**, we choose the y-axis in such a way that it is at right angles to **k**. The y-axis is normal to the plane defined by **k** and the z-axis. The x-axis will lie in the same plane.

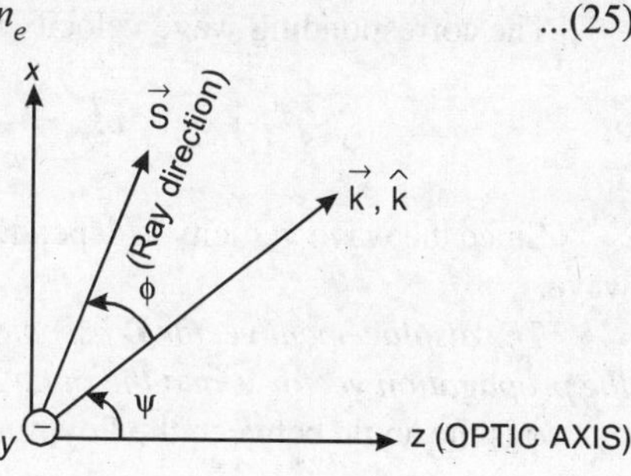

Fig. 33.2.

Let ψ be the angle that the **k** vector makes with the optic axis. Then

$$k_x = \sin\psi\,,\ k_y = 0, \text{ and } k_z = \cos\psi\,.$$

Eqs. (14), (15) and (16) become

$$\left(\frac{n_o^2}{n_w^2} - \cos^2\psi\right)E_x + \sin\psi\cos\psi\,E_z = 0 \qquad \text{...(26)}$$

$$\left(\frac{n_o^2}{n_w^2} - 1\right)E_y = 0 \qquad \text{...(27)}$$

$$\sin\psi\cos\psi\,E_x + \left(\frac{n_e^2}{n_w^2} - \sin^2\psi\right)E_z = 0 \qquad \text{...(28)}$$

One of the solutions of Eq. (27) is

$$n_w = n_{wo} = n_o \text{ with } E_x = 0 = E_z \text{ (y-polarized)} \qquad \text{...(29)}$$

The corresponding wave velocity is given by

$$v_w = v_{wo} = \frac{c}{n_o} \qquad \text{...(30)}$$

Since the wave velocity is independent of the direction of the wave, it is called the ordinary wave. *The electric vector associated with the ordinary wave is perpendicular to* **k** *and the optic axis.*

From Eqs. (26) and (28),

$$\frac{E_z}{E_x} = -\frac{\dfrac{n_o^2}{n_w^2} - \cos^2\psi}{\sin\psi\cos\psi} = -\frac{\sin\psi\cos\psi}{\dfrac{n_e^2}{n_w^2} - \sin^2\psi}$$

and $E_y = 0$

$$\therefore \qquad \frac{1}{n_w^2} = \frac{1}{n_{we}^2} = \frac{\cos^2\psi}{n_o^2} + \frac{\sin^2\psi}{n_e^2} \qquad \text{...(31)}$$

$$\frac{n_e^2 E_z}{n_o^2 E_x} = -\tan\psi$$

or $$\frac{\varepsilon_z E_z}{\varepsilon_x E_x} = \frac{D_z}{D_x} = -\tan\psi \text{ and } D_y = \varepsilon_y E_y = 0 \qquad ...(32)$$

The corresponding wave velocity is given by

$$v_{we}^2 = \frac{c^2}{n_{we}^2} = \frac{c^2}{n_o^2}\cos^2\psi + \frac{c^2}{n_e^2}\sin^2\psi \qquad ...(33)$$

Since the wave velocity is dependent on the direction of the wave, it is called the extraordinary wave.

The displacement vector **D** *associated with the extraordinary wave lies in the plane containing the propagation vector* **k** *and the optic axis and is normal to* **k**.

ϕ is the angle between the Poynting vector **S** and **k**.

$$\cos\phi = \frac{\mathbf{E}\cdot\mathbf{D}}{|\mathbf{E}||\mathbf{D}|} = \frac{E_x D_x + E_z D_z}{(E_x^2 + E_z^2)^{1/2}(D_x^2 + D_z^2)^{1/2}}$$

$$= \frac{1 + \frac{E_z}{E_x}\frac{D_z}{D_x}}{\left[1 + \left(\frac{E_z}{E_x}\right)^2\right]^{1/2}\left[1 + \left(\frac{D_z}{D_x}\right)^2\right]^{1/2}}$$

$$= \frac{1 + \frac{n_o^2}{n_e^2}\tan^2\psi}{\left[1 + \left(\frac{n_o^2\tan\psi}{n_e^2}\right)^2\right]^{1/2}\left(1 + \tan^2\psi\right)^{1/2}}$$

$$\therefore \quad \cos\phi = \frac{\left(n_e^2 + n_o^2\tan^2\psi\right)\cos\psi}{\left(n_e^4 + n_o^4\tan^2\psi\right)^{1/2}} \qquad ...(34)$$

33.5. Ray Velocity and Ray Refractive Index

The direction of energy propagation (or the ray propagation) is along the Poynting vector **S** (Fig. 33.3).

$$\mathbf{S} = \mathbf{E}\times\mathbf{H} \qquad ...(35)$$

S is perpendicular to both **E** and **H**. Therefore, **S** is in the same plane containing the vectors **k, E** and **D**.

Let $\hat{\mathbf{s}}$ be the unit vector along **S**. The direction of the propagation of the wave ($\hat{\mathbf{k}}$) is not along the direction of energy propagation ($\hat{s}$).

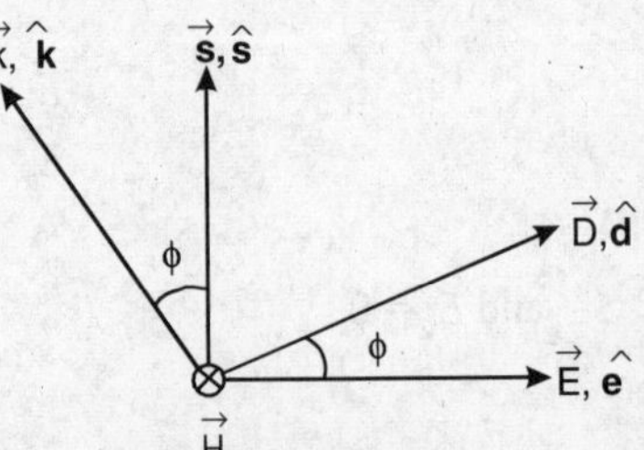

Fig. 33.3.

ϕ is the angle between $\hat{\mathbf{k}}$ and $\hat{s}$.

Let u be the energy density.

The ray velocity v_r is defined as

$$v_r = \frac{S}{u} \qquad ...(36)$$

$$u = \frac{1}{2}(\mathbf{D}\cdot\mathbf{E} + \mathbf{B}\cdot\mathbf{H}) = \frac{1}{2}(\mathbf{D}\cdot\mathbf{E} + \mu_o\mathbf{H}\cdot\mathbf{H}) \qquad ...(37)$$

The quantities $\frac{1}{2}\mathbf{D}\cdot\mathbf{E}$ and $\frac{1}{2}\mathbf{B}\cdot\mathbf{H}$ represent the electrical and magnetic energies per unit volume respectively.

Substituting for **H** and **D** from Eqs. (10) and (11), we get

$$u = \frac{1}{2\omega}[(\mathbf{H}\times\mathbf{k})\cdot\mathbf{E} + (\mathbf{k}\times\mathbf{E})\cdot\mathbf{H}]$$

$$= \frac{1}{2\omega}[\mathbf{k}\cdot(\mathbf{E}\times\mathbf{H}) + \mathbf{k}\cdot(\mathbf{E}\times\mathbf{H})]$$

$$\therefore \qquad u = \frac{1}{\omega}\mathbf{k}\cdot\mathbf{S} \qquad ...(38)$$

Substituting this value of u in Eq. (36),

$$v_r = \frac{\omega S}{\mathbf{k}\cdot\mathbf{S}} = \frac{\omega}{k\cos\phi} = \frac{v_w}{\cos\phi} \qquad ...(39)$$

The ray refractive index, n_r is defined as

$$n_r = \frac{c}{v_r} = \frac{c}{v_w}\cos\phi = n_w\cos\phi \qquad ...(40)$$

Let $\hat{\mathbf{e}}$ be a unit vector in the direction of **E**.

$$\mathbf{D} - (\mathbf{D}\cdot\hat{\mathbf{s}})\hat{\mathbf{s}} = (\mathbf{D}\cdot\hat{\mathbf{e}})\hat{\mathbf{e}} = (\mathbf{D}\cos\phi)\frac{\mathbf{E}}{E} \qquad ...(41)$$

Let $\hat{\mathbf{d}}$ be a unit vector in the direction of **D**.

$$\mathbf{E} = (\mathbf{E}\cdot\hat{\mathbf{d}})\hat{\mathbf{d}} + (\mathbf{E}\cdot\hat{\mathbf{k}})\hat{\mathbf{k}} \qquad ...(42)$$

$$\mathbf{E} = (\mathbf{E}\cdot\hat{\mathbf{k}})\hat{\mathbf{k}} = (\mathbf{E}\cdot\hat{\mathbf{d}})\hat{\mathbf{d}}$$

Substituting this value of $\mathbf{E} - (\mathbf{E}\cdot\hat{\mathbf{k}})\hat{\mathbf{k}}$ in Eq. (13), we get

$$\mathbf{D} = \frac{n_w^2}{\mu_0 c^2}(\mathbf{E}\cdot\hat{\mathbf{d}})\hat{\mathbf{d}}$$

or

$$\mathbf{D} = \frac{n_w^2}{\mu_0 c^2}\mathbf{E}\cos\phi \qquad ...(43)$$

Substituting in Eq. (41), we get

$$\mathbf{D} - (\mathbf{D}\cdot\hat{\mathbf{s}})\hat{\mathbf{s}} = \frac{n_w^2}{\mu_0 c^2}\cos^2\phi\mathbf{E}$$

From Eq. (40), $n_r = n_w\cos\phi$.

$$\mathbf{D} - (\mathbf{D}\cdot\hat{\mathbf{s}})\hat{\mathbf{s}} = \frac{n_r^2}{\mu_0 c^2}\mathbf{E} \qquad ...(44)$$

The x component of Eq. (44) is

$$D_x - (D_x s_x + D_y s_y + D_z s_z) s_x = \frac{n_r^2}{\mu_0 c^2} E_x = \frac{n_r^2}{\mu_0 c^2 \varepsilon_x} D_x$$

By using the relations

$$n_x^2 = \frac{\varepsilon_x}{\varepsilon_0}, c^2 = \frac{1}{\varepsilon_0 \mu_0} \text{ and } s_x^2 + s_y^2 + s_z^2 = 1$$

we get

$$\left(\frac{n_r^2}{n_x^2} - s_y^2 - s_z^2 \right) D_x + s_x s_y D_y + s_x s_z D_z = 0 \quad \text{...(45)}$$

Similarly

$$s_x s_y D_x + \left(\frac{n_r^2}{n_y^2} - s_x^2 - s_z^2 \right) D_y + s_z s_y D_z = 0 \quad \text{...(46)}$$

$$s_x s_z D_x + s_z s_y D_y + \left(\frac{n_r^2}{n_z^2} - s_x^2 - s_y^2 \right) D_z = 0 \quad \text{...(47)}$$

Eqs. (45), (46) and (47) form a set of three homogeneous equations.
For non-trivial solutions, we must have

$$\begin{vmatrix} \frac{n_r^2}{n_x^2} - s_y^2 - s_z^2 & s_x s_y & s_x s_z \\ s_x s_y & \frac{n_r^2}{n_y^2} - s_x^2 - s_z^2 & s_z s_y \\ s_x s_z & s_z s_y & \frac{n_r^2}{n_z^2} - s_x^2 - s_y^2 \end{vmatrix} \quad \text{...(48)}$$

For a given ray direction (*i.e.*, for given values of s_x, s_y and s_z), the solution of Eq. (48) gives us the two allowed values of n_r.

The two corresponding values of the ray velocities are given by Eq. (40),

$$n_r = \frac{c}{v_r}.$$

33.6. Ray Propagation along One of the Principal Axes

Consider ray propagation along one of the principal axes–say, along z-axis.

$$s_x = 0 = s_y \text{ and } s_z = 1.$$

Eqs. (45), (46) and (47) become

$$\left(\frac{n_r^2}{n_x^2} - 1 \right) D_x = 0$$

$$\left(\frac{n_r^2}{n_y^2} - 1 \right) D_y = 0$$

$$\frac{n_r^2}{n_z^2} D_z = 0$$

D_z is always zero.

The two possible values of the ray refractive index are

$$\left.\begin{aligned} n_r &= n_{r_1} = n_x \left[\text{with } D_x \neq 0 \text{ and } D_y = D_z = 0 \right] \\ n_r &= n_{r_2} = n_y \left[\text{with } D_y \neq 0 \text{ and } D_x = D_z = 0 \right] \end{aligned}\right\} \quad ...(49)$$

The two corresponding values of the ray velocities are

$$\left.\begin{aligned} v_r &= v_{r_1} = \frac{c}{n_{r_1}} = \frac{c}{n_x} \quad (x\text{-polarized wave}) \\ v_r &= v_{r_2} = \frac{c}{n_{r_2}} = \frac{c}{n_y} \quad (y\text{-polarized wave}) \end{aligned}\right\} \quad ...(50)$$

Comparing these with Eq. (21), we find that along the principal axes the ray and wave velocities are equal. They both propagate along the same direction.

33.7. Ray Propagation in a Uniaxial Crystal

Consider a uniaxial crystal with its optic axis along the z-direction (Fig. 33.4).

$$n_x = n_y = n_0 \text{ and } n_z = n_e.$$

The x-axis lies in the plane defined by **k** and the optic axis.

We choose the y-axis in such a way that the ray propagates in the x-z plane making an angle θ with the z-axis.

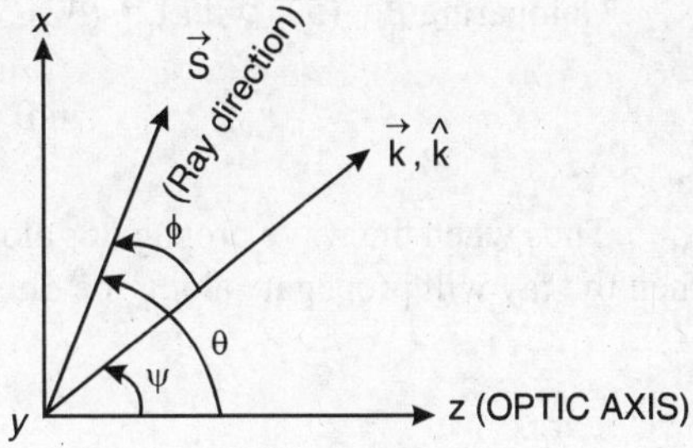

Fig. 33.4.

$\therefore \quad s_x = \sin\theta, \; s_y = 0 \text{ and } s_z = \cos\theta.$

Eqs. (45), (46) and (47) become

$$\left(\frac{n_r^2}{n_o^2} - \cos^2\theta\right) D_x + \sin\theta\cos\theta\, D_z = 0 \quad ...(51)$$

$$\left(\frac{n_r^2}{n_o^2} - 1\right) D_y = 0 \quad ...(52)$$

$$\sin\theta\cos\theta\, D_x + \left(\frac{n_r^2}{n_e^2} - \sin^2\theta\right) D_z = 0 \quad ...(53)$$

One of the roots is

$$n_r = n_{ro} = n_o \text{ with } D_x = 0 = D_z \; (y\text{-polarized}) \quad ...(54)$$

The corresponding ray velocity is given by

$$v_r = v_{ro} = \frac{c}{n_{ro}} = \frac{c}{n_o} \quad ...(55)$$

Since the ray velocity is independent of the direction of the ray, it is called the *ordinary ray*.

By using Eqs. (51) and (53), we get the other solution.

$$\frac{D_z}{D_x} = -\frac{\frac{n_r^2}{n_0^2} - \cos^2\theta}{\sin\theta\cos\theta} = -\frac{\sin\theta\cos\theta}{\frac{n_r^2}{n_e^2} - \sin^2\theta}$$

$$D_y = 0$$

$$\therefore \qquad n_r^2 = n_{re}^2 = n_o^2\cos^2\theta + n_e^2\sin^2\theta \qquad ...(56)$$

with
$$\frac{D_z / n_e^2}{D_x / n_o^2} = \frac{E_z}{E_x} = -\tan\theta,\ (D_y = 0) \qquad ...(57)$$

The corresponding ray velocity is given by

$$\frac{1}{v_r^2} = \frac{1}{v_{re}^2} = \frac{n_{re}^2}{c^2} = \frac{\cos^2\theta}{c^2 / n_o^2} + \frac{\sin^2\theta}{c^2 / n_e^2} \qquad ...(58)$$

It is called the *extraordinary ray*.

Eq. (58) represents an ellipse.

In three dimensions, Eq. (58) will represent an ellipsoid of revolution with the optic axis as the axis of rotation.

Comparing Eq. (57) with Eq. (32), we get

$$\tan\theta = -\frac{D_z / n_e^2}{D_x / n_o^2} = \frac{n_o^2}{n_e^2}\tan\psi \qquad ...(59)$$

Thus when the wave propagates along a direction which makes an angle ψ with the optic axis, then the ray will propagate along the direction

$$\theta = \tan^{-1}\left[\frac{n_o^2}{n_e^2}\tan\psi\right] \qquad ...(60)$$

CHAPTER

34 CHAUCHY'S DISPERSION FORMULA

34.1. Derivation of Cauchy's Dispersion Formula

In order to explain the variation of refractive index (n) with wavelength (λ) by the electromagnetic theory, we must take account of the molecular structure of matter.

When an electromagnetic wave is incident on an atom or a molecule, the periodic electric force of the wave sets the bound charges into vibratory motion. The frequency with which these charges are forced to vibrate is equal to the frequency of the wave. The phase of this motion relative to that of the impressed electric force will depend on the impressed frequency. It will vary with the difference between the impressed frequency and the natural frequency of the bound charges.

Dispersion can be explained with the concept of secondary waves that are produced by the induced oscillations of the bound charges. When a beam of light propagates through a transparent medium (solid or liquid), the amount of light scattered laterally is extremely small. The scattered waves travelling in a lateral direction produce destructive interference.

But the secondary waves travelling *in the same direction* as the incident beam superimpose on one another. The resultant vibration will depend on the phase difference between the primary and the secondary waves. This superimposition, changes the phase of the primary waves and this is equivalent to the change in the wave velocity. Wave velocity is defined as the speed at which a condition of equal phases is propagated. Hence, the variation in phase due to interference, changes the velocity of the wave through the medium. The phase of the oscillations and hence that of the secondary waves depends upon the impressed frequency. Therefore, the velocity of light in the medium varies with the frequency of light.

Also, refractive index depends upon the velocity of light in the medium. Therefore, the refractive index of the medium varies with the frequency (wavelength) of light.

This is the physical interpretation of dispersion.

The passage of electromagnetic wave produces polarisation of atoms of the medium. Hence the relative permittivity ε_r of the medium also gets altered. We assume that there is only one atomic frequency ε_0.

The relative permittivity of the medium is

$$\varepsilon_r = 1 + \chi_e = 1 + \frac{Ne^2}{\varepsilon_0 m\left(\omega_0^2 - \omega^2\right)} \quad ...(1)$$

Here, χ_e = electric susceptibility of the medium,

N = number of electrons per unit volume,

e = charge of the electron,

ε_0 = permittivity of free space,
m = mass of the electron,

$$\omega_0 = 2\pi\nu_0 = \frac{2\pi c}{\lambda_0} = \text{natural angular frequency of oscillation,}$$

$$\omega = 2\pi\nu = \frac{2\pi c}{\lambda} = \text{angular frequency of the incident wave,}$$

For transparent media which allow electromagnetic waves to pass through them,

$$n^2 = \varepsilon_r \qquad ...(2)$$

$$n^2 = 1 + \frac{Ne^2}{\varepsilon_0 m(\omega_0^2 - \omega^2)} \qquad ...(3)$$

$$n = \left[1 + \frac{Ne^2}{\varepsilon_0 m(\omega_0^2 - \omega^2)}\right]^{\frac{1}{2}} \qquad ...(4)$$

$$n = 1 + \frac{Ne^2}{2\varepsilon_0 m(\omega_0^2 - \omega^2)} \quad \text{(using the binomial expansion)}$$

$$= 1 + \left(\frac{Ne^2}{2\varepsilon_0 m\omega_0^2}\right)\left(1 - \frac{\omega^2}{\omega_0^2}\right)^{-1} = 1 + \left(\frac{Ne^2}{2\varepsilon_0 m\omega_0^2}\right)\left(1 + \frac{\omega^2}{\omega_0^2}\right)$$

Now, $\omega = \dfrac{2\pi c}{\lambda}$ and $\omega_0 = \dfrac{2\pi c}{\lambda_0}$;

$$\therefore \quad n = 1 + \frac{Ne^2}{2\varepsilon_0 m\left(\frac{2\pi c}{\lambda_0}\right)^2}\left(1 + \frac{\lambda_0^2}{\lambda^2}\right)$$

or

$$n = 1 + \frac{Ne^2\lambda_0^2}{8\pi^2\varepsilon_0 mc^2} + \frac{Ne^2\lambda_0^4}{8\pi^2\varepsilon_0 mc^2\lambda^2} \qquad ...(5)$$

Let $\quad 1 + \dfrac{Ne^2\lambda_0^2}{8\pi^2\varepsilon_0 mc^2} = A$ and $\dfrac{Ne^2\lambda_0^4}{8\pi^2\varepsilon_0 mc^2} = B$

$$\therefore \quad n = A + \frac{B}{\lambda^2} \qquad ...(6)$$

Eq. (6) represents Cauchy's dispersion formula.
A and B are called Cauchy's constants. The values of A and B depend on the medium.

EXERCISE

1. Derive Cauchy's formula for normal dispersion.

CHAPTER

35

OPTICAL ACTIVITY

35.1. Origin of Optical Activity in Inorganic Crystals

The explanation of optical activity in crystals like quartz lies in the *crystalline structure of the arrangement of atoms or molecules*. Reuch suggests that optically active crystals are made up of atomic layers which are twisted slightly from one another so that the atoms form a spiral path round the optic axis. In the right handed (dextro-rotatory) crystals, the layers are built up clockwise around the optic axis whereas in left handed (leavo-rotatory) crystals, they are anti-clockwise.

Fig. 35.1 shows the spiral arrangement of silicon and oxygen atoms in a quartz (SiO_2) crystal.

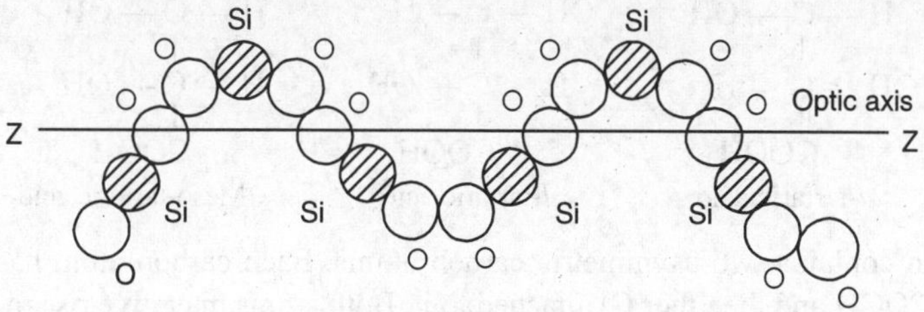

Fig. 35.1.

Si and *O* atoms are arranged in a spiral form along the optic axis *ZZ*. The characteristic property of rotatory polarisation in quartz arises on account of the fact that these spiral of atoms form planes which are twisted slightly from one another. This experimentally determined crystalline structure of quartz confirms the view point suggested by Reuch.

Reuch gave a demonstration of this theory. Mica is optically inactive. From a single cleaved thin sheet, several pieces were cut and then piled one over the other with their principal planes systematically turned in successive sheets. The pile showed optical rotation. An opposite arrangement produced the opposite rotation.

35.2. Origin of Optical Activity in Liquids and Solutions

In liquids and solutions, the optical activity is due to *molecular structure*. All the substances, which show optical activity in liquid state or in solutions, have in their molecules some asymmetry. The molecules of optically active compounds possess at least one carbon atom joined to four different atoms or groups. Such a carbon atom is known as asymmetric carbon atom. The four valencies of carbon atom are directed towards the corner of a tetrahedron. The carbon atom *C* is situated at its centre. Fig. 35.2 shows such a tetrahedral arrangement in which carbon atom *C* is surrounded by four different atoms *P*, *Q*, *R*, *S*.

In Fig. 35.2 (*a*), *Q, R and S* are arranged anticlockwise while in Fig. 35.2 (*b*), they are arrnaged clockwise as seen along line *PC*.

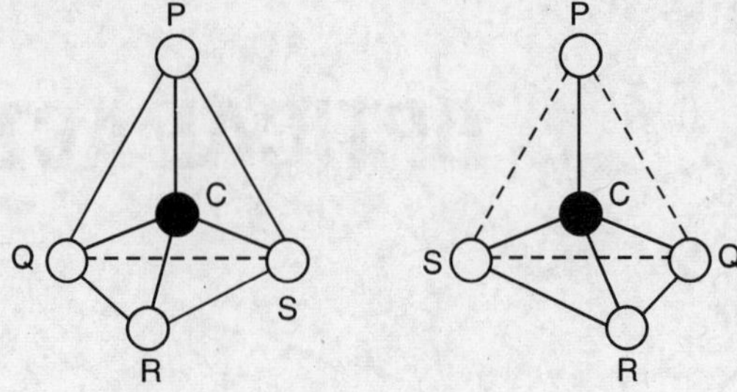

Fig. 35.2.

One structure is mirror image of the other about *PC*.

No rotation in space can make one structure coincide with the other. This suggests that the *optical activity is due to this mirror image asymmetry*. Thus the essential requirement for such asymmetry is that a tetravalent atom such as *C* is attached to four different atoms or groups. This explanation has been widely accepted.

Tartaric acid presents an interesting illustration. Structural formulae of dextro-(*d*–), laevo-(*l*–), and meso-tartaric acid are given below.

COOH	COOH	COOH
|	|	|
H — C — OH	OH — C — H	H — C — OH
|	|	|
OH — C — H	H — C — OH	H — C — OH
|	|	|
COOH	COOH	COOH
d-tartaric acid	*l*-tartaric acid	Meso-tartaric acid

Tartaric acid contains two asymmetric carbon atoms. Each carbon atom has four dissimilar groups (H, OH, COOH and the other C) attached to it. Both atoms may give rise to a laevo-rotatory (*l*-tartaric acid) or to a dextra-rotatory (*d*-tartaric acid). They may also give rise to a non-active meso-tartaric acid.

In *d*-tartaric acid and *l*-tartaric acid, both carbon atoms have H, COOH and OH groups in one cyclic order but the order is different in two cases. Hence the activities of the two groups add up in both the cases.

But in meso-tartaric acid the orders of H, COOH and OH in the two groups are opposite to one another. Hence the activities of two groups cancel out. Thus a solution containing equal numbers of left handed and right handed molecules is optically inactive.

EXERCISE

1. Give a qualitative explanation of optical rotation in molecular substances.

CHAPTER

36 GEOMETRICAL OPTICS II

THE THICK LENS

36.1. Introduction

A lens is referred to as a *thin lens* if its axial thickness is negligible compared with the radii of curvature of its two faces and the object and image distances from it. Its thickness t is then regarded to be *zero* and all distances from it are measured from its centre. If, on the other hand, the axial thickness of a lens be appreciable, it is regarded as a *thick lens*.

If the distance between the poles of two spherical refracting surfaces is not negligible in comparison to the radii of curvature of the spherical surfaces, the combination of two spherical refracting surfaces is called a thick lens.

Let an object OO' be placed on the axis XY of a thick lens (Fig. 36.1). The paraxial incident ray $O'P$ parallel to the principal axis starting from O', after refraction through the lens, passes through second focal point F_2 of the lens. Another paraxial ray $O'F_1Q$ starting from O' and passing through the first focal point F_1 of the lens, after refraction through the lens, emerges parallel to the principal axis. Both of these emergent rays intersect at I'. I' is the image of O'. The perpendicular II' drawn from I' on the axis is the complete image of object OO'.

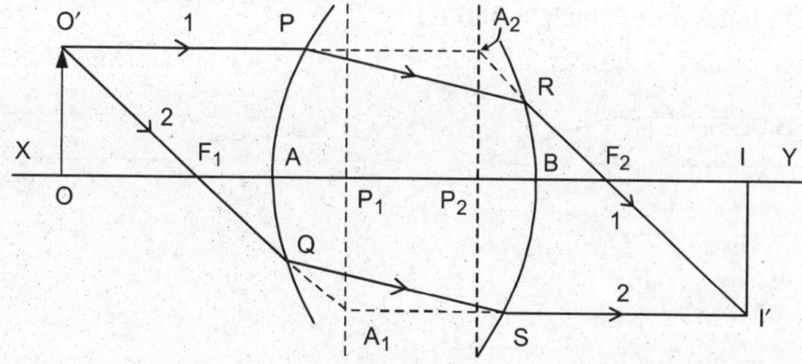

Fig. 36.1.

If the incident ray $O'P$ and the emergent ray RI' are produced, they meet at A_2. The plane passing through A_2 and normal to the principal axis is called the *second principal plane.* Its point of intersection P_2 with the principal axis is called the *second principal point.*

Similarly, the incident ray $O'F_1Q$ and the emergent ray SI', when produced, intersect at A_1. The plane passing through A_1 and normal to the principal axis is called the *first principal plane.* Its point of intersection P_1 with the principal axis is called the *first principal point.*

From Fig. 36.1 it is clear that the two consecutive deviations of ray '1' are equivalent to a single deviation taking place in the second principal plane. Similarly we may consider that the entire deviation produced in the ray '2' passing through first focal point has taken place in the first principal plane. The magnification of the first and second principal planes is unity. Therefore these planes are called *first and second unit planes* respectively. The distance of first focal point F_1 from the first principal point P_1 is called *first focal length* f_1. The distance of second focal point F_2 from the second principal point P_2 is called second focal length f_2. The planes passing through F_1 and F_2 and normal to the principal axis are called *first and second focal planes* respectively.

36.2. Cardinal Points of a Thick Lens

The *two focal points*, the *two principal points* and the *two nodal points,* all these six points are referred to as the *cardinal points of the thick lens.*

(*i*) **Focal points and planes of a thick lens:**

These are defined in exactly the same manner as in the case of a thin lens. The point on the axis of a lens, the image of which is formed at infinity is called the first focal point of the lens (Fig. 36.2 *a, b*). It is denoted by the letter F_1. The vertical plane passing through F_1, and perpendicular to the axis, is called the *first focal plane of* the lens.

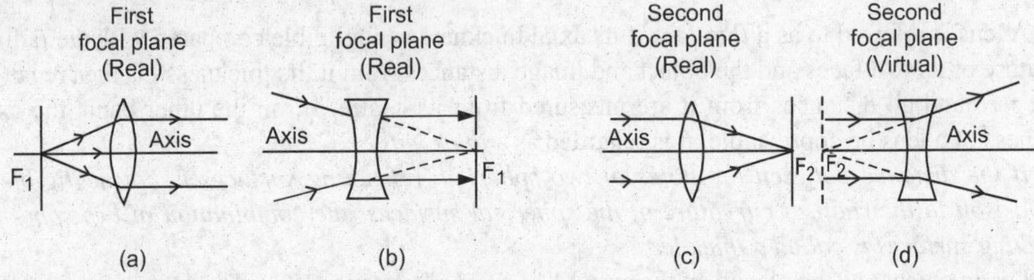

Fig. 36.2.

The point on the axis of a lens of which all incident rays parallel to the axis converge (in the case of a convex lens) or from which they appear to diverge (in the case of a concave lens) is called the *second focal point* of the lens. It is denoted by the letter F_2 (Fig. 36.2 *c, d*). The vertical plane passing through F_2, and perpendicular to the axis, is called the *second focal plane* of the lens.

(*ii*) **Principal planes and Principal points:**

Imagine a ray, parallel to the axis, incident on the lens placed in air (Fig. 36.3 *i*).

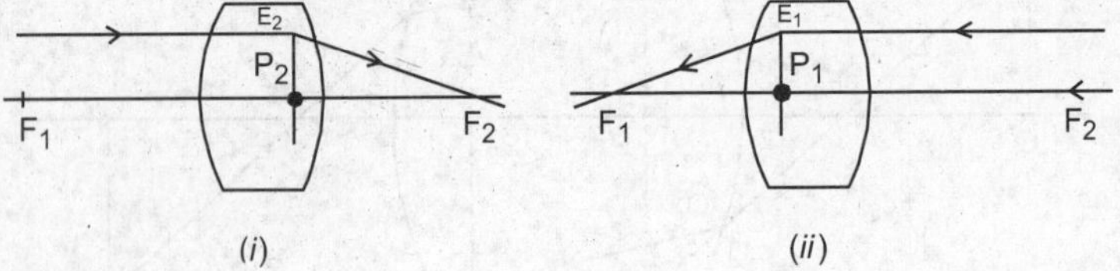

Fig. 36.3.

After refraction through the lens, it passes through the second focal point F_2 of the lens. On producing incident and the emergent rays forwards and backwards respectively, they intersect each other in E_2. The vertical plane through E_2, perpendicular to the axis, is called the *second principal plane.* The point P_2 where the plane cuts the axis is called the *second principal point* of the lens.

Similarly, if we imagine the ray, parallel to the axis, to be incident on the lens from the right side (Fig. 36.3 *ii*), the emergent ray will pass through the first focal point F_1 of the lens. The incident and emergent rays, produced as before, now meet in E_1. The vertical plane through E_1, perpendicular to the axis, is called the *first principal plane.* The point P_1, where the plane cuts the axis, is called the *first principal point* of the lens.

The distances of the first focal point and the object are measured from the first principal point and those of the second focal point and the image from the second principal point.

(*iii*) **Nodal points and Nodal planes**

These are two such points, on the axis, in the case of a lens, that a ray *incident on the lens and directed towards one (say N_1) emerges in a parallel direction through the other (N_2)* [Fig. 36.4]. In other words, *these are conjugate points of unit angular magnification*. The vertical planes, passing through them and perpendicular to the axis, are called *the first and the second nodal planes* respectively. They coincide with the two principal planes respectively, if the medium on either side of the lens be the same.

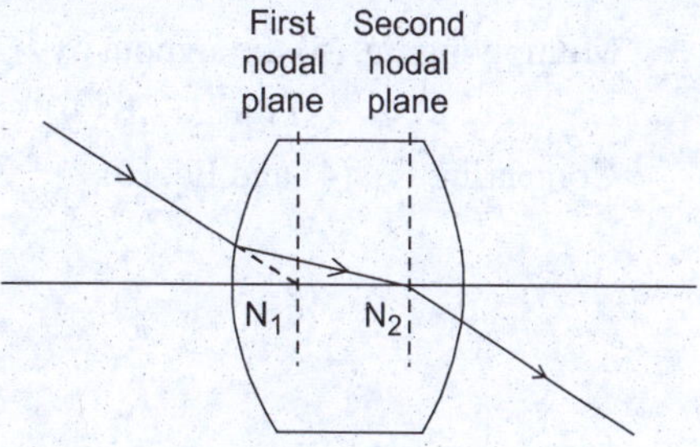

Fig. 36.4.

36.3. Refraction Through a Thick Lens

Assume that the medium on both sides of the lens is air. Consider a lens of thickness t and refractive index n. Let R_1 and R_2 be the radii of curvature of the first and second surfaces of the lens. Let O be an object placed at a distance u from the first surface of the lens (Fig. 36.5).

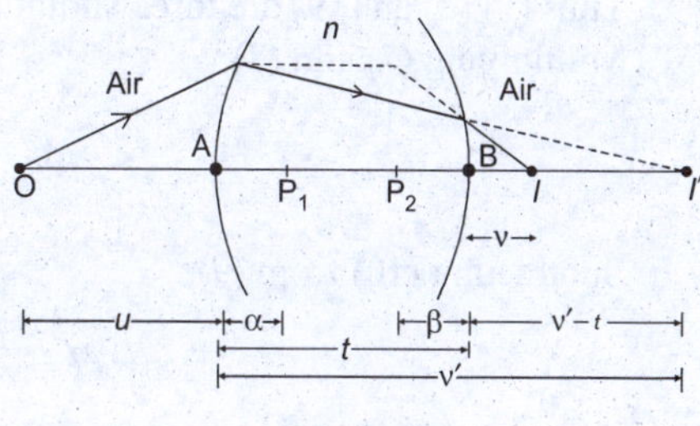

Fig. 36.5.

Then for refraction at the first surface,

$$\frac{n}{v'} - \frac{1}{u} = \frac{n-1}{R_1}$$

Here, v' is the distance of the image I' from the first surface

or $$v' = \frac{R_1 un}{R_1 + u(n-1)} \qquad ...(1)$$

The image I' formed by the first surface acts as the object for the second surface.

For refraction at the second surface, $u' = v' - t$.

The rays are now going from a denser to a rarer medium. Therefore

$$\frac{1/n}{v} - \frac{1}{v'-t} = \frac{(1/n)-1}{R_2}.$$

Here v is the distance of the final image (I) from the second surface.

or $$v' - t = \frac{R_2\, vn}{R_2 + v(n-1)} \qquad ...(2)$$

Substituting the value of v' from Eq. (1) in Eq. (2),

$$\frac{R_1\, un}{R_1 + u(n-1)} - t = \frac{R_2\, vn}{R_2 + v(n-1)}$$

Simplifying and collecting the terms in uv, u and v,

$$uv\,[n(n-1)(R_1 - R_2) - (n-1)^2 t] + u[n\,R_1\,R_2 - t\,R_2\,(n-1)]$$
$$+ v\,[-n\,R_1\,R_2 - t\,R_1\,(n-1)] - tR_1\,R_2 = 0 \qquad ...(3)$$

The equation is of the form,

$$uvA + uB + vC + D = 0$$

or $$uv + u\frac{B}{A} + v\frac{C}{A} + \frac{D}{A} = 0 \qquad ...(4)$$

The equation can be put in the form

$$\frac{1}{v-\beta}-\frac{1}{u-\alpha}=\frac{1}{f} \quad ...(5)$$

Multiplying Eq. (5) throughout by $f(u-\alpha)(v-\beta)$ and collecting terms in uv, u and v, we obtain

$$uv + u(-\beta - f) + v(-\alpha + f) + (-\beta f + \alpha f + \alpha\beta) = 0 \quad ...(6)$$

Comparing Eq. (4) and Eq. (6),

$$-\beta - f = \frac{B}{A} \quad ...(7)$$

$$-\alpha + f = \frac{C}{A} \quad ...(8)$$

$$-\beta f + \alpha f + \alpha\beta = \frac{D}{A} \quad ...(9)$$

Thus (7) (8) and (9) are three simultaneous equations to find the three unknowns α, β and f.
Multiplying (7) and (8),

$$-\beta f + \alpha f + \alpha\beta - f^2 = \frac{BC}{A^2} \quad ...(10)$$

Subtracting (10) from (9),

$$f^2 = \frac{D}{A} - \frac{BC}{A^2} = \frac{AD - BC}{A^2}$$

Substituting the values of A, B, C and D, from Eq. (3),

$$f^2 = \frac{[n(n-1)(R_1 - R_2) - (n-1)^2 t)](-t R_1 R_2) + [n R_1 R_2 - tR_2(n-1)][n R_1 R_2 + t R_1 (n-1)]}{[n(n-1)(R_1 - R_2) - (n-1)^2 t]^2}$$

$$f^2 = \frac{(n R_1 R_2)^2}{[n(n-1)(R_1 - R_2) - (n-1)^2 t]^2}$$

$$\therefore \qquad f = \frac{\pm n R_1 R_2}{n(n-1)(R_1 - R_2) - (n-1)^2 t} \quad ...(11)$$

[Here, only the –ve sign is to be retained].

This is the formula for the focal length of a thick lens in terms of the refractive index, thickness and radii of curvature of the two faces.

To determine the positions of the principal points of the lens, we proceed as follows:

Let α be the distance of the first principal point P_1 from the vertex A of the first surface of the lens. From Eq. (8),

$$\alpha = -\frac{C}{A} + f = \frac{t R_1}{n(R_1 - R_2) - t(n-1)} \quad ...(12)$$

Let β be the distance of the second principal point P_2 from the vertex B of the second surface of the lens. From Eq. (7),

$$\beta = -\left(\frac{B}{A} + f\right) = \frac{t R_2}{n(R_1 - R_2) - t(n-1)} \quad ...(13)$$

Position of focal points of a thick lens:

The distance of the second focal point F_2 from the pole of the second surface (Fig. 36.1),

$$BF_2 = f\left[1 - \frac{(n-1)t}{nR_1}\right]$$

The distance of the first focal point F_1 from the pole A of the first surface (Fig. 36.1),

$$AF_1 = -f\left[1 + \frac{(n-1)t}{nR_2}\right]$$

Power of a thick lens

$$P = \frac{1}{f} = \frac{-[n(n-1)(R_1 - R_2) - (n-1)^2 t]}{nR_1 R_2}$$

$$= \frac{(n-1)}{R_1} - \frac{(n-1)}{R_2} + \frac{(n-1)^2 t}{nR_1R_2}$$

$$= P_1 + P_2 - \frac{tP_1P_2}{n}.$$

Here, P_1 and P_2 are the powers of the two surfaces.

Example 1: *A thick double convex lens has the radii of curvature of its surfaces as 20 cm and 30 cm. The refractive index is 1.6 and the thickness is 6 cm. Calculate the focal length and the positions of the principal planes.*

Solution.

$$f = \frac{-nR_1 R_2}{n(n-1)(R_1 - R_2) - (n-1)^2 t}$$

$$= \frac{-1.6 \times (20) \times (-30)}{1.6 \times (1.6-1)[20-(-30)] - (1.6-1)^2 \times 6}$$

$f = 20.94$ cm.

$$\alpha = \frac{tR_1}{n(R_1 - R_2) - t(n-1)}$$

$$= \frac{6 \times 20}{1.6 \times [20-(-30)] - 6 \times (1.6-1)} = 1.571 \text{ cm.}$$

$$\beta = \frac{tR_2}{n(R_1 - R_2) - t(n-1)}$$

$$= \frac{6 \times (-30)}{1.6 \times [20-(-30)] - 6 \times (1.6-1)} = -2.356 \text{ cm.}$$

Example 2: *Calculate the focal length of a lens in the form of a sphere of glass. n = 1.5 and radius = 5 cm. Indicate the positions of the focal and principal planes on a diagram.*

Solution. The focal length f of a thick lens is

$$f = \frac{-nR_1 R_2}{n(n-1)(R_1 - R_2) - (n-1)^2 t}.$$

If R is the radius of the spherical lens, we have

$$R_1 = +R,\ R_2 = -R \text{ and } t = 2R.$$

$$\therefore \quad f = \frac{-nR(-R)}{n(n-1)\{R-(-R)\} - (n-1)^2\, 2R}$$

or
$$f = \frac{nR}{2(n-1)}$$

Here, $n = 1.5$ and $R = 5$ cm.

$$\therefore \quad f = \frac{1.5 \times 5}{2(1.5-1)} = 7.5 \text{ cm.}$$

Principal points: The distance of the first principal point P_1 from the vertex A of the first surface

$$\alpha = \frac{t R_1}{n(R_1 - R_2) - t(n-1)} = \frac{2R \times R}{n\{R - (-R)\} - 2R(n-1)}$$

$$= R = 5 \text{ cm.}$$

The distance of the second principal point P_2 from the vertex B of the second surface

$$\beta = \frac{t R_2}{n(R_1 - R_2) - t(n-1)} = \frac{2R \times (-R)}{n\{R - (-R)\} - 2R(n-1)}$$

$$= -R = -5 \text{ cm.}$$

Focal points: The distance of first focal point F_1 from the pole A of the first surface

$$AF_1 = -f\left[1 + \frac{(n-1)t}{nR_2}\right] = -7.5\left[1 + \frac{(1.5-1) \times 10}{1.5 \times (-5)}\right]$$

$$= -2.5 \text{ cm.}$$

The distance of second focal point F_2 from the pole B of the second surface

$$BF_2 = f\left[1 - \frac{(n-1)t}{nR_1}\right]$$

$$= 7.5\left[1 - \frac{(1.5-1) \times 10}{1.5 \times 5}\right]$$

$$= 2.5 \text{ cm.}$$

It is obvious from Fig. 36.6 that the principal points P_1, P_2 are coincident with the centre of the spherical lens of radius 5 cm. The focal points F_1 and F_2 are situated on both the sides of the lens at equal distances (i.e., 2.5 cm) from the poles A and B.

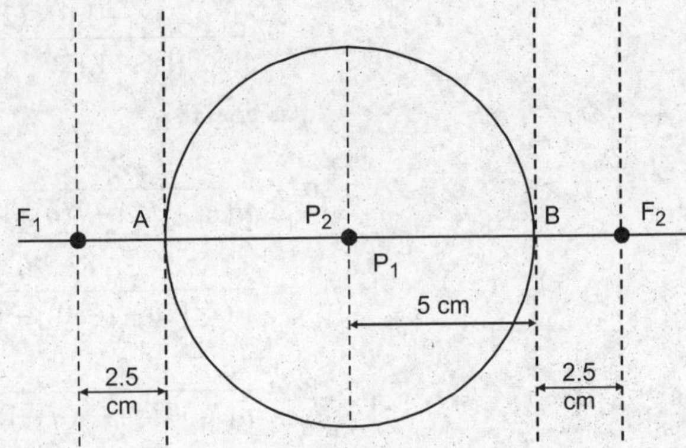

Fig. 36.6.

Example 3: *A convex lens 2 cm thick, has radii of curvature 6 cm and 8 cm. If the refractive index of the lens is 1.5, find the focal length and the position of the principal points of the lens.*

An object is situated at a distance of 10 *cm from the convex surface of radius of curvature 6 cm of the above lens. Find the position of the image.*

Solution. Here, $t = 2$ cm, $R_1 = 6$ cm, $R_2 = -8$ cm, $n = 1.5$.

$$f = \frac{-n_1 R_1 R_2}{n(n-1)(R_1 - R_2) - (n-1)^2 t}$$

$$= \frac{-1.5 \times 6 \times (-8)}{1.5 \times (1.5-1)\{6-(-8)\} - (1.5-1)^2 \times 2}$$

$$= 7.2 \text{ cm.}$$

$$\alpha = \frac{t R_1}{n(R_1 - R_2) - t(n-1)} = \frac{2 \times 6}{1.5\{6 - (-8)\} - 2(1.5 - 1)}$$

$$= 0.6 \text{ cm.}$$

$$\beta = \frac{t R_2}{n(R_1 - R_2) - t(n-1)} = \frac{2 \times (-8)}{1.5\{6 - (-8)\} - 2(1.5 - 1)}$$

$$= -0.8 \text{ cm.}$$

The distance of the object from the first principal plane

$$U = -10 - 0.6 = -10.6 \text{ cm.}$$

Applying the lens equation, $\frac{1}{V} - \frac{1}{U} = \frac{1}{f}$,

$$\frac{1}{V} - \frac{1}{-10.6} = \frac{1}{7.2}$$

$\therefore$ $V = 23.5$ cm from the second principal plane (Fig. 36.7).

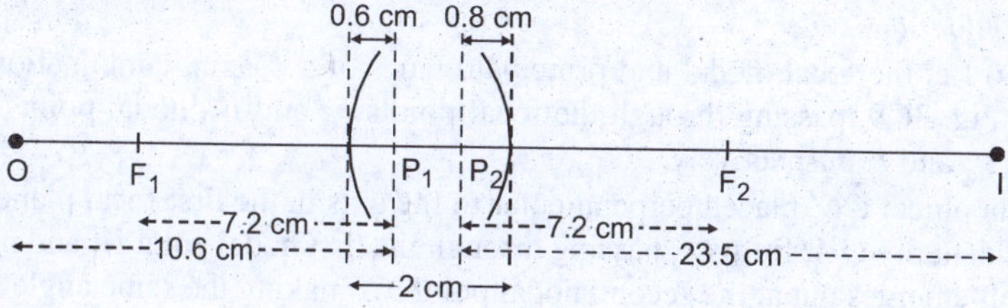

Fig. 36.7.

Example 4: *A convex lens of thickness 4 cm has radii of curvature 6 cm and 10 cm. Find the focal length and the positions of the focal points and the principal points.*

The refractive index = 1.5.

Solution. Here, $t = 4$ cm, $R_1 = 6$ cm, $R_2 = -10$ cm; $n = 1.5$.

$$f = \frac{-n R_1 R_2}{n(n-1)(R_1 - R_2) - (n-1)^2 t}$$

$$= \frac{-1.5 \times 6 \times (-10)}{1.5 \times (1.5 - 1)\{6 - (-10)\} - (1.5 - 1)^2 \times 4}$$

$$= 8.182 \text{ cm.}$$

$$\alpha = \frac{t R_1}{n(R_1 - R_2) - t(n-1)} = \frac{4 \times 6}{1.5 \times \{6 - (-10)\} - 4(1.5 - 1)}$$

$$= 1.091 \text{ cm.}$$

$$\beta = \frac{t R_2}{n(R_1 - R_2) - t(n-1)} = \frac{4 \times (-10)}{1.5 \times \{6 - (-10)\} - 4(1.5 - 1)}$$

$$= -1.818 \text{ cm.}$$

In Fig. 36.6, P_1 and P_2 show the positions of the principal points and F_1 and F_2 show the positions of the focal points. As the medium on the two sides of the lens is the same, the nodal points coincide with the principal points.

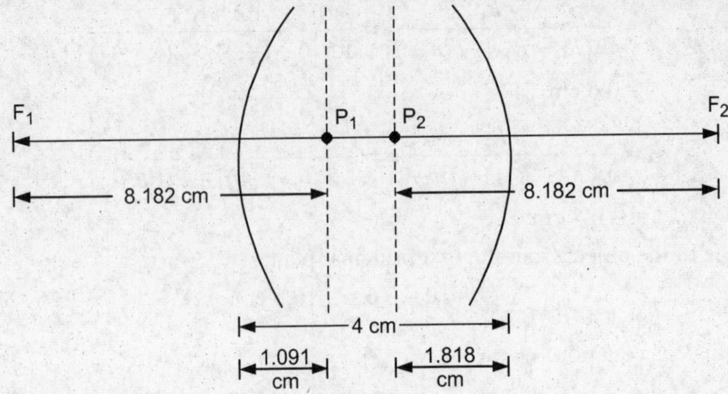

Fig. 36.8.

Example 5: *Show that in the case of a lens combination of two thin lenses, (i) the distance between the two principal points is equal to the distance between the two nodal points and (ii) if the medium on either side of the combination be the same, the principal points coincide respectively with the two nodal points.*

Solution. (*i*) Let the focal, nodal and principal planes of a coaxial combination of two lenses be as shown in Fig. 36.9, passing through the focal points F_1 and F_2, nodal points N_1 and N_2 and principal points P_1 and P_2 respectively.

Consider an object OO' placed perpendicular to the axis in the first focal plane of the system. Consider an incident ray $O'M$ from O', passing through the first nodal point N_1 and making an angle θ with the axis. It emerges out at the second nodal point N_2, making the same angle θ with the axis, i.e., parallel to $O'M$ such that it meets the second principal plane at N. Since the principal planes are conjugate planes of unit positive magnification, $P_2N = P_1M$.

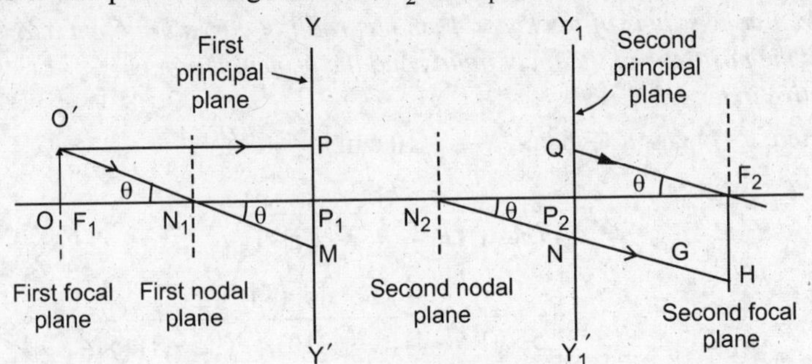

Fig. 36.9.

The right-angled Δs MP_1N_1 and NP_2N_2 are congruent, because $\angle MN_1P_1 = \angle NN_2P_2 = \theta$ and $P_1M = P_2N$.

$$\therefore \qquad P_1N_1 = P_1N_2.$$

Adding P_1N_2 to both sides, we have $N_1N_2 = P_1P_2$.

i.e., *the distance between the two nodal points is equal to the distance between the two principal points.*

(*ii*) In Fig. 36.9, consider another incident ray $O'P$ from O', meeting the first principal plane in P. It will emerge out at the same height from the axis at Q at the second principal plane, i.e., $P_2Q = P_1P$. It will pass through the second focal point F_2, in a direction parallel to the first emergent ray P_2G.

The right angled Δs $O'F_1N_1$ and QP_2F_2 are congruent ($\angle O'N_1F_1 = \angle QF_2P_2 = \theta$ and $F_1O' = P_2Q$)

$$\therefore \qquad P_2F_2 = N_1F_1 = P_1F_1 - P_1N_1 \text{ or } P_1N_1 = P_1F_1 - P_2F_2.$$

Since $P_1F_1 = -f_1$ and $P_2F_2 = f_2$, we have

$$P_1N_1 = -f_1 + f_2 = P_2N_2 \qquad (\because P_1N_1 = P_2N_2)$$

Thus, the distance between the corresponding principal and nodal points is equal to the sum of the two focal lengths of the system. Now, with the same medium on either side of the system, $f_1 = f_2$ (*numerically*).

$$\therefore \qquad P_1N_1 = P_2N_2 = -f_1 + f_2 = 0.$$

If the medium on either side of the system be the same, the two principal points coincide with the two nodal points respectively.

36.4. Abbe Refractometer

The instrument is used to measure the refractive index of liquids.

Apparatus. It consists of two dense flint glass prisms *ACB* and *ADB* (Fig. 36.10). The hypotenuse faces of the two prisms are in contact. The second prism *ADB* is placed below the first prism *ACB* and its face is left not much polished to prevent image formation.

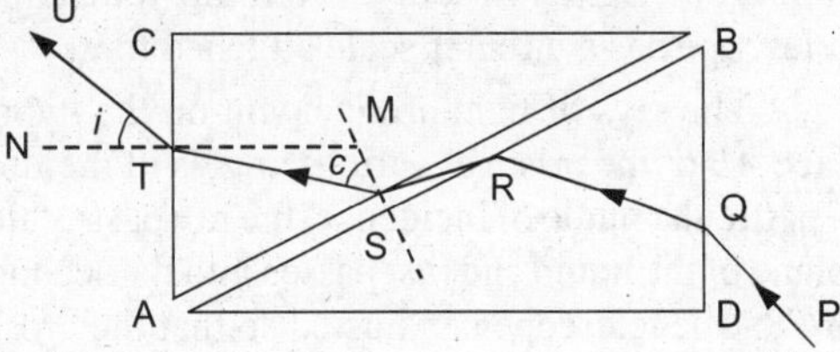

Fig. 36.10.

The liquid whose refractive index is to be determined is placed between the two prisms. The instrument is provided with an accurate scale.

The function of the lower prism *ADB* is to direct the ray at grazing incidence (direction *RS*) in the liquid film. The course of the ray is *PQRSTU* with the path *RS* grazing the standard prism surface. The refractive index of the material of this prism should be higher than that of any liquid to be examined.

Theory. Now *SM* is normal to face *AB* and $\angle TSM = C$, the critical angle.

Let n = refractive index of liquid and

n_g = refractive index of the prism.

$$\sin C = \frac{n}{n_g}. \qquad ...(1)$$

If $\angle TMS = \theta$ and $\angle NTU = i$, then $\angle MTS = 180 - (\theta + C)$.

Now $$\frac{\sin i}{\sin \angle MTS} = n_g$$

or $$\frac{\sin i}{\sin(\theta + C)} = n_g \qquad ...(2)$$

$$\frac{\sin i}{\sin\theta\cos C + \cos\theta\sin C} = n_g$$

or $$\frac{\sin i}{\sin\theta\sqrt{1-\sin^2 C} + \cos\theta . \frac{n}{n_g}} = n_g \qquad \left(\because \sin C = \frac{n}{n_g}\right)$$

or $$\frac{\sin i}{\sin\theta\sqrt{\left(1-\frac{n^2}{n_g^2}\right)} + \frac{n}{n_g}\cos\theta} = n_g$$

or $$\frac{\sin i}{\sin\theta\sqrt{n_g^2 - n^2} + n\cos\theta} = 1$$

or $$\sqrt{\left(n_g^2 - n^2\right)} = \frac{\sin i}{\sin \theta} - n \cot \theta .$$

Here we have eliminated C and a relation between n and i is determined.

Also $\theta = 180° - A$ where $A = \angle CAB$ of the prism.

Knowing i, θ and n_g, the refractive index of the liquid n can be calculated.

Experiment. Light is reflected from the ground glass surface into the prism ADB (Fig. 36.11). The emergent light is received by the telescope. The telescope has a circular scale attached to it.

The rays of light are incident on the hypotenuse face AB of the prism at various angles of incidence. At a particular angle of incidence, the ray passes along the plane of the liquid and just passes into the second prism ACB. The telescope is adjusted so that the cross wires lie on the dark edge of the field of view. This gives the position of the minimum angle of emergence i. The circular scale is calibrated to give the refractive index of the liquid directly.

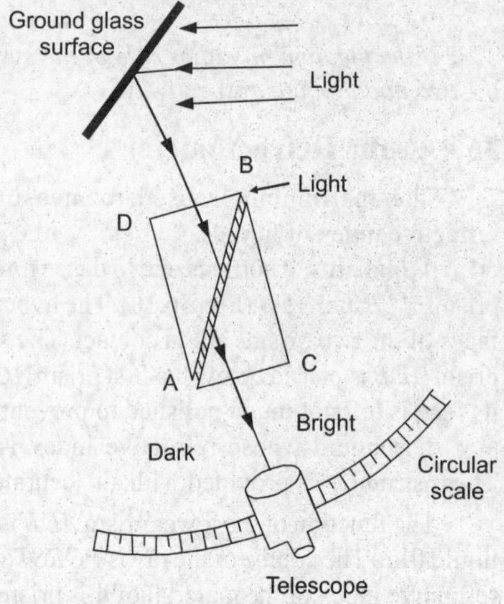

Fig. 36.11.

36.5. Irrational Dispersion

A beam of white light is passed through two prisms, one of crown glass and the other of flint glass of the same refracting angle (Fig. 36.12). A screen is placed at the same distance from each prism.

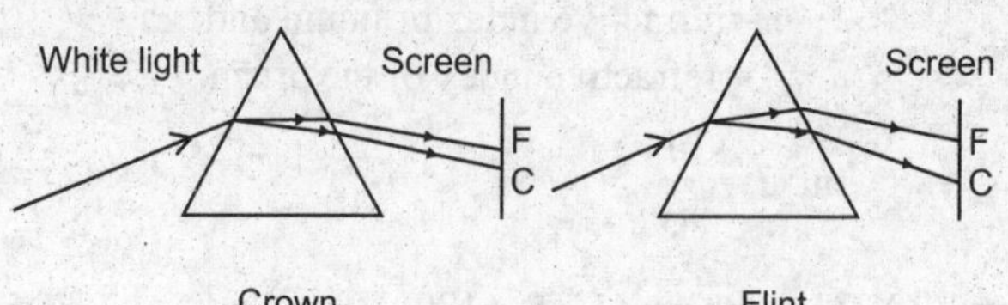

Fig. 36.12.

The width of the spectrum is different in the two cases. The width FC with a flint glass prism is more than that with a crown glass prism.

If the refracting angles of the two prisms are so chosen that the width of the spectra is the same in the two cases, then the intermediate colours are of different widths. The two do not exactly overlap with respect to colours.

(*i*) Let n and $n + dn$ be the refractive indices for two rays A, B in crown glass. Let v and $v + dv$ be the refractive indices for two other rays C, D in crown glass.

Then, the ratio of the dispersive powers of crown glass for those two pairs of rays is

$$\frac{dn/(n-1)}{dv/(v-1)} \qquad ...(1)$$

Similarly, let n', $n' + dn'$ and v', $v' + dv'$ be the refractive indices for the same two pairs of rays in flint glass. Then, the ratio of dispersive powers of flint glass for these two pairs of rays is

$$\frac{dn'/(n'-1)}{dv'/(v'-1)} . \qquad ...(2)$$

These two ratios are not equal. This is known as irrational dispersion.

36.6. Angular and Chromatic Dispersions

Definition of Angular Dispersion:

For a prism, angular dispersion is defined as the rate of change of angle of deviation θ with wavelength λ and is mathematically expressed as $\frac{d\theta}{d\lambda}$.

Here, $d\theta$ is the difference in the angle of deviation between two spectral lines of wavelengths λ and $\lambda + d\lambda$.

$\frac{d\theta}{d\lambda}$ is the *angular dispersion* between the wavelengths λ and $\lambda + d\lambda$.

Chromatic Dispersion.

The direct method to specify the dispersion of a material is given by the expression $\frac{dn}{d\lambda}$. Here, dn is the change in refractive index for a change in wavelength $d\lambda$.

Now we can write $\frac{d\theta}{d\lambda}$ as

$$\frac{d\theta}{d\lambda} = \frac{d\theta}{dn} \cdot \frac{dn}{d\lambda}$$

(*i*) The first term $\frac{d\theta}{dn}$ represents a quantity which is a geometrical factor of the prism.

(*ii*) The second terms $\frac{dn}{d\lambda}$ is a characteristic property of the prism material.

It is called *chromatic dispersion* of the material of the prism.

Determination of $\frac{d\theta}{d\lambda}$.

The refractive index n of the material of a prism is

$$n = \frac{\sin\left(\frac{A+\theta}{2}\right)}{\sin\left(\frac{A}{2}\right)} \qquad \text{...(1)}$$

Here, A = angle of the prism and θ = angle of minimum deviation.

$$\sin\left(\frac{A+\theta}{2}\right) = n \sin\left(\frac{A}{2}\right)$$

$$\therefore \qquad \cos\left(\frac{A+\theta}{2}\right) = \sqrt{1 - n^2 \sin^2 \frac{A}{2}} \qquad \text{...(2)}$$

Differentiating Eq.(1) with respect to n on the L.H.S., and θ on the R.H.S., we get

$$dn = \frac{\frac{1}{2}\cos\left(\frac{A+\theta}{2}\right)}{\sin\frac{A}{2}} . \, d\theta$$

or

$$\frac{dn}{d\lambda} = \frac{\cos\left(\frac{A+\theta}{2}\right)}{2\sin\frac{A}{2}} \cdot \frac{d\theta}{d\lambda} \qquad \text{...(3)}$$

or

$$\frac{d\theta}{d\lambda} = \frac{2 \cdot \sin\frac{A}{2}}{\cos\left(\frac{A+\theta}{2}\right)} \cdot \frac{dn}{d\lambda}.$$

Using Eq. (2), we get

$$\frac{d\theta}{d\lambda} = \frac{2 \sin (A/2)}{\sqrt{1 - n^2 \sin^2 (A/2)}} \frac{dn}{d\lambda} \quad ...(4)$$

The angular dispersion of the material of a prism depends on

(*i*) the angle of the prism A and

(*ii*) refractive index of the material of the prism, n.

Using a spectrometer and the given prism, a graph is drawn between n and λ (n along the Y-axis and λ along the X-axis). The tangent to the curve at any point measures the chromatic dispersion $\frac{dn}{d\lambda}$ for that particular wavelength. Substituting this value of $\frac{dn}{d\lambda}$ in Eq. (4), $\frac{d\theta}{d\lambda}$ is calculated.

36.7. The Kellner Eye-piece

It consists of two thin plano-convex lenses with their curved surfaces facing the incident beam [Fig. 36.13]. This minimises spherical aberration.

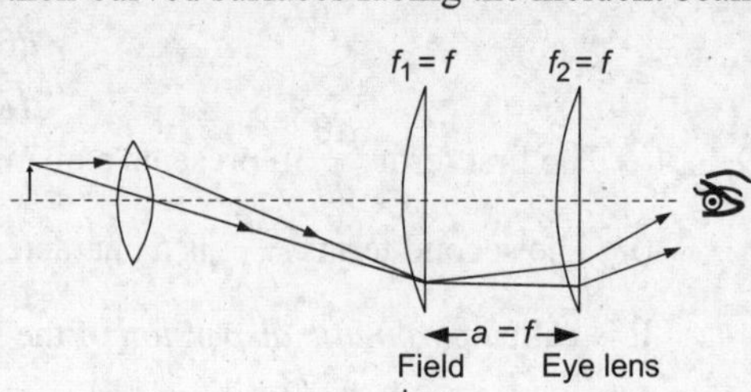

Fig. 36.13.

The two lenses are of the same focal length and the distance between them is equal to this focal length i.e., $f_1 = f, f_2 = f$ and $a = f$.

The condition $a = \frac{f_1 + f_2}{2}$ for minimum chromatic aberration is therefore satisfied in this case.

The distance of second principal point from second lens is

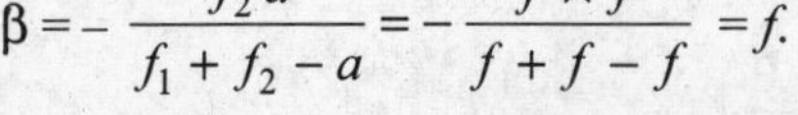

$$\beta = -\frac{f_2 a}{f_1 + f_2 - a} = -\frac{f \times f}{f + f - f} = f.$$

The distance of first principal point from first lens is

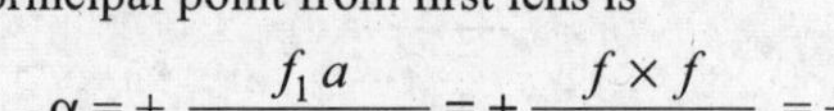

$$\alpha = +\frac{f_1 a}{f_1 + f_2 - a} = +\frac{f \times f}{f + f - f} = +f.$$

The focal length (F) of the equivalent lens is given by

$$\frac{1}{F} = \frac{1}{f_1} + \frac{1}{f_2} - \frac{a}{f_1 f_2} = \frac{1}{f} + \frac{1}{f} - \frac{f}{f^2} = \frac{1}{f}.$$

$$\therefore \quad F = f.$$

The above details show that the field lens is also the place of first focal plane of the system. The objective lens forms the image of the object along this plane and this is also the focal plane of eye lens. Therefore, any scratch or dust particle on the surface of the field lens is equally magnified. This is the chief defect of this eyepiece.

EXERCISE

1. Deduce an expression for the focal length of a thick lens in terms of the refractive index, thickness and radii of curvature of the two faces.
2. Obtain an expression for the focal length and the positions of the principal points of a thick lens in terms of its radii of curvature and thickness.
3. What is meant by a thick lens? Discuss the positions of cardinal points in the case of a thick lens.
4. What is a refractometer? Describe an Abbe refractometer. Describe an experiment to determine the refractive index of liquids using it.

CHAPTER 37

INTERFERENCE OF LIGHT-IV

37.1. Phase Change on Reflection

Principle of optical reversibility.

Statement. *In the absence of any absorption, a light ray that is reflected or refracted will retrace its original path if its direction is reversed.*

We investigate the reflection of light at an interface between two media using this principle.

(*i*) Consider a light ray *AO* incident on an interface of two media of refractive indices n_1 and n_2 (Fig. 37.1*a*).

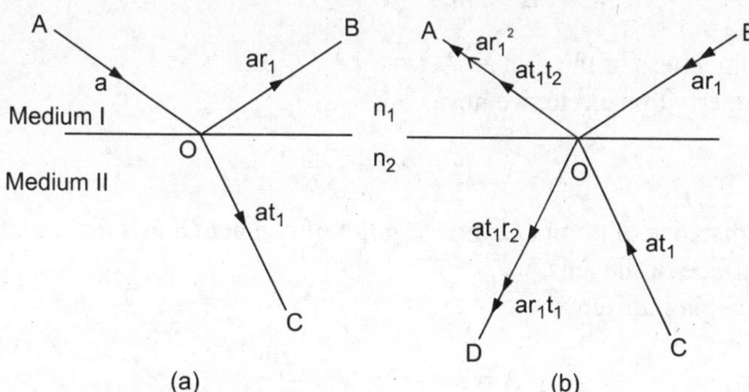

Fig. 37.1.

Let the amplitude reflection and transmission coefficients be r_1 and t_1 respectively.

The amplitude of the incident ray is *a*.

The amplitudes of the reflected and refracted rays are ar_1 and at_1 respectively.

(*ii*) Suppose that the directions of reflected and refracted rays are reversed (Fig. 37.1*b*).

Consider a ray *CO* of amplitude at_1 incident on medium 1 and a ray *BO* of amplitude ar_1 incident on medium 2.

The ray of amplitude at_1 gives rise to a reflected ray *OD* of amplitude at_1r_2 and a transmitted ray *OA* of amplitude at_1t_2.

Here, r_2 and t_2 are the amplitude reflection and transmission coefficients when a ray is incident from medium 2 on medium 1.

Similarly, the ray of amplitude ar_1 gives rise to a ray *OA* of amplitude ar_1^2 and a refracted ray *OD* of amplitude ar_1t_1.

According to the principle of optical reversibility, the two rays of amplitudes ar_1^2 and at_1t_2 must combine to give the incident ray of amplitude *a*, i.e.,

$$ar_1^2 + at_1t_2 = a$$

or $$t_1t_2 = 1 - r_1^2 \quad \text{...(1)}$$

The two rays of amplitudes at_1r_2 and ar_1t_1 must cancel each other, i.e.,

$$at_1r_2 + ar_1t_1 = 0$$

or $$r_2 = -r_1 \quad \text{...(2)}$$

Eqs. (1) and (2) are known as Stokes' relations.

The negative sign in Eq. (2) indicates a phase change of π either at reflection from rarer to denser medium or at reflection from denser to a rarer medium. In Lloyd's mirror experiment, where the interference takes place between direct waves from the source and reflected waves from optically denser medium (glass), it is observed that central fringe is dark instead of bright. This is only possible when light suffers a phase change of π on reflection at the mirror. This shows that a phase change of π is introduced when the reflection takes placc at the surface of denser medium.

No abrupt phase change occurs when light gets reflected by a rarer medium.

37.2. Newton's Rings by Transmitted Light

A wave train SP of monochromatic light is transmitted along PQA (Fig. 37.2). It is partly reflected at Q and then is refracted at H along HB The two wave trains '1' and '2' are coherent and are in a position to interfere.

The effective path difference between the two rays is

$$\Delta = 2nt - \frac{1}{2}\lambda + \frac{1}{2}\lambda = 2nt.$$

Here, t is the thickness of the film at P . (*i.e.*, $t = PQ$).

But by the property of a circle, we have

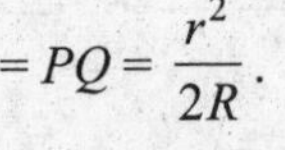

$$t = PQ = \frac{r^2}{2R}.$$

Here, r is the distance of point Q from the point of contact O. R is the radius of curved surface of the lens.

Fig. 37.2.

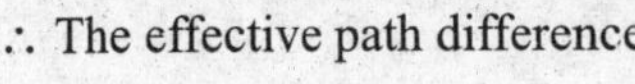

∴ The effective path difference

$$\Delta = 2nt = 2n\frac{r^2}{2R} = \frac{nr^2}{R}$$

∴ $$\Delta = \frac{nD^2}{4R}. \qquad \left(\because r = \frac{D}{2}\right)$$

For bright rings:

For *constructive interference,*

$$\frac{nD^2}{4R} = m\lambda, \qquad m = 1, 2, 3, \ldots$$

∴ Diameter of mth bright ring is given by

$$D_m^2 = \frac{4mR\lambda}{n}.$$

For air film $n = 1$.

∴ $$D_m^2 = 4m\lambda R$$

or $$D_m = \sqrt{4m\lambda R} = \sqrt{(4\lambda R)}\sqrt{m} \quad \text{...(1)}$$

The diameters of bright rings are proportional to the square roots of natural numbers. In this case, central spot is bright.

For dark rings:

For destructive interference,

$$\frac{nD^2}{4R} = (2m-1)\,\frac{\lambda}{2} \qquad m = 1, 2, 3, \ldots$$

Thus the diameter of mth dark ring is given by

$$D_m{}^2 = \frac{2\lambda R\,(2m-1)}{n}$$

For air film, $n = 1$

$$D_m{}^2 = 2\lambda R\,(2m-1)$$

or

$$D_m = \sqrt{(2\lambda R)}\,\sqrt{(2m-1)} \qquad \ldots(2)$$

The diameters of dark rings are proportional to the square roots of the odd natural numbers.

Thus we see that this is a case opposite to that with reflected light. Thus the system of rings by transmitted light is *complementary* to that by reflected light.

Fig. 37.3.

Fig. 37.3 shows Newton's rings formed by transmitted light.

37.3. Stationary Light Waves

37.3.1. Ives Experiment

Fig. 37.4 shows the experimental arrangement of Ives for studying stationary light waves. The emulsion side of a photographic plate was placed in contact with a film of mercury. A parallel beam of monochromatic light was allowed to fall normally on the glass plate. The beam was reflected on the mercury surface. Stationary waves were formed due to interference of the incident and reflected light waves. A section of the photographic film was cut along a plane normal to the surface. The cut section was viewed under a microscope.

Bright and dark bands (separated by regular intervals) were observed. By measuring the distance between two consecutive dark bands (which is equal to $\lambda/2$), we can calculate the wavelength of monochromatic light.

Because of the small wavelength of light ($\simeq 5 \times 10^{-7}\ m$), the distance between two consecutive dark (or bright) bands was extremely small and was, therefore, difficult to measure.

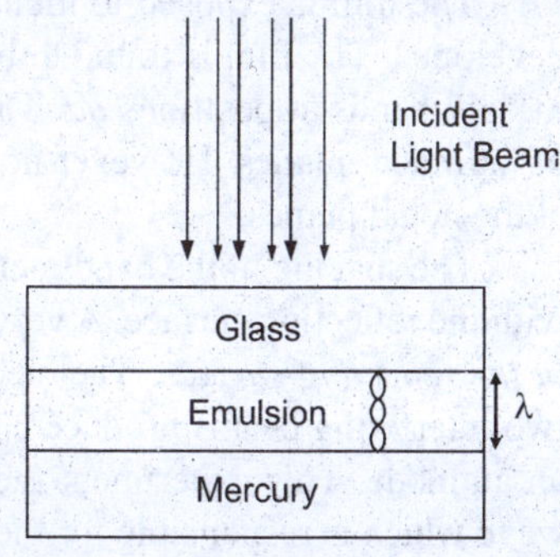

Fig. 37.4.

37.3.2. Wiener Experiment

Let a light wave be travelling along positive X-axis and reflected by a denser medium. The *incident* wave along X-axis is represented by

$$y_1 = a \sin 2\pi \left(\frac{t}{T} - \frac{x}{\lambda}\right) \qquad \ldots(1)$$

When it is reflected from the surface of a denser medium, it suffers a phase change of π. Therefore, *reflected wave* travelling along negative X-axis is represented by

$$y_2 = -\,a \sin 2\pi \left(\frac{t}{T} + \frac{x}{\lambda}\right) \qquad \ldots(2)$$

As a result of superposition of these two waves, a stationary wave is produced.

The resultant displacement y is given by

$$y = y_1 + y_2 = a \sin 2\pi \left(\frac{t}{T} - \frac{x}{\lambda}\right) - a \sin 2\pi \left(\frac{t}{T} + \frac{x}{\lambda}\right)$$

$$y = -2a \sin \frac{2\pi x}{\lambda} \cos \frac{2\pi t}{T} \qquad ...(3)$$

Here $2a \sin \frac{2\pi x}{\lambda}$ represents the amplitude of the resultant wave.

The *amplitude* varies with the position of x.

(*i*) **Nodes.** The amplitude is zero at points for which

$$x = 0, \frac{\lambda}{2}, \ldots\ldots, \frac{n\lambda}{2}, \qquad ...(4)$$

These points are the positions of *nodes*.

(*ii*) **Antinodes:** The amplitude is maximum at points for which

$$x = \frac{\lambda}{4}, \frac{3\lambda}{4}, \frac{5\lambda}{4} \ldots\ldots \frac{(2n+1)\lambda}{4}. \qquad ...(5)$$

These are the positions of *antinodes*.

From Eqs. (4) and (5) it is obvious that the resultant waves consist of alternate nodes and antinodes and hence are called *stationary waves of light.*

Fig. 37.5 shows the experimental arrangement of Wiener for studying stationary light waves. M is a highly polished plane mirror (reflecting surface). An extremely thin film of photographic emulsion coated on a glass plate is placed in front of the reflecting surface of the mirror in an inclined position such as AB.

The film is exposed to monochromatic light and developed. The film is found to be crossed by a series of dark bands at positions *aa*. These are the positions of antinodal planes. However no blackening is found at the nodal planes.

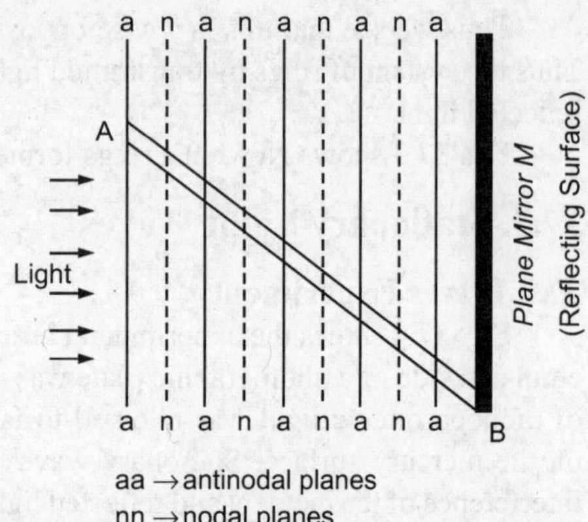

Fig. 37.5.

The spacing of the bands across AB increased on increasing the angle of inclination of the plate with the reflecting surface. A very significant result of this experiment is that *there is always a node at the reflecting surface.* The electromagnetic theory of light shows that the electric vectors of the two interfering beams produce node at the reflecting surface whereas the magnetic vectors produce an antinode. This experiment therefore proves that it is the electric vector of the electromagnetic wave which is responsible for the observed effects in optics.

37.3.3. Lippmann's Colour Photography

The *Lippmann* process of colour photography is based on the phenomenon of *stationary light waves*. Fig. 37.6 shows the *Principle of Lippmann Colour Photography.*

A photographic plate having very fine grain emulsion is placed in a camera. The glass side is to-wards the objective while the emulsion side is backed with a layer of mercury to form a reflecting surface.

Suppose that light of various wavelengths, corresponding to various colours present on the object to be photographed, is focussed on the plate. Each colour forms its own set of antinodal planes in the emulsion. *The distance between two antinodes* (= $\lambda/2$) *in each set will be different (being least for violet, maximum for red and in between for other colours)*. When such an exposed plate is developed, it contains a number of silver layers located at the antinodal planes.

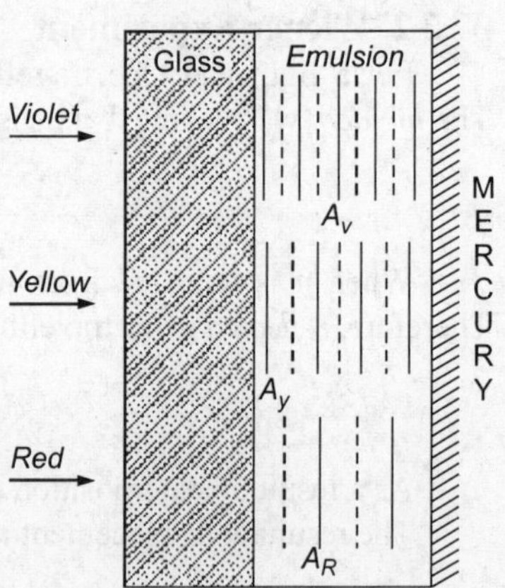

Fig. 37.6.

The separation of the silver layers is $\frac{\lambda}{2}$, where λ is wavelength of incident light. Since incident light contains a number of wavelengths, the silver layers are not uniformaly separated; but the layers from particular wavelength (colour) are uniformly separated.

When such a developed plate is illuminated by white light, each layer will reflect strongly only that particular colour by which it was originally formed. Each silver layer is thin enough to reflect only a small amount of light. As the white light penetrates the film, it is partially reflected at each successive silver layer. All radiations of the same wavelength (coherent) reflected from different layers due to which the layers were formed, are in phase with each other. Therefore they reinforce. The reflected light of that particular colour has large intensity. Thus a particular set of layers corresponds to a particular wavelength. For other colours, the path difference between successive reflected rays is not exactly one wavelength (not coherent). Therefore the rays of those colours will not reinforce and their intensity will be much less.

Thus an exposed Lippmann photographic plate will exhibit the object photographed in its *natural colours* when it is illuminated by white light and viewed by reflection. These colour photographs are extremely brilliant. But no prints can be taken from it. Moreover, the process to prepare Lippmann plates is tedius. Therefore, Lippmann process is not used commercially. The modern colour photography is based on *three colour process*.

37.3.4. Three Colour Process

Three colour process is based on the fact that by suitable mixture of three colours *viz.,* blue, green and red, all colours can be obtained. The two methods used are: (*i*) the additive method (*ii*) subtractive method. Here we shall briefly describe the *subtractive* method.

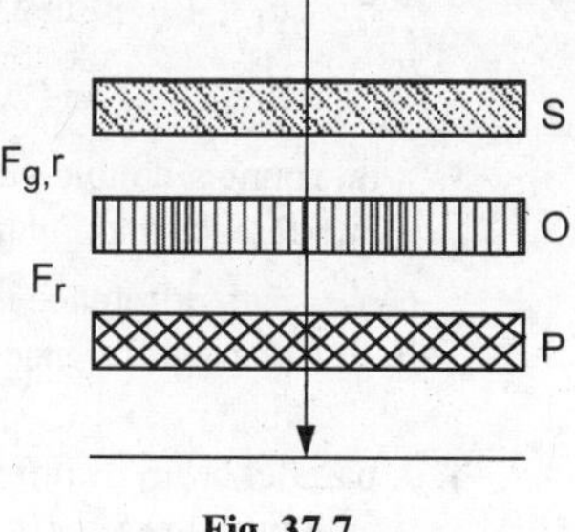

Fig. 37.7.

It consists of three layers of emulsions separated by two layers of filters (Fig. 37.7).

(*i*) The first (upper) emulsion is sensitive to blue and violet (mainly blue).

(*ii*) The second (orthochromatic) emulsion is sensitive to blue and green.

(*iii*) The third (lowest) emulsion (panchromatic) is sensitive to all colours.

These emulsions are painted on a celluloid base and a non-reflecting layer. Between the first and second emulsions is a filter $F_{g,r}$ which transmits green and red and no blue. Between the second and third emulsions is a filter F_r which transmits only red light.

When this film is exposed in a camera, three latent images are produced on the three emulsions by blue, green and red light from the object respectively.

Now by suitable operations the filters are bleached out and the film gives three negatives, each in a colour complementary to that colour of light by which the original image was produced.

(*i*) The first layer subtracts blue and violet from those parts on the object on which it was absent.

(*ii*) The second layer subtracts green from green missing parts of the object.

(*iii*) The third layer subtracts red light from red missing parts of the object.

Therefore if this film passes in front of a source of white light, the image produced is reproduced in natural colours of the object.

EXERCISE

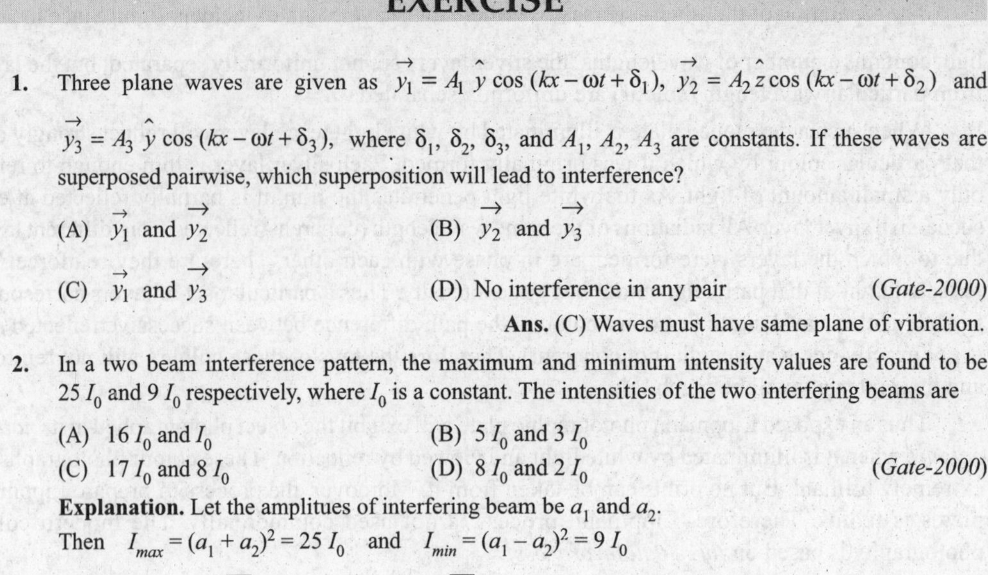

1. Three plane waves are given as $\vec{y_1} = A_1\, \hat{y} \cos (kx - \omega t + \delta_1)$, $\vec{y_2} = A_2\, \hat{z} \cos (kx - \omega t + \delta_2)$ and $\vec{y_3} = A_3\, \hat{y} \cos (kx - \omega t + \delta_3)$, where δ_1, δ_2, δ_3, and A_1, A_2, A_3 are constants. If these waves are superposed pairwise, which superposition will lead to interference?

(A) $\vec{y_1}$ and $\vec{y_2}$ (B) $\vec{y_2}$ and $\vec{y_3}$

(C) $\vec{y_1}$ and $\vec{y_3}$ (D) No interference in any pair *(Gate-2000)*

Ans. (C) Waves must have same plane of vibration.

2. In a two beam interference pattern, the maximum and minimum intensity values are found to be $25\, I_0$ and $9\, I_0$ respectively, where I_0 is a constant. The intensities of the two interfering beams are

(A) $16\, I_0$ and I_0 (B) $5\, I_0$ and $3\, I_0$

(C) $17\, I_0$ and $8\, I_0$ (D) $8\, I_0$ and $2\, I_0$ *(Gate-2000)*

Explanation. Let the amplitues of interfering beam be a_1 and a_2.

Then $I_{max} = (a_1 + a_2)^2 = 25\, I_0$ and $I_{min} = (a_1 - a_2)^2 = 9\, I_0$

$\Rightarrow \quad a_1 + a_2 = 5\sqrt{I_0}$ and $a_1 - a_2 = 3\sqrt{I_0}$

$\therefore \quad a_1 = 4\sqrt{I_0}$ and $a_2 = \sqrt{I_0}$

$I_1 = 16\, I_0$ and $I_2 = I_0$ **Ans.** (A)

3. In Young's double slit experiment with a helium-neon laser beam of wavelength 632 nm, the first interference maximum will occur when

(A) path difference is 948 nm (B) path difference is 316 nm

(C) phase difference is 2π radian (D) phase difference is π radian *(Gate-1997)*

Ans. (C)

4. A parallel beam of infrared radiation of wavelength of 1.01×10^{-6} m is incident normally on a screen with two slits 5×10^{-6} m apart and the resulting interference pattern is observed on a distant screen. What is the largest number of maxima that can be observed on the screen ?

(A) 4 (B) 9

(C) 13 (D) infinitely many *(Gate-2003)* **Ans.** (D)

CHAPTER

38

THE ECHELON GRATING AND ITS RESOLVING POWER

38.1. Principle

The resolving power of a grating is given by

$$R.P. = m \times N.$$

Here, m = order of the spectrum and
N = total number of lines on the grating.

The resolving power can be increased by increasing m and N.

(*i*) In practice, the diamond point ruling the grating fails when N approaches 100000. It thus sets a limit to the size of the surface which can be ruled.

(*ii*) On the other hand, with a ruled grating we cannot use an order of spectrum greater than the third, because the intensity is very poor in the higher orders.

Michelson, in 1898, devised a special type of grating in which N is kept fairly small (say 20 to 40) whereas the order of spectrum m is made very high, but with no loss in intensity. This is called an 'echelon grating'.

Michelson obtained enormously high resolving power with a fairly small number of diffracting elements (N = 20 *to* 40) *by accomplishing a very high order* (m = 200,000) *of interference between the diffracted rays from the corresponding points of the grating.*

38.2. Echelon Transmission Grating

Construction — It consists of a few plane, parallel and equally-thick plates of optically-worked glass or quartz (Fig. 38.1). The plates are arranged in a step formation such that each plate projects the same distance beyond the one below it. This distance (*i.e.,* the width of each step) is usually 1 mm. The width of the steps is made as accurately equal as possible.

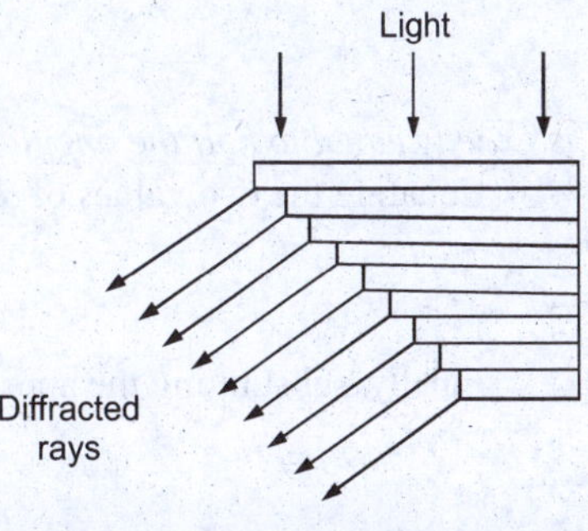

Fig. 38.1.

When a parallel beam of light falls normally on the grating, each point on the steps acts as a secondary source of disturbance sending out rays in all possible directions. A telescope (not shown) collects these diffracted rays. Since the width of each diffracting element (step) is very large compared with the wavelength of light, most of the incident light is diffracted through a small angle.

Theory. Let s be the width of each step, t the thickness of each plate and n the refractive index of the material of the plate (Fig. 38.2).

Let a parallel beam of monochromatic light of wavelength λ fall normally on the largest plate of the grating.

Consider two parallel rays diffracted through an angle θ from *corresponding points A* and *B*. The path difference between diffracted rays *AE* and *BF* is

$$\begin{aligned} \Delta &= BD \text{ in plate} - AC \text{ in air} \\ &= nBD - AC \\ &= nBD - (CN - AN) \\ &= nBD - (BM - AN) \\ &= nBD - (BD \cos\theta - AD \sin\theta) \\ &= nt - (t\cos\theta - s\sin\theta) \\ &= nt - t + s\theta \quad (\text{As } \theta \text{ is small, } \cos\theta = 1, \sin\theta = \theta) \\ &= (n-1)\,t + s\theta. \end{aligned}$$

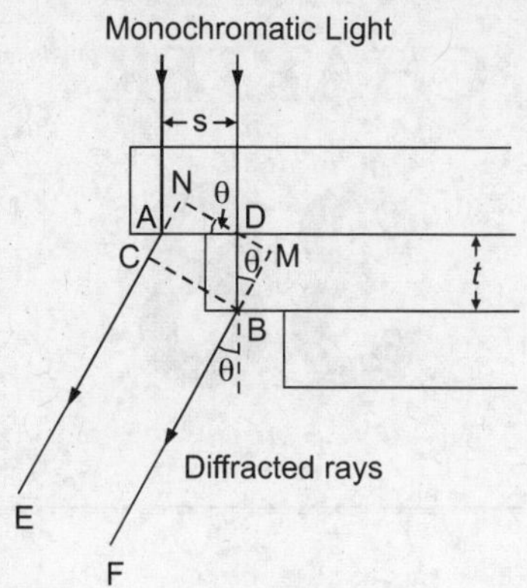

Fig. 38.2.

Therefore *the fundamental equation for principal maxima* is

Path difference, $\Delta = m\lambda$.

Here, m = order of the spectrum.

The grating equation giving the principal maximum is

$$(n-1)\,t + s\theta = m\lambda. \qquad ...(1)$$

The diffracted beam is collected by a telescope and it gives an interference pattern in the focal plane of the telescope.

Resolving Power of Transmission Echelon.

Let there be N plates in the echelon.

Multiplying Eq. (1) by N, we get

$$N(n-1)t + Ns\theta = mN\lambda \qquad ...(2)$$

The angle of diffraction $\theta \pm d\theta$ for the first minimum on either side of the principal maximum, as in the theory of plane grating, is given by the relation

$$N(n-1)t + Ns(\theta \pm d\theta) = (mN \pm 1)\lambda \qquad ...(3)$$

Substracting Eq. (2) from Eq. (3), we get

$$d\theta = \lambda / Ns \qquad ...(4)$$

This equation, therefore, expresses half the angular width of the principal maximum of mth order.

Lord Rayleigh defined two lines as just resolved when the principal maximum of one falls on the first minimum of the other i.e., when the angular separation between the principal maxima of two lines, given by the equation

$$d\theta = \frac{t}{s\lambda}\left[(n-1) - \lambda\frac{dn}{d\lambda}\right]d\lambda \qquad ...(5)$$

is exactly equal to half the angular width, given by Eq. (4) *of the principal maximum.*

Equating the two values of $d\theta$, one may get for the resolving power,

$$\frac{\lambda}{d\lambda} = \frac{Nt}{\lambda}\left[(n-1) - \lambda\frac{dn}{d\lambda}\right] \qquad ...(6)$$

Finally, substituting the approximate value of t, namely, $m\lambda / (n-1)$, in Eq. (6), we get

$$\frac{\lambda}{d\lambda} = Nm\,\frac{\{(n-1) - \lambda\,(dn/d\lambda)\}}{(n-1)} \qquad ...(7)$$

Due to very high resolving power of an echelon grating, it is used to study the hyperfine structure and Zeeman effect.

EXERCISE

1. How is the resolution improved in 'echelon grating'? Illustrate your answer by Michelson's echelon grating. (*Agra 81*)

CHAPTER

39 LASER IN INDUSTRY

39.1. Introduction

The beam which is coming out of a laser is usually a few millimeters (or more) in diameter. Hence for most material processing applications, we must use focussing elements (like lenses) to increase the intensity of the beam. There are numerous applications of laser when it is focussed to small, often diffraction limited, spots.

The beam from a laser has a well-defined wave front, which is either plane or spherical. When such a beam passes through a lens, then according to geometrical optics, the beam should get focussed to a point. When a plane wave of wavelength λ falls on a lens of radius a, then at the focal plane F of the lens we get an intensity distribution of the type shown in Fig. 39.1

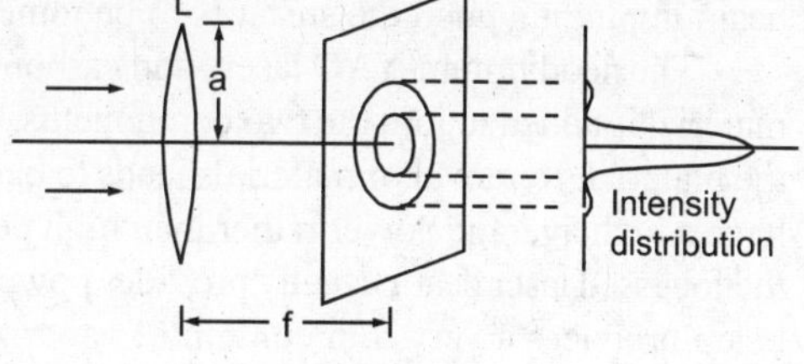

Fig. 39.1.

In actual practice, however, diffraction effects have to be taken into consideration.

If λ is the wavelength of the laser light, a is the radius of the beam, and f is the focal length of the lens, then the incoming beam will get focussed into a region of radius

$$b = \lambda f / a \qquad ...(1)$$

The dimension of this region is

(*i*) directly proportional to f and λ, and

(*ii*) inversely proportional to the radius a.

About 84% of the total energy is confined within a region of radius $\lambda f / a$.

If P represents the power of the laser beam, then the intensity I, obtained at the focussed region is

$$I \approx \frac{P}{\pi b^2} \approx \frac{Pa^2}{\pi \lambda^2 f^2} \qquad ...(2)$$

Aberrations increase the spot dimension, resulting in lower intensities. Thus if we focus a 1-W laser beam (with $\lambda = 1.06$ μm and having a beam radius of about 1 cm) by a lens of focal length 2 cm, then the intensity obtained at the focussed spot is

$$I \approx \frac{1}{3.14 \times (1.06 \times 10^{-4} \times 2)^2} \text{ W/cm}^2 \approx 7 \times 10^6 \text{ W/cm}^2.$$

The electric field strengths produced at the focus are of the order of 10^7 V/cm.

Such large intensities are produced in an extremely small region whose radius is $\approx 2 \times 10^{-4}$ cm.

Eq. (2) shows that, the larger the value of a, the greater will be the intensity. So we use a beam expander to increase the diameter of the beam. A beam expander consists of a set of two convex lenses as shown in Fig. 39.2.

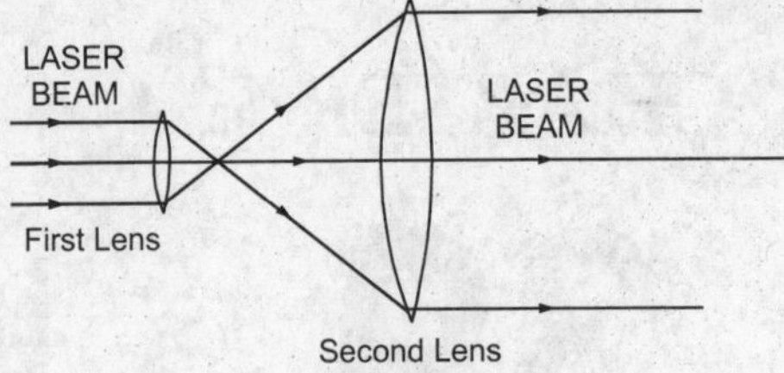

Fig. 39.2.

Depth of Focus

When we produce small focussed laser spots, the beam has a large divergence. Hence near the focussed region, the beam expands again within a very short distance. This distance (which may be defined as the distance over which the intensity of the beam drops to some percentage of that at the focus) defines the depth of focus. Thus, smaller focussed spots lead to a smaller depth of focus. This must also be kept in mind while choosing the parameters in a laser processing application.

39.2. Applications in Material Processing

39.2.1. Laser Welding

Welding is a materials joining process used in making welds. Laser beam welding produces coalescence of materials with the heat obtained from the application of a concentrated coherent light beam impinging upon the surfaces to be joined.

The neodymium-YAG lasers and carbon dioxide lasers are widely used in welding. In welding, material is added to join the two components. Thus the laser power must not be too high to evaporate the material. Removal of materials leads to bad welds. Thus the laser used in welding processes must have a high average power rather than high peak power. When used for *laser beam welding* (LBW), the focussed laser beam usually provides power intensities in excess of 10 kW/cm^2. The high-intensity beam produces a very thin column of vaporized metal with a surrounding liquid pool. As the laser advances, the liquid flows into the channel to produce a weld with depth-to-width ratio generally greater than 4:1. The narrow weld pool solidifies quickly, producing a very thin heat-affected zone and little thermal distortion.

Experiments have shown that the laser can be used effectively for welding. By passing the laser beam through a lens system, an extremely high energy density can be focussed in a very small area. The coherency of the light beam permits this high degree of focussing. The laser will be most useful for spot welding. Some of the advantages in using the laser for spot welding are:

(*a*) no physical contact with the material being welded,

(*b*) welding can be accomplished in air,

(*c*) very little rise in temperature in the area surrounding the weld, and

(*d*) a variety of metals have been successfully welded.

Welding thermocouples, electronic micro-modules are examples of the precise work which the laser can perform.

Laser welds can easily be performed between two dissimilar metals. Thus, a thermocouple may easily be welded to a substrate without much damage to adjacent material. We can simultaneously form the junction and attach the junction to the substrate. This method is used in attaching measuring probes to transistors, turbine blades, etc. Laser weld not only achieves welding between dissimilar metals but also allows precise location of the weld.

Another interesting aspect of welding with the laser is that it is possible to weld a joint even after the joint has been sealed inside a glass envelope.

Because of the sharply focussed beam and the short time (or high speed) of welding, laser welds are usually small, often less than 0.001 in. in width or diameter, and the total heat input is very low, often in the range 0.1 to 10 J. For this reason, laser welding has been adopted by the electronics industry for applications such as connecting wire leads to small electronic components. Lap, butt, tee, and cross-wire configurations can all be used. It is even possible to weld wires without removing the polyurethane insulation. The laser simply evaporates the insulation and completes the weld with the internal wire.

39.2.2. Hole Drilling

Drilling of holes in various substances is another interesting application of the laser. The laser has succeeded in cutting holes in diamond, one of the hardest materials.

Metals, Ceramics, and Polymers can be laser cut or drilled, irrespective of hardness. High power laser can be used for this purpose. Processing is easily automated for speed and accuracy, giving clean edges with a narrow kerf and heat affected zone.

The laser can be considered a direct replacement for many existing cutting techniques. Laser drilling is used in applications ranging from perforating cigarette papers to boring cooling channels in the components of jet engines.

39.2.3. Laser Cutting

The CO_2 laser generates high output powers of the order of several kilowatts. So it is commonly used in cutting processes.

In the cutting process, we essentially remove the materials along the cut. When cuts are obtained using pulsed lasers, then the repetition frequency of the pulse and the motion of the laser across the material is adjusted so that a series of partially overlapping holes are produced. The width of the cut should be as small as possible with due allowance to avoid any rewelding of the cut material. The efficiency of laser cutting can be increased by making use of a gas jet coaxial with the laser.

Fig. 39.3 shows a gas-jet-assisted laser cutting scheme. The gas jet assists in expelling the molten material from the cut.

(*i*) A highly reactive gas like oxygen is used to cut materials like stainless steel, low-carbon steel, titanium etc. When the laser heats up the materials, it interacts with the gas and gets burnt. The gas jet also helps in expelling the molten material.

(*ii*) Inert gases (like nitrogen or argon) are used to cut materials like wood, paper, plastic etc. Such a gas jet helps in expelling the molten materials. This technique is very efficient with materials which absorb most radiation at the laser wavelength.

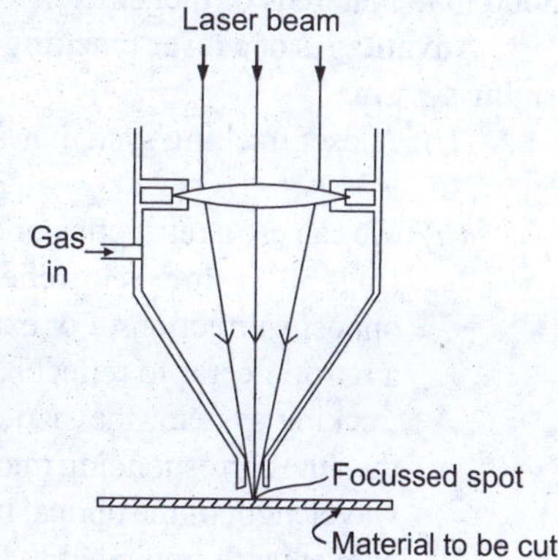

Fig. 39.3.

Laser cutting of stainless steel, nickel alloys, and other metals finds widespread application in the aircraft and automobile industries.

Laser cutting is used in the textile industry for cutting cloth.

Laser cutting is achieving prominence in the cutting of composite materials. The more uniform the thermal characteristics of the components, the better the cut and the less thermal damage to the materials. Kevlar-reinforced epoxy cuts easiest and gives a narrow heat-affected zone.

39.2.4. Other Applications

1. Lasers are used to vaporize materials for subsequent deposition on a substrate.

Advantages:

(*i*) No contamination occurs.

(*ii*) Some pre-selected areas of the source material may be evaporated.

(*iii*) The evaporant may be located very close to the substrate.

2. Brittle materials like rock, marble, etc., can be fractured using laser beams. This technique finds application in rock crushing and boring tunnels.
3. Lasers are used to remove microscopic quantities of material from balance wheels while in motion.
4. Lasers are used in trimming resistors to accuracies of 0.1%. Such micromachining processes find widespread use in semiconductor circuit processing.

Advantages:

(*i*) Small size of the focussed image with a precise control of energy.

(*ii*) Absence of any contamination, accuracy of positioning, and ease of automation.

5. Lasers are used in the opening of oysters.

A laser beam is focussed on that point on the shell where the muscle is attached. This results in detachment of the muscle, the opening of the shell, and leaving the raw oyster alive in the half shell.

39.3. Laser Tracking

By tracking we imply either the determination of the trajectory of a moving object like an aircraft or a rocket, or determining the daily positions of a heavenly object (like the Moon) or an artificial satellite.

Principle. We measure the time taken to travel to and fro for a sharp laser pulse sent by the observer to be reflected by the object and received back by the observer.

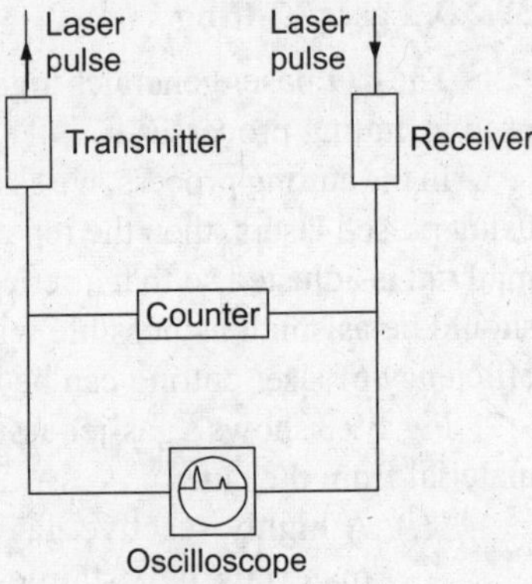

Fig. 39.4.

Fig. 39.4 gives a block diagram of a laser radar system for tracking of a satellite. A portion of the pulse that is sent is collected and is made to start an electronic counter. The counter stops counting as soon as the reflected pulse is received back. The counter may be directly calibrated in units of distance.

Typical ranges of distances that can be covered range from 5000 m to hundreds of megameters.

Advantages of a laser tracking system over a microwave radar system.

(*i*) A laser tracking system has a smaller size and its cost is less.

(*ii*) We can use a retroreflector on the object. A cube corner is often used to act as a retroreflector (Fig. 39.5). In a retroreflector, the incident and reflected rays are parallel and travel in opposite directions. For example, on the surface of the moon, or on a satellite, we can have a retroreflector to reflect back the incident radiation. For a laser tracking system, the size of the retroreflector is much smaller than the corresponding microwave reflector owing to the smaller wavelength of the optical beam. Hence the reflector can be more conveniently mounted in the system involving lasers.

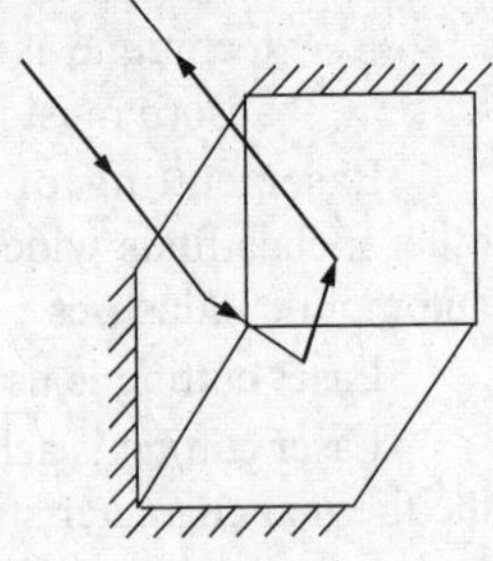

Fig. 39.5.

(*iii*) In a microwave radar system, we have to incorporate corrections because of the presence of the ionosphere and also because of the presence of water vapor in the troposphere. These corrections are much easier to incorporate in the case of an optical beam. As compared to a microwave radar system, the laser radar offers much higher spatial resolution.

Disadvantages

(*i*) When fog and snow are present in the atmosphere, it is very difficult to work at optical frequencies.

(*ii*) During daytime there is a large background noise.

(*iii*) The losses in the transmitter and receiver are also considerably larger in laser systems.

39.4. Lidar

Lidar (acronym for light detection and ranging) is used for monitoring the environment. Lidars study the laser beam scattered from the atmosphere.

Pulses of laser light are sent and the radiation that is scattered by various particles present in the atmosphere is picked up by the receiver. The background sunlight is removed by using filters. This scattered light gives information regarding the particles present in the atmosphere.

Working of Lidar system

Fig. 39.6 shows the block diagram of a pulsed lidar system to study the nature of aerosols present in the atmosphere.

A small fraction of the laser pulse is sent to the oscilloscope to record the zero time. We measure the time dependence of the intensity of the backscattered laser light using a photodetector. The backscattered light produces a current in the photodiode which is recorded on the oscilloscope. The time variation can be easily converted into the height from which the laser beam has been backscattered.

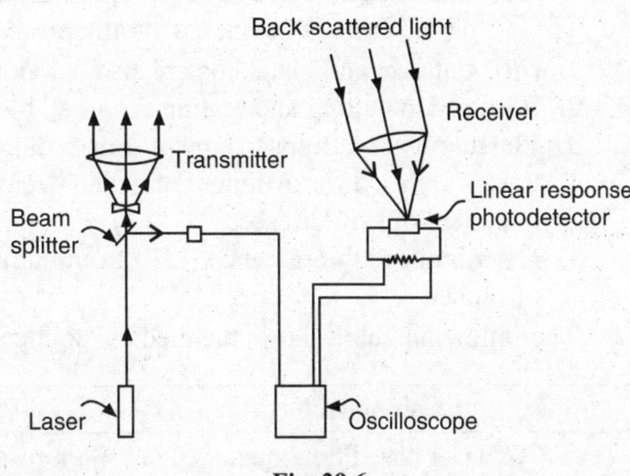

Fig. 39.6.

Fig. 39.7 (*a*) shows typical time dependence of the backscattered laser radiance as a function of height for a pure molecular atmosphere which has no aerosols. The backscattering is by pure molecular gases such as N_2, O_2, Ar, etc.

Fig. 39.7 (*b*) shows time dependence of the backscattered laser radiance as a function of height for an atmosphere containing aerosols. Kinks appear in the curve at points *X* and *Y* corresponding to the heights h_1 and h_2. Between the heights h_1 and h_2 there are aerosols which are responsible for a greater intensity (compared to that for a clear atmosphere) of the backscattered laser light.

The curve implies a haze which exists between the heights h_1 and h_2. Corresponding to the height h_2 the intensity is roughly the same as that from a pure molecular atmosphere. Thus, beyond the height h_2 we do not expect the presence of any aerosols.

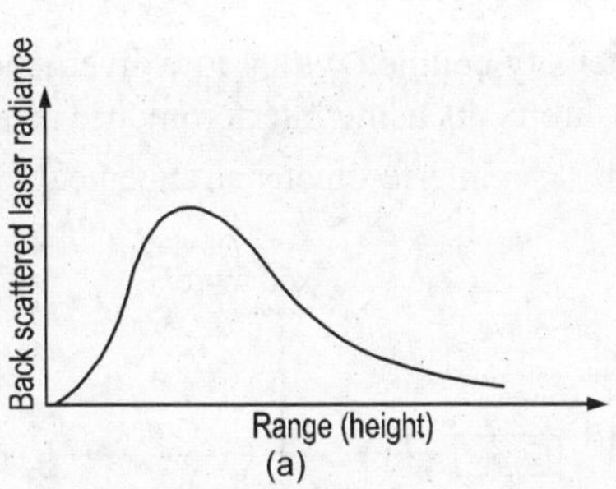

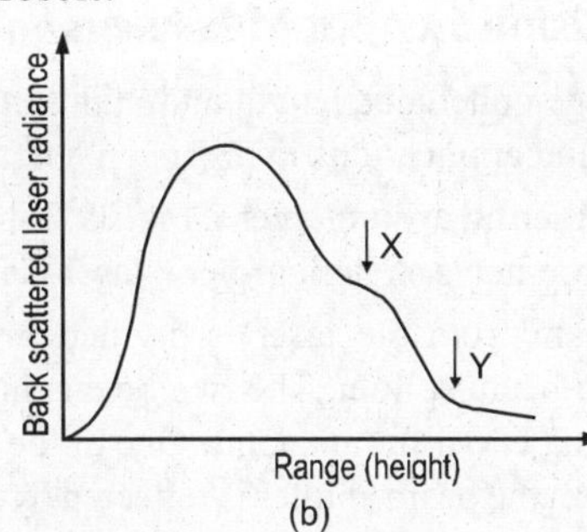

Fig. 39.7.

Other uses of Lidar

(*i*) We can study the concentrations and sizes of various particles present in the atmosphere, which are of extreme importance in pollution studies.

(*ii*) A lidar is used to study the visibility of the atmosphere, the diffusion of particulate materials (or gases released at a point) in the atmosphere, the presence of clouds, fog, etc.

The study of turbulence and winds and the probing of the stratosphere have also been carried out by lidar systems.

39.5. Lasers in Medicine

1. **Eye Surgery.** The most important use of lasers in the field of medicine is in eye surgery. The tremendous use of the laser in eye surgery is primarily due to the fact that the outer transparent regions of the eye allow light at suitable wavelengths to pass through for subsequent absorption by the tissues at the back of the eye.
   (*i*) The laser beam is used in delicate surgery like cornea grafting. Using laser beam, the surgical operation is completed in a much shorter time.
   (*ii*) In treating the detached retina, the laser beam is brought to a focus at the desired spot on the retina where it essentially produces a tiny spot weld.
   (*iii*) Cataract and glaucoma are also cured with lasers.
2. It is used in cutting and sealing the small blood vessels in brain operation.
3. **Plastic surgery.** Burns treatment, gynaecologic operations and treatment of festering wounds.
4. Lasers are used for treatment of dental decay, the destruction of malignant tumours and the treatment of skin diseases.
5. Laser induced fluorescence (LIF) from human tissues has recently been used for diagnosis of cancer.

The following Table shows the medical applications.

	Laser source	Treatment
1.	CO_2 laser based treatment	Skin resurfacing. TMR, Office based surgery.
2.	Erbium Laser	Dentistry, Skin resurfacing, Opthalmology.
3.	Ruby Laser	Hair removal, Tattoo removal, Pigmented lesions.
4.	Holmium Laser	Skin rejuvination opthalmology, TMR
5.	Excimer Laser	Glaucoma, PRK, LASIK, Pacemaker lead removal
6.	Dye Laser	Dermatology, PDT.
7.	Diode Laser	Surgery, Microsurgery, Dermatology, Hair removal, Opthalmology, Dentistry, Urology, Tissue welding
8.	Diode pumped Nd-YAG Laser	Photocoagulation Dermatology.

39.6. Precision Length Measurement

The large coherence length and high output intensity coupled with a low divergence, enables the laser to find applications in precision length measurements using interferometric techniques.

Experimental arrangement. Fig. 39.8 shows the laser interferometer arrangement for precision length measurements. A helium-neon laser is used.

The beam from the laser is divided into two portions by a beam splitter. The two beams interfere after traversing two different paths. One of the beams emerging from the beam splitter is reflected by a fixed reflector (mirror). The other beam is reflected by a retroreflector mounted on the surface whose position is to be monitored. The two reflected beams interfere to produce either constructive or destructive interference. Thus, as the reflecting surface is moved, we get alternatively constructive and destructive interference, which can be detected with the help of a photodetector.

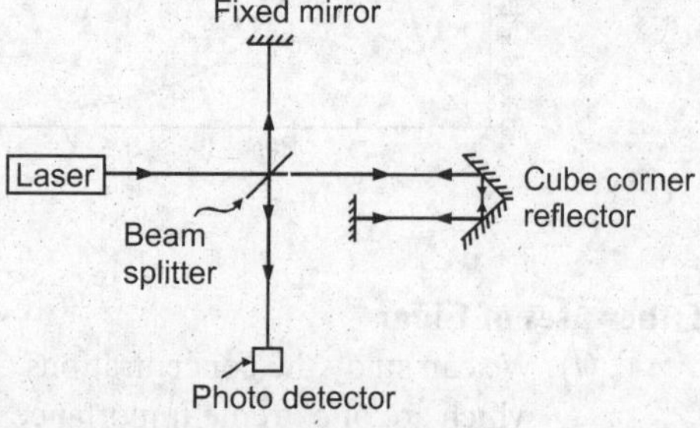

Fig. 39.8.

The change from a constructive to a destructive interference corresponds to a change of a distance of half a wavelength. So we can measure the distance traversed by the surface on which the reflector is mounted by counting the number of fringes which have crossed the photodetector. Accuracies up to 0.1 μm can be obtained by using this technique.

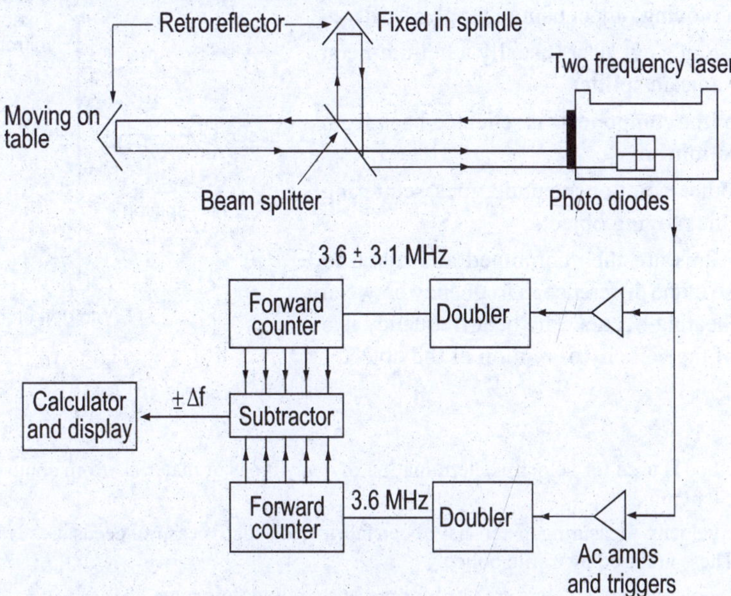

Fig. 39.9.

This technique is being used for accurate positioning of aircraft components on a machine tool, for calibration and testing of machine tools, for comparison with standards, etc.

Fig. 39.9 shows the components of a two frequency Hewlett-Packard *laser interferometer*.

It uses a helium-neon laser beam split into two beams, each of different frequency and polarized. When the beams are recombined, any relative motion between the optics creates a Doppler shift in the frequency. This shift is then converted into a distance measurement.

The company's first two-frequency interferometer calibration system was introduced in 1970 to overcome workplace contamination by thermal gradients, air turbulence, oil mist, and so on, which affect the intensity of light. Doppler laser interferometers are relatively insensitive to such problems. The system can be used to measure linear distances, velocities, angles, flatness, straightness, squareness, and parallelism in machine tools.

39.7. Velocity Measurement

Doppler Shift. When a light beam is scattered by a moving object, the frequency of the scattered wave is different from that of the incident wave. The shift in the frequency depends on the velocity of the object.

Let ν = light frequency,

v = velocity of the moving object which is moving at an angle θ with respect to the incident light beam,

c = velocity of light in free space.

The change in frequency $\Delta\nu$ between the incident and the reflected beams is given by

$$\frac{\Delta\nu}{\nu} = \frac{2v}{c}\cos\theta$$

Thus the change in frequency Δν is directly proportional to the velocity v of the moving object. This is known as the *Doppler shift*. Thus, by measuring the change in frequency suffered by a beam when scattered by a moving object, we can determine the velocity of the object.

Experimental Arrangement. Fig. 39.10 shows the experimental arrangement for measuring the velocity of a moving object using Doppler shift.

The beam from a cw laser (usually a helium-neon laser) is split by a beam splitter.

(*i*) One of the components is reflected back from a fixed mirror.

(*ii*) The other component undergoes scattering from the moving object.

The two beams are then combined and made to interfere. Because of the difference in frequency between the two beams, beating occurs. The beat frequency is a direct measure of the velocity of motion of the object.

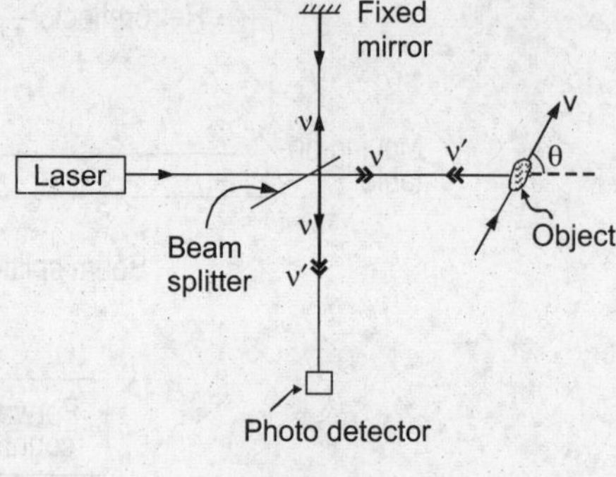

Fig. 39.10.

USES

1. This method is used for velocity determination of many types of materials from about 10 mm/min to about 150 m/min.
2. Portable velocity-measuring meters have been fabricated which measure speeds in the range of 15-130 km/hr. These are used by traffic police.
3. Laser Doppler velocity meters are used for measuring fluid flow rates.

CHAPTER 40

PHOTONICS

PHOTODETECTORS

40.1. Introduction

At the output end of an optical transmission line, there must be a receiving device which interprets the information contained in the optical signal. The first element of this receiver is a photodetector. The photodetector senses the luminescent power falling upon it and converts the variation of this optical power into a correspondingly varying electric current. Since the optical signal is generally weakened and distorted when it emerges from the end of the fiber, the photodetector must meet very high performance requirements.

The following criteria define the important performance and compatibility requirements for detectors:

1. *High sensitivity at the operating wavelengths.*
2. *High fidelity.* To reproduce the received signal waveform with fidelity, for analogy transmission the response of the photodetector must be linear with regard to the optical signal over a wide range.
3. *Large electrical response to the received optical signal.* The photodetector should produce a maximum electrical signal for a given amount of optical power, *i.e.,* the quantum efficiency should be high.
4. *Short response time to obtain a suitable bandwidth.*
5. *A minimum noise introduced by the detector.* Dark currents, leakage currents and shunt conductance must be low. Also the gain mechanism within either the detector or associated circuitry must be of low noise.
6. *Stability of performance characteristics.* Ideally, the performance characteristics of the detector should be independent of changes in ambient conditions.
7. *Low bias voltages.* Ideally the detector should not require excessive bias voltages or currents.
8. *High reliability.* The detector must be capable of continuous stable operation at room temperature for many years.
9. *Small size.* The physical size of the detector must be small for efficient coupling to the fiber and to allow easy packaging with the following electronics.

40.2. Optical Detection Principles

The basic detection process in an intrinsic absorber is illustrated in Fig. 40.1.

Operation of the *p–n* photodiode:

The *p–n* photodiode is reverse biased (Fig. 40.1 *b*). The electric field developed across the *p–n* junction sweeps mobile carriers (holes and electrons) to their respective majority sides (*p* and *n* type material). A depletion region or layer is therefore created on either side of the junction. This barrier has the effect of stopping the majority carriers crossing the junction in the opposite direction to the field. However, the field accelerates minority carriers from both sides to the opposite side of the junction, forming the reverse leakage current of the diode. Thus intrinsic conditions are created in the depletion region.

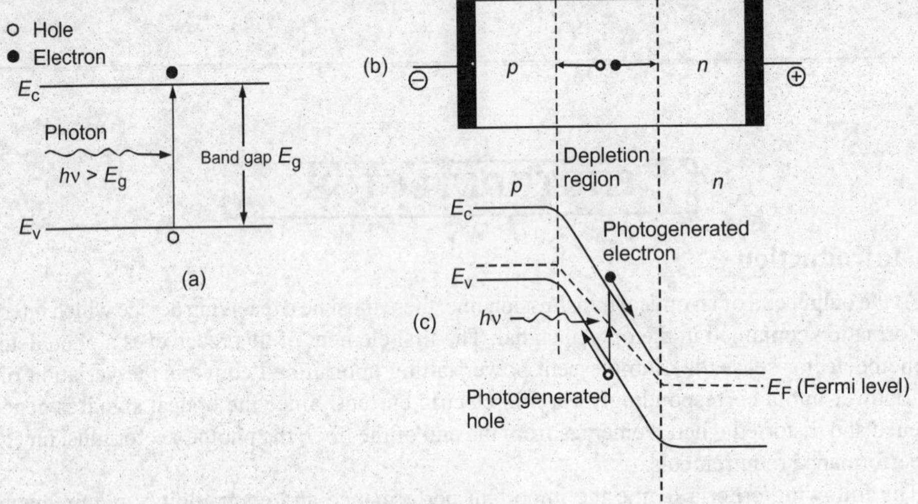

Fig. 40.1.

(*i*) Suppose a photon of light is incident in or near the depletion region.
If the energy of the incident photon ($h\nu$) is equal to or greater than the band gap energy (E_g) of the semiconductor material of the *p–n* junction, the photon will excite an electron in valence band to the conduction band. This process will generate an "electron-hole" pair (Fig. 40.1*a*), when

$$h\nu \geq E_g \qquad ...(1)$$

(*ii*) Fig. 40.1 (*b*) illustrates carrier drift in the depletion region. Carrier pairs generated near the junction are separated and swept (drift) under the influence of the electric field to produce a displacement by current in the external circuit in excess of any reverse leakage current.

(*iii*) Fig. 40.1 (*c*) shows the energy band diagram of the reverse biased *p–n* junction.
It shows photo-generation and the subsequent separation of an electron-hole pair in the depletion region.

The depletion region must be sufficiently thick to allow a large portion of the incident light to be absorbed in order to achieve maximum carrier-pair generation. Long carrier drift times in the depletion region restrict the speed of operation of the photodiode. So it is necessary to limit the width of the depletion region. Thus, a compromise between the number of photons absorbed (sensitivity) and the speed of response of the device should be made.

40.3. Absorption

40.3.1. Absorption coefficient

Optical radiation is absorbed in the semiconductor material according to the exponential law

$$P(x) = P_0\,(1 - e^{-\alpha_s(\lambda)x}) \qquad ...(1)$$

Here $\alpha_s(\lambda)$ is the *absorption coefficient* at a wavelength λ, P_0 is the incident optical power level, and $P(x)$ is the optical power absorbed in a distance x.

α_s depends strongly on the wavelength.

Fig. 40.2 shows optical absorption coefficient as a function of wavelength for silicon, germanium and gallium arsenide.

Thus a particular semiconductor material can only be used over a limited wavelength range.

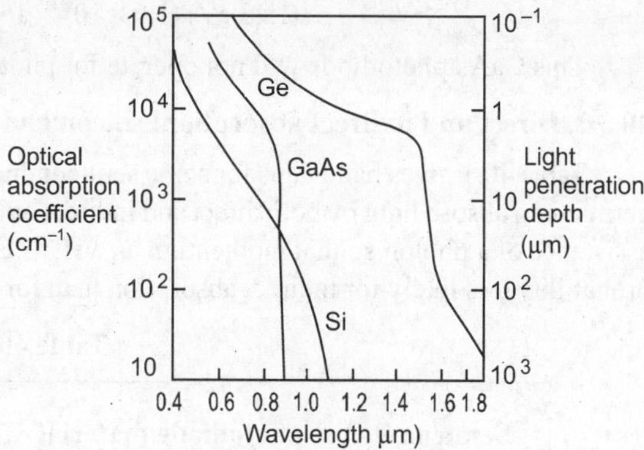

Fig. 40.2.

Long wavelength cutoff

The upper wavelength cutoff λ_c is determined by the band-gap energy E_g of the material. If E_g is expressed in units of electron volts (eV), then λ_c is given in units of micrometers (μm) by

$$\lambda_c\,(\mu m) = \frac{hc}{E_g} = \frac{1.24}{E_g\,(\text{eV})} \qquad \text{...(2)}$$

Derivation

When light having photon energies greater than or equal to the band gap energy of the semiconductor material is incident on a photodetector, the photons can give up their energy and excite electrons from the valence band to the conduction band.

Therefore, the photon energy

$$\frac{hc}{\lambda} \geq E_g \qquad \text{...(3)}$$

or

$$\lambda \leq \frac{hc}{E_g} \qquad \text{...(4)}$$

Thus the threshold for detection, commonly known as the long wavelength cutoff point λ_c, is:

$$\lambda_c = \frac{hc}{E_g} \qquad \text{...(5)}$$

The expression given in Eq. (5) allows the calculation of the longest wavelength of light to give photodetection for the various semiconductor materials used in the fabrication of detectors.

The cutoff wavelength is about 1.06 μm for Si and 1.6 μm for Ge. For longer wavelengths, the photon energy is not sufficient to excite an electron from the valence to the conduction band.

The absorption of photons in a photodiode to produce carrier pairs and thus a photocurrent, is dependent on the absorption coefficient α of the light in the semiconductor used to fabricate the device. At a specific wavelength and assuming only bandgap transitions (i.e., intrinsic absorber), the photocurrent I_p produced by incident light of optical power P_0 is given by

$$I_p = \frac{P_o e\,(1 - r)}{h\nu}\,[1 - \exp(-\alpha d)] \qquad \text{...(6)}$$

Here, e = charge on an electron,

r = Fresnel reflection coefficient at the semiconductor-air interface and

d = width of the absorption region.

Example 1. *A photodiode is constructed of GaAs, which has a band-gap energy of 1.43 eV at 300 K. Calculate the long-wavelength cutoff.*

Solution. $\lambda_c = \dfrac{hc}{E_g} = \dfrac{(6.625 \times 10^{-34}\ \text{J.s})\ (3 \times 10^8\ \text{m/s})}{(1.43\ \text{eV})\ (1.6 \times 10^{-19}\ \text{J/eV})} = 869\ \text{nm}.$

This GaAs photodiode will not operate for photons of wavelength greater than 869 nm.

40.3.2. Direct and indirect absorption: silicon and germanium

Table 40.1 gives band gaps for some semiconductor photodiode materials at 300 K. Silicon and germanium absorb light by both direct and indirect optical transitions. Indirect absorption requires the assistance of a photon so that momentum as well as energy are conserved. This makes the transition probability less likely for indirect absorption than for direct absorption where no photon is involved.

Table 40.1

Semiconductor photodiode material	Bandgap (eV) at 300 K	
	Indirect	Direct
Si	1.14	4.10
Ge	0.67	0.81
$In_{0.53}\ Ga_{0.47}As$	—	0.75

Silicon is only weakly absorbing over the wavelength band of interest in optical fiber communications (i.e., first generation 0.8 to 0.9 μm) [See Fig. 40.2]. This is because transitions over this wavelength band in silicon are due only to the indirect absorption mechanism. The threshold for indirect absorption occurs at 1.09 μm. The bandgap for direct absorption in silicon is 4.10 eV, corresponding to a threshold of 0.30 μm in the ultraviolet, and thus is well outside the wavelength range of interest.

Germanium is another semiconductor material for which the lowest energy absorption takes place by indirect optical transitions. However, the threshold for direct absorption occurs at 1.53 μm, below which germanium becomes strongly absorbing, corresponding to the kink in the characteristic shown in Fig. 40.2. Thus germanium may be used in the fabrication of detectors over the whole of the wavelength range of interest (i.e., first and second generation 0.8 to 1.6 μm), especially considering that indirect absorption will occur up to a threshold of 1.85 μm.

Since the bandgap of Ge is narrower than that of Si, Germanium photodiodes have relatively large dark currents. This is a major disadvantage with the use of germanium photodiodes, especially at shorter wavelengths (below 1.1 μm).

40.3.3. III-V alloys

Direct bandgap III–V alloys are superior to germanium because their bandgaps can be tailored to the desired wavelength by changing the relative concentrations of their constituents, resulting in lower dark currents. They may also be fabricated in heterojunction structures which enhances their high speed operations.

Ternary alloys such as InGaAs and GaAISb deposited on InP and GaSb substrates, respectively, have been used to fabricate photodiodes for the longer wavelength band. The alloy $In_{0.53}Ga_{0.47}As$ lattice matched to InP, which responds to wavelengths up to around 1.7 μm, has been extensively utilized in the fabrication of photodiodes for operation at both 1.3 and 1.55 μm. Quaternary alloys are also under investigation for detection at these wavelengths. InGaAsP grown on InP system finds significant application within advanced photodiode structures.

40.4. Quantum Efficiency

Definition. *The quantum efficiency* η *is defined as the fraction of incident photons which are absorbed by the photodetector and generate electrons which are collected at the detector terminals:*

$$\eta = \frac{\text{number of electrons collected}}{\text{number of incident photons}} \quad ...(1)$$

$$\therefore \qquad \eta = \frac{r_e}{r_p} \quad ...(2)$$

Here, r_p = incident photon rate (photons per second) and

r_e = corresponding electron rate (electrons per second).

The quantum efficiency η is an important photodetector performance parameter.

In practice, quantum efficiencies range from 30 to 95 percent.

In order to achieve a high value of quantum efficiency, the depletion layer must be made thick enough to allow a large portion of incident light to be absorbed. But when the depletion layer is made thicker, it takes a longer time for the photo generated carriers to drift across the reverse biased junction. Since the drift time determines the response speed of a photodetector, a compromise has to be made between the response speed and quantum efficiency.

Expression for Quantum Efficiency

Quantum efficiency
$$\eta = \frac{r_e}{r_p} \quad ...(1)$$

$$r_e = \frac{I_p}{e} \quad ...(2)$$

Here, I_p = average photo current generated by a steady average optical power P_o incident on the photo detector, and

e = charge on an electron.

Now,
$$r_p = \frac{P_o}{h\nu} \quad ...(3)$$

Substituting the values from Eqs. (2) and (3) in Eq. (1), we get

$$\eta = \frac{I_p / e}{P_o / h\nu} = \frac{I_p h\nu}{e P_o} \quad ...(4)$$

Example 1. *In a 100-ns pulse, 6×10^6 photons at a wavelength of 1300 nm fall on an InGaAs photodetector. On the average 3.9×10^6 electron-hole (e-h) pairs are generated. Calculate the quantum efficiency.*

Solution. $\eta = \dfrac{\text{number of } e\text{-}h \text{ pairs generated}}{\text{number of incident photons}} = \dfrac{3.9 \times 10^6}{6 \times 10^6} = 0.65.$

Thus the quantum efficiency at 1300 nm is 65 percent.

40.5. Responsivity

Definition. It is defined as:

$$R = \frac{I_p}{P_o} \; (AW^{-1}) \quad ...(1)$$

Here, I_p = output photocurrent in amperes and

P_o = incident optical power in watts.

Responsivity is defined as the ratio between the average current generated in the detector and the average optical power incident on the detector.

This parameter is quite useful. It gives the transfer characteristic of the detector (*i.e.*, photocurrent per unit incident optical power). Responsivity is related to the quantum efficiency by

$$R = \frac{I_p}{P_0} = \frac{\eta e}{h\nu} \qquad ...(2)$$

Expression for the responsivity of an intrinsic photodetector in terms of the quantum efficiency of the device and the wavelength of the incident radiation.

The energy of a photon $E = h$. ...(1)

The incident photon rate r_p may be written in terms of incident optical power and the photon energy as:

$$r_p = \frac{P_o}{h\nu} \qquad ...(2)$$

The electron rate (electrons per second) is given by:

$$r_e = \eta r_p \qquad ...(3)$$

Substituting from Eq. (2) we get

$$r_e = \frac{\eta P_o}{h\nu} \qquad ...(4)$$

Therefore, the output photocurrent is:

$$I_p = \frac{\eta P_o e}{h\nu} \qquad ...(5)$$

Here, e is the charge on an electron.

$$\therefore \qquad R = \frac{I_p}{P_o} = \frac{\eta P_o e/(h\nu)}{P_o}$$

or responsivity

$$R = \frac{\eta e}{h\nu} \qquad ...(6)$$

Let λ be the wavelength of the incident light.

The frequency ν of the incident photon is

$$\nu = \frac{c}{\lambda} \qquad ...(7)$$

Substituting this value of ν in Eq. (6), we get

$$R = \frac{\eta e\lambda}{hc} \qquad ...(8)$$

Eq. (8) shows that the responsivity is directly proportional to the quantum efficiency at a particular wavelength.

Fig. 40.3 shows the ideal responsivity against wavelength characteristic for a silicon photodiode with unit quantum efficiency. The responsivity of a typical practical photodiode is also shown. Fig. 40.3 shows that the responsivity decreases very rapidly beyond the cut off wavelength.

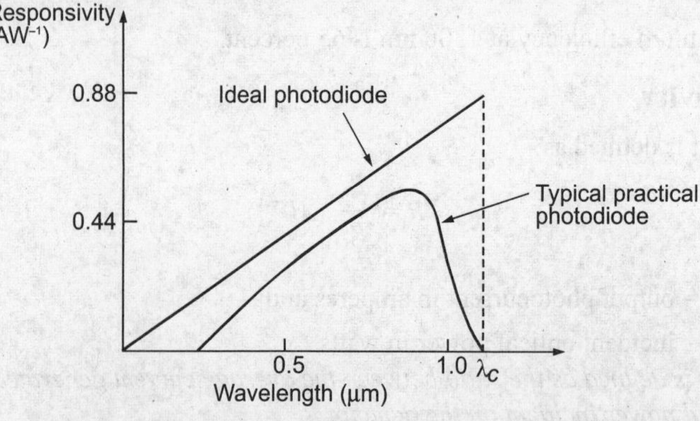

Fig. 40.3.

Example 1. *When 3×10^{11} photons each with a wavelength of 0.85 μm are incident on a photodiode, on average 1.2×10^{11} electrons are collected at the terminals of the device. Determine the quantum efficiency and the responsivity of the photodiode at 0.85 μm.*

Solution. $$\text{Quantum efficiency} = \frac{\text{number of electrons collected}}{\text{number of incident photons}} = \frac{1.2 \times 10^{11}}{3 \times 10^{11}} = 0.4$$

The quantum efficiency of the photodiode at 0.85 μm is 40%.

Responsivity $$R = \frac{\eta e \lambda}{hc}$$

$$= \frac{0.4 \times (1.602 \times 10^{-19}) \times (0.85 \times 10^{-6})}{(6.626 \times 10^{-34}) \times (2.998 \times 10^{8})}$$

$$= 0.274 \text{ AW}^{-1}$$

The responsivity of the photodiode at 0.85 μm is 0.27 AW^{-1}.

Example 2. *Photons of energy 1.53×10^{-19} J are incident on a photodiode which has a responsivity of 0.65 A/W. If the optical power level is 10 μW, calculate the photocurrent generated.*

Solution. $I_p = RP_0 = (0.65 \text{ A/W})\,(10\ \mu\text{W}) = 6.5\ \mu\text{A}$.

Example 3. *For the wavelength range 1100 nm < λ < 1600 nm, the quantum efficiency for InGaAs is about 60 percent, calculate the responsivity at 1300 nm.*

Solution. $$R = \frac{\eta e}{h\nu} = \frac{\eta e \lambda}{hc} = \frac{(0.60)\,(1.6 \times 10^{-19}\text{ C})\,(1.30 \times 10^{-6}\text{ m})}{(6.625 \times 10^{-34}\text{ J-s})\,(3 \times 10^{8}\text{ m/s})} = 0.63 \text{ A/W.}$$

Example 4. *A photodiode has a quantum efficiency of 65 percent when photons of energy 1.5×10^{-19} J are incident on it. (a) Calculate the wavelength at which the photodiode is operating and (b) incident optical power required when the current through the diode is 3μA.*

Solution. (*a*) $$E = h\nu = \frac{hc}{\lambda}$$

$$\therefore \quad \lambda = \frac{hc}{E} = \frac{(6.62 \times 10^{-34}) \times (3 \times 10^{8})}{(1.5 \times 10^{-19})} = 1.324 \times 10^{-6}\text{ m} = 1.324\ \mu\text{m.}$$

(*b*) Responsivity $$R = \frac{\eta e \lambda}{hc} = \frac{0.65 \times (1.602 \times 10^{-19}) \times (1.324 \times 10^{-6})}{(6.626 \times 10^{-34})\,(3 \times 10^{8})} = 0.694 \text{ A W}^{-1}$$

$$R = \frac{I_p}{P_o} \text{ or } P_o = \frac{I_p}{R} = \frac{3 \times 10^{-6}}{0.694} = 4.32\ \mu\text{W.}$$

40.6. *p–n* Photodiode

Principle. A *photodiode* consists of a single *p–n* junction, made from light sensitive semiconductor. If the energy of the incident photon ($h\nu$) is equal to or greater than the band gap energy (E_g) of the semiconductor material of the *p–n* junction, the photon will excite an electron in the valence band to the conduction band. This process will generate an "electron-hole" pair. The number of electron-hole pairs produced depends on the intensity of the incident photons. If a reverse bias is applied, the carriers are swept out of the transition region as a *photo-current.*

Fig. 40.4 shows a reverse biased p–n photodiode with both the depletion and diffusion regions.

The depletion region is formed by immobile positively charged donor atoms in the n type semiconductor material and immobile negatively charged acceptor atoms in the p type material, when the mobile carriers are swept to their majority sides under the influence of the electric field. The width of the depletion region is therefore dependent upon the doping concentrations for a given applied reverse bias.

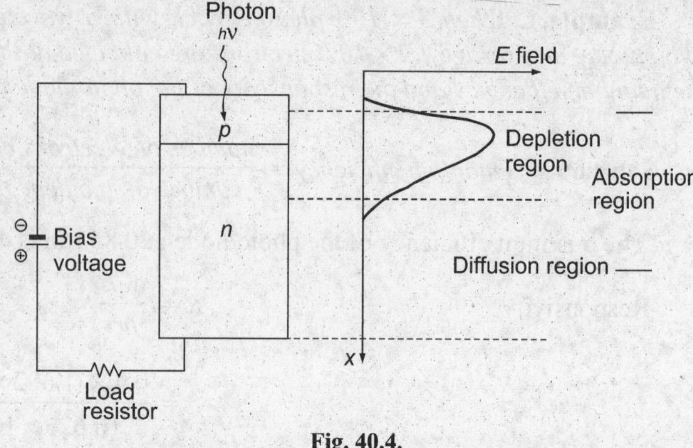

Fig. 40.4.

Photons may be absorbed in both the depletion and diffusion regions. The absorption region's position and width depends upon the energy of the incident photons and on the material from which the photodiode is fabricated. Thus in the case of the weak absorption of photons, the absorption region may extend completely throughout the device. Electron-hole pairs are therefore generated in both the depletion and diffusion regions. In the depletion region the carrier pairs separate and drift under the influence of the electric field, whereas outside this region the hole diffuses towards the depletion region in order to be collected. The diffusion process is very slow compared to drift and thus limits the response of the photodiode.

It is therefore important that the photons are absorbed in the depletion region. Thus it is made as long as possible by decreasing the doping in the n type material. The depletion region width in a p–n photodiode is normally 1 to 3 μm and is optimized for the efficient detection of light at a given wavelength. For silicon devices this is the visible spectrum (0.4 to 0.7 μm) and for germanium in the near infrared (0.7 to 0.9 μm).

Fig. 40.5 shows p–n photodiode output characteristics with light intensity as parameter.

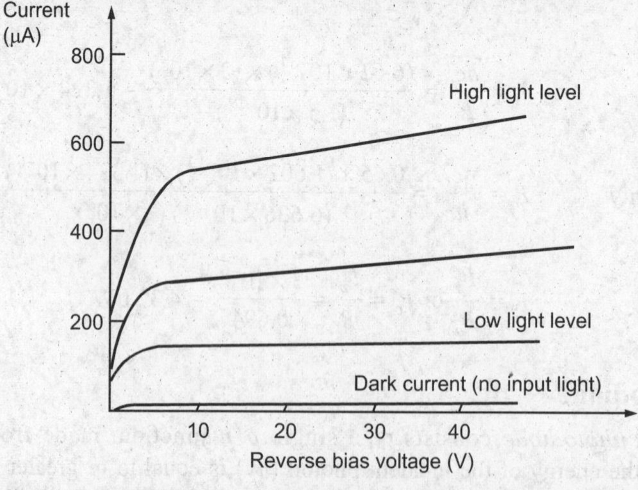

Fig. 40.5.

Curves with different illuminations are shown in Fig. 40.5. With zero illumination, the current equals the reverse saturation current shown as the dark current.

40.7. Avalanche Photodiodes

40.7.1. Avalanche Effect

Avalanche photodiodes (APDs) internally multiply the primary signal photocurrent before it enters the input circuitry of the following amplifier. This increases receiver sensitivity, since the photocurrent is multiplied before encountering the thermal noise associated with the receiver circuit. In order for carrier multiplication to take place, the photogenerated carriers must traverse a region where a very high electric field is present. In this high-field region, a photogenerated electron or hole can gain enough energy so that it ionizes bound electrons in the valence band upon colliding with them. This carrier multiplication mechanism is known as *impact ionization*. The newly created carriers are also accelerated by the high electric field, thus gaining enough energy to cause further impact ionization. This phenomenon is the *avalanche effect*.

40.7.2. Reach Through Avalanche Photodiode (RAPD)

Structure. A commonly used structure for achieving carrier multiplication with very little excess noise is the *reach-through* construction shown in Fig. 40.6.

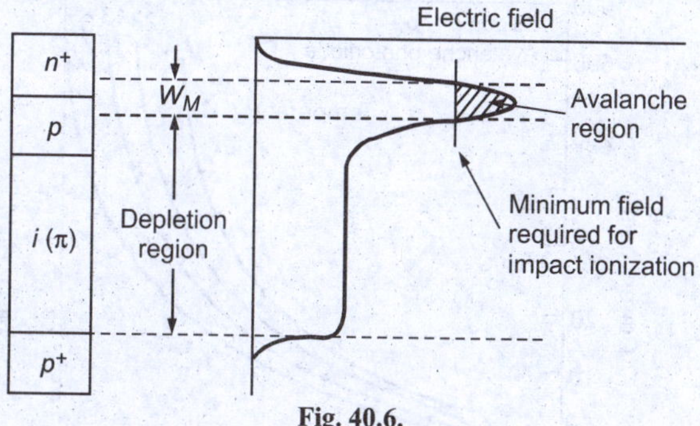

Fig. 40.6.

The RAPD is composed of a high-resistivity *p*-type material deposited as an epitaxial layer on a p^+ (heavily doped *p*-type) substrate. A *p*-type diffusion or ion implant is then made in the high-resistivity material followed by the construction of an n^+ (heavily doped *n*-type) layer. For silicon the dopants used to form these layers are normally boron and phosphorus, respectively. This configuration is referred to as a $p^+ \pi p n^+$ *reach-through* structure. The π layer is basically an intrinsic material that inadvertently has some *p* doping because of imperfect purification.

Operation. When a low reverse-bias voltage is applied, most of the potential drop is across the pn^+ junction. The depletion layer widens with increasing bias until a certain voltage is reached at which the peak electric field at the pn^+ junction is about 5 to 10 percent below that needed to cause avalanche breakdown. At this point the depletion layer just "reaches through" to the nearly intrinsic π region.

In normal usage the RAPD is operated in the fully depleted mode. Light enters the device through the p^+ region and is absorbed in the π material, which acts as the collection region for the photogenerated carriers. Upon being absorbed the photon gives up its energy, thereby creating electron-hole pairs.

The electron-hole pairs are then separated by the electric field in the π region. The photogenerated electrons drift through the π region to the pn^+ junction where a high electric field exists. It is in this high-field region that carrier multiplication takes place.

The average number of electron-hole pairs created by a carrier per unit distance travelled is called the *ionization rate*. Most materials exhibit different *electron ionization rates* α and *hole ionization rates* β.

The ratio $k = \beta/\alpha$ of the two ionization rates is a measure of the photodetector performance.

Multiplication factor. The multiplication factor *M* is a measure of the internal gain provided by the APD.

The multiplication *M* for all carriers generated in the photodiode is defined by

$$M = \frac{I_M}{I_P}.$$

Here, I_M = average value of the total multiplied output current and

I_P = primary unmultiplied photocurrent.

Fig. 40.7 shows the variation of the gain (multiplication factor) with temperature for a silicon RAPD operating at a wavelength of 0.825 μm. Thus temperature compensation is necessary to stabilize the operation of the device.

Typical room-temperature current gains for different wavelengths as a function of bias voltage for a silicon RAPD are shown in Fig. 40.8.

The dependence of the gain on the excitation wavelength is attributable to mixed initiation of the avalanche process by electrons and holes when most of the light is absorbed in the n^+p region close to the detector surface. This is especially noticeable at short wavelengths, where a larger portion of the optical power is absorbed close to the surface than at longer wavelengths. Since the ionization coefficient for holes is smaller than that for electrons in silicon, the total current gain is reduced at the short wavelengths.

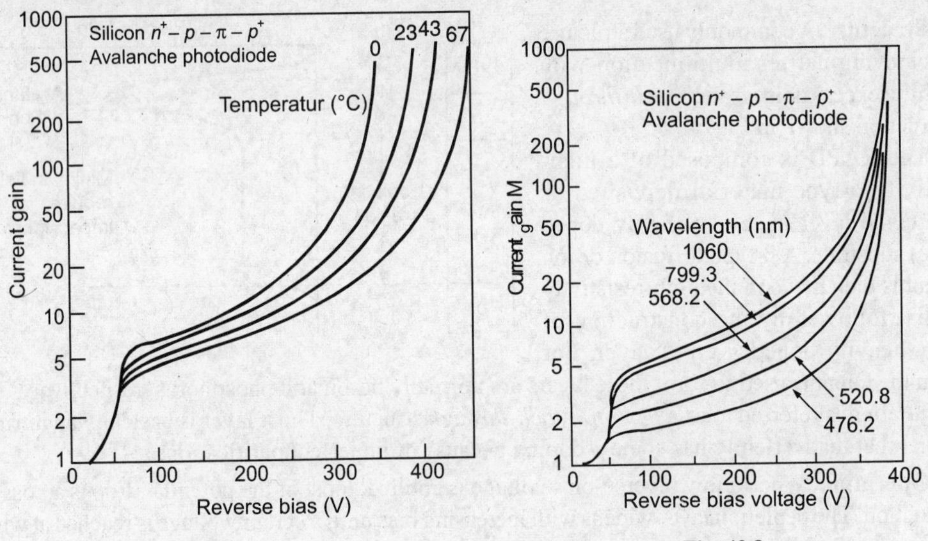

Fig. 40.7. Fig. 40.8.

Responsivity of Avalanche Photodiode

The performance of an APD is characterized by its "Responsivity" R_{APD}.

The responsivity R_{APD} is given by

$$R_{APD} = \frac{\eta e M}{h\nu} = R_o M.$$

Here, R_o is the unity gain responsivity.

Example 1. *The quantum efficiency of a particular silicon RAPD is 80% for the detection of radiation at a wavelength of 0.9 μm. When the incident optical power is 0.5 μW, the output current from the device (after avalanche gain) is 11 μA. Determine the multiplication factor of the photodiode under these conditions.*

Solution. The responsivity

$$R = \frac{\eta e\lambda}{hc} = \frac{0.8 \times (1.602 \times 10^{-19}) \times (0.9 \times 10^{-6})}{(6.626 \times 10^{-34}) \times (2.998 \times 10^{8})} = 0.581 \text{ A W}^{-1}$$

The primary photocurrent is

$$I_p = P_o R = (0.5 \times 10^{-6}) \times 0.581 = 0.291 \ \mu\text{A}.$$

The multiplication factor is

$$M = \frac{I_M}{I_p} = \frac{11 \times 10^{-6}}{0.291 \times 10^{-6}} = 37.8.$$

Thus the primary photocurrent is multiplied by a factor of 37.8.

40.8. Phototransistors

Principle. A phototransistor is capable of providing an internal gain of the photocurrent. This gain is achieved through the "transistor action".

Fig. 40.9(a) shows the symbolic representation of the n–p–n bipolar phototransistor.

The difference between a phototransistor and a conventional transistor is that in a phototransistor the base is unconnected. The base-collector junction is photosensitive and acts as a light gathering element. The absorbed light increases the base current. The amplification of the base current results in the increase of the primary photocurrent through the device.

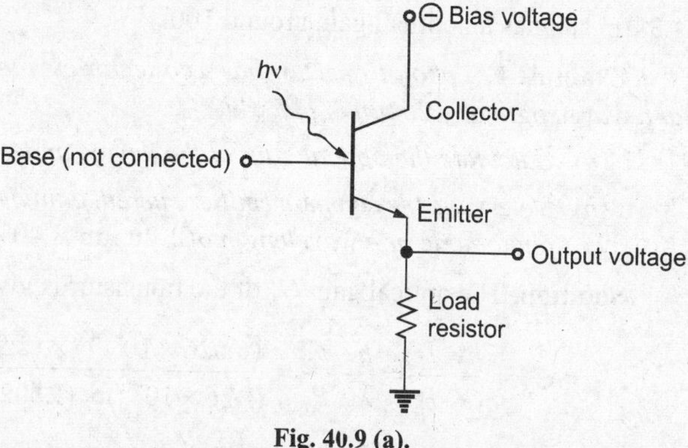

Fig. 40.9 (a).

Structure

Fig. 40.9 (b) shows the structure of a n–p–n InGaAsP/InP heterojunction phototransistor.

The three layer heterostructure is grown on an InP substrate using liquid-phase epitaxy (LPE). It consists of an n type InP collector layer followed by a thin (0.1 μm) p type InGaAsP base layer. The third layer is a wide bandgap n type InP emitter layer.

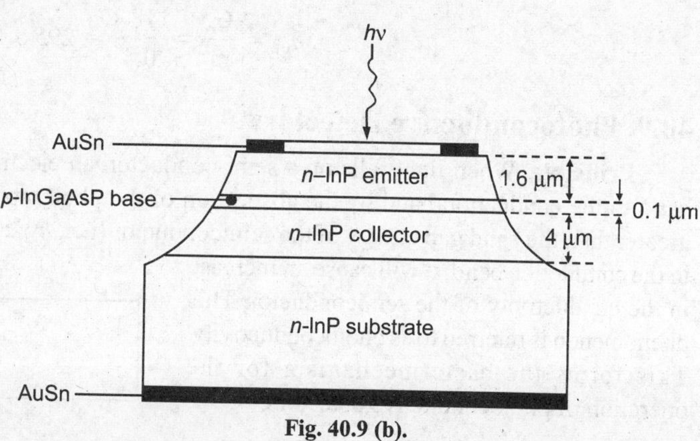

Fig. 40.9 (b).

Operation

Radiation incident on the phototransistor passes unattenuated through the wide bandgap emitter and is absorbed by the base layer, base collector depletion region and the collector. A large secondary photo current flowing between the collector and emitter is obtained as the photogenerated holes are swept into the base region increasing the forward bias on the device. The heterostructure permits a low value of capacitances between the emitter-base and the collector-base junction with a low value of base resistance. This is achieved through low emitter and collector doping levels coupled with heavy doping of the base, and allows large current gain. Moreover, the potential barrier created by the heterojunction at the emitter-base junction effectively eliminates hole injection from the base when the junction is forward biased. This gives good emitter base injection efficiency.

Optical Gain

The optical gain G_o of the device is given approximately by

$$G_0 \simeq \eta h_{FE} = \frac{h\nu}{e} \frac{I_c}{P_o} \qquad ...(1)$$

Here, η = quantum efficiency of the base-collector photodiode,

h_{FE} = common emitter current gain,

I_c = collector current,

P_o = incident optical power,

e = electronic charge and

$h\nu$ = photon energy.

The photo transistor of Fig. 40.9 (*b*) can operate over the range of wavelengths from 0.9 to 1.3 μm and has an optical gain around 100.

Example 1. *A phototransistor has a collector current of 15 mA when the incident optical power at a wavelength of 1.26 μm is 125 μW.*

(*a*) *Calculate the optical gain of the device under the above operating conditions.*

(*b*) *Determine the common emitter current gain if the quantum efficiency of the base–collector photodiode at a wavelength of* 1.26 μm *is* 40%.

Solution. The optical gain G_o of the transistor is given by

$$G_o = \frac{h\nu}{e}\frac{I_c}{P_o} = \frac{hc}{\lambda e}\frac{I_c}{P_o} = \frac{(6.626\times10^{-34})\times(2.998\times10^{8})\times(15\times10^{-3})}{(1.26\times10^{-6})\times(1.602\times10^{-19})\times(125\times10^{-6})} = 118.1$$

(*b*) The common emitter current gain is:

$$h_{FE} = \frac{G_o}{\eta} = \frac{118.1}{0.4} = 295.3$$

40.9. Photoconductive Detectors

Principle. When light falls on a semi-conductor, an electron may be raised from the valence band to the conduction band by the absorption of the photon provided the energy of the photon is greater than the band gap energy of the semiconductor (i.e., $h\nu \geq E_g$). As long as the electron remains in the conduction band, it will cause an increase in the conductivity of the semiconductor. This phenomenon is referred to as photoconductivity. This forms the basic mechanism for the operation of photoconductive detectors.

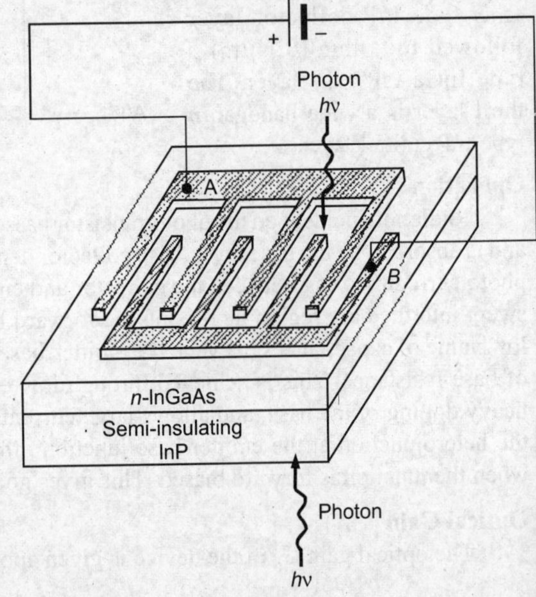

Fig. 40.10.

Structure

Fig. 40.10 shows the structure of the photoconducting device designed for operation in the longer wavelength region.

The conducting channel consists of a thin layer (1 to 2 μm) of *n*-type InGaAs which is capable of absorbing a considerable amount of incident light over the wavelengths of 1.1 to 1.6 μm. In particular, good sensitivities at reasonably high bandwidths have been achieved with a lightly doped (less than 5×10^{14} cm^{-3}) *n* type $In_{0.53}Ga_{0.47}As$ channel. Low resistance contacts are made to the conducting layer through the use of interdigital anodes and cathodes. A suitable bias voltage is applied across the contacts (A and B) to collect photo generated carriers. The optical coupling efficiency can also be improved by the application of an antireflection coating to the surface of the photoconductor facing the optical input.

Operation

The incident light on the channel region is absorbed, thereby generating additional electron-hole pairs. These photogenerated carriers increase the channel conductivity which results in an increased current in the external circuit. The optical receiver must therefore be sensitive to very small changes in resistance induced by the incident light.

Photo Conductive Gain *G*

When the fast electrons are collected at the anode, the corresponding photo generated holes are still moving across the channel. This creates an absence of electrons in the channel with a net positive charge in channel region. But the excess positive charge due to existence of holes is immediately neutralized by the injection of further electrons from the cathode to the channel. Thus further electrons may be generated from the absorption of a single photon. This mechanism creates what is known as the *photo conductive gain G.*

The photo conductive gain (G) may be defined as the ratio of the slow carrier transit time (or lifetime) t_s to the fast carrier transit time t_f. Hence:

$$G = \frac{t_s}{t_f} \quad ...(1)$$

Moreover, the photocurrent I_p produced by the photoconductor can be written as

$$I_p = \frac{\eta P_o e}{h\nu} G \quad ...(2)$$

Here, η = device quantum efficiency,
P_0 = incident optical power and
e = charge on an electron.

The current response in the photoconductor remains for a time t_s after the end of an incident optical pulse and exhibits an exponential decay with a time constant equal to this slow carrier transit time (or lifetime). So the maximum 3 dB bandwidth B_m for the device is given by:

$$B_m = \frac{1}{2\pi t_s} \quad ...(3)$$

The gain-bandwidth product of a photoconductor is

$$G \,.\, B_m = \frac{1}{2\pi t_f} \quad ...(4)$$

Eq. (4) shows that when t_f is fixed, photoconductive gain can only be obtained at the expense of the maximum bandwidth permitted by the device. There is therefore a compromise between gain (sensitivity) and the speed of response.

Noise

The sensitivity of a photoconductive detector is limited by the noise generated within the device, even though the quantum efficiency may be quite high. Noise arises from two sources:

(*i*) Johnson noise associated with the thermal noise from the bulk resistance of the photoconductor slab. This noise source, which is often dominant, results in a finite dark conductivity for the device which generates a randomly varying background dark current.

(*ii*) Generation–recombination noise caused by fluctuations in the generation and recombination rates of the photogenerated carrier pairs.

The signal to noise ratio *S/N* can be expressed as

$$\frac{S}{N} = \frac{\text{Conductivity when light falls}}{\text{Conductivity when no light falls}} = \frac{\text{(Conductivity) light}}{\text{(Conductivity) dark}}.$$

Thus in order to get a high value of *S/N,* the dark conductivity has to be reduced.

Example 1. *The electron transit time in an InGaAs photoconductive detector is 5 ps. Determine the maximum 3 dB bandwidth permitted by the device when its photoconductive gain is 70.*

Solution. $$B_{\text{m}} = \frac{1}{2\pi t_{\text{f}} G} = \frac{1}{2\pi \times (5 \times 10^{-12}) \times 70} = 454.7 \text{ MHz.}$$

EXERCISE

1. Explain the detection process in a *p–n* photodiode.
2. Define the quantum efficiency and the responsivity of a photodetector.
   Derive an expression for the responsivity of an intrinsic photodetector in terms of the quantum efficiency of the device and the wavelength of the incident radiation.
3. Explain what is meant by the long wavelength cutoff point for an intrinsic photodetector, deriving any relevant expressions.
4. Describe with relevant diagram the operation of the silicon RAPD.
5. Describe with necessary diagram the working principle of a phototransistor and give an expression for gain *G*.

CHAPTER

41 FIBRE OPTIC SENSORS

41.1. Measurement of Electric Field or Voltage Using Fibre Optic Sensor

Principle. Pockels Electro-optic Effect

When certain kind of birefringent crystals are placed in an electric field, their indices of refraction are altered by the presence of the field.

The variation of refractive index with respect to the electric field 'E' is written as

$$\frac{1}{n^2} = \frac{1}{n_0^2} + rE.$$

Here, n_0 = refractive index before the application of electric field,

r = linear electro-optic coefficient.

This effect is called Pockels effect.

For example potassium dihydrogen phosphate, KH_2PO_4 (KDP), crystal exhibits Pockels effect. In this crystal, when we apply the electric field or voltage along Z-axis, the light which is linearly polarised at an angle 45° with respect to X-axis undergoes a phase shift or phase retardation. The phase retardation 'Γ' is given by

$$\Gamma = \frac{2\pi n_0^3 \, r \, V_z}{\lambda}.$$

Here, r = electro-optic coefficient of the crystal,

V_z = applied voltage = $E_z \times l$,

l = length of the crystal,

E_z = applied electric field,

λ = wavelength of light.

Let I_0 be the incident light intensity.

The intensity of the transmitted light through the crystal is given by

$$I = I_0 \sin^2\left(\frac{\Gamma}{2}\right).$$

Thus the phase shift produced in the linearly polarised wave is directly proportional to the applied electric field or voltage.

Construction. Fig. 41.1 shows the optical assembly to measure the applied voltage or the applied electric field across the Pockels cell.

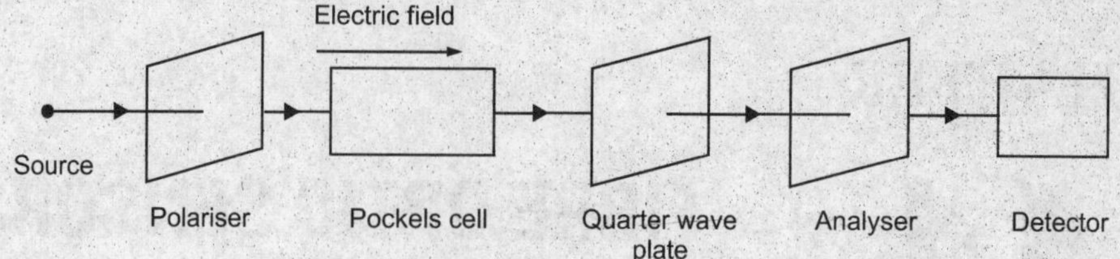

Fig. 41.1.

Working: (*i*) The polariser converts the incident ordinary light into a linearly polarised light.

(*ii*) When there is applied voltage across the Pockels cell, a phase shift is produced for the transmitted polarised beam.

(*iii*) The quarter wave plate produces a phase shift of $\frac{\pi}{2}$.

(*iv*) The transmitted light is analysed through an analyser.

The transmitted light intensity can be written as

$$I = \frac{1}{2} I_0 \left(1 + \sin \frac{V}{V_0}\right)$$

Here, I_0 = incident light intensity before transmission through the Pockels cell,

V_0 = halfwave voltage of the crystal *i.e.,* the value of the applied voltage which produces a phase shift of π radians.

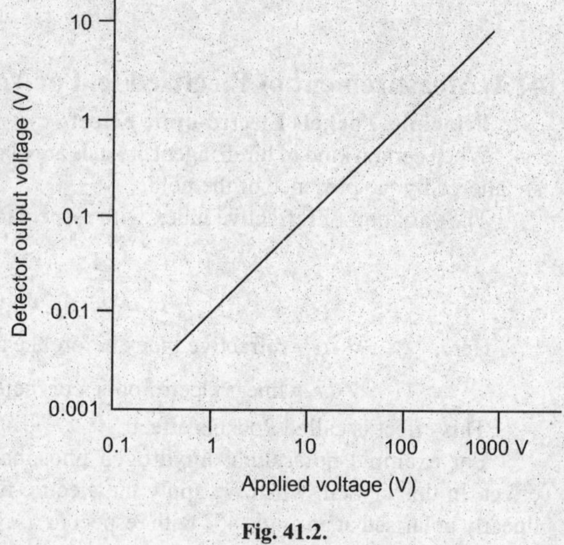

Fig. 41.2.

Fig. 41.2 shows the response curve of this Pockels cell sensor, made up of $BI_{12}Ge_{20}$ crystal. Even upto 1000 V, we get the linear response from this voltage sensor.

41.2. Intensity Modulated Temperature Sensor

Construction: Fig. 41.3 shows the intensity modulated temperature sensor based on reflective concept.

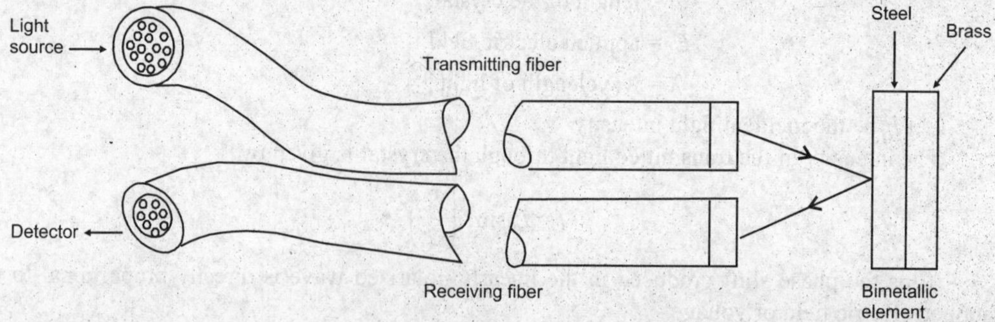

Fig. 41.3.

Here the bimetallic strip acts as the sensing element. It consists of steel and brass which are welded together to form a strip. The linear exansivity of brass is more than that of steel. So the strip bends into an arc when heated . The metal with higher linear expansivity will be on the outside of the arc when the strip is heated. The strip is attached to a bifurcated reflective fiber optic probe.

Working: The strip is designed to move continuously. The movement of the strip is directly proportional to temperature. The amount of the reflected light is converted into voltage by a photodiode.

Fig. 41.4 shows the linear response of the temperature sensor. The amount of light reflected decreases with the increase of temperature. So the output of the photodiode decreases with the increase of temperature.

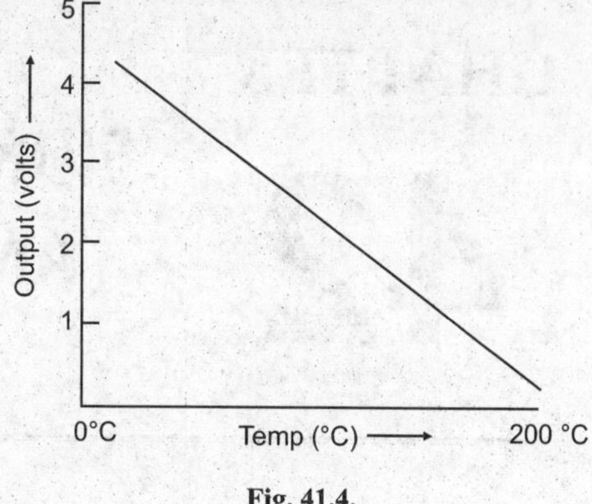

Fig. 41.4.

41.3. Phase Modulated Temperature Sensor

Principle: In the interferometric temperature sensor, the phase shift produced in the sensing fibre relative to the reference fiber is a function of temperature, *i.e.*,

$$\frac{\Delta\phi}{\Delta T} = \frac{2\pi L}{\lambda}\frac{\Delta n}{\Delta T}$$

Here, L is the fiber length and n is the refractive index.

Construction. Fig. 41.5 shows the Mach-Zehnder interferometer.

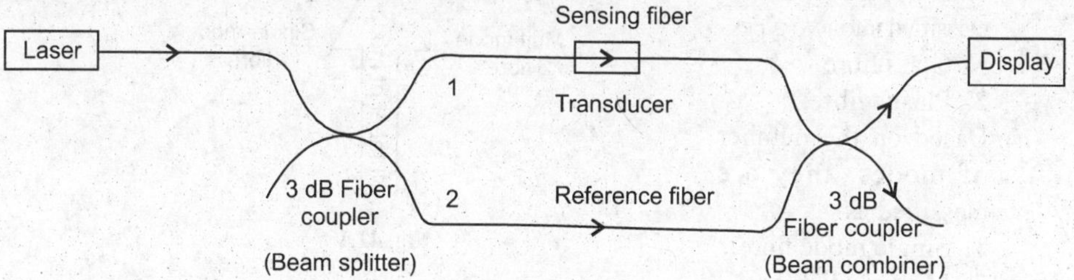

Fig. 41.5.

The light from a semiconductor laser is allowed to pass through a beam splitter. A 3 dB coupler acts as the beam splitter which sends the light through the sensing fiber (beam 1) and the reference fiber (beam 2). Another 3dB coupler acts as the combiner for these two beams. On the display screen, a series of bright and dark fringes are formed when the light from the two fibers interferes.

Working. A change in the phase of the light in the sensing fiber due to change of refractive index with respect to temperature with reference to the phase of the light in the reference fiber appears as a displacement of the fringe pattern. A phase change of 2 π radians causes a displacement of one fringe. By counting the fringe displacement, we can determine the magnitude of temperature.

Taking $\lambda = 0.633$ μm and $\frac{\Delta n}{\Delta T} = 10^{-6}$/K, we expect 2 fringes per Kelvin per metre length of the fiber. By increasing the length of the fiber, we can increase the sensitivity of the temperature sensor. Further a very small movement of fringe of the order of 10^{-13} m can be detected. So we can get high sensitivity.

CHAPTER

42 FIBRE CHARACTERISTICS AND CLASSIFICATION

42.1. Fibres: Classification

Optical fibres are classified into three major categories (Fig. 42.1) based on

(*i*) the type of material used;

(*ii*) the number of modes;

(*iii*) the refractive index profile.

(*i*) Based on the type of material used, they are classified into two types.
1. Glass fibre.
2. Plastic fibre.

(*ii*) Based on the number of modes, they are classified as:
1. Single mode fibre.
2. Multimode fibre.

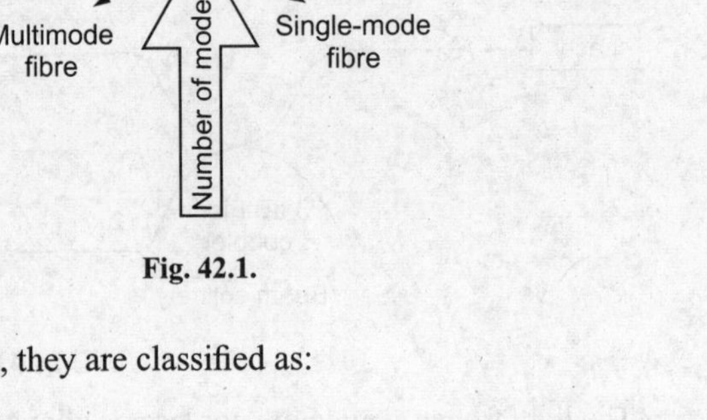

Fig. 42.1.

(*iii*) Based on the refractive index profile, they are classified as:
1. Step-index fibre.
2. Graded-index fibre.

Fig. 42.2 shows a single-fiber structure. The *core* of the fiber is a single solid dielectric cylinder of radius *a* and index of refraction n_1. The core is surrounded by a solid dielectric *cladding* having a refractive index n_2 that is less than n_1. The cladding reduces scattering loss resulting from dielectric discontinuities at the core surface, it adds mechanical strength to the fiber, and it protects the core from absorbing surface contaminants with which it could come in contact. An elastic, abrasion-resistant plastic buffer encapsulates the fiber.

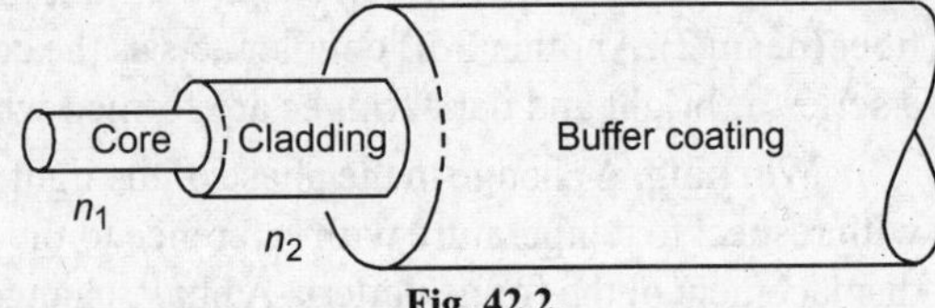

Fig. 42.2.

42.2. Fibre Materials

In selecting materials for optical fibers, the following requirements must be satisfied.

1. It must be possible to make long, thin, flexible fibers from the material.
2. The material must be transparent at a particular optical wavelength in order for the fiber to guide light efficiently.

3. Physically compatible materials having slightly different refractive indices for the core and cladding must be available.

Materials satisfying these requirements are glasses and plastics.

The majority of fibers are made of glass consisting either of silica (SiO_2) or a silicate. Plastic fibers are less widely used because of their substantially higher attenuation than glass fibers.

42.3. Single-mode Fibers

Single-mode fibers can have either a step index or graded index profile.

Fig. 42.3 shows the structure of a single-mode step index fiber.

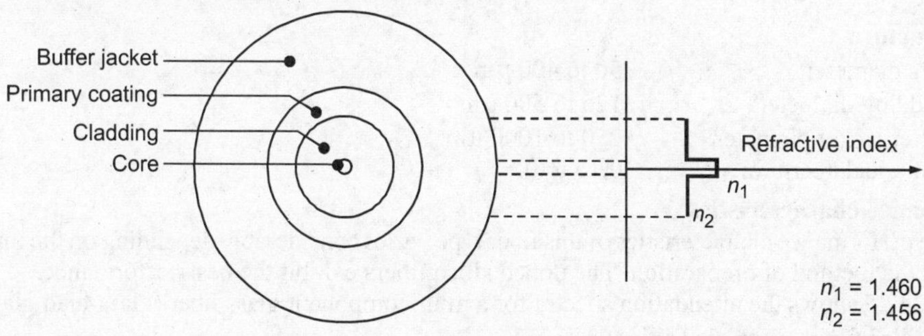

Fig. 42.3.

Structure

Core diameter	5 to 10 μm, typically around 8.5 μm
Cladding diameter:	generally 125 μm
Buffer jacket diameter:	250 to 1000 μm
Numerical aperture:	0.08 to 0.15, usually around 0.10.

Single-mode fibers have small core diameters to allow single-mode propagation.

The cladding diameter must be at least ten times the core diameter to avoid losses from the evanescent field. Single-mode step index fibres are high quality fibers for wideband, long-haul transmission. They are generally fabricated from doped silica (silica-clad silica) in order to reduce attenuation.

Bandwidth: Greater than 500 MHz km.

Applications: These fibers are ideally suited for high bandwidth very long-haul applications using single-mode injection laser sources.

The advantage of a singlemode step index fiber is that it has low value of intermodal dispersion because only one mode is transmitted.

42.4. Multimode Step Index Fibers

Multimode step index fibres are fabricated from either multicomponent glass compounds or doped silica. These fibres have large core diameters and large numerical apertures to facilitate efficient coupling to incoherent light sources such as light emitting diodes (LEDs).

Fig. 42.4 shows the structure of a glass multimode step index fiber.

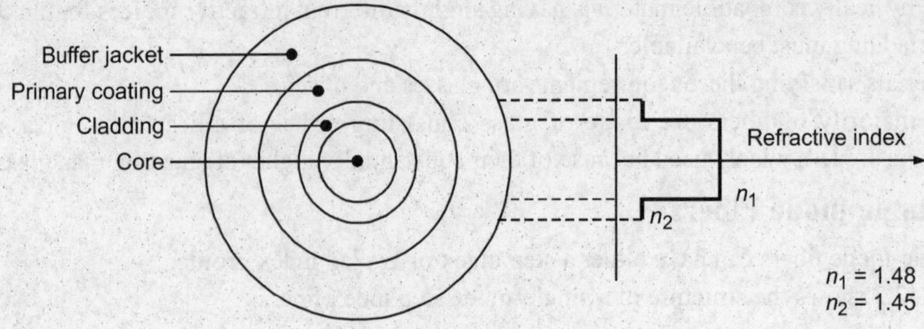

Fig. 42.4.

Structure

Core diameter:	50 to 400 μm
Cladding diameter:	125 to 500 μm
Buffer jacket diameter:	250 to 1000 μm
Numerical aperture:	0.16 to 0.5.

Performance characteristics

The performance characteristics of this fiber type varies considerably depending on the materials used and the method of preparation. The doped silica fibers exhibit the best performance.

Fig. 42.5 shows the attenuation spectra for a multicomponent glass fiber (glass-clad glass) and a doped silica fiber (silica-clad silica).

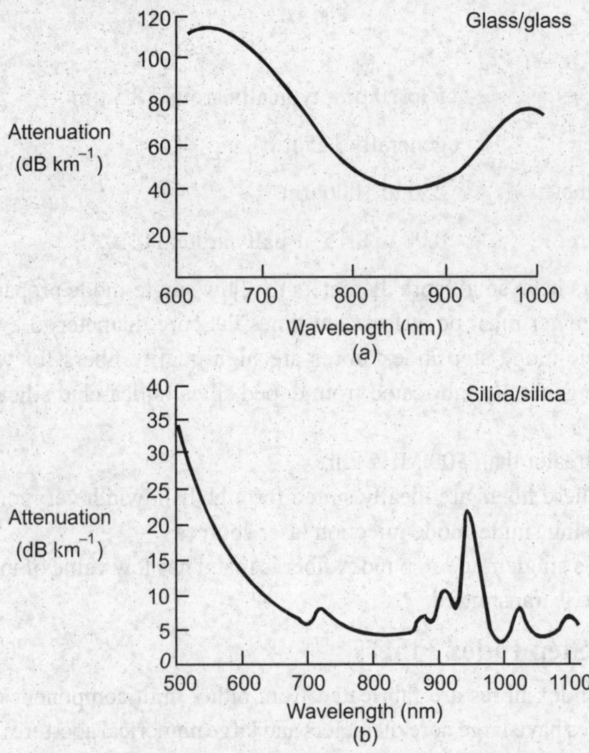

Fig. 42.5.

Bandwidth: 6 to 50 MHz km.

Applications: These fibers are best suited for short-haul, limited bandwidth and relatively low cost applications.

42.5. Multimode Graded Index Fibers

These fibers are fabricated using multicomponent glasses or doped silica. They are manufactured from materials with higher purity than the majority of multimode step index fibers in order to reduce fiber losses. Fig. 42.6 shows the structure for a glass multimode graded index fiber.

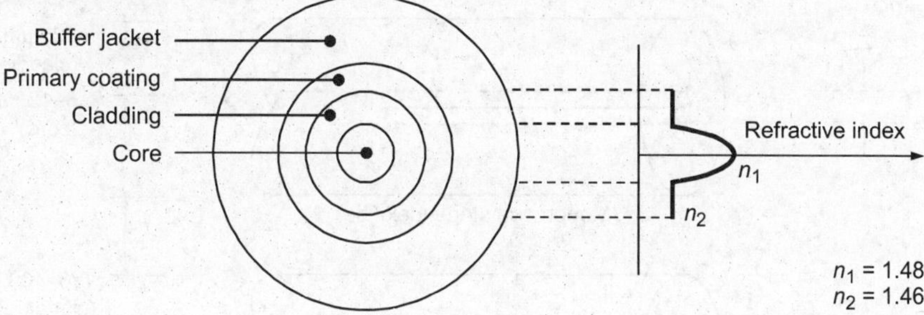

Fig. 42.6.

Structure

Core diameter:	30 to 100 μm.
Cladding diameter:	100 to 150 μm,
Buffer jacket diameter:	250 to 1000 μm.
Numerical aperture:	0.2 to 0.3.

Performance characteristics

The performance characteristics of multimode graded index fibers are better than those for multimode step index fibers due to the index grading and lower attenuation.

Bandwidth: 300 MHz km to 3 GHz km.

Applications: These fibers are best suited for medium-haul, medium to high bandwidth applications using incoherent and coherent multimode sources (i.e., LEDs and injection lasers respectively).

42.6. Comparison of Single-mode and Multimode Step-index and Graded-index Optical Fibers

(*i*) In the *step-index fiber,* the refractive index of the core is uniform throughout and undergoes an *abrupt* change (or step) at the cladding boundary.

In the *graded-index fiber*, the core refractive index is made to vary as a function of the radial distance from the center of the fiber.

Fig. 42.7 gives typical sizes of single-and multimode fibers, with ray paths.

A single-mode fiber sustains only one mode of propagation.

Multimode fibers contain many hundreds of modes.

Advantages of Multimode fibers over Single-mode fibres

1. The larger core radii of multimode fibers make it easier to launch optical power into the fiber and facilitate the connecting together of similar fibers.
2. Light can be launched into a multimode fiber using a light-emitting-diode (LED) source, whereas single-mode fibers must be excited with laser diodes.

A **disadvantage** of multimode fibers is that they suffer from *intermodal dispersion*

Intermodal dispersion can be described as follows. When an optical pulse is launched into a fiber, the optical power in the pulse is distributed over all (or most) of the modes of the fiber. Each of the modes that can propagate in a multimode fiber travels at a slightly different velocity. This means that the modes in a given optical pulse arrive at the fiber end at slightly different times, thus causing the pulse to spread out in time as it travels along the fiber. This effect is known as *intermodal dispersion.*

It can be reduced by using a graded-index profile in the fiber core. This allows graded-index fibers to have much larger band-widths (data rate transmission capabilities) than step-index fibers. Even higher bandwidths are possible in single-mode fibers, where intermodal dispersion effects are not present.

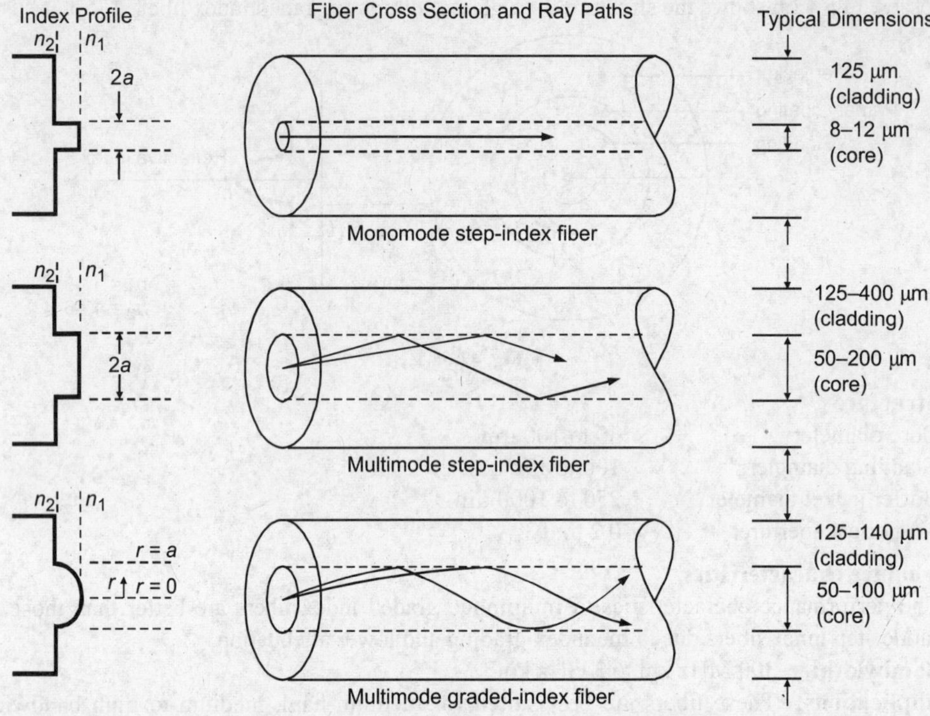

Fig. 42.7.

42.7. Plastic-clad Fibers

These fibers are composed of silica cores with the lower-refractive-index cladding being a polymer (plastic) material. These fibers are often referred to as PCS (plastic-clad silica) fibers.

Optical fibers constructed with glass cores and glass claddings are very important for long-distance applications where the very low losses achievable in these fibers are needed. For short-distance applications (up to several hundred meters), where higher losses are tolerable, the less expensive plastic-clad silica fibers can be used.

Fig. 42.8 shows the structure for a plastic-clad silica multimode step index fiber.

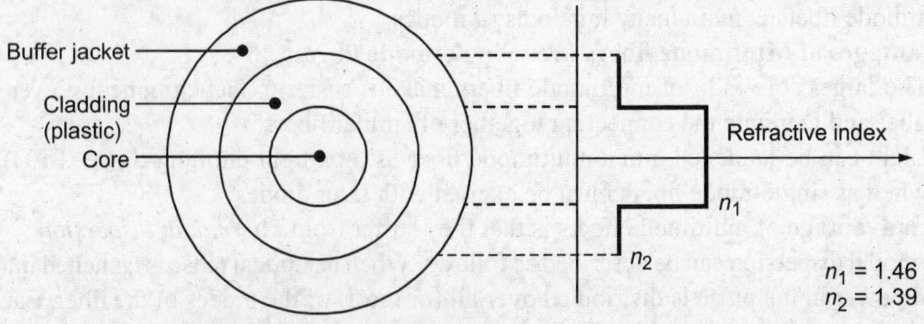

Fig. 42.8.

Structure

Core diameter:	Step index	100 to 500 μm
	Graded index	50 to 100 μm
Cladding diameter:	Step index	300 to 800 μm
	Graded index	125 to 150 μm
Buffer jacket diameter:	Step index	500 to 1000 μm
	Graded index	250 to 1000 μm
Numerical aperture:	Step index	0.2 to 0.5
	Graded index	0.2 to 0.3.

Performance characteristics

Attenuation:	Step index	5 to 50 dB km^{-1}
	Graded index	4 to 15 dB km^{-1}

The PCS fibers exhibit lower radiation-induced losses than silica-clad silica fibers and, therefore, have an improved performance in certain environments. Plastic-clad fibers are generally slightly cheaper than the corresponding glass fibers, but usually have more limited performance characteristics.

42.8. All-plastic Fibers

All-plastic or polymeric fibers are exclusively of the multimode step index type with large core and cladding diameters. Hence there is a reduced requirement for a buffer jacket for fiber protection and strengthening. These fibers are usually cheaper to produce and easier to handle than the corresponding silica-based glass variety.

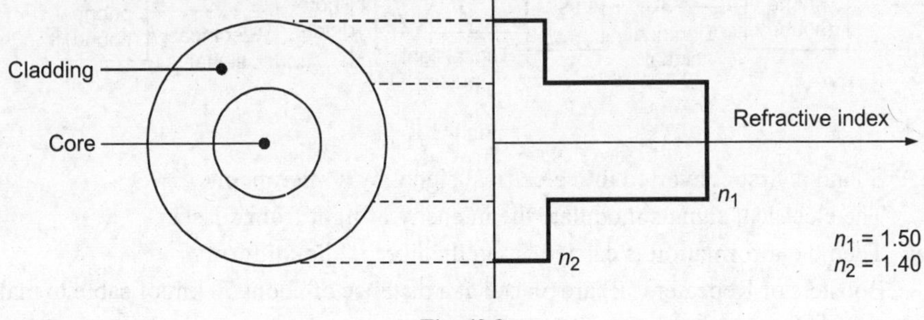

Fig. 42.9.

Fig. 42.9 shows the structure for an all-plastic fiber.

Structure

Core diameter:	200 to 600 μm
Cladding diameter:	450 to 1000 μm
Numerical aperture:	0.5 to 0.6.

Performance characteristics

Attenuation: 50 to 1000 dB km^{-1} at a wavelength of 0.65 μm.

Applications: All-plastic multimode step-index fibers are used for fairly short (up to about 100 m) and low-cost links. Although they exhibit considerably greater optical signal attenuations than glass fibers, the toughness and durability of plastic allow these fibers to be handled without special care.

EXERCISE

1. List thee various optical fibre types currently on the market indicating their important features. Hence, briefly describe the general areas of application for each type.
2. Describe the structures of different types of optical fibers with ray paths. What is the approximate diameter of an optical fiber in each case?
3. What are the advantages of cladding on a fiber?
4. What do you mean by single mode and multimode fibers?

CHAPTER

43 OPTICAL FIBRE AS A CYLINDRICAL WAVEGUIDE

43.1. Optical Fibre as an Optical Waveguide

Optical fibres are used as dielectric waveguides for electromagnetic signals at optical frequencies. Fig. 43.1 shows the block diagram of transmission of sound along an optical fibre and conversion again to sound at the other end.

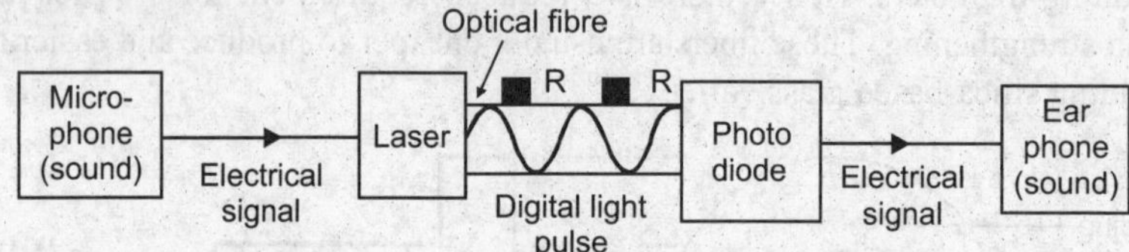

Fig. 43.1.

(*i*) Sound is first converted into electrical signal by a microphone.

(*ii*) The electrical signals modulate the intensity of light from a laser.

(*iii*) Then the information is carried along the fibre in digital form.
Boosters or Repeaters (R) are placed at a distance of about 50 km of cable to make up the signal losses occurring due to scattering and absorption.

(*iv*) At the receiving place, a photodiode converts the digital light pulses into corresponding electrical signals.

(*v*) The electrical signals are then converted into sound by an earphone (receiver).

Time division multiplexing system is used to transmit many thousands of telephone calls through a single optical fibre with the use of digital pulses.

43.2. Mode Theory for Circular Waveguides

To attain a more detailed understanding of the optical power propagation mechanism in a fiber, it is necessary to solve Maxwell's equations subject to the cylindrical boundary conditions of the fiber.

Maxwell's Equations

Assume that the fibre (waveguide) is made up of linear, isotropic dielectric material having no currents and free charges. Maxwell's equations are:

$$\nabla \times \mathbf{E} = -\frac{\partial \mathbf{B}}{\partial t} \qquad ...(1)$$

$$\nabla \times \mathbf{H} = \frac{\partial \mathbf{D}}{\partial t} \qquad ...(2)$$

$$\nabla \cdot \mathbf{D} = 0 \quad \text{...(3)}$$

$$\nabla \cdot \mathbf{B} = 0 \quad \text{...(4)}$$

Here, $\mathbf{D} = \varepsilon\mathbf{E}$ and $\mathbf{B} = \mu\mathbf{H}$.

The parameter ε is the permittivity (or dielectric constant) and μ is permeability of the medium. Taking the curl of Eq. (1) and making use of Eq. (2) yields

$$\nabla \times (\nabla \times \mathbf{E}) = -\mu \frac{\partial}{\partial t}(\nabla \times \mathbf{H}) = -\varepsilon\mu \frac{\partial^2 \mathbf{E}}{\partial t^2} \quad \text{...(5)}$$

Using the vector identity

$$\nabla \times (\nabla \times \mathbf{E}) = \nabla(\nabla \cdot \mathbf{E}) - \nabla^2\mathbf{E}$$

and using Eq. (3) (that is, $\nabla \cdot \mathbf{E} = 0$), Eq. (5) becomes

$$\nabla^2\mathbf{E} = \varepsilon\mu \frac{\partial^2 \mathbf{E}}{\partial t^2} \quad \text{...(6)}$$

Similarly, by taking the curl of Eq. (2), it can be shown that

$$\nabla^2\mathbf{H} = \varepsilon\mu \frac{\partial^2 \mathbf{H}}{\partial t^2} \quad \text{...(7)}$$

Equations (6) and (7) are the standard *wave equations*.

Waveguide Equations

Consider the propagation of electromagnetic waves along a cylindrical fiber (Fig. 43.2).

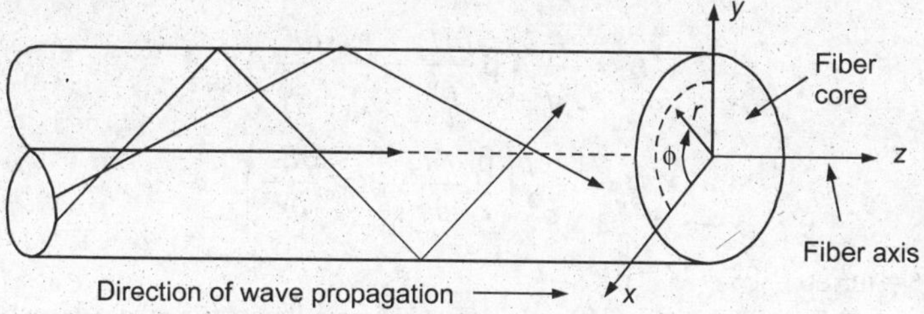

Fig. 43.2.

For this fiber, a cylindrical coordinate system (r, ϕ, z) is defined with the z axis lying along the axis of the waveguide.

When the electromagnetic wave propagates along the direction of Z-axis, the electric and magnetic field vectors have components (E_r, E_ϕ and E_z) and (H_r, H_ϕ and H_z) respectively. They have the functional dependence of the form:

$$\mathbf{E} = \mathbf{E}_0 (r, \phi)\, e^{j(\omega t - \beta z)} \quad \text{...(8)}$$

$$\mathbf{H} = \mathbf{H}_0 (r, \phi)\, e^{j(\omega t - \beta z)} \quad \text{...(9)}$$

which are harmonic in time t and coordinate z.

The parameter β is the z component of the propagation vector.

When Eqs. (8) and (9) are substituted into Maxwell's curl equations, we have, from Eq. (1),

$$\frac{1}{r}\left(\frac{\partial E_z}{\partial \phi} + jr\beta E_\phi\right) = -j\omega\mu\, H_r, \quad \text{...(10)}$$

$$j\beta E_r + \frac{\partial E_z}{\partial r} = j\omega\mu\, H_\phi, \quad \text{...(11)}$$

$$\frac{1}{r}\left[\frac{\partial}{\partial r}(rE_\phi) - \frac{\partial E_r}{\partial \phi}\right] = -j\mu\omega\, H_z, \quad \text{...(12)}$$

and from Eq. (2),

$$\frac{1}{r}\left(\frac{\partial H_z}{\partial \phi} + jr\beta H_\phi\right) = j\varepsilon\omega\, E_r, \quad \text{...(13)}$$

$$j\beta H_r + \frac{\partial H_z}{\partial r} = -j\varepsilon\omega E_\phi, \quad \text{...(14)}$$

$$\frac{1}{r}\left[\frac{\partial}{\partial r}(rH_\phi) - \frac{\partial H_z}{\partial \phi}\right] = -j\varepsilon\omega\, E_z, \quad \text{...(15)}$$

By eliminating variables these equations can be rewritten such that, when E_z and H_z are known, the remaining transverse components E_r, E_ϕ, H_r and H_ϕ can be determined.

For example, E_ϕ or H_r can be eliminated from Eqs. (10) and (14) so that the component H_ϕ or E_r, respectively, can be found in terms of E_z or H_z.

$$E_r = -\frac{j}{q^2}\left(\beta\frac{\partial E_z}{\partial r} + \frac{\mu\omega}{r}\frac{\partial H_z}{\partial \phi}\right) \quad \text{...(16)}$$

$$E_\phi = -\frac{j}{q^2}\left(\frac{\beta}{r}\frac{\partial E_z}{\partial \phi} - \mu\omega\frac{\partial H_z}{\partial r}\right) \quad \text{...(17)}$$

$$H_r = \frac{-j}{q^2}\left(\beta\frac{\partial H_z}{\partial r} - \frac{\omega\varepsilon}{r}\frac{\partial E_z}{\partial \phi}\right) \quad \text{...(18)}$$

$$H_\phi = \frac{-j}{q^2}\left(\frac{\beta}{r}\frac{\partial H_z}{\partial \phi} + \omega\varepsilon\frac{\partial E_z}{\partial r}\right) \quad \text{...(19)}$$

Here, $q^2 = \omega^2\,\varepsilon\mu - \beta^2 = k^2 - \beta^2$.

Substitution of Eqs. (18) and (19) into Eq. (15) results in the wave equation in cylindrical coordinates

$$\frac{\partial^2 E_z}{\partial r^2} + \frac{1}{r}\frac{\partial E_z}{\partial r} + \frac{1}{r^2}\frac{\partial^2 E_z}{\partial \phi^2} + q^2\, E_z = 0 \quad \text{...(20)}$$

Substituting Eqs. (16) and (17) into Eq. (12), we get

$$\frac{\partial^2 H_z}{\partial r^2} + \frac{1}{r}\frac{\partial H_z}{\partial r} + \frac{1}{r^2}\frac{\partial^2 H_z}{\partial \phi^2} + q^2\, H_z = 0 \quad \text{...(21)}$$

Eqs. (20) and (21) each contain either only E_z or H_z.

Eqs. (20) and (21) are wave equations for E_z and H_z respectively.

Different Modes of Propagation of Electromagnetic Waves

The electromagnetic waves propagate in different modes.

These are:

1. Transverse electric (TE) mode
2. Transverse magnetic (TM) mode
3. Hybrid EH mode
4. Hybrid HE mode.

1. TE Mode

When $E_z = 0$, the modes are called *transverse electric* or TE modes.

In the *TE* mode, the electric field vector lies entirely in transverse plane at right angle to the direction of propagation (which is Z-direction in this case) and the magnetic field vector has a component in the direction of propagation (Z-direction). Under this condition $E_z = 0$ but H_z is finite. The propagation takes place with the group velocity v_g, while the *E-H* plane moves at an angle normal to itself with the speed of light (Fig. 43.3).

2. TM Mode

When $H_z = 0$, the modes are called *transverse magnetic* or TM modes. In the transverse magnetic (TM) mode, the magnetic vector lies entirely in transverse plane (in a plane right angle to Z-direction in this example), while electric field components have both transverse component E_t and longitudinal component E_z so that $H_z = 0$ and E_z is finite. The propagation takes place with the group velocity v_g, while *E-H* wavefront travels at an angle normal to itself in the *Y-Z* plane with the speed of light (Fig. 43.4).

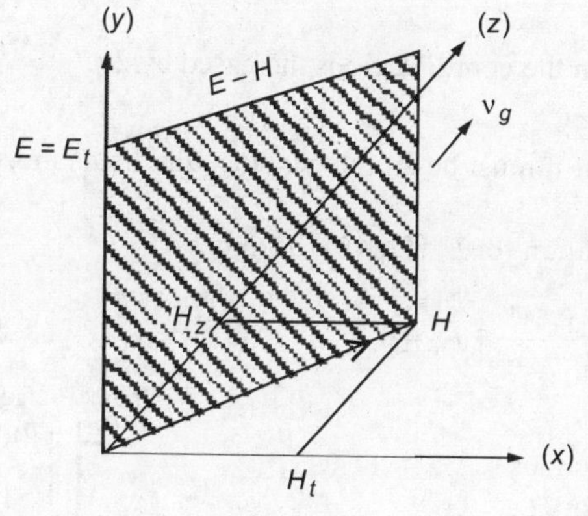

Fig. 43.3. **Fig. 43.4.**

3. Hybrid Modes: (EH and HE Modes)

Hybrid modes exist if both E_z and H_z are nonzero. These are designated as HE or EH modes, depending on whether H_z or E_z, respectively, makes a larger contribution to the transverse field.

(*i*) If $E_z > H_z$, the mode is called the *E-H* mode (Fig. 43.5).

(*ii*) If $H_z > E_z$, the mode is called the *HE* mode (Fig. 43.6).

The fact the hybrid modes are present in optical waveguides makes their analysis more complex than in the simpler case of hollow metallic waveguides where only TE and TM modes are found.

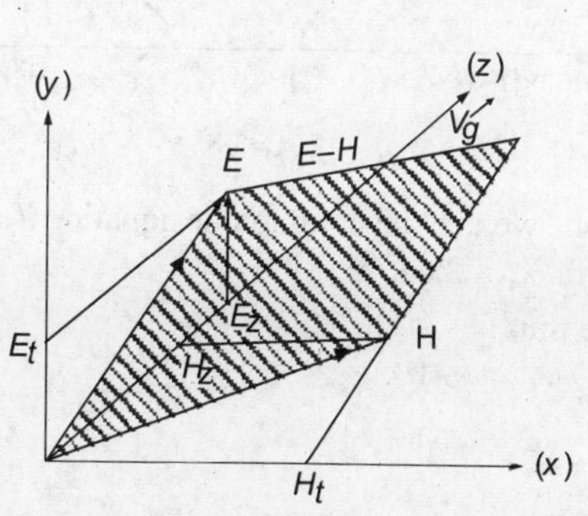

Fig. 43.5.

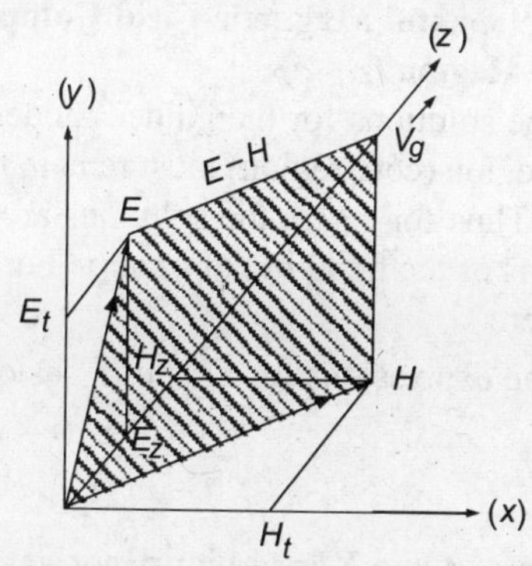

Fig. 43.6.

43.3. Wave Equations for Step-Index Fibers

The wave equations for E_z and H_z are

$$\frac{\partial^2 E_z}{\partial r^2} + \frac{1}{r}\frac{\partial E_z}{\partial r} + \frac{1}{r^2}\frac{\partial^2 E_z}{\partial \phi^2} + q^2 E_z = 0 \qquad ...(1)$$

$$\frac{\partial^2 H_z}{\partial r^2} + \frac{1}{r}\frac{\partial H_z}{\partial r} + \frac{1}{r^2}\frac{\partial^2 H_z}{\partial \phi^2} + q^2 H_z = 0 \qquad ...(2)$$

We solve Eq. (1) by the separation-of-variables method.

We assume a solution of the form

$$E_z = AF_1(r)\, F_2(\phi)\, F_3(z)\, F_4(t) \qquad ...(3)$$

The wave is sinusoidal in time and propagates in the z direction.

So the time-and z-dependent factors are given by

$$F_3(z)\, F_4(t) = e^{j(\omega t - \beta z)} \qquad ...(4)$$

The waveguide has circular symmetry.

So each field component must not change when the coordinate ϕ is increased by 2π.

$$\therefore \qquad F_2(\phi) = e^{j\nu\phi} \qquad ...(5)$$

The constant ν can be positive or negative, but it must be an integer since the fields must be periodic in ϕ with a period of 2π.

Substituting Eq. (5) into Eq. (3), the wave equation for E_z [Eq. (1)] becomes

$$\frac{\partial^2 F_1}{\partial r^2} + \frac{1}{r}\frac{\partial F_1}{\partial r} + \left(q^2 - \frac{\nu^2}{r^2}\right) F_1 = 0 \qquad ...(6)$$

Eq. (16) is differential equation for Bessel functions.

An exactly identical equation can be derived for H_z.

The step-index fiber has a homogeneous core of refractive index n_1 and radius a, which is surrounded by an infinite cladding of index n_2 (Fig. 43.7).

Fig. 43.7.

Fig. 43.8 shows the electric field distributions for several of the lower-order guided modes in a symmetrical slab waveguide.

The fields vary harmonically in the guiding region of refractive index n_1 and decay exponentially outside of this region.

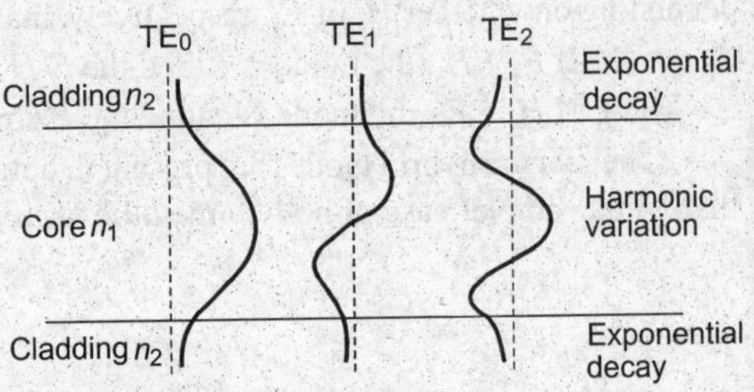

Fig. 43.8.

Eq. (6) must now be solved for the regions inside and outside the core.

(i) Electric and Magnetic Field Components in Core Region ($r < a$)

The solutions for the guided modes in the inside region (core region) must remain finite as $r \to 0$. Thus for $r < a$, the solutions are Bessel functions of the first kind of order ν. For these functions we use the common designation $J_\nu(ur)$.

Here $\qquad u^2 = k_1^2 - \beta^2$ with $k_1 = 2\pi n_1/\lambda$.

The expressions for E_z and H_z inside the core are thus

$$E_z(r < a) = AJ_\nu(ur)\, e^{j\nu\phi}\, e^{j(\omega t - \beta z)} \qquad ...(7)$$

$$H_z(r < a) = BJ_\nu(ur)\, e^{j\nu\phi}\, e^{j(\omega t - \beta z)} \qquad ...(8)$$

Here, A and B are arbitrary constants.

$J_\nu(ur)$ is an oscillating function.

(*ii*) Electric and Magnetic Field Components in Cladding Region ($r > a$)

The solutions for the guided modes in the cladding (outside the core) must decay to zero as $r \to \infty$.

Outside of the core, the solutions to Eq. (6) are given by modified Bessel functions of the second kind $K_\nu (wr)$, where $w^2 = \beta^2 - k_2^2$ with $k_2 = 2\pi n_2/\lambda$.

The expressions for E_z and H_z outside the core are

$$E_z (r > a) = CK_\nu (wr)\, e^{j\nu\phi}\, e^{j(\omega t - \beta z)} \quad ...(9)$$

$$H_z (r > a) = DK_\nu (wr)\, e^{j\nu\phi}\, e^{j(\omega t - \beta z)} \quad ...(10)$$

Here, C and D are arbitrary constants.

$K_\nu (wr)$ behaves like decaying exponential function.

Permissible range of β for Bound Solutions

First Condition. It is evident from the definition of the modified Bessel function that

$$K_\nu (wr) \to e^{-wr} \text{ when } wr \to \infty.$$

$K_\nu (wr)$ must go to zero when $r \to \infty$.

$$\therefore \qquad w > 0.$$

This indicates that $\beta \geq k_2$, which represents a cutoff condition.

The *cutoff condition* is the point at which a mode is no longer bound to the core region.

Second Condition. A second condition on β can be deduced from the behavior of $J_\nu(ur)$. Inside the core the parameter u must be real for F_1 to be real.

$$\therefore \qquad k_1 \geq \beta.$$

So for bound solutions, the allowed range of β is given by

$$n_2 k = k_2 \leq \beta \leq k_1 = n_1 k \quad ...(11)$$

Here, $k = 2\pi/\lambda$ is called the "free space propagation constant".

43.4. Modal Equations

The solutions for β must be determined from the boundary conditions.

The boundary conditions are:

(1) the tangential components E_ϕ and E_z of inside and outside the dielectric interface at $r = a$ must be the same.

(2) the tangential components H_ϕ and H_z of H inside and outside the dielectric interface at $r = a$ must be the same.

(*i*) The boundary conditions require that the tangential components E_ϕ and E_z of **E** inside and outside of the dielectric interface at $r = a$ must be the same.

The expression for E_z inside the core is

$$E_z (r < a) = AJ_\nu (ur)\, e^{j\nu\phi}\, e^{j(\omega t - \beta z)} \quad ...(1)$$

The expression for E_z outside the core is

$$E_z (r > a) = CK_\nu (wr)\, e^{j\nu\phi}\, e^{j(\omega t - \beta z)} \quad ...(2)$$

For the z component we have, from Eq. (1) at the inner core-cladding boundary ($E_z = E_{z1}$) and from Eq. (2) at the outside of the boundary ($E_z = E_{z2}$), that

$$E_{z1} - E_{z2} = AJ_\nu (ua) - CK_\nu (wa) = 0 \quad ...(3)$$

The ϕ component is found from the equation

$$E_\phi = -\frac{j}{q^2}\left(\frac{\beta}{r}\frac{\partial E_z}{\partial \phi} - \mu\omega \frac{\partial H_z}{\partial r}\right) \quad ...(4)$$

Inside the core the factor q^2 is given by

$$q^2 = u^2 = k_1^2 - \beta^2 \qquad ...(5)$$

where $k_1 = 2\pi n_1/\lambda = \omega\sqrt{\varepsilon_1\mu}$.

Outside the core

$$w^2 = \beta^2 - k_2^2 \qquad ...(6)$$

where $k_2 = 2\pi n_2/\lambda = \omega\sqrt{\varepsilon_2\mu}$.

The expressions for H_z inside the core and outside the core are

$$H_z\,(r < a) = BJ_\nu\,(ur)\, e^{j\nu\phi}\, e^{j(\omega t - \beta z)} \qquad ...(7)$$

and

$$H_z\,(r > a) = DK_\nu\,(wr)\, e^{j\nu\phi}\, e^{j(\omega t - \beta z)} \qquad ...(8)$$

Substituting Eqs. (1) and (7) into Eq. (4) to find $E_{\phi 1}$, and similarly using Eqs. (2) and (8) to determine $E_{\phi 2}$, yields at $r = a$

$$E_{\phi 1} - E_{\phi 2} = -\frac{j}{u^2}\left[A\frac{j\nu\beta}{a} J_\nu\,(ua) - B\omega\mu u\, J_\nu^r\,(ua)\right]$$

$$-\frac{j}{w^2}\left[C\frac{j\nu\beta}{a} K_\nu\,(wa) - D\omega\mu w\, K_\nu^r\,(wa)\right] = 0 \qquad ...(9)$$

Here, $J_\nu'\,(ua)$ and $K_\nu'\,(wa)$ indicate differentiation with respect to the argument.

Similarly, for the tangential components of **H** at $r = a$

$$H_{z1} - H_{z2} = BJ_\nu\,(ua) - DK_\nu\,(wa) = 0 \qquad ...(10)$$

and

$$H_{\phi 1} - H_{\phi 2} = - -\frac{j}{u^2}\left[B\frac{j\nu\beta}{a} J_\nu\,(ua) + A\omega\varepsilon_1 u\, J_\nu^r\,(ua)\right]$$

$$-\frac{j}{w^2}\left[D\frac{j\nu\beta}{a} K_\nu\,(wa) + C\omega\varepsilon_2 w\, K_\nu^r\,(wa)\right] = 0 \qquad ...(11)$$

Equations (3), (9), (10) and (11) are a set of four equations with four unknown coefficients *A*, *B*, *C* and *D*. A solution to these equations exists only if the determinant of these coefficients is zero:

$$\begin{vmatrix} J_\nu\,(ua) & 0 & -K_\nu\,(wa) & 0 \\ \frac{\beta\nu}{au^2} J_\nu\,(ua) & \frac{j\omega\mu}{u} J_\nu'\,(ua) & \frac{\beta\nu}{aw^2} K_\nu\,(wa) & \frac{j\omega\mu}{w} K_\nu'\,(wa) \\ 0 & J_\nu(ua) & 0 & -K_\nu\,(wa) \\ -\frac{j\omega\varepsilon_1}{u} J_\nu'\,(ua) & \frac{\beta\nu}{au^2} J_\nu\,(ua) & -\frac{j\omega\varepsilon_2}{w} K_\nu'\,(wa) & \frac{\beta\nu}{aw^2} K_\nu\,(wa) \end{vmatrix} = 0 \qquad ...(12)$$

Evaluating the above determinant, we get the following eigen value equation for β.

$$(\mathcal{J}_\nu + \mathcal{K}_\nu)\,(k_1^2\,\mathcal{J}_\nu + k_2^2\,\mathcal{K}_\nu) = \left(\frac{\beta\nu}{a}\right)^2 \left(\frac{1}{u^2} + \frac{1}{w^2}\right)^2 \qquad ...(13)$$

Here,

$$\mathcal{J}_\nu = \frac{J_\nu'\,(ua)}{uJ_\nu\,(ua)} \quad \text{and} \quad \mathcal{K}_\nu = \frac{K_\nu'\,(wa)}{wK_\nu\,(wa)}$$

If we solve Eq. (13) for β we find that only discrete values restricted to the range given by the equation

$$n_2 k = k_2 \le \beta \le k_1 = n_1 k$$

are allowed.

43.5. Modes in Step-Index Fibers

Fig. 43.9 shows the variation of the Bessel function $J_\nu(x)$ for the first three orders ($\nu = 0, 1, 2$) plotted as a function of x.

The J-type Bessel functions are similar to harmonic functions since they show oscillatory behavior for real values of wave vector k. Because of the oscillatory behavior of J_ν there will be m roots of the equation

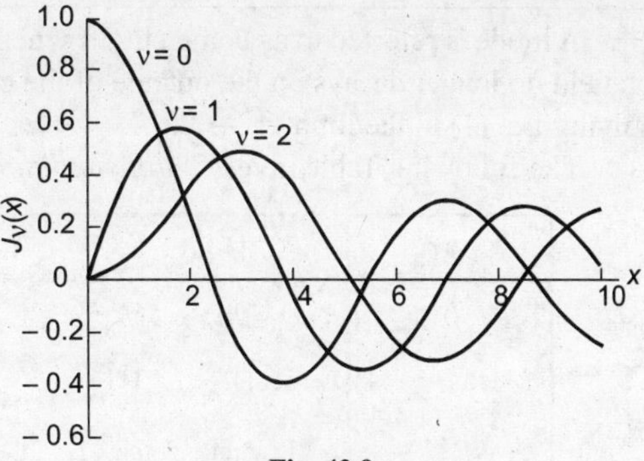

Fig. 43.9.

$$(\mathcal{J}_\nu + \mathcal{K}_\nu)\,(k_1^2\,\mathcal{J}_\nu + k_2^2\,\mathcal{K}_\nu) = \left(\frac{\beta\nu}{a}\right)^2 \left(\frac{1}{u^2} + \frac{1}{w^2}\right)^2 \qquad ...(1)$$

for a given ν value. These roots are designated by $\beta_{\nu m}$. The corresponding modes are either $TE_{\nu m}$, $TM_{\nu m}$, $EH_{\nu m}$, or $HE_{\nu m}$.

Fig. 43.10 shows cross-sectional views of the transverse electric field vectors for the four lowest-order modes in a step-index fiber.

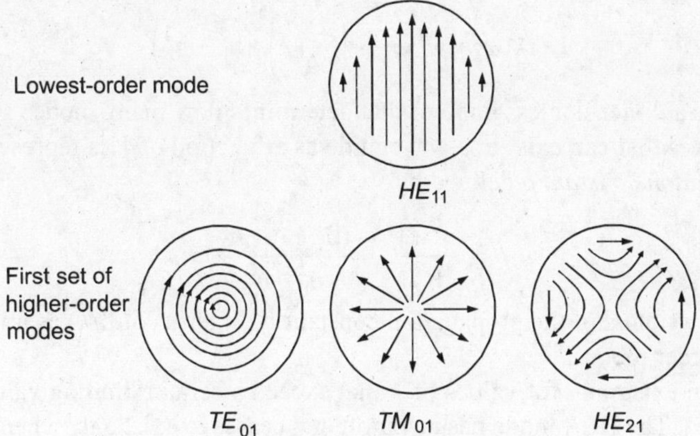

Fig. 43.10.

For the dielectric fiber waveguide all modes are hybrid modes except those for which $\nu = 0$. If we put $\nu = 0$ in Eq. (1), we get two different eigenvalue equations.

(*i*)

$$\mathcal{J}_0 + \mathcal{K}_0 = 0 \qquad ...(2)$$

or

$$\frac{J_1(ua)}{uJ_0(ua)} + \frac{K_1(wa)}{wK_0(wa)} = 0 \qquad ...(3)$$

For these modes $E_z = 0$ and these are called TE_{0m} modes.

(*ii*)

$$k_1^2\,\mathcal{J}_0 + k_2^2\,\mathcal{K}_0 = 0 \qquad ...(4)$$

or

$$\frac{k_1^2\,J_1(ua)}{uJ_0(ua)} + \frac{k_2^2\,K_1(wa)}{wK_0(wa)} = 0 \qquad ...(5)$$

For these modes $H_z = 0$ and these are called TM_{0m} modes.

Cutoff Conditions for Fiber Modes

Let us now study the cutoff conditions for fiber modes.

A mode is referred to as being cut off when it is no longer bound to the core of the fiber, so that its field no longer decays on the outside of the core. The cutoffs for the various modes are found by solving Eq. (1) in the limit $w^2 \to 0$.

The following Table gives *Cutoff conditions for some lower-order modes.*

ν	Mode	Cutoff condition
0	TE_{0m}, TM_{0m}	$J_0(ua) = 0$
1	HE_{1m}, EH_{1m}	$J_1(ua) = 0$
≥ 2	$EH_{\nu m}$	$J_\nu(ua) = 0$
	$HE_{\nu m}$	$\left(\frac{n_1^2}{n_2^2} + 1\right) J_{\nu-1}(ua) = \frac{ua}{\nu - 1} J_\nu(ua)$

V-parameters and Number of Modes

An important parameter connected with the cutoff condition is the *normalized frequency V* (also called the *V-number* or *V-parameter*) defined by

$$V^2 = (u^2 + w^2)a^2 = \left(\frac{2\pi a}{\lambda}\right)^2 \left(n_1^2 - n_2^2\right) \qquad \text{...(6)}$$

V-parameter is a dimensionless number that determines how many modes a fiber can support. The number of modes that can exist in a waveguide as a function of V is represented in terms of a *normalized propagation constant b* defined by

$$b = \frac{a^2 w^2}{V^2} = \frac{(\beta / k)^2 - n_2^2}{n_1^2 - n_2^2} \qquad \text{...(7)}$$

Fig. 43.11 shows plots of the propagation constant b (in terms of β/k) as a function of V for a few of the lowest-order modes.

Each mode can exist only for values of V that exceed a certain limiting value. The modes are cut off when $\beta/k = n_2$. The HE_{11} mode has no cutoff and ceases to exist only when the core diameter is zero. This is the principle on which the single-mode fiber is based.

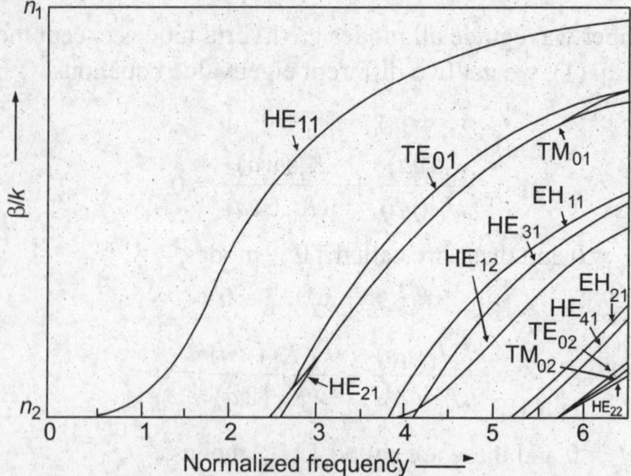

Fig. 43.11.

43.6. Design of a Single Mode Fiber

The basic principle on which a single mode fiber is designed is to choose the appropriate values of core radius a and the refractive indices n_1 and n_2 of the core and cladding respectively, so that

$$V = \frac{2\pi a}{\lambda}\left(n_1^2 - n_2^2\right)^{1/2} \leq 2.405 \qquad ...(1)$$

The value of $x = 2.405$ is the value at which the lowest-order Bessel function J_0 is zero. In this condition all modes except the HE_{11} mode are cut off.

Single-mode fibers are fabricated by making the dimensions of the core diameter to be a few wavelengths (usually 8 to 12) and by having a small index difference between the core and cladding indices. When $V = 2.4$ or less a single mode (HE_{11}) propagation is possible for fairly large variations in values of the physical core size a and the core-cladding index difference Δ.

$$\Delta = \frac{n_1 - n_2}{n_1} \qquad ...(2)$$

However, in practical designs of single-mode fibers, the core-cladding index difference varies between 0.2 and 1.0 percent, and the core diameter should be chosen to be just below the cutoff of the first higher-order mode.

That is, V should be slightly less than 2.4.

For example, a typical single-mode fiber has a core radius of 3 μm and a numerical aperture of 0.1 at a wavelength of 0.8 μm.

$$V = \frac{2\pi}{\lambda}\, a(NA) = \frac{2\pi \times (3.00 \times 10^{-6})}{(0.8 \times 10^{-6})} \times 0.1 = 2.356.$$

Example 1. *A step-index fiber has a normalized frequency* $V = 26.6$ *at a* 1300-nm *wavelength. If the core radius is* 25 μm, *find the numerical aperture.*

Solution.

$$V = \frac{2\pi a}{\lambda}\,\text{NA}$$

$$\text{NA} = V\,\frac{\lambda}{2\pi a} = 26.6\,\frac{(1300 \times 10^{-9})}{2\pi\,(25 \times 10^{-6})} = 0.22$$

43.7. Relationship between V and Number of Modes M in a Multimode Fiber

The parameter V is related to the number of modes M in a multimode fiber when M is large. An approximate relationship for step-index fibers can be derived from ray theory.

A ray congruence incident on the end of a fiber will be accepted by the fiber if it lies within an angle θ defined by the numerical aperture as given in the equation

$$\text{NA} = \sin\theta = \left(n_1^2 - n_2^2\right)^{1/2} \qquad ...(1)$$

For practical numerical apertures sin θ is small so that $\sin\theta \approx \theta$.

The solid acceptance angle for the fiber is

$$\Omega = \pi\theta^2 = \pi\left(n_1^2 - n_2^2\right) \qquad ...(2)$$

For electromagnetic radiation of wavelength λ emerging out from a laser or a waveguide, the number of modes per unit solid angle is given by $2A/\lambda^2$, where A is the area the mode is leaving or entering. The area A in this case is the core cross section πa^2. The factor 2 comes from the fact that the plane wave can have two polarization orientations.

The total number of modes M entering the fiber is

$$M \simeq \frac{2A}{\lambda^2}\Omega = \frac{2\pi^2 a^2}{\lambda^2}\left(n_1^2 - n_2^2\right) = \frac{V^2}{2} \qquad ...(3)$$

43.8. Linearly Polarized Modes

The exact analysis for the modes of a fiber is mathematically very complex. A highly accurate approximation is used, based on the principle that in a step-index fiber the difference between the indices of refraction of the core and cladding is very small, that is, $\Delta << 1$. This is the basis of the *weakly guiding fiber approximation.*

In this approximation the electromagnetic field patterns and the propagation constants of the mode pairs $HE_{v+1,m}$ and $EH_{v-1,m}$ are very similar. This holds likewise for the three modes TE_{0m}, TM_{0m} and HE_{2m}. This can be seen from Fig. 43.11 with (v, m) = (0, 1) and (2, 1) for the mode groupings {HE_{11}}, {TE_{01}, TM_{01}, HE_{21}}, {HE_{31}, EH_{11}}, {HE_{12}}, {HE_{41}, EH_{21}}, and {TE_{02}, TM_{02}, HE_{22}}. The result is that only four field components need to be considered instead of six. The field description is further simplified by the use of cartesian instead of cylindrical coordinates.

The eigenvalue equation for β is

$$(\mathcal{J}_v + \mathcal{K}_v)\,(k_1{}^2\,\mathcal{J}_v + k_2{}^2\,\mathcal{K}_v) = \left(\frac{\beta v}{a}\right)^2 \left(\frac{1}{u^2} + \frac{1}{w^2}\right)^2 \qquad ...(1)$$

When $\Delta \ll 1$, $k_1^2 \simeq k_2^2 \simeq \beta^2$.

Using these approximations, Eq. (1) becomes

$$\mathcal{J}_v + \mathcal{K}_v = \pm\frac{v}{a}\left(\frac{1}{u^2} + \frac{1}{w^2}\right) \qquad ...(2)$$

Thus Eq. [3 in Sec. 43.5] for TE_{0m} modes is the same as Eq. [5 in Sec. 43.5] for TM_{0m} modes. Using the recurrence relations for J_v' and K_v', we get two sets of equations for Eq. (2) for the positive and negative signs. The positive sign yields

$$\frac{J_{v+1}(ua)}{uJ_v(ua)} + \frac{K_{v+1}(wa)}{wK_v(wa)} = 0 \qquad ...(3)$$

The solution of this equation gives a set of modes called the EH modes.

The negative sign yields

$$\frac{J_{v-1}(ua)}{uJ_v(ua)} - \frac{K_{v-1}(wa)}{wK_v(wa)} = 0 \qquad ...(4)$$

Alternatively, taking the inverse of Eq. (4) and using the expressions for $J_v(ua)$ and $K_v(wa)$, we get

$$-\frac{uJ_{v-2}(ua)}{J_{v-1}(ua)} = \frac{wK_{v-2}(wa)}{K_{v-1}(wa)} \qquad ...(5)$$

The solution of this equation gives a set of modes called the HE modes. We define a new parameter

$$j = \begin{cases} 1 & \text{for TE and TM modes} \\ v+1 & \text{for EH modes} \\ v-1 & \text{for HE modes} \end{cases} \quad ...(6)$$

Eqs. [3 in Sec. 43.5], (3) and (5) can be written in the unified form

$$\frac{uJ_{j-1}(ua)}{J_j(ua)} = -\frac{wK_{j-1}(wa)}{K_j(wa)} \quad ...(7)$$

Equations (6) and (7) show that within the weakly guiding approximation all modes characterized by a common set of j and m satisfy the same characteristic equation. This means that these modes are degenerate. Thus, if an $HE_{v+1,m}$ mode is degenerate with an $EH_{v-1,m}$ mode (that is, if HE and EH modes of corresponding radial order m and equal circumferential order v form degenerate pairs), then any combination of an $HE_{v+1,m}$ mode with an $EH_{v-1,m}$ mode will likewise constitute a guided mode of the fiber.

Such degenerate modes are called *linearly polarized* (LP) modes.

They are designated LP_{jm} modes regardless of their TM, TE, EH, or HE field configuration.

Fig. 43.12 shows plots of the propagation constant b as a function of V for various LP_{jm} modes.

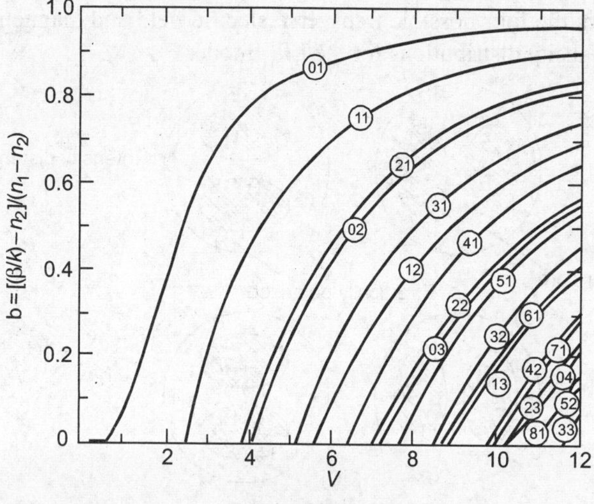

Fig. 43.12.

In general we have the following:

1. each LP_{0m} mode is derived from an HE_{1m} mode;

2. each LP_{1m} mode comes from TE_{0m}, TM_{0m} and HE_{2m} modes;

3. each LP_{vm} mode ($v \geq 2$) is from an $HE_{v+1,m}$ and an $EH_{v-1,m}$ mode.

The following Table gives the correspondence between the ten lowest LP modes (that is, those having the lowest cutoff frequencies) and the traditional TM, TE, EH, and HE modes.

This table also shows the number of degenerate modes.

A very useful feature of the LP-mode designation is the ability to readily visualize a mode. In a complete set of modes only one electric and one magnetic field component are significant. The electric field vector **E** can be chosen to lie along an arbitrary axis, with the magnetic field vector **H** being perpendicular to it. In addition there are equivalent solutions with the field polarities reversed. Since each of the two possible polarization directions can be coupled with either $a \cos j\phi$ or $a \sin j\phi$ azimuthal dependence, four discrete mode patterns can be obtained from a single LP_{jm} label.

Composition of the lower-order linearly polarized modes

LP-mode designation	Traditional-mode designation and number of modes	Number of degenerate modes
LP_{01}	$HE_{11} \times 2$	2
LP_{11}	TE_{01}, TM_{01}, $HE_{21} \times 2$	4
LP_{21}	$EH_{11} \times 2$, $HE_{31} \times 2$	4
LP_{02}	$HE_{12} \times 2$	2
LP_{31}	$EH_{21} \times 2$, $HE_{41} \times 2$	4
LP_{12}	TE_{02}, TM_{02}, $HE_{22} \times 2$	4
LP_{41}	$EH_{31} \times 2$, $HE_{51} \times 2$	4
LP_{22}	$EH_{12} \times 2$, $HE_{32} \times 2$	4
LP_{03}	$HE_{13} \times 2$	2
LP_{51}	$EH_{41} \times 2$, $HE_{61} \times 2$	4

Fig. 43.13 shows the four possible transverse electric field and magnetic field directions and the corresponding intensity distributions for the LP_{11} mode.

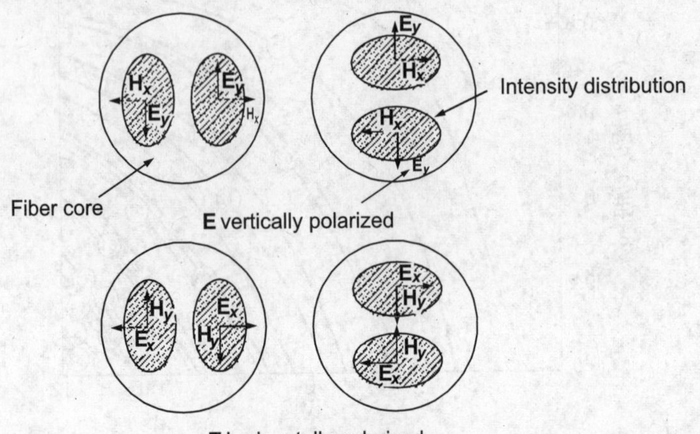

Fig. 43.13.

43.9. Power Flow in Step-Index Fibers

Let us study the fractional power flow in the core and cladding for a given mode for step-index fibers.

The electromagnetic field for a given mode does not decrease to zero at the core-cladding interface (Fig. 43.14).

The field changes from an oscillating form in the core to an exponential decay in the cladding. Thus the electromagnetic energy of a guided mode is carried partly in the core and partly in the cladding.

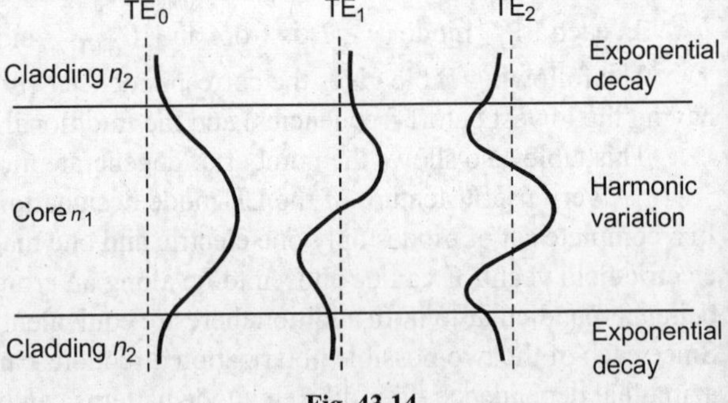

Fig. 43.14.

When the operating frequency is far away from the cut-off frequency of the concerned mode, the concentration of energy is more in the core.

The field penetrates more and more into the cladding region as cut-off is approached. Then, greater percentage of the electromagnetic energy travels in the cladding.

At the cut-off frequency, the field will not decay outside the core. The mode now becomes a fully radiating mode.

The relative amounts of power flowing in the core and the cladding is computed by integrating the Poynting vector in the axial direction, over the fibre cross-section.

$$S_z = \frac{1}{2} \operatorname{Re} (\mathbf{E} \times \mathbf{H}^*) \cdot \mathbf{e}_z \qquad \text{...(1)}$$

Here, $\mathbf{e}_z$ = unit vector in the *Z*-direction.

H^* is the complex conjugate of H.

The power in the core is given by

$$P_{\text{core}} = \frac{1}{2} \int_0^a \int_0^{2\pi} r(E_x H_y^* - E_y H_x^*)\, d\phi\, dr \qquad \text{...(2)}$$

Here, a = core radius.

The power in the cladding is

$$P_{\text{clad}} = \frac{1}{2} \int_a^\infty \int_0^{2\pi} r(E_x H_y^* - E_y H_x^*)\, d\phi\, dr \qquad \text{...(3)}$$

The relative core and cladding powers for a particular mode v are given by

$$\frac{P_{\text{core}}}{P} = \left(1 - \frac{u^2}{V^2}\right)\left[1 - \frac{J_v^2(ua)}{J_{v+1}(ua)\, J_{v-1}(ua)}\right] \qquad \text{...(4)}$$

and

$$\frac{P_{\text{clad}}}{P} = 1 - \frac{P_{\text{core}}}{P} \qquad \text{...(5)}$$

Here, P is the total power in the mode v.

Fig. 43.15 shows the variation of P_{core}/P and P_{clad}/P with V for various LP_{jm} modes.

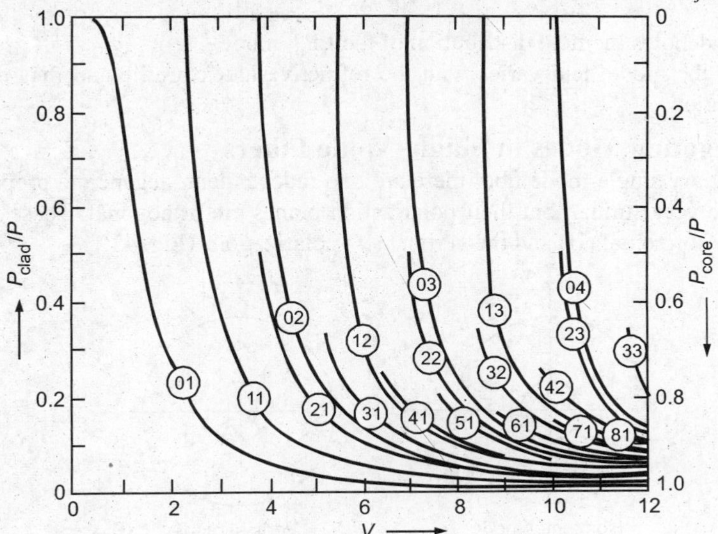

Fig. 43.15.

SINGLE-MODE FIBERS

43.10. Mode Field Diameter

A fundamental parameter of a single-mode fiber is the *mode field diameter* (MFD). This parameter can be determined from the mode field distribution of the fundamental LP_{01} mode.

Fig. 43.16 shows distribution of light in a single-mode fiber above its cutoff wavelength. For a gaussian distribution the MFD is given by the $1/e^2$ width of the optical power (Fig. 43.17).

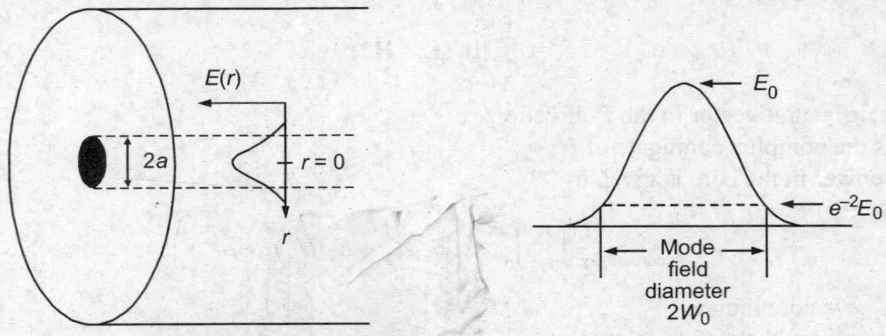

Fig. 43.16. **Fig. 43.17.**

Let us consider a model for characterizing and measuring the MFD. Assume the electric field distribution to be gaussian:

$$E(r) = E_0 \exp(-r^2 / W_0^2) \qquad ...(1)$$

Here, r is the radius, E_0 is the field at zero radius, and W_0 is the width of the electric field distribution.

Take the width $2W_0$ of the MFD to be twice the e^{-1} radius of the optical electric field (which is equivalent to the e^{-2} radius of the optical power) given in Eq. (1).

The MFD width $2W_0$ of the LP_{01} mode can then be defined as

$$2W_0 = \left[\frac{2\int_0^\infty r^3 E^2(r)\, dr}{\int_0^\infty r E^2(r)\, dr}\right]^{1/2} \qquad ...(2)$$

Here, $E(r)$ denotes the field distribution of the LP_{01} mode.

In general, the mode field varies with the refractive index profile and thus deviates from a gaussian distribution.

43.11. Propagation Modes in Single-Mode Fibers

In any ordinary single-mode fiber there are two independent, degenerate propagation modes. These modes are very similar, but their polarization planes are orthogonal. These may be chosen arbitrarily as the horizontal (H) and the vertical (V) polarizations (Fig. 43.18).

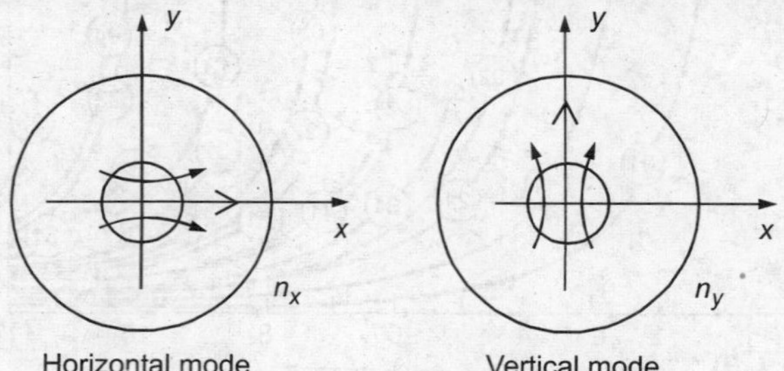

Fig. 43.18.

Either one of these two polarization modes constitutes the fundamental HE_{11} mode. In general, the electric field of the light propagating along the fiber is a linear superposition of these two polarization modes and depends on the polarization of the light at the launching point into the fiber.

Suppose we arbitrarily choose one of the modes to have its transverse electric field polarized along the x direction and the other independent orthogonal mode to be polarized in the y direction (Fig. 43.18). In ideal fibers with perfect rotational symmetry, the two modes are degenerate with equal propagation constants ($k_x = k_y$). Any polarization state injected into the fiber will propagate unchanged. In actual fibers there are imperfections such as asymmetrical lateral stresses, noncircular cores, and variations in refractive-index profiles. These imperfections break the circular symmetry of the ideal fiber and lift the degeneracy of the two modes. The modes propagate with different phase velocities. The difference between their effective refractive indices is called the fiber *birefringence,*

$$B_f = n_y - n_x \qquad ...(1)$$

The birefringence is also defined as

$$\beta = k_0 (n_y - n_x) \qquad ...(2)$$

Here, $k_0 = 2\pi/\lambda$ is the free-space propagation constant.

If light is injected into the fiber so that both modes are excited, then one mode is delayed in phase relative to the other as they propagate. When the phase difference between two modes is an integral multiple of 2π, the two modes will beat at this point and the input polarization state will be reproduced. The length over which this beating occurs is the *fiber beat length*

$$L_p = 2\pi/\beta \qquad ...(3)$$

43.12. Monomode and Multimode Step-index Fibers

Step index fiber has a core of refractive index n_1 and a cladding of slightly lower refractive index n_2. Mathematically, the refractive index profile of a step index fiber is expressed as:—

Refractive index $n(r) = n_1$ in the core where $r < a$ ...(1)

$n(r) = n_2$ in the cladding where $r > a$ ...(2)

Here, a is the radius of the fiber core and r is the radial distance from the centre of the core of the fiber.

Fig. 43.19 shows a multimode step index fiber of the core diameter of the order of 50 μm or more, which is sufficient to allow the propagation of signals in many modes in the fiber core.

There are many ray paths through the fiber core.

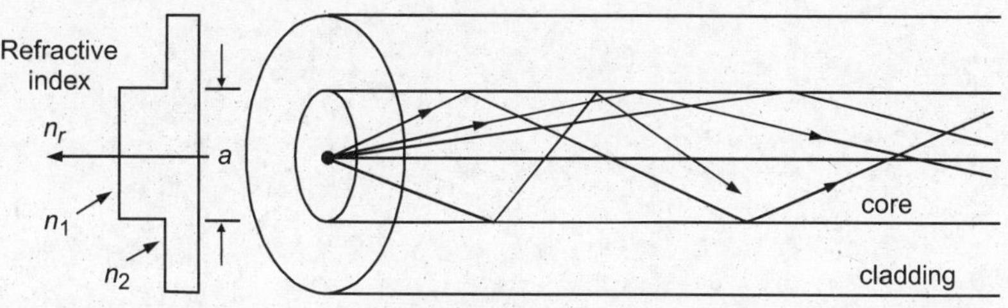

Fig. 43.19.

Fig. 4.20 shows a single mode step index fiber which only allows the propagation of only one transverse electromagnetic mode, typically HE_{11}. Its core diameter must be of the order of 1 to 10 μm. The propagation of signals in single mode corresponds to a single axial ray path only through the axis of the fiber core.

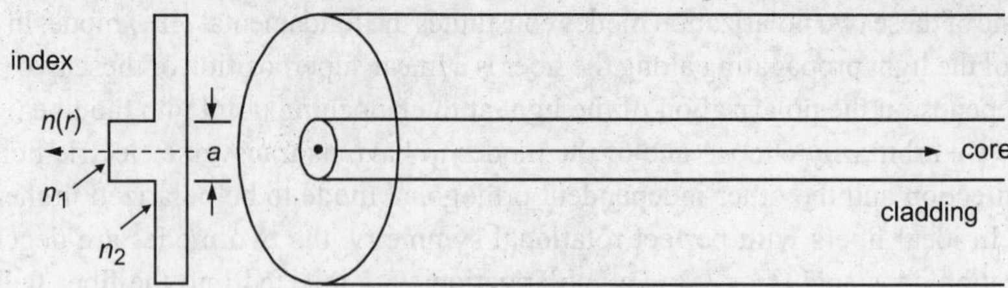

Fig. 43.20.

EXERCISE

1. Explain mode theory for circular waveguide.
2. Write down the Maxwell's field equations for a non dissipative, linear and isotropic dielectric medium with no free charge or current in it.
3. Express Maxwell's field equations in cylindrical coordinate system and find out the values of E_r, E_ϕ, H_r and H_ϕ in terms of E_z and H_z when the electromagnetic wave is propagating in the Z-direction.
4. Starting from the Maxwell's field equation, derive the wave equation for E_z and H_z for the wave travelling in the Z-direction.
5. Explain with relevant diagrams the different modes of an electromagnetic wave in an optical fiber. What do you mean by "hybrid mode"? Explain with ray diagram.
6. What are EH and HE hybrid modes?
7. Determine the solution for the wave equation for E_z and H_z in a step index fiber by the solution of partial differential equation method. Determine the value of E_z and H_z in the core and the cladding region.
8. Derive the wave equation in step-index fibre.
9. What do you mean by V-parameter? Give the expression for the same.
10. Derive expressions for the power flow through the core and cladding of a step index fiber.
11. Describe the propagation modes in single mode fiber.

CHAPTER

44

FIBRE LOSSES

44.1. Attenuation in Optical Fibres

Attenuation in fibres means 'loss of optical power' in the fibre itself. Attenuation of a light signal as it propagates along a fibre plays a major role in determining the maximum transmission distance between a transmitter and a receiver. The basic attenuation mechanisms in a fibre are (*i*) absorption, (*ii*) scattering and (*iii*) radiative losses of the optical energy.

(*i*) Absorption is related to the fibre material.

(*ii*) Scattering is associated both with the fibre material and with structural imperfections in the optical waveguide.

(*iii*) Attenuation owing to radiative effects originates from perturbations (both microscopic and macroscopic) of the fibre geometry.

44.2. Attenuation Units

Signal attenuation within optical fibers is usually expressed in the logarithmic unit of the decibel. The decibel is used for comparing two power levels. The decibel is defined for a particular optical wavelength as the ratio of the input (transmitted) optical power P_i into a fiber to the output (received) optical power P_o from the fiber as:

$$\textit{Number of decibels}\ (\text{dB}) = 10 \log_{10} \frac{P_i}{P_o} \qquad ...(1)$$

The numerical values are obtained using the relationship:

$$\frac{P_i}{P_o} = 10^{(\text{dB}/10)} \qquad ...(2)$$

In optical fiber communications, the attenuation is usually expressed in decibels per unit length (*i.e.*, dB km^{-1}).

$$\alpha_{\text{dB}} = \frac{10}{L} \log_{10}\left(\frac{P_i}{P_o}\right).$$

Here, α_{dB} = signal attenuation per unit length in decibels,

L = fiber length in km,

P_i = input (transmitted) optical power into the fibre and

P_o = output (received) optical power from the fiber.

α_{dB} is a function of the operating wavelength.

Definition. $$\alpha_{dB} = \frac{1}{L}\left[10\log_{10}\left(\frac{P_{in}}{P_{out}}\right)\right].$$

Explanation. An ideal fiber has no power loss so that $P_{out} = P_{in}$. This is equivalent to 0-dB attenuation. An ideal fiber is not found in practice. Suppose for a fiber $\alpha_{dB} = 3$ dB/km is average power loss. It means that the optical signal power will decrease by 50 percent over a path length of one kilometer.

Example 1. *When the mean optical power launched into an* 8 *km length of fiber is* 120 μW, *the mean optical power at the fiber output is* 3 μW.

Determine:

(*a*) *the overall signal attenuation or loss in decibels through the fiber assuming there are no connectors or splices;*

(*b*) *the signal attenuation per kilometer for the fiber.*

Solution. (*a*) The overall signal attenuation in decibels through the fiber is:

$$\textit{Signal attenuation} = 10\log_{10}\frac{P_i}{P_o} = 10\log_{10}\frac{(120\times10^{-6})}{(3\times10^{-6})}$$

$$= 10\log_{10}40 = 16.0\text{ dB}$$

(*b*) $$\alpha_{dB} = \frac{1}{L}\left[10\log_{10}\left(\frac{P_i}{P_o}\right)\right] = \frac{1}{8}\left[10\log_{10}\left(\frac{120\times10^{-6}}{3\times10^{-6}}\right)\right] = 2.0\text{ dB km}^{-1}$$

44.3. Absorption Losses

Material absorption is a loss mechanism related to the material composition and the fabrication process for the fiber, which results in the dissipation of some of the transmitted optical power as heat in the waveguide. The absorption of the light may be intrinsic (caused by the interaction with one or more of the major components of the glass) or extrinsic (caused by impurities within the glass).

1. Intrinsic absorption

Intrinsic absorption is the principal physical factor that defines the transparency window of a material over a specified spectral region.

Intrinsic absorption results (*i*) from electronic absorption bands in the ultraviolet region and (*ii*) from atomic vibration bands in the near infrared region.

(*i*) The electronic absorption bands are associated with the band gaps of the amorphous glass materials. Absorption occurs when a photon interacts with an electron in the valence band and excites it to a higher energy level.

Urbach's rule. *The ultraviolet edge of the electron absorption bands of both amorphous and crystalline materials follow the empirical relationship*

$$\alpha_{uv} = Ce^{E/E_0} \qquad ...(1)$$

This is called Urbach's rule.

Here C and E_0 are empirical constants and E is the photon energy.

Fig. 44.1 shows the optical fiber attenuation characteristics for a GeO_2-doped low-loss low-OH-content silica fiber.

Since E is inversely proportional to the wavelength λ, ultraviolet absorption decays exponentially with increasing wavelength. In particular, the ultraviolet loss contribution in dB/km at any wavelength can be expressed empirically as a function of the mole fraction x of GeO_2 as

$$\alpha_{uv} = \frac{154.2x}{46.6x+60} \times 10^{-2} \exp\left(\frac{4.63}{\lambda}\right) \quad ...(2)$$

The ultraviolet loss is small compared to scattering loss in the near infrared region.

(*ii*) In the near infrared region above 1.2 μm, the optical waveguide loss is predominantly determined by the presence of OH ions and the inherent infrared absorption of the constituent material. The inherent infrared absorption is associated with the characteristic vibration frequency of the particular chemical bond between the atoms of which the fiber is composed. An interaction between the vibrating bond and the electromagnetic field of the optical signal results in a transfer of energy from the field to the bond, thereby giving rise to absorption. This absorption is quite strong because of the many bonds present in the fiber.

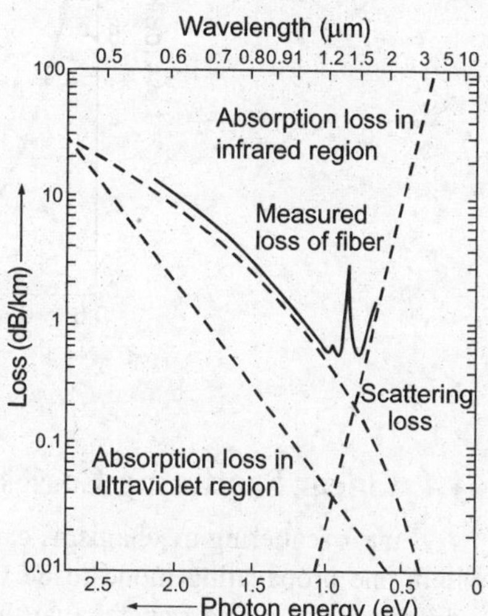

Fig. 44.1.

An empirical expression for the infrared absorption in dB/km for GeO_2-SiO_2 glass is

$$\alpha_{IR} = 7.81 \times 10^{11} \times \exp\left(\frac{-48.48}{\lambda}\right) \quad ...(3)$$

2. Extrinsic absorption

In practical optical fibers prepared by conventional melting techniques, a major source of signal attenuation is extrinsic absorption from transition metal element impurities. Some impurities like chromium and copper can cause attenuation in excess of 1 dB km^{-1} in the near-infrared region.

Fig. 44.2 shows absorption losses caused by Cr^{3+}, Cu^{2+}, Fe^{2+} and OH^- or hydroxyl ions as a function of wave length of an optical fiber.

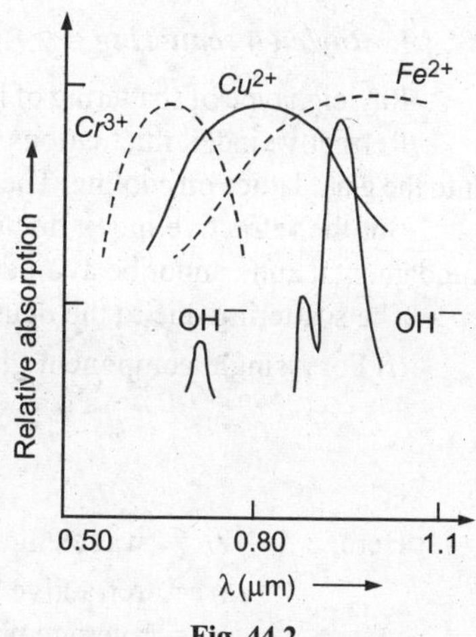

Fig. 44.2.

Transition element contamination may be reduced to acceptable levels (i.e., one part in 10^{10}) by glass refining techniques such as vapour-phase oxidation, which largely eliminates the effects of these metallic impurities.

Another major extrinsic loss mechanism is caused by absorption due to water (as the hydroxyl or OH ion) dissolved in the glass.

Significant progress has been made in reducing the residual OH content of fibers to less than 1 ppb.

For example, Fig. 44.3 shows the attenuation-versus-wavelength curve of a silica fiber prepared by the VAD method with an OH content of less than 0.8 ppb.

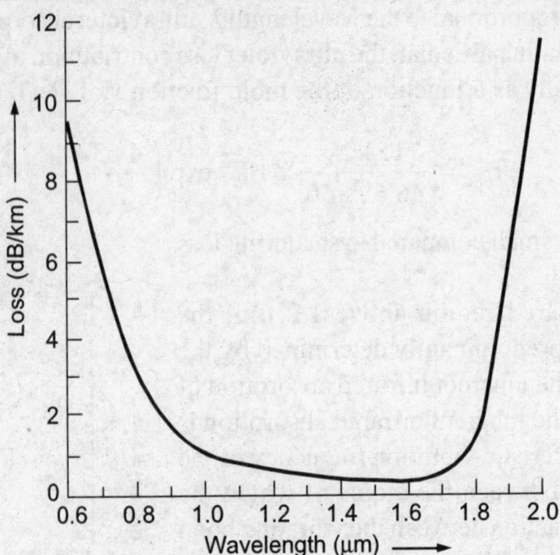

Fig. 44.3.

44.4. Linear Scattering Losses

Linear scattering mechanisms cause the transfer of some or all of the optical power contained within one propagating mode to be transferred linearly (proportionally to the mode power) into a different mode. The transfer may be to a leaky or radiation mode which does not continue to propagate within the fiber core, but is radiated from the fiber. So this process results in attenuation of the transmitted light. There is no change of frequency on scattering.

Linear scattering is categorized into two major types: Rayleigh and Mie scattering.

44.4.1. Rayleigh scattering

Rayleigh type of scattering of light takes place due to the variations in the refractive index in glass.

Refractive index fluctuations arise from density and compositional variations which are frozen into the glass lattice on cooling. The compositional variations may be reduced by improved fabrication.

But the refractive index fluctuations caused by the freezing-in of density inhomogeneities are fundamental and cannot be avoided.

The scattering due to the density fluctuations produces an attenuation proportional to $1/\lambda^4$.

(*i*) For a single component glass, the Rayleigh scattering coefficient γ_R is given by:

$$\gamma_R = \frac{8\pi^3}{3\lambda^4} n^8 p^2 \beta_c K T_F \qquad \text{...(1)}$$

Here, λ = wavelength of the optical radiation,

n = refractive index of the medium,

p = average photoelastic coefficient,

β_c = isothermal compressibility at a fictive temperature T_F, and

K = Boltzmann constant.

The fictive temperature is defined as the temperature at which the glass can reach a state of thermal equilibrium and is closely related to the anneal temperature.

The Rayleigh scattering coefficient γ_R is related to the transmission loss factor (transmissivity) of the fiber l by the following relationship:

$$l = \exp(-\gamma_R L) \qquad \text{...(2)}$$

Here, L is the length of the fiber.

It is apparent from Eq. (2) that the fundamental component of Rayleigh scattering is strongly reduced by operating at the longest possible wavelength.

(*ii*) For multicomponent glasses, the scattering is given by

$$\gamma = \frac{8\pi^3}{3\lambda^4}(\delta n^2)^2\,\delta V \qquad ...(3)$$

Here, the square of the mean square refractive-index fluctuation $(\delta n^2)^2$ over a volume of δV is

$$(\delta n^2)^2 = \left(\frac{\partial n}{\partial \rho}\right)^2 (\delta\rho)^2 + \sum_{i=1}^{m}\left(\frac{\partial n^2}{\partial C_i}\right)^2 (\delta C_i)^2 \qquad ...(4)$$

Here $\delta\rho$ is the density fluctuation and δC_i is the concentration fluctuation of the *i*th glass component.

Since Rayleigh scattering follows a characteristic λ^{-4} dependence, it decreases dramatically with increasing wavelength, as is shown in Fig. 44.1.

44.4.2. Mie scattering

Linear scattering may occur at inhomogeneities which are comparable in size to the guided wavelength. When the size of scattering inhomogeneities is greater than $\lambda/10$, the scattered intensity has an angular dependence and can be very large.

The scattering created by such inhomogeneities is mainly in the forward direction and is called Mie scattering. Depending upon the fiber material, design and manufacture, Mie scattering can cause significant losses. The inhomogeneities may be reduced by:

(*a*) removing imperfections due to the glass manufacturing process;

(*b*) carefully controlled extrusion and coating of the fiber;

(*c*) increasing the fiber guidance by increasing the relative refractive index difference between core and cladding.

By these means it is possible to reduce Mie scattering to insignificant levels.

44.5. Nonlinear Scattering Losses

Nonlinear scattering occurs at high optical power levels. This nonlinear scattering causes the optical power from one mode to be transferred in either the forward or backward direction to the same, or other modes, at a different frequency.

Nonlinear scattering is categorized into two major types:

(*i*) stimulated Brillouin scattering and (*ii*) Raman scattering.

They are observed at high optical power densities in long single-mode fibers.

The power density is defined as the power per unit area.

At normal power density, the nonlinear interactions are negligible.

44.5.1. Stimulated Brillouin scattering

Stimulated Brillouin scattering (SBS) may be regarded as the modulation of light through thermal molecular vibrations within the fiber.

The incident photons of light undergo nonlinear interaction to produce vibrational energy or "phonons" in the glass as well as the scattered light or "photons". The scattered light is found to be frequency modulated by the thermal energy, and both upward and downward frequency shifts are observed. The amount of the frequency shift and the strength of scattering vary as the function of the scattering angle. The frequency shift is a maximum in the backward direction reducing to zero in the forward direction.

Thus SBS occurs mainly in the backward direction.

The optical power level at which Brillouin scattering becomes significant in a single mode fiber is given by an empirical formula. The threshold power level P_B is given by:

$$P_B = 4.4 \times 10^{-3}\, d^2\, \lambda^2\, \alpha_{dB}\, v \text{ watts} \qquad ...(1)$$

Here, d and λ are the fiber core diameter and the operating wavelength, respectively, both measured in micrometres, α_{dB} is the fiber attenuation in decibels per kilometre and v is the source bandwidth (i.e., injection laser) in gigahertz.

Eq. (1) enables us to determine the threshold optical power which must be launched into a single-mode optical fiber before SBS occurs.

44.5.2. Stimulated Raman scattering

Stimulated Raman scattering (SRS) produces a high frequency optical phonon whereas low frequency acoustic phonons are produced in Brillouin scattering. SRS can occur in both the forward and backward directions in an optical fiber.

The threshold power level P_R for SRS to occur in a long single-mode fiber is given by:

$$P_R = 5.9 \times 10^{-2}\, d^2\, \lambda\, \alpha_{dB} \text{ watts.}$$

Here, d is the diameter of the fiber core in µm, λ is the wavelength emitted by the source in µm, α_{dB} is the fiber loss in dB/km and P_R is the threshold optical power.

44.6. Bending Losses

Whenever a fibre deviates from a straight-line path, radiative losses occur. Radiative losses occur whenever an optical fiber undergoes a bend of finite radius of curvature. Fibers can be subject to two types of bends:

1. Macroscopic bends have radii that are large compared to the fiber diameter. Such bends occur when a fiber cable turns a corner (Fig. 44.4.).

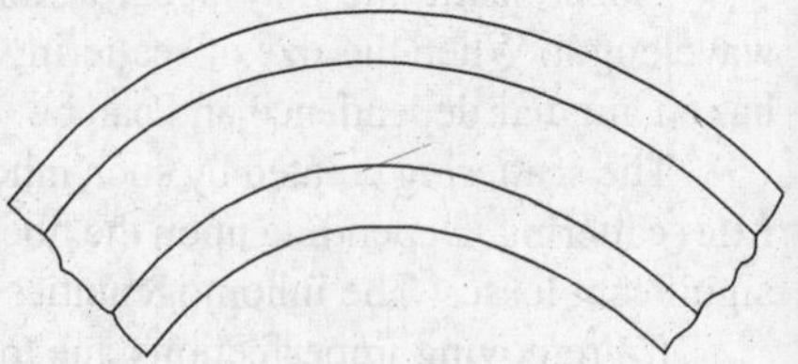

Fig. 44.4.

2. Microscopic bends are repetitive small-scale fluctuations in the radius of curvature of the fiber axis (Fig. 44.5).

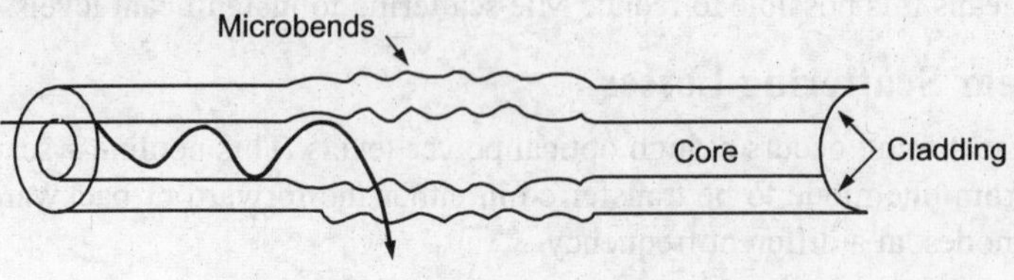

Fig. 44.5.

Microbending occurs due to nonuniform lateral pressure created during the cabling of the fiber.

***Macrobending Losses*:**

Large-curvature radiation losses are called *macrobending losses.* For slight bends the excess loss is extremely small and is unobservable. As the radius of curvature decreases, the loss increases exponentially until at a certain critical radius the curvature loss becomes observable. If the bend radius is made a bit smaller once this threshold point has been reached, the losses suddenly become extremely large.

Fig. 44.6 shows the fundamental mode field in a curved optical waveguide.

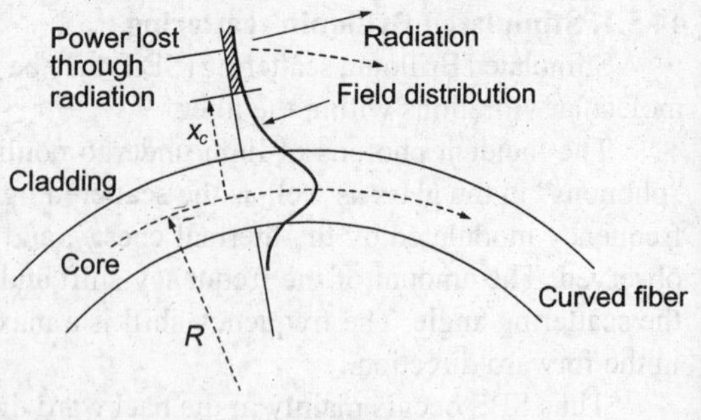

Fig. 44.6.

Any bound core mode has an evanescent field tail in the cladding which decays exponentially as a function of distance from the core. Since this field tail moves along with the field in the core, part of the energy of a propagating mode travels in the fiber cladding.

When a fiber is bent, the field tail on the far side of the center of curvature must move faster to keep up with the field in the core, for the lowest-order fiber mode. At a certain critical distance x_c from the center of the fiber, the field tail would have to move faster than the speed of light to keep up with the core field. Since this is not possible, the optical energy in the field tail beyond x_c radiates away.

The amount of optical radiation from a bent fiber depends on the field strength at x_c and on the radius of curvature R.

The radiation loss is represented by a radiation attenuation coefficient which has the form

$$\alpha_r = c_1 \exp(-c_2 R) \quad \text{...(1)}$$

Here, R is the radius of curvature of the fiber bend and

c_1, c_2 are constants which are independent of R.

Large bending losses occur in multimode fibers at a critical radius of curvature R_c which is calculated from the following relationship:

$$R_c = \frac{3 n_1^2 \lambda}{4\pi \left(n_1^2 - n_2^2\right)^{3/2}} \quad \text{...(2)}$$

Here, n_1 and n_2 are the refractive indices of the core and cladding of the fiber respectively.

Eq. (2) shows that potential macrobending losses may be reduced by :

(*a*) designing fibers with large relative refractive index differences;

(*b*) operating at the shortest wavelength possible.

Higher-order modes are bound less tightly to the fiber core than lower-order modes. So the higher-order modes will radiate out of the fiber first. Thus the total number of modes that can be supported by a curved fiber is less than in a straight fiber. Gloge has derived the following expression for the effective number of modes N_{eff} that are guided by a curved multimode fiber of radius a:

$$N_{eff} = N_\infty \left\{1 - \frac{\alpha+2}{2\alpha\Delta}\left[\frac{2a}{R} + \left(\frac{3}{2n_2 kR}\right)^{2/3}\right]\right\}. \quad \text{...(3)}$$

Here, α defines the graded-index profile, Δ is the core-cladding index difference, n_2 is the cladding refractive index, $k = 2\pi/\lambda$ is the wave propagation constant, and

$$N_\infty = \frac{\alpha}{\alpha+2}(n_1 ka)^2 \Delta \quad \text{...(4)}$$

is the total number of modes in a straight fiber.

Micro-bending Losses:

Microbends have small random deviations about a nominal straight line position. These are observed in cabled fibre. Microbends cause significant losses from cabled fiber.

An increase in attenuation results from microbending because the fiber curvature causes repetitive coupling of energy between the guided modes and the leaky or nonguided modes in the fiber (Fig. 44.7).

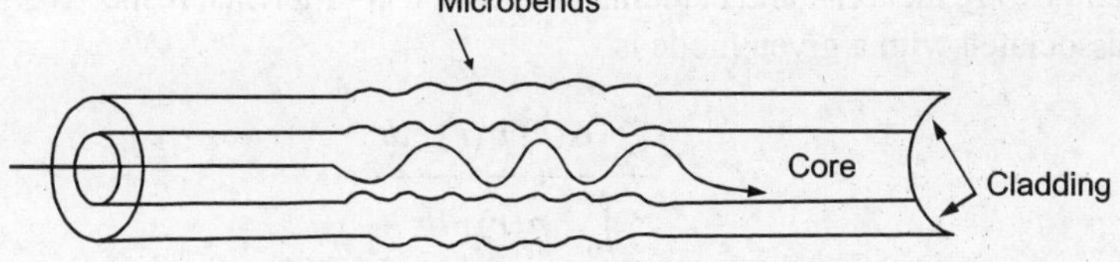

Fig. 44.7.

A compressible jacket extruded over the fiber minimizes microbending resulting from external forces. When external forces are applied to this configuration, the jacket will be deformed, but the fiber will tend to stay relatively straight as shown in Fig. 44.8.

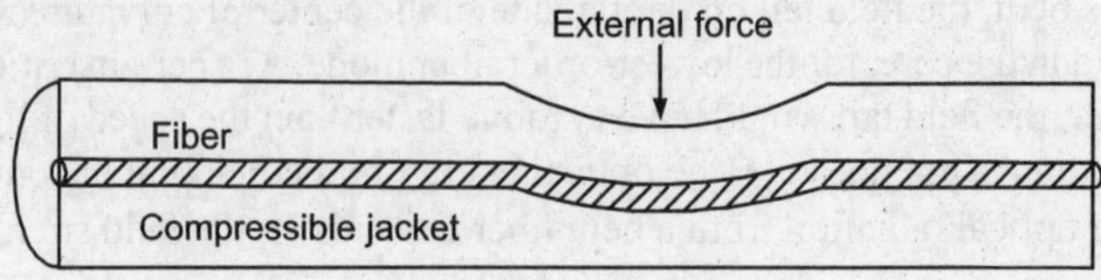

Fig. 44.8.

44.7. Core and Cladding Losses

In a fibre the core and cladding have different refractive indices, as they have different compositions. So the core and the cladding have different attenuation coefficients denoted α_1 and α_2, respectively. If we neglect the modal coupling, then the loss associated with a mode of order (v, m) for a step-index fibre is

$$\alpha_{vm} = \alpha_1 \frac{P_{\text{core}}}{P} + \alpha_2 \frac{P_{\text{clad}}}{P} \quad ...(1)$$

The fractional powers P_{core} / P and P_{clad} / P are shown in Fig. 44.9 for several low-order modes.

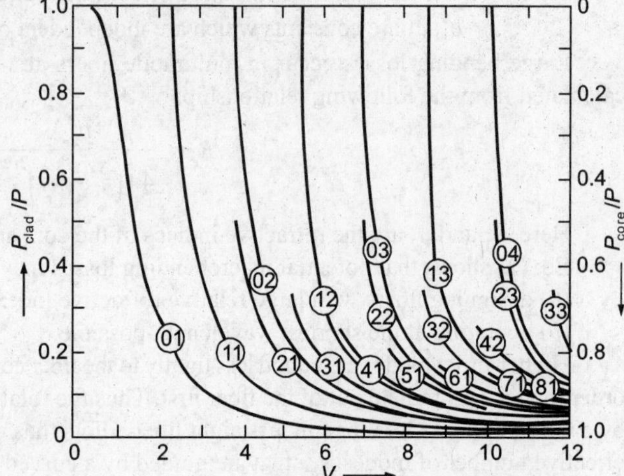

Fig. 44.9.

Using the equation

$$\frac{P_{\text{core}}}{P} = 1 - \frac{P_{\text{clad}}}{P},$$

Eq. (1) becomes

$$\alpha_{vm} = \alpha_1 + (\alpha_2 - \alpha_1)\frac{P_{\text{clad}}}{P} \quad ...(2)$$

The total loss of the fibre can be found by summing over all modes weighted by the fractional power in that mode.

But in case of graded index fibre, the relation is more complicated. In this case, both the attenuation coefficients and the modal power tend to be functions of the radial coordinates. So the loss at a distance r from the core axis is given by

$$\alpha(r) = \alpha_1 + (\alpha_2 - \alpha_1)\frac{n^2(0) - n^2(r)}{n^2(0) - n_2^2} \quad ...(3)$$

Here, α_1 and α_2 are the axial and cladding attenuation coefficients, respectively.

The loss associated with a given mode is

$$\alpha_{gi} = \frac{\int_0^\infty \alpha(r)\, p(r)\, r\, dr}{\int_0^\infty p(r)\, r\, dr} \quad ...(4)$$

Here $p(r)$ is the power density of that mode at r.

It has generally been observed that the loss increases with increasing mode number.

EXERCISE

1. Define signal attenuation. What is the unit of attenuation?
2. What are the mechanisms that cause absorption?
3. Discuss absorption losses in optical fibers, comparing and contrasting the intrinsic and extrinsic absorption mechanisms.
4. Briefly describe linear scattering losses in optical fibers with regard to:
   (*a*) Rayleigh scattering;
   (*b*) Mie scattering.
5. Compare stimulated Brillouin and stimulated Raman scattering in optical fibers, and indicate the way in which they may be avoided in optical fiber communicatiosns.
6. Explain what is meant by the critical bending radius for an optical fiber.
7. Explain how attenuation, absorption and scattering influence the properties of fiber.
8. Explain the bending losses.

   What is microbending? Illustrate with a diagram how the microbending is minimized by a compressible jacket.

CHAPTER

45

DISPERSION IN OPTICAL FIBRES

45.1. Introduction

(*i*) Dispersion

When data is sent along the communication fibre, information is generally contained in the form of pulses of intensity of light. Fig. 45.1 shows the input and output pulses across a fibre. It is observed that output pulse is wider than input pulse. Moreover the amplitude of output pulse is also decreased. This shows that the pulse becomes distorted as it propagates through the fibre. The reason of this distortion is dispersion. The term dispersion describes the pulse broadening effect in fibres. Consider an input pulse of width W_1 and output of width W_2 where $W_2 > W_1$. Then fibre dispersion D is defined as

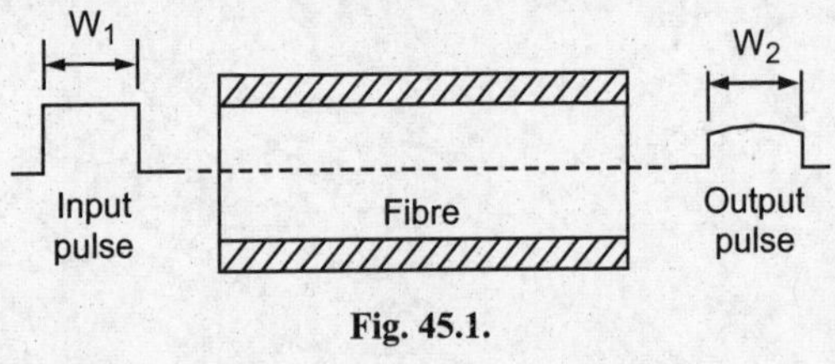

Fig. 45.1.

$$D = \sqrt{\left(W_2^2 - W_1^2\right)}\ .$$

Dispersion is measured in the unit of time either in nanoseconds (10^{-9}s) or picoseconds (10^{-12}s). The total dispersion produced by the fibre depends directly on its length. So, dispersion is typically measured in nanoseconds per kilometre (ns/km).

(*ii*) Intramodal dispersion

Intramodal dispersion is pulse spreading that occurs within a single mode. It is a result of the group velocity being a function of the wavelength λ. Since intramodal dispersion depends on the wavelength, its effect on signal distortion increases with the spectral width of the optical source.

The two main causes of intramodal dispersion are:

(*a*) Material dispersion and (*b*) Waveguide dispersion.

(*a*) Material dispersion

Material dispersion arises from the variation of the refractive index of the core material as a function of wavelength. This causes a wavelength dependence of the group velocity of any given mode; that is, pulse spreading occurs even when different wavelengths follow the same path.

(*b*) Waveguide dispersion

Waveguide dispersion occurs because a single-mode fiber confines only about 80 percent of the optical power to the core. Dispersion thus arises, since the 20 percent of the light propagating in the cladding travels faster than the light confined to the core. The amount of waveguide dispersion depends on the fiber design, since the modal propagation constant β is a function of a/λ (the optical fiber dimension relative to the wavelength λ; here a is the core radius.)

(*iii*) Intermodal dispersion

Light wave propagates through a fibre in different modes. Each of the modes in a multimode fibre travels at a slightly different velocity. So, there is a difference in propagation times for the different modes. The pulse spreads in time as it travels along the fibre. This is known as intermodal dispersion. This kind of dispersion is completely absent in single mode fibres. It also does not depend on the **spectral width** of the source.

45.2. Effect of Dispersion on Pulse Transmission

The term dispersion refers to spreading of light pulse as it propagates through the fiber. Dispersion of the transmitted optical signal causes distortion for both digital and analog transmission along optical fibers.

In digital optical communication system, the dispersion mechanisms within the fiber cause broadening of the transmitted light pulses as they travel along the channel.

Fig. 45.2 illustrates broadening of light pulses as they travel along a fiber using the digital bit pattern 1011.

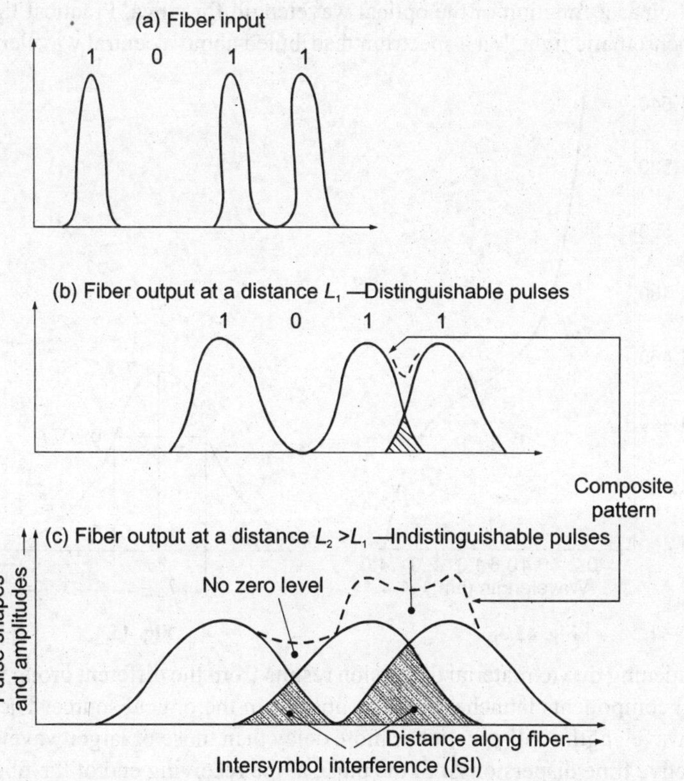

Fig. 45.2.

(*a*) Originally the pulses are separate;

(*b*) the pulses overlap slightly and are clearly distinguishable;

(*c*) eventually the pulses strongly overlap and are indistinguishable.

This leads to intersymbol interference and reduces the upperlimit on the pulse transmission rate. This distortion is a consequence of *intramodal dispersion* and *intermodal delay* effects.

For no overlapping of light pulses down on an optical fiber link the digital bit rate B_T must be less than the reciprocal of the broadened (through dispersion) pulse duration (2τ). Hence:

$$B_T \le \frac{1}{2\tau} \qquad ...(1)$$

This assumes that the pulse broadening due to dispersion on the channel is τ which dictates the input pulse duration which is also τ. Hence Eq. (1) gives the maximum bit rate that may be obtained on an optical fiber link as $1/2\tau$.

Another more accurate estimate of the maximum bit rate for an optical channel with dispersion is obtained by considering the light pulses at the output to have a Gaussian shape with an rms width of σ. The maximum bit rate is given approximately by

$$B_T\,(\max) \simeq \frac{0.2}{\sigma} \text{ bit s}^{-1}. \qquad ...(2)$$

45.3. Material Dispersion

The refractive index of core glass of fiber is not the same for lights of different wavelengths but it depends on the wavelength of light in a complicated way. Fig. 45.3 shows variations in the index of refraction as a function of the optical wavelength for silica. Practical light sources do not emit pure monochromatic light, but a spectrum distributed about a central wavelength λ_o (Fig. 45.4.)

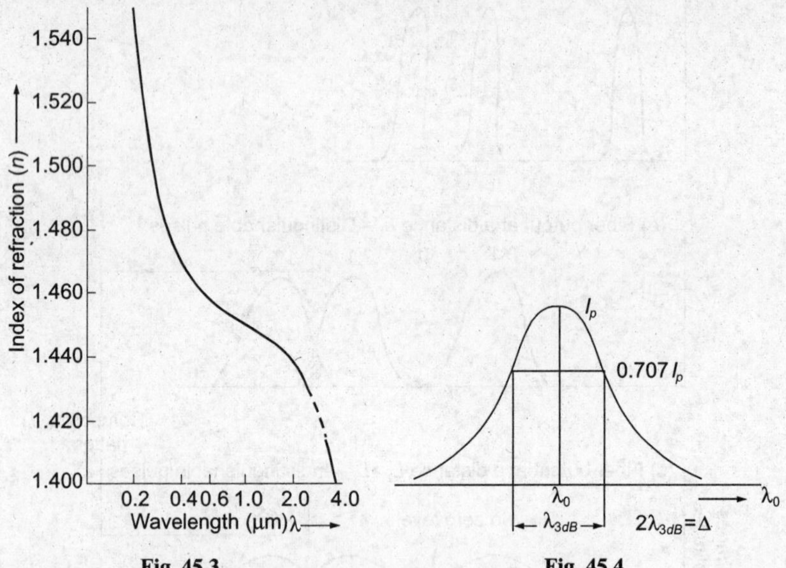

Fig. 45.3. **Fig. 45.4.**

Pulse broadening due to material dispersion results from the different group velocities V_g of the various spectral components launched into the fiber from the optical source. The pulse component having shorter wavelengths will experience more delay than those of larger wavelengths. As a result there is an effective time dispersion Δt of the pulse at the receiving end of the fiber.

A narrow bandwidth source will produce less dispersion than a wide band source. Material dispersion is, therefore, of particular importance for single-mode waveguides and for LED systems (since an LED has a broader output spectrum than a laser diode).

45.4. Relationship between Material Dispersion and Wavelength

Consider a signal that modulates an optical source. Assume that the modulated optical signal excites all modes equally at the input end of the fiber. Each mode thus carries an equal amount of energy through the fiber. Furthermore, each mode contains all of the spectral components in the wavelength band over which the source emits. The signal may be considered as modulating each of these spectral components in the same way.

As the signal propagates along the fiber, each spectral component travels independently. Each component undergoes a time delay or *group delay* per unit length in the direction of propagation given by

$$\frac{\tau_g}{L} = \frac{1}{V_g} = \frac{1}{c}\frac{d\beta}{dk} = -\frac{\lambda^2}{2\pi c}\frac{d\beta}{d\lambda} \qquad ...(1)$$

Here, L = distance travelled by the pulse,
β = propagation constant along the fiber axis,
k = $2\pi / \lambda$, and

$$V_g = c\left(\frac{d\beta}{dk}\right)^{-1} = \textit{group velocity}. \qquad ...(2)$$

Group velocity (V_g) is the velocity at which the energy in a pulse travels along a fiber. The delay difference per unit wavelength along the propagation path is approximately $d\tau_g / d\lambda$.

The total delay difference $\delta\tau$ over a distance L is

$$\delta\tau = \frac{d\tau_g}{d\lambda}\,\delta\lambda \qquad ...(3)$$

If the spectral width $\delta\lambda$ of an optical source is characterized by its root-mean-square (rms) value σ_λ, then the pulse spreading can be approximated by the rms pulse width

$$\sigma_g = \frac{d\tau_g}{d\lambda}\,\sigma_\lambda = -\frac{L\sigma_\lambda}{2\pi c}\left(2\lambda\frac{d\beta}{d\lambda} + \lambda^2\frac{d^2\beta}{d\lambda^2}\right) \qquad ...(4)$$

The factor

$$D = \frac{1}{L}\frac{d\tau_g}{d\lambda} \qquad ...(5)$$

is designated as the *dispersion.* It defines the pulse spread as a function of wavelength and is measured in picoseconds per kilometer per nanometer.

Consider a plane wave propagating in an infinitely extended dielectric medium that has a refractive index $n(\lambda)$ equal to that of the fiber core.

The propagation constant β is thus given by

$$\beta = \frac{2\pi n(\lambda)}{\lambda} \qquad ...(6)$$

Substituting this expression for β into Eq. (1) with $k = 2\pi / \lambda$, we get

$$\tau_{\text{mat}} = \frac{L}{c}\left(n - \lambda\frac{dn}{d\lambda}\right) \qquad ...(7)$$

τ_{mat} is the group delay due to material dispersion in a fiber of length L.

For a source with rms spectral width σ_λ and a mean wavelength λ, the rms pulse broadening due to material dispersion σ_{mat} is given by

$$\sigma_{\text{mat}} = \frac{d\tau_{\text{mat}}}{d\lambda}\,\sigma_\lambda \qquad ...(8)$$

Differentiating Eq. (7) with respect to wavelength, we get

$$\frac{d\tau_{\text{mat}}}{d\lambda} = \frac{L}{c}\left[\frac{dn}{d\lambda} - \lambda\frac{d^2n}{d\lambda^2} - \frac{dn}{d\lambda}\right] = \frac{-L\lambda}{c}\frac{d^2n}{d\lambda^2} \qquad ...(9)$$

Therefore, substituting the expression obtained in Eq. (9) into Eq. (8), the rms pulse broadening due to material dispersion is given by :

$$\sigma_{\text{mat}} = -\frac{L}{c}\lambda\frac{d^2n}{d\lambda^2}\sigma_\lambda \qquad ...(10)$$

Material dispersion can be reduced either by choosing sources with narrower spectral output widths (reducing σ_λ) or by operating at longer wavelengths.

A material is said to exhibit material dispersion when the second differential of the refractive index with respect to wavelength is not zero (*i.e.*, $d^2n/d\lambda^2 \neq 0$).

The material dispersion parameter M is defined as

$$M = \frac{1}{L}\frac{d\tau_{\text{mat}}}{d\lambda} = \frac{\lambda}{c}\left|\frac{d^2n}{d\lambda^2}\right|.$$

M is expressed in units of ps nm^{-1} km^{-1}.

Fig. 45.5 shows the variation of the material dispersion parameter M with wavelength for pure silica. Material dispersion tends to zero in the longer wavelength region around 1.3 μm.

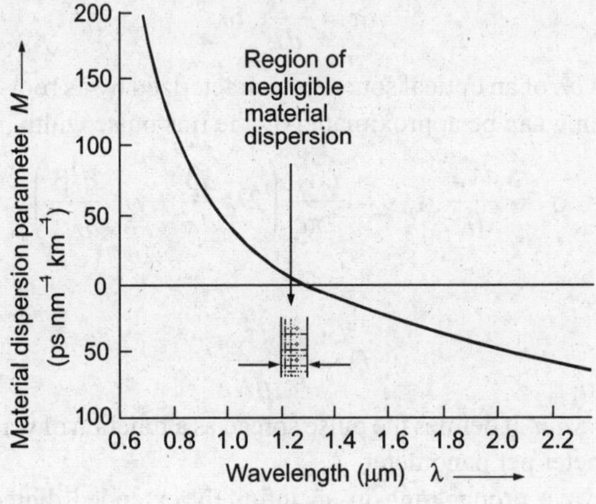

Fig. 45.5.

45.5. Waveguide Dispersion

Assume that the refractive index of the material is independent of wavelength.

The group delay τ_{wg}, that is, the time required for a mode to travel along a fiber of length L is given by

$$\tau_{wg} = \frac{L}{c}\frac{d\beta}{dk} \qquad ...(1)$$

Here, L = length of the fiber,

c = velocity of light in vacuum.

The group delay can be expressed in terms of the normalized propagation constant b defined by

$$b = 1 - \left(\frac{ua}{V}\right)^2 = \frac{\beta^2/k^2 - n_2^2}{n_1^2 - n_2^2}. \qquad ...(2)$$

Here, V is the cut-off parameter (normalized frequency) and a is the fiber radius.

When the refractive index difference between core and cladding is small, we can assume that

$$\beta/k + n_2 \approx n_1 + n_2.$$

$$\therefore \qquad b = \frac{\beta / k - n_2}{n_1 - n_2} \qquad ...(3)$$

or

$$\beta = n_2 k\,(b\Delta + 1) \qquad ...(4)$$

Here, $k = \dfrac{2\pi}{\lambda}$ and

$\Delta = (n_1 - n_2)/n_1$.

Substituting Eq. (4) in Eq. (1), we get

$$\tau_{wg} = \frac{L}{c}\frac{d\beta}{dk} = \frac{L}{c}\left[n_2 + n_2\Delta\frac{d(kb)}{dk}\right] \qquad ...(5)$$

We know that

$$V = ka\left(n_1^2 - n_2^2\right)^{1/2} \simeq kan_2\sqrt{2\Delta}\,.$$

Eq. (5) can be expressed in terms of *V*-parameter as

$$\tau_{wg} = \frac{L}{c}\left[n_2 + n_2\,\Delta\frac{d(Vb)}{dV}\right] \qquad ...(6)$$

The first term in Eq. (6) is a constant.

The second term represents the group delay arising from waveguide dispersion.

The factor $d(Vb)\,/\,dV$ can be expressed as

$$\frac{d(Vb)}{dV} = b\left[1 - \frac{2J_\nu^2\,(ua)}{J_{\nu+1}\,(ua)\,J_{\nu-1}\,(ua)}\right].$$

Fig. 45.6 shows a plot of $\dfrac{d(Vb)}{dV}$ versus V for various LP modes.

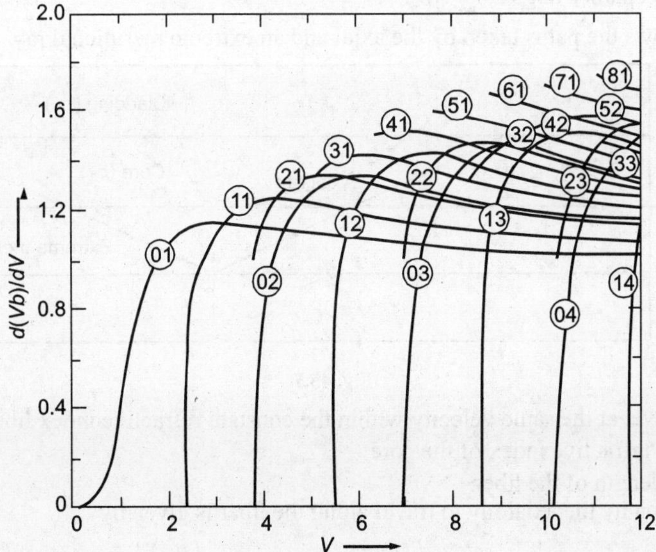

Fig. 45.6.

The plots show that, for a fixed value of V, the group delay is different for every guided mode. When a light pulse is launched into a fiber, it is distributed among many guided modes. These various modes arrive at the fiber end at different times depending on their group delay, so that a pulse spreading results.

45.6. Intermodal Dispersion

Pulse broadening due to intermodal dispersion results from the propagation delay differences between modes within a multimode fiber.

When a pulse of light is transmitted in a fiber, it will propagate over several different paths corresponding to the excited modes and come out of the far end at slightly different times. The received pulse at the far end is the summation of these mode pulses each delayed by different amount of time.

The shortest time delay will be suffered by the plane wave propagating straight along the axis of the fiber core. The longest time delay will be suffered by the wave propagating with highest order supported mode.

The pulse broadening is the difference between the travel time T_{max} of the longest ray congruence paths (the highest-order mode) and the travel time T_{min} of the shortest ray congruence paths (the fundamental mode). This is obtained from ray tracing and is given by

$$\sigma_{mod} = T_{max} - T_{min} = \frac{n_1 \Delta L}{c}.$$

Here, n_1 = refractive index of the core,
Δ = $(n_1 - n_2) / n_1$ and
L = length of the fiber.

45.6.1. Pulse broadening in multimode step index fiber

(*i*) The fastest mode propagating in the step index fiber is represented by the axial ray.

(*ii*) The slowest mode is represented by the extreme meridional ray (which is incident at the core-cladding interface at the critical angle ϕ_c).

In this analysis we consider only pulse broadening due to meridional rays and totally ignore skew rays with acceptance angles $\theta_{as} > \theta_a$.

Fig. 45.7 shows the paths taken by the axial and an extreme meridional ray.

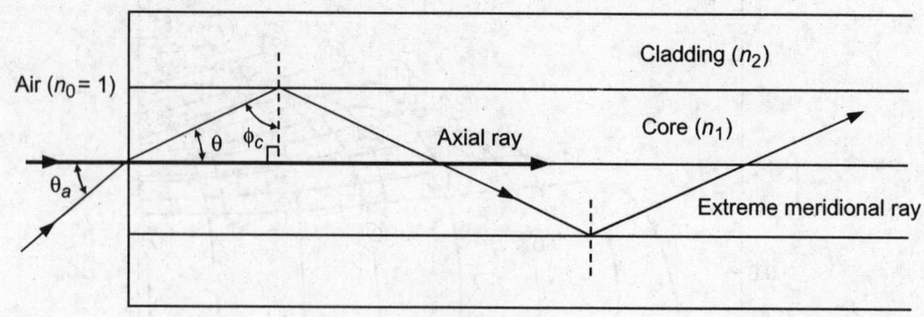

Fig. 45.7.

Both rays travel at the same velocity within the constant refractive index fiber core.
Let n_1 be the refractive index of the core.
Let L be the length of the fiber.
The time taken by the axial ray to travel along the fiber is given by

$$T_{min} = \frac{distance}{velocity} = \frac{L}{(c / n_1)} = \frac{L n_1}{c}. \qquad ...(1)$$

Here, c is the velocity of light in a vacuum.

The extreme meridional ray exhibits the maximum delay time T_{max}.

$$T_{max} = \frac{L/\cos\theta}{c/n_1} = \frac{Ln_1}{c\cos\theta} \quad \text{...(2)}$$

By Snell's law of refraction at the core-cladding interface

$$\sin\phi_c = \frac{n_2}{n_1} = \cos\theta \quad \text{...(3)}$$

Here, n_2 is the refractive index of the cladding.

Substituting into Eq. (2) for cos θ, we get

$$T_{max} = \frac{Ln_1^2}{cn_2} \quad \text{...(4)}$$

The delay difference δT_s between the extreme meridional ray and the axial ray is

$$\delta T_s = T_{max} - T_{min} = \frac{Ln_1^2}{cn_2} - \frac{Ln_1}{c} = \frac{Ln_1^2}{cn_2}\left(\frac{n_1 - n_2}{n_1}\right) \quad \text{...(5)}$$

$$\simeq \frac{Ln_1^2\Delta}{cn_2} \quad \text{when } \Delta \ll 1 \quad \text{...(6)}$$

The relative refractive index difference

$$\Delta \simeq \frac{n_1 - n_2}{n_2} \quad \text{...(7)}$$

Eq. (5) becomes $$\delta T_s = \frac{Ln_1}{c}\left(\frac{n_1 - n_2}{n_2}\right) \simeq \frac{Ln_1\Delta}{c} \quad \text{...(8)}$$

The relation between numerical aperture (NA) and Δ is

$$NA = n_1 (2\Delta)^{1/2} \Rightarrow \Delta = \frac{(\text{NA})^2}{2n_1^2} \quad \text{...(9)}$$

Substituting for Δ, Eq. (8) becomes $$\delta T_s \simeq \frac{L(NA)^2}{2n_1 c} \quad \text{...(10)}$$

Here, *NA* is the numerical aperture for the fiber. Eqs. (8) and (10) give the maximum pulse broadening in time due to intermodal dispersion in multimode step index fibers.

EXERCISE

1. Briefly explain the reasons for pulse broadening due to material dispersion in optical fibers.
2. Derive an expression for the rms pulse broadening due to material dispersion in an optical fiber and define the material dispersion parameter.
3. Describe the mechanism of intermodal dispersion in a multimode step index fiber.

   Show that the total broadening of a light pulse δT_s due to intermodal dispersion in a multimode step index fiber may be given by: $$\delta T_s \simeq \frac{L(NA)^2}{2n_1 c}$$

   where L is the fiber length, *NA* is the numerical aperture of the fiber, n_1 is the core refractive index and c is the velocity of light in a vacuum.

CHAPTER 46

OPTICAL FIBRE COMMUNICATION SYSTEM

46.1. Analog Optical Fiber Communication System

Fig. 46.1 shows the schematics of an analog optical fiber communication system.

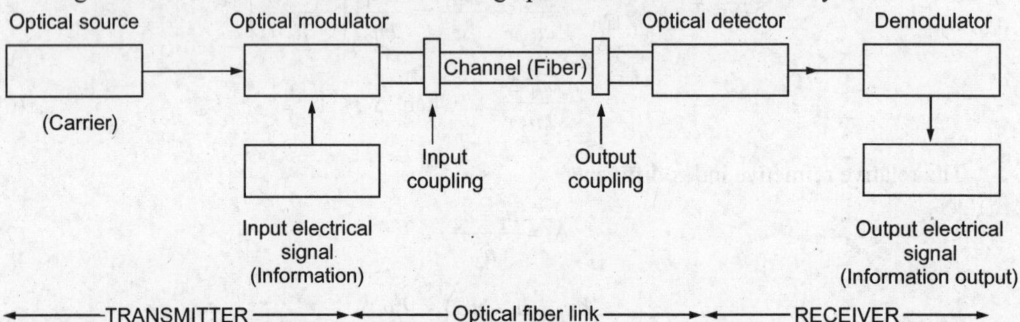

Fig. 46.1.

(*i*) An *analog signal* is continuous and reproduces the form of original message. For example, when a microphone picks up a sound wave, it produces the same shape as the wave itself (Fig. 46.2). The input electrical signal modulates the intensity of light from the optical source.

(*ii*) **Optical Source.** The function of optical source is to generate the carrier wave on which the information is transmitted. Light-emitting diode (LED) and laser diode (LD) are used as optical sources. These devices produce stable, single frequency waves with sufficient power for long distance propagation.

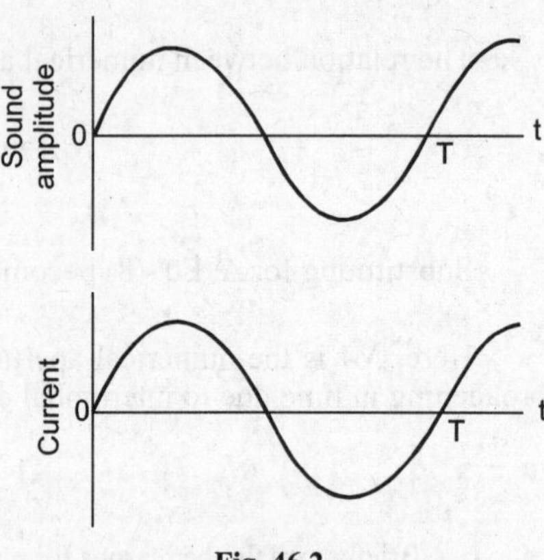

Fig. 46.2.

(*iii*) **Modulator.** The process of imposing information on a light stream is called *modulation.* The role of the modulator is to impress the information onto the continuous wave signal from the laser or LED source.

The optical carrier can be modulated internally or externally using an electro-optic modulator or acousto-optic modulator. Nowadays electro-optic modulators (KDP, $LiNbO_3$ or Beta Barium Borate) are widely used as external modulators.

(*iv*) The transmission channel is optical fiber link. The input coupler transfers the modulated light beam from the source to the optic fibre.

The input coupler is a microlens that focuses the optical signal onto the entrance plane of an optical fibre with the maximum possible efficiency.

After an optical signal is launched into a fiber, it will become progressively attenuated and distorted with increasing distance because of scattering, absorption, and dispersion mechanisms in the glass material.

Fig. 46.3 shows optical fiber attenuation as a function of wavelength.

Early technology made exclusive use of the 800-900-nm wavelength band, since, in this region, the fibers made at that time exhibited a local minimum in the attenuation curve. This region is called the *first window*.

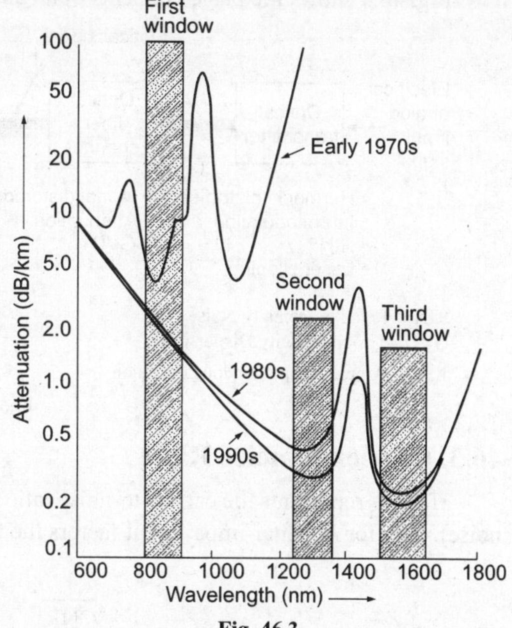

Fig. 46.3.

In the 1980s, optical fibres were fabricated with very low loss in the 1100-to-1600-nm region. This spectral band is called the *long-wavelength region.* The *second window* is centered around 1310 nm.

The *third window* is centered around 1550 nm.

(*v*) The output coupler directs the light emerging from the fibre onto the light detector. For this purpose, a simple butt connection is used.

(*vi*) At the receiver, a photo-diode detects the weakened optical signal emerging from the fiber end and converts it to an electric current (referred to as a *photocurrent*). Silicon photodiodes are used in the 800-to-900-nm region. The prime material in the 1100-to-1600-nm region is an InGaAs alloy.

(*vii*) The electrical signal is separated from the optical carrier by demodulator. Thus the original signal is obtained.

ANALOG SYSTEMS

46.2. Introduction

In certain areas of the telecommunication network, information transfer in *analog form* is advantageous. Some examples of this are microwave-multiplexed signals, subscriber services using hybrid fiber/coax (HFC), video distribution, antenna remoting, and radar signal processing. For most analog applications, laser diode transmitters are used.

An analog optical link consists of a transmitter, a fiber link and a receiver. The important parameters of an analog link are the carrier-to-noise ratio, bandwidth, and signal distortion.

The *carrier-to-noise ratio* is given by

$$\frac{C}{N} = \frac{\text{carrier power}}{\text{components noise power} + \text{intermodulation noises}}.$$

Fig. 46.4 shows the basic elements of an analog link and the major noise contributors.

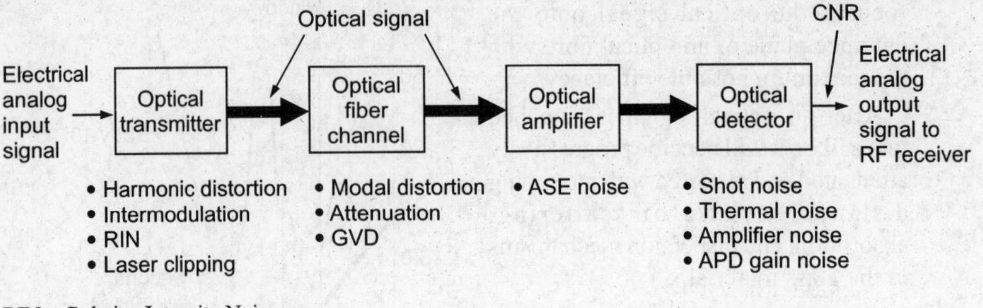

RIN – Relative Intensity Noise
GVD – Group Velocity Dispersion
ASE – Amplified spontaneous Emission

Fig. 46.4.

46.3. Carrier-to-noise Ratio

If CNR_i represents the carrier-to-noise ratio related to a particular signal contaminant (e.g., shot noise), then for N signal-impairment factors the total CNR is given by

$$\frac{1}{\text{CNR}} = \sum_{i=1}^{N} \frac{1}{\text{CNR}_i}.$$

For links in which only a single information channel is transmitted, the important signal impairments include laser intensity noise fluctuations, laser clipping, photo-detector noise, and optical-amplifier noise. When multiple message channels operating at different carrier frequencies are sent simultaneously over the same fiber, then harmonic and intermodulation distortions arise.

46.3.1. Carrier Power

Fig. 46.5 shows biasing conditions of a laser diode and its response to analog signal modulation.

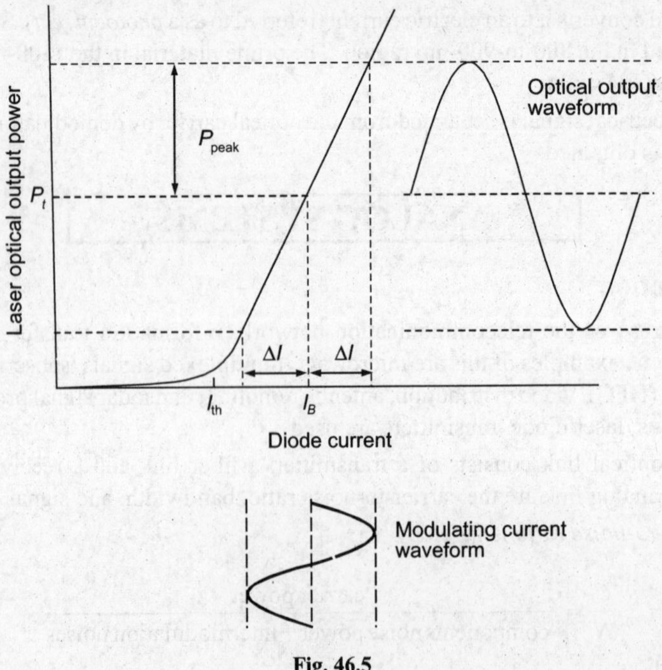

Fig. 46.5

The drive current through the optical source is the sum of the fixed bias current and a time-varying sinusoid. The source acts as a square-law device. So the envelope of the output optical power $P(t)$ has the same form as the input drive current.

Let $s(t)$ be the time-varying analog drive signal.

Let m be the *modulation index* defined as

$$m = \frac{\Delta I}{I'_B}$$

For laser diode, $I_B' = I_B - I_{th}$.

The parameter ΔI is the variation in current about the bias point.

Typical m values for analog applications range from 0.25 to 0.50.

The transmitted power

$$P(t) = P_t\,[1 + ms(t)] \quad \text{...(1)}$$

Here, P_t is the optical output power at the bias current level.

In terms of optical power, the modulation index is given by

$$m = \frac{P_{\text{peak}}}{P_t} \quad \text{...(2)}$$

Here, P_{peak} and P_t are defined in Fig. 46.5.

For a sinusoidal received signal, the carrier power C at the receiver output is given by

$$C = \frac{1}{2}(mR_0\, M\, \bar{P})^2 \quad \text{...(3)}$$

Here, R_0 = unity gain responsivity of the photodetector,
M = photodetector gain,
$\bar{P}$ = average received optical power.

46.3.2. Photodetector and Preamplifier Noises

The expression for the photodiode noise current is given by

$$\left\langle i_N^2 \right\rangle = \sigma_N^2 \approx 2q(I_p + I_D)M^2\, F(M)B \quad \text{...(1)}$$

Here, q = electronic charge,
I_P = $R_0 \bar{P}$ is the primary photocurrent,
I_D = detector bulk dark current,
M = photodiode gain with $F(M)$ being its associated noise figure, and
B = receiver bandwidth.

Then, the CNR for the photodetector only is

$$\text{CNR}_{\text{det}} = C/\sigma_N^2\,. \quad \text{...(2)}$$

The expression for the preamplifier noise current is given by

$$\left\langle i_T^2 \right\rangle = \sigma_T^2 = \frac{4k_B T}{R_{\text{eq}}}\, BF_t \quad \text{...(3)}$$

Here, R_{eq} is the equivalent resistance of the photodetector load and the pre-amplifier, and F_t is the noise factor of the preamplifier.

Then, the CNR for the preamplifier only is

$$\text{CNR}_{\text{preamp}} = C/\sigma_T^2. \quad \text{...(4)}$$

46.3.3. Relative Intensity Noise (RIN)

Within a semiconductor laser, fluctuations in the amplitude or intensity of the output produce optical intensity noise. These fluctuations could arise from temperature variations or from spontaneous emission contained in the laser output.

The resultant mean-square noise current is given by

$$\left\langle i_{\mathrm{RIN}}^2 \right\rangle = \sigma_{\mathrm{RIN}}^2 = \mathrm{RIN}\,(R_o\bar{P})^2\, B \qquad \text{...(5)}$$

Then, the CNR due to laser amplitude fluctuations only is $\mathrm{CNR}_{\mathrm{RIN}} = C/\sigma_{\mathrm{RIN}}^2$.

Here, RIN is the *relative intensity noise* of the laser.

The noise resulting from the random intensity fluctuations is called *relative intensity noise* (RIN). RIN is defined in terms of the mean-square intensity variations.

RIN is defined by the noise-to-signal power ratio

$$\mathrm{RIN} = \frac{\left\langle (\Delta P_L)^2 \right\rangle}{\bar{P}_L^2}\ (dB/Hz) \qquad \text{...(6)}$$

Here, $\left\langle (\Delta P_L)^2 \right\rangle$ is the mean-square intensity fluctuation of the laser output and

$\bar{P}_L$ is the average laser light intensity.

This noise decreases as the injection-current level increases according to the relationship

$$\mathrm{RIN} \propto \left(\frac{I_B}{I_{\mathrm{th}}} - 1\right)^{-3} \qquad \text{...(7)}$$

Here, I_B and I_{th} are the bias and threshold current of the laser, respectively.

For a single-channel AM system, the important signal impairments are

(*i*) laser intensity noise fluctuations, given by Eq. (5),

(*ii*) photo-detector noise, given by Eq. (1) and

(*iii*) optical-amplifier noise, given by Eq. (3).

The carrier power C at the receiver output is given by

$$C = \frac{1}{2}\,(m\,R_0\,M\,\bar{P})^2 \qquad \text{...(8)}$$

If CNR_i represents the carrier-to-noise ratio related to a particular signal contaminant (e.g., shot noise), then for N signal-impairment factors the total CNR is given by

$$\frac{1}{\mathrm{CNR}} = \sum_{i=1}^{N} \frac{1}{\mathrm{CNR}_i} \qquad \text{...(9)}$$

Substituting the CNRs resulting from Eqs. (8), (5), (1) and (3) into Eq. (9), we get

$$\frac{C}{N} = \frac{\frac{1}{2}(mR_0\,M\,\bar{P})^2}{\mathrm{RIN}\,(R_0\bar{P})^2\,B + 2q(I_p + I_D)\,M^2\,F(M)B + (4k_BT/R_{\mathrm{eq}})BF_t} \qquad \text{...(10)}$$

Eq. (10) gives the carrier-to-noise ratio for a single-channel AM system.

In analyzing the performance of analog systems, we calculate the ratio of rms carrier power to rms noise power at the input of the RF receiver following the photodetection process. This is known as the *carrier-to-noise ratio* (CNR). A widely used analog signal is a 525-line studio-quality television signal. Using amplitude modulation (AM) for such a signal requires a CNR of 56 dB, since the need for bandwidth efficiency leads to a high signal-to-noise ratio. Frequency modulation (FM), on the other hand, only needs CNR values of 15-18 dB.

46.4. Digital Optical Fiber Communication System

Fig. 46.6 shows the important components of a digital optical fiber communication system.

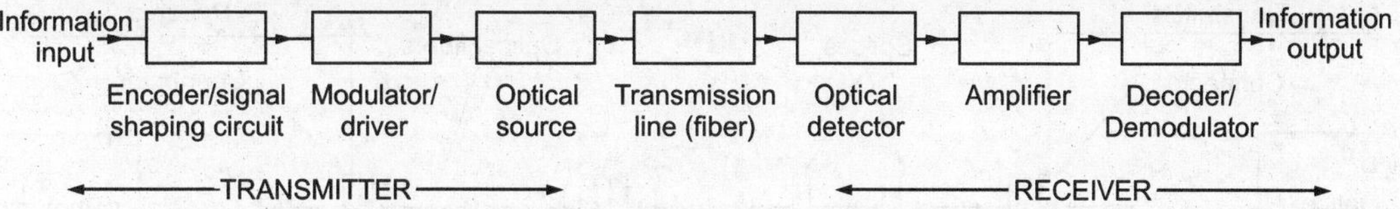

Fig. 46.6.

(*i*) Information encoder converts the information signal into coded pulses.

(*ii*) The modulator or electronic driver controls the operation of the optical source. The coded pulses modulate the intensity of light from the laser diode or LED by changing its switching current. The emitted light from the source is coupled to the fiber optic cable.

(*iii*) The light emerging from the other end of fiber optic cable is given to an optical detector like avalanche photodiode (APD) which converts the light pulses into electrical pulses.

(*iv*) Amplifier amplifies the detected electrical signal.

(*v*) The amplified signals are decoded to get the original information.

DIGITAL TRANSMISSION SYSTEMS

46.5. Introduction

A simple point-to-point digital optical communication link is illustrated in Fig. 46.7.

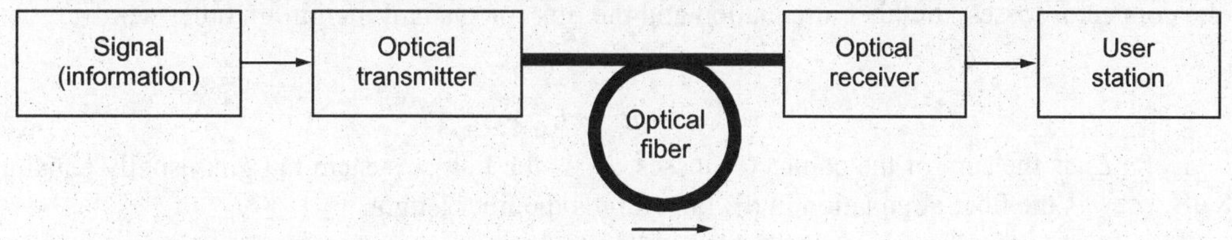

Fig. 46.7.

The key parameters characterizing the link are:

1. transmission distance,
2. data rate or channel bandwidth and
3. bit error rate (BER).

Choice of fibers, sources and detectors is made on the basis of the above three parameters.

(*i*) Important properties of the fiber are single-or multi-mode nature, dispersion, attenuation, core-size, and numerical aperture.

(*ii*) With respect to sources, the most important decision to be made is regarding the use of an LED or laser. Device properties such as output power, emission wavelength, spectral linewidth, and number of emitting modes are important considerations.

(*iii*) The transmission distance, bandwidth, and BER will decide the choice of detectors. The important parameters of these devices are:

bandwidth, sensitivity, responsivity, and operating wavelength.

The power loss and dispersion characteristics of an optical link are analyzed, respectively, by the *link power budget* and the *rise-time budget*.

46.6. Link Power Budget

Fig. 46.8 shows the optical power loss model for a point-to-point link.

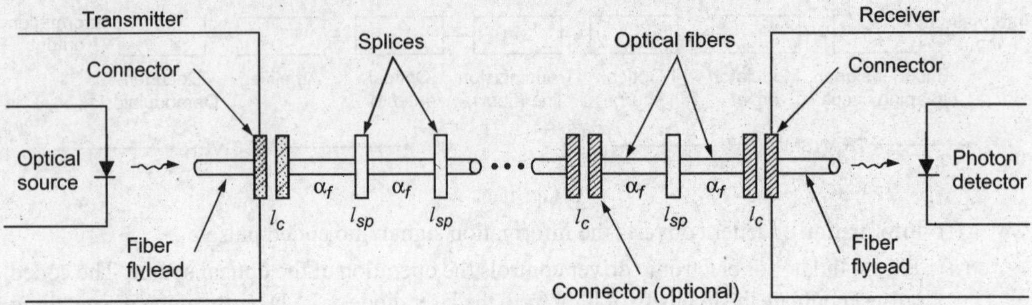

Fig. 46.8.

In the link power budget, the loss in each element of the link is expressed in decibels as

$$\text{loss} = 10 \log_{10} \frac{P_{\text{out}}}{P_{\text{in}}} \quad ...(1)$$

Here, P_{out} and P_{in} are the input and output powers of the element under consideration.

A link or system power margin is usually incorporated in the link power budget to take into account factors such as additional components, aging of components and other loss factors normally not accounted for.

The total allowed power loss P_L is then simply the difference between the power output at the fiber end of the transmitter, P_T, and the sensitivity of the photoreceiver, P_R. This, in turn, is the sum of the connector losses, the fiber attenuation and the link (or system) margin. In other words,

$$\begin{aligned} P_L &= P_T - P_R \\ &= L_C + L_M + \alpha_{\text{fiber}}\, l \end{aligned} \quad ...(2)$$

Here, L_C is the sum of the connector losses, L_M is the link or system margin, usually equal to 6-8 dB, α_{fiber} is the fiber attenuation in dB/km and l is the fiber length.

The splice loss is incorporated into the cable loss for simplicity.

46.7. Rise-Time Budget

The risetime budget accounts for the dispersion characteristics of the link.

The total rise time t_{sys} of the link is the root sum square of the rise times from each contributor t_i to the pulse rise-time degradation.

$$t_{\text{sys}} = \left(\sum_{i=1}^{N} t_i^2 \right)^{1/2} \quad ...(1)$$

The risetime of individual components limits the speed, or bandwidth, of the system. The important components whose risetimes need to be considered are the transmitter and receiver and the fiber itself. Again, the risetime of the fiber is usually the sum of the material dispersion and modal dispersion risetimes.

The four basic elements that may significantly limit system speed are the transmitter ris . time t_{tx}, the group-velocity dispersion (GVD) rise time t_{GVD} of the fiber, the modal dispersion rise time t_{mod} of the fiber, and the receiver rise time t_{rx}.

The total system rise time is given by

$$t_{\text{sys}} = \left[t_{tx}^2 + t_{\text{mod}}^2 + t_{\text{GVD}}^2 + t_{rx}^2 \right]^{1/2}$$

$$= \left[t_{tx}^2 + \left(\frac{440 L^q}{B_0} \right)^2 + D^2 \sigma_\lambda^2 L^2 + \left(\frac{350}{B_{rx}} \right)^2 \right]^{1/2} \quad ...(2)$$

Here, all the times are given in nanoseconds, L is the length of the link, the parameter q ranges between 0.5 and 1, B_0 is the bandwidth of a 1-km length of cable, D is the dispersion, σ_λ is the half-power spectral width of the source, and B_{rx} is the 3-dB electrical bandwidth of the receiver.

46.8. Line Coding

Digital data are transmitted as a stream of 0's and 1's.

Signal encoding uses a set of rules for arranging the signal symbols in a particular pattern. This process is called *channel* or *line coding.*

One of the principal functions of a line code is to introduce redundancy into the data stream for the purpose of minimizing errors that result from channel interference effects. Depending on the amount of redundancy introduced, any degree of error-free transmission of digital data can be achieved, provided that the data rate that includes this redundancy is less than the channel capacity. This is a result of the well-known Shannon channel-coding theory.

Line coding describes the manner in which the symbols are arranged in a pattern. Different codes, or formats, are followed to transmit two-level binary digital data. The two most commonly used formats are the *non-return-to-zero* (NRZ) and *return-to-zero* (RZ).

These are illustrated in Fig. 46.9.

In the NRZ format, the optical pulse remains on throughout the bit slot and its amplitude does not drop to zero between two or more successive 1 bits. As a result, pulse width varies depending on the bit pattern.

In the RZ format, each optical pulse representing bit 1 is shorter than the bit slot, and its amplitude returns to zero before the bit duration is over. The pulse width remains the same in the case of RZ format.

Other less popular line codes are the *phase encoded* (PE) format and the *multilevel binary* (MLB) format.

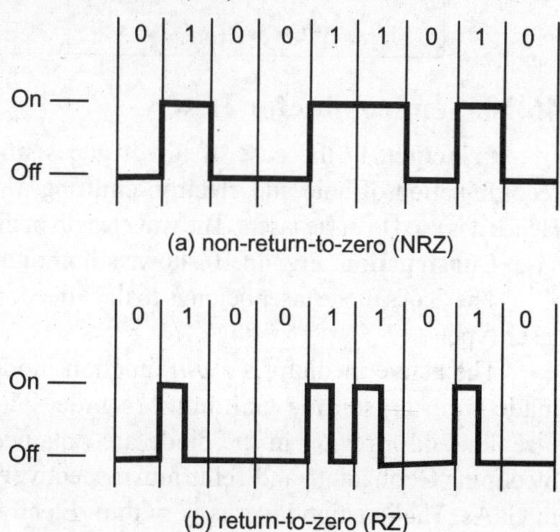

Fig. 46.9.

OPTICAL SOURCES

46.9. Requirements for Communication Light Sources

1. The source should emit light at wavelengths where the fiber has low losses and low dispersion and where the detectors are efficient.
   It must emit the required wavelengths of 1.31 μm and 1.55 μm in the case of silica fibers.
2. The source should have a very narrow spectral bandwidth (linewidth) in order to minimize dispersion in the fiber.
3. It should be possible to modulate the source at high speeds (more than several Gb/s).
4. The light source should have compact size and high efficiency.

5. The source must be capable of maintaining a stable optical output which is largely unaffected by changes in ambient conditions (e.g., temperature).
6. It is essential that the source is comparatively cheap and highly reliable in order to compete with conventional transmission techniques.

The principal light sources used for fiber optic communications applications are heterojunction-structured semiconductor *laser diodes* and *light-emitting diodes* (LEDs).

These devices are suitable for fiber transmission systems because:

(*i*) they have adequate output power for a wide range of applications,
(*ii*) their optical power output can be directly modulated by varying the input current to the device,
(*iii*) they have a high efficiency, and
(*iv*) their dimensional characteristics are compatible with those of the optical fiber.

Advantages of LEDs and Semiconductor Lasers

Characteristic	LED	Laser
Output Power	Lower	Higher
Spectral Width	Wider	Narrower
Numerical Aperture	Larger	Smaller
Rise Time	Slower	Faster
Cost	Less	More
Ease of Operation	Easier	More Complicated.

46.10. Semiconductor Laser

Principle. In the case of direct gap semiconductors, there is a large possibility for direct recombination of hole and electron emitting a photon. GaAs is a direct band gap semiconductor. Hence it is used to make lasers. The wavelength of the emitted light depends on the band gap of the material.

Construction. Fig. 46.10 shows a homojunction GaAs semiconductor diode laser.

The GaAs diode laser belongs to the injection laser type.

The active medium is a *p-n* junction diode made from crystalline Gallium Arsenide. The *p*-region and *n*-region in the diode are obtained by doping Germanium and Tellurium respectively in GaAs. The junction layer is very thin. Electric current is applied to the crystal through an electrode strip fixed to its upper surface. At the junction, light is emitted. The sides through which emitted light comes out are well polished and parallel to each other.

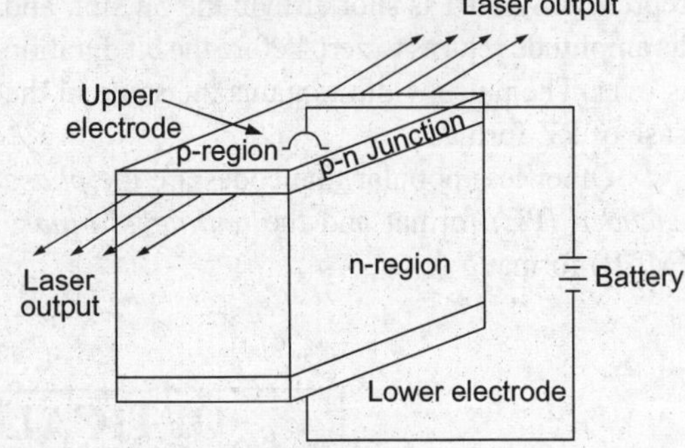

Fig. 46.10.

Working. When the *pn* junction diode is forward biased, electrons are injected across the junction from the (*n*) to (*p*) region. Thus a population inversion is achieved. This is efficient if a large current of the order of 10^4 A/cm^2 is passed through the junction. The electrons and holes recombine and produce radiations in the form of light.

The emitted photons also increase the rate of recombination of injected electrons. Thus more number of photons are produced. This provides stimulated emission. Hence the emitted photons from induced recombinations are having same phase and frequency as that of original inducing photons. The efficiency of laser emission increases when we cool the GaAs diode.

Explanation of the operation of a laser diode based on band theory

(*i*) **Absorption.** An electron in the valence band can make an upward transition to the conduction band by absorbing a photon of energy $(E_c - E_v)$. Here, E_v is the maximum energy of the valence band and E_c is the minimum energy of the conduction band.

(*ii*) **Spontaneous emission.** An electron in the conduction band can recombine with a hole in the valence band by spontaneously emitting a photon of energy $(E_c - E_v)$.

(*iii*) **Stimulated emission.** An electron in the conduction band can recombine with a hole in the valence band in the presence of a stimulating or inducing photon whose energy also equals to $(E_c - E_v)$ giving stimulated emission. The emitted photon and inducing photon travel in the same direction with the same energy. The emitted photon is fully coherent with the inducing photon.

Advantages

1. Semiconductor lasers are compact, efficient and can be fabricated with ease.
2. They can have a continuous wave output or pulsed output.

Disadvantages

Their monochromaticity, coherence and directionality are inferior to those of other lasers.

Example 1. *The band gap of GaAs is 1.44 eV. Calculate the wavelength of emitted radiation.*

Solution.
$$E_g = h\nu = \frac{hc}{\lambda}$$

$$\lambda = \frac{hc}{E_g} = \frac{(6.626\times10^{-34})\times(3\times10^8)}{1.44\times(1.6\times10^{-19})} \text{ m}$$

$$= 8.628 \times 10^{-7} \text{ m} = 862.8 \text{ nm.}$$

46.11. Hetero-Junction Lasers (Modern Laser Diodes)

Heterojunction means that the material on one side of the junction differs from that on the other side of the junction. In modern GaAs diode lasers, a hetero junction is formed between GaAs and Ga AI As.

Construction of hetero junction laser diode:

Fig. 46.11 shows a heterojunction GaAs diode laser.

A layer of GaAs is sandwiched between two layers of Ga AI As which have a wider energy gap and a lower refractive index than Ga As on GaAs substrate. Table below shows the different Laser materials for GaAs diode.

Layer Number	*Name of the Layer*	*Material*
1.	Contact layer	GaAIAs
2.	P-type wide band gap confinement	GaAIAs
3.	p-type narrow band gap active layer	GaAs
4.	N-type wide band gap confinement	GaAIAs
5.	n-type substrate	GaAs

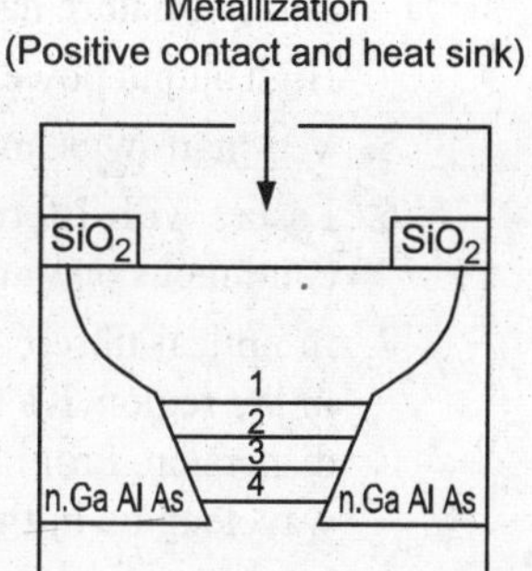

Fig. 46.11.

The Ga As diode has N-p-P structure. N and P represent the wider band gap semiconductors. p represents the narrow band gap semiconductor.

The front and the back faces of the active layer are dielectric coated so that the reflections at the active layer-air interface provide sufficient feedback for laser oscillation.

Working:

(*i*) The basic principle of working of hetero junction laser diode and homojunction laser diode is the same.

We have seen that the threshold current density for homojunction lasers is very large due to poor optical and carrier confinement. But threshold current density for heterojunction lasers is very low. Using multilayers in the heterojunction laser structure, carriers are confined to a narrow region so that population is built up at lower current levels.

(*ii*) Charge injection takes place into the active layer. As a result, spontaneous emission of photons is produced. Some of the injected charges are stimulated to emit by other photons. A large number of injected charges are available for stimulated recombination and they create population inversion, provided the current density is sufficiently high and optical gain is also high.

(*iii*) The band gap difference prevents the diffusion of injected charge carriers from the active layer.

(*iv*) Stimulated radiation gives coherent laser radiation.

(*v*) The step change in refractive index provides an efficient waveguide structure. Thus the injected charges and radiation are confined mainly to the active layer. This leads to get a lower value of threshold current. Further the active region is few microns wide so that the threshold current is further reduced.

(*vi*) The stripe geometry provides the confinement of charges in the lateral direction and the longer life for the laser diodes.

(*vii*) The output wavelength is 0.8 μm when the band gap of active layer is 1.55 eV.

(*viii*) Thus high power output, narrow spectral width, high efficiency and high coherence can be achieved through the double hetero junction stripe laser diodes.

Advantages of hetero-junction lasers :

1. Threshold current density is very low (10 A/mm^2) at room temperature.
2. Continuous wave operation is possible.
3. Carrier and optical confinement can be achieved simultaneously. Thus more efficient wave guide structure has been achieved and hence the beam divergence is reduced.
4. High output power (10 mW) can be achieved even with low operating current (< 500 mA).
5. Very narrow beam with high coherence and monochromacity is achieved.
6. Lasers with lifetimes in excess of 40,000 hours are now available corresponding to continuous operation over a five-year period.
7. In optical fibre communications, it is desirable to have a laser emitting at wavelengths in the region 1.1 to 1.6 μm, where present optical fibers have minimum attenuation and dispersion. From lasers fabricated with quaternary compounds such as $Ga_x\ In_{1-x}\ As_{1-y}\ P_y$, wavelengths in this range can be obtained. Hence these hetero-junction laser diodes are used as optical sources in optical fibre communication.

46.12. Light Emitting Diode (LED)

A *light* emitting diode (LED) is a specially made forward biased P-N junction diode which emits visible light when energized. The energy emitted from visible LEDs is spontaneous or noncoherent.

Theory

When a junction diode is forward biased, electrons from *n*-side and holes from *p*-side move towards the depletion region and they recombine. During this process, energy is released because electrons make transition from conduction band (higher energy level) to valence band (lower energy level) [Fig. 46.12].

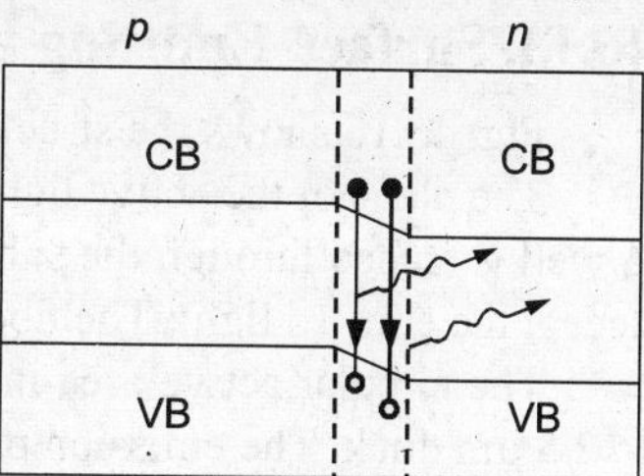

Fig. 46.12.

If E_g is semiconductor band gap, then the energy $E_g = h\nu = \frac{hc}{\lambda}$ is emitted in the form of radiation.

The corresponding emission wavelength is given by

$$\lambda = \frac{hc}{E_g}.$$

In case of Gallium Arsenide Phosphide (Ga As P), band gap $E_g = 1.9$ eV.

$$\lambda = \frac{hc}{E_g} = \frac{(6.62\times10^{-34})\times(3\times10^8)}{1.9\times(1.6\times10^{-19})} m = 655.3\, nm\, (red).$$

Construction

At first an *N*-type layer is grown on a substrate and then a *P*-type layer is deposited on it by the process of diffusion (Fig. 46.13). Metal contacts (Anode) are made at the outer edge of the *P*-layer so that more upper surface is left free for light to escape. For making Cathode connections, a metal film (preferably gold) is coated at the bottom of the substrate. This film also reflects as much light as possible to the surface of the device.

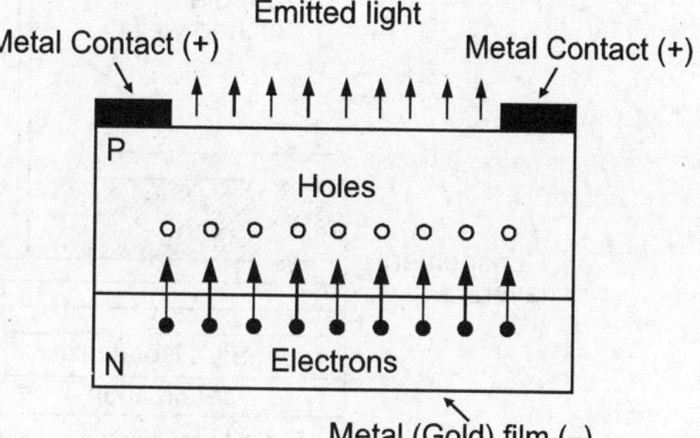

Fig. 46.13.

Fig. 46.14 shows the schematic symbol of LED.

Ga As P LED emits red light when forward biased (Fig. 46.15).

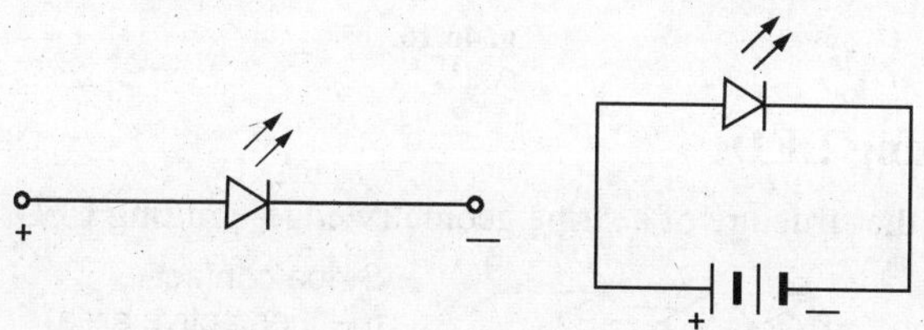

Fig. 46.14. Fig. 46.15.

Fiber Optical Communication based LEDs:

LED material	Band gap energy (eV)	Wavelength range (μm)
In Ga As P	0.73 – 1.35	0.9 – 1.7
In Ga As	0.95 – 1.24	1 – 1.3
Ga Al As	1.40 – 1.55	0.8 – 0.9
Ga As	1.4	0.9

The two basic LED configurations being used for fiber optics are *surface emitters* (also called *Burrus* or *front emitters*) and *edge emitters*.

46.13. Surface-Emitting LED (Burrus type)

Fig. 46.16 shows the structure of a high-radiance surface-emitting LED.

The plane of the active light-emitting region is oriented perpendicularly to the axis of the fiber. A well is etched through the substrate of the device, into which a fiber is then cemented in order to accept the emitted light. The fiber is properly aligned to optimize coupling of the emitted radiation.

The circular active area in practical surface emitters is nominally 50 μm in diameter and up to 2.5 μm thick. The emission pattern is essentially isotropic with a 120° half-power beam width.

This isotropic pattern from a surface emitter is called a *lambertian pattern.* In this pattern, the source is equally bright when viewed from any direction, but the power diminishes as cos θ, where θ is the angle between the viewing direction and the normal to the surface (this is because the projected area one sees decreases as cos θ). Thus, the power is down to 50 percent of its peak when θ = 60°, so that the total half-power beam width is 120°.

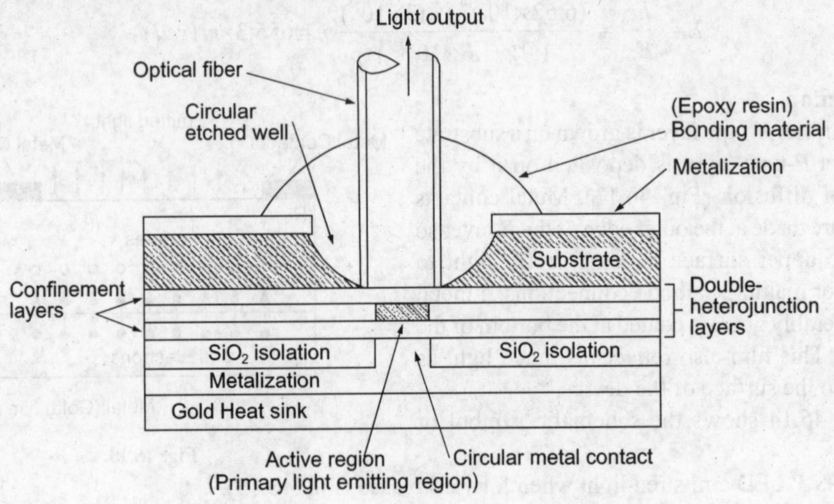

Fig. 46.16.

46.14. Edge-Emitting LED

Fig. 46.17 shows the structure of a stripe geometry edge-emitting LED.

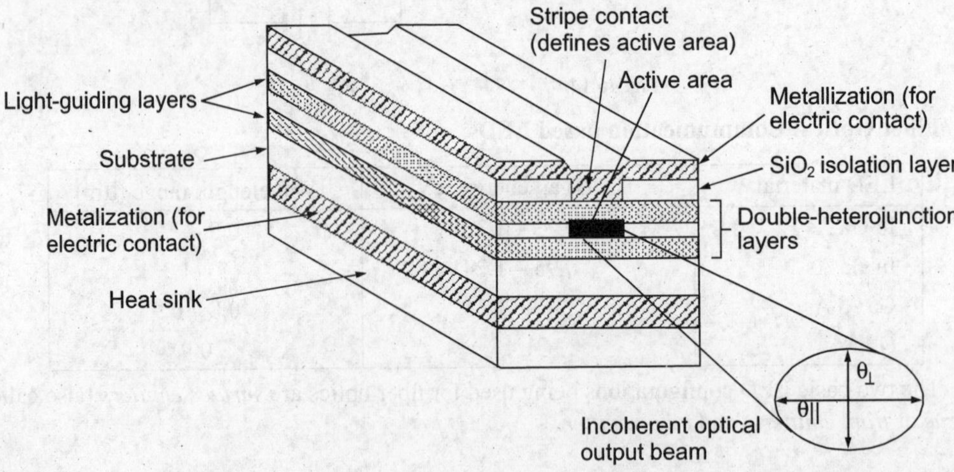

Fig. 46.17.

It consists of an active junction region, which is the source of the incoherent light, and two guiding layers. The guiding layers both have a refractive index which is lower than that of the active region but higher than the index of the surrounding material. This structure forms a waveguide channel that directs the optical radiation toward the fiber core.

To match the typical fiber-core diameters (50–100 μm), the contact stripes for the edge emitter are 50-70 μm wide. The stripe geometry, made by selective metallization on the top surface through a window opened in an SiO_2 layer, allows higher carrier injection densities for the same drive current. In other words, the power conversion efficiency is improved.

Lengths of the active regions usually range from 100 to 150 μm.

The output beam is lambertian in the plane of the *pn* junction ($\theta_{\parallel} = 120°$) and highly directional perpendicular to the *pn* junction ($\theta_{\perp} \approx 30°$).

The emission pattern of the edge emitting diode is more directional than that of the surface emitting LED. Drive current is about 100 mA with output optical power about 0.5 μW to 10 μW. Manufacturing of edge emitting LEDs is more expensive. Modulation bandwidth is about 500 kHz with rise time of 600 to 800 ns. In Ga As P and Ga Al As are used as LED materials at 1.3 μm. Ga Al As is used as a LED material at 0.85 μm having smaller rise time about 2 to 20 ns and bandwidth is greater than 200 MHz. The spectral width is about 50 nm at 0.85 μm wavelength and 70 nm at 1.3 μm wavelength.

MODULATION

46.15. Introduction

Modulation is the process by which the waveform of a high-frequency carrier wave is modified suitably to transmit information. Usually, the information signal, at a much lower frequency, is impressed on the high-frequency carrier wave.

Modulation is defined as the process by which some characteristic property of a high frequency wave is varied in accordance with the instantaneous value of the signal.

Modulation is the process of varying one of the parameters such as amplitude, intensity, frequency, phase and polarisation associated with a carrier wave to enable it to carry information. But the optical detectors respond only to the intensity or irradiance of the light. So we have only intensity modulators at optical frequencies.

46.16. Analog and Digital Modulation

(*i*) In **analog modulation** the information signal or wave varies the light from the source, or the high-frequency signal, in a continuous manner. Thus, both could be sinusoidal, as shown Fig. 46.18.

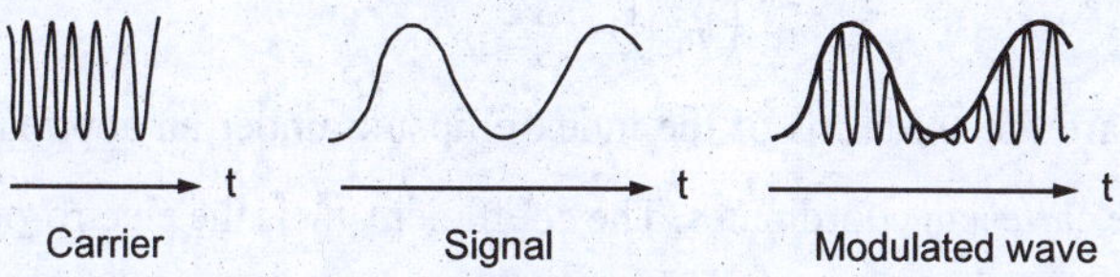

Fig. 46.18.

There is always a one-to-one correspondence between the information signal and the magnitude of the modulated carrier.

For example, in the case of intensity modulation, there is a one-to-one relationship between the signal amplitude and the intensity of laser light.

Disadvantages. Analog modulation requires a higher signal-to-noise ratio at the receiver or detector. For large bandwidth applications the laser is driven at high-current levels, at which the light-current characteristics are very nonlinear. Analog modulation may be more suited for low modulation frequencies. Digital modulation is more suited for large bandwidth optical transmission and reception.

(*ii*) In **digital modulation,** discrete changes (on-off) in the intensity of the carrier are caused by the information signal. Information is then transmitted by the high-frequency signal as a series of discrete pulses (0 and 1), as shown in Fig. 46.19.

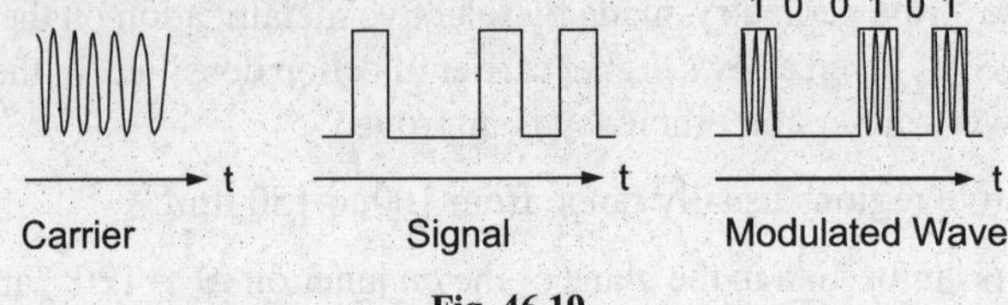

Fig. 46.19.

46.17. Electro-optic Modulators

There is a change in the refractive index of the crystal [like KH_2PO_4(KDP)] in the directions of the ordinary and extraordinary rays due to the application of an electric field. This is called *electro-optic effect.*

The field-dependent change in the refractive index can be expressed by the equation

$$\Delta\left(\frac{1}{n_r^2}\right) = r^l\mathbf{E} + s^q\mathbf{E}^2 \qquad \text{...(1)}$$

(*i*) Here r^l is called the *linear electro-optic coefficient.*

If the value of r^l is very large, the corresponding electro-optic effect is called the *Pockels effect.*

(*ii*) s^q is called the *quadratic electro-optic coefficient.*

If s^q is large and makes the quadratic term dominant, then the corresponding electro-optic effect is called the *Kerr effect.*

Most of the crystals exhibit Pockels effect.

Pockels effect refers to the change in the refractive index of the medium which is linearly proportional to the applied field or voltage.

For KDP crystal, *linear electro-optic coefficient,*

$$r_{ij}^l = 10.6 \text{ pm V}^{-1}.$$

The Pockels effect is accurately defined by the equation

$$\Delta\left(\frac{1}{n_r^2}\right)_i = \sum_{j=1}^{3} r_{ij}^l E_j \qquad \text{...(2)}$$

Here, $i(= 1 - 6)$ are the six terms of the index ellipsoid under an applied field and $j(= 1 - 3)$ correspond to the three cartesian coordinates. The coefficient r_{ij}^l is the electro-optic tensor described by a (6 × 3) matrix. In crystals that have inversion symmetry, $r_{ij}^l = 0$. Symmetry considerations determine the zero or nonzero values of the different r_{ij}^l coefficients. For example, in materials such as GaAs, which crystallize in the zincblende structure, the only three and equal nonzero linear coefficients are $r_{41}^l = r_{52}^l = r_{63}^l$.

Theory of Pockels effect

Consider a KDP crystal which has tetragonal structure.

Fig. 46.20 shows the electro-optic modulator and its components.

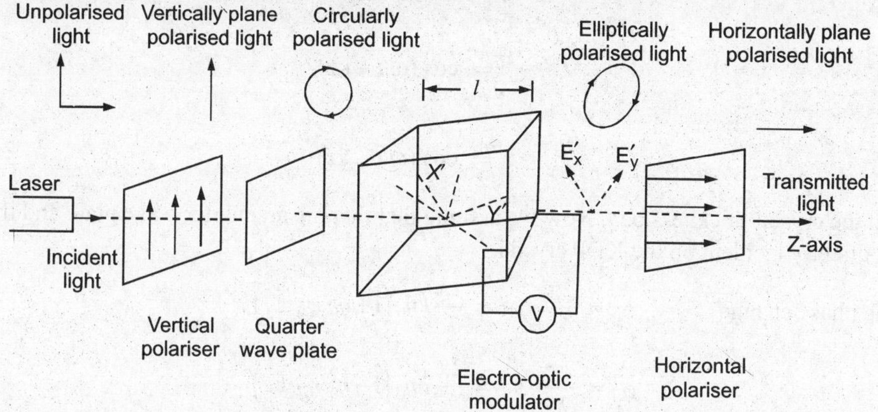

Fig. 46.20.

Light propagates through the crystal in the *z*-direction.

The equation of the optical indicatrix of the crystal is written as

$$\left(\frac{X}{n_x}\right)^2 + \left(\frac{Y}{n_y}\right)^2 + \left(\frac{Z}{n_z}\right)^2 = 1 \qquad ...(3)$$

Here, *X, Y* and *Z* are crystallographic axes;

n_x, n_y and n_z are the refractive indices along *X, Y* and *Z* axes.

Optical indicatrix is a three dimensional figure representing the values of refractive index of the crystalline solid along *X, Y* and *Z* axes.

When an electric field E_z is applied in the *z*-direction, then the principal axes *X* and *Y* are rotated through 45° which induces the new principal axes *X′* and *Y′*.

The new equation for optical indicatrix after applying the electric field E_z can be written as

$$\left(\frac{X'}{n'_x}\right)^2 + \left(\frac{Y'}{n'_y}\right)^2 + \left(\frac{Z}{n'_z}\right)^2 = 1 \qquad ...(4)$$

In terms of initial crystallographic axes *X, Y, Z*, Eq. (4) can be written as

$$\left(\frac{X}{n_0}\right)^2 + \left(\frac{Y}{n_0}\right)^2 + \left(\frac{Z}{n_e}\right)^2 + 2\, r_{63}\, E_z XY = 1 \qquad ...(5)$$

Here n_0 are n_e are the ordinary ray refractive index and extra ordinary ray refractive index respectively.

The new values of refractive indices along *X′* and *Y′* directions are given by

$$n_x' = n_0 + \frac{n_0^3}{2} r_{63}\, E_z$$

$$n_y' = n_0 - \frac{n_0^3}{2} r_{63}\, E_z$$

Thus

$$\Delta\left(\frac{1}{n^2}\right) = -\frac{2\Delta n}{n^3} = r_{63}\, E_z \qquad ...(6)$$

Thus r_{63} is the effective electro-optic coefficient for KDP crystal.

Let the plane polarised light from the vertical polariser be written as

$$E = E_0 \cos(\omega t - kz) \qquad \text{...(7)}$$

$$\therefore \qquad E_x' = \frac{E_0}{\sqrt{2}} \cos(\omega t - kz) \qquad \text{...(8)}$$

$$E_y' = \frac{E_0}{\sqrt{2}} \cos(\omega t - kz) \qquad \text{...(9)}$$

Let the crystal thickness be l. Now we assume that there is no quarter wave plate and the light beam is directly incident on the KDP crystal.

The phase change $\qquad \phi_{x'} = \frac{2\pi}{\lambda} n_{x'}\, l = \frac{2\pi}{\lambda} l\, n_0 \left(1 + n_0^2\, r_{63}\, E_z\right)$

$$\phi_{y'} = \frac{2\pi}{\lambda} n_{y'}\, l = \frac{2\pi}{\lambda} l\, n_0 \left(1 - n_0^2\, r_{63}\, E_z\right)$$

Let $\qquad \phi_{x'} = \phi_0 + \Delta\phi$

$$\phi_{y'} = \phi_0 - \Delta\phi$$

Net phase shift ϕ or total retardation between E_x' and E_y' is given by $\phi = \phi_y' = 2\Delta\phi$

$$\therefore \qquad 2\Delta\phi = 2\,\frac{\pi}{\lambda} l\, r_{63}\, n_0^3\, E_z = 2\frac{\pi}{\lambda} r_{63}\, n_0^3\, V.$$

In terms of intensity,

$$I = \frac{\omega}{2\pi} \int_0^{2\pi/\omega} E^2\, dt$$

Here, $E = \frac{E_0}{2} [\cos(\omega t + \Delta\phi) - \cos(\omega t - \Delta\phi)]$

$= - E_0 \sin\Delta\phi \sin\omega t.$

Thus $\qquad I = I_0 \sin^2 \Delta\phi = I_0 \sin^2\left(\frac{\phi}{2}\right)$

or $\qquad \frac{I}{I_0} = \sin^2\left(\frac{\pi}{\lambda} r_{63}\, n_0^3\, V\right) \qquad \text{...(10)}$

Let $V_\pi = \frac{\lambda}{2\, r_{63}\, n_0^3}$ = the voltage required for the net phase shift of π or maximum transmission. (i.e., $I = I_0$). This voltage is called **Half wave voltage** which produces a retardation of π between two component waves.

$$\therefore \qquad \frac{I}{I_0} = \sin^2\left(\frac{\pi}{2}\frac{V}{V_\pi}\right).$$

The effect of introducing quarter wave plate will produce an additional phase shift of $\pi/2$. Thus

$$\phi = \pi/2 + 2\Delta\phi = \frac{\pi}{2} + \pi\frac{V}{V_\pi}$$

$$\therefore \quad \frac{I}{I_0} = \sin^2\left(\frac{\pi}{4}+\frac{\pi}{2}\frac{V}{V_\pi}\right) = \frac{1}{2}\left(1+\sin\frac{\pi V}{V_\pi}\right).$$

Thus for small applied voltages, the change of intensity of light is nearly linear with respect to the applied voltage. To reduce the half wave voltage, it is better to use a long thin Pockels cell.

Fig. 46.20 explains *Electro-optic phase modulation.*

(*i*) The emitted laser light is unpolarised.

(*ii*) Unpolarised light is converted into vertically plane polarised light by a vertical polariser.

(*iii*) The quarter wave plate converts the plane polarised light into circularly polarised light. The purpose of the quarterwave plate is to get linearity between the applied voltage and intensity of transmitted light.

(*iv*) Circularly polarised light is incident on the electro-optic crystal plate subjected to a voltage *V*. Now it is resolved into two components along *X*′ and *Y*′, the induced principal (axes) directions. Due to net phase shift or retardation between two component waves, the emergent light from the crystal is elliptically polarised.

(*v*) The horizontal polariser converts elliptically polarised light into horizontally plane polarised light which will act as output light. Thus the intensity of the output light is the function of voltage applied across the width of the crystal.

DEMODULATION

Demodulation means the reverse process of modulation, i.e., extraction of the original signal from the modulated carrier, detected at the receiver.

46.18. Demodulation schemes

(*i*) Heterodyne detection scheme

Fig. 46.21 shows receiver configuration for optical heterodyne receiver.

A beat-note signal between the incoming optical signal and the local oscillator signal produces the IF signal which is obtained using the square law optical detector. The IE signal is amplified by an intermediate frequency amplifier. The IF signal, which generally has a frequency of between three and four times the transmission rate, is then demodulated into the baseband using either a synchronous or nonsynchronous detection technique.

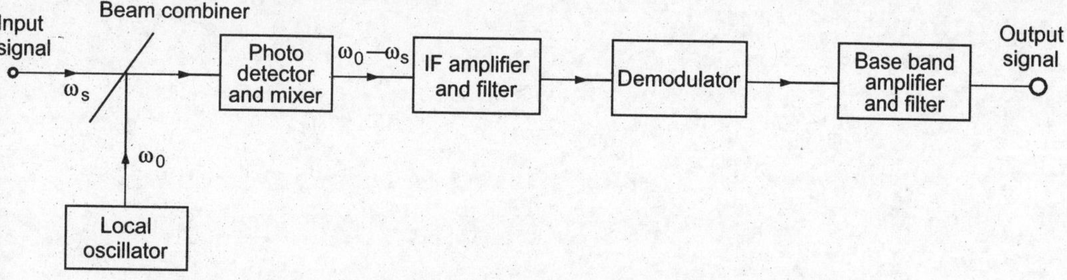

Fig. 46.21

(*ii*) Homodyne detection scheme

Fig. 46.22 shows the block diagram of a homodyne detection scheme.

Here the phase of the local oscillator signal is locked to the incoming signal. Therefore, a synchronous detection scheme must be employed. Moreover, the result of the mixing process in the optical detector produces an information signal which is in the baseband and thus requires no further demodulation.

The local oscillator frequency and modulated laser frequency should be equal and stabilised.

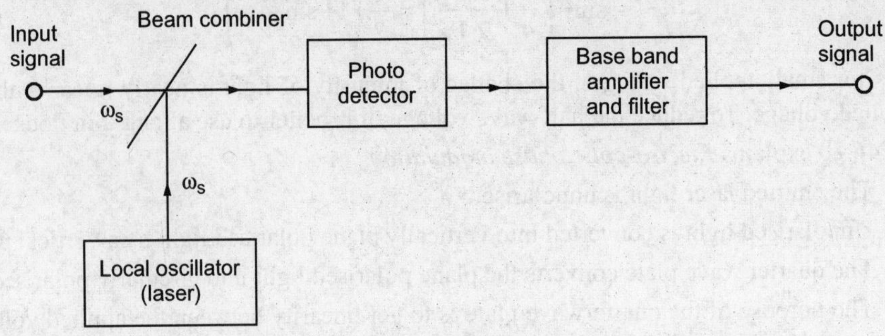

Fig. 46.22

EXERCISE

1. Explain analog optical fibre communication system with a block diagram.
2. Explain digital optical fibre communication system with a block diagram.
3. Describe any one of the heterojunction laser diodes used in the optical fibre communication.

CHAPTER

47

OPTICAL FIBRE FABRICATION

47.1. Double Crucible Method

Fig. 47.1 shows **Double-Crucible fibre drawing process.**

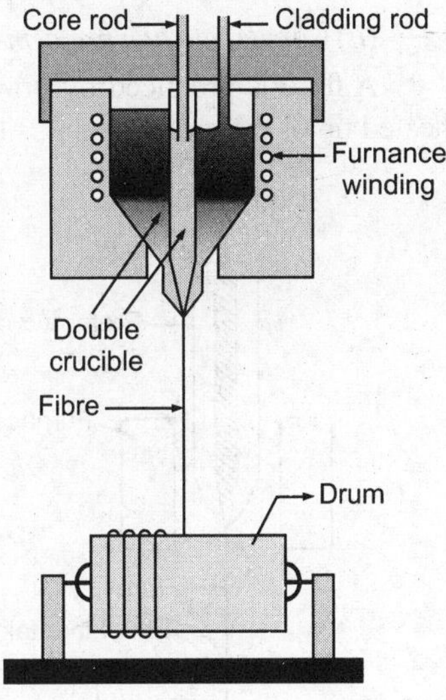

Fig. 47.1.

At first, two very high purity glass rods, with different refractive indices, are prepared. For example, these rods consist of soda boro-silicate (Na_2O –B_2 O_3 – SiO_2) with different compositions for core and cladding. To draw cladded-core fibres from these glass rods, two cylindrical crucibles are arranged concentrically.

The rods are fed slowly downward into the heated crucibles where they are melted, filling up the crucibles with glass-melt. The inner crucible contains the molten core. The outer crucible contains the molten cladding. The height of melt must be constant. The crucible temperature is maintained between 850°C to 1100°C, depending on the choice of glass compositions.

At the base, the crucibles taper down to concentric nozzles, where the glasses are pulled to a fibre. The pulling speed is given by the rotating drum on to which the fibre is wound.

With this method, a fibre is continuously fabricated from the molten glass in a single fabrication step.

The disadvantage of this method is the possibility of introducing contaminants during the melting process.

47.2 Modified Chemical Vapor Deposition (MCVD) Method

Principle:

A doped silica layer is deposited onto the inner surface of a pure silica tube. The deposition occurs as a result of a chemical reaction taking place between the vapor constituents that are being passed down the tube. Vapors used are $SiCl_4$, $GeCl_4$ and O_2. The following reactions take place:

$$SiCl_4 + O_2 \rightarrow SiO_2 + 2Cl_2 \quad ...(1)$$

$$GeCl_4 + O_2 \rightarrow GeO_2 + 2Cl_2 \quad ...(2)$$

(*i*) Production of fibre preform

Stage I. The reactants are introduced into one end of a silica tube. [Fig. 47.2 (a)].

The zone where the reaction takes place is moved along the tube by locally heating the tube to a temperature in the range 1200 – 1600°C with a traversing oxy-hydrogen flame. If the process is repeated with different input concentrations of the dopant vapors, layers of different impurity concentrations may be built up sequentially. This technique thus allows a much greater control over the index profile than does the double crucible method.

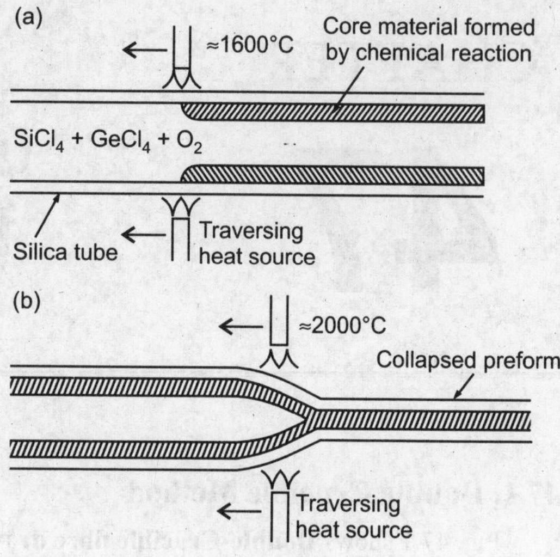

Fig. 47.2.

Stage II. Once the deposition process is complete, the tube is heated to its softening temperature, ≈ 2000°C [Fig. 47.2(b)]. Surface tension effects then cause the tube to collapse into a solid rod called a *preform.*

(*ii*) *Conversion of a preform rod into a fibre*

A fibre is produced by drawing from the heated tip of the preform as it is lowered into a furnace (Fig. 47.3).

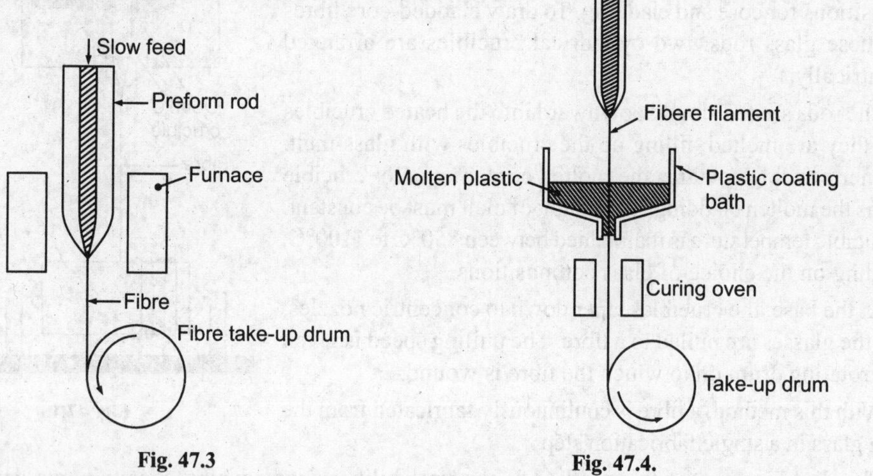

Fig. 47.3

Fig. 47.4.

To exercise tight control over the fibre diameter, a thickness monitoring gage is used before the fibre is drawn onto the take-up drum, and feedback is applied to the drum take-up speed.

- The fiber is usually coated with a layer of plastic for protection before being wound on the drum. Protective plastic coating is applied to the outside of the fibre by passing it through a bath of molten plastic and a curing oven (Fig. 47.4).

CHAPTER 48

ELECTROMAGNETIC WAVES-II

48.1. Maxwell's Equations in Integral Form

By means of Gauss' and Stoke's theorems we can put the field equations in integral form and hence obtain their physical significance.

1. Maxwell's first equation is $\nabla \cdot \mathbf{D} = \rho$.

Integrating this over an arbitrary volume V, we get

$$\int_V \nabla \cdot \mathbf{D}\, dV = \int_V \rho\, dV$$

But from Gauss theorem, we get

$$\int_S \mathbf{D} \cdot d\mathbf{S} = \int_V \rho\, dV = q$$

Here, q is the net charge contained in volume V. S is the surface bounding volume V. Therefore, Maxwell's first equation signifies that:

The total electric displacement through the surface enclosing a volume is equal to the total charge within the volume.

2. Maxwell's second equation is $\nabla \cdot \mathbf{B} = 0$.

Integrating this over an arbitrary volume V, we get

$$\int_V \nabla \cdot \mathbf{B} = 0$$

Using Gauss divergence theorem to change volume integral into surface integral, we get

$$\int_S \mathbf{B} \cdot d\mathbf{S} = 0$$

Maxwell's second equation signifies that:

The total outward flux of magnetic induction **B** *through any closed surface S is equal to zero.*

3. Maxwell's third equation is $\nabla \times \mathbf{E} = -\dfrac{\partial \mathbf{B}}{\partial t}$.

Integrating this equation over a surface S, bounded by a curve C, we get

$$\int_S (\nabla \times \mathbf{E}) \cdot d\mathbf{S} = -\int_S \frac{\partial B}{\partial t} \cdot d\mathbf{S}$$

Converting the surface integral of L.H.S. into line integral by Stoke's theorem, we get

$$\oint_C \mathbf{E} \cdot d\mathbf{l} = -\int_S \frac{\partial \mathbf{B}}{\partial t} \cdot d\mathbf{S}$$

Maxwell's third equation signifies that:

The electromotive force $\left(e.m.f.\ e = \int_C \mathbf{E} \cdot d\mathbf{l}\right)$ *around a closed path is equal to negative rate of change of magnetic flux linked with the path* $\left(\text{since magnetic flux } \phi = \int_S \mathbf{B} \cdot d\mathbf{S}\right)$.

4. Maxwell's fourth equation is

$$\nabla \times \mathbf{H} = \mathbf{J} + \frac{\partial \mathbf{D}}{\partial t}$$

Taking surface integral over surface S bounded by curve C, we obtain

$$\int_S \nabla \times \mathbf{H} \cdot d\mathbf{S} = \int_S \left(\mathbf{J} + \frac{\partial \mathbf{D}}{\partial t} \right) \cdot d\mathbf{S}$$

Using Stoke's theorem to convert surface integral on L.H.S. of above equation into line integral, we get

$$\oint_C \mathbf{H} \cdot d\mathbf{l} = \int_S \left(\mathbf{J} + \frac{\partial \mathbf{D}}{\partial t} \right) \cdot d\mathbf{S}$$

Maxwell's fourth equation signifies that:

The magnetomotive force $\left(m.m.f. = \oint_C \mathbf{H} \cdot d\mathbf{l} \right)$ *around a closed path is equal to the conduction current plus displacement current through any surface bounded by the path.*

48.2. Poynting's Theorem

Statement:

The work done on the charges by the electromagnetic force is equal to the decrease in energy stored in the field, less the energy that flowed out through the surface.

Derivation:

Suppose we have some charge and current configuration which, at time t, produces fields **E** and **B**. In the next instant, dt, the charges move around a bit. According to the Lorentz force law, the work done on a charge q is

$$dW = F \cdot d\mathbf{l} = q\,(\mathbf{E} + \mathbf{v} \times \mathbf{B}) \cdot \mathbf{v}\, dt = q\mathbf{E} \cdot \mathbf{v}\, dt.$$

Here, $q = \rho d\tau$ and $\rho \mathbf{v} = \mathbf{J}$.

The rate at which work is done on all the charges in a volume V is

$$\frac{dW}{dt} = \int_V (\mathbf{E} \cdot \mathbf{J})\, d\tau \quad \text{...(1)}$$

$\mathbf{E} \cdot \mathbf{J}$ is the work done per unit time, per unit volume = *power* delivered per unit volume.

Ampere's law with Maxwell's correction is

$$\mathbf{J} = \frac{1}{\mu_0} (\nabla \times \mathbf{B}) - \epsilon_0 \frac{\partial \mathbf{E}}{\partial t}$$

$$\mathbf{E} \cdot \mathbf{J} = \frac{1}{\mu_0} \mathbf{E} \cdot (\nabla \times \mathbf{B}) - \epsilon_0\, \mathbf{E} \cdot \frac{\partial \mathbf{E}}{\partial t} \quad \text{...(2)}$$

Eq. (2) expresses power delivered per unit volume in terms of the fields alone.

We have the vector identity

$$\nabla \cdot (\mathbf{E} \times \mathbf{B}) = \mathbf{B} \cdot (\nabla \times \mathbf{E}) - \mathbf{E} \cdot (\nabla \times \mathbf{B})$$

Using Faraday's law $(\nabla \times \mathbf{E} = -\,\partial \mathbf{B}/\partial t)$, we get

$$\nabla \cdot (\mathbf{E} \times \mathbf{B}) = -\,\mathbf{B} \cdot \frac{\partial B}{\partial t} - \mathbf{E} \cdot (\nabla \times \mathbf{B})$$

$$\therefore \qquad \mathbf{E} \cdot (\nabla \times \mathbf{B}) = -\,\mathbf{B} \cdot \frac{\partial B}{\partial t} - \nabla \cdot (\mathbf{E} \times \mathbf{B}) \quad \text{...(3)}$$

We know the self-evident equalities

$$\mathbf{B} \cdot \frac{\partial \mathbf{B}}{\partial t} = \frac{1}{2} \frac{\partial}{\partial t} (B^2), \text{ and } \mathbf{E} \cdot \frac{\partial \mathbf{E}}{\partial t} = \frac{1}{2} \frac{\partial}{\partial t} (E^2) \quad \text{...(4)}$$

Eq. (2) becomes

$$\mathbf{E}\cdot\mathbf{J} = \frac{1}{2}\frac{\partial}{\partial t}\left(\epsilon_0 E^2 + \frac{1}{\mu_0}B^2\right) - \frac{1}{\mu_0}\nabla\cdot(\mathbf{E}\times\mathbf{B})$$

Putting this into Eq. (1), we get

$$\frac{dW}{dt} = \int_{\text{v}}\left[-\frac{1}{2}\frac{\partial}{\partial t}\left(\epsilon_0 E^2 + \frac{1}{\mu_0}B^2\right)\right]d\tau - \int_{\text{v}}\left[\frac{1}{\mu_0}\nabla\cdot(\mathbf{E}\times\mathbf{B})\right]d\tau$$

Using gauss divergence theorem to the second term, we get

$$\int_{\text{v}}\left[\frac{1}{\mu_0}\nabla\cdot(\mathbf{E}\times\mathbf{B})\right]d\tau = \oint_S\left[\frac{1}{\mu_0}(\mathbf{E}\times\mathbf{B})\right]\cdot d\mathbf{a}$$

Here, S is the surface bounding the volume V.

$$\therefore \quad \frac{dW}{dt} = -\frac{d}{dt}\int_{\text{v}}\frac{1}{2}\left(\epsilon_0 E^2 + \frac{1}{\mu_0}B^2\right)d\tau - \frac{1}{\mu_0}\oint_S(\mathbf{E}\times\mathbf{B})\cdot d\mathbf{a} \qquad \text{...(6)}$$

This is **Poynting's theorem**.

- The first integral on the right is the total energy stored in the fields.

$$U_{\text{em}} = \frac{1}{2}\int\left(\epsilon_0 E^2 + \frac{1}{\mu_0}B^2\right)d\tau$$

- The second term represents the rate at which energy is carried out of v, across its boundary surface, by the electromagnetic fields.

Poynting's theorem says, that the work done on the charges by the electromagnetic force is equal to the decrase in energy stored in the field, less the energy that flowed out through the surface.

The energy per unit time, *per unit area*, transported by the field is called the **Poynting vector**:

$$\text{S} = \frac{1}{\mu_0}(\mathbf{E}\times\mathbf{B}) \qquad \text{...(7)}$$

$\mathbf{S}\cdot d\mathbf{a}$ is the energy per unit time crossing the infinitesimal surface $d\mathbf{a}$– the energy *flux*.

We can use Poynting vector to express Poynting's theorem more compactly:

$$\frac{dW}{dt} = -\frac{dU_{em}}{dt} - \oint_S S\cdot d\mathbf{a} \qquad \text{...(8)}$$

Differential version of Poynthing's theorem

The work W done on the charges will increase their mechanical energy (kinetic, potential, or whatever).

Let u_{mech} denote the mechanical energy density of the fields,

$$\frac{dW}{dt} = \frac{d}{dt}\int_{\text{v}} u_{mech}\, d\tau \qquad \text{...(9)}$$

Let u_{em} denote the energy density of the fields,

$$u_{\text{em}} = \frac{1}{2}\left(\epsilon_0 E^2 + \frac{1}{\mu_0}B^2\right) \qquad \text{...(10)}$$

Eq. (6) becomes

$$\frac{d}{dt}\int_{\text{v}}(u_{\text{mech}} + u_{\text{em}})d\tau = -\oint_S \boldsymbol{S}\cdot d\mathbf{a} = -\int_{\text{v}}(\nabla\cdot\boldsymbol{S})d\tau$$

$$\therefore \quad \frac{\partial}{\partial t}(u_{\text{mech}} + u_{\text{em}}) = -\nabla\cdot\mathbf{S} \qquad \text{...(11)}$$

Eq. (11) represents Ponthing's theorem in differential form.

48.3. Scalar and Vector Potentials

In the case of a magnetic field,

$$\nabla \cdot \mathbf{B} = 0$$

Since the divergence of any curl is zero,

$$\mathbf{B} = \nabla \times \mathbf{A} \qquad ...(1)$$

Here **A** is called the *magnetic vector potential.*

The differential form of Faraday's law is

$$\nabla \times \mathbf{E} = -\frac{\partial \mathbf{B}}{\partial t}$$

or

$$\nabla \times \mathbf{E} = -\frac{\partial}{\partial t}(\nabla \times \mathbf{A})$$

or

$$\nabla \times \left(\mathbf{E} + \frac{\partial \mathbf{A}}{\partial t}\right) = 0$$

Here is a quantity whose curl *does* vanish. It can therefore be written as the gradient of a scalar potential V.

$$\mathbf{E} + \frac{\partial \mathbf{A}}{\partial t} = -\nabla V$$

V is called the *scalar potential.*

In terms of V and **A**, then,

$$\mathbf{E} = -\nabla V - \frac{\partial \mathbf{A}}{\partial t} \qquad ...(2)$$

The potential formulation [Eqs. (1) and (2)] satisfies the following two homogeneous Maxwell equations.

$$\nabla \cdot \mathbf{B} = 0$$

$$\nabla \times \mathbf{E} = -\frac{\partial \mathbf{B}}{\partial t}$$

Now consider the Maxwell's equation

$$\nabla \cdot \mathbf{E} = \frac{1}{\epsilon_0}\rho$$

Substituting for **E** from Eq. (2), we get

$$\nabla \cdot \left[-\nabla V - \frac{\partial \mathbf{A}}{\partial t}\right] = -\frac{1}{\epsilon_0}\rho$$

or

$$\nabla^2 V + \frac{\partial}{\partial t}(\nabla \cdot \mathbf{A}) = -\frac{1}{\epsilon_0}\rho \qquad ...(3)$$

Now consider the Maxwell's equation

$$\nabla \times \mathbf{B} = \mu_0 \mathbf{J} + \mu_0 \epsilon_0 \frac{\partial \mathbf{E}}{\partial t}$$

Putting Eqs. (1) and (2) into the above equation, we get

$$\nabla \times (\nabla \times \mathbf{A}) = \mu_0 \mathbf{J} - \mu_0 \epsilon_0 \nabla\left(\frac{\partial V}{\partial t}\right) - \mu_0 \epsilon_0 \frac{\partial^2 \mathbf{A}}{\partial t^2}$$

Using the vector identity $\nabla \times (\nabla \times \mathbf{A}) = \nabla(\nabla \cdot \mathbf{A}) - \nabla^2 \mathbf{A}$ and rearranging terms, we get

$$\left(\nabla^2 \mathbf{A} - \mu_0 \epsilon_0 \frac{\partial^2 \mathbf{A}}{\partial t^2}\right) - \nabla\left(\nabla \cdot \mathbf{A} + \mu_0 \epsilon_0 \frac{\partial V}{\partial t}\right) = -\mu_0 \mathbf{J} \qquad ...(4)$$

Equations (3) and (4) are Maxwell's field equations in terms of electromagnetic potentials, V and **A**. In this formulation four Maxwell's equations have been reduced to two.

48.4. Non-uniquencess of Electromagnetic Potentials and Gauge Transformations

In terms of electromagnetic potentials field vectors are given by

$$\mathbf{B} = \nabla \times \mathbf{A} \quad \text{...(1)}$$

and
$$\mathbf{E} = -\nabla V - \frac{\partial \mathbf{A}}{\partial t} \quad \text{...(2)}$$

From equations (1) and (2) it is clear that for a given **A** and V, each of the field vectors **B** and **E** has only one value *i.e.*, **A** and V determine **B** and **E** uniquely. However the converse is not true *i.e.*, field vectors do not determine the potentials **A** and V completely. This in turn implies that for a given **A** and V there will be only one **E** and **B** while for a given **E** and **B** there can be an infinite number of **A**'s and V's.

Suppose we have two sets of potentials, (V, **A**) and (V', **A**′), which correspond to the *same* electric and magnetic fields.

Let
$$\mathbf{A}' = \mathbf{A} + \alpha \text{ and } V' = V + \beta$$

Since the two **A**'s give the same **B**, their curls must be equal [Eq. (1)].

∴
$$\nabla \times \alpha = 0$$

The curl of the gradient of any scalar vanishes identically. Hence we can write α as the gradient of some scalar λ:

$$\alpha = \nabla\lambda$$

The two potentials also give the same **E**. So Eq. (2) requires that

$$\nabla\beta + \frac{\partial \alpha}{\partial t} = 0$$

or
$$\nabla\left(\beta + \frac{\partial \lambda}{\partial t}\right) = 0$$

The term in parentheses is therefore independent of position. However, it could be a function of time, say $k(t)$. Then,

$$\beta = -\frac{\partial \lambda}{\partial t} + k(t)$$

Actually, we may as well absorb $k(t)$ into λ, defining a new λ by adding $\int_0^t k(t')\,dt'$ to the old one. This will not affect the gradient of λ. It just adds $k(t)$ to $\partial\lambda/\partial t$.

∴
$$V' = V - \frac{\partial \lambda}{\partial t}$$

Also
$$\mathbf{A}' = \mathbf{A} + \nabla\lambda$$

The magnetic field **B** remains unchanged by the transformation

$$\mathbf{A} \to \mathbf{A}' = \mathbf{A} + \nabla\lambda$$

Likewise, the electric field **E** remains unchanged by the transformation

$$V \to V' = V - \frac{\partial \lambda}{\partial t}$$

Here, λ is an arbitrary scalar function. The addition of an arbitrary scalar function to the potential function such that electric and magnetic fields are not affected is known as *Gauge transformation.* The *invariance of fields under gauge transformation is known as Gauge invariance.*

Thus we can add ∇λ to **A**, provided we simultaneously subtract $\partial\lambda/\partial t$ from V. None of this will affect the physical quantities **E** and **B**. Such changes in V and **A** are called *Gauge transformations*.

48.5. Coulomb Gauge

Suppose we are assuming a field satisfying the condition

$$\nabla \cdot \mathbf{A} = 0 \qquad ...(1)$$

This conforms to a *gauge transformation* which is known as *Coulomb gauge.*
The Coulomb gauge restricts the divergence of **A** as

$$\text{div } \mathbf{A} = \nabla \cdot \mathbf{A} = 0$$

The equation for scalar potential V is

$$\nabla^2 V + \frac{\partial}{\partial t}(\nabla \cdot \mathbf{A}) = -\frac{1}{\epsilon_0}\rho \qquad ...(2)$$

Using Eq. (1), Eq. (2) reduces to *Poisson's equation*

$$\nabla^2 V = -\frac{1}{\epsilon_0}\rho \qquad ...(3)$$

The *advantage* of the Coulomb gauge is that the *scalar* potential is particularly simple to calculate.
Using Coulomb gauge condition Eq. (1), equation of the vector potential [Eq. (4) of Section 48.3] reduces to

$$\left(\nabla^2 \mathbf{A} - \mu_0 \epsilon_0 \frac{\partial^2 \mathbf{A}}{\partial t^2}\right) = -\mu_0 \mathbf{J} + \mu_0 \epsilon_0 \nabla\left(\frac{\partial V}{\partial t}\right) \qquad ...(4)$$

48.6. Lorentz Gauge

Maxwell's equation in terms of electrodynamic potentials are expressed as

$$\left(\nabla^2 \mathbf{A} - \mu_0 \epsilon_0 \frac{\partial^2 \mathbf{A}}{\partial t^2}\right) - \nabla\left(\nabla \cdot \mathbf{A} + \mu_0 \epsilon_0 \frac{\partial V}{\partial t}\right) = -\mu_0 \mathbf{J} \qquad ...(1)$$

$$\nabla^2 V + \frac{\partial}{\partial t}(\nabla \cdot \mathbf{A}) = -\frac{1}{\epsilon_0}\rho \qquad ...(2)$$

If we put

$$\nabla \cdot \mathbf{A} = -\mu_0 \epsilon_0 \frac{\partial V}{\partial t} \qquad ...(3)$$

Eqs. (1) and (2) become

$$\nabla^2 \mathbf{A} - \mu_0 \epsilon_0 \frac{\partial^2 \mathbf{A}}{\partial t^2} = -\mu_0 \mathbf{J} \qquad ...(4)$$

$$\nabla^2 V - \mu_0 \epsilon_0 \frac{\partial^2 V}{\partial t^2} = -\frac{1}{\epsilon_0}\rho \qquad ...(5)$$

The condition given by relation (3) is known as *Lorentz condition* and expresses a relationship between V and **A**. When the vector and scalar potential satisfy the Lorentz condition, the gauge is known as *Lorentz gauge*.

Eqs. (4) and (5) are written more compactly by defining the *d'Alembertian operator* as

$$\nabla^2 - \mu_0 \epsilon_0 \frac{\partial^2}{\partial t^2} \equiv \Box^2 \qquad ...(6)$$

Then we get

$$\left.\begin{aligned} \Box^2 \mathbf{A} &= -\mu_0 \mathbf{J} \\ \Box^2 V &= -\frac{1}{\epsilon_0}\rho \end{aligned}\right\} \qquad ...(7)$$

Lorentz gauge treats V and **A** on an equal footing. The same differential operator occurs in both equations. In special relativity, the d' Alembertian plays somewhat the same role as the Laplacian in classical physics, and Eq. (7) may be regarded as a four-dimensional version of Poisson's equation.

48.7. Monochromatic Plane Waves in Vacuum

The electric and magnetic field vectors **E** and **B** in empty space, satisfy the three-dimensional wave equation

$$\nabla^2 \mathbf{E} = \frac{1}{c^2}\frac{\partial^2 \mathbf{E}}{\partial t^2}, \nabla^2 \mathbf{B} = \frac{1}{c^2}\frac{\partial^2 \mathbf{B}}{\partial t^2} \quad ...(1)$$

Here, $c = 1/\sqrt{\epsilon_0 \mu_0}$ is the speed of light in vacuum.

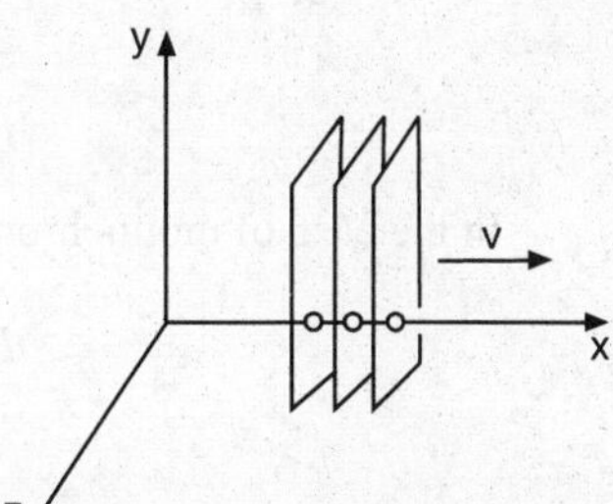

Fig. 48.1.

We consider sinusoidal waves of frequency ω. Such waves are called *monochromatic*. Suppose the waves are travelling in the x-direction and have no y-or z-dependence. These are called *plane waves*, because the fields are uniform over every plane perpendicular to the direction of propagation (Fig. 48.1).

$$\tilde{\mathbf{E}}(x,t) = \tilde{\mathbf{E}}_0\, e^{i(\kappa x - \omega t)}, \tilde{\mathbf{B}}(x,t) = \tilde{\mathbf{B}}_0\, e^{i(\kappa x - \omega t)} \quad ...(2)$$

κ is called the *wave number*.

$\tilde{\mathbf{E}}_0$ and $\tilde{\mathbf{B}}_0$ are the *complex amplitudes* of the electric and magnetic fields. The *physical* fields are the real parts of $\tilde{\mathbf{E}}$ and $\tilde{\mathbf{B}}$.

Since $\nabla \cdot \mathbf{E} = 0$ and $\nabla \cdot \mathbf{B} = 0$, it follows that

$$(\tilde{E}_0)_x = (\tilde{B}_0)_x = 0 \quad ...(3)$$

This implies that *electromagnetic waves are transverse*.

The electric and magnetic fields are perpendicular to the direction of propagation.

Faraday's law, $\nabla \times \mathbf{E} = -\partial \mathbf{B}/\partial t$, implies a relation between the electric and magnetic amplitudes.

$$-\kappa(\tilde{E}_0)_z = \omega(\tilde{B}_0)_y, \ \kappa(\tilde{E}_0)_y = \omega(\tilde{B}_0)_z \quad ...(4)$$

or

$$\tilde{\mathbf{B}}_0 = \frac{\kappa}{\omega}(\hat{i} \times \tilde{\mathbf{E}}_0) \quad ...(5)$$

E and **B** are *in phase* and *mutually perpendicular*. The real amplitudes of **E** and **B** are related by

$$B_0 = \frac{\kappa}{\omega} E_0 = \frac{1}{c} E_0 \quad ...(6)$$

48.8. Monochromatic Plane Waves Travelling in an Arbitrary Direction

Let $\vec{\kappa}$ be the *propagation* (*wave*) *vector* pointing in the direction of propagation. The magnitude of $\vec{\kappa}$ is the wave number κ. Then, $(\kappa \cdot \mathbf{r})$ is the generalization of κx (Fig. 48.2).

Considering the vector nature of the electromagneic field, we express the spatial and temporal dependence of field vectors as

$$\tilde{\mathbf{E}}(\mathbf{r}, t) = \tilde{\mathbf{E}}_0\, e^{i(\kappa \cdot r - \omega t)}\, \hat{n} \quad ...(1)$$

$$\tilde{\mathbf{B}}(\mathbf{r},t) = \frac{1}{c}\tilde{E}_0\, e^{i(\kappa \cdot r - \omega t)}(\hat{\kappa} \times \hat{n}) = \frac{1}{c}\hat{\kappa} \times \tilde{\mathbf{E}} \quad ...(2)$$

Here, $\hat{n}$ is the polarization vector.

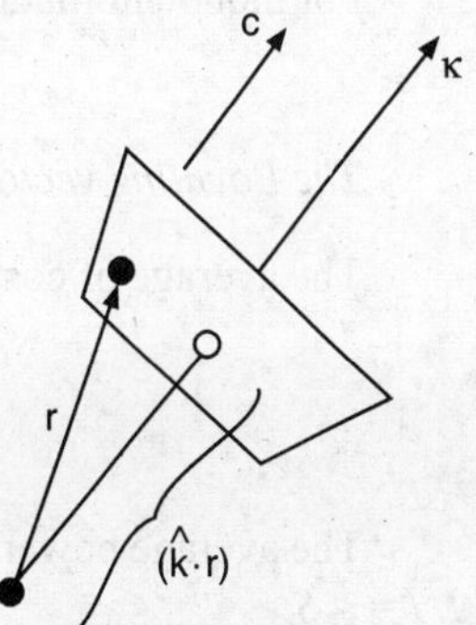

Fig. 48.2

The actual (real) electric and magnetic fields in a monochromatic plane wave with propagation vector κ and polarization $\hat{n}$ are

$$\mathbf{E}(\mathbf{r}, t) = E_0 \cos(\kappa \cdot \mathbf{r} - \omega t + \delta)\, \hat{n} \quad ...(3)$$

$$\mathbf{B}(\mathbf{r}, t) = \frac{1}{c} E_0 \cos(\kappa \cdot \mathbf{r} - \omega t + \delta)\, (\hat{\kappa} \times \hat{n}) \quad ...(4)$$

48.9. Energy and Momentum of Electromagnetic Waves

(i) Energy of e.m. waves

The energy per unit volume stored in an electromagnetic field is

$$U = \frac{1}{2}\left(\epsilon_0 E^2 + \frac{1}{\mu_0} B^2\right) \quad ...(1)$$

In the case of monochromatic plane wave,

$$B^2 = \frac{1}{c^2} E^2 = \mu_0 \epsilon_0 E^2 \quad ...(2)$$

$$\therefore \quad U = \frac{1}{2}\left(\epsilon_0 E^2 + \frac{1}{\mu_0} \mu_0 \epsilon_0 E^2\right) = \epsilon_0 E^2$$

If **E** points in the y-direction, then

$$\mathbf{E}(x, t) = E_0 \cos(\kappa x - \omega t + \delta)\hat{j},$$

$$\mathbf{B}(x, t) = \frac{1}{2} E_0 \cos(\kappa x - \omega t + \delta)\hat{k},$$

$$\therefore \quad U = \epsilon_0 E^2 = \epsilon_0 E_0^2 \cos^2(\kappa x - \omega t + \delta) \quad ...(3)$$

The average of $\cos^2$ over a complete cycle is $\frac{1}{2}$.

$$\therefore \quad \langle U \rangle = \frac{1}{2} \epsilon_0 E_0^2 \quad ...(4)$$

As the wave propagates, it carries this energy along with it.

The rate at which energy is transmitted through unit area perpendicular to the direction of propagation of energy is called Poynting vector and is represented by **S**.

$$\mathbf{S} = \frac{1}{\mu_0} (\mathbf{E} \times \mathbf{B}) \quad ...(5)$$

For monochromatic plane waves,

$$\mathbf{S} = c\epsilon_0 E_0^2 \cos^2(\kappa x - \omega t + \delta)\hat{i} = cU\hat{i} \quad ...(6)$$

The Poynting vector **S** *is c times the energy density* (*U*) *of the field.*

The average of $\cos^2$ over a complete cycle is $\frac{1}{2}$.

$$\therefore \quad \langle \mathbf{S} \rangle = \frac{1}{2} c\epsilon_0 E_0^2 \hat{i} = c \langle U \rangle \hat{i}$$

The average power per unit area transported by an electromagnetic wave is called the *intensity I*. $I = \langle S \rangle$.

(ii) Momentum of e.m. waves

Electromagnetic fields carry *momentum*.
The momentum density stored in the fields is given by

$$\rho = \frac{1}{c^2}\mathbf{S} \qquad ...(7)$$

For monochromatic plane waves,

$$\rho = \frac{1}{c} \epsilon_0 E_0^2 \cos^2 (\kappa x - \omega t + \delta)\hat{i} = \frac{1}{c} U\hat{i} \qquad ...(8)$$

$$\therefore \qquad U = \rho c \qquad ...(9)$$

The average of cos² over a complete cycle is $\frac{1}{2}$.

$$<\rho> = \frac{1}{2}\frac{1}{c} \epsilon_0 E_0^2 \hat{i} = \frac{1}{c} <U> \hat{i}$$

48.10. Propagation Through Linear Media

Consider plane monochromatic electromagnetic waves traveling through *any linear medium*. The energy density is

$$U = \frac{1}{2}\left(\epsilon E^2 + \frac{1}{\mu} B^2\right)$$

The Poynting vector is

$$\mathbf{S} = \frac{1}{\mu} (\mathbf{E} \times \mathbf{B})$$

The *intensity* of the wave is

$$I = \frac{1}{2} \epsilon v E_0^2$$

EXERCISE

1. State and prove Poynting theorem.
2. Show that for any scalar and vector potentials representing a particular EM field, we can always find another set of potentials representing the same field. What is Lorentz gauge condition?
3. Obtain expressions for the average energy and momentum of an electromagnetic wave. What is intensity of the wave?

CHAPTER

49

SPECTROSCOPY-II

Microwave Spectroscopy

49.1. The Rotation of Molecules

Microwave spectroscopy deals with the pure rotational motion of molecules. The rotation of a three-dimensional body is quite complex. It is convenient to resolve it into rotational components about three mutually perpendicular directions through the centre of mass - the principal axes of rotation. Thus a body has three principal *moments of inertia*, one about each axis. They are usually denoted by I_a, I_b, and I_c.

We can classify molecules in terms of the three moments of inertia according to the following Table.

Moment of inertia	Type of rotator	Example
$I_a = 0$ $I_b = I_c$	Linear	HCl, OCS
$I_b = I_c > I_a$	prolate } symmetric	CH_3F
$I_b = I_c < I_a$	oblate } top	BCl_3
$I_a = I_b = I_c$	Spherical top	CH_4
$I_a \neq I_b \neq I_c$	Asymmetric top	H_2O

A spherical top molecule does not possess a permanent dipole moment and therefore is microwave inactive.

49.2. Rotational Spectra

Rotational energy, along with all other forms of molecular energy is quantized. It means that a molecule cannot have any arbitrary amount of rotational energy. Its energy is limited to certain definite values depending on the shape and size of the molecule concerned. The permitted energy values are called rotational energy *levels*. The rotational energy levels may be calculated for any molecule by solving the Schrodinger equation for the system represented by that molecule. For simple molecules, the mathematics involved is simple. For complex molecules, the mathematics involved is tedious and gross approximations have to be used. We shall discuss the rigid diatomic molecule in detail.

49.3. The Rigid Diatomic Molecule

Consider a diatomic molecule (Fig. 49.1). Masses m_1 and m_2 are joined by a rigid bar (the bond) whose length is

$$r_0 = r_1 + r_2 \qquad ...(1)$$

The molecule rotates end-over-end about the centre of mass *C*.

The moment of inertia about *C* is defined by

$$I = m_1 r_1^2 + m_2 r_2^2 \qquad ...(2)$$

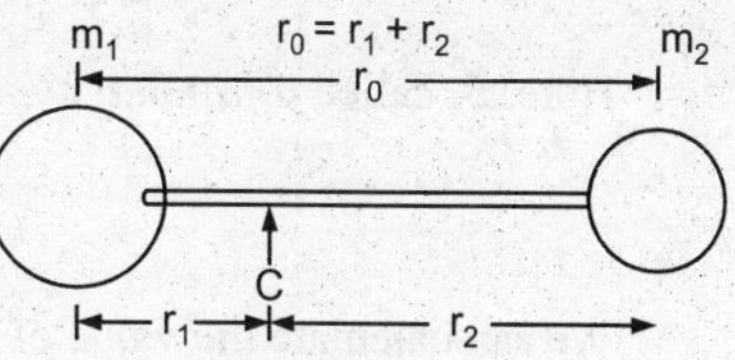

Fig. 49.1.

Here, r_1 and r_2 are the distances of m_1 and m_2 from the axis is of rotation through *C*.

As the system is balanced about its centre of mass *C*, we have

$$m_1 r_1 = m_2 r_2 \qquad ...(3)$$

From Eqs. (1) and (3), we have

$$r_1 = \frac{m_2}{m_1 + m_2} r_0$$

and.

$$r_2 = \frac{m_1}{m_1 + m_2} r_0$$

Putting these values of r_1 and r_2 in Eq. (2), we get

$$I = \frac{m_1 m_2^2}{(m_1 + m_2)^2} r_0^2 + \frac{m_2 m_1^2}{(m_1 + m_2)^2} r_0^2$$

$$= \frac{m_1 m_2}{(m_1 + m_2)} r_0^2$$

or

$$I = \mu r_0^2 \qquad ...(4)$$

Here, the *reduced mass* μ is defined as

$$\mu = \frac{m_1 m_2}{(m_1 + m_2)} \qquad ...(5)$$

Eq. (4) defines the moment of inertia in terms of the atomic masses and the bond length.

By the use of the Schrodinger equation it may be shown that the rotational energy levels allowed to the rigid diatomic molecule are given by the expression

$$E_J = \frac{h^2}{8\pi^2 I} J(J+1) \qquad ...(6)$$

J is called the *rotational quantum number*. $J = 0, 1, 2, 3, ...$

Eq. (6) expresses the allowed energies in joules.

We are interested in the wave number

$$\bar{\nu} = \frac{\Delta E}{hc}$$

of the radiation emitted or absorbed as a consequence of changes between energy levels. So we consider energies in these units. Thus, we write

$$\varepsilon_J = \frac{E_J}{hc} = \frac{h}{8\pi^2 Ic} J(J+1) \qquad ...(7)$$

ε_J is called *rotational term* (unit m^{-1}). Eq. (7) can be written as

$$\varepsilon_J = BJ(J+1)\ m^{-1},\ J = 0, 1, 2, \ldots \qquad \ldots(8)$$

Here, B, called *rotational constant*, is given by

$$B = \frac{h}{8\pi^2 Ic}\ m^{-1} \qquad \ldots(9)$$

We can calculate the value of ε_J for $J = 0, 1, 2, \ldots$ using Eq. (8).

The allowed rotational energy levels of a rigid diatomic molecule are shown in Fig. 49.2.

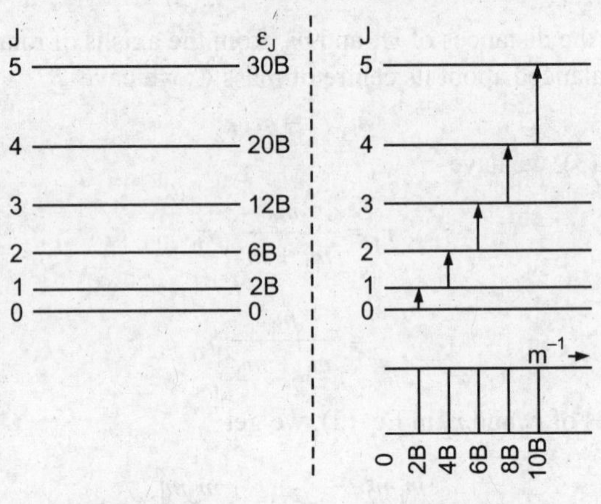

Fig. 49.2. Fig. 49.3.

Frequency of Rotational Spectral Lines

Rotational spectra are always obtained in absorption. Hence for a transition from some initial state of quantum number J to the next higher state of quantum number $J+1$, the wave number of absorbed photon is

$$\bar{\nu}_{J\to J+1} = B(J+1)(J+2) - BJ(J+1) = 2B(J+1) \qquad \ldots(10)$$

If we imagine the molecule to be in the $J = 0$ state, we can let incident radiation be absorbed to raise it to the $J = 1$ state. Then,

$$\bar{\nu}_{J=0\to J=1} = 2B$$

An absoprtion line will occur at $2\,B\ m^{-1}$.

If now the molecule is raised from the $J = 1$ state to the $J = 2$ state by the absorption of more energy, then

$$\bar{\nu}_{J=1\to j=2} = 4\,B$$

From Eq. (10), it is clear that a stepwise raising of the rotational energy results in an absorption spectrum consisting of lines at $2B$, $4B$, $6B$, ... m^{-1}. This is shown in Fig. 49.3. Therefore, on the frequency (wave number) scale, the lines are equidistant.

Selection Rule

Rotational transitions occur only in those molecules which possess a permanent electric dipole moment. Thus molecules such as HCl and CO will show a rotational spectrum, while N_2 and O_2 will not.

The selection rule for rotational transitions is $\Delta J = \pm 1$ ...(11)

From the above rule, it is evident that only those transistions are permitted in which there is an increase or decrease by unity in the rotational quantum number.

It means that $J = 0 \rightarrow J = 2 \rightarrow J = 4$ ... transitions are not possible. These transitions are spectroscopically *forbidden*. Thus Eq. (10) gives the *whole* spectrum to be expected from such a molecule.

Infra-red Spectroscopy

49.4 The Energy of a Diatomic Molecule

Consider a diatomic molecule consisting of two atoms A and B separated by equilibrium distance r_e. Fig. 49.4 shows how the potential energy of a diatomic molecule varies with the internuclear distance r. The minimum of this curve corresponds to the normal configuration of the molecule. The zero of curve is found at $r = r_e$. Any energy in excess of this, for example, ε_1, arises because of extension or compression of the bond. If one atom (A) is considered to be stationary on the $r = 0$ axis, the other will oscillate between B' and B''. If the energy is increased to ε_2, the oscillation will become more vigorous.

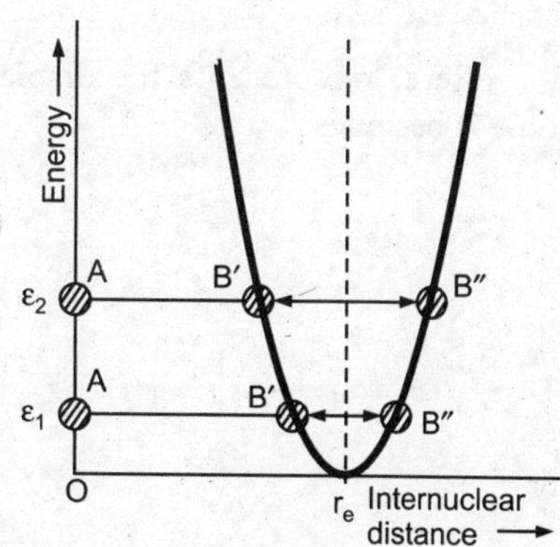

Fig. 49.4.

Suppose the bond is distorted from its equilibrium length r_e to new length r. The bond, like a spring, obeys Hooke's law. Then,

$$f = -k(r - r_e) \qquad \text{...(1)}$$

Here, f is the restoring force, k the force constant, and r the internuclear distance.

In this case the energy curve is parabolic and has the form

$$E = \frac{1}{2}k\,(r - r_e)^2 \qquad \text{...(2)}$$

This model of a vibrating diatomic molecule is called "simple harmonic oscillator" model.

49.5. Vibrating Diatomic Molecule as a Harmonic Oscillator

Frequency of Vibrations

In the diatomic molecule, we have two bodies of masses m_1 and m_2 joined by a spring. Suppose the bond is distorted from its equilibrium length r_e to a new length r. Then, restoring forces on each atom of a diatomic molecule are

$$m_1 \frac{d^2 r_1}{dt^2} = -k(r - r_e) \qquad \text{...(1)}$$

and

$$m_2 \frac{d^2 r_2}{dt^2} = -k(r - r_e) \qquad \text{...(2)}$$

Here, k represents the *force constant*. It is a measure of the stiffness of the bond. r_1 and r_2 are the positions of atoms 1 and 2 relative to the centre of mass of the molecule. From Art. 49.3, we have,

$$r_1 = \frac{m_2 r}{m_1 + m_2}$$

$$r_2 = \frac{m_1 r}{m_1 + m_2}$$

Putting the value of r_1 in Eq. (1), we get

$$\frac{m_1 m_2}{m_1 + m_2} \frac{d^2 r}{dt^2} = -k(r - r_e) \qquad \text{...(3)}$$

Since r_e is constant, $$\frac{d^2r}{dt^2} = \frac{d^2(r - r_e)}{dt^2}$$

$$\therefore \qquad \frac{m_1 m_2}{m_1 + m_2}\frac{d^2(r - r_e)}{dt^2} = -k(r - r_e) \qquad ...(4)$$

Put $$r - r_e = x, \text{ and } \frac{m_1 m_2}{m_1 + m_2} = \mu.$$

Here, x represents the displacement of the bond length from its equilirium position. Eq. (4) becomes

$$\mu\frac{d^2x}{dt^2} = -kx$$

or $$\frac{d^2x}{dt^2} + \frac{k}{\mu}x = 0$$

or $$\frac{d^2x}{dt^2} + \omega^2 x = 0$$

where $\omega^2 = k/\mu$.

This is the equation of simple harmonic motion with frequency of vibration,

$$\nu = \frac{1}{2\pi}\sqrt{\left(\frac{k}{\mu}\right)}.$$

In m^{-1} unit (wave number),

$$\bar{\nu} = \frac{1}{2\pi c}\sqrt{\left(\frac{k}{\mu}\right)}.$$

Energy levels

Vibrational energies, like other molecular energies, are quantised. The permitted vibrational energies for any particular system can be calculated from Schrodinger equation. The eigen values for the energy (in joules) of a linear harmonic oscillator are given by

$$E_v = \left(v + \frac{1}{2}\right)h\nu.$$

Here v is the *vibrational quantum number*, equal to zero or an integer.

Converting to the spectroscopic units, m^{-1}, we have,

$$\varepsilon_v = \frac{E_v}{hc} = \left(v + \frac{1}{2}\right)\bar{\nu}.$$

Here, ε_v is called *term value*. Its unit is m^{-1}.

$\bar{\nu}$ is the vibrational frequency of the oscillator in wave numbers.

Zero-point energy in $m^{-1} = \frac{1}{2}\bar{\nu}$.

The *selection rule* for the harmonic oscillator undergoing vibrational changes is

$$\Delta v = \pm 1.$$

Vibrational energy changes will only give rise to an observable spectrum if the vibration can interact with radiation, *i.e.*, if the vibration involves a change in the dipole moment of the molecule. Thus vibrational spectra will be observable only in heteronuclear diatomic molecules (like HF, HCl, HBr) since homonuclear molecules (like H_2, N_2 and O_2) have no dipole moment.

Applying the selection rule, we have

$$\varepsilon_{v+1 \rightarrow v} = (v + 1 + 1/2)\,\bar{\nu} - (v + 1/2)\,\bar{\nu} = \bar{\nu}$$

for emission and

$$\varepsilon_{v \rightarrow v+1} = \bar{\nu}$$

for absorpiton, whatever the initial value of v.

It means that all the vibrational lines obtained from harmonic oscillator are of the same frequency. The allowed vibrational energy levels and transitions between them for a diatomic molecule undergoing simple harmonic motion are shown in Fig. 49.5.

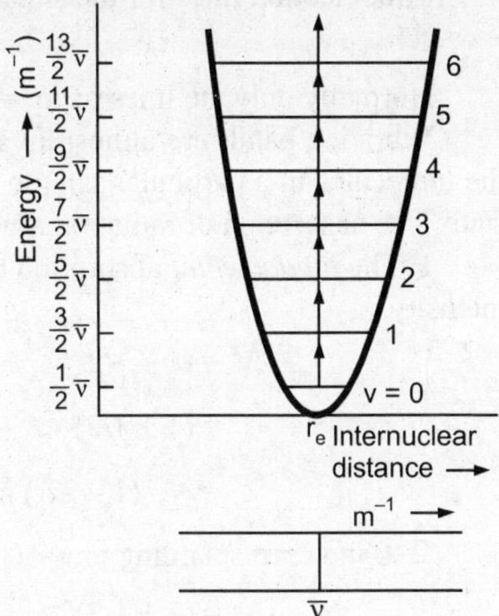

Fig. 49.5.

Pure vibrational spectra are observed only in liquids. This is because interactions between neighbouring molecules prevent their rotational motion.

49.6. The Anharmonic Oscillator

Real molecules do not obey exactly the laws of simple harmonic motion. Real molecules dissociate at large values of r and acquire very large amounts of energy as r decreases. Fig. 49.6 shows the energy curve for a typical diatomic molecule undergoing anharmonic extensions and compressions together with (dotted) the ideal simple harmonic parabola.

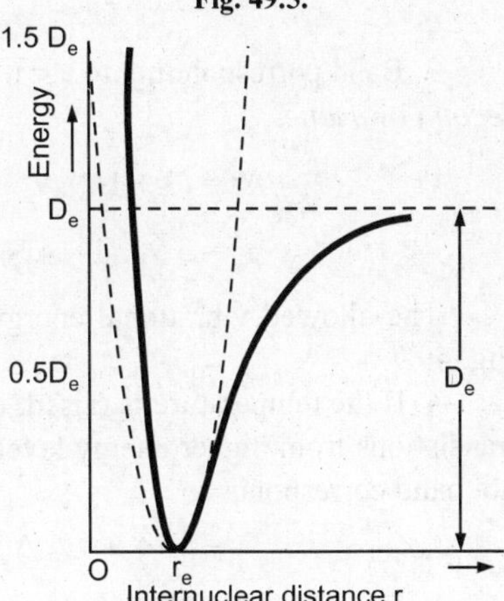

Fig. 49.6.

A purely empirical expression which fits this curve to a good approximation was derived by P.M. Morse, and is called the Morse function:

$$E = D_e\,[1 - \exp.\ \{a\,(r_e - r)\}]^2 \qquad ...(1)$$

where D_e is the dissociation energy, a is constant for a particular molecule, r_e is the value of internuclear distance r that corresponds to the minimum of Morse curve.

Energy Levels

When Eq. (1) is used in the Schrodinger equation, the allowed vibrational energy levels are found to be

$$\varepsilon_v = (v + 1/2)\,\bar{\nu}_e - (v + 1/2)^2\,\bar{\nu}_e\,x_e\ \mathrm{m}^{-1}$$

$$(v = 0, 1, 2, ...) \qquad ...(2)$$

Here, $\bar{\nu}_e$ is called the *equilibrium oscillation frequency* of the molecule expressed in wave numbers and is the value for small displacements.

x_e is the corresponding anharmonicity constant. $x_e \approx 0.01$.

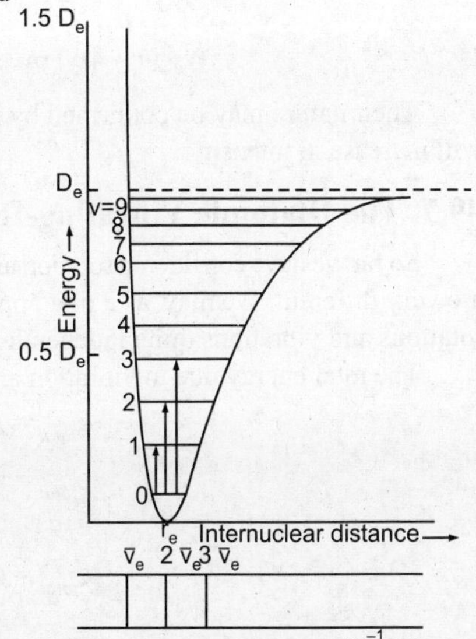

Fig. 49.7.

The selection rules for the anharmonic oscillator are found to be

$$\Delta v = \pm 1, \pm 2, \pm 3, \ldots \quad (3)$$

Normally only the lines of $\Delta v = \pm 1, \pm 2$, and ± 3, have observable intensity.

Vibration bands are almost invariably observed in absorption. At room temperature nearly all the molecules in a particular sample have only the zero point vibrational energy and exist in $v = 0$ state. The absorption of radiation must, therefore, result in transitions starting from $v = 0$.

1. The *fundamental* absorption band corresponds to $v = 0 \rightarrow v = 1$, $\Delta v = +1$, with considerable intensity.

$$\begin{aligned}\Delta\varepsilon &= \varepsilon_{v=1} - \varepsilon_{v=0} \\ &= (1 + 1/2)\,\bar{\nu}_e - x_e\,(1 + 1/2)^2\,\bar{\nu}_e - \{1/2\,\bar{\nu}_e - (1/2)^2 x_e \bar{\nu}_e\} \\ &= \bar{\nu}_e\,(1 - 2x_e)\ \text{m}^{-1}\end{aligned} \quad \ldots(4)$$

2. Band corresponding to $v = 0 \rightarrow v = 2$, $\Delta v = +2$ with small intensity, is called *first overtone.*

$$\begin{aligned}\Delta\varepsilon &= (2 + 1/2)\,\bar{\nu}_e - x_e(2 + 1/2)^2\,\bar{\nu}_e - \{1/2\,\bar{\nu}_e - (1/2)^2 x_e \bar{\nu}_e\} \\ &= 2\,\bar{\nu}_e\,(1 - 3x_e)\ \text{m}^{-1}\end{aligned} \quad \ldots(5)$$

3. Band corresponding to $v = 0 \rightarrow v = 3$, $\Delta v = +3$, with normally negligible intensity is called *second overtone.*

$$\begin{aligned}\Delta\varepsilon &= (3 + 1/2)\,\bar{\nu}_e - (3 + 1/2)^2\,\bar{\nu}_e x_e - \{1/2\,\bar{\nu}_e - (1/2)^2 x_e \bar{\nu}_e\} \\ &= 3\,\bar{\nu}_e\,(1 - 4x_e)\ \text{m}^{-1}\end{aligned} \quad \ldots(6)$$

The allowed vibrational energy levels and these three absorption transistions are shown in Fig. 49.7.

4. If the temperature is raised, the population of the $v = 1$ state may become appreciable. The transitions from higher energy levels with the selection rule $\Delta v = 1$ are called *hot bands*. The first hot band corresponds to

$v = 1 \rightarrow v = 2$, $\Delta v = +1$, normally very weak.

$$\begin{aligned}\Delta\varepsilon &= 2\frac{1}{2}\bar{\nu}_e - 6\frac{1}{4}x_2\bar{\nu}_e - \left\{1\frac{1}{2}\bar{\nu}_e - 2\frac{1}{4}x_e\bar{\nu}_e\right\} \\ &= \bar{\nu}_e\,(1 - 4x_e)\ \text{m}^{-1}.\end{aligned}$$

Their nature may be confirmed by raising the temperature of the sample when a true hot band will increase in intensity.

49.7. The Diatomic Vibrating-Rotator

So far we have considered rotation and vibration quite separately. The energies of the two motions are very different. We may, as a first approximation, consider that a diatomic molecule can execute rotations and vibrations quite independently. This is called the Born-Oppenheimer approximation.

The total energy due to vibration and rotation is given by

$$E_{total} = E_r + E_v \text{ (joules)}$$

$$\varepsilon_{total} = \varepsilon_r + \varepsilon_v \text{ (m}^{-1}\text{)} \quad \ldots(1)$$

$$\varepsilon_{total} = \varepsilon_{J,v} = BJ(J+1) + \left(v + \frac{1}{2}\right)\bar{\nu}_e - x_e\left(v + \frac{1}{2}\right)^2 \bar{\nu}_e \quad \ldots(2)$$

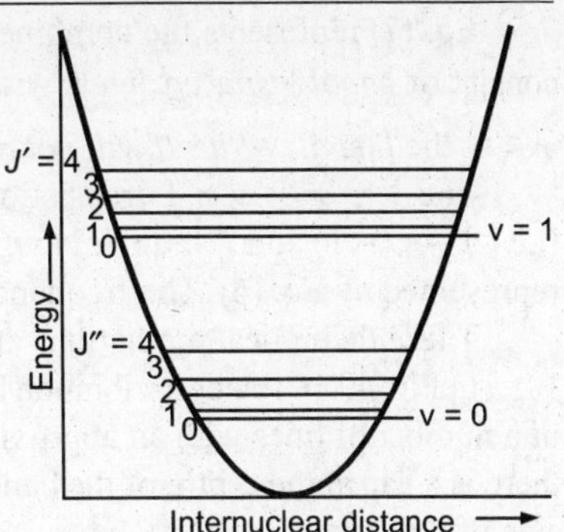

Fig. 49.8.

The rotational levels are sketched in Fig. 49.8 for the two lowest vibrational levels, $v = 0$ and $v = 1$.

It may be shown that the selection rules for the combined motions are the same as those for each separately. Therefore,

$$\Delta v = \pm 1, \pm 2, \text{ etc.} \quad ...(3)$$

$\Delta J = \pm 1$. The transition with $\Delta J = 0$ is not allowed.

In Fig. 49.9 are shown relevant energy levels and transistions, designating rotational quantum number in $v = 0$ state as J'' and in $v = 1$ state as J' (single prime is used for upper state and double prime for lower state).

An analytical expression for the spectrum can be obtained by applying the selection rules to the energy levels. Considerating only $v = 0$ to $v = 1$ transistion we have,

$$\Delta\varepsilon_{j.v} = \varepsilon_{j'.v=1} - \varepsilon_{j'.v=0}$$

$$= BJ'(J'+1) + 1\frac{1}{2}\bar{\nu}_e - 2\frac{1}{4}x_e\bar{\nu}_e$$

$$- \{BJ''(J''+1) + \frac{1}{2}\bar{\nu}_e - \frac{1}{4}x_e\bar{\nu}_e\}$$

$$= \bar{\nu}_0 + B(J'-J'')(J'+J''+1) \text{ m}^{-1} \quad ...(4)$$

Here, for $\bar{\nu}_e(1-2x_e)$, we have written $\bar{\nu}_0$.

We have assumed that rotation is unaffected by vibrational changes. Hence B will remain identical in the upper and lower vibrational states.

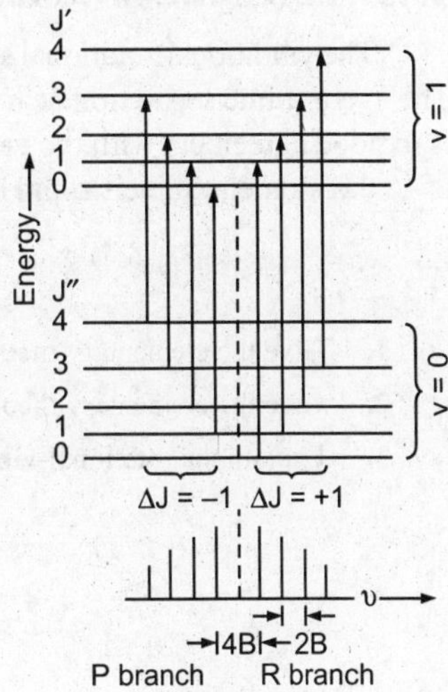

Fig. 49.9.

Now we can have:

1. $\Delta J = +1$, *i.e.*, $J' = J'' + 1$ or $J' - J'' = +1$; hence

$$\Delta\varepsilon_{J,v} = \bar{\nu}_0 + 2B(J''+1) \text{ m}^{-1} \qquad J'' = 0, 1, 2, ... \quad ...(5)$$

2. $\Delta J = -1$, *i.e.*, $J'' = J' + 1$ or $J' - J'' = -1$; and

$$\Delta\varepsilon_{J,v} = \bar{\nu}_0 - 2B(J'+1) \text{ m}^{-1} \qquad J' = 0, 1, 2, ... \quad ...(6)$$

Eqs. (5) and (6) can be combined in the form

$$\Delta\varepsilon_{J,v} = \bar{\nu}_{\text{spect}} = \bar{\nu}_o + 2Bm \text{ m}^{-1} \qquad m = \pm 1, \pm 2, ... \quad ...(7)$$

m cannot be zero since this would imply values of J' or J'' to be –1.

The frequency $\bar{\nu}_o$ is usually called the *band origin or band centre.*

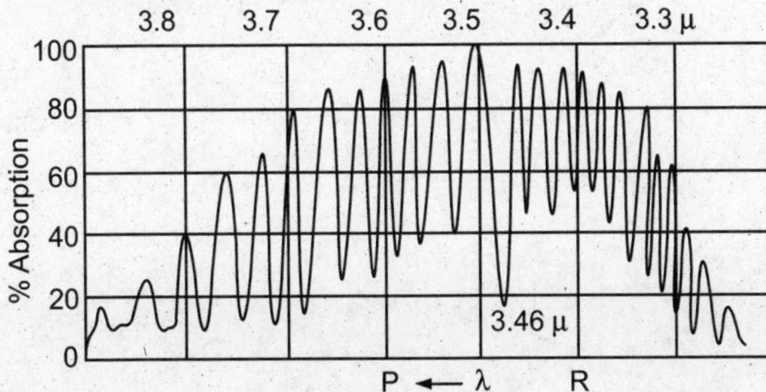

Fig. 49.10.

Eq. (7) represents the combined vibration-rotation spectrum. Such a spectrum will obviously consist of *equally spaced lines*, with a spacing $2B$, on each side of the band origin $\bar{\nu}_0$ but, since $m \neq 0$, the *line at* $\bar{\nu}_0$ *itself will not appear.*

The $v = 0 \rightarrow v = 1$ transistions fall into two categories, the P branch in which $\Delta J = -1$ (*i.e.*, $J \rightarrow J - 1$) and R branch in which $\Delta J = +1$ (*i.e.*, $J \rightarrow J + 1$). The frequencies of P branch are represented by Eq. (6). The frequencies of R branch are represented by Eq. (5).

These theoretical predictions are in agreement with the experiment.

In Fig. 49.10 rotation-vibration bands of hydrogen chloride (HCl) are shown. It is found to consist of a number of lines with an approximately constant frequency separation of $2B$ cm^{-1}. As expected, there is a gap at the centre of the band, and therefore spacing between the lines immediately on each side of the centre is thus $4B$ cm^{-1}.

49.8 The Vibration-Rotation Spectrum of Carbon Monoxide

The vibrational-rotational spectrum of gaseous CO includes an intense band centred at 2143 cm^{-1}. The average line separation near the centre is 3.83 cm^{-1}. Hence $2B = 3.83$ cm^{-1} or $B = 1.915$ cm^{-1}. This is in good agreement with the value $B = 1.92118$ cm^{-1} derived by microwave studies. Using the value of B, the moment of inertia and bond length can be calculated from infra-red data alone.

EXERCISE

1. Give the elementary theory of the origin of pure rotational spectrum of a molecule.
2. Give the elementary theory of the origin of the vibrational spectrum of a molecule.
3. Explain the rotational-vibrational spectrum of diatomic molecules. (*Osmania* 1992)

APPENDIX

A1. To Find the Thickness or Refractive Index of a Thin Transparent Sheet Using Michelson's Interferometer

Suppose a thin film of a transparent substance of thickness t and refractive index n is introduced into the path of one of the interfering beams in the interferometer (Fig. A1). The optical path in this beam is increased from t to nt. Since the beam traverses the film twice, this will introduce an extra path difference of $2(n-1)t$ between the two beams. Hence the fringe system will shift.

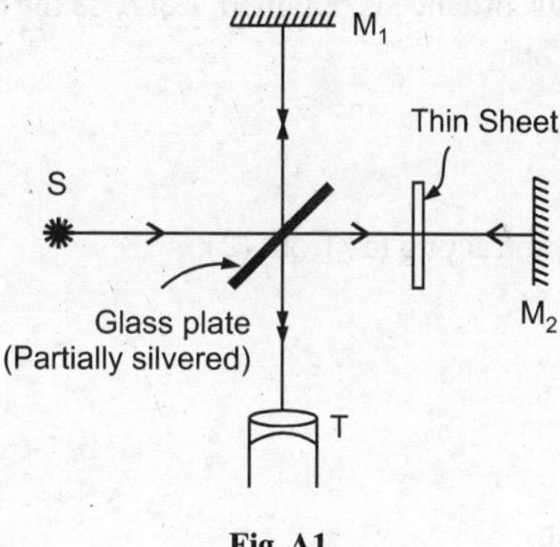

Fig. A1.

Let m be the number of fringes by which the fringe system is displaced and λ the wavelength of the light used. Then,

$$2(n-1)t = m\lambda,$$

or

$$n = \left[\frac{m\lambda}{2t} + 1\right].$$

Since with the monochromatic light all the fringes are exactly similar, it is not possible to distinguish one fringe from the other. To overcome this difficulty, white light fringes are super-imposed over the monochromatic light fringes and the central white fringe serves as the reference. The cross-wires in the telescope are focused on the central white fringe without the film. After inserting the film, the number of fringes of the monochromatic light between the cross-wires and the new position of the central white fringe is counted. This gives the value of m. If the thickness t of the plate is known, n may be calculated. If n is known, t may be calculated. This method is suitable for thin plates.

For thick plates, two identical plates are inserted in the two beams. One of the plates is rotated through a small angle and the number of fringes shifted as a result of this is noted. Knowing the angle turned, its refractive index can be calculated. This method can also be employed to determine the variation of refractive index of a gas with its pressure and temperature.

Example. *A thin film of glass of refractive index 1.50 is inserted in one arm of a Michelson's interferometer. Find the thickness of the film if a shift of 10 fringes is obtained, assuming the wavelength of light to be* 5890×10^{-8} *cm.*

Solution. $2(n-1)t = m\lambda$

Here, $n = 1.5, m = 10, \lambda = 5.890 \times 10^{-7}$ metre;

$$t = \frac{m\lambda}{2(n-1)} = \frac{10 \times \left(5.890 \times 10^{-7}\right)}{2(1.5-1)} = 5.890 \times 10^{-6} \text{ metre}.$$

A2. To determine the Wavelength Difference Between Spectral Lines of a Close Doublet Using Fabry Perot Interferometer

The difference in wavelength by Fabry Perot interferometer can be found by *coincidence method.*

Let λ_1 and λ_2 be two very close wavelengths in the incident light (say $\lambda_1 > \lambda_2$). At first, the movable plate is brought in contact with the other. Now the rings due to λ_1 and λ_2 partially coincide. Now by moving movable plate t is increased, the fringe systems due to λ_1 and λ_2 separate and the maximum discordance occurs when the rings due to λ_2 are half way between those due to λ_1.

Let for maximum discordance the separation between plates be t_1 We have,

$$2t_1 = n_1\lambda_1 = \left(n_1 + \frac{1}{2}\right)\lambda_2 \qquad \text{...(1)}$$

By further increasing t, the rings coincide again and then separate out and second maximum discordance is obtained. Let t_2 be the distance between two plates for second maximum discordance. Then,

$$2t_2 = n_2\lambda_1 = \left(n_2 + \frac{3}{2}\right)\lambda_2 \qquad \text{...(2)}$$

Subtracting (1) from (2),

$$2(t_2 - t_1) = (n_2 - n_1)\lambda_1 \qquad \text{...(3)}$$

$$= (n_2 - n_1)\lambda_2 + \lambda_2$$

$$\therefore \quad (n_2 - n_1)(\lambda_1 - \lambda_2) = \lambda_2$$

or

$$n_2 - n_1 = \frac{\lambda_2}{\lambda_1 - \lambda_2}$$

Substituting for $(n_2 - n_1)$ in Eqn. (3), we get

$$2(t_2 - t_1) = \frac{\lambda_2}{\lambda_1 - \lambda_2}.\lambda_1$$

or

$$\lambda_1 - \lambda_2 = \frac{\lambda_1\lambda_2}{2(t_2 - t_1)} = \frac{\lambda^2}{2(t_2 - t_1)} \qquad \text{...(4)}$$

Here, λ is the average wavelength. $(t_2 - t_1)$ is the distance moved by the movable plate in between the two maximum discordance. The value of $t_2 - t_1$ may be found from the difference in the micrometer screw attached to movable plate.

Example. *A Fabry-Perot interferometer is used to determine the small difference in wavelength of the two components of a radiation of average wavelength 640 nm. It is found that at a given point near the centre, the two sets of rings coincide for a given distance between the two plates and again when the distance is increased by 1.024 mm. Find the difference in wavelength.*

Solution.

$$\delta\lambda = \lambda_1 - \lambda_2 = \frac{\lambda^2}{2(t_2 - t_1)}.$$

Here, $\lambda = 640 \times 10^{-9}$ m, $t_2 - t_1 = 1.024 \times 10^{-3}$ m.

$$\therefore \qquad \delta\lambda = \frac{\left(640 \times 10^{-9}\right)^2}{2\left(1.024 \times 10^{-3}\right)} = 0.2 \times 10^{-9} \text{ m} = 0.2 \text{ nm}.$$

A3. Superiority of Fabry Perot Interferometer over that of Michelson's Interferometer.

(*i*) The *resolving power of F.P. interferometer is much higher than that of Michelson interferometer.* Hence if incident light consists of two or more close wavelengths (as D_1 and D_2 lines in sodium light), then, in F.P. instrument, each wavelength produces its own pattern of fringes clearly separated by pattern of other wavelength, while in Michelson's interferometer separate patterns are not produced. Therefore *this interferometer is very useful for study of fine structure of spectral lines.*

(*ii*) In Michelson's interferometer, two waves of equal intensities produce interference. Therefore, the intensity distribution is given by

$$I = I_{max} \cos^2 \frac{\delta}{2}.$$

If

$$I = \frac{I_{max}}{2}, \cos^2 \frac{\delta}{2} = \frac{1}{2}.$$

$$\therefore \qquad \cos \frac{\delta}{2} = \frac{1}{\sqrt{2}} = \cos\left(\frac{\pi}{4}\right)$$

$$\therefore \qquad \frac{\delta}{2} = \frac{\pi}{4}$$

or

$$\delta = \frac{\pi}{2} \text{ rad} = 1.57 \text{ rad}.$$

In case of Fabry Perot interferometer, if $R = 0.9$, $\delta = 0.105$ rad. Fringes of F.P. interferometer have comparatively very much sharpness than Michelson's fringes. This is the reason that F.P. interferometer is more suited for the study of hyperfine structure.

A4. Fraunhofer Diffraction by a Rectangular Aperture

If the length and breadth of the slit are comparable to each other, the slit is termed as a rectangular aperture. Consider a rectangular aperture of length *l* and width *a*. The analytical treatment of diffraction through a rectangular aperture will involve double integration over both the dimensions of the wavefront. A comprehensive analytical treatment yields the expression for intensity distribution as

$$I = a^2 l^2 \frac{\sin^2 \alpha}{\alpha^2} \cdot \frac{\sin^2 \gamma}{\gamma}.$$

Here, α is phase difference between two rays arising from two points on two extremities of width *a*.

γ is the phase difference between two rays arising from two points on the two extremities of the slit.

$$\alpha = \frac{\pi}{\lambda} a \sin\theta \text{ and } \gamma = \frac{\pi}{\lambda} l \sin\phi$$

where θ and ϕ are the angles measured in two mutually perpendicular planes containing the two sides of the rectangular aperture.

A5. Spherical Aberration in Mirrors

When the rays from an object to a spherical mirror make large angles with the principal axis, the images formed are confused. The rays from a point source on the axis do not come to a focus at a common point. The rays reflected from the outer parts of a spherical mirror, called the *marginal rays*, come to a focus nearer to the mirror than the paraxial rays (Fig. A2). This imperfection is called the *spherical aberration*.

Consecutive rays moving away from the axis intersect at points lying on a curved line. This line is called the *caustic curve*. Mirror surfaces may be specially "corrected" in such a way that rays from all zones come to a common focus. The surface is then not spherical and is called "aspherical". Such arrangements are used in all expensive telescopes and cameras using large aperture.

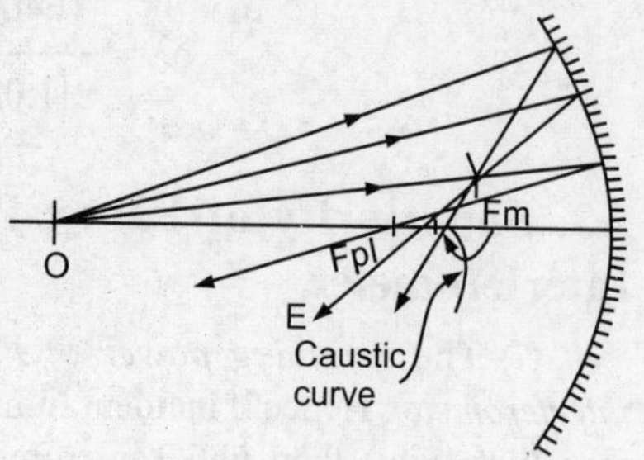

Fig. A2.

Schmidt Corrector Plate

The Schmidt *corrector plate* (Fig. A3) eliminates the spherical aberration of the mirror.

The rays falling on outer parts (larger h) come to focus earlier than those falling on the central part of the mirror. The Schmidt plate effectively compensates this by acting as a concave lens for rays at larger h and as a convex lens for rays at smaller h. Its positioning is at the centre of curvature of the reflector to ensure that the effect is independent of the inclination of the chief ray.

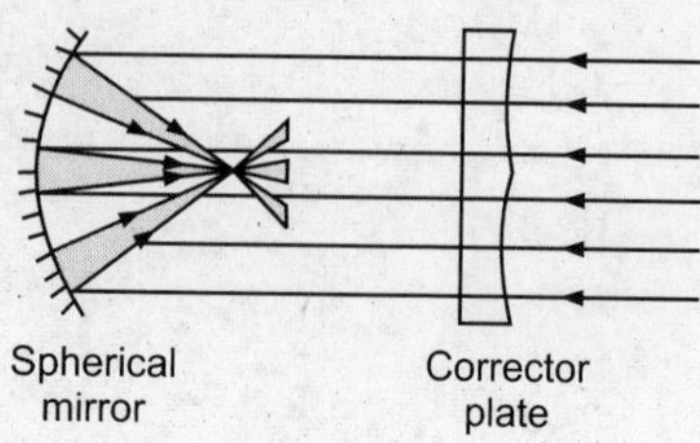

Fig. A3.

The Schmidt corrector plate is used widely with large aperture reflector telescopes.

A6. Blazed Grating

The dispersive power of a diffraction grating is directly proportional to n, the order of the spectrum. So higher order spectra are desirable for good dispersion. But the intensity is maximum in the zero order and falls as n increases. Blazed grating is designed to shift the intensity maximum towards a desired high order spectrum. Grooves with inclined reflecting faces are made in the grating surface (Fig. A4). The inclination γ is called the *blaze angle*.

R. W. Wood has been able to produce gratings which concentrate as much as 90 percent of the light of a particular wavelength in a single order on one side. Thus one of the chief disadvantages of gratings compared with prisms–the presence of multiple spectra, none of which is very intense–is overcome.

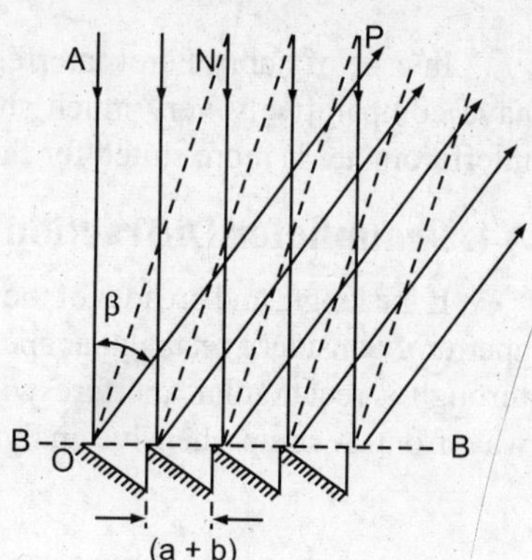

Fig. A4.

Suppose light is incident along normal AO to the grating surface BB. The *central diffraction maximum* now falls not along OA, but at an angle 2γ from OA. The spectrum seen at angle β from OA corresponds to

$$(a + b)\sin\beta = n\lambda.$$

If β is in the neighbourhood of 2γ, then the spectrum has large intensity, despite n being high. Suppose we want to choose a grating blazed for a given order n for a given region γ. Then we calculate β and set $\gamma = \frac{1}{2}\beta$ as the blaze angle.